T0191664

Lecture Notes in Artificial Intelligence   11645

Subseries of Lecture Notes in Computer Science

Series Editors

Randy Goebel
University of Alberta, Edmonton, Canada
Yuzuru Tanaka
Hokkaido University, Sapporo, Japan
Wolfgang Wahlster
DFKI and Saarland University, Saarbrücken, Germany

Founding Editor

Jörg Siekmann
DFKI and Saarland University, Saarbrücken, Germany

De-Shuang Huang · Zhi-Kai Huang ·
Abir Hussain (Eds.)

Intelligent Computing Methodologies

15th International Conference, ICIC 2019
Nanchang, China, August 3–6, 2019
Proceedings, Part III

Springer

Editors
De-Shuang Huang
Tongji University
Shanghai, China

Zhi-Kai Huang
Nanchang Institute of Technology
Nanchang, China

Abir Hussain
Liverpool John Moores University
Liverpool, UK

ISSN 0302-9743   ISSN 1611-3349 (electronic)
Lecture Notes in Artificial Intelligence
ISBN 978-3-030-26765-0   ISBN 978-3-030-26766-7 (eBook)
https://doi.org/10.1007/978-3-030-26766-7

LNCS Sublibrary: SL7 – Artificial Intelligence

This Springer imprint is published by the registered company Springer Nature Switzerland AG
The registered company address is: Gewerbestrasse 11, 6330 Cham, Switzerland

Preface

The International Conference on Intelligent Computing (ICIC) was started to provide an annual forum dedicated to the emerging and challenging topics in artificial intelligence, machine learning, pattern recognition, bioinformatics, and computational biology. It aims to bring together researchers and practitioners from both academia and industry to share ideas, problems, and solutions related to the multifaceted aspects of intelligent computing.

ICIC 2019, held in Nanchang, China, during August 3–6, 2019, constituted the 15th International Conference on Intelligent Computing. It built upon the success of previous ICIC conferences held in Wuhan, China (2018), Liverpool, UK (2017), Lanzhou, China (2016), Fuzhou, China (2015), Taiyuan, China (2014), Nanning, China (2013), Huangshan, China (2012), Zhengzhou, China (2011), Changsha, China (2010), Ulsan, Republic of Korea (2009), Shanghai, China (2008), Qingdao, China (2007), Kunming, China (2006), and Hefei, China (2005).

This year, the conference concentrated mainly on the theories and methodologies as well as the emerging applications of intelligent computing. Its aim was to unify the picture of contemporary intelligent computing techniques as an integral concept that highlights the trends in advanced computational intelligence and bridges theoretical research with applications. Therefore, the theme for this conference was "Advanced Intelligent Computing Technology and Applications." Papers focusing on this theme were solicited, addressing theories, methodologies, and applications in science and technology.

ICIC 2019 received 609 submissions from 22 countries and regions. All papers went through a rigorous peer-review procedure and each paper received at least three review reports. Based on the review reports, the Program Committee finally selected 217 high-quality papers for presentation at ICIC 2019, included in three volumes of proceedings published by Springer: two volumes of *Lecture Notes in Computer Science* (LNCS), and one volume of *Lecture Notes in Artificial Intelligence* (LNAI). This volume of *Lecture Notes in Artificial Intelligence* (LNAI) includes 72 papers.

The organizers of ICIC 2019, including Tongji University, Nanchang Institute of Technology, and East China Institute of Technology, as well as Shandong University at Weihai, made an enormous effort to ensure the success of the conference. We hereby would like to thank the members of the Program Committee and the referees for their collective effort in reviewing and soliciting the papers. We would like to thank Alfred Hofmann, executive editor from Springer, for his frank and helpful advice and guidance throughout, and for his continuous support in publishing the proceedings. In particular, we would like to thank all the authors for contributing their papers. Without the high-quality submissions from the authors, the success of the conference would not have been possible. Finally, we are especially grateful to the International Neural Network Society, and the National Science Foundation of China for their sponsorship.

June 2019   De-Shuang Huang
                                                                 Zhi-Kai Huang
                                                                 Abir Hussain

ICIC 2019 Organization

General Co-chairs

De-Shuang Huang, China
Shengqian Wang, China

Program Committee Co-chairs

Kang-Hyun Jo, South Korea
Phalguni Gupta, India

Organizing Committee Co-chairs

Chengzhi Deng, China
Zhikai Huang, China
Yusen Zhang, China

Organizing Committee Members

Shumin Zhou, China
Wei Tian, China
Yan Li, China
Keming Liu, China
Shaoquan Zhang, China
Liling Zhang, China

Award Committee Chair

Vitoantonio Bevilacqua, Italy

Tutorial Chair

M. Michael Gromiha, India

Publication Chair

Ling Wang, China

Special Session Chair

Abir Hussain, UK

Special Issue Chair

Kyungsook Han, South Korea

International Liaison Chair

Prashan Premaratne, Australia

Workshop Co-chairs

Jair Cervantes Canales, Mexico
Michal Choras, Poland

Publicity Co-chairs

Valeriya Gribova, Russia
Laurent Heutte, France
Chun-Hou Zheng, China

Exhibition Contact Chair

Di Wu, Tongji University, China

Program Committee Members

Abir Hussain	Dah-Jing Jwo	Tianyong Hao
Khalid Aamir	Shaoyi Du	Mohd Helmy Abd Wahab
Kang-Hyun Jo	Dunwei Gong	Hao Lin
Angelo Ciaramella	Xiaoheng Deng	Hongmin Cai
Wenzheng Bao	Meng Joo Er	Xinguo Lu
Binhua Tang	Eros Pasero	Hongjie Wu
Bin Qian	Evi Syukur	Jianbo Fan
Bingqiang Liu	Fengfeng Zhou	Jair Cervantes
Bin Liu	Francesco Pappalardo	Junfeng Xia
Li Chai	Gai-Ge Wang	Juan Carlos
Chin-Chih Chang	LJ Gong	Figueroa-Jiangning Song
Wen-Sheng Chen	Valeriya Gribova	Joo M. C. Sousa
Michal Choras	Michael Gromiha Maria	Ju Liu
Xiyuan Chen	Naijie Gu	Ka-Chun Wong
Jieren Cheng	Guoliang Li	Kyungsook Han
Chengzhi Liang	Fei Han	Seeja K. R.

Yoshinori Kuno
Laurent Heutte
Xinyi Le
Bo Li
Yunxia Liu
Zhendong Liu
Hu Lu
Fei Luo
Haiying Ma
Mingon Kang
Marzio Pennisi
Gaoxiang Ouyang
Seiichi Ozawa
Shaoliang Peng
Prashan Premaratne
Boyang Qu
Rui Wang
Wei-Chiang Hong
Xiangwei Zheng
Shen Yin
Sungshin Kim

Surya Prakash
Tar Veli Mumcu
Vasily Aristarkhov
Vitoantonio Bevilacqua
Ling Wang
Xuesong Wang
Waqas Haider Khan
   Bangyal
Bing Wang
Wenbin Liu
Weidong Chen
Wei Jiang
Wei Wei
Weining Qian
Takashi Kuremoto
Shitong Wang
Xiao-Hua Yu
Jing Xiao
Xin Yin
Xingwen Liu
Xiujuan Lei

Xiaoke Ma
Xiaoping Liu
Xiwei Liu
Yonggang Lu
Yongquan Zhou
Zu-Guo Yu
Yuan-Nong Ye
Jianyang Zeng
Tao Zeng
Junqi Zhang
Le Zhang
Wen Zhang
Qi Zhao
Chunhou Zheng
Zhan-Li Sun
Zhongming Zhao
Shanfeng Zhu
Quan Zou
Zhenran Jiang

Additional Reviewers

Huijuan Zhu
Yizhong Zhou
Lixiang Hong
Yuan Wang
Mao Xiaodan
Ke Zeng
Xiongtao Zhang
Ning Lai
Shan Gao
Jia Liu
Ye Tang
Weiwei Cai
Yan Zhang
Yuanpeng Zhang
Han Zhu
Wei Jiang
Hong Peng
Wenyan Wang
Xiaodan Deng
Hongguan Liu
Hai-Tao Li

Jialing Li
Kai Qian
Huichao Zhong
Huiyan Jiang
Lei Wang
Yuanyuan Wang
Biao Zhang
Ta Zhou
Wei Liao
Bin Qin
Jiazhou Chen
Mengze Du
Sheng Ding
Dongliang Qin
Syed Sadaf Ali
Zheng Chenc
Shang Xiang
Xia Lin
Yang Wu
Xiaoming Liu
Jing Lv

Lin Weizhong
Jun Li
Li Peng
Hongfei Bao
Zhaoqiang Chen
Ru Yang
Jiayao Wu
Dadong Dai
Guangdi Liu
Jiajia Miao
Xiuhong Yang
Xiwen Cai
Fan Li
Aysel Ersoy Yilmaz
Agata Giełczyk
Akila Ranjith
Xiao Yang
Cheng Liang
Alessio Ferone
José Alfredo Costa
Ambuj Srivastava

Mohamed Abdel-Basset
Angelo Ciaramella
Anthony Chefles
Antonino Staiano
Antonio Brunetti
Antonio Maratea
Antony Lam
Alfredo Pulvirenti
Areesha Anjum
Athar Ali Moinuddin
Mohd Ayyub Khan
Alfonso Zarco
Azis Ciayadi
Brendan Halloran
Bin Qian
Wenbin Song
Benjamin J. Lang
Bo Liu
Bin Liu
Bin Xin
Guanya Cai
Casey P. Shannon
Chao Dai
Chaowang Lan
Chaoyang Zhang
Chuanchao Zhang
Jair Cervantes
Bo Chen
Yueshan Cheng
Chen He
Zhen Chen
Chen Zhang
Li Cao
Claudio Loconsole
Cláudio R. M. Silva
Chunmei Liu
Yan Jiang
Claus Scholz
Yi Chen
Dhiya AL-Jumeily
Ling-Yun Dai
Dongbo Bu
Deming Lei
Deepak Ranjan Nayak
Dong Han
Xiaojun Ding

Domenico Buongiorno
Haizhou Wu
Pingjian Ding
Dongqing Wei
Yonghao Du
Yi Yao
Ekram Khan
Miao Jiajia
Ziqing Liu
Sergio Santos
Tomasz Andrysiak
Fengyi Song
Xiaomeng Fang
Farzana Bibi
Fatih Adıgüzel
Fang-Xiang Wu
Dongyi Fan
Chunmei Feng
Fengfeng Zhou
Pengmian Feng
Feng Wang
Feng Ye
Farid Garcia-Lamont
Frank Shi
Chien-Yuan Lai
Francesco Fontanella
Lei Shi
Francesca Nardone
Francesco Camastra
Francesco Pappalardo
Dongjie Fu
Fuhai Li
Hisato Fukuda
Fuyi Li
Gai-Ge Wang
Bo Gao
Fei Gao
Hongyun Gao
Jianzhao Gao
Jianzhao Gao
Gaoyuan Liang
Geethan Mendiz
Geethan Mendiz
Guanghui Li
Giacomo Donato
    Cascarano

Giorgio Valle
Giovanni Dimauro
Giulia Russo
Linting Guan
Ping Gong
Yanhui Gu
Gunjan Singh
Guohua Wu
Guohui Zhang
Guo-Sheng Hao
Surendra M. Gupta
Sandesh Gupta
Gang Wang
Hafizul Fahri Hanafi
Haiming Tang
Fei Han
Hao Ge
Kai Zhao
Hangbin Wu
Hui Ding
Kan He
Bifang He
Xin He
Huajuan Huang
Jian Huang
Hao Lin
Ling Han
Qiu Xiao
Yefeng Li
Hongjie Wu
Hongjun Bai
Hongtao Lei
Haitao Zhang
Huakang Li
Jixia Huang
Pu Huang
Sheng-Jun Huang
Hailin Hu
Xuan Huo
Wan Hussain Wan Ishak
Haiying Wang
Il-Hwan Kim
Kamlesh Tiwari
M. IkramUllah Lali
Ilaria Bortone
H. M. Imran

Ingemar Bengtsson
Izharuddin Izharuddin
Jackson Gomes
Wu Zhang
Jiansheng Wu
Yu Hu
Jaya Sudha
Jianbo Fan
Jiancheng Zhong
Enda Jiang
Jianfeng Pei
Jiao Zhang
Jie An
Jieyi Zhao
Jie Zhang
Jin Lu
Jing Li
Jingyu Hou
Joe Song
Jose Sergio Ruiz
Jiang Shu
Juntao Liu
Jiawen Lu
Jinzhi Lei
Kanoksak
 Wattanachote
Juanjuan Kang
Kunikazu Kobayashi
Takashi Komuro
Xiangzhen Kong
Kulandaisamy A.
Kunkun Peng
Vivek Kanhangad
Kang Xu
Kai Zheng
Kun Zhan
Wei Lan
Laura Yadira
Domínguez Jalili
Xiangtao Chen
Leandro Pasa
Erchao Li
Guozheng Li
Liangfang Zhao
Jing Liang
Bo Li

Feng Li
Jianqiang Li
Lijun Quan
Junqing Li
Min Li
Liming Xie
Ping Li
Qingyang Li
Lisbeth Rodríguez
Shaohua Li
Shiyong Liu
Yang Li
Yixin Li
Zhe Li
Zepeng Li
Lulu Zuo
Fei Luo
Panpan Lu
Liangxu Liu
Weizhong Lu
Xiong Li
Junming Zhang
Shingo Mabu
Yasushi Mae
Malik Jahan Khan
Mansi Desai
Guoyong Mao
Marcial Guerra de
 Medeiros
Ma Wubin
Xiaomin Ma
Medha Pandey
Meng Ding
Muhammad Fahad
Haiying Ma
Mingzhang Yang
Wenwen Min
Mi-Xiao Hou
Mengjun Ming
Makoto Motoki
Naixia Mu
Marzio Pennisi
Yong Wang
Muhammad Asghar
 Nadeem
Nadir Subasi

Nagarajan Raju
Davide Nardone
Nathan R. Cannon
Nicole Yunger Halpern
Ning Bao
Akio Nakamura
Zhichao Shi
Ruxin Zhao
Mohd Norzali Hj Mohd
Nor Surayahani Suriani
Wataru Ohyama
Kazunori Onoguchi
Aijia Ouyang
Paul Ross McWhirter
Jie Pan
Binbin Pan
Pengfei Cui
Pu-Feng Du
Kunkun Peng
Syed Sadaf Ali
Iyyakutti Iyappan
 Ganapathi
Piyush Joshi
Prashan Premaratne
Peng Gang Sun
Puneet Gupta
Qinghua Jiang
Wangren Qiu
Qiuwei Li
Shi Qianqian
Zhi-Xian Liu
Raghad AL-Shabandar
Rafał Kozik
Raffaele Montella
Woong-Hee Shin
Renjie Tan
Rodrigo A. Gutiérrez
Rozaida Ghazali
Prabakaran
Jue Ruan
Rui Wang
Ruoyao Ding
Ryuzo Okada
Kalpana Shankhwar
Liang Zhao
Sajjad Ahmed

Sakthivel Ramasamy
Shao-Lun Lee
Wei-Chiang Hong
Hongyan Sang
Jinhui Liu
Stephen Brierley
Haozhen Situ
Sonja Sonja
Jin-Xing Liu
Haoxiang Zhang
Sebastian Laskawiec
Shailendra Kumar
Junliang Shang
Wei-Feng Guo
Yu-Bo Sheng
Hongbo Shi
Nobutaka Shimada
Syeda Shira Moin
Xingjia Lu
Shoaib Malik
Feng Shu
Siqi Qiu
Boyu Zhou
Stefan Weigert
Sameena Naaz
Sobia Pervaiz
Somnath Dey
Sotanto Sotanto
Chao Wu
Yang Lei
Surya Prakash
Wei Su
Qi Li
Hotaka Takizawa
FuZhou Tang
Xiwei Tang
Li-Na Chen
Yao Tuozhong
Qing Tian
Tianyi Zhou
Junbin Fang
Wei Xie
Shikui Tu
Umarani Jayaraman
Vahid Karimipour
Vasily Aristarkhov

Vitoantonio Bevilacqua
Valeriya Gribova
Guangchen Wang
Hong Wang
Haiyan Wang
Jingjing Wang
Ran Wang
Waqas Haider Bangyal
Pi-Jing Wei
Wei Lan
Fangping Wan
Jue Wang
Minghua Wan
Qiaoyan Wen
Takashi Kuremoto
Chuge Wu
Jibing Wu
Jinglong Wu
Wei Wu
Xiuli Wu
Yahui Wu
Wenyin Gong
Wu Zhang
Zhanjun Wang
Xiaobing Tang
Xiangfu Zou
Xuefeng Cui
Lin Xia
Taihong Xiao
Xing Chen
Lining Xing
Jian Xiong
Yi Xiong
Xiaoke Ma
Guoliang Xu
Bingxiang Xu
Jianhua Xu
Xin Xu
Xuan Xiao
Takayoshi Yamashita
Atsushi Yamashita
Yang Yang
Zhengyu Yang
Ronggen Yang
Xiao Yang
Zhengyu Yang

Yaolai Wang
Yaping Yang
Yue Chen
Yongchun Zuo
Bei Ye
Yifei Qi
Yifei Sun
Yinglei Song
Ying Ling
Ying Shen
Yingying Qu
Lvjiang Yin
Yiping Liu
Wenjie Yi
Jianwei Yang
Yu-Jun Zheng
Yonggang Lu
Yan Li
Yuannong Ye
Yong Chen
Yongquan zhou
Yong Zhang
Yuan Lin
Yuansheng Liu
Bin Yu
Fang Yu
Kumar Yugandhar
Liang Yu
Yumin Nie
Xu Yu
Yuyan Han
Yikuan Yu
Yong Wang
Ying Wu
Ying Xu
Zhiyong Wang
Shaofei Zang
Chengxin Zhang
Zehui Cao
Tao Zeng
Shuaifang Zhang
Yan Zhang
Liye Zhang
Zhang Qinhu
Sai Zhang
Sen Zhang

Shan Zhang
Shao Ling Zhang
Wen Zhang
Wei Zhao
Bao Zhao
Zheng Tian
Sijia Zheng
Zhenyu Xuan
Fangqing Zhao
Zhao Fangqing

Zhipeng Cai
Xing Zhou
Xiong-Hui Zhou
Lida Zhu
Ping Zhu
Qi Zhu
Zhong-Yuan Zhang
Ziding Zhang
Junfei Zhao
Zhe Li

Juan Zou
Quan Zou
Qian Zhu
Zunyan Xiong
Zeya Wang
Yatong Zhou
Shuyi Zhang
Zhongyi Zhou

Contents – Part III

Flower Species Recognition System Combining Object Detection and Attention Mechanism

Wei Qin[1($\boxtimes$)], Xue Cui[1], Chang-An Yuan[2], Xiao Qin[2], Li Shang[3], Zhi-Kai Huang[4], and Si-Zhe Wan[1]

[1] Tongji University, Shanghai, China
qinwei118114@163.com
[2] Science Computing and Intelligent Information Processing of GuangXi Higher Education Key Laboratory, Nanning Normal University, Nanning, Guangxi, China
[3] Department of Communication Technology, College of Electronic Information Engineering, Suzhou Vocational University, Jiangsu, China
[4] College of Mechanical and Electrical Engineering, Nanchang Institute of Technology, Nanchang, Jiangxi, China

Abstract. In this paper, a flower species recognition system combining object detection and Attention Mechanism is proposed. In order to strengthen the ability of the model to process images under complex backgrounds, we apply the method of object detection to locate flowers that we want to recognize. For less time of training, object detection and classification are Integrate into an end-to-end network stacked attention modules to generate attention-aware features.

Experiments are conducted on Flower 102, our method can recognize flower species against a complex background. With model owning attention module and transfer learning, we increase mAP from 73.8% to 74.7% and training time is reduced by about 15%.

Keywords: Flower species recognition · Object recognition · Attention mechanism

1 Introduction

In recent years, Flower Species Recognition enjoys great popularity in Mobile phone camera and related mobile application. However, the present model can only identify flower species in relatively simple images where the flower usually at the center of the photo with a single target object.

In previous work, feature engineering was widely used to extract features from images. Features such as color (HS, CbCrCg, and RGB), texture (grey level co-occurrence matrix) and shape (zero crossing rate) [5]. The downside of this approach is that the limit of low robustness of the model. People usually don't know how to select suitable features to hit the best performance of the model.

In recent years, Deep learning is developing rapidly. Many scholars apply the method of deep learning to complete the task. Gogul and Kumar [7] introduced a Deep learning approach using Convolutional Neural Networks (CNN) is to recognize flower

D.-S. Huang et al. (Eds.): ICIC 2019, LNAI 11645, pp. 1–8, 2019.
https://doi.org/10.1007/978-3-030-26766-7_1

species Guan [6] proposed a Residual Network to extract features. However, these model parameters are huge so it is hard to train with unbearable training time.

Chai [8] divided this task into two steps: segmentation and classification. The segmentation itself can be divided into three parts: first, then the original image is partitioned into small segments (superpixels); second, the segments are individually classified as foreground (flower) or background; in the final stage, the color information within the whole image is taken into account using GrabCut initialized from the classified region.

During the classification stage, a Bag-of-Words approach is used in combination with a kernelized SVM. The SVM consists of 1-vs-rest classifiers for each species. However, this network is not an end-to-end network.

In this paper, we proposed an end-to-end network. In order to implement the model the ability to process images against a complex background, we use an object detection mechanism like yolov2 to locate the flower we want to recognize. For less training time, we stack an attention module into the layer of feature extraction. With the help of the attention module, we can get more discriminative feature representation. Our model also exhibits the following appealing properties:

(1) Our model can process images against complex backgrounds. Our model is less demanding on picture quality, our model can recognize flower species in images with more objects and noise.
(2) Our mode can recognize several flowers in a single image. Furthermore, our model can give the location of the flower and the confidence in our prediction.
(3) Our model uses transfer training in the training process and attention module is stacked in our model which further reduce training time.

2 Object Detection

Object detection is a technology that deals with detecting instances of semantic objects of a certain class. Deep learning is widely used in this field such as R-CNN, Fast R-CNN, and Faster R-CNN. Compared with other deep learning methods, YOLOv2 is very fast at detecting objects, because it integrated bounding box prediction and classification into a single network. We also use a network based on Darknet-19 like YOLOv2 to get feature maps.

It is very difficult to predict the coordinates of bounding boxes directly. Faster R-CNN predicts bounding boxes with the help of hand-picked priors, Faster R-CNN using RPN predicts offsets for anchor boxes. We use a feature map to predict offsets instead of coordinates, which simplifies the problem and makes it easier for the network to learn. Instead of hand-priors, we use a method of k-mean clustering to find good priors like YOLOV2. Then, we get 5 anchors.

Each image is converted to 416 * 416. Our system divides the input image into an $S \times S$ grid. In Yolo, each bounding box consists of 5 predictions: x, y, w, h, and confidence. The (x, y) coordinates represent the center of the box relative to the bounds of the grid cell, w and h represent the width and height predicted relative to the whole image. The confidence prediction represents the IOU between the predicted box and any ground truth box. Redmon and Farhadi [2] pointed out that the most instability of the model comes from predicting the (x, y) locations for the bounding box. In our model, we use (t_x, t_y) instead of the (x, y) coordinates. The (x, y) coordinates are calculated by formula 1

$$
\begin{aligned}
x &= (t_x * w_a) - x_a \\
y &= (t_y * w_a) - y_a
\end{aligned}
\tag{1}
$$

(x_a, y_a) represents the center of an anchor box. (w_a, h_a) means the width and height of an anchor box.

We select 50 categories of Flower 102 and we manually label the image as our dataset Flower 50. In our model, each grid can generate B bounding box x, y, w, h. Our final prediction is an $S \times S \times (5 + C) \times B$ tensor. We use $S = 13$, $C = 50$, $B = 5$. Our final prediction is a $13 \times 13 \times 55 \times 5$ tensor. The network of feature extraction is shown in Table 1.

During training, we optimize the following multi-part loss function shown in formula 2.

$$
\begin{cases}
loss = \lambda_{obj}^{coord} \sum_i^{S^2} \sum_j^B l_{ij}^{responsible\_obj} (x_{ij}^{pred} - x_{ij}^{obj})^2 + (y_{ij}^{pred} - y_{ij}^{obj})^2 + (w_{ij}^{pred} - w_{ij}^{obj})^2 \\
\qquad\qquad\qquad\qquad\quad +(h_{ij}^{pred} - h_{ij}^{obj})^2 \\[4pt]
+ \lambda_{noobi}^{coord} \sum_i^{S^2} \sum_j^B l_{ij}^{no\_responsible\_obj} (x_{ij}^{pred} - x_{ij}^{anchor\_center})^2 + (y_{ij}^{pred} - y_{ij}^{anchor\_center})^2 \\
\qquad\qquad\qquad\qquad\quad +(w_{ij}^{pred} - w_{ij}^{anchor\_default})^2 + (h_{ij}^{pred} - h_{ij}^{anchor\_default})^2 \\[4pt]
+ \lambda_{obj}^{conf} \sum_i^{S^2} \sum_j^B l_{ij}^{responsible\_obj} \{conf_{ij}^{pred} - iou(box_{ij}^{pred}, box_{ij}^{truth})\}^2 \\[4pt]
+ \lambda_{noobj}^{conf} \sum_i^{S^2} \sum_j^B l_{ij}^{no\_responsible\_obj} \{conf_{ij}^{pred} - 0\}^2 \\[4pt]
+ \sum_i^{S^2} \sum_j^B l_{ij}^{responsible\_obj} \{p_{ij}^{pred}(c) - p_{ij}^{truth}(c)\}^2 \\[4pt]
x_{ij}^{anchor\_center} = y_{ij}^{anchor\_center} = 0.5 \\[4pt]
w_{ij}^{anchor\_default} = h_{ij}^{anchor\_default} = 1.0 \\[4pt]
\lambda_{obj}^{coord} = 1.0 \\[4pt]
\lambda_{obj}^{coof} = 5.0 \\[4pt]
\lambda_{noobj}^{conf} = \begin{cases} 0.0, & (iou > thresh) \\ 1.0, & (iou \leq thresh) \end{cases}
\end{cases}
\tag{2}
$$

3 Attention Module

In recent years, efforts are made to introduce an attention mechanism into a deep neural network. Attention mechanism has the ability to select a focused location and enhances different representations of objects at that location. We stack an attention module into the layer of feature extraction.

Table 1. The network of feature extraction. The network has 28 layers. Include 15 convolution layers, 5 pooling layer, and 1 residual module.

Type	Filters	Size/Stride	Input	Output
Convolutional	32	3 x 3/1	416 x 416 x 3	416 x 416 x 32
Maxpool		2 x 2/2	416 x 416 x 32	208 x 208 x 32
Convolutional	64	3 x 3/1	208 x 208 x 32	208 x 208 x 64
Maxpool		2 x 2/2	208 x 208 x 64	104 x 104 x 64
Convolutional	128	3 x 3/1	104 x 104 x 64	104 x 104 x 128
Convolutional	64	1 x 1/1	104 x 104 x 128	104 x 104 x 64
Convolutional	128	3 x 3/1	104 x 104 x 64	104 x 104 x 128
Maxpool		2 x 2/2	104 x 104 x 128	52 x 52 x 128
Convolutional	256	3 x 3/1	52 x 52 x 128	52 x 52 x 256
Convolutional	128	1 x 1/1	52 x 52 x 256	52 x 52 x 128
Convolutional	256	3 x 3/1	52 x 52 x 128	52 x 52 x 256
Maxpool		2 x 2/2	52 x 52 x 256	26 x 26 x 256
Convolutional	512	3 x 3/1	26 x 26 x 256	26 x 26 x 512
Convolutional	256	1 x 1/1	26 x 26 x 512	26 x 26 x 256
Convolutional	512	3 x 3/1	26 x 26 x 256	26 x 26 x 512
Convolutional	256	1 x 1/1	26 x 26 x 512	26 x 26 x 256
Convolutional	512	3 x 3/1	26 x 26 x 256	26 x 26 x 512
Maxpool		2 x 2/2	26 x 26 x 512	13 x 13 x 512
Convolutional	1024	3 x 3/1	13 x 13 x 512	13 x 13 x 1024
Convolutional	512	1 x 1/1	13 x 13 x 1024	13 x 13 x 512
Convolutional	1024	3 x 3/1	13 x 13 x 512	13 x 13 x 1024
Convolutional	512	1 x 1/1	13 x 13 x 1024	13 x 13 x 512
Convolutional	1024	3 x 3/1	13 x 13 x 512	13 x 13 x 1024
Convolutional	1024	3 x 3/1	13 x 13 x 1024	13 x 13 x 1024
Convolutional	1024	3 x 3/1	13 x 13 x 1024	13 x 13 x 1024
Route 16				
Convolutional	1024	3 x 3/1	13 x 13 x 2048	13 x 13 x 1024
Convolutional	275	1 x 1/1	13 x 13 x 1024	13 x 13 x 275

Each Attention Module has two branches: mask branch and trunk branch. The trunk branch is responsible to feature processing. The trunk layer in our model the 16th to the 26th layer of the network shown in Fig. 1. Given trunk-branch output $T(x)$ with input x, the mask branch uses bottom-up top-down structure to learn same size mask $M(x)$. The output is shown in formula 3.

$$H_{i,c}(x) = M_{i,c}(x) * T_{i,c}(x) \tag{3}$$

Where represent spatial positions and $c \in \{1, \ldots, C\}$ is the index of the channel.

However, according to previous research, ordinary Attention Modules, the attention mechanism will degrade the performance of the model. Two factors contribute to the problem. Firstly, the activation function of mask branch is sigmoid function and $T_{i,c}(x)$ is normalized between 0 and 1, which will weaken the features that we extracted.

Secondly, simple attention module will destruct the advantages of the original model. For example, because of the existing of Residual Network, neural networks can reach great depths because the residual network overcomes the problem of gradient disappearance. If we add the attention module roughly, the backpropagation gradient of a residual network will be influenced. To overcome the first disadvantage, we change the output of our feature extraction layer shown as formula 4.

$$H_{i,c}(x) = (M_{i,c}(x) + 1) * T_{i,c}(x) \qquad (4)$$

Experiments show that we achieved a better result. In order to solve the second problem, we design a mask branch of attention module as the frame of the residual network so that mask branch will not bring bad influence to the original model.

Figure 1 shows the structure diagram of the attention module.

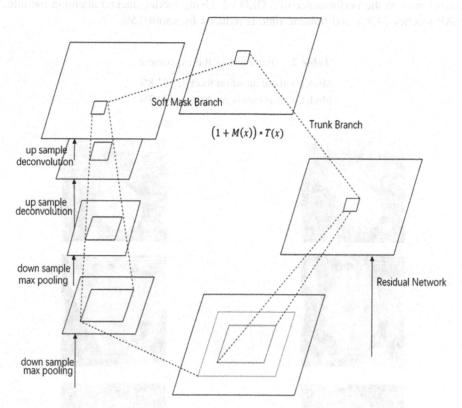

Fig. 1. The structure diagram of the attention module. Trunk branch is a residual network. Soft mask branch is consist of two down-sample layers and two up-sample layers. Down-sample layer is also a residual network. The dimension of the feature map is recovered by the method of deconvolution in the up-sample layer.

4   Experiment

4.1   Transfer Learning

Reasonable initialization strategy of model parameters can greatly accelerate the training speed of the model. Transfer learning is a method that transfers weights of a model into another model. Transfer learning can greatly reduce the training time and avoid the model falling into local optimization. In our model in the stage of training, we use the weights of Yolov2 to partly initialize the parameters of our model.

4.2   Results

Experiments show that attention mechanisms do work. As is shown in Table 2, we get 73.8% of mAP using the model without the attention module, which is basically consistent with the performance of YOLO v2. Using model stacked attention module, mAP reaches 74.8% and training time is reduced by about 15%.

Table 2. The results of the experiment

Model without attention module	73.8%
Model with attention module	74.7%

Fig. 2. The results of our system. The left image shows there types of images in our model from top to bottom: an image has a single target object, an image has two target objects, an image with more than two images. The right image is the output of our model.

The visualization results are shown in Fig. 2. From Fig. 2, we can see that our model can recognize flower species against the complex background and can identify several target objects at the same time. The accuracy of our model classification is very high.

5   Conclusion

We introduce an end-to-end network stacked attention module. Our model can identify several flowers in a single image and at the same time, our model can return the coordinates of the flowers. An effective attention module is used to enhance different representations of objects not destroying the performance of the original network.

In the experiment, our model is very fast in training and prediction. It still has some limitations:

Firstly, our model has a limited ability to recognize small objects. Secondly, as we can see in Fig. 2, our model has a strong ability to identify the front of a flower and weaker ability to identify the side.

For future work, we hope to further study the effect of attention mechanism to on YOLO so as to improve the accuracy and speed of object detection.

Acknowledgements. This work was supported by the grants of the National Science Foundation of China, Nos. 61672203, 61572447, 61772357, 31571364, 61861146002, 61520106006, 61772370, 61702371, 61672382, and 61732012, China Post-doctoral Science Foundation Grant, No. 2017M611619, and supported by "BAGUI Scholar" Program and the Scientific & Technological Base and Talent Special Program, GuiKe AD18126015 of the Guangxi Zhuang Autonomous Region of China.

References

1. Redmon, J., et al.: You Only Look Once: Unified Real-Time Object Detection (2015)
2. Redmon, J., Farhadi, A.: IEEE 2017 IEEE Conference on Computer Vision and Pattern Recognition (CVPR) - Honolulu, HI (2017.7.21-2017.7.26). In: 2017 IEEE Conference on Computer Vision and Pattern Recognition (CVPR) - YOLO9000: Better, Faster, Stronger, pp. 6517–6525 (2017)
3. Ren, S., et al.: Faster R-CNN: towards real-time object detection with region proposal networks. IEEE Trans. Pattern Anal. Mach. Intell. **39**(6), 1137–1149 (2017)
4. Wang, F., et al.: Residual Attention Network for Image Classification (2017)
5. Hong, K.S., Lee, H.H., Kim, J.H.: Mobile-based flower species recognition in the natural environment. Electron. Lett. **51**(11), 826–828 (2015)
6. Guan, Y.: Flower species recognition system based on residual network transfer learning. Comput. Eng. Appl. **55**(1), 174–179 (2019)
7. Gogul, I., Kumar, V.S.: Flower species recognition system using convolution neural networks and transfer learning. In: 2017 Fourth International Conference on Signal Processing, Communication and Networking (ICSCN). IEEE (2017)
8. Chai, Y.: Recognition between a large numbers of flower species. Masters' thesis at the Swiss Federal Institute of Technology Zurich (2011)

9. Mnih, V., Heess, N., Graves, A., Kavukcuoglu, K.: Recurrent models of visual attention. In: NIPS (2014)
10. Neubeck, A., Gool, L.J.V.: Efficient Non-Maximum Suppression (2006)
11. Huang, D.S.: Systematic Theory of Neural Networks for Pattern Recognition. Publishing House of Electronic Industry of China, May 1996. (in Chinese)
12. Huang, D.S., Du, J.-X.: A constructive hybrid structure optimization methodology for radial basis probabilistic neural networks. IEEE Trans. Neural Netw. 19(12), 2099–2115 (2008)
13. Huang, D.S.: Radial basis probabilistic neural networks: model and application. Int. J. Pattern Recognit. Artif. Intell. 13(7), 1083–1101 (1999)
14. Huang, D.S., Jiang, W.: A general CPL-AdS methodology for fixing dynamic parameters in dual environments. IEEE Trans. Syst. Man Cybern. Part B 42(5), 1489–1500 (2012)
15. Wang, X.F., Huang, D.S., Xu, H.: An efficient local Chan-Vese model for image segmentation. Pattern Recogn. 43(3), 603–618 (2010)
16. Wang, X.-F., Huang, D.S.: A novel density-based clustering framework by using level set method. IEEE Trans. Knowl. Data Eng. 21(11), 1515–1531 (2009)
17. Huang, D.S., Ip, H.H.S., Chi, Z.-R.: A neural root finder of polynomials based on root moments. Neural Comput. 16(8), 1721–1762 (2004)
18. Huang, D.S.: A constructive approach for finding arbitrary roots of polynomials by neural networks. IEEE Trans. Neural Netw. 15(2), 477–491 (2004)
19. Huang, D.S., Ip, H.H.S., Law, K.C.K., Chi, Z.: Zeroing polynomials using modified constrained neural network approach. IEEE Trans. Neural Netw. 16(3), 721–732 (2005)
20. Zhao, Z.Q., Glotin, H., Xie, Z., Gao, J., Wu, X.: Cooperative sparse representation in two opposite directions for semi-supervised image annotation. IEEE Trans. Image Process. (TIP) 21(9), 4218–4231 (2012)
21. Zhu, L., Huang, D.S.: Efficient optimally regularized discriminant analysis. Neurocomputing 117, 12–21 (2013)
22. Zhu, L., Huang, D.S.: A rayleigh–ritz style method for large-scale discriminant analysis. Pattern Recogn. 47(4), 1698–1708 (2014)

A Person-Following Shopping Support Robot Based on Human Pose Skeleton Data and LiDAR Sensor

Md. Matiqul Islam[(✉)], Antony Lam, Hisato Fukuda,
Yoshinori Kobayashi, and Yoshinori Kuno

Graduate School of Science and Engineering, Saitama University, Saitama, Japan
{matiqul,antonylam,fukuda,
kuno}@cv.ics.saitama-u.ac.jp,
yosinori@hci.ics.saitama-u.ac.jp

Abstract. In this paper, we address the problem of real-time human pose-based robust person tracking for a person following shopping support robot. We achieve this by cropping the target person's body from the image and then apply a color histogram matching algorithm for tracking a unique person. After tracking the person, we used an omnidirectional camera and ultrasonic sensor to find the target person's location and distance from the robot. When the target person is fixed in front of shopping shelves the robot finds the fixed distance between the robot and target person. In this situation our robot finds the target person's body movement orientation using our proposed methodology. According to the body orientation our robot assumes a suitable position so that the target person can easily put his shopping product in the basket. Our proposed system was verified in real time environments and it shows that our robot system is highly effective at following a given target person and provides proper support while shopping the target person.

Keywords: Ultrasonic sensor · Histogram matching ·
Omnidirectional camera · Person following

1 Introduction

In the field of computer vision, research in human detection and tracking is a challenging area with decades of efforts. This area has numerous applications in robot vision [1]. A prime example is in human following robots, where such types of robots must localize the walking person and avoid obstacles. These human following robots also are used in many applications such as wheel-chairs [2] that carry belongings when travelling, automated shopping cart robots [3], nurse following carts [4] and a line following intelligent shopping cart controlled by a smartphone that can guide the cart to the locations of the wanted items to be purchased [5].

Our goal of this research is to track a target person based on a pose skeleton model and a color histogram matching algorithm and compute the target person's body movement orientation for shopping support. Specifically, we use the Open-Pose model [6] implemented on a 360^0 camera and crop the body image of target person using the

© Springer Nature Switzerland AG 2019
D.-S. Huang et al. (Eds.): ICIC 2019, LNAI 11645, pp. 9–19, 2019.
https://doi.org/10.1007/978-3-030-26766-7_2

left shoulder, right shoulder, left hip and right hip key points, then further apply a color histogram matching algorithm in each frame to track the target person. After tracking the target person, we take the left or right ankle key points and calculate the angle value ($0°$ to $270°$) of the person's position in the image frame according to Fig. 4. When we get the angle value of person's position, we take the rotation and translation value of the robot using LiDAR (Hokuyo UTM-30LX) sensors. Figure 1 shows our person following mini cart robot.

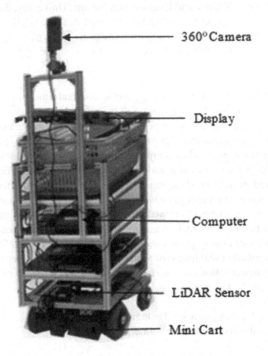

Fig. 1. Our person following mini cart robot.

2 Related Work

Person-tracking is currently an active research topic in robotics and these person-trackers are typically used for a robot to follow the person being tracked. Camera face, color blob detection or contour tracking are common methods of tracking people [7]. Other blob tracking systems have used Kalman filter prediction to track positions of blobs between frames [8] and features such as adaptive templates update each new frame [9]. Now-a-days, many researchers combine both camera-based and laser-based tracking methods. In particular, many researchers will use a laser to find legs and a camera either to detect faces [10] or to track other visual regions of interest [11]. For face detection, the person always needs be in a face-to-face position to the robot. This is inconvenient for the robot to follow from behind and or even move next to a person.

Other methods for person tracking require attaching a tracking device (sensors) to the person [12]. These sensors provide accurate location information for the robot to follow behind the person. However, the need for the person to wear sensors limits applications.

Color based tracking is one of the most simple and robust methods of object tracking. Many algorithms exist for color-based object tracking. Histogram intersection techniques for foreground extraction [13] use color histograms as the main tracking tool. In [14], the authors apply color and edge histograms that are compared using Chi-squared and Intersection histogram comparison measures. The Earth Movers Distance (EMD) of color histograms is computed for object tracking in [15]. The weighted combination of more than one feature histogram was applied for tracking in [16]. Histogram comparison based on Euclidian distance along with region correspondence was developed in [17].

In this paper the system proposed is a shopping support robot that implements a person following behavior. The system is robust and simple specially for older people and it represents a relatively low cost compared to other approaches. To follow the target person, he or she does not need to wear any extra sensor to follow him or her.

3   Design Approach

The major issue of person-following is how to localize the target person by tracking. A common approach is to localize the target person over time and identify them based on motion continuity. Figure 2 shows our system model. In this model, we first detect the person's pose using a 360^0 camera. This wide-angle camera allows us to detect different people for all 360^0 viewing angles. Among these detected people, we apply the color histogram intersection algorithm to track the target person. After we get the target person, we record that person's ankle x coordinate points. Using these points, we calculate the angle value (0^0 to 270^0) of the person's position in the image frame according to Fig. 4. By this angle and LiDAR sensor data, we finally calculate the robot rotation and translation values to follow the target person.

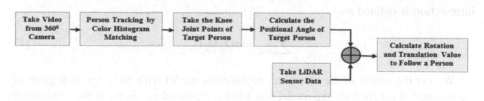

Fig. 2. Block diagram of person following procedure.

Our paper is organized as follows: Sect. 3.1 describes the person tracking procedure. Section 3.2 describes how we calculate the person's positional angle value. In Sect. 3.3, we describe the procedure for following the person based on calculated rotation angles and translation values. Section 3.4 illustrates and finds the target person's body orientation angle. In Sect. 4, the results and discussion are analyzed.

3.1   Person Tracking

In our system, we use the histogram intersection algorithm to track the target person. From the human pose skeleton data, we crop the body using the left shoulder, right shoulder, left hip and right hip key points as shown in Fig. 3. Then in each frame we compare each person's body image to the cropped image using the histogram intersection algorithm.

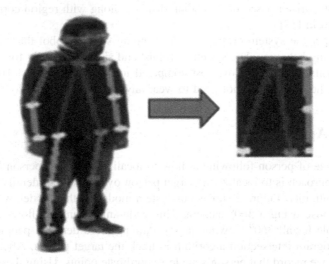

Fig. 3. Cropped image from skeletal key points information

The histogram intersection algorithm was proposed by Swain and Ballard in [18]. Here the author considered two histograms I and M, each containing n bins, and the intersection is defined as,

$$\sum_{j=1}^{n} min(I_j, M_j) \tag{1}$$

We get the output of the histogram intersection model with an image histogram as the number of pixels from the model that have corresponding pixels of the same color in the image. The match value is given by,

$$\frac{\sum_{j=1}^{n} min(I_j, M_j)}{\sum_{j=1}^{n} M_j} \tag{2}$$

When an unknown object image is given as input, we compute the histogram intersection for all the stored models, the highest value is the best match.

3.2 Positional Angle Value of Tracked Person

We calculate the positional angle value of the tracked person based on the 360^0 wide angle camera. However, in practice, we find it sufficient to calibrate the total width of the image to be from 0^0 to 270^0 as shown in Fig. 4. Given the width of the image is W and the x coordinate value, A then we can easily calculate the ankle angle value by the following equation,

$$AnkleAngle = \frac{270}{W} * (W - A) \tag{3}$$

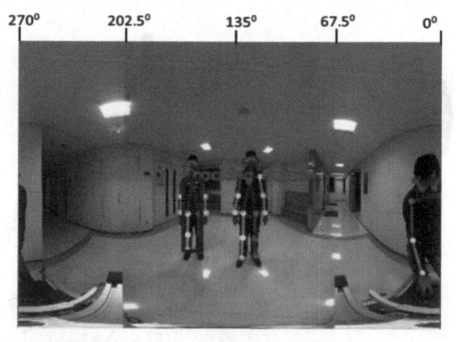

Fig. 4. Calibrating the width of the image 0^0 to 270^0.

3.3 Calculate the Rotation and Translation Value

Figure 5 shows how we calculate the rotation and translation values of our mini cart robot. The whole procedure is given below,

Step 1: Track the target person among multiple people using the histogram intersection algorithm described in Sect. 3.1

Step 2: Take the tracked person's ankle x coordinate value.

Step 3: Calculate the tracked person's ankle angle using Eq. (3) as described in Sect. 3.2

Step 4: If the person's angle value is not in the range $(100–135)^0$ of the LiDAR sensor we calculate the rotation angle θ needed to be in the range (100–135), as illustrated by the example in Fig. 5

Step 5: After rotation of our mini cart at the angle θ we calculate the distance between the tracked person's ankle points to the mini cart robot. We also set the threshold value such that if the distance is greater than $1m$, we operate the translation operation. We calculate the distance using the Euclidean distance as,

$$Distance = \sqrt{(0-x)^2 + (0-y)^2} \qquad (4)$$

Where (x, y) is the coordinate value of the left or right ankle from the sensor output. Using the above procedure, the robot follows its target person effectively.

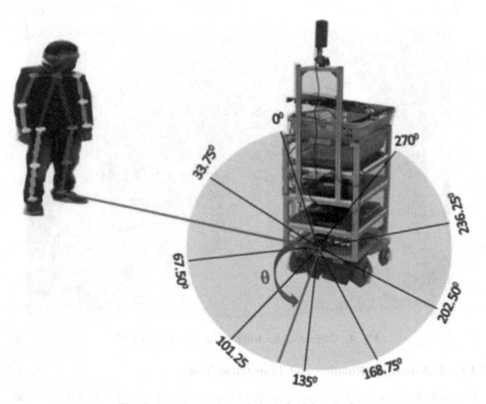

Fig. 5. Rotation and translation of mini cart robot.

3.4   Body Orientation Angle of Target Person

In this paper, the body orientation of the target person is estimated in eight directions as shown in Fig. 6(b). We take four angle values of the target person to predict body orientations. As for example to calculate the angle of EAC as shown in Table 1, we use the following formula,

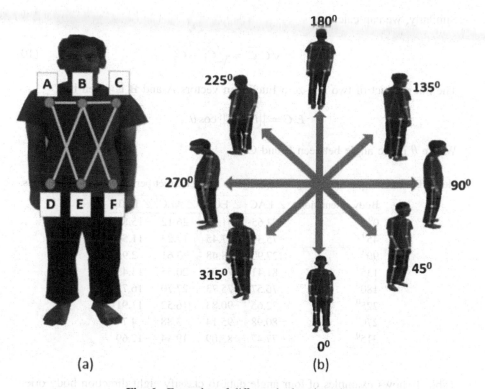

(a)   (b)

Fig. 6. Examples of different body orientations

First, we generate the vector E and C as,

$$E = EA = (e1 - a1, e2 - a2) = [E1, E2] \qquad (5)$$

$$C = CA = (c1 - a1, c2 - a2) = [C1, C2] \qquad (6)$$

The dot product of vector E and C is defined as,

$$E.C = E1C1 + E2C2 \qquad (7)$$

The magnitude of a vector E is denoted by $\|E\|$. The dot product of vector A with itself is

$$E.E = \|E\|^2 = E_1^2 + E_2^2 \qquad (8)$$

Which gives,

$$\|E\| = \sqrt{E.E} = \sqrt{E_1^2 + E_2^2} \qquad (9)$$

Similarly, we can calculate

$$\|C\| = \sqrt{C.C} = \sqrt{C_1^2 + C_2^2} \qquad (10)$$

The dot product of two non-zero Euclidean vectors A and B is given by

$$E.C = \|E\|\|C\| \cos \theta \qquad (11)$$

Where θ is the angle between E and C.

Table 1. Different angles according to Fig. 6(a) to predict target person's body orientations.

Body orientations	$\angle$ EAC	$\angle$ ECA	$\angle$ AEC	$\angle$ DBF
0^0	−72.65	81.22	26.12	−15.83
45^0	−75.31	87.45	17.23	−11.50
90^0	127.90	−48.48	−3.61	2.95
135^0	81.41	−78.05	−20.55	14.45
180^0	76.57	−75.73	−27.70	16.72
225^0	72.65	−90.83	−16.52	11.91
270^0	−80.98	95.14	3.88	−4.74
315^0	−77.47	83.09	19.44	−12.69

Table 1 shows examples of four angle data to classify eight direction body orientations. We take different person's four angle data according to Fig. 6(a) and make a dataset for SVM classifier to classify person's body orientations.

4   Results and Discussion

Figure 7 shows our histogram intersection algorithm-based person tracking results in different frames. Using this tracking result, we take the tracked person's left and right ankle points and calculate his or her positional angle value.

Our LiDAR sensor (Hokuyo UTM-30LX) can estimate the distance by calculating the phase difference with in a range of 30 m. The LiDAR sensor gives us 1080 distance points. We calibrate these 1080 distance points to the $(0–270)^0$ range. According to Eq. 4 we take the target person's ankle angle and find the distance between robot and target person.

In Fig. 8 the red dots give the distance per degree. The blue and green solid circles are the plot from the tracked person's left and right ankle angle using the 360^0 camera. The black square shows the actual blob of the left and right ankle angle detected by the sensor. We see that both solid circles and the detected black square are very close.

Table 2 shows the person's positional left ankle angle value using the LiDAR sensor and 360^0 camera. We see that the detected angles of the LiDAR sensor and 360^0 camera are almost the same and varies only by $(2–4)^0$ which is an acceptable amount of variation. The cause of this variation is mainly due to the variation of detected left ankle skeleton data.

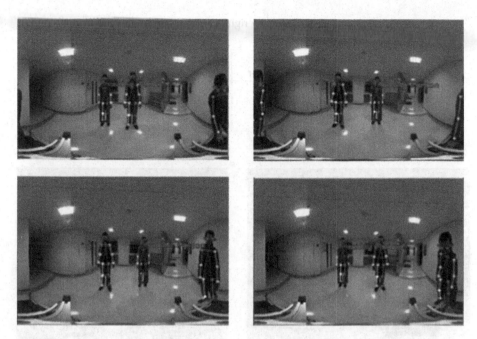

Fig. 7. Result of tracking a person.

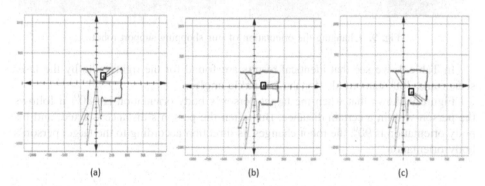

| (a) | (b) | (c) |

Fig. 8. The LiDAR sensor output for calculating the target person's positional angle and corresponding distance. (Color figure online)

Table 2. Positional angle value of the left ankle by laser sensor and 360^0 camera.

Fig. no.	Left ankle angle detected by laser sensor	Left ankle angle detected by 360^0 camera	Error
Fig. 8(a)	167^0	165^0	2^0
Fig. 8(b)	136^0	138^0	2^0
Fig. 8(c)	107^0	111^0	4^0

Table 3. Positional angle value of right ankle by laser sensor and 360^0 camera.

Fig. no.	Right ankle angle detected by laser sensor	Right ankle angle detected by 360^0 camera	Error
Fig. 8(a)	169^0	167^0	2^0
Fig. 8(b)	140^0	138^0	2^0
Fig. 8(c)	111^0	115^0	4^0

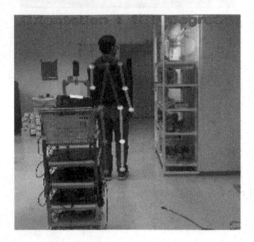 

Fig. 9. Changing the orientation of our shopping support robot.

In Table 3, we see that identical results are found for the right ankle by the laser sensor and 360^0 camera like for the left ankle angles.

Figure 9 shows that when the target person's body orientation is 180^0, it follows the person with-in a certain distance. But when the person's distance is fixed, and his body orientation is 90^0, the robot changes its direction according to the target person's body orientation.

5   Conclusions and Future Work

A person following shopping cart was developed and tested. We developed the system using human pose data, a wide-angle camera, color histogram matching algorithm and 2D LiDAR sensor. The best applications for this type of robot is in warehouses, shopping malls, airports, hospitals and more. Its performance is dependent on accurate person tracking and accurately identified human key point detections. In the future, we will continue to improve the accuracy of our tracking algorithm so that the robot can follow a target person in settings such as a shopping mall so that the robot can carry the target person's goods and recognize the target person's shopping behaviors.

Acknowledgement. This work was supported by JSPS KAKENHI Grant Number JP26240038.

References

1. Moeslund, T.B., Hilton, A., Krüger, V., Sigal, L.: Visual Analysis of Humans. Springer, London (2011). https://doi.org/10.1007/978-0-85729-997-0
2. Ahmad, M.F., Alhady, S.S.N., Kaharuddin, S., Othman, W.A.F.W.: Visual based sensor cart follower for wheelchair by using microcontroller. In: ICCSCE (2015)
3. Rawashdeh, N.A., Haddad, R.M., Jadallah, O.A., To'ma, A.E.: A person-following robotic cart controlled via a smartphone application: design and evaluation. In: REM (2017)
4. Ilias, B., Nagarajan, R., Murugappan, M., Helmy, K., Awang Omar, A.S., Abdul Rahman, M.A.: Hospital nurse following robot: hardware development and sensor integration. Int. J. Med. Eng. Inf. **6**(1), 1–13 (2014)
5. Ng, Y.L., Lim, C.S., Danapalasingam, K.A., Tan, M.L.P., Tan, C.W.: Automatic human guided shopping trolley with smart shopping system. Jurnal Teknologi **73**(3) (2015)
6. Cao, Z., Simon, T., Wei, S.E., Sheikh, Y.: Realtime multi-person 2D pose estimation using part affinity fields. In: CVPR (2017)
7. Montemerlo, M., Thrun, S., Whittaker, W.: Conditional particle filters for simultaneous mobile robot localization and people-tracking. In: Proceedings 2002 IEEE International Conference on Robotics and Automation (Cat. No. 02CH37292) (2002)
8. Grimson, W.E.L., Stauffer, C., Romano, R., Lee, L.: Using adaptive tracking to classify and monitor activities in a site. In: Proceedings of the 1998 IEEE Computer Society Conference on Computer Vision and Pattern Recognition (Cat. No. 98CB36231) (1998)
9. Beymer, D.J.: Real-time tracking of multiple people using stereo. In: FRAME-RATE: Frame-rate Applications, Methods and Experiences with Regularly Available Technology and Equipment (1999)
10. Kleinehagenbrock, M., Lang, S., Fritsch, J., Lomker, F., Fink, G.A., Sagerer, G.: Person tracking with a mobile robot based on multi-modal anchoring. In: Proceedings. 11th IEEE International Workshop on Robot and Human Interactive Communication (2002)
11. Kobilarov, M., Sukhatme, G., Hyams, J., Batavia, P.: People tracking and following with mobile robot using an omnidirectional camera and a laser. In: International Conference on Robotics and Automation (2006)
12. Bianco, R., Caretti, M., Nolfi, S.: Developing a robot able to follow a human target in a domestic environment. In: Proceedings of the First RoboCare Workshop (2003)
13. Lu, W., Tan, Y.P.: A color histogram-based people tracking system. In: International Symposium on Circuits and Systems (ISCAS) (2001)
14. Mason, M., Duric, Z.: Using histograms to detect and track objects in color video. In: Applied Imagery Pattern Recognition Workshop (AIPR) (2001)
15. Wojtaszek, D., Laganiere, R.: Using color histograms to recognize people in real time visual surveillance. In: International Conference on Multimedia, Internet and Video Technologies (2002)
16. Bajramovic, F., Deutsch, B., Grabl, C., Denzler, J.: Efficient adaptive combination of histograms for real-time tracking. EURASIP J. Image Video Process. **2008** (2008). Article ID 528297
17. Fang, Y., Wang, H., Mao, S., Wu, X.: Multi-object tracking based on region corresponding and improved color-histogram matching. In: International Symposium on Signal Processing and Information Technology (2007)
18. Swain, M.J., Ballard, D.H.: Color indexing. Int. J. Comput. Vis. **7**(1), 11–32 (1991)

A New Hybrid Calibration Method for Robot Manipulators by Combining Model–Based Identification Technique and a Radial Basis Function–Based Error Compensation

Phu-Nguyen Le[1] and Hee-Jun Kang[2(✉)]

[1] Graduate School of Electrical Engineering, University of Ulsan,
Ulsan 680-749, South Korea
phunguyen07dl@gmail.com
[2] School of Electrical Engineering, University of Ulsan,
Ulsan 680-749, South Korea
hjkang@ulsanac.kr

Abstract. Though the kinematic parameters had been well identified, there are still existing some non-negligible non-geometric error sources such as friction, gear backlash, gear transmission, temperature variation etc. They need to be eliminated to further improve the accuracy of the robotic system. In this paper, a new hybrid calibration method for improving the absolute positioning accuracy of robot manipulators is proposed. The geometric errors and joint deflection errors are simultaneously calibrated by robot model identification technique and a radial basis function neural network is applied for compensating the robot positions errors, which are caused by the non-geometric error sources. A real implementation was performed with Hyundai HH800 robot and a laser tracker to demonstrate the effectiveness of the proposed method.

Keywords: Radial basis function · Robot calibration · Robot accuracy

1 Introduction

In the offline programming (OLP), the kinematic model parameters are used to achieve a given Cartesian position of the arm's tip. However, a nominal geometric model does not include the errors arising in manufacturing or assembly. The model-based robotic calibration is developed based on error model which represents the relationship between the geometric parameter error and the end effector positioning error [1, 2]. A major advantage of using a model-based method is that it is possible to calibrate the model parameters accurately, so long as the error model properly represents the error sources. Many works have addressed the modeling of the error sources of robots for calibration, such as the Denair–Hardenberg (DH) model [3–5], the zero-reference position model [6], the complete and parametrically continuous (CPC) model [7, 8], and the product of exponentials (POE) error models [9–11]. Some errors can be modeled easily and correctly, for example, the robot link geometric errors and the robot joint compliance are easily modeled [12–14]. However, it is virtually impossible to consider all the sources contributing to the positioning errors while forming the

© Springer Nature Switzerland AG 2019
D.-S. Huang et al. (Eds.): ICIC 2019, LNAI 11645, pp. 20–31, 2019.
https://doi.org/10.1007/978-3-030-26766-7_3

kinematic identification model of a robot manipulator. Recently, non-geometric calibration has been developed the extended Kalman filter (EKF) algorithm [15], genetic algorithm [16], maximum likelihood estimation [17, 18], particle filter (PF) algorithm [19], genetic algorithm [18], particle swarm optimization [20, 21], and radial basis function (RBF) [22, 23]. Although, these methods were successful used in robot calibration, they do not supply knowledge of error sources in robot structure, has slow convergence, and the calibration space is limited.

In this study, a new calibration method is proposed to combine the model–based identification technique and a radial basis function–based error compensation. At first, model-based robotic calibration is used to simultaneous identify the kinematic parameters and joint stiffness parameters. This work is the extension to closed chain manipulator for the previous work [12]. Then, in order to compensate the errors that causes by the un-model sources (such as friction, gear backlash, gear transmission, temperature variation and so on), the radial basis function is applied for compensating the residual position errors [24–26] by describing the relationship of end-effector positions and corresponding position errors. Finally, experimental studies on an industrial Hyundai HH800 robot with carrying heavy loads are performed to demonstrate the effectiveness and correctness of the method.

This paper is organized in six sections as below. Following this introduction, the kinematic model of the HH 800 robot is described in Sect. 2. Section 3 present the simultaneous identification of joint stiffness and kinematic parameters. Section 4 constructs RBF to compensate for the robot position errors. Section 5 describes experimental calibration and validation results. Finally, some conclusions are presented in Sect. 6.

2 Kinematic Model of the HH800 Robot

The 6 degrees of freedom (dof) HH800 robot, which is described in Fig. 1, consists of a main open kinematic chain (6 dof) and a closed-loop mechanism (2 dof). The open chain is connected by the revolute joints 1, 2, 3p, 4, 5, and 6. The closed-loop PQRS is connected by the revolute joints 2, 3, Q, R, and S (both joints' axes 2 and 3 are coincident at point P). The frames are fixed at the open chain's links based on the Denavit–Hartenberg (D-H) convention.26 The nominal D-H parameters of the open chain and the closed loop are shown in Table 1.

The homogeneous transformation matrix from the robot base frame to end effector frame can be computed by:

$$ {}_E^0T = {}_1^0T(\theta_1){}_2^1T(\theta_2){}_{3p}^2T(\theta_{3p}){}_4^{3p}T(\theta_4){}_5^4T(\theta_5){}_6^5T(\theta_6){}_E^6T \tag{1} $$

The calculation process of transformation matrixs between the tool frame and the frame 6 are shown in Eq. 2:

$$ {}_T^6T = Tr_X(a_6)Tr_Y(b_6)Tr_Z(d_7) \tag{2} $$

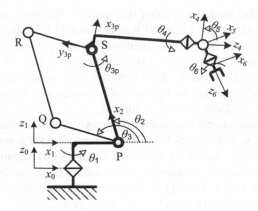

Fig. 1. Kinematic model of Hyundai robot HH800.

The passive joint θ_{3p} is obtained by solving a system of constraint equations with the given input joint angles θ_2 and θ_3. The closed loop actuating mechanism PQRS can be considered as a planar mechanism as follows:

$$a_2 c\theta_2 + L_3 c\theta_{2,3p} - L_5 c\theta_3 - L_4 c\theta_{3,4p} = 0$$
$$a_2 s\theta_2 + L_3 s\theta_{2,3p'} - L_5 s\theta_3 - L_4 s\theta_{3,4p} = 0 \tag{3}$$

The closed-chain is a parallelogram, therefore, $a_2 = L_4$, $L_3 = L_5$, and the passive joint position θ_{3p} can be solved from (2) as follows $\theta_{3p} = \theta_{3p'} - \delta$, where $\theta_{3p'} = \theta_3 - \theta_2$ and $\delta = 90^0$ is a constant angle.

Table 1. Nominal D-H parameters of Hyundai robot HH800.

i	$\alpha_{i-1}(deg)$	$a_{i-1}(m)$	$\beta_{i-1}(deg)$	$b_{i-1}(m)$	$d_i(m)$	$\theta_i(deg)$
D-H parameters of the main open chain						
1	0	0	0	0	0	θ_1
2	90	0.515	–	–	0	θ_2
3	0	1.6	0	–	0	θ_3
4	90	0.35	–	–	1.9	θ_4
5	–90	0	–	–	0	θ_5
6	90	0	–	–	0.445	θ_6
T	–	−0.45	–	0.11	0.930	–

Link lengths of the closed loop					
$L_5(m)$	0.8	$L_4(m)$	1.6	$L_3(m)$	0.8

3 Robot Joint Compliance Model

3.1 Computing Effective Torques Applied on Robot Joints

The total effective torque in joint i-th is:

$$\tau_i = \sum_{j=i}^{n+1} \tau_{i,j} = \sum_{j=i}^{n+1} J_{\theta_{i,j}}{}^T F_j \tag{4}$$

$F_j = [0 \quad 0 \quad -M_j g]^T$ is a gravity force vector of link j, M_j is the link mass, $g = 9.81 \frac{m}{s^2}$ is the gravity of Earth, n is the number of degree of freedom (DOF) of the robot, and F_{n+1} is the gravity force vector (or external force) related to the payload. The transpose of the Jacobian matrix, $J_{\theta_{i,j}}$ is the relationship between forces and torques at the equilibrium point.

$$J_{\theta_{i,j}} = [z_i \times P_{i,j}] \tag{5}$$

P_{ij} is the 3×1 vector between the origin of the frame $\{i\}$ and the mass center of link j-th. The effective torques through the closed-chain mechanism is given [12]

$$\begin{bmatrix} \tau_1 \\ \tau_2 \\ \tau_3 \\ \tau_4 \\ \tau_5 \\ \tau_6 \end{bmatrix} = \begin{bmatrix} 1 & 0 & 0 & 0 & 0 & 0 \\ 0 & 1 & 0 & 0 & 0 & 0 \\ 0 & J_{\theta_2}^{3p} & J_{\theta_3}^{3p} & 0 & 0 & 0 \\ 0 & 0 & 0 & 1 & 0 & 0 \\ 0 & 0 & 0 & 0 & 1 & 0 \\ 0 & 0 & 0 & 0 & 0 & 1 \\ 1 & 0 & 0 & 0 & 0 & 0 \\ 0 & 0 & 1 & 0 & 0 & 0 \\ 0 & J_{\theta_2}^{4p} & J_{\theta_3}^{4p} & 0 & 0 & 0 \end{bmatrix} \begin{bmatrix} \tau_{1,1} \\ \tau_2 \\ \tau_{3p} \\ \tau_4 \\ \tau_5 \\ \tau_6 \\ \tau_{1,2} \\ \tau_3 \\ \tau_{4p} \end{bmatrix} \tag{6}$$

3.2 Simultaneous Joint Stiffness and Kinematic Parameter

Assuming the robot joint can be modeled as a linear torsional spring, joint stiffness value of the ith joint is represented by a constant value k_i. We also assume the link of robot is much stiffer than its actuated joint, and most compliance errors are due to the flexibility of robot joints under the link self-gravity and external payload. The deflection of the ith joint due to the effective torques can be obtained by:

$$\Delta\theta_{c_i} = \frac{\tau_i}{k_i} = \tau_i c_i \tag{7}$$

The Cartesian position errors due to small joint deflections can be modeled as:

$$\Delta P_c = J_\theta \Delta\theta_c = (J_\theta \tau)C \tag{8}$$

where $C = [\begin{array}{ccc} c_1 & c_2 & \cdots & c_n \end{array}]^T$ is the joint compliance vector, $\Delta\theta_c = [\Delta\theta_{c_1} \ \Delta\theta_{c_2} \ \ldots$ $\Delta\theta_{c_n}]$ is the joint deflection vector and $\tau = diag(\tau_1, \tau_2, \cdots, \tau_n)$ is the effective torque in the robot joints at the balance position. The real position vector of the robot end-effector can be expressed as:

$$P_{real} = P_{kin} + \Delta P_{kin} + \Delta P_c + \Delta P_{extra} \tag{9}$$

where P_{kin} is the results of forward kinematics based on the current kinematic parameters. ΔP_{kin} is position errors due to kinematic parameter error, ΔP_c is position errors due to joint deflections, and ΔP_{extra} is position errors due to other sources. Because of the error caused by the other sources (such as friction, gear backlash, gear transmission, temperature variation and so on) we cannot model correct and completely. Therefore, in this subsection, the error model or kinematic and stiffness parameters will be considered first, and the compensation for other sources of errors will be presented in the next section. Zhou [12] presented a method that can be used for simultaneous identification of joint stiffness and kinematic parameters. The comprehensive error model of kinematic and stiffness parameters can be expressed as:

$$\begin{aligned}
\Delta X &= \Delta P_{kin} + \Delta P_{stiffness} \\
&= J\Delta\phi + J_\theta\Delta\theta_{stiffness} \\
&= J\Delta\phi + J_\theta C\tau \\
&= [\begin{array}{cc} J & J_\theta\tau \end{array}] \begin{bmatrix} \Delta\phi \\ C \end{bmatrix} = J_\Phi\Delta\Phi
\end{aligned} \tag{10}$$

Thus, kinematic parameters and joint compliances can be identified simultaneously with utilizing the iterative least-square method

4   Radial Basis Function Error Compensation Technique

After the simultaneous calibration of both the link geometry and the joint compliance error is applied, the radial basis function (RBF) is employed for the additional compensation of the existing position errors. The error compensation method with RBF can be stated as follows.

Given a set of N different points x_i and N real numbers, y_i, construct a smooth function F satisfying the following two equations:

$$F(x_i) = y_i, i = 1, 2, 3, \ldots, N, \tag{11}$$

$$F(p) = \sum_{i=1}^{N} \emptyset(||w_i - p||b) \tag{12}$$

where W_i is weight vector of the node i, p is the input vector and b is the bias. The function $\emptyset$ is a basis function that is continuous:

$$\emptyset(||w_i - p||) = \exp\left(-||w_i - p||^2\right) \qquad (13)$$

It is proven that the RBF networks could have the capability of universal approximation [26]. The radial basis function is applied for compensating the residual position errors [24–26] by describing the relationship of end-effector positions and corresponding position errors.

In this study, a RBF that has three layers is constructed as in Fig. 2. The input layer consists of six nodes. These nodes represent six robot joint positions $[\theta i]$, $i = 1, \ldots, 6$. The hidden layer consists of 20 nodes. The output layer consists of three nodes representing three elements of a robot position error vector.

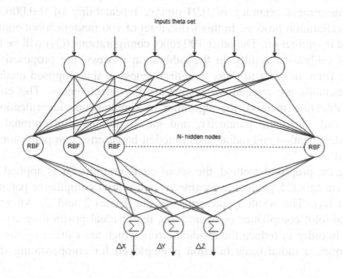

Fig. 2. Radial basis function network employed for error compensation

The correction process for the robot position is depicted as follow. First, the comprehensive error model of kinematic and stiffness parameters is identified by Eq. (10). After that the residual robot position errors after compensation for link geometry and joint compliance errors are used for training the RBF. The residual position errors are supposedly caused by other un-modeled non-geometric error sources such as robot link deformation and gear backlash. The modified robot model and the trained RBF are used to compute the robot end-effector positions for a given robot joint position reading.

5 Experiment and Validation Results

In this section, the proposed calibration method is applied to the experimental calibration of a 6 DOF closed chain Hyundai robot HH800. In addition, the calibration results are also validated at the other robot configurations that are not used in calibration. Furthermore, to demonstrate the effectiveness of the proposed method, the results of both calibration and validation are compared with those obtained by the conventional kinematic calibration, the algorithm of simultaneous identification of joint compliance and kinematic parameter, and the combining conventional kinematic parameters identification and radial basis function-based error compensation technique.

5.1 Experimental Calibration Results

A set of 200 robot configurations (moves overall workspace) of the Hyundai robot HH800 carrying a heavy load (745 kg) are measured by a laser tracker (API Laser Tracker, measurement accuracy of 0.01 mm/m, repeatability of ± 0.006 mm/m) to perform the calibration process. In this work, a set of 100 random robot configurations (Q_1) are used in calibration. The other 100 robot configurations (Q_2) will be used in the validation of calibration results. In the calibration process, the proposed method is applied first. Then, in order to show the effectiveness of the proposed methods, some calibration methods are applied for comparing the effectiveness. The conventional kinematic calibration method, the algorithm of simultaneous identification of joint compliance and kinematic parameter, and the combining conventional kinematic parameters identification and radial basis function-based error compensation technique are employed.

By using the proposed method, the set of configurations Q_1 is applied for identifying 29 parameters (25 geometric parameters and 4 joint compliance parameters, K_2, K_3, K_4, and K_5). The result is presented in the Tables 2 and 3. After robot link geometry and joint compliance compensation, the residual positioning errors are still fairly large. In order to reduce the residual errors which are caused by the other non-kinematic errors, a radial basis function is employed for compensating the residual errors.

For comparison, the Q_1 is also apply for conventional kinematic calibration. This method is well-studied and presented (3, 4, 5). Also, the algorithm of simultaneous identification of joint compliance and kinematic parameter (31) is used. At the end, in comparing to the proposed method, a radial basis function is employed for compensating the residual errors after the conventional kinematic calibration. The calibration result is shown in Tables 2, 3 and Fig. 2. By applying the simultaneous calibration geometric and joint stiffness parameter method, the kinematic parameters together with 4 joint values are identified and shown in Tables 2 and 3 respectively. The result of the proposed method and the others are shown in the Table 4. A RBF is generated based on the residual position errors (the output) and the robot joint position data (input). The Matlab RBF toolbox (newrb) is employed for training the network. The goal is set to 0.4 for overcoming the overfitting problem. After the training, the result is a RBF network with 20 hidden nodes.

For the combining conventional kinematic parameters identification and radial basis function-based error compensation technique. The residual position errors after conventional calibration method are used at the output and the robot joint position data is the input, the result is another RBF network with 20 hidden nodes.

The calibration result shows that the proposed calibration method has the ability to reduce the mean of error to 0.2612 (mm) from 4.0654 mm (before calibration) (performance increase 93.57%). The maximum error is 0.7661 (mm) and standard deviation is 0.142 (mm), before calibration, the maximum error is 6.3291 (mm) (performance increase 87.89%). It is easy to see from the Table 4 that the proposed method had advantages over the others method. For conventional kinematic calibration method, it reduces the mean error 68.14% (from 0.8199 mm to 0.2612 mm) and the maximum error 78.45% (from 3.5552 mm to 0.7661 mm). The proposed method had advantage of compensating the non-geometric error and has the better result in comparing to the algorithm of simultaneous identification of joint compliance and kinematic parameters. It reduces the mean error 63.97% (from 0.7250 mm to 0.2612 mm) and the maximum error 70.16% (from 2.5671 mm to 0.7661 mm). At last, the proposed method had advantage of well identified the geometric and joints compliance parameters over the combining conventional kinematic parameters identification and radial basis function-based error compensation technique. As a result, the proposed method has a better performance. It reduces the mean error 20.8% (from 0.3298 mm to 0.2612 mm) and the maximum error 19.03% (from 0.9462 mm to 0.7661 mm). The result also shows that the proposed method has the better standard deviation.

Table 2. Stiffness identification (without measurement noise).

	K_2	K_3	K_4	K_5
Simultaneous calibration	6.1590 * 10^7	4.388 * 10^6	3.151 * 10^6	2.220 * 10^6

Table 3. Identified D-H parameters of HH800 robot ("–": unavailable, "X": un-identifiable).

Identified D-H parameters of the main open chain					
i	$\alpha_{i-1}(deg)$	$a_{i-1}(m)$	$\beta_{i-1}(deg)$	$b_{i-1}(m)$	$d_i(m)$	$\theta_i(deg)$
1	0.8752	0.0003	0.006	0.0001	0.0976	0.3468
2	89.9412	0.5157	–	–	0(X)	−0.8836
3	0.0133	1.5998	0.001	–	−0.0014	−1.2385
4	90.1172	0.3545	–	–	1.8862	3.3033
5	−90.038	0.0002	–	–	4.087e−05	2.5786
6	90.0371	0.0003	–	–	0.445(X)	0(X)
T	–	−0.4511	–		0.0111	0.9279	–

Identified link lengths of the closed loop					
$L_5(m)$	0.7996	$L_4(m)$	1.601	$L_3(m)$	0.8(X)

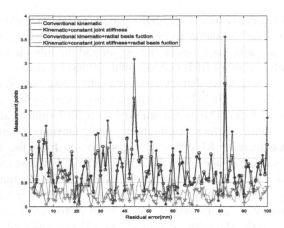

Fig. 3. Absolute position error of the HH800 robot after calibration.

Table 4. Absolute position accuracy of the HH800 robot (calibration).

	Mean (mm)	Maximum (mm)	Standard deviation (mm)
Before calibration (nominal robot model)	4.0654	6.3291	0.8803
After kinematic calibration	0.8199	3.5552	0.5422
After robot link geometry and joint compliance compensation	0.7250	2.5671	0.4107
After kinematic calibration and RBF compensation	0.3298	0.9462	0.1898
After proposed calibration	0.2612	0.7661	0.142

5.2   Experimental Validation Results

To validate the correctness of the proposed method, a set robot configurations Q_2 which not used in identification process are performed to validate for the robot accuracy after calibration.

The validation results show in Fig. 3 and Table 5. The proposed method shows its effectiveness by reducing the mean error to 0.4186(mm) (form 4.0629(mm) before calibration) and the maximum value is 1.234(mm) (form 6.1681(mm) before calibration). In comparing to the others method, the robot average position accuracy after conventional kinematic calibration is reduced to 0.8491(mm) and the maximum error is 3.7468 (mm).

By simultaneously calibrate the link geometry and compensation the joint compliance, the mean error is 0.767 (mm) and the maximum error is reduced to 2.7578 (mm). It is recorded that by combining the conventional kinematic calibration and RBF compensation the mean error is 0.5016 (mm) and the maximum value is 2.097 (mm) (Fig. 4).

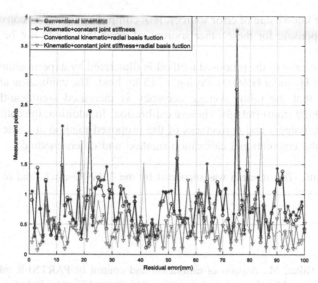

Fig. 4. Absolute position error of the HH800 robot after validation.

Table 5. Absolute position accuracy of the HH800 robot (validation).

	Mean (mm)	Maximum (mm)	Standard deviation (mm)
Before calibration (nominal robot model)	4.0629	6.1681	0.8451
After kinematic calibration	0.8491	3.7486	0.5099
After robot link geometry and joint compliance compensation	0.7670	2.7578	0.4142
After kinematic calibration and RBF compensation	0.5016	2.0970	0.3551
After proposed calibration	0.4186	1.234	0.2401

It is noted form the validation results that the position accuracy of the robot error is improved by a rate of 50.7% ((0.8491 − 0.4186)/0.8491) when the proposed calibration method is applied in comparing to the conventional calibration method. For the others two methods, the rates of improving are 45.42% (comparing to the simultaneously calibrate the link geometry and compensation the joint compliance method) and 16.55% (comparing to the combining the conventional kinematic calibration and RBF compensation method). Also, the standard deviation of the proposed method (0.2410 (mm)) and the maximum error are the best in comparing to the other methods.

6 Conclusions

The study has presented a new practical calibration method for enhancing robot position accuracy. It combines the techniques of model-based and radial basis function compensation method. The advantages of the method could be seen clearly by the result: accurate knowledge of error sources, less computing time, fast convergence, and ability to compensate for these other sources of error that could not be modeled by using a RBF.

The effectiveness of the proposed method is illustrated by experimental study of the industrial robot Hyundai HH800 carrying a 745 kg load. The calibration and validation results shown that the robot average accuracy is increased significantly to 0.4186 (mm) from 4.0629 (mm) (89.69%) before calibration. In addition, the calibration result shows the effectiveness and correctness of the proposed method at a rate of 50.7% in comparing to the conventional calibration method and others method.

Acknowledgment. This research was supported by the 2019 Research fund of University of Ulsan, Ulsan, Korea.

References

1. Brisan, C., Hiller, M.: Aspects of calibration and control of PARTNER robots. In: 2006 IEEE International Conference on Automation, Quality and Testing, Robotics, Cluj-Napoca, Romania, pp. 272–277. IEEE (2006)
2. Manne, R.: Analysis of two PLS algorithms for multivariate calibration. Chemometr. Intell. Lab. Syst. **2**, 187–197 (1987)
3. Denavit, J., Hartenberg, R.S.: A kinematic notation for lower-pair mechanisms based on matrices. ASME J. Appl. Mech. **77**, 215–221 (1955)
4. Hayati, S., Mirmirani, M.: Improving the absolute positioning accuracy of robot manipulators. J. Field Robot. **2**, 397–413 (1985)
5. Hayati, S., Tso, K.S., Roston, G.: Robot geometry calibration. In: Proceedings. 1988 IEEE International Conference on Robotics and Automation, Philadelphia, PA, USA. IEEE (1988)
6. Mooring, B.W., Tang, G.R.: An improved method for identifying the kinematic parameters in a six-axis robot. In: Computers in Engineering Proceedings of the International Computers in Engineering Conference and Exhibit, pp. 79–84. ASME, Las Vegas, Nevada (1984)
7. Zhuang, H., Roth, Z.S., Hamano, F.: A complete and parametrically continuous kinematic model. IEEE Trans. Robot. Autom. **8**, 451–463 (1992)
8. Zhuang, H., Wang, L.K., Roth, Z.S.: Error-model-based robot calibration using a modified CPC model. Robot. Comput.-Integr. Manuf. **10**, 287–299 (1993)
9. Okamura, K., Park, F.: Kinematic calibration using the product of exponentials formula. Robotica **14**(4), 415–421 (1996)
10. Chen, G., Kong, L., Li, Q., Wang, H., Lin, Z.: Complete, minimal and continuous error models for the kinematic calibration of parallel manipulators based on POE formula. Mech. Mach. Theory **121**, 844–856 (2018)
11. Chen, G., Wang, H., Lin, Z.: Determination of the identifiable parameters in robot calibration based on the POE formula. IEEE Trans. Robot. **30**(5), 1066–1077 (2014)
12. Zhou, J., Nguyen, H.N., Kang, H.J.: Simultaneous identification of joint compliance and kinematic parameters of industrial robots. Int. J. Precis. Eng. Manuf. **15**(11), 2257–2264 (2014)
13. Dumas, C., Caro, S., Garnier, S., Furet, B.: Joint stiffness identification of six-revolute industrial serial robots. Robot. Comput.-Integr. Manuf. **27**(4), 881–888 (2011)

14. Lightcap, C., Hamner, S., Schmitz, T., Banks, S.: Improved positioning accuracy of the PA10-6CE robot with geometric and flexibility calibration. IEEE Trans. Robot. **24**(2), 452–456 (2008)
15. Martinelli, A., Tomatis, N., Tapus, A., Siegwart, R.: Simultaneous localization and odometry calibration for mobile robot. In: Proceedings 2003 IEEE/RSJ International Conference on Intelligent Robots and Systems, Las Vegas, NV, USA, pp. 1499–1504. IEEE (2004)
16. Song, Y., Zhang, J., Lian, B., Sun, T.: Kinematic calibration of a 5-DoF parallel kinematic machine. Precis. Eng. **45**, 242–261 (2016)
17. Renders, J.M., Rossignol, E., Becquet, M., Hanus, R.: Kinematic calibration and geometrical parameter identification for robots. IEEE Trans. Robot. Autom. **7**(6), 721–732 (1991)
18. Zhou, J., Kang, H.J.: A hybrid least-squares genetic algorithm–based algorithm for simultaneous identification of geometric and compliance errors in industrial robots. Adv. Mech. Eng. **7**(6), 1687814015590289 (2015)
19. Jiang, Z., Zhou, W., Li, H., Mo, Y., Ni, W., Huang, Q.: A new kind of accurate calibration method for robotic kinematic parameters based on the extended Kalman and particle filter algorithm. IEEE Trans. Ind. Electron. **65**(4), 3337–3345 (2018)
20. Barati, M., Khoogar, A.R., Nasirian, M.: Estimation and calibration of robot link parameters with intelligent techniques. Iran. J. Electr. Electron. Eng. **7**(4), 225–234 (2011)
21. Xie, X., Li, Z., Wang, G.: Manipulator calibration based on PSO-RBF neural network error model. In: AIP Conference Proceedings, Malang City, Indonesia, p. 020026. AIP Publishing (2018)
22. Jang, J.H., Kim, S.H., Kwak, Y.K.: Calibration of geometric and non-geometric errors of an industrial robot. Robotica **19**(3), 311–321 (2001)
23. Tao, P.Y., Yang, G.: Calibration of industrial robots with product-of-exponential (POE) model and adaptive neural networks. In: 2015 IEEE International Conference on Robotics and Automation (ICRA), Seattle, Washington, USA, pp. 1448–1454. IEEE, May 2015
24. Kluk, P., Misiurski, G., Morawski, R.Z.: Total least squares versus RBF neural networks in static calibration of transducers. In: IEEE Instrumentation and Measurement Technology Conference Sensing, Processing, Networking. IMTC Proceedings, Ottawa, Ontario, Canada, pp. 424–427. IEEE (1997)
25. Liao, W.H., Aggarwal, J.K.: Curve and surface interpolation using rational radial basis functions. In: Proceedings of 13th International Conference on Pattern Recognition, Vienna, Austria, pp. 8–13. IEEE (1996)
26. Park, J., Sandberg, I.W.: Universal approximation using radial-basis-function networks. Neural Comput. **3**(2), 246–257 (1991)

Smart Wheelchair Maneuvering Among People

Sarwar Ali[✉], Antony Lam, Hisato Fukuda, Yoshinori Kobayashi,
and Yoshinori Kuno

Graduate School of Science and Engineering, Saitama University, Saitama, Japan
{sarwar_ali,antonylam,fukuda,yoshinori,
kuno}@cv.ics.saitama-u.ac.jp

Abstract. The advancement of technology is increasing with several applications on robotics like smart wheelchairs providing autonomous functions for severely impaired users with less need of caregiver support. One of the main issues in robotic wheelchair research is autonomous pedestrian avoidance for safety and smooth maneuvering. However, this is difficult because most of the pedestrians change their motion abruptly. Thus we need a fully autonomous smart wheelchair that can avoid collisions with individual or multiple pedestrians safely and with user comfort in mind for crowded environments like train stations. This paper presents a method for our smart wheelchair to maneuver through individual and multiple pedestrians by detecting and analyzing their interactions and predicted intentions with the wheelchair. Our smart wheelchair can obtain head and body orientations of pedestrians by using OpenPose. Using a single camera, we infer the walking directions or next movements of pedestrians by combining face pose and body posture estimation with our collision avoidance strategy in real-time. For an added layer of safety, we also use a LiDAR sensor for detection of any obstacles in the surrounding area to avoid collisions in advance.

Keywords: Wheelchair · OpenPose · Pedestrians · Interaction · Intention

1 Introduction

Nowadays, there are a vast number of technologies such as robotic wheelchairs, which improve the quality of life for many by providing independent mobility for individuals like the elderly, physically disabled, and mentally handicapped [1–5]. On the other hand, a significant number of such people have found difficulties in controlling their wheelchairs [6]. For some wheelchair users, traveling by using wheelchairs may be a difficult task, especially when they are in dynamic environments where pedestrians move around in dense public areas such as airport terminals, shopping malls, train stations, and so forth. Therefore, integration of smart wheelchairs that have high maneuverability and navigational intelligence with autonomy, will provide safety and ease of operation to both users and pedestrians. For this to happen, one of the important functions of autonomous wheelchairs is to detect and avoid pedestrians smoothly.

© Springer Nature Switzerland AG 2019
D.-S. Huang et al. (Eds.): ICIC 2019, LNAI 11645, pp. 32–42, 2019.
https://doi.org/10.1007/978-3-030-26766-7_4

Researchers have developed many ways to detect people and avoid them for wheelchairs. For example, the approaches in [7, 8], detect pedestrians as static obstacles but only in indoor environments and a 2D grid safety map was given for navigation. Other approaches have assumed simple independent motions for pedestrians (e.g. constant velocity) [9] and avoid them using simultaneous localization and mapping (SLAM). In actual human occupied environments, pedestrian motions can exhibit a lot of variation. One of the main drawbacks of these methods is that their smart wheelchairs are not familiar with pedestrian behaviors and movements in real-world environments. In our work, we take into account pedestrian behaviors in the real-world. One observation we have made is that when a wheelchair encounters pedestrians that are aware of it, the pedestrians try to make way for the wheelchair. Hence, when pedestrians are aware of the wheelchair, the wheelchair can generally maintain its current trajectory. Therefore consideration of how pedestrians interact with the wheelchair is essential to determining how the wheelchair should plan its path for navigating crowds.

In our previous works [10–12], we considered the awareness levels of pedestrians by detecting face/head orientations towards the wheelchair and devised a smart wheelchair navigation strategy where head detection and the orientations of pedestrians was used for collision avoidance in indoor environments, using both RGBD and laser sensors. The experimental results in those studies showed that the wheelchair could detect and avoid a single pedestrian by observing his/her awareness of the wheelchair. However, there are also situations in real-world environments where some pedestrians might not be aware of the wheelchair even though they are facing towards wheelchair. Therefore, to drive smoothly and successfully in dynamic environments among multiple pedestrians, using only awareness from face detection and orientations is insufficient for obtaining an acceptable amount of information for human avoidance. There are two ways to obtain information on the pedestrians' movements. The first way is through interactions that indicate pedestrian awareness of the wheelchair and another is through pedestrian intentions, which can be obtained from body orientations predicting the next movements of the pedestrian. When people walk in any given environment, the personal spaces (PS) proxemics concept [13] needs to be considered. The PS proxemics concept indicates that a certain amount of boundary around people exists and people use this boundary space to protect themselves from collisions. Thus as long as people are aware of other pedestrians (or wheelchairs) in the environment, they will make an attempt at avoiding collisions by reorienting their bodies and moving along a new direction. We are interested in exploring such important circumstances, where we consider pedestrian awareness of the wheelchair along with predicting their intended future movements so that the wheelchair maneuvers smoothly and safely in an effective manner in crowded real-world environments, where pedestrians and the wheelchair can collaboratively pass each other.

In this paper, we propose a method to enhance the movements of the smart wheelchairs for severely impaired users by analyzing and detecting the intentions and interactions of individual and multiple pedestrians. In our experiments, we employ a system that is mostly based on one RGB camera for frontal observations. We use a Tensorflow based OpenPose [14] system to detect and obtain human skeleton data of the full body. With the human skeleton data, we calculate the pose of different body parts, and from those parts, we can estimate any given pedestrian's body and face orientations. From this, we create a model for estimating the pedestrians' awareness and next movements. Our study consists of three steps for analyzing both individual and multiple moving pedestrians. In the first step, we determine the head orientations to see if any given pedestrian is aware of the wheelchair or not. In the second step, we determine the body orientations for estimating the intended future movement directions of pedestrians. Finally, we combine these two orientations (head and body) to determine a strategy for which direction our smart wheelchair should move in to avoid collisions with pedestrians in real-time. Moreover, we determine the relative distance between pedestrians and the wheelchair using [15] to check if the pedestrians are too close to the wheelchair, to maintain a safe distance away from pedestrians. During its maneuvers, for added safety, we also use a laser sensor that scans for surrounding objects that might collide with the wheelchair and generate steering commands for avoiding them by using our previous method described in [11]. The system architecture, methods, and results are given below.

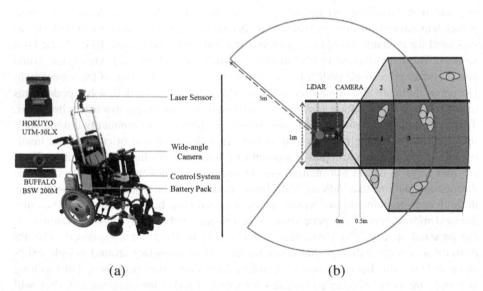

                        (a)   (b)

Fig. 1. (a) Our wheelchair system and (b) wheelchair sensor detection ranges and our defined "zones" to determine potential collisions. The LiDAR's range is shown as the orange arcs. (Color figure online)

2   System Overview

In our proposed system, the smart wheelchairs needs to have a wide enough range to sense the environment so that it can detect the various types of hazards in real environments. We use a combination of an RGB camera and laser sensor and evaluate its effectiveness in perceiving all possible types of obstacles that exist in outdoor environments and developed our robotic wheelchair as shown in Fig. 1(a). We use a BUFFALO BSW 200M series camera as a major sensing unit for a wide-viewing angle of 120° and mount it 75 cm above the ground in front of the wheelchair with a 0° inclination to get an aligned picture with the horizontal plane. This camera has a focal length of f = 1.8 mm with a max resolution of 1920 × 1080 at 30 fps. For our experiments, we use 800 × 600 resolution image frames. Moreover, the wheelchair is equipped with a Laser Range Sensor (UTM-30LX by Hokuyo) with a coverage angle of 270° installed on the rear frame at 1.05 m from the ground. In our experiment, we use our previous detection method in [11] as an added layer of safety to avoid unexpected obstacles and consider the maximum range of the laser sensor to be 5 m and set a buffer for free space with a minimum range of 0.5 m around the wheelchair to avoid collisions like in Fig. 1(b).

In Fig. 1(b), we divided the field of view of our vision camera equally into three sections as middle (1 and 3), right (2 and 3) and left (2 and 3). Here, area number 1 is considered as the area with the most potential for collisions, while area number 2, is considered as the "moderate zone". Area number 3 is considered less prone to collisions. Our smart wheelchair mainly runs in the forward, left, and right directions.

3 Methodology

Our smart wheelchair system primarily uses two procedures to control its speed along with avoidance of obstacles. At first, our system estimates the pedestrians' head and body orientations and with the help of those estimates, we determine their intentions and interactions with the wheelchair. Secondly, we use our direction finding technique to select the direction at which the wheelchair will move autonomously to reduce the user's workload.

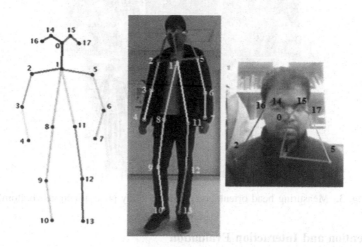

Fig. 2. Human keypoints extracted by OpenPose based on the COCO dataset model.

3.1 Pedestrians' Pose Estimation

We used a pose detection library, the "Tensorflow based OpenPose" [14] for pose estimation. The original OpenPose library runs on the Caffe deep learning framework, which is computationally intensive. On the other hand, Tensorflow is a very fast framework for real-time performance. The Tensorflow port of OpenPose allows for high-speed performance at the cost of some loss in accuracy. However, we have found that this slight decrease in accuracy is a good tradeoff for our real-world application. We used the pre-trained COCO dataset model [15] of OpenPose in our program with Tensorflow in Python on a Notebook GPU (NVIDIA GTX1070), which gives us detections speeds of around 10fps. The result is a network with a depth of 57 layers including 18 layers for body parts localization as shown in Fig. 2, 1 layer for the background, and 19 layers for limbs information in each of the x and y directions.

We use the output data of detected body parts to determine head and body orientations. For head orientations, the detection step generates possible regions of the head using the neck, nose, two eyes, and two ears. In our detections, we take the nose, eyes, and neck to estimate various angles and make the observation that when a person turns his head left, the left eye angle is greater than the right eye angle with the connecting line from the left shoulder to left ear also vanishing and vice versa as in Fig. 3 (top). Whereas, Fig. 3 (bottom) shows the body orientations, where the left, middle, and right images show that the person's body is posing in the left, forward, and right directions respectively. We use the coordinates of the neck, two shoulders, and two hips to estimate body orientations.

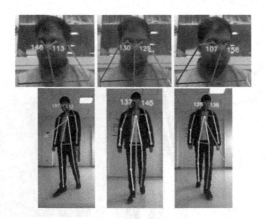

Fig. 3. Measuring head orientation (top) and body pose directions (bottom).

3.2 Intention and Interaction Evaluation

In our system, we have found it effective to first divide the frame equally into three sections, namely, the middle, right, and left based on our defined zones in front of the camera in Fig. 1(b). We then determine human awareness of the wheelchair using head orientations and the wheelchair's movement direction is determined by combining the pedestrians' awareness with their body positions and orientations.

Whenever a wheelchair encounters pedestrians, the wheelchair can most likely maintain its path because the pedestrians usually give way for the wheelchair. However, this may depend on the pedestrians' awareness of the wheelchair. If we assume a person does not tilt his/her head towards the ground or is not using a phone while walking, we can interpret that he/she may be clearly aware of the wheelchair's presence. We used this strategy for all persons detected by OpenPose and with their head orientations estimated by our method (Sect. 3.1). We point out some basic interactions of pedestrians to simplify our problem and assume the types of interactions with the wheelchair based on each frame and for how long any given pedestrian interacts with the wheelchair. The basic concept of our algorithm for determining the pedestrians' awareness of the wheelchair depending upon the position in the frame and head orientations of pedestrians can be stated as:

- Three conditions, when (1) the pedestrian's position is in the center of the frame and the pedestrian is facing straight, (2) the pedestrian's position is in the right of the frame and facing left, and (3) the pedestrian's position is in the left of the frame and facing right, then our wheelchair upon detecting these conditions, considers the pedestrian as looking at the wheelchair i.e. aware of it and that the pedestrian will try to avoid the wheelchair.
- For the rest of the positions and head orientations of the pedestrians, the wheelchair will assume that pedestrians are not looking at the wheelchair and unaware of it.
- If the pedestrian is looking at the wheelchair for a fraction of a second, the wheelchair will assume the pedestrian is unaware of it and will stop and process the next frame to get the head orientations to try to determine awareness.

Next, we compute body orientations from skeleton data and also count the pedestrians' positions in the frame, for determining human intentions. Basically, using the combination of pedestrian awareness and body positions in the frame and orientations, we determine the probable locations that pedestrians will move to in order to determine where the wheelchair should move as shown in Fig. 4.

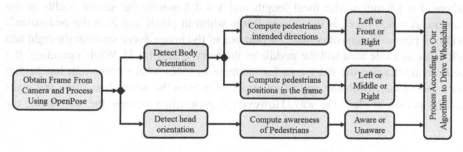

Fig. 4. Detections process diagram of our smart wheelchair.

3.3 Autonomous Navigation

Our smart wheelchair uses discrete actions (forward, right, left, slightly right, slightly left, slow down, and stop): at each time step, in order to avoid collisions and navigate efficiently and smoothly towards the goal. After detecting the awareness of the wheelchair and body orientations of the pedestrians with their position in the frame we determine which direction the wheelchair should move according to the decisions Table 1.

Table 1. Decision matrix of wheelchair steering procedure for different scenarios.

	Body orientation	Person location in frame		
		Right	Middle	Left
Aware	Right	Move forward	Move slightly left	Slow down
	Straight		Slow down	
	Left	Slow down	Move slightly right	Move forward
Unaware	Right	Move forward	Move left	Stop
	Straight		Stop	Move forward
	Left	Stop	Move right	

We also calculate the partially relative distance by how much the pedestrian is covering the frame, for some special cases where the pedestrian is unaware of the wheelchair and his/her position in the camera frame appears to be too close to the wheelchair. By that we can determine the closest potential collision and avoid the collision by stopping the wheelchair. To determine the distance (D) from the camera to a person, we consider the usual width of a pedestrian (Y), which is around 400 mm and the number of pixels the person covers in the image frame (X) as in Fig. 5. Then, the distance equation is:

$$D = fYA/sX \tag{1}$$

where, f = 1.8 mm is the focal length and s = 4.5 mm is the sensor width of the camera. A = 800 pixel, is the camera frame width in pixels and X is the pedestrian's width in pixels. We consider the 25% portions of the image frame on both the right and left sides as a safe area and the middle as the risky area (Fig. 5). While operating, if a pedestrian occupies the risky area for example in cases where pedestrian has his/her back facing the wheelchair and is not aware of it, then the wheelchair stops and waits for the person to clear the way. However, if pedestrians occupy the safe areas, the wheelchair will continue to move while performing our detection and maneuvering algorithms. Finally, we use the laser sensor detection process from [11] for additional safety, to avoid any pedestrians or objects that are out the camera's range as shown in Fig. 1. Experimental results of our method and discussions are given below.

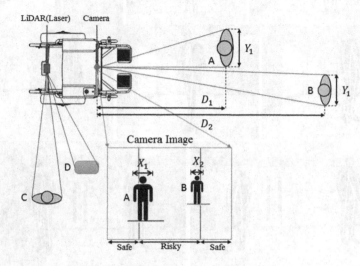

Fig. 5. Distance measurement using the camera.

4   Experimental Results and Discussion

We first tested our method inside our lab as in Fig. 6. We tested our method for a single person and attempted to estimate the interactions and intentions of his/her next move. Then, we conducted our tests for two persons in a group and analyzed their behaviors as in Fig. 6. To test the prediction accuracy, we extracted 2D trajectories from real-world pedestrian videos shot in a train station. In Fig. 6, the estimated human trajectory is represented as the yellow colored arrows in the frames and the wheelchair moving direction is represented in red. Head directions are represented in blue. For planned slight movements of the wheelchair, we denote the arrow as curved otherwise, the movements are denoted as straight arrows. Black arrows indicate the slowing down or stopping of the wheelchair.

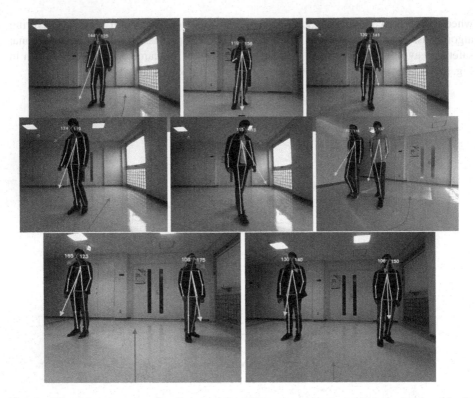

Fig. 6. Wheelchair maneuvering directions for different combinations of human head and body orientations. (Color figure online)

We now show what happens when a pedestrian is too close to the wheelchair. For example, the measurement of distance is given in Fig. 7. Here, the distance of the pedestrian is too close, almost 0.7 m in front of the wheelchair. In this situation, the wheelchair would stop and process the next frame and get directions until it sees a clear path. Figure 8 shows the detections by a laser sensor and our method respectively.

Fig. 7. Distance measurement using camera when a pedestrian is too close.

We have proposed a robotic wheelchair that observes pedestrians and the environment. It can recognize the pedestrians' intentions from his/her behaviors and environmental information. Experimental results show our approach is promising. We collected data from a train station and also from our campus to evaluate our methods.

Fig. 8. In a train station, detection of pedestrians with a laser sensor (left) and with our method (right). In the left image, the light blue arrows represent pedestrian directions, yellow denotes the detected pedestrian, and red is for static obstacles like walls. (Color figure online)

5 Conclusions

In outdoor environments, there is an uncertainty of the pedestrians' motions and for constantly changing environments; it is quite difficult for autonomous smart wheelchairs to detect the pedestrians' interactions and intentions. In this paper, we have developed a wheelchair maneuverability approach for severely impaired users. We have found that our system, which utilizes Tensorflow based OpenPose data can detect pedestrian interactions with the wheelchair and their intended next movements.

In summary, we investigated the combination of two possible orientations of pedestrian (head and body) and evaluated its effectiveness in perceiving various types of wheelchair movement directions in response to pedestrian behaviors in crowded outdoor environments. The final system results in auto-navigation that generates wheelchair movements that are safe and comfortable to wheelchair user and other people in real-time multi-human scenarios.

For future work we will build a more sophisticated method for very busy environments to achieve robust performance with more precise estimates of the pedestrians next move and free space for our wheelchair to move through. We are also planning to develop this method using a neural network structure to train the wheelchair's decisions to attain more robust and better performance.

Acknowledgement. This work was supported by JSPS KAKENHI Grant Number JP26240038.

References

1. Miller, D.P., Slack, M.G.: Design and testing of a low-cost robotic wheelchair prototype. Auton. Robots **2**(1), 77–88 (1995)
2. Simpson, R.C.: Smart wheelchairs: a literature review. J. Rehabil. Res. Dev. **42**(4), 423 (2005)
3. Gomi, T., Griffith, A.: Developing intelligent wheelchairs for the handicapped. In: Mittal, V. O., Yanco, H.A., Aronis, J., Simpson, R. (eds.) Assistive Technology and Artificial Intelligence. LNCS, vol. 1458, pp. 150–178. Springer, Heidelberg (1998). https://doi.org/10. 1007/BFb0055977
4. Mamun, S.A., Lam, A., Kobayashi, Y., Kuno, Y.: Single laser bidirectional sensing for robotic wheelchair step detection and measurement. In: Huang, D.-S., Hussain, A., Han, K., Gromiha, M.M. (eds.) ICIC 2017. LNCS (LNAI), vol. 10363, pp. 37–47. Springer, Cham (2017). https://doi.org/10.1007/978-3-319-63315-2_4
5. Ali, S., Al Mamun, S., Fukuda, H., Lam, A., Kobayashi, Y., Kuno, Y.: Smart robotic wheelchair for bus boarding using CNN combined with hough transforms. In: Huang, D.-S., Gromiha, M.M., Han, K., Hussain, A. (eds.) ICIC 2018. LNCS (LNAI), vol. 10956, pp. 163–172. Springer, Cham (2018). https://doi.org/10.1007/978-3-319-95957-3_18
6. Fehr, L., Langbein, W.E., Skaar, S.B.: Adequacy of power wheelchair control interfaces for persons with severe disabilities: a clinical survey. J. Rehabil. Res. Dev. **37**(3), 353–360 (2000)
7. Bauer, A., et al.: The autonomous city explorer: towards natural human-robot interaction in urban environments. Int. J. Soc. Robot. **1**(2), 127–140 (2009)
8. Tomari, M.R.M., Kobayashi, Y., Kuno, Y.: Development of smart wheelchair system for a user with severe motor impairment. In: IRIS (2012)
9. Kümmerle, R., Ruhnke, M., Steder, B., Stachniss, C., Burgard, W.: Autonomous robot navigation in highly populated pedestrian zones. J. Field Robot. **32**(4), 565–589 (2015)
10. Tomari, R., Kobayashi, Y., Kuno, Y.: Enhancing wheelchair maneuverability for severe impairment users. Int. J. Adv. Rob. Syst. **10**, 92–105 (2013). https://doi.org/10.5772/55477
11. Tomari, R., Kobayashi, Y., Kuno, Y.: Socially acceptable smart wheelchair navigation from head orientation observation. Int. J. Smart Sens. Intell. Syst. **7**(2), 630–643 (2014)
12. Murakami, Y., Kuno, Y., Shimada, N., Shirai, Y.: Collision avoidance by observing pedestrians' faces for intelligent wheelchairs. J. Robot. Soc. Jpn **20**(2), 206–213 (2002)
13. Hall, E.T.: Proxemics. Curr. Anthropol. **9**(2–3), 83–108 (1968)
14. Human pose estimation using OpenPose with TensorFlow. https://arvrjourney.com/human-pose-estimation-using-openpose-with-tensorflow-part-1-7dd4ca5c8027. Accessed 22 Jan 2019
15. COCO dataset. http://cocodataset.org. Accessed 5 Oct 2018

A Human-Robot Interaction System Based on Calling Hand Gestures

Aye Su Phyo$^{(\boxtimes)}$, Hisato Fukuda$^{(\boxtimes)}$, Antony Lam$^{(\boxtimes)}$,
Yoshinori Kobayashi$^{(\boxtimes)}$, and Yoshinori Kuno$^{(\boxtimes)}$

Graduate School of Science and Engineering, Saitama University, Saitama, Japan
{ayesuphyo, fukuda, antonylam,
kuno}@cv.ics.saitama-u.ac.jp,
yosinori@hci.ics.saitama-u.ac.jp

Abstract. Human-Robot Interaction (HRI) is one of the most important aspects of development in social service robots. Interacting with social robots via nonverbal cues allows for natural and efficient communication with humans. This paper presents on going work in developing service robots that provide assisted-care to the elderly. The major goal of this system is to recognize natural calling gestures from people in an interaction scenario where the robot continuously observes the behavior of a humans. In our approach, firstly, the robot moves amongst people. At that time, when the person calls the robot by a hand gesture, the robot detects the person who is calling the robot from among the crowd. While approaching to the potential caller, the robot observes whether the person is actual calling the robot or not. We tested the proposed system at a real elderly care center. Experiment results validate the practicality and effectiveness of the system.

Keywords: Human-Robot Interaction · Calling gesture · Icart-mini

1 Introduction

Over the past decade, many different types of human-friendly robots have been developed. Depending on the objectives, some robots are developed for helping humans in industrial environments and some are designed to function in indoor environments. As technology gets more sophisticated and advanced, the focus has been shifted to social service robots. The goal of these robots is to communicate with humans in a human-like manner and perform different tasks as instructed by human users. This leads us to social behavior in robots. These social robots should recognize humans, their verbal communication and gestures in order to realize natural communication [1].

Social behavior in robots generally depend upon efficient human-robot interaction (HRI). The most common modes of human interaction are either by vocal communication or by body gestures. According to [2], 65% of our communication consists of human gestures and only 35% consists of verbal content. This two-thirds of our mode of communication shows the significance of gestures. For this purpose, recognition of nonverbal content becomes essential for HRI. Human gestures are an important form of nonverbal content, which is used with or without verbal communication in expressing

© Springer Nature Switzerland AG 2019
D.-S. Huang et al. (Eds.): ICIC 2019, LNAI 11645, pp. 43–52, 2019.
https://doi.org/10.1007/978-3-030-26766-7_5

the intended meaning of the speech. Such gestures may include hand, arm, or body gestures and it may also include use of the eyes, face, head and more.

In particular, gesture recognition has been a popular topic in computer vision. The topic has been studied numerous times because of its important applications in surveillance systems, elderly care, in the field of medicine (e.g. gait analysis, surgical navigation), in the field of sports, augmented reality, sign language for hearing impaired people and human behavior analysis. Hand gestures are also critical in face-to-face communication scenarios. Especially during discussions, hand gestures become more animated. They emphasize points and convey the enthusiasm of the speaker. Recent years have seen the increasing popularity of gesture-based applications, where users use the movements of their hands, fingers, head, face and other parts of the body to interact with virtual objects. Furthermore, studies have been carried out to investigate how older users use gesture inputs in their interactions with information technologies.

Most existing gesture recognition systems are designed to be used in face-to-face situation. This is not the case for our service robot application. The robot should find people calling by hand gestures. But there may be various hand movements in multiple people environments. To solve this problem, we propose a real-time human robot interaction system based on natural calling gestures. The system may be perfect when we ask users to use complicated specified gestures, but we would like to use natural gestures not specified in advance. When the robot moves and finds people calling it amongst other people, the robot may detect many calling candidates. At that time, the robot observes the calling candidates while approaching to them to verify whether the initially detected calling gestures were false positives. We validate our findings using our experimental setup, which is composed of a humanoid robot (Aldebaran's NAO) and an i-Cart mini (T-frog) that carries the NAO humanoid and a webcam.

The remainder of the paper is organized as follows: Sect. 2 discusses hand gesture recognition systems in Human-Robot Interaction. Section 3 presents the system configuration for our calling gesture to assist elderly people. Experiments and results are discussed in Sect. 4, and the final section presents conclusions and future work.

2 Related Work

Hand gesture recognition is commonly used to control a robot in close proximity of the user. In recent years, the scientific literature on the development of hand gesture recognition systems for HRI has expanded substantially. Many papers presented the use of Kinect sensors [3–11] and they select several features and different classification methods to build gesture models in several applicative contexts. This section will overview those studies mainly focused on hand gesture recognition approaches in the HRI context [12]. The ability of the OpenNI framework to easily provide the position and segmentation of the hand has stimulated many approaches to the recognition of hand gestures [6, 13, 14]. The hand's orientation and four hand gestures (open hand, fist, pointing index, and pointing index and thumb) are recognized in [13] for interacting with a robot, which uses the recognized pointing direction to define its goal on a map.

In [14], static hand gestures are also recognized to control a hexagon robot by using the Kinect and the Microsoft SDK library. In [15], the authors also propose a system that recognizes seven different hand gestures for interactive navigation by a robot, although in contrast to the previously cited works, the gestures are dynamic. This involves the introduction of a start-/end-point detection method for segmenting the 3D hand gesture from the motion trajectory. First, an interactive hand-tracking strategy based on a Camshift algorithm and which combines both color and depth data is designed. Next, the gestures classified by a Hidden Markov Model (HMM) are used to control a dual-arm robot. Dynamic gestures based on arm tracking are instead recognized in [16]. The proposed system is intended to support natural interaction with autonomous robots in public places, such as museums and exhibition centers. The proposed method uses the arm joint angles as motor primitives (features) that are fed to a neural classifier for the recognition process. In [17], the movement of the left arm is recognized to control a mobile robot. The joint angles of the arm with respect to the person's torso are used as features. A preprocessing step is first applied to the data in order to convert feature vectors into finite symbols as discrete HMMs are considered for the recognition phase. Human motion is also recognized by using an algorithm based on HMMs in [18]. A service robot with an on-board Kinect receives commands from a human operator in the form of gestures. Six different gestures executed by the right arm are defined and the joint angles of the elbow are used as features.

The authors in [19] have collaborated in the development of a robot that can look towards and then approach a person who waves a hand at the robot in a laboratory setting, there has not yet been a robot designed that can respond to a person's request through hand waving or other embodied means in naturally occurring multiparty settings. In [20], they reported the results of interaction between caregivers and the elderly, then discussed the implications for developing robotic system. Based on these findings, [21] developed a system that can initiate interaction with a human through nonverbal behaviors such as head gestures (e.g. head turning) and gaze. From these works, we focus on natural calling gesture recognition in crowded scenes using gaze.

In our system, when the robot moves and finds people calling it amongst people, the robot may detect many calling candidates. At that time, the robot approaches the calling candidates to verify whether they really are calling the robot. Specifically, the robot first detects a person in the scene and the hand-wrist to localize the calling gesture. Next, by using a heuristic approach, the system recognizes the hand gestures that are calling commands for the robot [22].

3 System Configuration

3.1 Service Robot Setup

The current prototype of our service robot consisting of an i-Cart mini (T-frog), USB webcam, a humanoid robot (Aldebaran's NAO) and a mobile workstation, is shown in Fig. 1. The i-Cart mini (T-frog) is a small, light weighted robot that can easily move. To control the robot's hardware, we used the Robot Operating System (ROS), which is a framework with software libraries and tools designed specifically to develop robot applications. The USB wide-angle webcam allows 120 degrees wide coverage to capture the whole body of each person in the scene. Hokuyo Laser sensor is used to measure the distance between the caller and robot. The NAO robot will be used for our future work to interact with the caller once the robot has arrived the caller's location.

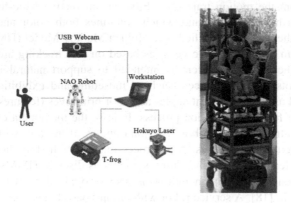

Fig. 1. System architecture showing interactions between components

The system is configured as follows: The USB wide camera, mounted on top of the robot, obtains a subject's full-body position data and Hokuyo laser sensor's data is used to calculate the distance between the caller and robot. Such data becomes available to all devices connected throughout the Robot Operating System (ROS) network and subsequently, it is processed to recognize the body gestures that will indicate the robot's actions.

3.2 Process Model of Human-Robot Interaction Based on Gesture

Fig. 2. A four-stage model of information processing

To recognize gestures in HRI, it is beneficial to investigate into a generic and simplified information processing model. As shown in Fig. 2, the generalized human information processing is broken into a four-stage model. Based on this model, we propose a specific model for gesture recognition in HRI.

Figure 3 shows the process model of the proposed service robot with gesture recognition system. Firstly, let us say there is a natural calling gesture within the camera's field of view. Then it is detected in real time and the key-points information of the whole body are obtained, i.e., positions of nodes in the skeleton model detected by OpenPose [23]. And, gaze and hand detections are performed based on the key-points information of individual people. After detecting the gaze and hand, the detailed key-points of the fingertips are extracted by zooming into the hand-wrist part. After detecting potential calling gestures, the robot faces and approaches to the candidates of calling gestures by transmitting data to the motion control system. At that time, the robot approaches the potential caller and further verifies whether they the caller was really calling to reduce false positives. After determining whether there was an actual call or not from the hand gesture's command, the robot responds to the action by moving to the ones who are calling the robot. We then map the recognized gesture to the appropriate command – so that if the caller continues the calling hand gesture while approaching, the robot will keep coming to the caller. And, if the caller still holds gaze to the robot while approaching, the robot will continue coming action. Like this action, service robots can help the elderly.

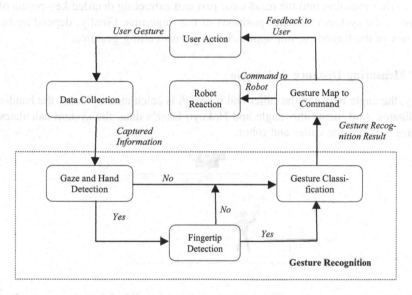

Fig. 3. A process model of the proposed service robot with gesture recognition system.

Algorithm 1 Procedure for human-robot interaction

```
1: open human-robot interaction switch;
2: loop:
3: capture skeleton data and feed data into system, and
   move around the people;
4: if no calling gesture then
5:     goto loop;
6: else
7:     turn toward direction of caller and move towards
       caller;
8:     if user keeps calling or holding gaze then
9:         continue moving towards caller;
10:    else
11:        goto loop;
12:    end if
13: end if
```

3.3 Gesture Recognition

In this step, the system uses the skeleton key-points of individuals among the crowd using OpenPose that as in our previous work [22]. Firstly, we extract the body and hand fingertip key-points of individuals in the scenes. The gaze of people that are looking at the robot is detected to find candidate people that are more likely to be calling the robot. We then localize the hand-wrist positions. Specifically, the system looks for positions of the wrist that are higher than the position of the below and at the same time, the position of the elbow is either higher or lower than the position of the shoulder. In the final step, the system looks at the configuration of the fingertips for classification. So as to not to miss fine details of the hand in crowded scenes, we also zoom into the hand part of the candidate person based on the locations of body key-points. After zooming into the hand-wrist part and extracting detailed key-points of the fingertips, the system calculates positions of the fingertips. Finally, depending on the positions of the fingertips, our approach recognizes calling gestures.

3.4 Measuring Distance and Angle

Firstly, the angle between the caller and robot, θ, is calculated by using the hand-wrist coordinates. And, using this angle and Hokuyo laser's data, the system calculates the distance between the caller and robot.

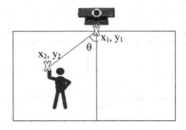

$$\theta = \tan^{-1} \frac{|(x_1 - x_2)|}{(y_2 - y_1)} \tag{1}$$

θ = the angle between the caller and robot
x_1 = half of image horizontal dimension (e.g. if 640 × 480, x_1 = 320)
y_1 = 0
x_2 = x coordinate of detected calling hand-wrist
y_2 = y coordinate of detected calling hand-wrist

In Eq. (1), x_1 is defined as half of image horizontal dimension and y_1 is 0 because the camera is mounted on the center of the i-Cart mini. The direction of the angle, θ, depends on the sign of $(x_1 - x_2)$. If the sign is '+', the direction is right, otherwise, the direction is left.

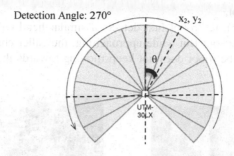

Detection Angle: 270°   x₂, y₂

The system calculates the distance between the caller and robot by using the angle, θ, x, y coordinates of detected calling hand-wrist and Hokuyo laser's data. Hokuyo laser sensor can detect range from 0.1 m to 30 m within 270° wide angle. When the system implements on ROS, the obstacles for navigation can be avoided.

4 Experiments and Results

Fig. 4. The robotic system designed for this study in example scene.

In order to evaluate the performance of human-robot interaction based on gesture recognition, we prepared an indoor test scenario at a real elderly care center in several specific situations and evaluated our system on them. Human test subjects were asked to call the robot by hand. In our approach, firstly, the robot moves around the people as shown in Fig. 4. At that time, if the person within the camera's field of view calls the robot, the robot moves to that person. If two or more different person within the camera views start to call the robot at a time, the robot will move first detect first move.

Figure 5 shows example cases of our experiment. In Fig. 5(a) and (b), when the robot recognizes the hand calling gesture, the robot moves to that person. The robot keeps moving as the caller continues the hand calling gesture.

In Fig. 5(c) and (d), when the robot detects the calling gesture, the robot approaches the caller. Moving towards the caller, the robot observes whether there is an actual call or not. The robot approaches the caller because of the caller maintaining consistent gaze towards the robot.

In Fig. 5(e) and (f), the robot first detects a similar hand calling gesture and the robot moves to that person. But while approaching, the caller changes hand direction and gaze. Therefore, the robot does not keep moving towards that person.

Fig. 5. Example results of our human-robot interaction system.

In our experiments, the robot's main computer is a tablet PC with Intel(R) Core (TM) i7-7700HQ CPU @ 2.8G Hz*4, RAM of 32.00 GB, GPU Geforce GTX 1070. In the process of experiments, the system needs a few seconds to execute the behaviors after users signal with hand gestures. If another signal is sent in the same period of time, it takes some time to perform. And, sometimes, gestures are missed as the robot moves around. Based on these cases, the program is set to sleep a short period after one signal is sent. This proves efficient in solving the problem. Another way to overcome our limitation is by using multiple or powerful GPUs to enhance the temporal performance of our system.

5 Conclusion and Future Work

This paper proposes a service robot system that provides assisted-care to the elderly. This system recognizes natural calling gestures in an interaction scenario where the robot visually observes the behavior of humans. After detecting the calling gesture, the robot moves to the caller. While approaching, the robot observes whether the user is actually calling or not. From this result, the interaction between humans and robot more effective. In multiple callers' case at the same time, the robot will move first detect first move. We expect that many service robot systems will make effective use of the proposed system for human robot interaction.

The use of multiple GPUs for OpenPose can enhance the temporal performance of our system. It can be extended for a more natural and cooperative experience between humans and robots. Multiple object detection, their localization in the scene and handling objects depending on the elderly's actions will be added in future work.

Acknowledgement. This work was supported by JSPS KAKENHI Grant Number JP26240038.

References

1. Breazeal, C., Takanishi, A., Kobayashi, T.: Social robots that interact with people. In: Siciliano, B., Khatib, O. (eds.) Springer Handbook of Robotics. Springer, Heidelberg (2008). https://doi.org/10.1007/978-3-540-30301-5_59
2. Chen, W.: Gesture-based applications for elderly people. In: Kurosu, M. (ed.) HCI 2013. LNCS, vol. 8007, pp. 186–195. Springer, Heidelberg (2013). https://doi.org/10.1007/978-3-642-39330-3_20
3. Cruz, L., Lucio, F., Velho, L.: Kinect and RGBD images: challenges and applications. In: XXV SIBGRAPI IEEE Conference and Graphics, Patterns and Image Tutorials (2012)
4. Di Paola, D., Milella, A., Cicirelli, G., Distante, A.: An autonomous mobile robotic system for surveillance of indoor environments. Int. J. Adv. Robot. Syst. 7(1), 8 (2010)
5. D'Orazio, T., Guaragnella, C.: A graph-based signature generation for people re-identification in a multi-camera surveillance system. In: International Conference on Computer Vision Theory and Applications (VISAPP), Rome, Italy, pp. 414–417, February 2012
6. D'Orazio, T., Cicirelli, G.: People re-identification and tracking from multiple cameras: a review. In: IEEE International Conference on Image Processing (ICIP 2012), Orlando, Florida, September 2012
7. Fasola, J., Mataric, M.J.: Using socially assistive human-robot interaction to motivate physical exercise for older adults. Proc. IEEE 100(8), 2512–2526 (2012)
8. Fujii, T., Lee, J.H., Okamoto, S.: Gesture recognition system for human-robot interaction and its application to robotic service task. In: Proceedings of the International Multi-Conference of Engineers and Computer Scientists (IMECS), Hong Kong, vol. 1, March 2014

9. Goodrich, M.A., Schultz, A.C.: Human-robot interaction: a survey. Found. Trends Hum.-Comput. Inter. **1**(3), 203–275 (2007)
10. Gu, Y., Do, H., Ou, Y., Sheng, W.: Human gesture recognition through a kinect sensor. In: IEEE International Conference on Robotics and Biomimetics (ROBIO), pp. 1379–1384 (2012)
11. Hachaj, T., Ogiela, M.R.: Rule-based approach to recognizing human body poses and gestures in real time. Multimedia Syst. **20**, 81–99 (2013)
12. Itauma, I.I., Kivrak, H., Kose, H.: Gesture imitation using machine learning techniques. In: Proceedings of 20th IEEE Signal Processing and Communications Applications Conference, Mugla, Turkey, April 2012
13. Jacob, M.G., Wachs, J.P.: Context-based hand gesture recognition for the operating room. Pattern Recogn. Lett. **36**, 196–203 (2014)
14. Lai, K., Konrad, J., Ishwar, P.: A gesture-driven computer interface using kinect. In: IEEE Southwest Symposium on Image Analysis and Interpretation (SSIAI), pp. 185–188 (2012)
15. Miranda, L., Vieira, T., Martinez, D., Lewiner, T., Vieira, A.W., Campos, M.F.M.: Real-time gesture recognition from depth data through key poses learning and decision forests. In: 25th SIBGRAPI Conference on Graphics, Patterns and Images (SIBGRAPI), pp. 268–275 (2012)
16. Oh, J., Kim, T., Hong, H.: Using binary decision tree and multiclass SVM for human gesture recognition. In: International Conference on Information Science and Applications (ICISA), pp. 1–4 (2013)
17. Kun, Q., Jie, N., Hong, Y.: Developing a gesture based remote human-robot interaction system using kinect. Int. J. Smart Home **7**(4), 203–208 (2013)
18. Sigalas, M., Baltzakis, H., Trahanias, P.: Gesture recognition based on arm tracking for human-robot interaction. In: Proceedings of IEEE/RSJ International Conference on Intelligent Robots and Systems (IROS), Taipei, Taiwan, October 2010
19. Miyauchi, D., Sakurai, A., Nakamurai, A., Kuno, Y.: Bidirectional eye contact for human-robot communication. IEICE Trans. Inf. Syst. **E88-D**(11), 2509–2516 (2005)
20. Yamazaki, K., et al.: Prior-to-request and request behaviors within elderly day care: implications for developing service robots for use in multiparty settings. In: Bannon, L.J., Wagner, I., Gutwin, C., Harper, R.H.R., Schmidt, K. (eds.) ECSCW 2007, pp. 61–78. Springer, London (2007). https://doi.org/10.1007/978-1-84800-031-5_4
21. Quan, W., Niwa, H., Ishikawa, N., Kobayashi, Y., Kuno, Y.: Assisted-care robot based on sociological interaction analysis. In: Proceedings of the ICIC 2009 (2009)
22. Phyo, A.S., Fukuda, H., Lam, A., Kobayashi, Y., Kuno, Y.: Natural calling gesture recognition in crowded environments. In: Huang, D.-S., Bevilacqua, V., Premaratne, P., Gupta, P. (eds.) ICIC 2018. LNCS, vol. 10954, pp. 8–14. Springer, Cham (2018). https://doi.org/10.1007/978-3-319-95930-6_2
23. Hidalgo, G., Cao, Z., Simon, T., Wei, S.-E., Joo, H., Sheik, Y.: Openpose. https://github.com/CMU-Perceptual-Computing-Lab/openpose

A Welding Defect Identification Approach in X-ray Images Based on Deep Convolutional Neural Networks

Yuting Wang, Fanhuai Shi[✉], and Xuefeng Tong

College of Electronics and Information Engineering, Tongji University,
Shanghai 201804, China
fhshi@tongji.edu.cn

Abstract. Welding defect detection and identification is a crucial task in the industry that has a significant economic impact and can pose safety risks if left unnoticed. Currently, an inspector visually evaluates the condition of welds to guarantee reliability in industrial processes. This way is time-consuming and subjective. This paper proposes a deep learning-based approach to identify automatically multiple welding defect types and locations in X-ray images by adopting a pre-trained RetinaNet-based convolutional neural network. To realize this, a dataset including 6714 images labeled for three types of welding defect— blowhole, under fill or incomplete penetration, and tungsten inclusion—is developed. Then, the RetinaNet-based model is designed and trained using this database. The proposed approach can not only directly detect welding defect on the original input images without any pro-processing, but also identify the types of defect. Moreover, the experimental tests show that the proposed approach works well on the images even in low resolution and mean average precision (mAP) ratings are 0.76, 0.79, and 0.92 respectively for three defect types.

Keywords: Convolutional neural network · Deep learning · Welding defect identification · RetinaNet

1 Introduction

For welded joints, the welding quality largely determines the service lifetime of the structural component. However, due to the instability of parameters in the welding process and the deformation in the structural component, there may be some inherent defects in the welded joints. Hence, the non-destructive testing (NDT) is performed to confirm the presence of any defect, its type, extent, and location in the component during the welding process.

At present, there are various NDT methods available, such as visual inspection, radiographic testing (RT), ultrasonic testing (UT) and so on. Among them, the non-destructive testing based on X-ray digital radiography has the advantage of clean, convenient and easy to view, and its inspection result has high judgment value.

There are two ways to evaluate the inspection result for X-ray images: manual assessment and computer-aided assessment. The manual assessment is labor intensive and inefficient. Additionally, the assessment result is so subjective that can cause too

D.-S. Huang et al. (Eds.): ICIC 2019, LNAI 11645, pp. 53–64, 2019.
https://doi.org/10.1007/978-3-030-26766-7_6

many false positives and false negatives. Therefore, it is necessary to reduce the work intensity of the assessors and realize the scientific, objective and standardized evaluation process by computer-aided automatic assessment.

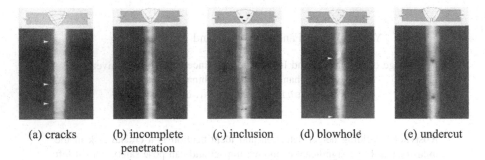

   (a) cracks   (b) incomplete   (c) inclusion   (d) blowhole   (e) undercut
                   penetration

Fig. 1. Some common types of welding defect.

Traditional image processing techniques (IPTs) for detecting welding defect mostly use the geometric features of defect, but the IPTs methods are highly susceptible to different environments. Machine learning algorithms (MLAs) [8] is a possible solution that is more appropriate to the real-world situation, so it is also widely used in the detection of welding defect. However, MLAs are not performing well in accuracy. During the past few decades, deep learning (DL) has shown a significant performance owing to the capabilities of parallel computing using graphics processing units (GPUs) and large well-labeled databases like ImageNet [13].

Feature learning by using convolutional neural networks (CNNs) has achieved excellent performance but not obviously in the field of NDT. There are two reasons. First, welding defect datasets labeled by experts are too small to train DL networks. Second, most methods detect only specific types of defect, such as blowhole or crack. Actually, in the industry, under fill and incomplete penetration are also the major types of welding defect that are strictly not allowed in the welded joints. However, expertise and experience of professionals are strongly required to identify under fill and incomplete penetration, while even non-professionals can observe some blowhole and crack from images. Some common types of welding defect are shown in Fig. 1.

To deal with the two challenges, a deep learning-based approach to identify automatically multiple welding defect types and locations in X-ray images by adopting a pre-trained RetinaNet-based convolutional neural network is proposed in this paper. To realize this, a dataset including 6714 images labeled for three types of welding defect—blowhole, under fill or incomplete penetration, and tungsten inclusion—is developed. Then, the RetinaNet-based network is trained and tested using the proposed dataset. The proposed approach can not only directly detect welding defects on the original input images without any pro-processing, but also identify the types of defects. Moreover, the experimental tests show that the proposed approach works well on the images even in low resolution and mean average precision (mAP) ratings are 0.76, 0.79, and 0.92 respectively for three defect types. The main contribution of this paper is to achieve effective automatic identification of welding defects by using deep learning-based approach, rather than distinguish the differences among different network architectures.

The organization of this paper is as follows. Section 2 discusses the related work of this paper. Next Sect. 3 describes the methodology for this paper, where welding defect identification based on deep convolutional neural networks is discussed and the proposed approach is described in details. In Sect. 4, datasets and experimental results are presented. Finally, Sect. 5 concludes this paper.

2 Related Work

The computer-aided assessment transforms the manual assessment work into the computer process using computer vision-based techniques. Computer-aided automatic assessment for welding defects makes the evaluation more objective and stable. Besides, the speed and efficiency of computer-aided assessment are much higher than the manual assessment. Hence, the automatic detection and identification of welding defects is gradually becoming an important developing direction in the future.

Previous studies on welding defect identification are mainly categorized into three parts: image processing techniques (IPTs), machine learning algorithms (MLAs), and convolutional neural networks (CNNs). In the IPTs methods, Tang et al. [1] analyzed the characteristics of X-ray images and proposed an edge detection algorithm based on the multi-scale and multi-resolution of wavelet transform. Chen et al. [2] implemented a new automatic detection approach based on the EMD (Empirical Modal Decomposition) method. Felisberto et al. [3] described a genetic algorithm to search for suitable parameters values. In another work [4], the author proposed a more effective method based on fuzzy theory, and images were filtered by applying fuzzy reasoning using local characteristics. Kasban et al. [5] extracted the Mel-Frequency Cepstral Coefficients (MFCCs) and polynomial coefficients. Tong et al. [6] described a weld defect segmentation algorithm based on mathematical morphology theories and iterative threshold method. And Hassan et al. [7] tended to localize defects with maximum interclass variance and minimum intra-class variance. However, traditional IPTs methods are highly susceptible to different environments and have poor robustness.

Machine learning algorithms (MLAs) [8] is a possible solution that is more appropriate to the real-world situation, so it is also widely used in welding defect identification. In both of these work [9, 10], the authors both used support vector machine (SVM) to achieve welding defects classification. And Sikora et al. [11] described three approaches for defect detection, including fuzzy logic, artificial neural networks (ANNs). However, MLAs are not performing well in accuracy.

In recent years, convolutional neural networks (CNNs) have been achieving great performance on image tasks such as object recognition [12]. CNNs can learn the appropriate features that were designed artificially in traditional algorithms. Owing to the excellent performance of CNNs, Yu et al. [13] described a localization method for casting defect based on feature matching, but the resolution was too low to detect small defect in the castings. In another work [14], the author constructed a classification model based on deep neural network, but the approach cannot classify different types of welding defects. And Liu et al. [15] achieved welding defect classification with VGG16-based neural network, but the network was trained and tested on defect regions manually cropped instead of the whole image.

To overcome the limitation of previous studies, a deep learning-based approach to identify automatically multiple welding defect types and locations in X-ray images by adopting a pre-trained deep CNN, called RetinaNet, is proposed.

3   Welding Defect Identification Based on RetinaNet

3.1   Overview of the Proposed Approach

The overall flowchart of the proposed approach is shown in Fig. 2. The welding X-ray images are used as an input for RetinaNet-based network. The locations and types of welding defect are identified at the output of network. As shown in Fig. 2, the proposed approach does not require any pre-processing and post-processing.

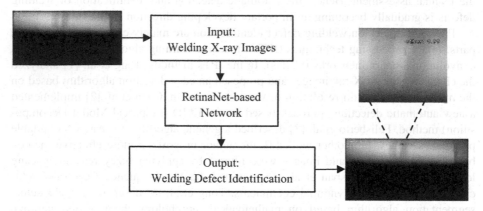

Fig. 2. The proposed approach for welding defect identification based on RetinaNet.

3.2   Model Architecture Based on RetinaNet

The networks architectures for object detection are usually divided into two categories, namely one-stage and two-stage object detectors. In two-stage detectors like R-CNN (Region-based CNN) [16], Fast R-CNN [17], Faster R-CNN [18] and Mask R-CNN [19], a Region Proposal Network (RPN) is performed to generate the region of interests (ROI) proposals in the first stage. Then, these ROI proposals are used for object classification and bounding box regression in the second stage. The two-stage methods are generally very slow but have high accuracy. On the other hand, one-stage detectors such as YOLO [20, 21] and SSD [22, 23] do not have a pre-selection step to generate foreground candidates and they simply view object detection as a simple regression problem. Therefore, the one-stage detectors can be faster and simpler but yield a lower accuracy. However, RetinaNet can reach the speed of one-stage detectors while surpassing the accuracy of two-stage detectors by addressing foreground-background class imbalance [24]. Class imbalance is also a serious problem in welding defect images, because a single welding image generally contains only one or several defects. In order to meet the requirements of both speed and accuracy of welding defect identification, the proposed approach uses the one-stage object detector based on RetinaNet. The model architecture based on RetinaNet is shown in Fig. 3 [24].

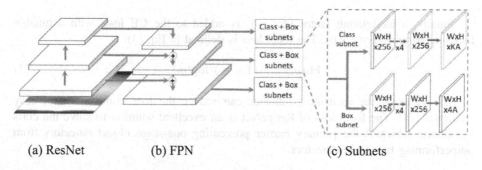

(a) ResNet   (b) FPN   (c) Subnets

Fig. 3. The model architecture based on RetinaNet generate a rich, multi-scale convolutional feature pyramid using a (b) feature pyramid net (FPN) on the top of (a) residual network (ResNet) architecture. To the (b) FPN, RetinaNet attaches (c) two subnets, class subnet for classifying welding defect (top) and box subnet for regressing from anchor boxes to ground-truth object boxes (bottom).

RetinaNet is a single, one-stage network made up of a backbone network and two subnets. Feature Pyramid Network (FPN) from [25] is adopted as the backbone network of RetinaNet, and the network efficiently generates a rich, multi-scale convolutional feature pyramid from a single resolution input image, see Fig. 3(b). Each level of the feature pyramid can be used for detecting objects at a different scale. According to [25], FPN is built on top of the ResNet architecture [26], see Fig. 3(a). In order to reduce the calculating quantity of network, the number of layers in ResNet is set to 50. The classification subnet is a small fully convolutional network (FCN) attached to each level of FPN, and the parameters of classification subnet are shared across all levels of FPN. In contrast to region proposal networks (RPN) [18], the classification subnet is deeper and does not share parameters with the box regression subnet. Box regression subnet is another small FCN attached to each level of FPN and used for regressing from anchor boxes to ground-truth object boxes. The box regression subnet has a common structure with the classification subnet but uses separate parameters, see Fig. 3(c).

RetinaNet implements a new loss function called focal loss (FL) to deal with foreground-background class imbalance, which is more effective than previous approaches [24]. The focal loss is a dynamically scaled cross entropy loss, and the scaling factor can automatically focus the model on hard negative examples during training. First, the standard cross-entropy (CE) loss for binary classification is in Eq. (1):

$$CE(p, y) = \begin{cases} -\log(p) & if \ y = 1 \\ -\log(1-p) & otherwise. \end{cases} \quad (1)$$

In the above, $y \in \{\pm 1\}$ defines the ground-truth class, and $p \in \{0, 1\}$ is the estimated probability for the class with label $y = 1$. Then, p_t is defined to represent the probability of easy examples in Eq. (2):

$$p_t = \begin{cases} p & \textit{if } y = 1 \\ 1 - p & \textit{otherwise.} \end{cases} \tag{2}$$

Finally, a modulating factor $(1 - p_t)^\gamma$ is added to the CE loss, with a tunable focusing parameter $\gamma (\gamma \geq 0)$. The focal loss is defined in Eq. (3):

$$\mathrm{FL}(p_t) = -(1 - p_t)^\gamma \log(p_t). \tag{3}$$

The value of the focusing parameter can reduce the loss contribution of easy examples [24]. The focal loss of RetinaNet is an excellent solution to solve the class imbalance, which is the primary barrier preventing one-stage object detectors from outperforming two-stage detectors.

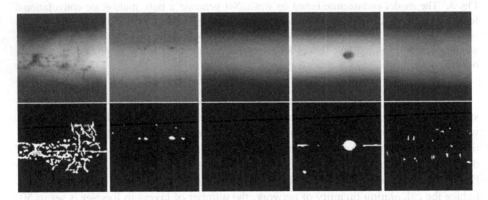

Fig. 4. Some images of group welds (upper row) and corresponding binary segmentation images (lower row).

4 Experimental Test and Discussions

4.1 Dataset Description

In order to verify the effectiveness of the proposed approach, a large number of well-labeled welding X-ray images are required for training and testing. The proposed dataset includes a public dataset and a self-labeled dataset. The details of these datasets are presented as follows.

Dataset from GDXray. The public dataset is from GDXray images [27]. The dataset includes five groups of X-ray images: Castings, Welds, Baggage, Natural Objects, and Settings. Besides the welds dataset, the castings dataset is also used in this paper. The reason for this is that the castings dataset has more than 3000 labeled defects and can be trained to detect blowhole of welding defect. Some examples of welds and castings images are respectively shown in Figs. 4 and 5 [27].

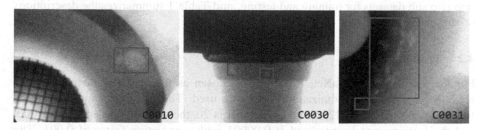

Fig. 5. Some annotated images of group castings showing bounding boxes of defects.

Fig. 6. Some images of the self-labeled dataset and bounding box.

Self-labeled Dataset. Another dataset used in the experiments is from a domestic welding company. By working with professionals, about 800 welding X-ray images are labeled with the locations and classes of defects. According to the welding images from the company, three main defect types are labeled, including blowhole, under fill or incomplete penetration, and tungsten inclusion. Under fill and incomplete penetration are classified into one category because they are similar in appearance. After labeling the images, 822 images are obtained. Each image may be free of defect or have one or several types of defects. Some examples of the self-labeled welding defect dataset are shown in Fig. 6.

Table 1. Descriptions of the proposed dataset.

Datasets	Sub datasets	Number of images	Total
Castings	Train	2662	3328
	Test	666	
Welds	Train	2051	2564
	Test	513	
Self-labeled dataset	Train	658	822
	Test	164	

Generally, large datasets are required to train deep CNNs for excellent performance. Therefore, data augmentation is used in this paper to increase datasets by flipping images horizontally, vertically and both. The dataset of all images is divided into two sub datasets for training and testing, and Table 1 summarizes the descriptions of sub datasets.

4.2   Training Process

For the training of RetinaNet, ResNet50 is chosen as the backbone network and a method for stochastic optimization (Adam) is used. The training parameters are as follows: the batch size is 12, the epoch number is 20, the steps of each epoch are 5000, and the learning rate is initialized at 0.00001 with a reduction factor of 0.001. The training is performed using a server with NVIDIA GeForce GTX 1080 graphic card. The algorithm is implemented by Keras-Tensorflow on the Ubuntu 16.04 operating system. More specifically, version of python is 3.5.2, version of Tensorflow-GPU is 1.4.1, NVIDIA CUDA toolkit version is 8.0, and NVIDIA CUDA deep neural network library (cuDNN) version is 6.0. The loss in the training process is shown in Fig. 7.

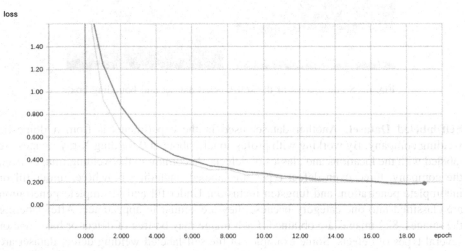

Fig. 7. The loss of the experiment (The translucent curve represents the original loss and the solid curve represents the smoothed loss)

4.3   Experimental Results

After training, testing is performed with the same configuration as the training environment. In the testing process, the mean average precision (mAP) of each category is calculated as shown in Table 2.

Table 2. The mAP of the proposed approach.

Category	Blowhole	Under fill or incomplete penetration	Tungsten inclusion
mAP	0.76	0.79	0.92

The comparisons between the proposed approach and existing approaches, including traditional image processing techniques (IPTs), machine learning algorithms (MLAs), and deep learning (DL), are summarized in Table 3.

Table 3. Comparisons between the proposed approach and existing approaches for detection and identification of Welding defect.

Category	Method	Detection	Classification		
			Blowhole	Inclusion	Under fill
IPTs	EMD [2]	✓	✗	✗	✗
	GA [3]	✓	✗	✗	✗
	FR [4]	✓	✗	✗	✗
MLAs	SVM [9, 10]	✓	✗	✗	✗
	ANN [11]	✗	✓	✓	✓
DL	DNN [14]	✓	✗	✗	✗
	VGG16 [15]	✗	✓	✓	✗
Proposed	Retina-based	✓	✓	✓	✓

The IPTs methods for welding defect detection, such as Empirical Modal Decomposition (EMD) [2], genetic algorithm (GA) [3], and fuzzy reasoning (FR) [4], mainly detected whether there were defects in the welding area without classification for different defect types.

As for machine learning algorithm (MLAs) shown in Table 3, support vector machine (SVM) and artificial neural networks (ANN) are commonly used for welding defect detection and identification. The SVM classifier for welding defect [9, 10] was used to distinguish non-defects from defects, while ANN [11] can classify different defect types but not performing well in accuracy. In the [11], the MSE (mean square errors) of under fill was 1.312 that was not satisfying. The poor effects revealed the difficulty in distinguishing multiple defect types.

In the DL methods, deep neural network (DNN) [14] learned the intrinsic feature of welding defect and can detect the whole images by utilizing sliding-window approach but unable to classify defect types. In addition, in the work [15], VGG16-based convolutional neural networks achieved a high accuracy of welding defect classification, including blowhole and inclusion. Nevertheless, the VGG16-based network was trained and tested on defect regions manually cropped instead of the whole image.

The proposed approach can realize both detection and identification of welding defect on original input images, while other methods only achieved detection or classification of defect in the welding regions. The average time for the proposed approach to identify a welding image is about 0.19 s. Moreover, the main contribution of this paper is to realize effective identification of welding defect by deep learning-based approach, rather than distinguish the differences among different network architectures.

The correct identification cases from the proposed approach are shown in Fig. 8, which proves the approach can work well not only on the public dataset but also on the self-labeled dataset. As seen in Fig. 8(a), castings defect can be correctly detected and classified even in low resolution. In Fig. 8(b), the proposed approach can also correctly detect and identify welding defect directly on the original input image rather than only in welding regions. For a better observation, the identified welding defect area among the whole image is enlarged in Fig. 8(c).

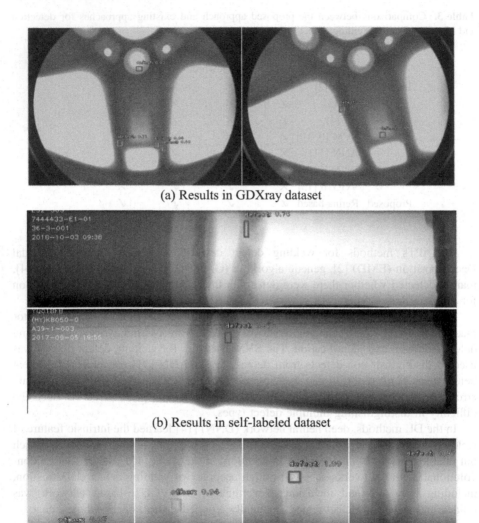

(a) Results in GDXray dataset

(b) Results in self-labeled dataset

(c) The enlarged welding defect area

Fig. 8. Testing results show that the method can work well both on (a) public dataset and (b) self-labeled dataset. (c) The detection areas of self-labeled dataset are enlarged for a better observation. (Detection boxes are in red and ground-truth bounding boxes are in green). (Color figure online)

5 Conclusion

In this paper, a practical welding defect detection and identification approach based on deep learning is presented. The approach can automatically recognize types and locations of defects by adopting a pre-trained RetinaNet. The proposed dataset includes three types of weld defect—blowhole, under fill or incomplete penetration, and tungsten inclusion. The experimental results show that the proposed approach works well on welding defect identification and mean average precision (mAP) ratings are 0.76, 0.79, and 0.92 respectively for the three defect types. The results also show that the proposed approach can detect and identify defects accurately on the original welding image instead of small region segment. The main contribution of this paper is to achieve effective automatic identification of welding defects by using deep learning-based approach, rather than distinguish the differences among different network architectures. Moreover, the proposed approach can not only directly detect welding defects on the original images without any pro-processing, but also identify the types of defects. The experimental results may have a greater improvement with more advanced network structures and hardware environments.

References

1. Tang, G., Zhong, X., Zhang, F., et al.: X-ray image edge detection based on wavelet transform and lipschitz exponent. In: International Symposium on Intelligent Information Technology and Security Informatics, pp. 195–197. IEEE (2009)
2. Chen, F.L., Wang, L.M., Han, Y.: X-ray detection of tiny defects in strongly scattered structures using the EMD method. In: International Congress on Image and Signal Processing, pp. 1033–1037. IEEE (2010)
3. Felisberto, M.K., Lopes, H.S., Centeno, T.M., et al.: An object detection and recognition system for weld bead extraction from digital radiographs. Comput. Vis. Image Underst. 102(3), 238–249 (2006)
4. Lashkia, V.: Defect detection in X-ray images using fuzzy reasoning. Image Vis. Comput. 19(5), 261–269 (2001)
5. Kasban, H., Zahran, O., Arafa, H., et al.: Welding defect detection from radiography images with a cepstral approach. NDT E Int. 44(2), 226–231 (2011)
6. Tong, T., Cai, Y., Sun, D.: Defects detection of weld image based on mathematical morphology and thresholding segmentation. In: 2012 8th International Conference on Wireless Communications, Networking and Mobile Computing (WiCOM), pp. 1–4. IEEE (2012)
7. Hassan, J., Awan, A.M., Jalil, A.: Welding defect detection and classification using geometric features. In: 2012 10th International Conference on Frontiers of Information Technology (FIT), pp. 139–144. IEEE (2012)
8. LeCun, Y., Bottou, L., Bengio, Y., Haffner, P.: Gradient-based learning applied to document recognition. Proc. IEEE 86, 2278–2324 (1998)
9. Wang, Y., Guo, H.: Weld defect detection of X-ray images based on support vector machine. IETE Tech. Rev. 31(2), 137–142 (2014)
10. Shao, J., Shi, H., Du, D., Wang, L., Cao, H.: Automatic weld defect detection in real-time X-ray images based on support vector machine. In: 2011 4th International Congress on Image and Signal Processing (CISP), vol. 4, pp. 1842–1846. IEEE (2011)

11. Sikora, R., Baniukiewicz, P., Chady, T., et al.: Detection and classification of weld defects in industrial radiography with use of advanced AI methods. In: Nondestructive Evaluation/Testing: New Technology & Application (2013)
12. Schmidhuber, J.: Deep learning in neural networks: an overview. Neural Netw. **61**, 85–117 (2015)
13. Yu, Y., Du, L., Zeng, C., Zhang, J.: Automatic localization method of small casting defect based on deep learning feature. Chin. J. Sci. Instrum. **37**, 1364–1370 (2016)
14. Hou, W., Wei, Y., Guo, J., et al.: Automatic detection of welding defects using deep neural network, p. 012006 (2018)
15. Liu, B., Zhang, X., Gao, Z., Chen, L.: Weld defect images classification with VGG16-based neural network. In: Zhai, G., Zhou, J., Yang, X. (eds.) IFTC 2017. CCIS, vol. 815, pp. 215–223. Springer, Singapore (2018). https://doi.org/10.1007/978-981-10-8108-8_20
16. Girshick, R., Donahue, J., Darrell, T., Malik, J.: Rich feature hierarchies for accurate object detection and semantic segmentation. In: Proceedings of the IEEE Conference on Computer Vision and Pattern Recognition, Columbus, OH, USA, 23–28 June 2014, pp. 580–587 (2014)
17. Girshick, R.: Fast R-CNN. In: Proceedings of the IEEE International Conference on Computer Vision, pp. 1440–1448 (2015)
18. Ren, S., He, K., Girshick, R., et al.: Faster R-CNN: towards real-time object detection with region proposal networks. In: Advances in Neural Information Processing Systems, pp. 91–99 (2015)
19. He, K., Gkioxari, G., Dollár, P., et al.: Mask R-CNN. In: Proceedings of the IEEE International Conference on Computer Vision, pp. 2961–2969 (2017)
20. Redmon, J., Divvala, S., Girshick, R., Farhadi, A.: You only look once: unified, real-time object detection. In: CVPR (2016)
21. Redmon, J., Farhadi, A.: YOLO9000: better, faster, stronger. In: CVPR (2017)
22. Liu, W., et al.: SSD: Single Shot MultiBox Detector. In: Leibe, B., Matas, J., Sebe, N., Welling, M. (eds.) ECCV 2016. LNCS, vol. 9905, pp. 21–37. Springer, Cham (2016). https://doi.org/10.1007/978-3-319-46448-0_2
23. Fu, C.-Y., Liu, W., Ranga, A., Tyagi, A., Berg, A.C.: DSSD: deconvolutional single shot detector. arXiv: 1701.06659 (2016)
24. Lin, T.Y., Goyal, P., Girshick, R., et al.: Focal loss for dense object detection. In: Proceedings of the IEEE International Conference on Computer Vision, pp. 2980–2988 (2017)
25. Lin, T.-Y., Dollár, P., Girshick, R., He, K., Hariharan, B., Belongie, S.: Feature pyramid networks for object detection. In: CVPR (2017)
26. He, K., Zhang, X., Ren, S., Sun, J.: Deep residual learning for image recognition. In: CVPR (2016)
27. Mery, D., et al.: GDXray: the database of X-ray images for nondestructive testing. J. Nondestruct. Eval. **34**(4), 1–12 (2015)

Person Re-identification Based on Feature Fusion

Qiang-Qiang Ren[✉], Wei-Dong Tian, and Zhong-Qiu Zhao

College of Computer and Information, Hefei University of Technology,
Hefei, China
18505691311@163.com

Abstract. Person re-identification is a matching task of person images captured from different camera views. In the real scene, This task is extremely challenging due to changes in pedestrian poses, camera angles and lighting. How to extract robust pedestrian features has become a key step in Person re-identification. In this paper, we propose an person re-identification model based on combined visual features. Our features consist of traditional visual features and convolutional neural network (CNN) features. We extract the CNN feature of person image and fuse them with robust Local Maximal Occurrence (LOMO) features. This fused feature has better performance. Before extracting features, we use the Retinex algorithm to preprocess person images. Finally adopt a random sampling softmax loss to effectively train the model. We experimentally show the effectiveness and accuracy of the proposed method on the VIPeR and PRID450s datasets.

Keywords: Person re-identification · Convolutional neural network ·
Local maximal occurrence

1 Introduction

Person re-identification is a research hotspot in intelligent video surveillance systems [1] and has received extensive attention in recent years. The purpose of pedestrian re-identification is to realize the association of the same target in the multi-camera network [2], to find out whether a specific target appears under the camera, to capture the motion trajectory of the target in the monitoring network, and to achieve continuous tracking of the moving target. This method of finding the target trajectory by the person re-identification technique does not require coverage of a large-area camera, and can effectively reduce the cost, and thus is more in line with the needs of the actual monitoring system. However, due to various factors such as illumination changes, occlusion, and shooting angle, the recognition performance may be affected [3]. (See in Fig. 2) How to improve the recognition rate of person re-identification technology is still a challenging problem [4–8].

In recent years, image features extracted using convolutional neural networks have proven to be very successful for person re-identification [9–12]. However, CNN-based features focus on the global look and lack of powerful representation of viewpoint changes. The Local Maximum Occurrence (LOMO) feature analyzes the local features

© Springer Nature Switzerland AG 2019
D.-S. Huang et al. (Eds.): ICIC 2019, LNAI 11645, pp. 65–73, 2019.
https://doi.org/10.1007/978-3-030-26766-7_7

and maximizes them to provide a stable representation of the change in viewpoint. Therefore, we extract the LOMO function, which is designed for re-recognizing and mining more local information of the character image for the pedestrians with the same viewpoint. These low-level hand-crafted features combine the advanced learning CNN functionality in the post-training and processing phases to obtain a more powerful feature representation that simulates the appearance of characters acquired in different states image enhancement is an effective means to improve the quality of feature extraction. Therefore, we recommend applying the Retinex algorithm [13–15] to pre-process the character image. The Retinex algorithm is designed to produce color images with sharper colors. The restored image usually contains vivid color information and enhanced detail in the shaded area. Finally, we use the random sampling softmax loss to effectively train the model (Fig. 1).

(a) VIPeR   (b) CUHK01   (c) PRID450s

Fig. 1. Sample images from VIPeR, CUHK01 and PRID450s datasets. Images on the same column represent the same person.

Based on the considerations above, we using an identification model to get the person identification. And propose fuses The Local Maximal Occurrence (LOMO) feature with the CNN features to improve representation ability of the person image feature. And in a nutshell, the main contributions of this paper are summarized as follows:

1. We propose an person re-identification model based on combined visual features and fuses The Local Maximal Occurrence (LOMO) feature with the CNN features.
2. We use the Retinex algorithm to preprocess person images. Mining vivid color information for feature extraction.

3. We evaluate the proposed model on two person re-ID benchmark datasets (VIPeR and PRID450s) and the experiment results show that the effectiveness of the proposed method.

2 Related Work

2.1 Person Re-identification Based on Convolutional Neural Network

With the rapid development of deep learning, convolutional neural network have become an effective method to solve the problem of person re-identification. A large number of person re-identification frameworks based on convolutional neural networks have been proposed.

In order to solve the problems caused by pose and viewpoint variants. The Deep Filter Paired Neural Network (FPNN) was first proposed [16], it uses the patch matching layer and the maximum out pool. And F-PNN was the first method to use deep learning to solve the problem of person re-identification. [17] Improve the deep learning architecture, they specifically designed the cross input neighborhood difference layer. Later, the "Siamese" [18] deep neural structure and cosine layer were used to deal with large changes in the image of the character. In addition, [19] transferred cross-domain visual knowledge to the target dataset, and on this basis, proposed Deep Transfer Metric Learning.

2.2 Person Re-identification Based on Traditional Features

The person re-identification algorithm based on traditional features is the earliest and most basic method. The purpose is to establish a robust and well-discriminating feature descriptor [20]. Researchers often use pedestrian local and global features to characterize person, including colors and textures. Color feature is one of the most widely used visual features, and has good immutability to changes in viewing angle and person attitude, such as RGB, HsV, LAB, Ycbcr. Texture features are generally used to reflect the homogenous phenomena existing in the image, and have good invariance to the effects of illumination. Commonly used are LBP, SILTP [19], HOG. Generally, the statistical histogram is used to obtain the color and texture features, and the extraction method is simple, and can be used as an effective basis for pedestrian matching. In order to improve the matching rate of person re-identification, most researchers have combined these features in different forms and degrees.

3 Proposed Method

3.1 Network Architecture

The architecture of the person re-identification method proposed in this paper is shown in Fig. 2. The framework consists of two parts. One part is the traditional convolutional neural network, which is mainly composed of a convolutional layer and a pooling layer. This part can extract the convolution neural network feature of pedestrian images. The other part extracts the Local Maximal Occurrence (LOMO) feature of the viewpoint. This is an efficient and effective visual feature representation that is specifically used to mine local information of images and solve problems caused by viewpoint changes, and to achieve the most advanced results [3]. After the LOMO features are extracted, the LOMO features are processed using principal component analysis (PCA). There are two purposes for doing this: first, reduce the size of the LOMO feature vector, reduce the amount of computation and unnecessary information. Second, the CNN feature is kept consistent with the LOMO feature size, which combines the information represented in the two feature vectors. Finally, the random sampling softmax loss is used to effectively train the model.

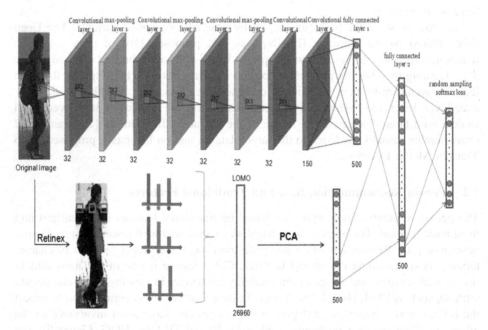

Fig. 2. The structure of our person re-identification model.

3.2 CNN Features

Our proposed CNN model can extract the deep feature of pedestrians in the image. This feature provides a powerful distinguishing ability because we learn these features directly from the big dataset. The part of our extraction of deep feature is in the upper part of Fig. 2: CNN consists of five convolutional layers and three layers of pooling layers. The size of the first convolution kernel is 5 × 5. Other convolution kernel sizes are 3 × 3. The kernel size of max-pooling is 2 × 2. In addition, we set the number of channels for the convolution and pooling layers to 32. In the last two convolutional layers we use two 1D horizontal convolutions of size 3 × 1, which reduces the feature map to a single column vector. The 1D convolution can extract texture features with high resolution from the pedestrian image. In the last convolutional layer, we set the number of channels to 150 and use these feature maps as fully connected inputs to generate an output vector of size 500. In addition, batch normalization and ReLU activation are followed by all convolutional layers.

3.3 Local Maximal Occurrence

Color is one of the important features that describe the appearance of a person. However, if the camera is in different lighting conditions or camera settings are different, the acquired image may be perceived differently in color. Therefore, the perceived color of the same person will change as the view of the camera changes. As shown in Fig. 3(a) shows some sample images in the VIPeR dataset [9]. It can be seen that the images of the same person in the two camera views vary greatly in color appearance and illumination.

In this paper, we apply the Retinex algorithm to preprocess the character image before training the model. Retinex focuses on the brightness and color perception of the image. It can produce sharper, brighter color images. The restored image usually contains vivid color information, especially enhanced detail in the shaded area. Figure 3(b) shows the pedestrian image processed by the Retinex algorithm.

(a)   **(b)**

Fig. 3. Image enhancement by the Retinex algorithm

We use a sliding window to extract the local detail features of the character image. The specific method is as follows: the sub-window size is set to 10×10, and the overlapping step size is 5 pixels to locate the partial patch in the 128×48 image. In each sub-window, we extract two SILTP histograms and an $8 \times 8 \times 8$-bin joint HSV histogram. Each histogram bin represents the probability of occurrence of a pattern in the child window. To resolve viewpoint changes, we check all child windows in the same horizontal position and maximize the local occurrence of each pattern in those child windows. The resulting histogram achieves some invariance of viewpoint changes while capturing the local region features of the person. The bottom of Fig. 2 shows the proposed LOMO feature extraction process.

3.4   Random Sampling Softmax Loss

We replaced the original softmax loss with a more efficient random sampling softmax (RSS) loss. The traditional softmax loss is suitable for small batches classification problems. In monitoring video, a large number of pedestrians may need to be identified, so it is not suitable for solving pedestrian re-identification problems. We use a random sampled softmax loss to solve this problem by randomly selecting a subset of softmax neurons for each input sample to calculate the loss and gradient. The detailed calculation method is as follows.

Suppose the number of target classes is $C + 1$, where the $C + 1$ class is the background and the other classes are the pedestrian's identities. Each data sample is represented by $\{x, t\}$, where $x \in R^{C+1}$ is the input of softmax, which is the classifier score, and t is the 1-of-$(C + 1)$ binary vector representing the label. Then the original softmax loss can be written as:

$$l = -\sum_{i=1}^{C+1} t_i \log y_i, where\ y_i = \frac{e^{x_i}}{\sum_{j=1}^{C+1} e^{x_j}}$$

During training, the random sampling softmax loss will randomly select the K $(K \ll C + 1)$ dimension from x and t to calculate the loss and gradient. Assuming the selected index is $i_1, i_2, \ldots, i_K$, the sample classifier score. Then define the RSS loss function as:

$$\tilde{l} = -\sum_{i=1}^{K} \tilde{t}_i \log \tilde{y}_i, where\ \tilde{y}_i = \frac{e^{\tilde{x}_i}}{\sum_{j=1}^{K} e^{\tilde{x}_j}}$$

4   Experiments

4.1   Datasets

CUHK01 [15], VIPeR [8] and PRID450s are three publicly available datasets that we test based on this data set. Each pair of pedestrian images in the dataset is presented in two disjoint camera views with significant illumination changes, posture changes, and body part distortion. We briefly describe these three datasets in Table 1. During training, we propose to apply the Retinex algorithm to preprocess person images.

Table 1. Re-identification datasets used in our experiments.

	CUHK01	VIPeR	PRID450s
No. of images	3884	1264	900
No. of identities	971	632	450
No. of images in training set	485	316	225
No. of camera views	2	2	2
No. of images per view per ID	2	1	1

4.2 Parameters Setting

The weight of the neural network is initialized using a Gaussian distribution. Set the mean to 0 and the standard deviation to 0.01. The deviation is set to 0. The learning rate is set to 0.01. We used a fully connected layer with dropout at a rate of 0.6.

The randomly initialized convolutional neural network is pre-trained on the CUHK01. Because VIPeR has only 632 training images, it can't get good training results. Therefore, the network is pre-trained using random sampling softmax on the CUHK01 data set, and then the CNN part is fine-tuned using the VIPeR data set. The learning rate is set to 0.0005.

4.3 Evaluation on Datasets

In order to verify the performance of our proposed model, we tested on VIPeR, PRID450s. As can be seen from Tables, our method is superior to most comparison methods, which proves the effectiveness of the proposed method.

4.3.1 Evaluation on VIPeR
See Table 2.

Table 2. Results (Rank1, Rank5 and Rank10 matching accuracy in %) on the VIPeR dataset

Rank	1	5	10
Our model	42.06	64.01	73.39
Deep feature learning	**40.50**	60.80	70.40
LOMO + XQDA	**40.00**	67.40	80.51
mFilter	**29.11**	52.10	67.20
SalMatch	**30.16**	52.31	65.54
LFDA	*24.18*	52.85	67.12
LADF	*29.34*	61.04	75.98
RDC	*15.66*	38.42	53.86
KISSME	*24.75*	53.48	67.44
LMNN-R	*19.28*	48.71	65.49
L2 − norm	*10.89*	22.37	32.34

4.3.2 Evaluation on PRID450S
Since LOMO focuses on the characteristics of HSV and SILTP histograms, it performs better on PRID450, which is undergoing specific lighting conditions. So the fused feature have a better performance on the PRID450 dataset (Table 3).

Table 3. Results (Rank1, Rank5 and Rank10 matching accuracy in %) on the PRID450s

Rank	1	5	10
Our model	60.62	82.84	90.84
Mirror KMFA	**55.42**	79.29	87.82
Ahmed's Deep Re-id	**34.81**	63.72	76.24
ITML	**24.27**	47.82	58.67
LFDA	**36.18**	61.33	72.40
KISSME	*36.31*	65.11	75.42
LMNN-R	*28.98*	55.29	67.64
L2 − norm	*11.33*	24.50	33.22

5 Conclusion

In this paper, we proposed an effective person re-identification model. The model uses CNN features and local maximum occurrence features together. It automatically adjusts the weight of this information through the neural network's backpropagation process. We conducted experiments on two challenging human re-identification data sets (VIPeR, PRID450s). Experiments show that our proposed pedestrian re-identification model has good performance.

Acknowledgment. This research was supported by the National Natural Science Foundation of China (No. 61672203) and Anhui Natural Science Funds for Distinguished Young Scholar (No. 170808J08).

References

1. Khedher, M.I., El-Yacoubi, M.A., Dorizzi, B.: Fusion of appearance and motion-based sparse representations for multi-shot person re-identification. Neurocomputing **248**, 94–104 (2017)
2. Su, C., Zhang, S., Yang, F., et al.: Attributes driven tracklet-to-tracklet person re-identification using latent prototypes space mapping. Pattern Recognit. **66**, 4–15 (2017)
3. Zhao, R., Oyang, W., Wang, X.: Person re-identification by saliency learning. IEEE Trans. Pattern Anal. Mach. Intell. **39**(2), 35–46 (2017)
4. Cao, J., Pang, Y., Li, X.: Pedestrian detection inspired by appearance constancy and shape symmetry. IEEE Trans. Image Process. **25**(12), 15–28 (2016)
5. Chen, D., Yuan, Z., Chen, B., et al.: Similarity learning with spatial constraints for person re-identification. In: Computer Vision and Pattern Recognition, pp. 1268–1277. IEEE (2016)
6. Cho, Y.J., Yoon, K.J.: Improving person re-identification via pose-aware multi-shot matching. In: CVPR, pp. 1354–1362 (2016)
7. Liu, H., Tian, Y., Wang, Y., et al.: Deep relative distance learning: tell the difference between similar vehicles. In: CVPR, pp. 2167–2175. IEEE (2016)
8. Matsukawa, T., Okabe, T., Suzuki, E., et al.: Hierarchical gaussian descriptor for person re-identification. In: CVPR, pp. 1363–1372 (2016)

9. Li, W., Zhao, R., Xiao, T., Wang, X.: DeepReID: deep filter pairing neural network for person re-identification. In: Proceedings of the IEEE Conference on Computer Vision and Pattern Recognition (CVPR) (2014)
10. Ahmed, E., Jones, M., Marks, T.K.: An improved deep learning architecture for person re-identification. In: Proceedings of the IEEE Conference on Computer Vision and Pattern Recognition (CVPR) (2015)
11. Cheng, D., Gong, Y., Zhou, S., Wang, J., Zheng, N.: Person re-identification by multi-channel parts-based cnn with improved triplet loss function. In: Proceedings of the IEEE Conference on Computer Vision and Pattern Recognition (CVPR) (2016)
12. Varior, R.R., Haloi, M., Wang, G.: Gated siamese convolutional neural network architecture for human re-identification. In: Leibe, B., Matas, J., Sebe, N., Welling, M. (eds.) ECCV 2016. LNCS, vol. 9912, pp. 791–808. Springer, Cham (2016). https://doi.org/10.1007/978-3-319-46484-8_48
13. Land, E.H., McCann, J.: Lightness and retinex theory. JOSA **61**(1), 1–11 (1971)
14. Jobson, D.J., Rahman, Z.-U., Woodell, G.A.: Properties and performance of a center/surround retinex. IEEE Trans. Image Process. **6**(3), 451–462 (1997)
15. Li, W., Zhao, R., Xiao, T., Wang, X.: DeepReID: deep filter pairing neural network for person re-identification. In: IEEE CVPR (2014)
16. Ahmed, E., Jones, M., Marks, T.K.: An improved deep learning architecture for person re-identification. In: IEEE CVPR (2015)
17. Yi, D., Lei, Z., Liao, S., Li, S.Z.: Deep metric learning for person re-identification. In: IEEE ICPR (2014)
18. Hu, J., Lu, J., Tan, Y.-P.: Deep transfer metric learning. In: IEEE CVPR (2015)
19. Liao, S., Hu, Y., Zhu, X., Li, S.Z.: Person re-identification by local maximal occurrence representation and metric learning. In: Proceedings of the IEEE International Conference on Computer Vision and Pattern Recognition (CVPR), pp. 2197–2206 (2015)
20. Liao, S., Zhao, G., Kellokumpu, V., et al.: Modeling pixel process with scale invariant local patterns for background subtraction in complex scenes. In: Computer Vision and Pattern Recognition, pp. 1301–1306. IEEE (2010)

Exploiting Local Shape Information for Cross-Modal Person Re-identification

Md. Kamal Uddin$^{(\boxtimes)}$, Antony Lam, Hisato Fukuda,
Yoshinori Kobayashi, and Yoshinori Kuno

Graduate School of Science and Engineering, Saitama University, Saitama, Japan
{kamal,antonylam,fukuda,kuno}@cv.ics.saitama-u.ac.jp,
yosinori@hci.ics.saitama-u.ac.jp

Abstract. In computer vision, person re-identification (Re-id) is an important problem, aiming to match people across multiple camera views. Most of the existing Re-id systems widely use RGB-based appearance cues, which is not suitable when lighting conditions are very poor. However, for many security reasons, sometimes continued surveillance via camera in low lighting conditions is inevitable. To overcome this problem, we take advantage of the Kinect sensor based depth camera (e.g., Microsoft Kinect), which can be installed in dark places to capture video, while RGB based cameras can be installed in good lighting conditions. Such types of heterogeneous camera networks can be advantageous due to the different sensing modalities available but face challenges to recognize people across depth and RGB cameras. In this paper, we propose a body partitioning method and novel HOG based feature extraction technique on both modalities, which extract local shape information from regions within an image. We find that combining the estimated features on both modalities can sometimes help to better reduce visual ambiguities of appearance features caused by lighting conditions and clothes. We also propose an effective metric learning approach to obtain a better re-identification accuracy across RGB and depth. Experimental results on two publicly available RGBD-ID datasets show the effectiveness of our proposed method.

Keywords: Person re-identification · Cross-modality matching ·
Heterogeneous camera network

1 Introduction

Nowadays, person re-identification (Re-id) has gained increasing attention in the research community due to its importance in various surveillance and intelligent applications, such as forensic search, multi-camera tracking, pedestrian detection in autonomous driving and access control. Recent Re-id research has mainly focused on RGB-RGB matching, which is the most common scenario where there is only a single-modality. However, RGB based Re-id systems have limitations in surveillance when lighting is either very poor, since RGB based cameras cannot capture sufficient information in dark environments. In comparison to RGB cameras, depth cameras can capture video even in low lighting conditions. So, it is possible to extract depth

© Springer Nature Switzerland AG 2019
D.-S. Huang et al. (Eds.): ICIC 2019, LNAI 11645, pp. 74–85, 2019.
https://doi.org/10.1007/978-3-030-26766-7_8

information and the body skeleton using depth cameras [1] (e.g., Microsoft Kinect) in dark environments. The Kinect sensor in particular, can capture the depth information of each pixel by using an infrared sensor, regardless of the pedestrian's color appearance and illumination in indoor environments (see Fig. 1).

Fig. 1. Examples of RGB and depth images captured in indoor environments. In Row 1, columns 1, 4, 5 and 6 show the RGB images in good illumination conditions, with columns 2 and 3 in poor illumination conditions accordingly. Row 2 shows the depth images of all RGB images.

Most existing works in person Re-id emphasize on either RGB camera networks [9–14] or depth camera networks [1, 2]. Very few recent works utilize RGB-Infrared heterogeneous camera network [3, 4]. While less sensitive than RGB, infrared cameras can still be affected by illumination changes from real-world environments, as well as temperature changes in the environment. Zhuo *et al.* [5] proposes a dictionary learning based method on heterogeneous camera networks that contain RGB and depth images. Specifically, the authors in [5] proposed two kinds of edge gradient features for RGB images, which are the classic Histogram of Oriented Gradient (HOG) [21] and Scale Invariant Ternary Patterns (SILTP) [13]. Both of them can describe the body's shape coarsely. For depth images, they extract Eigen-depth features from 3D point clouds of segmented torso and head parts only.

In this work, we propose a body partitioning method and HOG based feature extraction technique on both modalities because it captures edge or gradient structures which represent local shapes in scenes. In [5], they extract features only from segmented regions (head and torso part) from the depth domain. However, to the best of our knowledge, our work is the first attempt to extract edge gradient features on both RGB and depth modalities at the same time. To learn discriminant features, we first apply Principle Component Analysis (PCA) for dimensionality reduction, then we exploit the beneficial properties of Linear Discriminant Analysis (LDA) within the PCA subspace to find the low intra-class variation and high inter-class variation of the data. This allows us to gain good improvements for the task of person re-identification on heterogeneous camera networks.

We tested our methods on two publicly available datasets, the BIWI RGBD-ID [7] and IAS-Lab RGBD-ID [8]. Our contributions can be summarized as follows:

1. We propose a body partitioning method and HOG based feature extraction technique on both modalities, RGB and Depth domains, which extract local shape information from the image. To the best of our knowledge, this is the first attempt to extract edge gradient features on both modalities.
2. We utilize PCA and LDA based metric learning to increase re-identification accuracy.
3. Extensive experiments show the effectiveness of the proposed method over two RGB-D benchmark re-identification datasets.

2   Related Work

In this section, we first review some related works in person re-identification, especially for the single modal and multi-modal cases, then we introduce cross-modal approaches which are the most related to our work.

Most of the existing Re-id systems work with RGB-based camera networks which use color cues as appearance features [9–14]. As the feature matching process is RGB-RGB, this is a single-modality problem. These approaches have been effective but in the situations where people may change their clothing in long-term monitoring or in dark environments, these RGB-based appearance features tend to fail.

There are some works that address multi-modal person re-identification [1, 6, 15–18] by combining RGB and depth information in order to extract robust discriminative features. In [1], a depth-shape descriptor called Eigen-depth has been proposed to extract features from the depth domain. The authors also proposed a kernelized implicit feature transfer scheme to estimate Eigen-depth features implicitly from RGB images when depth information is not available. Though, the methodology is in principle applicable in cross-modal re-identification, the authors did not perform any evaluations in this scenario. In [6], a depth based segmentation technique is used to extract the features from the foreground body parts. In [15], the authors fused clothing appearance features with anthropometric measures extracted from depth information. In [16], a multi-modal uniform deep learning method was proposed to extract the anthropometric and appearance features from RGB and depth images. The authors in [17], proposed a tri-modal based person re-identification method by combining RGB, depth and thermal data. In [18], the authors proposed a height-based gait feature that integrated an RGB based height histogram and gait feature from the depth domain.

Besides the single and multi-modal cases, very few works [3–5, 19, 20] have investigated with cross-modal person re-identification. Among them, [3, 4, 19, 20] were published concerning cross-modal person re-identification between RGB and Infrared images. As noted earlier, infrared can be affected by temperature changes in the environment. At the same time, it can still be affected by illumination changes in real-world settings (though to a lesser extent than RGB). Only [5] performed cross-modal person re-identification between depth and RGB on heterogeneous camera networks. In [5], the authors proposed a dictionary learning based method to encode different-modality body shape features such as edge gradient feature and Eigen-depth feature which are extracted from the RGB and depth domain respectively.

In contrast to the above works on cross-modal re-identification, we propose extracting local shape information from partitioned regions of the body for the RGB and depth domains on heterogeneous camera networks, and also propose an effective metric learning approach.

3   Proposed Method

Our re-identification approach has three distinct phases: (1) Feature extraction, (2) Metric learning, and (3) Feature matching. The overall system is illustrated in Fig. 2.

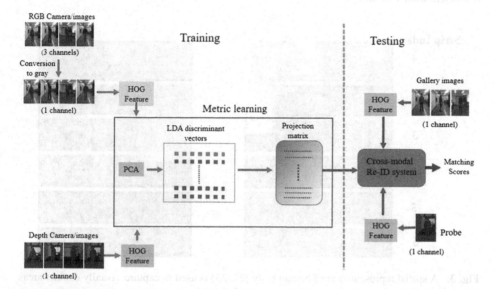

Fig. 2. Overview of our proposed approach. In the training stage, labeled image pairs from RGB and depth cameras are used to jointly learn the discriminative features by LDA. After dimensionality reduction, the projected features are matched by using Euclidean distance in the testing stage.

3.1   Feature Extraction

In this section, we give the details of feature extraction using HOG [21], which extracts features from both camera images. HOG has been widely accepted as one of the best features to capture edges or local shape information. Though HOG can extract features from a true color (RGB and LAB color spaces) or grayscale images, we find extracting features from grayscale images works best in our RGB-depth setup. According to our proposed method (see Fig. 2), RGB images are captured by an RGB camera and depth images captured by a depth camera (e.g. Kinect) on the heterogeneous camera network. To facilitate cross-modal learning, our aim is to first make the images from the RGB and depth domains as similar as possible. Since RGB images have three color channels, we need to convert it to a single channel for convenience because the Kinect sensor depth images are 16-bit depth monochrome images with 65,536 levels of sensitivity [6–8].

In this work, we divide a person image into six horizontal stripes (see Fig. 3). This is a generic human body partitioning method that is widely used in existing methods [22, 23] to capture distinct areas of interest. For HOG features, each strip is further divided into 2×2 blocks of 8×8 pixel cells with 50% overlapping blocks, and each cell contains 9 orientation bins. Each strip returns the features as a 1-by-v_1 vector. Finally, the feature vectors of all 6 strips are concatenated to construct a final feature vector of 1-by-d, where $d = v_1 + \ldots + v_6$, is inside an image window. Since the histograms are computed for regions of a given size within a window, HOG is robust to some location variability of body parts. HOG is also invariant to rotations smaller than the orientation bin size.

Fig. 3. A spatial representation of human body [22, 23] is used to capture visually distinct areas of interest. The representation employs six equal-sized horizontal strips in order to capture approximately the head, upper and lower torso and upper and lower legs.

3.2 Metric Learning

In the metric learning approaches, we first extract features for each image, and then learn a metric with which the training data have strong inter-class differences and intra-class similarities. In such a case, we employ linear discriminant analysis (LDA) to determine a set of projection vectors maximizing the between-class scatter matrix (S_b) while minimizing the within-class scatter matrix (S_w) in the projective space. However, LDA often suffers from issues such as small sample size and high dimensionality (so there are too many variables). When there are not enough training samples and/or the dimensionality is too high, (S_w) may become singular, and it is difficult to compute the LDA vectors. In our work, we use a two-stage approach PCA + LDA [24] to address this problem. First, we reduce the feature dimensionality using Principal Component

Analysis (PCA), and then LDA is applied on the reduced PCA subspace, in which (S_w) is non-singular.

LDA tries to find the projection matrix W maximizing the ratio of the determinant of S_b to S_w,

$$W = \arg \max \left| \frac{W^T S_b W}{W^T S_w W} \right| \tag{1}$$

Consider that the training set contains C classes which are taken from the RGB camera and depth camera, and each class X_i has n_i samples. S_b and S_w are defined as,

$$S_b = \sum_{i=1}^{C} n_i (\mu_i - \mu)(\mu_i - \mu)^T \tag{2}$$

$$S_w = \sum_{i=1}^{C} \sum_{\bar{x}_k \in X_i} (\bar{x}_k - \mu_i)(\bar{x}_k - \mu_i)^T \tag{3}$$

where μ is the mean of all data, μ_i is the mean for the class X_i, and $\bar{x}_k$ is the sample belonging to class X_i. W can be computed from the eigenvectors of $S_w^{-1} S_b$ [25]. The eigenvectors corresponding to the first m largest eigenvalues are used to construct the projection matrix.

3.3 Feature Matching/Classification

After obtaining the projection matrix, we aim to recognize a certain person on heterogeneous camera network. The goal of our cross-modal person re-identification system now is to find a person image that has been selected in the depth camera (probe image) in all images from the RGB camera (gallery images). This is obtained by calculating the Euclidean distances between the probe image and all gallery images using the learned metric, and returning those gallery images with the smallest distances as potential feature matches.

4 Experiments

In this section, we evaluate the performance of our approach by performing experiments on two RGB-D person re-identification datasets BIWI RGBD-ID [7] and IAS-Lab RGBD-ID [8] recorded by Microsoft Kinect cameras. Both datasets target long-term people re-identification from RGB-D cameras. In our work, besides the HOG based feature extraction technique, we also experimented on two well-known local shape descriptors SILTP [13] and LBP [26, 27] for both datasets. The SILTP descriptor is an improved operator over LBP [13].

4.1 Datasets

BIWI RGBD-ID [7]. This dataset has three groups of sequences, namely "Training", "Still" and "Walking", each of which contains groups of 50, 28 and 28 people respectively with different clothes, and collected on different days and in a different scenes. Each person is associated with about 300 sequence of frames of depth images, RGB images and skeletons. The BIWI dataset consists of RGB images with a resolution of 1280 × 960 and depth images with a resolution of 640 × 480.

IAS-Lab RGBD-ID [8]. In the IAS-Lab RGBD-ID dataset, there are 11 different people. This dataset contains three groups of sequences "Training", "TestingA" and "TestingB", and each person performs out-of-plane rotations on himself and walks in the recordings. There are about 500 frames of depth images, RGB images and skeletons for each person. The first (Training) and second (TestingA) sequences were acquired when same person was wearing different clothes, while the third one (TestingB) was collected in a different room, but with the same clothing as in the first group (Training). Some sequences in "TestingB" were recorded under low lighting. The IAS-Lab dataset consists of RGB images with a resolution of 640 × 480 and depth images with a resolution of 640 × 480.

Data Pre-processing. As depth images are single channel, so we convert all RGB images to gray scale images. Before HOG feature extraction, all depth and RGB images are resized to 256 × 192 to maintain the original aspect ratio of the images, which retain edge gradient shape without distortions.

4.2 Evaluation Metrics

We show the results in terms of recognition rate as a cumulative matching characteristic (CMC) curve and rank-k accuracy, which are common practice in the Re-id literature [1, 11–14]. Rank-k accuracy is the cumulative recognition rate of correct matches at rank k. The CMC curve represents the cumulative recognition rates at all ranks. The evaluation is repeated 10 times and the average results are reported. For quantitative evaluation, the average rank 1, 5 and 10 accuracy performance measures are reported.

4.3 Compared Methods

To evaluate the effectiveness of our approach, we compare our method with a recently proposed cross-modal re-identification approach on a heterogeneous camera network [5]. In [5], the authors performed the Re-id task across the depth and RGB modalities and proposed a dictionary learning based method to encode different-modality body shape features including an edge gradient feature and the Eigen-depth feature for the BIWI RGBD-ID and RGBD-ID datasets. In our work, we use PCA and LDA based metric learning method for edge gradient feature extraction on both modalities. Besides HOG features, we also investigate two local body shape descriptors including SILTP [13] and LBP [26, 27] on our proposed approached. These feature descriptors are extracted using same algorithm for both modalities. LBP has a nice invariant property under monotonic gray-scale transforms, but it is not robust to image noise. SILTP improves on LBP by introducing a scale invariant local comparison tolerance and robustness to image noise [13].

4.4 Evaluation on BIWI RGBD-ID

We use the complete "Training" and "Still" groups in our experiment, hence there are 78 video sequences (samples) in total. And then we randomly select five frames each from the RGB and depth video sequences for each sample. By convention, we randomly choose about half of the samples, 40 pedestrians for training and the remaining for testing. Each experiment is carried out in two cases. For the first case, we select RGB images as the gallery and depth images as the probe, and in the second case, we use depth images as the gallery and RGB images as the probe. In our experiment, the single-shot setting is used, where one image per sample is randomly selected as the gallery. The results are shown in Table 1.

Table 1. Average recognition rates of our approach for different scenarios on the BIWI dataset. This table also shows the comparison with [5], where in [5] no detailed information on the evaluation procedure was given.

Approach	Gallery-RGB, Probe-Depth			Gallery-Depth, Probe-RGB		
	Rank-1 (%)	Rank-5 (%)	Rank-10 (%)	Rank-1 (%)	Rank-5 (%)	Rank-10 (%)
Eigen-depth HOG, CCA [5]	6.31	27.63	40.79	6.31	24.21	40.79
Eigen-depth SILTP, CCA [5]	6.58	27.37	45.00	8.42	26.32	41.58
Eigen-depth HOG, LSSCDL [5]	7.11	28.42	41.32	8.42	27.11	46.05
Eigen-depth SILTP, LSSCDL [5]	7.37	29.47	50.26	9.47	24.21	40.26
Eigen-depth SILTP, diction. learning [5]	9.21	26.32	46.05	12.11	26.32	41.58
Eigen-depth HOG, diction. learning [5]	11.32	30.26	48.16	11.84	28.42	44.47
LBP, PCA + LDA metric learn. (Ours)	35.01	82.51	95.08	34.30	82.53	95.21
SILTP, PCA + LDA metric learn. (Ours)	36.89	84.20	96.52	36.14	83.34	95.21
HOG, PCA + LDA metric learn. (Ours)	41.43	82.51	94.36	36.52	79.73	92.38

These results outperform the results from [5] for the Eigen-depth features combined with HOG/SILTP. In [5], the authors also compared their method with Least Square Semi-Coupled Dictionary Learning (LSSCDL) [28] and Canonical Correlation Analysis (CCA) [29]. In the results, we also see that when LBP and SILTP features are extracted from both modalities and we apply our metric learning approach, then results also outperform the method proposed by [5] on the heterogeneous camera network. The average results of the three local shape descriptors with our metric learning approach is shown in Fig. 4 using a CMC curve over 10 trials.

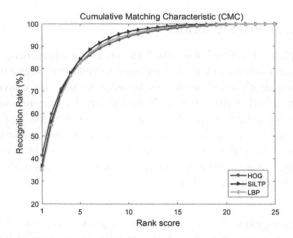

Fig. 4. Performance on BIWI RGBD-ID (single-shot) dataset for three local shape descriptors with our approach, where we set Gallery-RGB and Probe-Depth images.

4.5 Evaluation on IAS-Lab RGBD-ID

On this dataset, the evaluation also follows the same settings as with the BIWI dataset with one exception. In this experiment, we randomly select ten frames from the RGB and depth images to avoid singularity problem with LDA. We use the complete "Training" and "TestingA" groups in our experiment, hence there are 22 samples in total. By convention, we randomly choose exact half of the samples, 11 pedestrians for training and the remaining for testing. The average rank-1, rank-5 and rank-10 accuracies over 10 trials of evaluation are reported in Table 2. The performance of the tested methods is shown in Fig. 5 using a CMC curve over 10 trials.

Table 2. Average recognition rates of our approach for different scenarios on IAS-Lab dataset.

Approach	Gallery-RGB, Probe-Depth			Gallery-Depth, Probe-RGB		
	Rank-1 (%)	Rank-5 (%)	Rank-10 (%)	Rank-1 (%)	Rank-5 (%)	Rank-10 (%)
LBP, PCA + LDA metric learn. (**Ours**)	34.11	94.38	99.94	33.71	95.45	99.74
SILTP, PCA + LDA metric learn. (**Ours**)	37.20	95.33	100	35.24	96.04	99.78
HOG, PCA + LDA metric learn. (**Ours**)	38.93	96.28	99.89	38.21	96.80	99.99

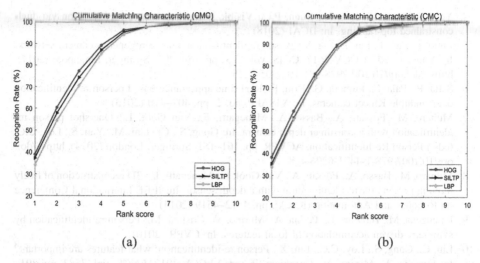

Fig. 5. Performance on IAS-Lab RGBD-ID (single-shot) dataset for three local shape descriptors with our approach. (a) Gallery-RGB and Probe-Depth images, (b) Gallery-Depth and Probe-Depth images.

5 Conclusions and Future Work

In this paper, we have presented a cross-modal re-identification system for RGB and depth heterogeneous camera networks. This is in contrast to most existing camera networks, which are based on RGB cameras only. Such RGB only camera networks tend to fail in poor lighting conditions or dark environments. To the best of our knowledge, ours is the first attempt at cross-modal person re-identification where edge gradient features for local shape descriptors are used the same for both modalities. We have also proposed an effective metric learning approach to obtain a better re-identification matching score across the RGB and depth modalities. Experimental results on two benchmark RGB-D person re-identification datasets show the effectiveness of our proposed approach for the cross-modal re-identification problem.

In future work, we plan to extend the work by combining depth information and RGB appearance cues in order to extract more discriminative features, and exploit DNN based approaches to improve the matching accuracy across multi-modal features.

References

1. Wu, A., Zheng, W.S., Lai, J.H.: Robust depth-based person re-identification. IEEE Trans. Image Process. **26**(6), 2588–2603 (2017)
2. Haque, A., Alahi, A., Fei-Fei, L.: Recurrent attention models for depth-based person identification. In: CVPR (2016)
3. Wu, A., Zheng, W.S., Yu, H.X., Gong, S., Lai, J.: RGB-infrared cross-modality person re-identification. In: ICCV (2017)

4. Ye, M., Wang, Z., Lan, X., Yuen, P.C.: Visible thermal person re-identification via dual-constrained top-ranking. In: IJCAI (2018)
5. Zhuo, J., Zhu, J., Lai, J., Xie, X.: Person re-identification on heterogeneous camera network. In: Yang, J. (ed.) CCCV 2017. CCIS, vol. 773, pp. 280–291. Springer, Singapore (2017). https://doi.org/10.1007/978-981-10-7305-2_25
6. Satta, R., Pala, F., Fumera, G., Roli, F.: Real-time appearance-based person re-identification over multiple Kinect cameras. In: VISAPP, no. 2, pp. 407–410 (2013)
7. Munaro, M., Fossati, A., Basso, A., Menegatti, E., Van Gool, L.: One-shot person re-identification with a consumer depth camera. In: Gong, S., Cristani, M., Yan, S., Loy, C.C. (eds.) Person Re-Identification. ACVPR, pp. 161–181. Springer, London (2014). https://doi.org/10.1007/978-1-4471-6296-4_8
8. Munaro, M., Basso, A., Fossati, A., Van Gool, L., Menegatti, E.: 3D reconstruction of freely moving persons for re-identification with a depth sensor. In: IEEE International Conference on Robotics and Automation (ICRA), pp. 4512–4519 (2014)
9. Farenzena, M., Bazzani, L., Perina, A., Murino, V., Cristani, M.: Person re-identification by symmetry-driven accumulation of local features. In: CVPR (2010)
10. Liu, C., Gong, S., Loy, C.C., Lin, X.: Person re-identification: what features are important? In: Fusiello, A., Murino, V., Cucchiara, R. (eds.) ECCV 2012. LNCS, vol. 7583, pp. 391–401. Springer, Heidelberg (2012). https://doi.org/10.1007/978-3-642-33863-2_39
11. Bhuiyan, A., Perina, A., Murino, V.: Person re-identification by discriminatively selecting parts and features. In: Agapito, L., Bronstein, Michael M., Rother, C. (eds.) ECCV 2014. LNCS, vol. 8927, pp. 147–161. Springer, Cham (2015). https://doi.org/10.1007/978-3-319-16199-0_11
12. Gray, D., Tao, H.: Viewpoint invariant pedestrian recognition with an ensemble of localized features. In: Forsyth, D., Torr, P., Zisserman, A. (eds.) ECCV 2008. LNCS, vol. 5302, pp. 262–275. Springer, Heidelberg (2008). https://doi.org/10.1007/978-3-540-88682-2_21
13. Liao, S., Hu, Y., Zhu, X., Li, S.Z.: Person re-identification by local maximal occurrence representation and metric learning. In: CVPR (2015)
14. Panda, R., Bhuiyan, A., Murino, V., Roy-Chowdhury, A.K.: Unsupervised adaptive re-identification in open world dynamic camera networks. In: CVPR (2017)
15. Pala, F., Satta, R., Fumera, G., Roli, F.: Multimodal person re-identification using RGB-D cameras. IEEE Trans. Circ. Syst. Video Technol. 26(4), 788–799 (2016)
16. Ren, L., Lu, J., Feng, J., Zhou, J.: Multi-modal uniform deep learning for RGB-D person re-identification. Pattern Recogn. 72, 446–457 (2017)
17. Mogelmose, A., Bahnsen, C., Moeslund, T., Clapes, A., Escalera, S.: Tri-modal person re-identification with rgb, depth and thermal features. In: CVPR (2013)
18. John, V., Englebienne, G., Krose, B.: Person re-identification using height-based gait in colour depth camera. In: ICIP (2013)
19. Ye, M., Lan, X., Li, J., Yuen, P.C.: Hierarchical discriminative learning for visible thermal person re-identification. In: AAAI (2018)
20. Dai, P., Ji, R., Wang, H., Wu, Q., Huang, Y.: Cross-modality person re-identification with generative adversarial training. In: IJCAI (2018)
21. Dalal, N., Triggs, B.: Histograms of oriented gradients for human detection. In: CVPR (2005)
22. Prosser, B.J., Zheng, W.S., Gong, S., Xiang, T., Mary, Q.: Person re-identification by support vector ranking. In: BMVC (2010)
23. Zheng, W.S., Gong, S., Xiang, T.: Re-identification by relative distance comparison. IEEE Trans. Pattern Anal. Mach. Intell. 35(3), 653–668 (2013)

24. Belhumeur, P.N., Hespanha, J.P., Kriegman, D.J.: Eigenfaces vs. fisherfaces: recognition using class specific linear projection. IEEE Trans. PAMI **19**(7), 711–720 (1997)
25. Webb, A.R.: Statistical Pattern Recognition. Wiley, Hoboken (2003)
26. Ojala, T., Pietikäinen, M., Harwood, D.: A comparative study of texture measures with classification based on featured distributions. Pattern Recogn. **29**(1), 51–59 (1996)
27. Zhang, Y., Li, S.: Gabor-LBP based region covariance descriptor for person re-identification. In: Sixth International Conference on Image and Graphics, pp. 368–371 (2011)
28. Zhang, Y., Li, B., Lu, H., Irie, A., Ruan, X.: Sample-specific SVM learning for person re-identification. In: CVPR (2016)
29. An, L., Kafai, M., Yang, S., Bhanu, B.: Reference-based person re-identification. In: AVSS (2013)

Single and Multi-channel Direct Visual Odometry with Binary Descriptors

Brendan Halloran[(✉)] , Prashan Premaratne , Peter James Vial,
and Inas Kadhim

University of Wollongong, Northfields Avenue, Wollongong,
NSW 2522, Australia
bh294@uowmail.edu.au

Abstract. Visual odometry is a popular area of computer vision that has seen a paradigm shift towards *direct methods*, where whole image alignment is used to determine camera poses. Current methods not robust to lighting changes in the scene and rely on standard feature-based methods for multi-camera systems. Binary descriptors are an option for alleviating both problems, but current methods do not scale well to larger and more robust descriptors. We present a method for performing direct tracking with binary descriptors of any size by approximating the gradient and descent direction with Hamming weights. We also present alternative methods that approximate the entire descriptor by its Hamming weights. Our results show improved accuracy compared to tracking on intensity alone, and our primary method improves significantly upon similar methods.

Keywords: Lucas-Kanade · Direct visual odometry · Binary descriptors

1 Introduction

Visual Odometry (VO) is one of the most active areas of research in computer vision. It is the process of estimating the relative poses of subsequent frames from a camera and forms the core of Visual Simultaneous Localisation and Mapping (V-SLAM) seen in algorithms such as LSD-SLAM [1], ORB-SLAM [2], SVO [3] and DSO [4].

There are two main types of VO – *indirect* and *direct* methods. Indirect methods use robustly matched feature correspondences to estimate poses [5], such as in MonoSLAM [6] and ORB-SLAM [2]. Direct methods perform whole-image alignment on intensity to estimate poses, generally using the Lucas-Kanade (LK) algorithm [7–9] and recently the more efficient Inverse Compositional (IC-LK) formulation [10]. Such direct VO algorithms include DTAM [11], LSD-SLAM [1] and DSO [4]. A middle ground between these methods is to use densely evaluated features in a direct VO framework. Standard direct methods rely on the *Brightness Consistency Assumption* (BCA) which is not robust to illumination changes seen in real image sequences. Feature-base direct methods alleviate this using a *Descriptor Consistency Assumption* (DCA) instead [12].

Our work explores direct tracking with binary descriptors such as BRIEF [13], ORB [14] and BRISK [15] which are efficient to compute and match. Many existing

D.-S. Huang et al. (Eds.): ICIC 2019, LNAI 11645, pp. 86–98, 2019.
https://doi.org/10.1007/978-3-030-26766-7_9

feature-based direct methods either use expensive features such as SIFT [16], or simpler binary descriptors such as the Census transform [17]. Our work involves a simple single-channel method for direct tracking with IC-LK and binary descriptors which performs extremely well on much larger descriptors and could be extended to multi-view systems.

1.1 Related Work

There has been extensive work on improving the robustness of the LK algorithm, particularly for use in VO and SLAM. Some algorithms still operate directly on intensity values but also model changes in brightness, such as the pixel-wise lighting and shadow model of Silveira and Malis [18], and the affine lighting model of Engel et al. [4].

Feature-based LK has been used in areas including VO and optical flow. The SIFT Flow algorithm uses dense SIFT descriptors with LK on each channel [16]. Bristow and Lucey showed that while dense descriptors are poor predictors of the error surface they still have very good performance for gradient based methods [19]. Sevilla-Lara and Learned-Miller proposed *distribution fields* which 'explode' a single-channel image into many channels based on the intensity values, preventing smoothing from erasing small details [20]. They describe this form of DCA as *channel consistency* [21]. Crivellaro and Lepetit use a similar idea where they separate the image into channels for first and second order gradients which are further separated by sign [22].

Binary features, mainly the Census transform, have also seen use in LK. Although binary features are non-convex and non-differentiable, Alismail et al. showed that they can be used in LK by splitting up each bit of the descriptor into a different channel [23]. Their method had good performance particularly in low-light scenes but does not scale well to larger descriptors which could require as many as 512 channels. Recently, Park et al. showed that these Census-based methods were more effective than other descriptor-based methods on real-world images [24].

1.2 Contributions

We present a method for performing LK-based VO with binary features which operates only on a single channel, allowing the use of robust descriptors which are suitable for wide baseline matching and challenging lighting conditions. Our contributions are:

- Considering the Census transform and its related Rank transform, we present an LK-based algorithm that compares the binary features using their Hamming distance but determines gradient and descent direction by comparing Hamming weight.
- We show that this algorithm achieves excellent performance when applied to BRIEF [13], ORB [14], and BRISK [15] descriptors, still operating on only a single channel.
- We demonstrate an alternative multi-channel channel form, with still fewer channels than other methods, by using the Hamming weight of segments of the descriptor.

The remainder of the paper is organised as follows: Sect. 2 covers the background information on the LK algorithm. Section 3 discusses our methods for tracking on binary descriptors. Section 4 presents our test results and analyses the performance. Finally, Sect. 5 concludes the paper and presents our direction for future work.

2 Background

2.1 Direct Visual Odometry

Consider two images I and I'. The intensity of each image at pixel coordinate $p = (u, v)^T$ is given by $I(p) \in \mathbb{R}$ and $I'(p) \in \mathbb{R}$ respectively. Direct visual odometry seeks to find the camera motion parameters $\theta \in \mathbb{R}^n$ which minimise the photometric error,

$$\theta^* = \mathrm{argmin}_\theta \sum_{p \in \Omega} \| I'(w(p; \theta + \Delta\theta)) - I(p) \|^2, \tag{1}$$

where $w(\cdot)$ is a function that warps a point from image I to I' by θ, and Ω is the set of points selected from image I. The pose between the two images is generally represented by a rigid body transformation $G \in SE(3)$ with a minimal representation using the element of the associated Lie-algebra $\theta \in \mathfrak{se}(3)$ related by the exponential map in (2).

$$G = \begin{bmatrix} R & t \\ 0 & 1 \end{bmatrix} \quad \text{with} \quad R \in SO(3), \quad t \in \mathbb{R}^3$$
$$G = \exp_{\mathfrak{se}(3)}(\theta) \tag{2}$$
$$\theta = \log_{SE(3)}(G)$$

With the representation of (2), we have optimisation parameters $\theta \in \mathbb{R}^6$. From this, our warp function is the reprojection from one camera view to another by

$$w(p; d, \theta) = \pi\big(R\pi^{-1}(p; d) + t\big)$$

where $\pi(\cdot)$ projects a 3D point into the image plane and $\pi^{-1}(\cdot)$ is the inverse projection finding the 3D point from the image point p and inverse depth d. We assume known intrinsic parameters of the projection. For improved efficiency, the IC-LK algorithm exchanges the roles of the template image, I, and tracked image, I' [10], giving

$$\theta^* = \mathrm{argmin}_\theta \sum_{p \in \Omega} \| I(w(p; d, \Delta\theta)) - I'(w(p; d, \theta)) \|^2. \tag{3}$$

This allows the Jacobian matrix J used in each iteration to remain constant. At each iteration we calculate the parameter update which tends towards the optimal θ,

$$\Delta\theta = \sum_{p \in \Omega} \big(J(p, \theta)^T J(p, \theta)\big)^{-1} J(p, \theta)^T r(p, \theta)$$
$$\text{where} \quad r(p, \theta) = I'(w(p; d, \theta)) - I(p)$$
$$\text{and} \quad J(p, \theta) = \nabla I(p) \frac{\partial w(p; d, \theta)}{\partial \theta} \Big|_{\theta = 0}.$$

Each iteration composes the inverse of the update to the current estimate,

$$\theta \leftarrow \theta \circ (\Delta\theta)^{-1} = \log_{\mathrm{SE}(3)}\left(\exp_{\mathfrak{se}(3)}(\theta) \cdot \exp_{\mathfrak{se}(3)}(-\Delta\theta)\right).$$

2.2 Multi-channel and Descriptor-Based Lucas-Kanade

The LK algorithm can operate on an arbitrary number of image channels $\mathcal{N}_c$ by finding the L_2 or similar norm, between each channel separately. This allows us to perform alignment of multi-channel images, such as RGB images, or even using dense descriptor images, such as densely evaluated SIFT descriptors. This gives us the form

$$\theta^* = \operatorname{argmin}_\theta \sum_{p\in\Omega} \sum_{i\in\mathcal{N}_c} \left\| \Phi_i(w(p;d,\Delta\theta)) - \Phi_i'(w(p;d,\theta)) \right\|^2 \qquad (4)$$

where $\Phi_i(\cdot)$ is the value of the i th channel. This gives us a different update, given by

$$\Delta\theta = \sum_{p\in\Omega} \sum_{i\in\mathcal{N}_c} \left(J_i(p,\theta)^T J_i(p,\theta)\right)^{-1} J_i(p,\theta)^T r_i(p,\theta)$$
$$where \quad r_i(p,\theta) = \Phi_i'(w(p;d,\theta)) - \Phi_i(p)$$
$$and \quad J_i(p,\theta) = \frac{\partial\Phi_i}{\partial p} \cdot \frac{\partial w(p;d,\theta)}{\partial\theta}\Big|_{\theta=0}.$$

3 Direct Tracking with Binary Features

3.1 The Census and Rank Transforms

We firstly consider the Census and Rank transforms which are single channel descriptors invariant to global monotonically increasing rescaling of the image [17]. The Census transform compares a target pixel to each other pixel in a local patch, for example 3×3, and sets a corresponding bit in a bit string to one if it is greater than or equal.

$$\Phi_{census}(p) := \{\mathbb{1}_{(p \geq p+\Delta p_0)}, \dots, \mathbb{1}_{(p \geq p+\Delta p_{N_b})}\}$$

$$with \quad \mathbb{1}_{(x)} = \begin{cases} 1 & \text{if } x \text{ is true,} \\ 0 & \text{otherwise} \end{cases}$$

where $\{\Delta p_i\}_{i=0}^{\mathcal{N}_b}$ is the set of neighbouring pixels within the patch and $\mathcal{N}_b$ is the number of bits in the descriptor. The Rank transform is simply the sum of all set bits in the corresponding Census transform. This is essentially the *Hamming weight*, given by

$$\Phi_{rank}(p) := \|\Phi_{census}(p)\|_H$$
$$where \quad \|\Phi\|_H = \sum_{i=0}^{N_b}(\Phi)_i$$

with $(\cdot)_i$ indicating the i th bit. We also consider two additional descriptors. The Complete Census computes a Census transform for each pixel in the image patch and the Complete Rank takes the Hamming weight of each of those Census transforms [12]. An example of these four descriptors is shown in Fig. 1.

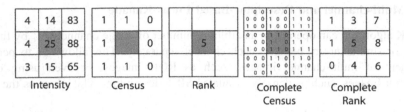

Intensity   Census   Rank   Complete Census   Complete Rank

Fig. 1. The Census transform uses comparison operations over intensity to encode the local structure into a bit string. The Rank transform is the sum of set bits in the Census transform.

3.2 Estimating Gradients and Descent Direction

Binary descriptors are matched using Hamming *distance* norm rather than L_2 norm. This counts the number of bits that differ between two binary descriptors, given by

$$\|\Phi_1, \Phi_2\|_H = \sum_{i=0}^{N_b} (\Phi_1 \oplus \Phi_2)_i$$

where $\oplus$ is the bitwise exclusive OR operator. This norm results in a non-convex and non-differentiable cost surface, preventing us from solving (4). Bitplanes [23] resolves this by separating each bit of the descriptors into a different channel and performing LK on each binary image, however, this method becomes computationally expensive with larger descriptors which could have as many as 512 channels.

For a single channel solution, we consider the relationship between Rank and Census transforms. To minimise (4) for Census descriptors we are minimising the Hamming *distance*, and for Rank descriptors we are minimising the difference in Hamming *weights*. The main issue with the former is gradient and descent direction. Consider two Census descriptors, $\Phi_1 = \{1, 0, 1, 1, 1, 0, 1, 1\}$ and $\Phi_2 = \{1, 1, 0, 0, 1, 0, 0, 1\}$. Their Hamming distance is 4 and Hamming weights are 6 and 4 respectively, giving a difference of 2. Moving from Φ_2 to $\Phi_2' = \{1, 1, 0, 0, 1, 0, 1, 1\}$, the Hamming distance is now 3 and the difference in Hamming weights is 1. Therefore, in this case, following the Rank descent also improved Census descriptor cost. Figure 2 shows the percentage of movements where reducing difference in Hamming weights also reduces Hamming distances. The Rank transform alone is less discriminative than the Census transform. Therefore, we propose to get residual size from the Hamming distance of Census descriptors, but gradient and decent direction from the Rank descriptors. That is,

$$r(p, \theta) = \mathbb{1}^{\epsilon}_{\left(\Phi'_{rank}(w(p;d,\theta)), \Phi_{rank}(p)\right)} \cdot \left\| \Phi'_{census}\big(w(p;d,\theta)\big), \Phi_{census}(p) \right\|_{H} \quad (5)$$

$$J(p, \theta) = \frac{\partial \Phi_{rank}}{\partial p} \cdot \left. \frac{\partial w(p; d, \theta)}{\partial \theta} \right|_{\theta=0} \quad (6)$$

where $\mathbb{1}^{\epsilon}_{(x_1, x_2)}$ is the signed indicator function,

$$\mathbb{1}^{\epsilon}_{(x_1, x_2)} = \begin{cases} 1 & \text{if } x_1 - x_2 > \epsilon \\ -1 & \text{if } x_2 - x_1 > \epsilon \\ 0 & \text{otherwise.} \end{cases}$$

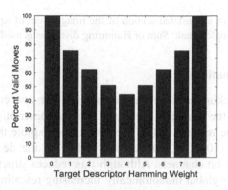

Fig. 2. For pairs of Census descriptors Φ_1 and Φ_2, the percentage of changes from Φ_2 to some other Φ'_2 where a reduction in difference of Hamming weights also results in a reduction in Hamming distance, separated by weight of Φ_1.

3.3 Extending to Arbitrary Binary Descriptors

The residuals and Jacobians in (5) and (6) can be extended to a binary descriptor of any size by using Hamming weights and distances in general. That is,

$$r(p, \theta) = \mathbb{1}^{\epsilon}_{\left(\|\Phi'(w(p;d,\theta))\|_{H}, \ \|\Phi(p)\|_{H} \right)} \cdot \left\| \Phi'\big(w(p; d, \theta)\big), \Phi(p) \right\|_{H} \quad (7)$$

$$J(p, \theta) = \frac{\partial \|\Phi\|_{H}}{\partial p} \cdot \left. \frac{\partial w(p; d, \theta)}{\partial \theta} \right|_{\theta=0} \quad (8)$$

For dense descriptor images to work with LK they need good convergence basins, which are the region around the correct pixel match resulting in convergence. To demonstrate a reasonable convergence basin for the descriptors we are using we show in Fig. 3 the sum of squared difference and sum of Hamming distance costs over a translated window, which show a clear basin leading to the correct translation at $(0, 0)$.

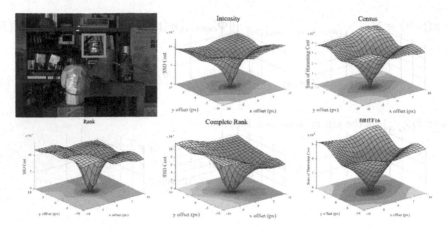

Fig. 3. The cost surface of a translated section of the image. Sum of squared difference is used for intensity, rank and complete rank. Sum of Hamming distances is used for Census and BRIEF.

3.4 Rank Approximations

In the same vein as tracking on Rank transform descriptors, we consider the possibility of tracking entirely on the Hamming weight of a binary descriptor. This would use the Jacobian of (8) with the residual of (9). We could also explore this in a multi-channel manner, expressed in (10)–(11), where each n bits of the descriptor is a separate channel, for example 8-bit channels. Although this reduces structural discrimination it maintains invariance to global monotonically increasing rescaling of intensity.

$$r(p, \theta) = \left\| \Phi'(w(p; d, \theta)) \right\|_H - \left\| \Phi(p) \right\|_H \qquad (9)$$

$$r_i(p, \theta) = \left\| \Phi'_i(w(p; d, \theta)) \right\|_H - \left\| \Phi_i(p) \right\|_H \qquad (10)$$

$$J_i(p, \theta) = \frac{\partial \left\| \Phi_i \right\|_H}{\partial p} \cdot \frac{\partial w(p; d, \theta)}{\partial \theta} \bigg|_{\theta=0}. \qquad (11)$$

3.5 Implementation Details

The binary descriptor-based IC-LK tracking was implemented in C++ using the OpenCV and Eigen libraries. Points were selected in each image using local maximums within a 3×3 patch with non-zero gradients. We tracked on a four level Gaussian pyramid and built a depth map for the point warping from small baseline epipolar searches between images of known relative pose. For intensity, we used the method of LSD-SLAM [1], searching the epipolar line for the minimum SSD over a patch. For the binary features we replaced the patch-based SSD with Hamming distance between descriptors. This process is shown in Fig. 4. This method requires bootstrapping for the first depth map or pose, as each require the other. We used a standard feature-based method to estimate the first pose then proceeded as described [5].

Fig. 4. Example stereo pair from New Tsukuba dataset [25]. Middle column has dense depth maps built from scanning epipolar lines for minimum SSD over a pixel patch (top), and scanning epipolar lines for best BRIEF16 descriptor match (bottom). Right column shows the sparse depth map with BRIEF16 descriptors (top) and successful points overlaid on image (bottom).

4 Evaluations

We evaluated our proposed methods using the KITTI Visual Odometry dataset seen in Fig. 5 [26]. We tested the combined Census and Rank method of Sect. 3.2 as well as using Rank and Compete Rank alone. Then, we tested BRIEF16, BREIF32, ORB and BRISK descriptors with the method of Sect. 3.3, and the single and multi-channel forms for BRIEF16 from Sect. 3.4. We also compared this to an implementation of Bitplanes [23]. We compare this to ground truth and intensity in Figs. 6 and 7.

Fig. 5. Example images from the KITTI dataset used for evaluation. Frame 0 (top) and frame 200 (bottom) from sequence 00 [26].

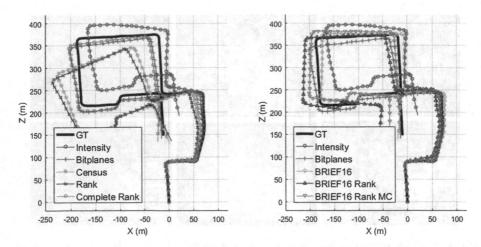

Fig. 6. Estimated paths for the first 1500 frames of the test dataset, using intensity, Bitplanes, Census, Rank, Complete Rank, BRIEF16, and Hamming weight approximations of BRIEF16.

Tracking on intensity alone does not perform very well, however Bitplanes performs much better. Our single-channel Census method is slightly better than a Rank transform approach but performs below Bitplanes. Interestingly, the Complete Rank transform performs almost as well as Bitplanes. The more complex descriptors perform much better, and all similarly. However, approximating the entire descriptor with its Hamming weight, even if a multi-channel case, sees poorer performance.

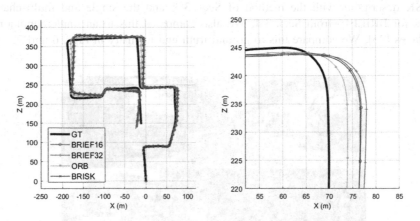

Fig. 7. Estimated paths for the first 1500 frames of the test dataset, using intensity, BRIEF16, BRIEF32, ORB, and BRISK descriptors. A zoomed in section is shown on the right.

Comparing the frame-by-frame error of BRIEF16 to Bitplanes in Fig. 8, we can see the very similar but slightly improved accuracy. This is because, although our descent direction is only an approximation, BRIEF descriptors provide much more information to exploit compared to Census descriptors.

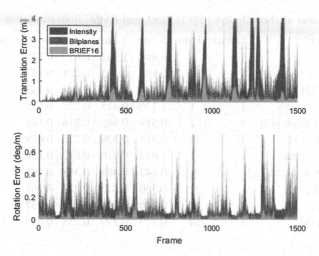

Fig. 8. Error per frame in estimations of relative poses, shown for intensity, Bitplanes, and BRIEF16 descriptors.

Looking at the ranges of errors for each descriptor type across all frames in Fig. 9, we can see that our methods for BRIEF, ORB and BRISK out perform the other methods quite substantially, with ORB being marginally better than BRIEF16.

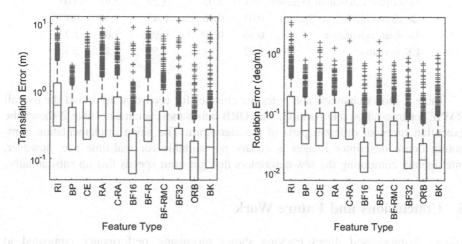

Fig. 9. Translation and rotation errors for relative pose estimations between frames, for raw intensity, bitplanes, census, rank, complete rank, BRIEF16 with single and multi-channel approximations, BRIEF32, ORB, and BRISK descriptors.

Table 1. Number of channels, bytes per channel, and RMSE from translation and rotation.

Descriptor	Channels	Bytes	X (m)	Y (m)	Z (m)	Rotation (deg/m)
Intensity	1	1	0.814	1.342	0.655	0.272
Bitplanes [23]	8	1	0.508	0.413	0.231	0.147
Census [17]	1	1	0.966	0.291	0.551	0.133
Rank [17]	1	1	1.136	0.415	0.725	0.115
Complete Rank [12]	9	1	0.519	0.845	0.318	0.193
BRIEF16 [13]	1	16	0.311	0.330	0.236	0.059
BRIEF16 Rank	1	1	1.044	0.910	0.617	0.118
BRIEF16 Rank MC	16	1	0.643	0.409	0.318	0.105
BRIEF32 [13]	1	32	0.360	0.373	0.273	0.068
ORB [14]	1	32	**0.270**	**0.304**	**0.173**	**0.057**
BRISK [15]	1	64	0.379	0.399	0.258	0.078

Table 2. Execution time for different parts of the algorithm for raw intensity and four descriptor types. Times given in milliseconds (ms).

	RI	BRIEF16	BRIEF32	ORB	BRISK
Pyramid construction	0.195				
Descriptor calculation (Dense)	N/A	120.93	265.67	58.74	636.80
Gradient calculation (Dense)	0.25	13.87	13.27	13.88	13.41
Descriptor calculation (Sparse)	N/A	3.03	3.27	2.97	85.05
Gradient calculation (Sparse)	0.03	0.46	0.43	0.48	0.43
Jacobian calculation	0.46	0.70	0.99	1.09	1.23
LK Iteration	0.52	1.11	1.59	1.87	2.58

Table 1 shows channels and bytes per channel for each descriptor and the overall RMSE for translation and rotation, with ORB being most accurate. Table 2 shows the execution times of different parts of the algorithm. The naïve implementation computing dense descriptor images is clearly not suitable for real-time use, however, intelligently computing the few descriptors that get used speeds this up substantially.

5   Conclusions and Future Work

Binary feature-based direct tracking shows promising performance compared to intensy. Our method approximates gradient and descent direction with Hamming weights, allowing the descriptors to still be compared using Hamming distance. We derived this by considering the relationship between Census and Rank transforms and extending to more complex descriptors. Compared to alternative approaches for binary descriptors LK, ours uses a single channel regardless of descriptor size. We also described further single- and multi-channel approximations using entirely Hamming weights. On Census transforms, our method has similar performance to intensity, but on BRIEF, ORB and BRISK descriptors we have much improved performance. Our alternative approximations also perform better than intensity but worse than our primary method.

We plan to integrate this type of binary descriptor tracking into existing VO and SLAM methods to thoroughly demonstrate its performance. Also, we seek to improve the speed of the algorithm by intelligently computing the minimum number of sparse descriptors require, which our results showed is necessary for real-time execution. We can further improve speed with single-instruction-multiple-data (SIMD) optimisations. Finally, as the features we use are suitable for wide-baseline matching, our method could be extended into a multi-camera VO system.

Acknowledgements. This research has been conducted with the support of the Australian Government Research Training Program Scholarship.

References

1. Engel, J., Schöps, T., Cremers, D.: LSD-SLAM: large-scale direct monocular SLAM. In: Fleet, D., Pajdla, T., Schiele, B., Tuytelaars, T. (eds.) ECCV 2014. LNCS, vol. 8690, pp. 834–849. Springer, Cham (2014). https://doi.org/10.1007/978-3-319-10605-2_54
2. Mur-Artal, R., Tardós, J.D.: ORB-SLAM2: an open-source SLAM system for monocular, stereo, and RGB-D cameras. IEEE Trans. Rob. **33**(5), 1255–1262 (2017)
3. Forster, C., et al.: SVO: semidirect visual odometry for monocular and multicamera systems. IEEE Trans. Rob. **33**(2), 249–265 (2017)
4. Engel, J., Koltun, V., Cremers, D.: Direct sparse odometry. IEEE Trans. Pattern Anal. Mach. Intell. **40**(3), 611–625 (2018)
5. Torr, P.H.S., Zisserman, A.: Feature based methods for structure and motion estimation. In: Triggs, B., Zisserman, A., Szeliski, R. (eds.) IWVA 1999. LNCS, vol. 1883, pp. 278–294. Springer, Heidelberg (2000). https://doi.org/10.1007/3-540-44480-7_19
6. Davison, A.J., et al.: MonoSLAM: real-time single camera SLAM. IEEE Trans. Pattern Anal. Mach. Intell. **29**(6), 1052–1067 (2007)
7. Lucas, B.D., Kanade, T.: An iterative image registration technique with an application to stereo vision, p. 674 (1981)
8. Premaratne, P., Ajaz, S., Premaratne, M.: Hand gesture tracking and recognition system for control of consumer electronics. In: Huang, D.-S., Gan, Y., Gupta, P., Gromiha, M.M. (eds.) ICIC 2011. LNCS (LNAI), vol. 6839, pp. 588–593. Springer, Heidelberg (2012). https://doi.org/10.1007/978-3-642-25944-9_76
9. Premaratne, P., Ajaz, S., Premaratne, M.: Hand gesture tracking and recognition system using Lucas-Kanade algorithms for control of consumer electronics. Neurocomputing **116**, 242–249 (2013)
10. Baker, S., Matthews, I.: Lucas-kanade 20 years on: a unifying framework. Int. J. Comput. Vis. **56**(3), 221–255 (2004)
11. Newcombe, R.A., Lovegrove, S.J., Davison, A.J.: DTAM: dense tracking and mapping in real-time. In: 2011 International Conference on Computer Vision. IEEE (2011)
12. Demetz, O.: Feature invariance versus change estimation in variational motion estimation. Dissertation, Saarländische Universitäts-und Landesbibliothek (2015)
13. Calonder, M., et al.: BRIEF: computing a local binary descriptor very fast. IEEE Trans. Pattern Anal. Mach. Intell. **34**(7), 1281–1298 (2012)

14. Rublee, E., et al.: ORB: an efficient alternative to SIFT or SURF. In: ICCV, vol. 11. no. 1 (2011)
15. Leutenegger, S., Chli, M., Siegwart, R.: BRISK: binary robust invariant scalable keypoints. In: 2011 IEEE International Conference on Computer Vision (ICCV). IEEE (2011)
16. Liu, C., Yuen, J., Torralba, A.: Sift flow: dense correspondence across scenes and its applications. IEEE Trans. Pattern Anal. Mach. Intell. **33**(5), 978–994 (2011)
17. Zabih, R., Woodfill, J.: Non-parametric local transforms for computing visual correspondence. In: Eklundh, J.-O. (ed.) ECCV 1994. LNCS, vol. 801, pp. 151–158. Springer, Heidelberg (1994). https://doi.org/10.1007/BFb0028345
18. Silveira, G., Malis, E.: Real-time visual tracking under arbitrary illumination changes. In: 2007 IEEE Conference on Computer Vision and Pattern Recognition. IEEE (2007)
19. Bristow, H., Lucey, S.: In defense of gradient-based alignment on densely sampled sparse features. In: Hassner, T., Liu, C. (eds.) Dense Image Correspondences for Computer Vision, pp. 135–152. Springer, Cham (2016). https://doi.org/10.1007/978-3-319-23048-1_7
20. Sevilla-Lara, L., Learned-Miller, E.: Distribution fields for tracking. In: 2012 IEEE Conference on Computer Vision and Pattern Recognition. IEEE (2012)
21. Sevilla-Lara, L., Sun, D., Learned-Miller, E.G., Black, M.J.: Optical flow estimation with channel constancy. In: Fleet, D., Pajdla, T., Schiele, B., Tuytelaars, T. (eds.) ECCV 2014. LNCS, vol. 8689, pp. 423–438. Springer, Cham (2014). https://doi.org/10.1007/978-3-319-10590-1_28
22. Crivellaro, A., Lepetit, V.: Robust 3D tracking with descriptor fields. In: Proceedings of the IEEE Conference on Computer Vision and Pattern Recognition (2014)
23. Alismail, H., et al.: Direct visual odometry in low light using binary descriptors. IEEE Rob. Autom. Lett. **2**(2), 444–451 (2017)
24. Park, S., Schöps, T., Pollefeys, M.: Illumination change robustness in direct visual slam. In: 2017 IEEE International Conference on Robotics and Automation (ICRA). IEEE (2017)
25. Peris, M., et al.: Towards a simulation driven stereo vision system. In: Proceedings of the 21st International Conference on Pattern Recognition (ICPR 2012). IEEE (2012)
26. Geiger, A., Lenz, P., Urtasun, R.: Are we ready for autonomous driving? The kitti vision benchmark suite. In: 2012 IEEE Conference on Computer Vision and Pattern Recognition. IEEE (2012)

Performance Evaluation of Faster R-CNN for On-Road Object Detection on Graphical Processing Unit and Central Processing Unit

Tanvir Ahmad[✉] and Yinglong Ma

School of Control and Computer Engineering,
North China Electric Power University, Beijing, China
ahmdtanvir@yahoo.com, yinglongma@ncepu.edu.cn

Abstract. On road object detection is very active research area for autonomous cars driving, pedestrian detection etc. Despite recent momentous enhancements, on road object detection is still a challenge that calls for more accuracy. In this study, we present the implementation of Faster R-CNN training for on road object detection and recognition. We have trained the model with our own dataset categorized into three classes such as car, cycle, pedestrian and test against three different datasets, such as KITTI dataset, video of Beijing road and also tested on our dataset to check the performance of the Faster R-CNN on GPU and CPU. we used this data to utilize Faster R-CNN and to analyze the impact of several factors like training datasets size, pre training model, iteration time, and training methods on the detection results of vehicle and pedestrians. Training the Faster R-CNN model by our own dataset on GPU has took 12 h while CPU took 11 days and 9 h to complete 50000 iterations. The GPU processed the video with a frame rate of 8 fps while CPU processed with 4 fps. The result shows that Faster R-CNN on GPU has higher mean Average Precision than CPU.

Keywords: Faster R-CNN · CNN · On-road object detection · GPU and CPU

1 Introduction

Computer vision has been used generously in the domain of object detection recognition. From last decade, several research scholars were attracted by convolutional neural network because of the tremendous advantages in object detection and recognition [1].

For Human beings it takes just a few glimpses to distinguish people and objects, recognize events and distinguish possibly dangerous circumstances. In order to achieve very multipart tasks such as playing sports or driving a vehicle, it is very important for human to have exact interpretation of different graphical stimuli. Correspondingly, localizing and recognizing numerous objects in an image is still challenging for machines to achieve. However, in the recent years a significant work has been done on object detection with Convolutional neural network (CNN) [2].

CNN works just like the normal neural network, which initially has convolution layers. Good performance in numerous jobs such as classification image, digit

© Springer Nature Switzerland AG 2019
D.-S. Huang et al. (Eds.): ICIC 2019, LNAI 11645, pp. 99–108, 2019.
https://doi.org/10.1007/978-3-030-26766-7_10

recognition and face recognition has been recorded by CNN [3] and is considered as state of the art [4]. Object detection and recognition is the latest invention in deep neural network (DNN) area, which makes the task simple for image recognition. Algorithms on Deep learning are the subset of machine learning algorithm, that are admirable in recognizing patterns, but usually it needs more data.

The main contribution behind CNN's is to learn automatically a hard model that can retrieve visual features from image, manipulating a sequence of operations such as filtering, local contrast normalization, local pooling, non-linear activation. Handcrafted features are used by traditional methods, that result in the feature extraction pipeline of human intuitions and the raw data understanding [5, 6].

In this study, we implemented faster R-CNN on CPU and GPU mode to detect car, pedestrian and cycle in images, recorded video as well as in real time.

The main contribution of the study as follows

- To make a detector, know and test it for low and high computational, to see how much computational power is necessary for a system that can accurately detect and classify objects of interest in images, on recorded video as well as in real time in many real-world applications. The input to our detector is an unprocessed recorded video or real time video.
- We implemented faster R-CNN to detect and classify car, pedestrian and cycle in given input.

This paper is arranged in the following order, CNN background is described in Sect. 2 in Sect. 3 Faster RCNN architecture is described, Sect. 4 describes experimental setting, Sect. 5 describe implementation of CPU and GPU mode, performance and comparison, Sect. 6 describe the paper conclusion.

2 Related Work

Detecting and identifying multiple object in an image is hard for machines to recognize and classify. However, a noteworthy effort has been carried out in the past years on detection of objects using convolutional neural networks (CNNs).

In object detection recognition field, Neural Networks are in use from a decade but became prominent due to improvement of hardware's new techniques for training these networks on large dataset [7]. The powerful deep features helped in building up recent successes [8], educated from the job of ImageNet classification [9, 10]. R-CNN. OverFeat [11] has been achieved good results on object detection on PASCAL VOC dataset [12]. Recently, more computationally efficient versions of neural networks have developed that can be trained on larger datasets such as COCO [13]. For example, to speed-up the processing, researchers in [14] proposed Fast-RCNN, which is based on sharing the convolutions across different region proposals. In Dai et al., and Ren et al., [11, 15], the researchers proposed R-FCN and Faster-RCNN respectively which is based on the idea of merge region proposal generation framework to search for object instances, which makes a complete back-to-back version. The Overfeat detector sliding-window paradigm build up other computationally-efficient approaches like YOLO [16], SSD [17] and Dense Box [18]. Faster R-CNNs performed well compared to earlier R-CNNS, that is due to the RPN addition which decreases the bottleneck by finding the region and their scores at the same time on a region.

The frame rate of VGG-16 Faster R-CNN model is 5fps, though attaining state-of-the-art object detection accuracy on MS COCO and PASCAL VOC (2007, 2012) datasets with only 300 proposals per image [19, 20]. On the other hand, the images processing time for YOLO model is 45 fps. The object detection mAP value for Faster R-CNN is 73.2 mAP and is better than YOLO that is 63.4 mAP [15].

Faster R-CNN is divided into two units, fully convolutional network and Faster R-CNN detector [11]. The whole system for object detection is a unified single network, [21], while RPN unit direct Faster R-CNN unit where to look.

3   An Overview of Faster-RCNN Architecture

The key perception of Faster R-CNN was to replace the slow search algorithm with faster one, for that's they introduced region proposal network (RPN).

Input to the Faster R-CNN is an entire image. Base on this input, it classifies and detect objects by using many regions which are expected to have objects. Faster R-CNN can be distributed into three parts: (1) the image regions are used to extract features for classification by CNN. (2) Object anticipated regions are gathered by region proposal network (RPN). (3) Region-based convolution neural network, which categories all the regions.

The whole image is first processed by the network with some convolutional (convl) and max pooling layers, for creating convl feature maps. Then, a region of interest (ROI) for each object is proposed using pooling layer and it retrieve a static-length feature vector from the feature maps. Figure 1 shows general overview of faster R-CNN.

Each feature map is given as an input into a particular order of fully connected (fc) layers that lastly divides into two output layers. Finally, bounding boxes are gathered with scores. The bounding boxes are classified and regressed by Multi-Scale anchors. The changing scales and aspect ratios report multiple sizes and scales of the images. Each anchor is assigned a label during training. Positive label is assigned to the anchor if the value > 0.7, representing detection of an object. if the IoU ratio is < 0.3 then anchors are assigned negative label for all ground-truth boxes, indicating there is no object detected.

Here's how the RPN works:

- an initial CNN last layer of, a sliding window of 3×3 changes along the feature map and plots it to a lesser measurement (e.g. 256-d)
- using k fixed-ratio anchor boxes, it produces multiple possible regions for each sliding-window location.
- Each proposal region contains an "abjectness'" score for proposed area and the 4 coordinates signifying the leaping box of the area.

However, the RPN training is carried out by fully Conv network which can be optimized using back propagation and stochastic gradient descent [22]. For training network, all mini batch outcome from a particular image that contains many positive

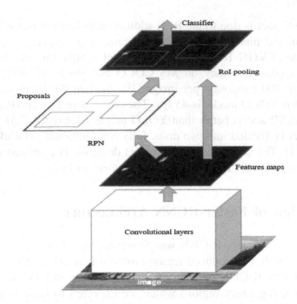

Fig. 1. Architecture flow chart of Faster RCNN [11]

and negative instance anchors for computing the loss function and to acquire a near 1:1 ratio. Additionally, we gathered bounding boxes for each class using non-maximum suppression to side-step duplicate detection. Figure 2 depicts RPN structure.

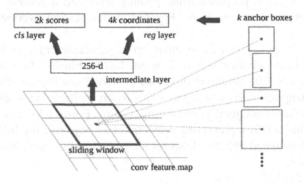

Fig. 2. RPN idea [21]

On every single sliding-window, at the same time it predicts multiple regions. Here k denotes maximum number of possible proposals for every single location. By this way the reg layers has 4k outputs and classification layers has 2k to evaluate the presences of the objects for every proposal. k proposals are parameterized relative to k reference boxes, called anchors. At use 3 scales and 3 aspect ratios by default, generating k = 9 anchors at each sliding position.

4 Experimental Setup

We carried out experiment by using our own dataset to gauge the performance of faster RCNN on GPU and CPU mode and see how much the computational power is important and can affect the accuracy and speed of algorithm in real time object detection.

4.1 Dataset

We perform experiment using our own dataset, which entails of 14000 images, divided it by a ratio of 70% training and 30% testing. We had kept the testing dataset isolated and had never showed to the system. Every picture in dataset is labeled and having no clamor. Collecting and labeling traffic images data manually is troublesome, we used this data to utilize faster-rcnn and to analyze the impact of several factors like training datasets size, pre training model, iteration time, and training methods on the detection results of vehicle and pedestrians. Our dataset contains images (1080 × 680) from different circumstances like partially occluded, night scene images and daytime images. In our experiment we use only three classes for detection, car, pedestrian and cycle class. Below Fig. 3 shows some example pictures from our dataset. For testing the system, we used KITTI dataset (1392 × 512) and Beijing road recoded video (1000 × 800), KITTI datasets are recorded by driving around in Karlsruhe Germany, by using station wagon equipped with two high resolution color and gray scale cameras.

4.2 System Dataset Flow Architecture

Machine learning algorithm performs objects detection and classification by stimulating human's artificial intelligence; input to the system is some unprocessed image, system performs some operation and gives its category is an output.

5 Implementation of Model

We implemented Object detection and classification on two different GPU and CPU systems with windows 10 operating systems. The GPU GTX 1070 with NVIDIA GeForce having 8 GB graphic card memory and 1920 CUDA cores. The second system CPU is Intel(R) i3-4130 with 3.4 GHz processor and 8 GB RAM with 64-bit operating system.

For training and testing faster RCNN we used Tensorflow with python and for user interface we used OpenCV for displaying, obtaining and saving images. The TensorFlow has many inbuild libraries which makes the work very easy for convolutional neural network to perform matrices, classification and other such kind of operations. Computer has the incapability to characterize the images and convert it into numbers, for which it needs platforms like Tensorflow, Caffe etc., to perform such operations for deep learning content.

Fig. 3. Images from our datasets

5.1 Detection Results

The model is trained on our own dataset which consists of 14000 images and tested for performance evaluation on a variety of pictures from Kitti dataset and recorded video of Beijing road, the pictures are in different climatic conditions.

5.2 Performance Evolution (mAP)

The model performance is estimated in terms of detection precision. The detection accuracy of system is calculated by mean Average Precision (mAP). This is measured by dividing the number of true detections for all the classes over the sum of number of true detection and number of false detections as given in below Eqs. (1), (2) and (3) respectively.

$$P = \frac{\text{No. of True detection}}{\text{No. of True detection} + \text{no of false detection}} \tag{1}$$

$$AP = \frac{\sum P \; \forall \; \text{True detection}}{\text{True detection}} \tag{2}$$

$$mAP = \frac{\sum AP}{\text{No. Classes}} \tag{3}$$

We have achieved a mAP of 81.0% for Kitti_drive0005 on GPU, which is better than 71.7% in [22] (Figs. 4, 5, 6 and Tables 1, 2).

Fig. 4. GPU Car, Cyc, Ped detection on Kitti_drive0005

Fig. 5. CPU Car, Cyc, Ped detection on Kitti_drive0005

Here is some short coming in the system as well, it is getting confused in cycle and pedestrian class very often and misidentifying it, some time it detects cycle is a pedestrian and pedestrian is a cycle specially in CPU mode, at the moment we are also unaware that why there is difference in precision, although the testing images provided to the systems, Number of iteration and architecture are the same. We will investigate this issue in our future work and will train the systems with other various models.

Fig. 6. Example of Car, Cyc, Ped on Beijing road video

Table 1. Mean average precession for CPU

No of classes	Kitti (AP)	Beijing (AP)	Our data (AP)	Kitti mAP%	Beijing mAP%	Our data mAP%	FPS
Car	0.95	0.753	0.73	65	76.1	77.7	4
Cycle	0	0.71	0.759				
Pedestrian	1	0.821	0.843				

Table 2. Mean average precession for GPU

No of classes	Kitti (AP)	Beijing (AP)	Our data (AP)	Kitti mAP%	Beijing mAP%	Our data mAP%	FPS
Car	0.95	0.78	0.921	81	84.3	87.4	8
Cycle	0.82	0.881	0.87				
Pedestrian	0.5	0.87	0.832				

6   Conclusion

In this experiment, we trained and tested Faster R-CNN for object detection and recognition on road objects. We have trained the model with our own dataset and tested for three classes such as car, cycle, pedestrian, by using three different datasets, KITTI datasets images, video of Beijing road to check the performance of the system on GPU and CPU. Training the Faster R-CNN model by our own dataset on GPU has took 12 h while CPU took 11 days and 9 h to complete 50000 iterations. From the experiment our finding is that, computational power affects the detection and recognition rate and not suitable for real time applications. The CPU also takes more time in detecting the object during testing. Thus, computing resources plays a major role in detecting the on-road objects for real time applications.

In our future work, we will be focusing on improving the detection performance by fine-tuning to achieve better results.

Acknowledgement. This work is partially supported by the National Key R&D Program of China (2018YFC0831404, 2018YFC0830605).

References

1. Yao, B., Fei-Fei, L.: Recognizing human object interactions in still images by modeling the mutual context of objects and human poses. IEEE Trans. Pattern Anal. Mach. Intell. **34**(9), 1691–1703 (2012)
2. Qi, Q., Zhang, K., Tan, W., Huang, M.: Object detection with multi-RCNN detectors. In: Proceedings of the 2018 10th International Conference on Machine Learning and Computing, pp. 193–197. ACM (2018)
3. Tomè, D., Monti, F., Baroffio, L., Bondi, L., Tagliasacchi, M., Tubaro, S.: Deep convolutional neural networks for pedestrian detection. Sig. Process. Image Commun. **47**, 482–489 (2016)
4. Szegedy, C., et al.: Going deeper with convolutions. In: Proceedings of the IEEE Conference on Computer Vision and Pattern Recognition, pp. 1–9 (2015)
5. Benenson, R., Omran, M., Hosang, J., Schiele, B.: Ten years of pedestrian detection, what have we learned? In: Agapito, L., Bronstein, M.M., Rother, C. (eds.) ECCV 2014. LNCS, vol. 8926, pp. 613–627. Springer, Cham (2015). https://doi.org/10.1007/978-3-319-16181-5_47
6. Hosang, J., Omran, M., Benenson, R., Schiele, B.: Taking a deeper look at pedestrians. In: Proceedings of the IEEE Conference on Computer Vision and Pattern Recognition, pp. 4073–4082 (2015)
7. Zheng, Y., Zhu, C., Luu, K., Bhagavatula, C., Le, T.H.N., Savvides, M.: Towards a deep learning framework for unconstrained face detection. In: 2016 IEEE 8th International Conference on Biometrics Theory, Applications and Systems (BTAS), pp. 1–8, September 2016
8. Krizhevsky, A., Sutskever, I., Hinton, G.E.: Imagenet classification with deep convolutional neural networks. In: Advances in Neural Information Processing Systems, pp. 1097–1105 (2012)
9. Girshick, R., Donahue, J., Darrell, T., Malik, J.: Rich feature hierarchies for accurate object detection and semantic segmentation. In: Proceedings of the IEEE Conference on Computer Vision and Pattern Recognition, pp. 580–587 (2014)
10. Sermanet, P., Eigen, D., Zhang, X., Mathieu, M., Fergus, R., LeCun, Y.: Overfeat: integrated recognition, localization and detection using convolutional networks. arXiv preprint arXiv: 1312.6229 (2013)
11. Ren, S., He, K., Girshick, R., Sun, J.: Faster R-CNN: towards real-time object detection with region proposal networks. In: Advances in Neural Information Processing Systems, pp. 91–99 (2015)
12. Lin, T.-Y., et al.: Microsoft COCO: common objects in context. In: Fleet, D., Pajdla, T., Schiele, B., Tuytelaars, T. (eds.) ECCV 2014. LNCS, vol. 8693, pp. 740–755. Springer, Cham (2014). https://doi.org/10.1007/978-3-319-10602-1_48
13. Girshick, R.: Fast R-CNN. In: ICCV (2015)
14. Dai, J., Li, Y., He, K., Sun, J.: R-FCN: object detection via region-based fully convolutional networks. In: Advances in Neural Information Processing Systems, pp. 379–387 (2016)

15. Redmon, J., Divvala, S., Girshick, R., Farhadi, A.: You only look once: unified, real-time object detection. In: Proceedings of the IEEE Conference on Computer Vision and Pattern Recognition, pp. 779–788 (2016)
16. Liu, W., et al.: SSD: single shot MultiBox detector. In: Leibe, B., Matas, J., Sebe, N., Welling, M. (eds.) ECCV 2016. LNCS, vol. 9905, pp. 21–37. Springer, Cham (2016). https://doi.org/10.1007/978-3-319-46448-0_2
17. Huang, L., Yang, Y., Deng, Y., Yu, Y.: DenseBox: unifying landmark localization with end to end object detection. arXiv preprint arXiv:1509.04874 (2015)
18. Deng, J., Dong, W., Socher, R., Li, L.-J., Li, K., Fei-Fei, L.: Imagenet: a large-scale hierarchical image database. In: IEEE Conference on Computer Vision and Pattern Recognition, CVPR 2009, pp. 248–255. IEEE (2009)
19. Everingham, M., Van Gool, L., Williams, C.K., Winn, J., Zisserman, A.: The pascal visual object classes (VOC) challenge. Int. J. Comput. Vis. **88**(2), 303–338 (2010)
20. Bell, S., Lawrence Zitnick, C., Bala, K., Girshick, R.: Inside-outside net: detecting objects in context with skip pooling and recurrent neural networks. In: Proceedings of the IEEE Conference on Computer Vision and Pattern Recognition, pp. 2874–2883 (2016)
21. https://towardsdatascience.com/deep-learning-for-object-detection-a-comprehensive-review-73930816d8d9
22. Prabhakar, G., Kailath, B., Natarajan, S., Kumar, R.: Obstacle detection and classification using deep learning for tracking in high-speed autonomous driving. In: IEEE Region 10 1 Symposium (TENSYMP), pp. 1–6. IEEE (2017)
23. Chorowski, J.K., Bahdanau, D., Serdyuk, D., Cho, K., Bengio, Y.: Attention-based models for speech recognition. In: Advances in Neural Information Processing Systems, pp. 577–585 (2015)

Research on Full Homomorphic Encryption Algorithm for Integer in Cloud Environment

Lijuan Wang[1], Lina Ge[1,2,3]($\boxtimes$), Yugu Hu[1], Zhonghua He[1],
Zerong Zhao[1], and Hangui Wei[1]

[1] College of Information Science and Engineering,
Guangxi University for Nationalities, Nanning, China
66436539@qq.com
[2] Key Laboratory of Network Communication Engineering,
Guangxi University for Nationalities, Nanning, China
[3] Key Laboratory of Guangxi High Schools Complex System
and Computational Intelligence, Nanning, China

Abstract. With the continuous development of computer networks, cloud storage is now the mainstream way for people to store information, but some important information leaks in the process of storage, and information security issues have become the most concerned issue at present. Homomorphic encryption not only has the property of data encryption of traditional encryption algorithm, but also has the result of ciphertext operation equivalent to the corresponding plaintext operation. This paper proposes an improved scheme based on DGHV, it (MDGHV) is mainly implemented from two points: changing the encryption formula $c = pq + 2r + m$ to $c = pqr + m$, the domain of clear text m is expanded to a custom value, which can be used as a parameter; without acquiring sensitive information, data owners can directly calculate multiple ciphertexts and ensure high efficiency to meet the security requirements of cloud environment. An improved scheme based on DGHV, which expands the plaintext domain and enable search encryption operation to encrypt larger values, thus improving the efficiency. The security of this algorithm still depends on the similar GCD problem, which can solve some practical security problems in the cloud environment.

Keywords: Cloud storage · Full homomorphic encryption · Retrieval · Confidentiality

1 Introduction

In today's information age, data presents an explosive growth [1]. Cloud computing is an information technology mode, which can quickly allocate resources and advanced services to users and reduce users' management operations [2]. The development of cloud computing has not only changed the development mode of the computer industry, but also changed the life habits of users and the operation mode of enterprises. With the continuous development of network technology, cloud computing plays an important role in the future development process. In recent years, China's Internet giants such as Alibaba and Tencent have launched their own cloud plans and provided

© Springer Nature Switzerland AG 2019
D.-S. Huang et al. (Eds.): ICIC 2019, LNAI 11645, pp. 109–117, 2019.
https://doi.org/10.1007/978-3-030-26766-7_11

their own cloud service platforms. And low quality, flexible fees, convenient and rapid advantages of the domestic masses of recognition. Therefore, more and more service platforms and websites are built on the cloud platform. In the ideal expectation of users, cloud service should be the most trusted third-party participant. Under no circumstances will it reveal any information, nor cheat any users. But in practice cloud services are not always infallible, or they are semi-trustworthy. It will fulfill its relevant obligations, implement relevant agreements and provide good services for users, but it may retain intermediate results or even relevant data information in the process of work. In the era of big data, a large amount of data is like a gold mine to be mined. Under the pressure of business and regulation, cloud services often use or directly steal users' data to gain great benefits.

Traditional data encryption can not effectively solve the above problems, and even brings about the complexity of the problem, especially the encryption of huge data and local computing resources are relatively small. In fact, the current full homomorphic encryption technology can solve the above problems, which can meet the requirements of data encryption function, but also can meet the requirements of the data has been encrypted in the case of data computing, very suitable for the storage of information in the cloud environment. The unique feature of identical encryption is that after the ciphertext is calculated, the result of decryption is exactly the same as that of plaintext corresponding to the ciphertext $f(Enc(m)) = Dec(f(m))$. (Enc is the encryption calculation, Dec is the decryption calculation, and f is the addition or multiplication calculation).

This paper focuses on how to apply the full homomorphic encryption scheme on more plaintext integers to the cloud computing environment. Meaning is that:

The unique property of full homomorphic encryption enables it to fundamentally solve the problem of cloud computing data security, which will promote the development of cloud computing.

Studying and researching on relevant schemes have promoted the development of full homomorphic encryption technology.

The further learning and improvement of schemes on related integers promotes the continuous development of fully homomorphic encryption schemes.

This paper mainly analyzes the research status of full homomorphic encryption, describes the algorithm of full homomorphic encryption (DGHV) based on integer, and proposes the improved algorithm in this paper. Then the algorithm description, correctness analysis, security analysis, as well as the algorithm in different stages of efficiency analysis, and finally draw a conclusion.

2 Research Status of Full Homomorphic Encryption

With the development of computing networks. In order to ensure the security and integrity of data in the process of storage and transmission, it is necessary to encrypt the stored and transmitted data to avoid data theft and modification by illegal users in the process of storage and transmission. Full homomorphic encryption is one of the more popular encryption algorithms from the historical shift method to the current full homomorphic encryption algorithm development. The characteristic of the algorithm is to calculate the ciphertext without knowing the key. The result of decryption is the same as that of corresponding operation on the plaintext corresponding to the ciphertext, $f(Enc(m)) = Enc(f(m))$.

The development of encryption technology is divided into three stages. The first phase is from 1978 to 2009. The encryption schemes proposed during this period are all known as classical public key encryption schemes, which cannot satisfy homomorphic addition and homomorphic multiplication at the same time. They are partial homomorphisms and can only carry out finite times of calculation. In the second stage, until 2009, Gentry et al. [3] proposed the first fully homomorphic encryption scheme (a fully homomorphic encryption scheme based on ideal lattice over polynomial ring). For example, DGHV [4] scheme and CAFDE [5] scheme can both satisfy the addition homomorphism and multiplication homomorphism, as well as perform countless ciphertext calculations, thus achieving true complete homomorphism. Brakerski and Vaikuntanathan proposed a fully homomorphic encryption scheme based on fault tolerant Learning (LWE) With Errors [6].

3 Full Homomorphic Encryption Algorithm

Full homomorphic encryption technology is able to perform direct operations on the ciphertext, and perform algebraic operations on the ciphertext in the encrypted state, and the results after the operations are also the encrypted state. Ciphertext calculation and plaintext calculation based on full homomorphic encryption are carried out on polynomial rings and there is a certain mapping relationship between them. Both plaintext and ciphertext data are decomposed into polynomials, and there is always a mapping relation in the polynomial ring, so that the ciphertext algebraic operation corresponds to the plaintext algebraic operation relation.

Specific operations of homomorphic encryption algorithm can be described as follows: Assuming that the encryption operation is Enc and the decryption operation is Dec; The homomorphic addition operator is, and the homomorphic multiplication operator is; m1 and m2 respectively represent two plaintext to be calculated; c1 and c2 represent the ciphertext corresponding to the plaintext m1 and m2 encryption.

Homomorphism plus properties [4]: $Enc(m1 + m2) = Enc(m1) + E(m2) = c1 + c2$, or $Dec(c1 + c2) = m1 + m2$.
Homomorphic multiplication property [4]: $Enc(m1 * m2) = Enc(m1) * E(m2) = c1 * c2$, or $Dec(c1 * c2) = m1 * m2$.

If an algorithm satisfies the properties of homomorphic addition and homomorphic multiplication, the algorithm is called full homomorphic encryption algorithm. If an algorithm can only satisfy one of the two properties of homomorphic multiplication and homomorphic addition, it is called a partial homomorphic encryption algorithm.

The following case is used to analyze and explain the full homomorphic encryption algorithm: Given that p is the public key, q is the private key, r is the noise in encryption, c is the ciphertext, m is the plaintext, c1 and c2 are the ciphertext corresponding to the encryption of m1 and m2 respectively. The encryption operation is

$c = pq + 2r + m$; The decryption operation is $m = (c \bmod p) \bmod 2$. The encryption operations of plaintext m1 and m2 are as follows:

$$c_1 = pq_1 + 2r_1 + m_1 \tag{1}$$

$$c_2 = pq_2 + 2r_2 + m_2 \tag{2}$$

The homomorphic encryption operation is shown in formula 2:

$$
\begin{aligned}
\mathrm{Enc}(m_1 + m_2) &= c_1 + c_2 = p(q_1 + q_2) + 2(r_1 + r_2) \\
\mathrm{Dec}(c_1 + c_2) &= (c_1 + c_2)\bmod p \bmod 2 \\
&= (2(r_1 + r_2) + (m_1 + m_2))\bmod 2 \\
&= m_1 + m_2
\end{aligned}
\tag{3}
$$

According to formula 2, the result of decryption after ciphertext addition is the same as that of plaintext addition, which satisfies the property of homomorphism addition.

The homomorphic multiplication operation is shown in formula 3:

$$
\begin{aligned}
c_1 * c_2 &= (pq_1 + 2r_1 + m_1) * (pq_2 + 2r_2 + m_2) \\
&= p(2pq_1q_2 + 2q_1r_2 + m_2q_1 + 2q_2r_1 + m_1q_2) + \\
&\quad 2(r_1r_2 + r_1m_2 + r_2m_1) + m_1m_2 \\
&= \mathrm{Enc}(m_1 * m_2) \\
\mathrm{Dec}(c_1 * c_2) &= (c_1 * c_2)\bmod p \bmod 2 \\
&= [2(r_1r_2 + r_1m_2 + r_2m_1) + m_1m_2]\bmod 2 \\
&= m_1 * m_2
\end{aligned}
\tag{4}
$$

Formula 3 shows that the result of decryption after ciphertext multiplication is the same as that of plaintext multiplication.

4 Improvement of DGHV - Based Scheme (MDGHV)

The fully homomorphic encryption algorithm of DGHV scheme is $c = pq + 2r + m$, and the decryption algorithm is $m = (c \bmod p) \bmod 2$. These two encryption, decryption algorithms are generally some of the integer on the ordinary operation method, is relatively easy to understand.

The most important feature of the integer identical encryption scheme is that it is easy to understand. It is realized through the heavy encryption technology. However, DGHV is more concise than Gentry's first scheme, so later scholars have tried to improve DGHV in various aspects. It can be clearly seen from the encryption formula $c = pq + 2r + m$ and the decryption formula $m = (c \bmod p) \bmod 2$ that the improvement direction of the scheme mainly lies in the module and public key. Coron et al. [7] converted the public key from linear to binary form, which reduced the size of the public key and improved the performance and efficiency of the algorithm. Later, Sun et al. [8] expanded the plaintext space of the algorithm in the scheme from {0,1} to {0,1}k, so that the module expanded from 2 to 2 k. Xiong et al. [9] expanded from 1 bit to n bit plaintext space, simultaneously encrypted n bits into a ciphertext, and optimized the scheme's heavy encryption.

In addition to the above scheme, there are other scholars who rapidly compress the size of the public key and improve the linearization of the public key to quadratic or even to the form of k, or double the number of bits in clear text. Both of these methods can reduce the size of the key and expand the plaintext space, but due to the constraint of their low efficiency, the improved efficiency is still low and far below the demand. To solve this problem, this paper will improve the homomorphic encryption algorithm on integer:

(1) change the encryption formula $c = pq + 2r + m$ to $c = pqr + m$, the domain of clear text m is expanded to a custom value, which can be used as a parameter;
(2) Without acquiring sensitive information, data owners can directly calculate multiple ciphertexts and ensure high efficiency to meet the security requirements of cloud environment.

The improved algorithm is composed of six parts:

Setup(1^λ):The user requests the key control center to generate security parameters
PreGen(λ): Private key sk is generated based on security parameters
PubGen(λ,s): Generate p public key according to λ,s
Enc(p,s,m): The user encrypts the clear text $c = PSR + m$. Upload ciphertext c to the cloud
Eva(f,c_1,c_2,...,c_n): The user issues the instruction f to add or multiply ciphertext ci, and the cloud performs Eva() according to f, and then returns the result c*.
Dec(s,c*): According to s, decrypt $m = c*modp$. I get the clear text m.

4.1 MDGHV Correctness

Suppose c is the ciphertext, p is the public key, q is the private key, r is the encryption noise, and m is the plaintext.

From $c_1 = pq_1 + m_1$ and $c_2 = pqr_2 + m_2$:

$$c_1 + c_2 = pqr_1 + pqr_2 + m_1 + m_2$$
$$c_1 * c_2 = p^2q^2r_1r_2 + pqr_1m_2 + pqr_2m_1 + m_1 * m_2 \tag{5}$$

Decrypting c1 + c2:

$$(c_1 + c_2)modp = (pqr_1 + pqr_2 + m_1 + m_2)modp = m_1 + m_2 \tag{6}$$

Decrypting C1*C2

$$(c_1 * c_2) \bmod p$$
$$= (p^2q^2r_1r_2 + pqr_1m_2 + pqr_2m_1 + m_1 * m_2) \bmod p \qquad (7)$$
$$= m_1 * m_2$$

The improved scheme decrypts the encrypted ciphertext and the result is consistent with the calculation result of plaintext. So the improved algorithm is correct.

4.2   Performance Analysis of MDGHV

The improved DGHV algorithm was tested and analyzed, and the MDGHV secret key generation time, encryption time and decryption time were compared with the DGHV. In the DGHV scheme, only one bit-ciphertext $\{0,1\}$ can be encrypted at a time. Therefore, we operate in the laboratory under the condition of 1-bit ciphertext. In fact, the improved algorithm MDGHV can encrypt data in a larger space.

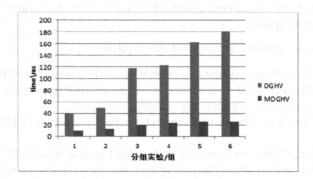

Fig. 1. The Setup operation time for scheme

It can be seen from Fig. 1 that the time difference in the Setup stage is relatively large, and it can be seen that MDGHV takes precedence over DGHV. DGHV depends on the problem of sparse subset sum in the phase of secret key generation. During execution, a subset sequence needs to be constructed first, which wastes a lot of time and is random and uncontrollable. It may take a long time to generate a sequence that meets the requirements.

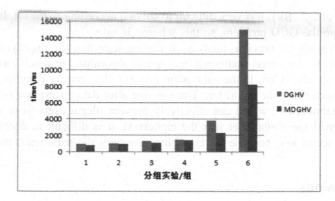

Fig. 2. The encryption operation time for scheme

As can be seen from Fig. 2, in the process of plaintext encryption, although MDGHV has one more multiplication operation in the form, the efficiency of MDGHV execution is still higher than that of MDGHV. Because MDGHV in each encryption process, need to be a re-encryption, to ensure the range of noise, to ensure that the next ciphertext calculation can be carried out smoothly, in this process consumes a very long time.

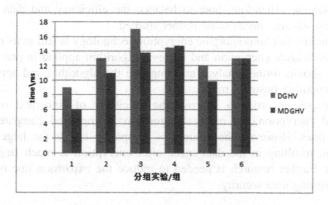

Fig. 3. The decryption operation time for scheme

It can be seen from Fig. 3 that the data difference between MDGHV and DGHV in decryption time, although extremely small, is still improved. Because the decryption phase is modular, which is expected.

In conclusion, the overall performance of MDGHV is better than that of DGHV.

4.3　MDGHV Security Analysis and Efficiency Analysis

The security of MDGHV comes down to the problem of approximate GCD. During the execution of the algorithm, k large integer q are randomly generated, approximate multiples {b1, b2 ... Bk}, it is very difficult to solve q according to the set. It is difficult to approximate the GCD problem, so this scheme is safe.

Compared with the previous analysis, it can be seen that the improved MDGHV algorithm and the improved traditional encryption algorithm are much more efficient than the original algorithm. In the encryption process, the security parameters are also randomly generated by the algorithm. The file size after data encryption is larger than the size of plaintext file, which can effectively prevent illegal users from stealing the analysis. Even if the illegal user gets the ciphertext, it is difficult to decrypt without knowing the secret key, because the algorithm generates the parameters randomly.

5   Conclusions

Aiming at the data security problems faced by cloud storage at present, this paper studies and improves the existing homomorphic encryption and proposes a new encryption algorithm and retrieval algorithm.

(1) The improvement of DGHV, the all-homomorphic encryption scheme based on integer in this paper, is a relatively novel way to expand the plaintext domain and improve the efficiency, which is applicable to some simple computing problems in cloud storage. However, the security of this solution depends on the GCD. With the continuous development of network technology, the size of secret keys will surely increase. Therefore, how to balance the efficiency and data security of encryption schemes needs to be further studied.

(2) At present, the full homomorphic encryption technology is still in its infancy, and there is no mature encryption and retrieval algorithm applied in practice. Therefore, we should further analyze and improve the algorithm and apply it to the cloud computing business model.

(3) In this paper, in order to improve the efficiency of ciphertext retrieval, the improved encryption and retrieval algorithm is improved compared with the existing ones. However, the volume of the public key is too large during the operation, resulting in the data space after encryption is much larger than the plaintext. Further research is needed to reduce the expansion rate of ciphertext while ensuring data security.

Acknowledgements. This work is supported by the National Science Foundation of China under Grant No. 61862007, Guangxi Natural Science Foundation under Grant No. 2018GXNSFAA138147. Postgraduate research and innovation project of Guangxi University for nationalities No. gxun-chxzs2017114, and Teaching Reform Project of Guangxi University for Nationalities under Grant No. 2016XJGY33 and No. 2016XJGY34.

References

1. Smart, N.P., Vercauteren, F.: Fully homomorphic encryption with relatively small key and ciphertext sizes. In: Nguyen, Phong Q., Pointcheval, D. (eds.) PKC 2010. LNCS, vol. 6056, pp. 420–443. Springer, Heidelberg (2010). https://doi.org/10.1007/978-3-642-13013-7_25
2. Mell, P., Grance, T.: The NIST definition of cloud computing. Commun. ACM 53(6), 1–3 (2011)
3. Gentry, C.: Fully homomorphic encryption using ideal lattices. In: Proceedings of the Annual ACM Symposium on Theory of Computing vol. 9, no. 4, pp. 169–178 (2009)
4. Namjoshi, J., Gupte, A.: Service oriented architecture for cloud based travel reservation software as a service. In: Proceedings of the 2009 IEEE International Conference on Cloud Computing, Bangalore, India, 21–25 September 2009, Los Alamitos, CA, USA, pp. 147–150. IEEE Computer Society (2009)
5. Fellows, M., Koblitz, N.: Combinatorial cryptosystems galore. In Finites: Theory, Applications and Algorithms, Contemporary Mathematics, Las Vegas, vol. 168, pp. 51–61 (1993)
6. Xu, P., Liu, C., Xue, S.: Full homomorphic encryption algorithm based on integer polynomial ring. Comput. Eng. 38(24), 1–4 (2012)
7. Coron, J.-S., Mandal, A., Naccache, D., Tibouchi, M.: Fully homomorphic encryption over the integers with shorter public keys. In: Rogaway, P. (ed.) CRYPTO 2011. LNCS, vol. 6841, pp. 487–504. Springer, Heidelberg (2011). https://doi.org/10.1007/978-3-642-22792-9_28
8. Ni-gang, S., Haoran, Z., Weixin, W.: A fully homomorphic encryption scheme for integers with n bits. Comput. Appl. Res. 04, 55–68 (2018)
9. Xiong, W., Wei, Y., Wang, H.: An improved scheme of full homomorphic encryption based on integer. Comput. Appl. Res. (03), 67–78 (2016)

Implementation of a Centralized Scheduling Algorithm for IEEE 802.15.4e TSCH

Jules Pascal Gilles Rugamba, Dinh Loc Mai,
and Myung Kyun Kim[✉]

University of Ulsan, Daehak-ro 93, Nam-Gu, Ulsan 44610, South Korea
mkkim8404l@gmail.com

Abstract. IEEE 802.15.4e Time Slotted Channel Hopping (TSCH) is an important standard in the actual scalable Internet of Things (IoT), to habilitate deterministic low-power mesh networking. In TSCH, scheduling can be distributed or centralized scheduling. In this paper we describe the implementation of a centralized link scheduling for IEEE 802.15.4e TSCH based industrial Low Power Wireless Networks, known as Path Conflict free Least Laxity First (PC-LLF) algorithm in Contiki OS, a real time operating system for wireless sensor by using RPL protocol for network topology construction. PC-LLF prioritizes the assignment of the communications resources for each transmission. For each node to know its assigned cell in the schedules, we implemented a method to broadcast the schedule information packets to all transmissions generator (nodes) with the aim to reduce packets collision and loss. We compared our method with the broadcast of the same packet by Minimum Schedule Function (MSF) and the results show that our method performs better than the MSF and use less number of packets messages than other CoAP and OCARI broadcasting method. The implementation devices are Texas instruments CC 2650 Launch-pad. The implementation aim is to make a real IoT network running a centralized scheduling algorithm of TSCH which can be effectively applied in the industrial networks and the results proved that.

Keywords: TSCH · IEEE 802.15.4e · Scheduling algorithm implementation · Contiki · RPL

1 Introduction

The use of Wireless Sensor Networks (WSNs) in industrial area is increasing and it is found in different industrial application such as monitoring applications, radiation check, leakage detection, distributed and process control [1]. Those real-time applications are sensitive to the delays and demand high communication reliability and a satisfaction of a high scalability. The end-to-end delay are constrained by upper bounds which varies according to the field of application [2]. To overcome those challenges, different wireless technology approaches which reduces the cost of wired installation, have been considered such as the IEEE802.15.4e with a MAC protocol named Time Slotted Channel Hopping (TSCH). With the time slotted and multi-channel frequency hopping features, TSCH reduces latency and ensures high resistance against interferences as well as multi path fading.

© Springer Nature Switzerland AG 2019
D.-S. Huang et al. (Eds.): ICIC 2019, LNAI 11645, pp. 118–129, 2019.
https://doi.org/10.1007/978-3-030-26766-7_12

Although the IEEE802.15.4e standard defines the process to execute communication schedule, it does not define the process of how the schedule is built, maintained, updated as well as the design of the entity in charge of performing such tasks. Scheduling on TSCH networks has been studied widely. Two types can be distinguished: 1. *Distributed scheduling* where there is no central manager (or coordinator) to handle the schedules but nodes in the network agree on the schedule to be used; 2. *Centralized scheduling* where a scheduling is built by a central entity or Coordinator by collecting the network information. Our implementation will be based on a realistic centralized scheduling algorithm known as PC-LLF.

Path Conflict free Least Laxity First (PC-LLF) [3] is a centralized scheduling algorithm which assigns priority to each packet transmission dynamically based on its laxity (the remaining time to the end-to-end deadline) and the amount of latency imposed by the collisions throughout its path to the final destination node (root node).

With the aim to give a contribution to the study of implementation of centralized scheduling in TSCH networks in real environment, we implemented PC-LLF in real hardware, a low-power microcontroller of a Texas instruments TI CC2650 LaunchPad kit using Contiki OS an open-source implementation of a fully finished protocol stack based on Internet of Things standards. As the coordinator node makes the schedules to be used in a given network, the principal key during implementation of centralized scheduling, is the reception by all nodes in the network of the schedule table information packet i.e. each node has to know through that packet, at which timeslot and channel it has to wake up for transmission or reception. To avoid the loss of such essential packets, we designed a method to send it in broadcasting manner by assigning to each node its own timeslot to broadcast the packets, which eliminates competition between nodes, reduces collision and decrease considerably the loss.

The remainder of this paper is organized as follows. Some backgrounds are described in Sect. 2. A brief description of the PC-LLF will be given in Sect. 3. In Sect. 4 the implementation process will be described, in Sect. 5 we will show the evaluation of our implementation and the conclusion will be given in Sect. 6.

2   Background

2.1   IEEE 802.15.4e (TSCH) and RPL

TSCH builds a globally synchronized mesh network where nodes can join the network after hearing a beacon from another node. The time synchronization flows from the PAN coordinator to the leaf nodes along the network.

In TSCH the time is divided into a fixed number and equal-sized timeslots N which form a repeatable and periodic slotframe. One timeslot is long enough to contain the transmission of a data frame and its acknowledgment. Based on a schedule, a node is set in transmitting, reception or sleeping mode at a specific timeslot. TSCH defines the Absolute Slot Number (ASN) which is the total number of timeslots that has elapsed from the start of the network or a randomly determined start time by the coordinator.

It is used as the slot counter and it increments globally in the network on the beginning of each time slot.

In TSCH, 16 or less different channels can be used for communication. Multiple nodes can communicate at same time slots using different channel offset. The communication between 2 nodes in TSCH occurs in cell which is a pair of a timeslot in the slotframe and channel offset in that timeslot. A communication cell can be shared or dedicated i.e. contention-based with CSMA back-off or contention-free respectively. In the timeslot, the sender begins the transmission TsTxOffset after the beginning of timeslot, and the receiver listens to the channel guardTime before. Thereby, it implies that the devices cannot be desynchronized more than guardtime μs. In this paper, the implemented centralized scheduling algorithm allows only the use of dedicated cells.

To form a mesh network or any topology in TSCH, there exist different used protocol such as RPL [4] which is a routing protocol for constrained Low power and Lossy network standardized by the IETF group and designed to meet some requirements described in [5] such as latency, reliability, scalability among others. The routers are organized through the destination oriented directed acyclic graph (DODAG) rooted at the edge router. It is used to get an accurate vision of the network status and thus to compute the best path for downward and upward traffic. Those path are computed based on an Objective Function (OF) and its set of metrics and constraints.

Designed for IPv6 networks, RPL uses the features of ICMP6 neighbor discovery messages for its auto-configuration.

The building process of a DODAG is triggered by a root node and four types of control message *DODAG Information Object (DIO), DODAG Information Solicitation (DIS), Destination Advertisement Object (DAO), Destination Advertisement Object Acknowledgment (DAO-ACK)* are used between nodes during that process to exchange multiple information.

2.2 Contiki OS

Written in C, Contiki operating system [6] was designed for resource constrained devices like Low-Power wireless IoT devices and includes in its latest version [7] the implementation of TSCH based on the IEEE 802.15.4e standards. Its library has a full TCP/IP stack for radio network communication and it supports different variety of radio enabled platforms. Contiki network stack includes diverse protocol at each layer and some modules are specific to each protocol and some are common modules. The *queuebuf.c* module which is common for example provides a way to manage multiple packets at a time. Each other module to access the *queuebuf.c* needs to maintain pointers to it. 6LoWPAN at adaptation layer would take advantage of *queuebuf.c* for fragmenting IPv6 datagrams into multiple packets and TSCH at MAC layer use it for queues transmission. From IETF (Internet Engineering Task Force) known networks and applications standards, Contiki supports in its library TCP, 6LowPAN, UDP, RPL, IPv6 and CoAP. Contiki OS includes an event driven kernel used to load and unload process at run time.

3 PC-LLF: Path Conflict Aware Least Laxity First Algorithm

The PC-LLF algorithm [3] is a centralized algorithm designed over IEEE 802.15.4e TSCH with the aim to reduce latency of messages by finding feasible schedule for all transmissions of each flow along their designated path. As described previously, the transmission in TSCH occurs at a certain time slot t and channel ch, which compose a cell (t, ch).

Designed for TSCH networks, PC-LLF algorithm obeys to some constraints of that technology described below. Firstly, network forms a topology mesh network (Fig. 1.) as a Directed Acyclic Graph (DAG) $G = (N, L)$ where $N = \{N_1, N_2, ..., Nn\}$ represents all devices in the considered TSCH networks and arcs in L are communication link. NpNq represents a link where Np is the transmitter node and Nq is the receiver node. Interfering links cannot be scheduled on the same cell and the duplex conflicting links cannot share the same time slots. The duplex conflicting links are all the links that have same receiver or sender nodes. From the Fig. 1, duplex conflicting links of the link N_6N_2 are CNF $(N_6N_2) = \{N_5N_2, N_4N_2, N_2N_1\}$.

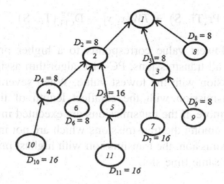

Fig. 1. Example of network $G = (N, L)$, $N = \{N_1... N_{11}\}$ and D the Period assigned to each transmission from its source node; N_1 is the root node

All transmissions are upward and considered periodic. The transmission's period which is equal to its deadline, is chosen randomly by each node from a power of two value $[2^3, 2^9]$. A packet's transmission miss deadline when it arrives to the destination later than its release time added its deadline considered in time slot value. To calculate priority used by PC-LLF to make feasible schedule, some parameters are considered such as time window denoted $TW[x, y]$ for each transmission which is the time phase from the earliest time slot (x) and the latest time slot (y) that the transmission can occur. The earliest time slot is the actual time slot and thus it is dynamically recomputed at each timeslot. The latest time slot has a fixed value and it depends on the deadline value. The second considered parameter is the average delay due to collisions that each remaining transmissions may deal with at the time slot S. The considered collisions are duplex conflicting collisions which happen due to a transmission existing in a conflicting link with a time window *TW* that overlaps with the time window of the actual considered transmission. Let denote the amount of conflict transmissions with a certain transmission T_i as N_{CFL} (T_i). The next step is to get the average latency of a packet due to the collisions through the remaining path at a certain time slot.

The remaining path is considered to be composed by the actual transmission and the future transmission of the packet in order to reach the root nodes. Let denote the average latency due to duplex conflicting transmission at a certain time slot S as D_{CFL}^{avg} (T_i, S) and its value change dynamically at each time slot because the amount of duplex conflicting transmissions vary due to the fact that some transmissions may be scheduled previously or others may be released later. Let denote k, the number of hop count away for a transmission to reach the root node. Thus the average latency is given by the sum of all conflicting links divided by the number of hop count as in the equation below:

$$D_{CFL}^{avg}(T_i, S) = \sum N_{CFL}(T_i)/k \qquad (1)$$

From the Eq. (1), the priority of a transmission at a certain time slot S is computed by the difference between the time window of the packet's transmission and its average latency as described in the equation below:

$$Pr(T_i, S) = TW[x, y] - D_{CFL}^{avg}(T_i, S) \qquad (2)$$

From Eq. (2), the lowest value corresponds to a higher priority. Based on the computed priorities for all transmissions, PC-LLF algorithm assigns the time slot and channel to the transmission with the lowest value. When several transmissions have same priority, the transmission with the smallest length of time window will be assigned firstly. Furthermore, if the transmissions are executed in an environment with more than one channel, among the transmissions which are not in duplex conflict with the firstly assigned transmission, the transmission with highest priority is assigned to a different channel in the same time slot.

4   Implementation

The PC-LLF algorithm was implemented using Contiki OS into the low-power microcontroller of a Texas instruments TI CC2650 LaunchPad kit. Its microcontroller is based on ARM Cortex-M3 architecture which operates at 48 MHz, with 128 KB of flash memory and 20 KB of SRAM.

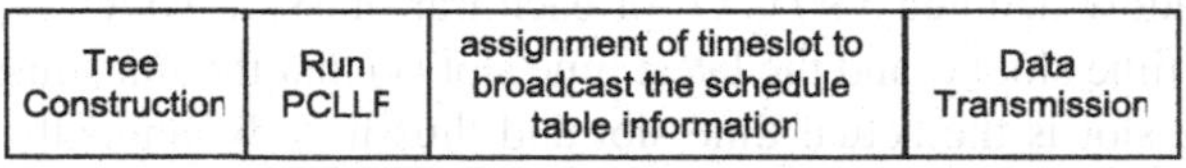

Tree Construction	Run PCLLF	assignment of timeslot to broadcast the schedule table information	Data Transmission

Fig. 2. Implementation process

Figure 2 gives a big picture of the implementation process divided into four parts. Firstly is the tree construction where nodes make a network topology using the RPL routing protocol. Secondly is the execution by root node of the implemented PC-LLF algorithm. At third, root node assigns to each node a timeslot to broadcast the schedule table information packet, broadcasts the resultant schedule table to all nodes and that part includes also the broadcast of the schedule table information packets. Fourthly all nodes will start data transmission following the PC-LLF cells assignment.

4.1   Tree Construction

Contiki OS includes the RPL routing protocol in storing and non-storing mode. We chose to use RPL non-storing mode as it allows the root node to collect all network information from each node in the network through the exchanged packets. We assume a fix number of nodes in network and known by root node in the beginning and thus, it can check through defined function if it had received all the information from all nodes in order to start running the PC-LLF algorithm.

The nodes form a network by RPL non-storing mode and run Minimum Scheduling Function [8] in the tree construction period. The root nodes broadcast first the DIO packet to inform of the existence of a network and each receiver node try to join by following the process describes in [4]. Once joined, the node may also act as a router and broadcast the DIO at its turn.

Type: 0x0A	Option length	Period
1 byte	2 bytes	2 bytes

Fig. 3. Structure of new Period Option in RPL DAO packet

There exist different DAO option used for carrying routing information such as target option which carry the information of DAGid of the new joiner node and the transit option which carry multiple information such as the joiner node's parent information and thus the links information are forwarded to the root node. To send to the root node the period value of each node's transmission needed to run PC-LLF, a new option named *Period Option* as shown in Fig. 3 was added in the DAO packet by modifying the *rpl-icmp6.c* module in Contiki OS. The period option has three field: *Type* to identify the option type, *Option length* to specify the length and *Period* which contains the period for the transmission.

4.2   Implementation of PC-LLF Algorithm

From the link and transmission's period information collected by root node during the tree construction period, the root node executes the PC-LLF algorithm to schedule all node's transmissions. In case of an unschedulable network topology, the root node has to stop the execution and the change in the topology is required.

The input parameter for the execution of algorithm are G the links information built from RPL previously, R which is the list of released transmission and in the beginning each node is considered to have a transmission to be forwarded. The next parameter is P, each node's transmission period which is equal to its deadline and Nch the number of channel used in data transmission and we considered two channel (Nch = 2). The output is the schedule table of all transmissions (Fig. 4) which is forwarded to all nodes from root node.

The PC-LLF algorithm was written in C as it has to run under Contiki OS and was added as new created library in Contiki. The library access the routing layer handled by RPL to get network topology with routing links and period of each transmission as input.

Timeslot	1	2	3	4	5	6	7	8	9	10	11	12	13	14	15	16
Ch0	T(2-1)-1	T(4-2)-1	T(4-1)-1	T(6-2)-1	T(6-1)-1	T(11-2)-1	T(11-1)-1	T(10-2)-1	T(10-1)-1	T(2-1)-2	T(4-2)-2	T(4-1)-2	T(5-2)-1	T(5-1)-1	T(6-2)-2	T(6-1)-2
Ch1	T(9-3)-1	T(9-1)-1	T(11-5)-1	T(3-1)-1	T(10-4)-1	T(8-1)-1	T(7-2)-1	T(7-1)-1	T(9-3)-2		T(9-1)-2		T(3-1)-2		T(8-1)-2	

Fig. 4. Example of scheduling table information from a network topology in Fig. 1

The algorithm must finish to schedule all transmissions no longer than the Least Common Multiplier (LCM = 16 in the example) of all periods value in the network as seen in example Fig. 4. At timeslot one (Ts = 1, ch = 0), the transmission from node 2 to root node (1) was assigned. The transmission are noted as $T(N_{id} - M_{id}) - I$ where N_{id} is the source node id of the packet, M_{id} is the receiver node id and I is the instance of the transmission which specifies the current period of the transmission, as all transmission are periodic. In the example (Fig. 1) nodes N_2, N_3, N_8, N_4, N_6 and N_9 for example have two period of transmission within the LCM. The number of period of transmission is defined by the LCM divided by the period value of the sender node's transmission.

4.3 Assignment of Timeslot to Broadcast the Schedule Table Information

This process is divided into three parts. Firstly, we assigned timeslot for each node to broadcast the schedule table information and secondly is to send the assignment information to all nodes. Last one is the broadcast of schedule table information packets by all nodes as assigned.

4.3.1 Assignment of Timeslot to Broadcast the Schedule Table Information

After execution of PC-LLF algorithm, each node in the network has to receive the schedule information. For that, the packets have to flow from the root node towards all the nodes in the network. To ensure the minimum loss of schedule table information packets, we assigned a dedicated timeslot to each node for broadcasting that packet.

Based on the constructed network topology and the gathered network information by root node, the assignment was done by Breadth-First Search Algorithm (BFS) [9]. Each node in the network is assigned a unique identifier (*node_id*) used to assign timeslot to the nodes to broadcast the schedule table information packet. By following the BFS standard, we assign firstly the master node (root node) in the first timeslot and then the algorithm traverse the tree from the adjacent node of the root node and assigns timeslot to the node with the smallest *node_id* value until all the adjacent nodes to the root nodes are assigned. At next hop-count, the algorithm firstly traverses the adjacent nodes to the firstly assigned nodes from the adjacent nodes to the root and assign timeslot based on the value of *node_id*. The process continues until all nodes are assigned. The result of assignment is shown in Fig. 5.

Timeslot	1	2	3	4	5	6	7	8	9	10	11
	1→	2→	3→	8→	4→	5→	6→	7→	9→	10→	11→

Fig. 5. Result of assignment of Timeslot to each node in Fig. 1 to send schedule table information

This assignment reduces the number of collision firstly when transmitting as only one node is assigned for transmitting and there is no competition between nodes, and secondly when receiving the packet because one node (or more) can receive only from one transmitter, the assigned node.

4.3.2 Sending the Assignment Information to All Nodes

The assigned timeslots must be known by all node in the network and thus the root node will, after creating the assignment information packet, broadcast it to each node. To realize that, we used the Contiki OS implemented module named *tsch-queue* which defines a queue of outgoing packets for each neighbor and we added others queues neighbor beside the existing one, for the purpose of our implementation. The *n_assign* is used for queuing the assignment information packet when it is broadcasted and *n_data* for handling the outgoing data packet used specifically for data transmission period.

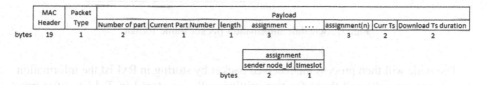

Fig. 6. Packet frame format of assignment information packet

The root node triggers the download of assignment information packet by broadcasting the packet. When a node receive the assignment packet, it activates its assignment information download period. Each node is assigned a number of timeslot delay between two transmissions. A larger number of node in the network implies a large packet and hence it has to be divided into parts to be transmitted. Each assignment information packet in node's queue is transmitted after the delay. The use of *MSF* i.e.

one timeslot and one channel offset as specified in the standard during that period may cause collisions in the packet traffic. To reduce the collision, we implemented a competition scheme within a timeslot which allow each node to access the communications resources by reducing contention time based on the sequence of the assignment information packets received and by checking if the channel is idle or not.

When broadcasting the assignment information packet, we took in consideration the maximum packet size originally support in TSCH which is 127 KB. Figure 6 shows the packet frame format and the size we used for the assignment information packet. Based on the frame format, one packet can include up to 30 assignment information which make a total size of 118 KB. This allow to send a large number of assignment information packet while the allowed maximum size of packet is not exceeded.

Taking into consideration that the assignment information sent in one packet are considerable and that one assignment information contains only one node information, we assumed that the needed number of timeslot to send all packets in a given network and to be received by all nodes is equal to the number of nodes in the network multiply by two to avoid loss.

4.3.3   Broadcast of the Schedule Table Information
To transmit data packets, each node needs to know in which timeslot it has to wake up for transmitting or receiving the data packet. That means the broadcast of the result of the PC-LLF algorithm scheduling (Fig. 4) which is done by following the assigned timeslot for each node to broadcast the schedule information packet. After the reception of the schedule information packet, the node adds it to the queue via *n_schedule* queue neighbor used for queuing the schedule table information packet and created through the *tsch_queue* module described above.

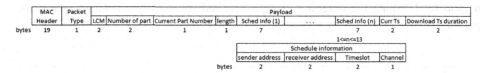

Fig. 7. Schedule information packet frame format

The node will then process the received packet by storing in RxList the information of its reception cells and those for transmitting cells are stored in TxList. After processing the packet, the node get the packet from the queue and broadcast at its turn in its assigned timeslot.

Figure 7 shows the schedule information packet constitutes by *MAC header, packet type* to identify the packet. A large number of nodes in the network implies the fragmentation of the full schedule information packet and thus the payload is divided into different fields to handle that. It has *LCM* which is the least common multiplier for all period value of all transmission, *Number of part* which contains the number of packets to transmit all the schedule information packet, the *current part number* for identifying the currently being transmitted packet, *length* with the value of total transmission information in the current packet, *sched info* containing the assigned cell for a transmission as well as the sender and receiver node address. It contains also the *curr Ts* which informs on the current used timeslot and the *Download Ts duration* which is the number of timeslot to be used for broadcasting the schedule information packet. Considering that each node has its own timeslot to broadcast the packet, we considered the maximum duration to broadcast the packet to be equal to the number of node multiply the number of packet used to transmit all the full packet (*Number of Part*). Based on the packet size and the allowed size packet to be transmitted in TSCH (127 KB), one packet can contain at most 13 scheduled transmission information (or cells) which make the size equal to 121 KB.

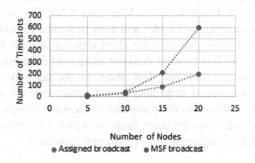

Fig. 8. Reception delay in timeslot value of the schedule table information packet

4.4 Data Transmission Period

During the data transmission period, all nodes transmit their data towards to the root node based on the schedule computed by the PC-LLF algorithm. After the broadcast process, each node knows from the RxList and TxList exactly at which cell it has to transmit or to receive data packet. The Absolute slot Number (ASN) described above, gives the possibility to know exactly the current used timeslot and hence, each node can trace and know its timeslot to wake up.

The data transmission period duration or number of timeslot for data transmission is equal to the LCM value of all node's periods in the network. During data transmission, when a node wakes up at its associated cell, it checks first if the current cell is for reception mode then the node listen for the packet and after reception, it adds the packet in its queue and if is for transmission mode, the node firstly checks if any packet is available in its queue buffer, get the packet from the queue and send it; after transmission the packet is removed from the queue. In case the queue buffer is empty, it switches into sleep mode. Once the packet is ready, the 6LoWPAN of Adaptation Layer, passes it to the MAC layer for being transmitted. Likewise, when receiving packet, the MAC layer passes it to adaptation Layer using the *packetbuf.c* module. As each node wake up at the specific cell for transmitting or receiving, the data packet transmission is done by broadcasting from its source node and only the node which woke up at that cell for reception can receive the packet.

5 Implementation Evaluation

The PC-LLF algorithm was implemented using Contiki OS real hardware of Texas Instruments TI CC2650 LaunchPad kit in which one acts as root node.

We compared the schedule table information packet reception delay in the network when the packet is broadcasted by Minimum scheduling function (MSF) and when it is broadcasted by our assigned broadcast method described in Sect. 4.3 by considering the number of timeslot used in relation to the number of nodes in the network. The more nodes are in network, the more packets of schedule information as the full packet is fragmented into multiple packet to avoid oversizing the maximum allowed packet size. We also compared the number of packets messages used by our method and those used by CoAP and OCARI to broadcast the schedule packets as described in [10].

From Fig. 8, we can see the variation of the reception delay of the schedule table information packet when broadcasted by MSF and when our assigned broadcast is used. We made a test using firstly 5 nodes and then 10 nodes, 15 nodes and 20 nodes in the network. The result show that with the short number of node in the network, the broadcast using MSF uses less timeslot than our assigned broadcast. This is due to less competition process as there are only few nodes in the network. As the number of nodes increases, we can see a considerable difference and the assigned broadcast outperforms the MSF broadcast as the competition becomes though when then number of node increase and collisions are more frequent which can cause considerable loss and make MSF vulnerable compare to our assigned broadcast.

Table 1. Table of each Flow and their period and delay time.

Flow	Period	End-to-end delay	Flow	Period	End-to-end delay
2 → 1	8	1	7 → 1	16	8
3 → 1	8	4	8 → 1	8	6
4 → 1	8	3	9 → 1	8	2
5 → 1	16	14	10 → 1	16	9
6 → 1	8	5	11 → 1	16	7

The period of each transmission equals to its deadline and that value is expressed in timeslot number. The end-to-end delay of each transmission flow (represented in first column with sender node id → destination node id), expressed in timeslot number as shown by Table 1, is less than its period value and thus all the flow didn't exceed the deadline for transmission.

Table 2. Number of packets required to broadcast the schedule packets.

Number of node	Number of cells	Our method (packets)	CoAP (packets)	OCARI (packets)
5	12 × 7(84 bytes)	1 × 5	3 × 5	2 × 5
10	28 × 7(196 bytes)	3 × 5	7 × 5	3 × 5
15	60 × 7(420 bytes)	5 × 5	14 × 5	6 × 5
20	78 × 7(546 bytes)	6 × 5	18 × 5	7 × 5

In Table 2, the *number of cells* column describes the existing cells in one schedule table and obviously increase when the number of nodes increases. Each cell is 7 bytes as shown in Fig. 7. Our method allows at most 91 bytes over the 127 bytes max packet as the remaining is used for other system and configuration information.

As described in [10], CoAP and OCARI allow respectively 32 bytes and 80 bytes. Considering the topology in Fig. 1, the nodes that can broadcast the schedule packets i.e. with at least one children node, are 5 and thus we multiply by 5 to get the total number of packets to broadcast. As result, our method use less packets which has a positive impact on node's duty cycle as well as on the duration of broadcasting the schedule.

6 Conclusion

In this paper, a centralized schedule for TSCH was implemented using Contiki OS and RPL for the network topology. The PC-LLF component was integrated in the existing stack structure of Contiki OS. To ensure the reception by all nodes of the schedules table information, we successfully used our assigned broadcast method which shown better results than broadcasting with Minimum Scheduling Function (MSF) algorithm and use less packet than CoAP or OCARI. The result of implementation show that our method to broadcast schedule table and the PC-LLF algorithm can run in real device which is helpful for an implementation in real environment with a centralized schedule.

Acknowledgements. This work was supported by the 2019 Research Fund of University of Ulsan.

References

1. Sinopoli, B., Sharp, C., Schenato, L., Schaffert, S., Sastry, S.S.: Distributed control applications within sensor networks. In: Proceedings of IEEE, vol. 91, no. 8, pp. 1235–1246 (2003)
2. Zurawski, R.: Networked embedded systems: an overview. In: Zurawski, R. (ed.) Networked Embedded Systems, Ch. 1, pp. 1.11–1.16. CRC Press, Boca Raton (2009)
3. Darbandi, A., Kim, M.K.: Path collision-aware real-time link scheduling for TSCH wireless networks. KSII Trans. Internet Inf. Syst. (2019). Accepted to be published
4. Winter, T., Thubert, P., Brandt, A., Hui, J., Kelsey, R.: RPL: IPV6 "Routing protocol for Low Power and Lossy Networks". IETF request for comments 6550, March 2012
5. Pister, E.K., Thubert, E.P., Dwars, S., Phinney, T.: Industrial routing requirements in Low-Power and Lossy Networks. IETF request for comments:5673, March 2009
6. Dunkels, A., Gronvall, B., Voigt, T.: Contiki - a lightweight and flexible operating system for tiny networked sensors. In: 29th Annual IEEE international conference on Local Computer Networks, pp. 455–462, November 2004. 17
7. Contiki 3.x. https://github.com/contiki-os/contiki.Online. Accessed 2 June 2016. 17
8. Minimum Scheduling Function (MSF). https://tools.ietf.org/html/draft-ietf-6tisch-minimal-21
9. Breadth-First Searching (BFS) Algorithm. https://en.wikipedia.org/wiki/Breadth-first_search
10. Livolant, E., Minet, P., Watteyne, T.: The cost of installing a 6TiSCH schedule. In: Mitton, N., Loscri, V., Mouradian, A. (eds.) ADHOC-NOW 2016. LNCS, vol. 9724, pp. 17–31. Springer, Cham (2016). https://doi.org/10.1007/978-3-319-40509-4_2

Learning Automata-Based Solutions to the Multi-Elevator Problem

O. Ghaleb[1] and B. John Oommen[1,2(⊠)]

[1] School of Computer Science, Carleton University, Ottawa K1S 5B6, Canada
omar.ghaleb@carleton.ca, oommen@scs.carleton.ca
[2] University of Agder, Grimstad, Norway

Abstract. In the last century, AI has been the topic of interest in many areas, where the focus was on mimicking human behaviour. It has been researched to be incorporated into different domains, such as security, diagnosis, autonomous driving, financial prediction analysis and playing games such as chess and Go. They also worked on different subfields of AI such as machine learning, deep learning, pattern recognition and other relevant subfields. Our work in a previous paper [1] focused on a problem that has not been tackled using AI before, which is the elevator-problem. In which we try to find the optimal parking floor for the elevator for the single elevator problem. In this paper, our work extends the model by solving the more complicated scenario, which is the multi elevator problem (MEP) using Learning Automata (LA). This problem can then be generalized to be applied to a variety of problems that share the same characteristics with the elevator problem. We refer to these problems as Elevator-Like Problems (ELPs). For the extended version (MEP) we try to find the optimal parking floors for the set of elevators so as to minimize the passengers' Average Waiting Time (AWT). Apart from proposing benchmark solutions, we have provided two different novel LA-based solutions for the multi-elevator scenario. The first solution is based on the well-known LRI scheme, and the second solution incorporates the *Pursuit* concept to improve the performance and the convergence speed of the first solution, leading to the $P L_{RI}$ scheme. The simulation results presented demonstrate that our solutions performed better than those used in modern-day elevators, and provided results that are near-optimal, yielding a performance increase of up to 91%.

Keywords: Learning Automata (LA) · Learning systems ·
Multi Elevator Problem (MEP) · Elevator-like problem (ELP) · Parking problem

1 Introduction

Directly from its early stages, the objective of the field of Artificial Intelligence (AI) has been to influence PCs to react smartly in situations that are normal or challenging. Turing, in his Turing test, said that a machine could be considered to have AI if a human spectator would not have possessed the capacity to recognize whether its behaviour was the machines behaviour or a real human. The objective, despite the fact that it has been elevated, has been accomplished to a remarkable degree. Computer AI-based programs have tested and beaten the best players in a variety of games, including

© Springer Nature Switzerland AG 2019
D.-S. Huang et al. (Eds.): ICIC 2019, LNAI 11645, pp. 130–141, 2019.
https://doi.org/10.1007/978-3-030-26766-7_13

Chess, Go and recently StarCraft. It is impressive to record the regions in which AI has been utilized. In to- day's world, machine learning, pattern recognition, medical diagnosis and voice- operated systems are commonplace. AI has been utilized in essentially every application domain.

This paper considers an area in which AI has not been utilized much. Consider a tennis player who is moving inside his side of the court. The question that he encounters after he hits the ball is to know where he should place himself in order to best counter the response of his opponent. This problem can be modelled as a parking problem, i.e., where should the player "park" himself such that the ball is in his vicinity of where he "parked" when his opponent hits the ball. These types of problems could be referred to as "elevator-like" problems. They are, in fact, in a unidimensional domain, related to the problem of where an elevator within a building should be parked. In other words, if there is a building with n floors and if the elevator car is requested at floor i, where should the car move to after the passenger has been dropped off? The answer to this question, of course, depends on the function of the criterion, but it is appropriate to work with the understanding that the car should be parked near the next passenger call. The same analogy can be seen if we extend the problem area to know where police vehicles or emergency vehicles should be parked in order to be available for the next call as soon as possible.

Although the above problems generally involve transport problems, their common aspect can also be extended to other areas. For example, one might consider where the read/write disk head should be placed in a memory bank so that data can be accessed more quickly. In fact, by determining where the underwater sensors should be located, our problem model can be extended to consider improving underwater communication systems.

In this paper, all these problems are referred to as "Elevator-like" problems, and this is the main focus of the research. However, in order to make the problem non-trivial, we assume that the "world" in which we operate is stochastic and the underlying distributions are unknown.

We will discuss the Multi Elevator Problem (MEP) and explain the solutions by which researchers have tackled it in the past. In MEP, we have a building with n floors and has more than a single elevator, and we are to design a policy for the elevator in order to operate in such a way so as to save energy, reduce the travel time or the waiting time for passengers. Moreover, the elevator policy should decide how the elevator should operate in order to achieve its goal. For example, one possible policy could require that the elevator picks the best route in order to pick passengers, while another could be to serve the longest queue first or to decide where the elevator should wait for the next call. We will initially submit the solutions used as benchmarks and show their results. After that, we will introduce our LA-based solutions and display their simulation results. We will then compare our findings with the benchmark results to demonstrate that we have obtained similar or better results.

In a previous paper [1], we discussed the SEP and how researchers have attempted to solve it. In this paper, we will consider an extension for the SEP which is the MEP. In the MEP, the problem is to try to optimize a set of elevators in a building instead of a single elevator. This extension, obviously, introduces more complexity to the problem and adds more constraints. The problem can be solved in two different types of environments. In the first case, we deal with a stationary environment, where the environment does not change with time. In the second scenario, the environment is non-stationary, and it changes with time. Here, the search space changes while one attempts to find the optimal solution to the problem. The solution could consist of a set of movements for the elevator or finding the optimal parking floors where the elevators should be parked to minimize the service time or the waiting time for passengers.

In Crites and Barto's paper [2], they proposed a Reinforcement Learning (RL)-based solution to the elevator dispatching problem. At the time of their work, there were a variety of challenges related to this problem that had not been addressed earlier. In the elevator system that they examined, they simulated a building with 10-stories and four elevators. They used a Poisson distribution for the arrival of passengers at each floor. They then modelled the system as a discrete event system where the events happened at discrete time instances. They also used a different RL agent for each elevator. Such a design made them encounter a couple of complications such as the noise that the agents obtained when they received a reinforcement signal because of the effects of other agents. After they conducted their experiments, they compared their results with eight previous algorithms and heuristics, and were able to demonstrate, by their simulations, that their algorithm noticeably outperformed the previous algorithms.

In [3, 4], the authors presented an algorithm to calculate estimates for the average waiting time for passengers in group elevator systems. They modelled the system as a discrete-state Markov chain for the moving elevator car. They thus used dynamic programming to analyze the Markov Chain and to compute the expected measures for their proposed metrics for the performance of the elevators. These metrics were, for example, the expected average waiting time for the passengers. They then introduced a linear-time controller that was based on the method that they had earlier specified, referred to as the ESA-DP algorithm.

In their paper, they first introduced the "base" ESA-DP scheme after which they extended the algorithm to deal with specific cases. Their base algorithm involved some assumptions such as an unlimited elevator capacity, no future passenger arrivals, full state information and a known marginal distribution for the destination floors. In the subsequent work, they extended the algorithm to deal with the non-uniform distributions for the destination floors. They then showed, in their results, that this algorithm outperformed other algorithms that were used as benchmarks and that they were able to reduce the waiting times by 30%–40% in heavy traffic scenarios. Besides, their algorithm was applicable and computationally efficient for real-life elevator systems.

In [5], which might be considered to be the closest representation and approach to the model and solution that we propose, the authors presented a paper concerning the problem of optimally parking elevator cars in a building with an attempt to minimize the waiting time for passengers. This was done by anticipating the arrival of the passengers. They studied two traffic patterns of passengers which involved the down-peak and up-peak settings. These patterns represent the cases when the frequency of passengers going down to the lobby is high, and alternatively when the frequency of going up the building from the lobby was high, respectively. The authors of [5] modelled the system as a Markov Decision Process whose states corresponded to possible parking locations or floors. Using this, they proposed two dynamic programming solutions for the elevators for selecting which location they should park the cars for the up-peak and down-peak traffic patterns. The results that they obtained demonstrated that their algorithm was able to acquire the dependency between the number of cars parked that needed to be parked at the lobby and the arrival rate of passengers. It also yielded good results for the medium and low arrival patterns.

1.1 Learning Automata (LA)

We now concentrate on the field that we shall work in, namely, that of LA. The concept of LA was first introduced in the early 1960's in the pioneering research done by Tsetlin [6]. He proposed a computational learning scheme that can be used to learn from a random (or stochastic) Environment which offers a set of actions for the automaton to choose from. The automaton's goal is to pick the best action that maximizes the reward received from the Environment and minimizes the penalty. The evaluation is based on a function that permits the Environment to stochastically measure how good an action is, and to thereafter send an appropriate feedback signal to the automata.

After the introduction of LA, different structures of LA, such as the deterministic and the stochastic schemes, were introduced and studied by the famous researchers Tsetlin, Krinsky and Krylov in [6] and Varshavskii in [7].

Like many of the Reinforcement Learning techniques in a variety of problems, mainly optimization problems and in many AI fields, the LA field has been used. It has been used in neural network adaptation [8], solving communication and networking problems [9], [10] and [11], in distributed scheduling problems [12], and in the training of Hidden Markov Models [13].

We will cover the relevant background to help the reader understand the basic concepts for our proposed work in this section[1].

In Fig. 1, we have the general stochastic learning model associated with LA. The components of the model are the *Random Environment*, the *Automaton*, the *Feedback* received from the environment and the *Actions* chosen by the automaton.

1.2 Environment

First, we define the stochastic Random Environment[2] with which the automaton interacts. Initially, the automaton selects and communicates to the environment an action from the set of actions available to it. The environment assesses this action according to a random function and sends a feedback signal back to the automaton, depending on whether the action resulted in a reward or a penalty.

---

[1] We will not go through irrelevant details and/or the proofs of the LA-related claims.

[2] We will only present what each part of LA does. Full definitions are mentioned in [14].

$$\underline{c} = \{c_i\}$$

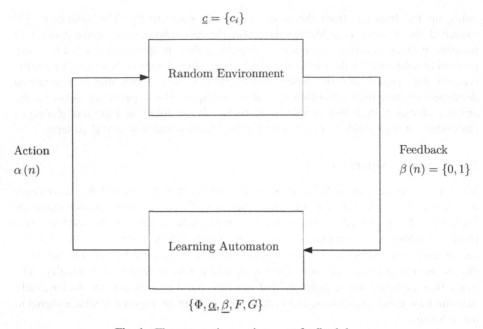

Action
$\alpha(n)$

Feedback
$\beta(n) = \{0,1\}$

$$\{\Phi, \underline{\alpha}, \underline{\beta}, F, G\}$$

Fig. 1. The automation-environment feedback loop.

1.3 Automaton

As an iterative operation based on the Environment and the interaction between them, the LA achieves its learning process. This process consists of two main steps, policy assessment, which is how the selected action is assessed by the environment. In addition, the second step involves policy improvement, in which the LA improves the likelihood of picking an action that maximizes the environmental reward.

1.4 Estimator and Pursuit Algorithms

In [18, 19], Thathachar und Sastry introduced the concept of estimator algorithms when they realized that the Absolutely Expedient algorithms were absorbing and had little chance not to converge to the better action. Initial estimator algorithms were based on Maximum Likelihood estimates (ML), which were used later in the Bayesian estimates to update the probabilities of the actions. This concept was achieved by monitoring the number of rewards received for the selected actions. In doing so, the LA converged quicker to the actions with higher reward estimates.

The original Pursuit Algorithms are the simplest versions of those using the Estimator paradigm introduced by Thathachar and Sastry in [15]. These algorithms are based on the pursuit strategy where the idea is to have the algorithm pursuing the best-known action based on the corresponding reward estimates. The pursuit algorithms were proven to be ϱ-optimal. After the initial family of Pursuit algorithms, the concept of designing discretized Pursuit LA was introduced by Oommen and Lanctot in [16].

1.5 Contributions of This Paper

The novel contributions of this paper are:

- We have surveyed a subfield of AI, namely the field of Learning Automata (LA), and have concluded that it has not been used previously to solve the MEP.
- We were able to identify two different models of computations for the elevator problem, where the first requires the calling distribution to be known *a priori*, and the second does not need such information.
- We were able to model the elevator problem in such a way that it can be solved using LA approaches.
- We introduced a Linear Reward-Inaction (L_{RI})-based solution to tackle the multi-elevator problem. It has been referred to as MEP3.
- We also presented an improvement on MEP3 by including the so-called pursuit phenomenon into the LA solution. This led to the P L_{RI}-based solution, referred to as MEP4, and this yielded better results and faster convergence than MEP3.
- To summarize, we have shown that LA-based solutions can solve the MEP without requiring any knowledge of the underlying distributions. Amazingly enough, the results and solutions that they yielded are near-optimal.

2 The Multi-Elevator Problem (MEP)

The problem we are trying to tackle can be stated as follows: We have a specific building with n floors and e elevators. The floors and passengers are characterized by distributions $C = \{c_1, \ldots, c_n\}$ and $D = \{d_1, \ldots, d_n\}$, where c_i is the probability of receiving a call from floor i, and d_j is the probability that the elevator drops the passenger off at floor j. These distributions are unknown to the decision-making algorithm, and our goal is to design LA-based solutions such that they adaptively determine a set of floors for the e elevators to park at during the idle period so as to minimize the passengers' waiting time.

The metric used as a performance measure in our study is the Average Waiting Time (AWT) of the passengers. This is clearly a function of the number of floors, and of also how close the converged solution is to the optimal floor.

3 Competitive Solutions

3.1 Do Nothing Policy: MEP1

The first policy, called the "Do Nothing" policy, is referred to as MEP1, and is presented in [14]. As one can observe, the simulations start by having the elevators parked at initial random floors, selected in an equiprobable manner. Thereafter, a call is issued according to the distribution C. The main controller then checks which elevator is the closest to the calling floor and sends that elevator to pick up the passenger. Once the elevator drops off the passenger at the destination floor, which is determined according to the distribution D, the elevator waits at that drop-off floor and stops for the next call.

To evaluate the average waiting time, we calculated the waiting time for each call which, as mentioned, is used to evaluate the performance of the policy. This is done by using the following equation:

$$WT = \theta * |calling\,floor - parking\,floor|, \tag{1}$$

where θ is a parameter characterizing the pace variable, and which differs from one system to another because of different shaft speeds and the associated accelerations/decelerations of the elevator shafts. Since this is constant for all the simulations, we ignored this pace parameter and focused on the travel distance from the parked location to the floor where the call is made, $|calling\,floor - parking\,floor|$, which does not change for different elevator systems even if their paces are different. By doing this, the model would be generalized so that it can be applied to different elevator systems since it will be system independent. This yields the final waiting time equation to be:

$$WT = |calling\,floor - parking\,floor|, \tag{2}$$

which is used to calculate the waiting time for all the policies studied in the paper.

3.2   Myopic Policy: MEP2

The second policy, MEP2, is an extension to the previous solution for the single elevator scenario, SEP2, that was introduced in the previous paper [1], but modified to work with the multi-elevator setting, as explained in [17]. The model uses the principle of using the call distributions, C, to calculate the best possible parking floors for the number of available elevators.

The simulation followed the same procedure as in the previously presented model, MEP1, where the elevators received a passenger call and then decided which elevator was to serve this passenger based on how far the elevators were, and where the closest one was, which was the one chosen to serve the current passenger. Subsequently, the elevator dropped-off the passenger at the destination location, but instead of staying at the destination floor as a parking floor, the elevators moved to predetermined parking floors that are associated with the elevators after every service. The corresponding algorithm is given in [14]. This was done by exhaustively searching across the combination of floors which minimized a function representing the expected waiting time. The extended equation for this is as follows:

$$T(f_1,\ldots,f_e) = \sum_{y=1}^{mp_1} |y - f_1| * g(y) + \ldots + \sum_{y=mp_{e-1}}^{n} |y - f_e| * g(y) \tag{3}$$

where $f1$ to fe are the floors selected as parking floors for the set of available elevators, n is the number of floors in the building, $g(y)$ is the probability of receiving a call from the floor y, m_{pe-1} is the midpoint between the two consecutive parking floors f_{e-1} and fe. In our study, as earlier, we employed 2 and 3-elevator models.

4 LA-Based Solutions

In this section, we are going to present our proposed LA-based solutions for the MEP. First, we will show how we have modelled the problem, and thereafter we present an L_{RI}-based solution to the problem. Subsequently, we submit an enhancement on the L_{RI} solution, in which we use the pursuit concept for the L_{RI}, and this yielded the second and even better solution, which is the P L_{RI}- based solution.

4.1 Problem Modelling

As mentioned in Sect. 1.1, any LA-based structure consists of an Environment and the LA itself. Unlike the SEP, in the MEP we have to take into consideration the fact that we are having multiple elevators, and this renders the problem to be more complicated. To model the problem, we have to define what would constitute the LA, and what the Environment would do. The Environment acts as the evaluator for what the LA picks as the parking floor for the elevators.

On the other hand, the LA design differs in as much as we could employ two models. In the first model, the actions are specified as a tuple of a combination of floors, and the issue of selecting the floors as seen as a single decision. However, this method requires a considerable amount of memory as the number of elevators increases, because it involves an exponentially increasing quantity in order to store each combination of floors. The second model requires having different LA for each elevator. This paradigm saves memory as it requires a linearly increasing amount of space. This is the paradigm used in our thesis.

Each LA acts as an elevator's controller, which picks one of the available floors at which it will park at so as to wait for the next call. Moreover, all the LAs (i.e., the controller), will share the same Environment as they all operate in the same building.

4.2 L_{RI}-Based Solution: MEP3

Our first proposed solution is based on the LRI scheme where the LA updates the probabilities of the actions based on the feedback received from the Environment. It updates them when it receives a reward feedback and keeps them unchanged when it receives a penalty feedback.

The corresponding algorithm of MEP3 is presented in [14]. The simulation starts by having the elevator randomly selecting initial parking floors in the building. Each elevator acts independently from the other, and thus, each one has its own action probability for the parking floors, and both are initially created in an equiprobable manner. The Environment generates a call floor and a destination floor using the associated distributions C and D respectively. There- after, the closest elevator to the calling floor will be dispatched to service that passenger, and its waiting time is used to determine whether the parking floor that the elevator came from was good or not. The corresponding feedback from the Environment is presented in [14].

Once the Environment sends the feedback to the elevator, only the active elevator's action probabilities are updated. After the active elevator drops off the passenger, all elevators pick a parking floor, each according to its action or parking probabilities[3]. The corresponding updating algorithm is presented in [14].

After this, the procedure repeats itself until the LA converges to a set of floors that correspond to the number of elevators in the building.

4.3   P L_{RI}-Based Solution: MEP4

The second LA-based solution, referred to as MEP4, is the PLRI-based model. This is based on applying the pursuit concept, where we keep track of the ratios of rewards to the number of times an action was selected, and we use this to enhance the previous MEP3 solution. We observed for the single elevator case [1], that applying the pursuit concept often superior results to the non-pursuit LRI-based solution.

The corresponding algorithm for MEP4 is presented in [14]. As before, the simulation started by selecting each action, which represents the selection of a combination of parking floors for the set of elevators, for a small number of calls, i.e., in our case, 10 times. Thereafter, the Environment evaluated the selected actions and recorded the ratio of the number of times an action was rewarded to the number of times it was chosen. The simulation then proceeded in the same manner as MEP3.

The Environment generated a number of calls and destinations according to the distributions C and D, and for each call, the closest elevator served the passenger. It then requested the feedback from the Environment, which evaluated the current parking floors using the feedback algorithm mentioned in [14]. Subsequently, the active elevator dropped the passenger off at the destination floor, and then, if the LA of the active elevator received a reward feedback from the Environment, it updated its parking probabilities. On the other hand, it did nothing if it received a penalty feedback. The corresponding updating algorithm is presented in [14].

After the active elevator updated the parking probabilities, the LA updated its rewards' estimates for each elevator as per its parking probabilities.

5   Results and Discussion

We now comparatively discuss the results of the simulations obtained from the above solutions. In Table 1, we submit the results of the four discussed solutions, namely, MEP1, MEP2, MEP3 and MEP4. From the table, one can see that MEP1, which we believe is the most popular policy currently used in buildings, performed very poorly in comparison to our proposed solutions, MEP3 and MEP4. The improvement in the average waiting time was more than 40% and up to 86%. This policy, MEP1, serves as a lower bound for our benchmark, as no solution should perform worse than this.

---

[3] This is done to prevent any of the elevators from being "stuck" at one position without being called during the whole simulation.

MEP2 was able to provide the set of optimal parking floors from the beginning, but it required the a priori knowledge of C for each floor. On the other hand, our LA-based solutions were able to achieve a close-to-optimal AWT and parked at floors close to the optimal floors, without the knowledge of C.

MEP3 showed that it was better than MEP1 and recorded an AWT improvement of 85%, 86%, 42% and 45% for the *Exp, InveExp, Gaussian and BiModal* distributions respectively. Moreover, the results were close to the values achieved by MEP2. Further, the increase in the performance was more than for the SEP models, which demonstrates the effect of having multiple intelligent elevators especially when the number of elevators increases.

We attempted to improve MEP3 by proposing MEP4 that applied the Pursuit concept. MEP4 was able to achieve slightly better results than MEP3. The algorithm helped the system to converge faster to the optimal locations in most of the cases. It also resulted in a superior AWT.

One result that we found from the simulations in the SEP and the MEP is that having a single intelligent elevator could result in a better performance than having multiple elevators where none of them used intelligent policies. The scenario for the 12-floor building system with two elevators is one such example.

While in MEP1, the AWT was between 5 and 6 floors, the SEP3 resulted in an average of about only 2 floors. Also, just as in the case of the SEP, the more equally-distributed the calls were, the higher was the value of the AWT. On the other hand, the importance of having a good policy becomes evident when the calling distribution is skewed toward a specific region or a specific floor.

In Fig. 2, we present the performance of each algorithm and how our proposed LA-based algorithms, MEP3 and MEP4, were able to achieve an AWT that is very close to the optimal solution, MEP2, and how it significantly outperforms MEP1.

Table 1. Simulation results for the previous and newly proposed solutions for the MEP for an ensemble of 200 experiments for the 12-floor settings. The results are given here as a tuple (α, β) where the first field, $\alpha = (\alpha_1, \alpha_2)$, is the best optimal parking floors for each elevator and the second field, β is the AWT in terms of number of floors travelled for the elevator to reach the passenger from the parked location.

Dist	MEP1	MEP2	MEP3	MEP4
Exp	(−, 5.3604)	((1, 3), 0.3597)	((1, 2), 0.493)	((1, 2), 0.4334)
InvExp	(−, 5.3697)	((10, 12), 0.3568)	((11, 12), 0.508)	((11, 12), 0.446)
Gaussian	(−, 2.6252)	((4, 8), 1.2511)	((6, 7), 1.658)	((6, 8), 1.464)
Bimodal	(−, 2.6770)	((4, 9), 1.0651)	((5, 9), 1.715)	((4, 9), 1.448)

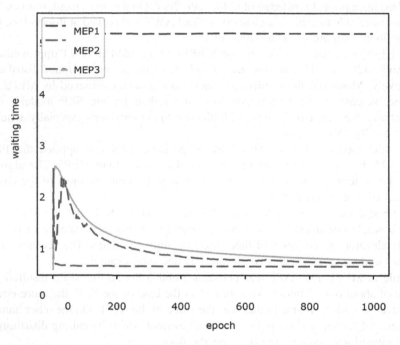

Fig. 2. The Average Waiting Time for MEP1, MEP2, MEP3 and MEP4 for an ensemble of 200 experiments for the case of the Exp distribution.

6   Conclusions

In this paper, we reviewed different parking policies that were used in various building settings and for previously reported solutions. We discussed two different solutions in the multi-elevator scenario and tabulated the experimental results that correspond to the various simulations. We then introduced our LRI-based solution, MEP3, and it achieved the optimal parking floor without knowing C a priori.

We further introduced an improvement on MEP3 to incorporate the Pursuit concept. This yielded our second solution, MEP4, a P L_{RI}-based solution. From the simulation results for MEP4, we demonstrated that it performed even better than MEP3.

This paper included our findings and compared the results of our proposed solutions to the previous solutions. We showed that our LA-based solutions performed better than MEP1 and converged to the optimal floor. They also reduced the AWT to be close to the optimal value with the advantage that they did not require us to know the distributions to determine the best floor. The case in which the environment is acting in a non-stationary manner and combining multiple objective functions to achieve better results are to be open for future work.

References

1. Ghaleb, O., John Oommen, B.: Learning automata-based solutions to the single elevator problem. In: MacIntyre, J., Maglogiannis, I., Iliadis, L., Pimenidis, E. (eds.) AIAI 2019. IAICT, vol. 559, pp. 439–450. Springer, Cham (2019). https://doi.org/10.1007/978-3-030-19823-7_37
2. Barto, A., Crites, R.H.: Improving elevator performance using reinforcement learning. In: Proceedings of the 8th International Conference on Neural Information Processing Systems, pp. 1017–1023 (1995)
3. Nikovski, D., Brand, M.: Decision-theoretic group elevator scheduling. In: 13th International Conference on Automated Planning and Scheduling, pp. 133–142 (2003)
4. Zerz, E., et al.: Exact calculation of expected waiting times for group elevator control. IEEE Trans. Automatic Control **49**(10), 2002–2005 (2004)
5. Brand, M., Nikovski, D.: Optimal parking in group elevator control. In: IEEE International Conference on Robotics and Automation, Proceedings. ICRA 2004, vol. 1, pp. 1002–1008 (2004)
6. Tsetlin, M.: On behaviour of finite automata in random medium. Avtomat. i Telemekh **22** (10), 1345–1354 (1961)
7. Varshavskii, V., Vorontsova, I.P.: On the behavior of stochastic automata with a variable structure. Avtomatika i Telemekhanika **24**(3), 353–360 (1963)
8. Meybodi, M.R., Beigy, H.: New learning automata-based algorithms for adaptation of backpropagation algorithm parameters. Int. J. Neural Syst. **12**(01), 45–67 (2002)
9. Misra, S., Oommen, B.J.: GPSPA: A new adaptive algorithm for maintaining shortest path routing trees in stochastic networks. Int. J. Commun Syst **17**(10), 963–984 (2004)
10. Obaidat, M.S., Papadimitriou, G.I., Pomportsis, A.S., Laskaridis, H.S.: Learning automata-based bus arbitration for shared-medium ATM switches. IEEE Trans. Syst. Man Cybern. Part B (Cybern.) **32**(6), 815–820 (2002)
11. Oommen, B.J., Roberts, T.D.: Continuous learning automata solutions to the capacity assignment problem. IEEE Trans. Comput. **49**(6), 608–620 (2000)
12. Seredyński, F.: Distributed scheduling using simple learning machines. Eur. J. Oper. Res. **107**(2), 401–413 (1998)
13. Kabudian, J., Meybodi, M.R., Homayounpour, M.M.: Applying continuous action reinforcement learning automata (carla) to global training of hidden Markov models. In: Proceedings of International Conference on Information Technology: Coding and Computing, ITCC 2004, vol. 2, pp. 638–642. IEEE (2004)
14. Ghaleb, O., Ooommen, B.: Novel Solutions and Applications to Elevator-like Problems By (2018)
15. Thathachar, M.A.L., Sastry, P.S.: Estimator algorithms for learning automata. In: Platinum Jubilee Conference on Systems and Signal Processing (1986)
16. Oommen, B.J., Lanctot, J.K.: Discretized pursuit learning automata. IEEE Trans. Syst. Man Cybern. **20**(4), 931–938 (1990)
17. Parlar, M., Sharafali, M., Ou, J.: Optimal parking of idle elevators under myopic and state-dependent policies. Eur. J. Oper. Res. **170**(3), 863–886 (2006)
18. Thathachar, M.A.L., Sastry, P.S.: A new approach to the design of reinforcement schemes for learning automata. IEEE Trans. Syst. Man Cybern. SCM **15**(1), 168–175 (1985)
19. Thathachar, M.A.L., Sastry, P.S.: A class of rapidly converging algorithms for learning automata. In: IEEE International Conference on Systems, Man and Cybernetics. IEEE (1984)

Developing System from Low-Cost Devices to Build a Security and Fire System as a Part of IoT

Štefan Koprda, Zoltán Balogh$^{(\boxtimes)}$, and Martin Magdin

Department of Informatics, Faculty of Natural Sciences,
Constantine the Philosopher University in Nitra, Tr. A. Hlinku 1,
949 74 Nitra, Slovakia
{skoprda, zbalogh, mmagdin}@ukf.sk

Abstract. Nowadays electronic security systems are not only about the alarm going off with a loud beeping when an object has been disturbed. They cover multiple other useful features that ensure the owner's safety and his properties. These features can be the basic elements such as fire monitoring, gas leakage, flooding, assault or health emergency. Or more sophisticated features such as an interlinked net of motion detectors, pressure and temperature change in the room, a camera system with a facial and fingerprint recognition, retinal scan, etc. Each critical situation is evaluated and automatically reported via a text message or via the Internet to the user application. The paper describes a complex system of securing people and their properties with the use of innovative technologies. The paper is dealing with the creation of a secure and fire protection system using the microcontroller Arduino, and it deals with the creation of a web interface for system control and management.

Keywords: Innovative technologies · Arduino · Security system ·
Fire system · Web interface

1 Introduction

Currently, the security devices in IoT started to create a big complex system where each devices communicate with the external peripherals that ensure the support. These systems have a redesigned net of components that the user can control and monitor from work via his phone. It is not just only the sensors, electric locks, but also the lighting in the home, heating and blind control, pool heating, etc.

There are various devices supporting IoT such as Arduino, Raspberry PI [1] and other microelectronic devices. IoT is able to use the Internet and the wireless technology to create an environment of a remote-control system to create a domestic automated system at home/in the office to monitor/control [2].

The wireless network of sensors is a type of a network that consists of knots that collectively gather the information about the physical parameters in real time [3]. The network can have about thousands of sensor knots, collectors and gates.

Each knot is equipped with multiple sensors, microcontrollers and transceivers. The knot scans and then transfers the scanned data into the communication network through

© Springer Nature Switzerland AG 2019
D.-S. Huang et al. (Eds.): ICIC 2019, LNAI 11645, pp. 142–154, 2019.
https://doi.org/10.1007/978-3-030-26766-7_14

the gate. The knots are the devices limited by the means related to memory and computing capabilities and the energetic capacity [4].

The communication networks with the Internet technologies support a significant base of the communication system that enables the thought of linking anytime and anywhere in the architecture of the IoT [5, 6].

The paper deals with the current state of the security technologies at the market and with the creation of a complex system to secure people and their properties with the use of innovative low-cost technologies. A security fire system has been created with the use of microcontroller Arduino. The system has been then tested in practice and its features have been evaluated with appropriate statistical methods.

2 Related Work

The concept of "smart" is appearing more and more in various contexts with technological solutions that enable people to actively and mutually communicate with these devices. At the same time, they provide possibilities for a more comfortable life and such the lowering the electric energy usage and the total efficiency form the point of view of saving time [7]. For this reason, many opinions come up that the concept of a smart home itself does not gain any scientific content and for the scientific community it has a minimal significance [8, 9]. It was not only the obsolete technology of communication with remote control used in the past representing the problem but the missing possibility of connecting to the client interface [10, 11]. The expansion of communication technologies has caused that the majority of the smart homes have been focused on the use of wireless communication to connect home appliances. The researchers started to deal with this field since with the development of the Internet and multiple devices that can be connected actively a new concept has been born which is IoT [12, 13]. The new communication technologies such as WIFI (Wireless Fidelity), ZigBee (wireless smart home technology), LoraWan, Sigfox or RFID (Radio-frequency identification) have become rather popular in the mutual communication of the particular sensors [14].

In the last ten years, various authors of scientific publications have presented their visions and their practical solutions about how and what a smart home should cover in their opinion.

Chung et al. [13] presented a smart home that uses the technology of Bluetooth. Each device and sensor is controlled with a smart phone with OS Android without the possibility of connecting to the Internet. Despite that, the expanses on the system are rather high, and the user cannot influence the system at a distance.

In the last ten years, the demand for smart homes has grown quickly mainly because of IoT [15]. The latest progress in the telecommunication technologies has helped the developers and researchers to design homes with regard to health, safety and energetic efficiency [16, 17]. It has been reached by smart watches that monitored the basic physiological functions of a person or smart phones connected to IoT. Thus, it is possible to monitor and to control various parts of the smart home according to the requirements of the user [16]. Raja et al. [18] designed a simple security system where the control unit is an Arduino Uno. PIR sensor was used to detect the unauthorized motion of a person inside of the secured object while the magnetic sensor was placed on the doors and the windows. Then it enables to evaluate the data whether the doors or the windows were open. The temperature sensor LM35 is used to detect the temperature increase above the designated limit as a precaution of fire.

Gulve et al. [19] proposed the use of an intelligent monitoring and saving the video recording via their own application based on the principles of IoT. The designed system is able to detect the presence of any person in the area. For detection and recognition, OpenCV software is used (for video monitoring). To send messages, GSM module is used for the network 900 MHz (for the SMS and the e-mail notification). Kader et al. [20] states other ways of securing the access in a smart home system. As entry into the secured area analyses the fingerprint of the person and if it matches with the fingerprint saved in the database, according to the categorization it enables the way of access. The microcontroller Arduino ATMega328P has been used as a control unit. Santoso et al. [21] have designed a smart home system that uses a microcontroller Arduino as a control unit to process data from the RFID reader to secure the access and the PIR sensors to detect motion. The control of the control unit is secured with a web application and the communication runs through WI-FI. Prasetyo et al. [22] designed a smart office system where they used the control unit Arduino ATMega 2560. The proposed system with the help of the camera connected to Raspberry Pi is able to find out the threat in the case of a violent intrusion into the secured area using dangerous objects made of metal, breaking the glass, fire or unauthorized entry of a person. The entry of the user is secured with the keypad or with the help of RFID.

3 The Creation Proposition of a Security System

An alarm security and distress system were created from three parts – hardware, software and a protective case. The hardware part was created from an Arduino platform and its components. When creating the protective case, modeling programs Solidworks and PrusaControl were used. To create diagrams, the web interface draw.io was used and Friting was used for schemes.

The software part of the system consisted of a code for Arduino and a code for the web page. To create a program for Arduino, the development environment of Arduino (IDE) was used that is a multiple platform application programmed in Java. The programs for Arduino were written in C or C++. IDE contained the library of functions that made the writing of the most basic operations with the hardware easier. The user had to define only two functions to gain a working program.

3.1 Solution Design and Creation Description

The security system solution was divided into three parts. When creating the system, a hardware design solution was made for each system separately. This contained connecting all the used modules, shields for Arduino, described in previous chapters. When all parts were functional and tested, a complex solution design was created where the relevant components were found. The design contained a basic board of

Arduino Mega, Ethernet shield, a keyboard, an LCD display, an RFID reader, an SD car and a board to simulate the circuit.

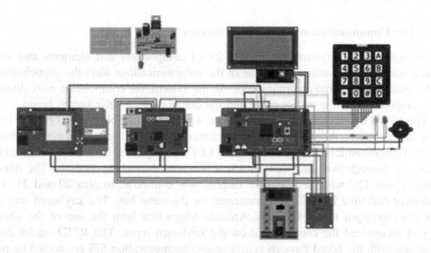

Fig. 1. The diagram of security system connection

Each component had to have its own library defined in the program that enabled the correct function and communication with the motherboard. The first version of the program was supposed to load the card, to login the user, to record the change of the sensors' state and to record into the database. The database was created on the web page examine.sk and it contained a chart recording the sensor and security state, a chart for the users, and a chart recording the new disruption or other alert. To communicate with the database, the relevant forms had to program in PHP code to record in the database, and the program Notepad++ was used. At a sensor change or an alert, a part of Arduino has called the PHP file and it has sent the given value, evaluated by the form and recorded into the database.

The next part was dealing with programming the web interface for the user to control and check the state of the system. It was done in the environment Notepad++. First, the basic HTML and the CSS layout were designed. In the layout, the visual environment to signalize each sensor was developed, shown in Fig. 1. A side panel was created where the last five records, done by the system, were written.

The web interface contained the subpages Control and Statement that were displayed in the upper right menu. The subpage Control contained a panel to turn off and on the security and the fire protection. Another element of the interface was a total text statement of the records from the database found in the subpage Statement. The records were created by Arduino every hour at any change as already mentioned. All the elements and the records were written into the layout via PHP forms that connected the database and the web interface. The creation of a login system on the web interface for the users was an important step. The interface was adjusted so that to access the data, a login name and a password was needed that was saved in the database. The password was protected by a hash function which is a function for the input data string transfer on a short output string.

3.2   The Communication and the Connection

The created security system contained a lot of components and elements that used several ways of communication. Some of the communications were the asynchronous or the synchronous serial communication. In the system, the components were divided into components for the access panel and the components for the switch board.

The switch board consisted of a keyboard 4 × 4, LCD display, RFID reader, LED diode and a speaker. All the components communicated with the control panel Arduino Mega. To communicate with Arduino, the LCD display used the I2C bus that communicated through two data wires and was powered through two others. The library LiquidCrystal_I2C was used, and the display was connected to pins 20 and 21. The module of real time DS1307 communicated on the same bus. The keyboard sent the data from its input to the board of Arduino Mega that with the use of the library Keypad recognized the character hit on the keyboard input. The RFID reader communicated with the board through synchronous communication SPI connected on pins SS (Slave Select) 5, RST 6. It used the library MFRC522 that was able to recognize RFID readers and cards. Two LED diodes were connected on digital pins 45 and 43. The speaker was connected on the digital pin 49 (Fig. 2).

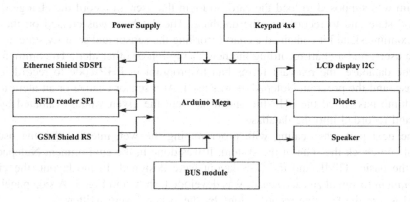

Fig. 2. The diagram of the communication between the peripheries and the board of Arduino

The components used for the switch board were Arduino Mega, Ethernet Shield, GSM shield, the module of real time DS1307, the bus module and the power supply. The communication between the Ethernet shield and Arduino took place on the bus SPI and was connected on pins ICSP and SSpin 10 that served for turning on and off the module. The library Ethernet was used with which it was possible to connect to the given server and to send data to it. The GSM shield communicated with an asynchronous communication (pins 7 and 8) and it used the LibrarySoftware Serial that enabled to send SMS texts. The sensor module used the communication through one-wire bus to find out the temperature and was connected to pin 2. It contained two more balanced current loops sending the given data about the sensors to pins A11 and A13.

3.3   The Programs of the Security System

Two programs were created in the security and fire protection system. The first control program was created for Arduino. It served to control and to operate the whole system and also to operate the access panel. It was recorded on the board of Arduino Mega. The second program enabled the user a distant access to the security system through the web interface. It served to inform the state of the system and its operation.

The Control Program

The control program served to operate the whole security and fire protection system. It was programmed in the development environment Arduino IDE (Fig. 3).

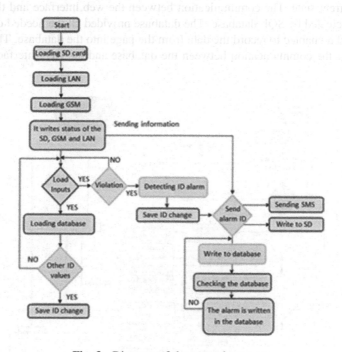

Fig. 3. Diagram of the control program

It was divided into two circuits that were connected to each other. One part served for the user access through the access panel where the user could physically manage and operate the system. The other part of the control program worked in the background and served as a communication among the sensors and web interface. It also served as an SMS notification and ensured a secured protection. If the user wanted to access the panel, he had to load the RFID card and enter the correct password. On the access panel, the security system would turn off and the program would turn on that operated the menu in the access panel and enabled multiple functions. The programs for each circuit like the access panel, the subprogram menu and the base of the whole system were placed on one board of Arduino Mega 2560. Each circuit communicated via Void methods. The methods were called after each action and enabled that the relevant part of the code was always carried out. After launching, the program loaded and addressed the relevant modules for communication. Then it evaluated their states and set the system according to which part was available. The sensors' state and the inputs were loaded. Every 30 s, the system checked the data from the database and in the case of a change that happened in the web interface; the system would set itself according to these values. In the case there was an input disruption, the algorithm would evaluate which input was disrupted and then sent to the ID problem evaluated in the functions of sending. The functions were set that a relevant text message was sent to ID.

Web Interface
The web interface enabled the user to operate the security system and informed him about its current state. The communication between the web interface and the security system was created by SQL database. The database provided all the needed data for the interface and it enabled to record the data from the page into the database. The diagram below shows the communication between the database and the web interface (Fig. 4).

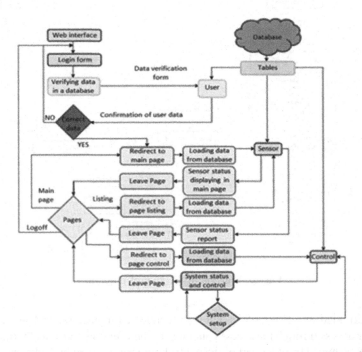

Fig. 4. Diagram of communication between the database and the web interface

To enter the page, the user needed to login. The user filled in the access form that was sent into the database which then verified the data. After verification of the correct data, it enabled access to the user to the page. The main page requested information about the sensors and a report statement from the database. After loading the data from the database, the page would show up.

4 Results and Error Analysis

The reliability and the function were the basic requirements that a security and a fire system should fulfil. That is the reason why these features were examined together with the error rate of its behaviour. In the given time interval, the system was creating records about its activity and stored the records in a database. Data preparation was done based on Munk et al. [23, 24]. The records were obtained from the database and saved to log files in a standard log format. Data preparation consisted of obtaining the raw log files from the database and data transformation. The data transformation phase consisted of creating a dataset from these records. The dataset described how the system behaved. Error analysis was done using an expected dataset that was created using false data that were generated for every hour. The sets were then compared and it then showed in what extent was the system reliable. The result was expressed in a graph and in percentage.

The reliability of the system influenced several factors that might have happened in the system during its run. These factors could be a sudden error, an external intervention into the system, or sensor disruption. In the case of an external intervention, the system would have to submit to a superior command, creating a different record that might affect the records. An example of that could be the switching off of the system through web interface. A sudden error could be caused by an unwanted phenomenon that might spoil the correct behaviour of the system.

The sensors' disruption was another option. In the case of such disruption, they were recorded, and they affected the given results. In order to a correct research, the real set of data had to be cleared of such unwanted records and to keep only the required data.

4.1 Behaviour Testing and Analysis

The security and fire system were tested in a family house. The system's reliability was tested, whether it turned on and off in the given time interval and whether it created reports about its state. The testing took place in a range of 22 days where the system was launched with two independent protection circuits. The interval was set via web interface so that the security circuit would turn on at seven o'clock in the morning and would turn off at three o'clock in the afternoon. The security circuit was turned on in this period and each hour it recorded the state of the system. The fire system circuit was turned on continuously 24 h a day and it recorded the state each hour. The records were saved into the table Sensor in the database.

These states informed the user about the activities of the security system and possible disruptions. The attribute Name reported which circuit or sensor had been disrupted, turned on or off. The next attribute Value reported the state of being turned on and off of the circuits and sensors.

The states about being turned on and off in the system were recorded in the database only when the system was turning on and off automatically or manually. The records about the states were written every hour automatically, but in the case of a disruption, the system recorded the state about the disruption in that very moment automatically.

In order to process the data correctly, a data matrix was created into which the data were placed. The matrix contained a real set of data from the security system, but these data needed to be cleared of all unwanted states.

The data from the field Value were showing the real values about the system being turned on or off. These values were converted into binary data that represented states 1 if it was turned on and 0 if it was turned off recorded in LogValue. A support field Supposition was created in the data matrix. This field expressed the expected values, how the system was supposed to behave. The values of the field were converted into the binary field Log-Supposition. The next step was a comparison of the binary field with real data and the field with expected data. If the data equalled in time, the state 1 was entered into the field Comparison, and if the data did not equal, state 0 was entered. The field were then expressed in percentage as shown in Table 1.

Table 1. The calculation of conformity and discrepancy for the system

Values	Number of records	Conformity	Discrepancy
Numbers	754	750	4
Percentage	100.00%	99.47%	0.53%

The table showed that there were altogether 754 records and it represented the 100%. There were 750 identical records, and it represented 99.47% of successful records. There were only four non-identical records and the represented 0.53% of unsuccessful records. The percentage was created for each security system separately and their success rate was examined. In an excel data matrix, the data were filtered from the field Status only for the fire system. Then the filtered results were shown on the field Comparison.

Table 2. The calculation of conformity and discrepancy for the fire system

Values	Number of records	Conformity	Discrepancy
Numbers	518	516	2
Percentage	100.00%	99.61%	0.39%

Table 2 showed that the number of all records for the fire system was 518, representing 100%. There were 516 identical records that the represented 99.61%. There were only two non-identical records, representing 0.39%.

A similar comparison was made for the security system. In the data matrix, the data was filtered for the security system, and the percentage was created that is shown of Table 3.

Table 3. The calculation of conformity and discrepancy for the security system

Values	Number of records	Conformity	Discrepancy
Numbers	236	234	2
Percentage	100.00%	99.15%	0.85%

In both security and fire system, there were two discrepancies that represented 0.85%. There were 234 identical records representing 99.15%. The total number of records was 236. The research showed that the security system was 99% reliable under the terms of function and reliability in a time interval. This means that the system has turned on at the time to which it had been set to.

5 Discussion

In the theoretical part, the current state of the field of security technologies had been analysed. It turned out that the quality of the system depended on its reliability and functionality. The functionality of the systems was rather high, it included the resistance to false alarms, power failure, central support or wireless communication of the sensors. The paper dealt with a detailed description of a well-known security system in the foreign market. The authors discovered that the system was rather reliable and had a wide use. Harmfulness of the wireless sensors had been taken into account as well and a research, that took place, showed that the radiation of the wireless sensors was not harmful for the human body. A security system can be created with the use of the microcontroller Arduino that is cheaper and its reliability and functionality is high. Our own security system has been compared with the Jablotron 100 security system.

Table 4. Price comparison of components

Products/Sensors	Our security system	Jablotron 100
Control panel	70 €	410 €
Access module	10 €	86 €
Power supply	10 €	9 €
Backup unit	17 €	56 €
Magnetic contact	4 €	3 €
PIR sensor	7 €	22 €
Glass break sensor	40 €	32 €
Laser motion sensor	20 €	Not available
Humidity sensor	5 €	17 €
Gas sensor	2 €	37 €
Temperature sensor	2 €	In the gas sensor
Total price:	187 €	672 €

By comparison (Table 4), it was found that the Arduino microcontroller's security system and its modules, sensors are much cheaper than the Jablotron 100 professional security system. Jablotron 100 offers various pricing packages for protection, home, apartment and business, making it difficult to compare with its own security system but in terms of components it is a more cost-effective Arduino-based security system.

In the practical part, the authors dealt with the creation of a security system with the use of the microcontroller Arduino. A complex security system has been designed that had a simple control and tested in practice. The system has been tested in a family house for three weeks. During this period, the system recorded its activities that were evaluated in the practical part. The reliability of the system was 99%.

The advantage of the system was to secure fire protection and safety that were not dependent. When detecting an alert, the system evaluated the state and then offered three options of notifying the alert: 1. sending the record to the web interface, 2. sending an SMS text to the provided phone number, 3. sending the record to the SD card.

6 Conclusion

The paper dealt with the design and the creation of a security system with the use of the microcontroller Arduino. The knowledge in the theoretical part was used and a high-quality security fire system was created. It was designed so that it would contain a switchboard connecting the access panel to the physical control and web interface for the distant access. The sensor module served to connect the sensors that were divided into two independent safety loops – the fire protection loop and the securing loop. The alert notification was set in the system via Internet mobile connection and recording it on an SD card. The functionality of the security system was verified through a data matrix that analysed the records of the system and evaluated them. The analysis evaluated the behavior and the error rate of the system. It proved the system to be 99% reliable and fully functional for the user. The result of the practical part was an automated and a functional security system. The main advantage was that it could be used in a family house or a flat. The user could control the system and would not need any technical support.

References

1. De Sousa, M.: Internet of Things with Intel Galileo. Packt Publishing Ltd., Birmingham (2015)
2. Reddy, V.M., Vinay, N., Pokharna, T., Jha, S.S.K.: Internet of things enabled smart switch. In: 2016 Thirteenth IEEE and IFIP International Conference on Wireless and Optical Communications Networks. IEEE, New York (2016)
3. Akyildiz, I.F., Su, W.L., Sankarasubramaniam, Y., Cayirci, E.: A survey on sensor networks. IEEE Commun. Mag. 40(8), 102–114 (2002). https://doi.org/10.1109/mcom.2002.1024422
4. Xu, J., Yang, G., Chen, Z.Y., Wang, Q.Q.: A survey on the privacy-preserving data aggregation in wireless sensor networks. China Commun. 12(5), 162–180 (2015). https://doi.org/10.1109/cc.2015.7112038

5. Ali, A., Hamouda, W., Uysal, M.: Next generation M2M cellular networks: challenges and practical considerations. IEEE Commun. Mag. **53**, 18–24 (2015)
6. Holler, J., Tsiatsis, V., Mulligan, C., Avesand, S., Karnouskos, S., Boyle, D.: From Machine-To-Machine to the Internet of Things (2014). https://doi.org/10.1016/c2012-0-03263-2
7. Cottone, P., Gaglio, S., Lo Re, G., Ortolani, M.: User activity recognition for energy saving in smart homes. Pervasive Mob. Comput. **16**, 156–170 (2015). https://doi.org/10.1016/j.pmcj.2014.08.006
8. Mowad, M.A.E.L., Fathy, A., Hafez, A.: Smart home automated control system using android application and microcontroller. Int. J. Sci. Eng. Res. **5**(5), 935–939 (2014)
9. Kamelia, L., Alfin Noorhassan, S., Sanjaya, M., WS, E.M.: Door-automation system using Bluetooth-based android for mobile phone. ARPN J. Eng. Appl. Sci. **9**(10), 1759–1762 (2014)
10. Coskun, I., Ardam, H.: A remote controller for home and office appliances by telephone. IEEE Trans. Consumer Electron. **44**(4), 1291–1297 (1998). https://doi.org/10.1109/30.735829
11. Patil, M., Reddy, S.: Design and implementation of home/office automation system based on wireless technologies. Int. J. Comput. Appl. **79**(6), 19–22 (2013)
12. Hussein, A., Adda, M., Atieh, M., Fahs, W.: Smart home design for disabled people based on neural networks. In: Shakshuki, E.M. (ed.) 5th International Conference on Emerging Ubiquitous Systems and Pervasive Networks/the 4th International Conference on Current and Future Trends of Information and Communication Technologies in Healthcare, vol. 37, pp. 117+ https://doi.org/10.1016/j.procs.2014.08.020. Procedia Computer Science
13. Chung, C.-C., Huang, C.Y., Wang, S.-C., Lin, C.-M.: Bluetooth-based Android interactive applications for smart living. In: 2011 Second International Conference on Innovations in Bio-inspired Computing and Applications, pp. 309–312. IEEE (2011)
14. Hsu, C.-L., Yang, S.-Y., Wu, W.-B.: Constructing intelligent home-security system design with combining phone-net and Bluetooth mechanism. In: 2009 International Conference on Machine Learning and Cybernetics, pp. 3316–3323. IEEE (2009)
15. Yuan, D., Fang, S., Liu, Y.: The design of smart home monitoring system based on WiFi electronic trash. J. Softw. **9**(2), 425–429 (2014)
16. Piyare, R.: Internet of things: ubiquitous home control and monitoring system using android based smart phone. Int. J. Internet Things **2**(1), 5–11 (2013)
17. Panth, S., Jivani, M.: Home automation system (HAS) using android for mobile phone. Int. J. Electron. Comput. Sci. Eng. **3**(1), 1–11 (2013)
18. Raja, K.B., Saranya, M., Shahana, R., Sreeraj, S., Vishnu, R.: Assorted security system to defend intrusion. Int. J. Adv. Res. Sci. Eng. (IJARSE) **7**(2), 762–774 (2018)
19. Gulve, S.P., Khoje, S.A., Pardeshi, P.: Implementation of IoT-based smart video surveillance system. In: Behera, H.S., Mohapatra, D.P. (eds.) Computational Intelligence in Data Mining. AISC, vol. 556, pp. 771–780. Springer, Singapore (2017). https://doi.org/10.1007/978-981-10-3874-7_73
20. Kader, M., Haider, M.Y., Karim, M.R., Islam, M.S., Uddin, M.M.: Design and implementation of a digital calling bell with door lock security system using fingerprint. In: 2016 International Conference on Innovations in Science, Engineering and Technology (ICISET), pp. 1–5. IEEE (2016)
21. Santoso, F.K., Vun, N.C.H.: Securing IoT for smart home system. In: 2015 IEEE International Symposium on Consumer Electronics (2015)
22. Prasetyo, T., Zaliluddin, D., Iqbal, M.: Prototype of smart office system using based security system. J. Phys.: Conf. Ser. **1**, 012189 (2018)

23. Munk, M., Benko, L., Gangur, M., Turcani, M.: Influence of ratio of auxiliary pages on the pre-processing phase of web usage mining. E & M Ekonomie Manag. **18**(3), 144–159 (2015). https://doi.org/10.15240/tul/001/2015-3-013

24. Munk, M., Pilkova, A., Benko, L., Blazekova, P.: Pillar3: market discipline of the key stakeholders in CEE commercial bank and turbulen times. J. Bus. Econ. Manag. **18**(5), 954–973 (2017). https://doi.org/10.3846/16111699.2017.1360388

Full-Order Sliding Mode Control Algorithm for Robot Manipulators Using an Adaptive Radial Basis Function Neural Network

Anh Tuan Vo[1], Hee-Jun Kang[2(✉)], and Tien Dung Le[3]

[1] Graduate School of Electrical Engineering, University of Ulsan,
Ulsan, South Korea
voanhtuan2204@gmail.com
[2] School of Electrical Engineering, University of Ulsan, Ulsan, South Korea
hjkang@ulsan.ac.kr
[3] The University of Danang - University of Science & Technology,
Danang, Vietnam
ltdung@dut.udn.vn

Abstract. In this paper, a full-order sliding mode tracking control system is developed for industrial robots. First, to dismiss the effects of perturbations and uncertainties, while to improve faster response time and to eliminate the singularity, a full-order sliding function is selected. Next, to reach the prescribed tracking path and to remove the chattering, a control method is designed for robot manipulators by using a combination of full-order sliding function and a continuous adaptive control term. Additionally, the unknown dynamic model of the robot is estimated by adopting a radial basis function neural network. Due to the combination of these methodologies, the proposed controller can run free of exact robot dynamics. The suggested controller provides strong properties of high tracking accuracy and quick response with minimum tracking errors. In simulation analysis, the simulated performances verify high effectiveness of the proposed controller in trajectory tracking control of a 3-DOF robot manipulator.

Keywords: Full-order sliding mode control ·
Radial basis function neural network · Adaptive control ·
Industrial robot manipulators

1 Introduction

Literature concerning industrial robots has proposed many tracking control schemes focused on attaining the desired performance against various uncertainties, including external perturbations. Sliding mode control (SMC) has been validated to provide high robustness against uncertainties and perturbation for nonlinear systems. Accordingly, the SMC has been usually applied in practical systems [1, 2]. Nonetheless, the conventional SMC still possesses obstacles such as requiring an exact dynamic model, singularity, a chattering occurrence, and finite-time convergence. Some study attempts have focused on treating these weaknesses. For the system state variables to reach the desired sliding mode function within a finite-time, the use of the terminal sliding mode control (TSMC), based on the nonlinear sliding mode function, has been stated in the

© Springer Nature Switzerland AG 2019
D.-S. Huang et al. (Eds.): ICIC 2019, LNAI 11645, pp. 155–166, 2019.
https://doi.org/10.1007/978-3-030-26766-7_15

literature [3, 4]. However, the TSMC convergence time is slower than the conventional SMC convergence time, and still encompasses a singularity. In order to resolve convergence time and singularity glitch, several fast TSMC (FTSMC) [5, 6] and nonsingular TSMC (NTSMC) [7, 8] systems have been reported. Practically, private methodologies, such as FTSMC or NTSMC, have only handled an individual weakness or neglected to handle the other weaknesses of the traditional SMC. Accordingly, the nonsingular fast TSMC (NFTSMC) has been developed [9–13]. Here, NFTSMC can deal with many drawbacks of the traditional SMC or other control systems based on TSMC. Nonetheless, chattering has not been eliminated by using a high-frequency switching control rule to the control input of the above systems, which include TSMC, FTSMC, NTSMC, and NFTSMC. Consequently, some useful approaches have been proposed to treat this issue by application of the full-order sliding mode control (FOSMC) [14, 15], or high-order sliding mode control (HOSMC) [16].

One of the major challenges in the design of the control system according to SMC or TSMC is to calculate an exact dynamic model of the robot, which one does not readily know in advance for practical robot systems. To approximate this unknown model, several computing attempts have been proposed such as neural networks [17–19] and fuzzy logic systems [20, 21] due to their approximation abilities.

Whilst each drawback of the conventional SMC and TSMC has been considered individually, this work focuses on simultaneous resolution of the drawbacks of SMC and TSMC, including the condition for an exact dynamic model, the existence of a singularity, chattering occurrence, and finite-time convergence.

Therefore, the objective is to design a FOSM controller for industrial robots using an adaptive radial basis function neural network (ARBFNN) scheme. The important advantages of the suggested method include: (1) The inheritance of FOSMC benefits in provisions of non-singularity, finite-time convergence, speedy transient response, low steady-state errors, and high path tracking accuracy; (2) The attainment of smooth control inputs with chattering abolition; (3) The elimination of demand for an exact dynamic model by utilizing an ARBFNN to obtain an unknown robot function; (4) Better tracking performance and less impact by external perturbations and uncertainties compared other control schemes based on SMC; (5) Enhanced stability and robustness of the robot system, as confirmed by Lyapunov theory.

The rest of the work is organized as follows. Following the introduction, the preliminaries and problem formulations are presented, succeeded by the design method for the suggested controller, where the proposed algorithm is used to allow joint path tracking control simulation for a 3-degree of freedom (DOF) industrial robot. Here, its tracking performance is compared with ANN-SMC to evaluate the effectiveness of the proposed control system. Finally, conclusions are stated.

2   Preliminaries and Problem Formulation

2.1   Radial Basis Function Neural Network

RBFNNs have benefits, including simple design, good generalization, strong tolerance to input disturbance and online adjusting capability. Compared with multiplayer neural network, the RBFNN has a simpler and converges faster. An RBFNN contains three layers: input layer, hidden layer and output layer which is illustrated in Fig. 1.

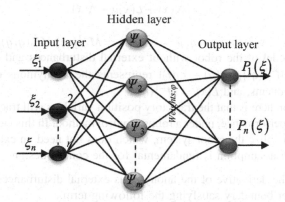

Fig. 1. The architecture of RBFNN.

The RBFNN output is calculated as follows:

$$P(\xi) = \varphi^T \Psi(\xi) + e(\xi), \tag{1}$$

where $\xi \in R^n$ and $P(\xi)$ correspond to the RBFNN input and output. $\varphi^T \in R^{n \times m}$ represents the weight matrix linking the hidden layer and the output layer, $\Psi(\xi)$ represents the nonlinear function of the hidden nodes, and $e(\xi) \in R^n$ represents an approximation error.

A Gaussian function is defined for the nonlinear function as follows:

$$\Psi(\xi) = \exp\left(\frac{-(\xi - \eta_l)^T(\xi - \eta_l)}{d_l^2}\right), l = 1, 2, \ldots, m, \tag{2}$$

where, d and η correspond to the width and center of Gaussian function.

2.2 Dynamic Model of Robot Manipulator

The robot dynamic model is described as (refer to [22, 23])

$$M(q)\ddot{q} + C(q, \dot{q})\dot{q} + G(q) + F(\dot{q}) + \tau_d = \tau, \tag{3}$$

where q, $\dot{q}$ and $\ddot{q} \in \mathbb{R}^n$ correspond to the position, velocity, and acceleration at each joint of the robot. $M(q) \in R^{n \times n}$ is inertia matrix, $C(q, \dot{q}) \in R^{n \times 1}$ is the matrix from the centrifugal force and Coriolis, $G(q) \in \mathbb{R}^{n \times 1}$ represents the gravity force matrix, $F(\dot{q}) \in R^{n \times 1}$ represents friction force matrix, $\tau \in R^{n \times 1}$ is the designed control input of actuators, $\tau_d \in R^{n \times 1}$ represents an external disturbance matrix.

Equation (3) can be transformed into the following expression:

$$\begin{cases} \dot{x}_1 = x_2 \\ \dot{x}_2 = N(x) + L(x)u + \Delta(x) \end{cases} \tag{4}$$

where $x = [x_1, x_2]^T$, $x_1 = q$, $x_2 = \dot{q}$, $u = \tau$, $N(x) = M^{-1}(q)[-C_m(q,\dot{q})\dot{q} - G(q)]$ is the nominal robot model of the robot without external disturbances and dynamic uncertainties, $\Delta(x) = M^{-1}(q)[-F_r(\dot{q}) - \tau_d(t)]$ represents the anonymous disturbance and uncertainty components, and $L(x) = M^{-1}(q)$.

The suggestion here is that the trajectory position variables will track the prescribed paths, with good performance, under a robust control method. In this case, the proposed control method stabilize the robot system, which does not need an exact robot model.

The following assumption is fundamental for the control design.

Assumption 1: The derivative of the anonymous external disturbances and uncertainties have an upper-boundary satisfying the following term,

$$|\dot{\Delta}(x)| \le \bar{\nabla}, \tag{5}$$

where $\bar{\nabla}$ is known positive constant.

3 Design Procedure of Tracking Control Method

As a similar design approach to the classic SMC, in this section, an adaptive neural network full-order sliding mode control algorithm is developed for robot system (3), which is described by two following major stages.

For the first stage, a sliding function termed as FOSM function is prior defined with the following expression (refer to [14, 15]):

$$s = \dot{\varepsilon}_2 + \lambda_2 |\varepsilon_2|^{\phi_2} \mathrm{sgn}(\varepsilon_2) + \lambda_1 |\varepsilon_1|^{\phi_1} \mathrm{sgn}(\varepsilon_1), \tag{6}$$

where $\varepsilon_i = x_i - x_{ir}$, $(i = 1, 2, \cdots, n)$ is the tracking positional error, x_{ir} is described as the prescribed trajectory value, λ_1, λ_2, ϕ_1, and ϕ_2 are positive constants, which are assigned as in [15].

In order to attain the prescribed control performance, the control input system is designed for robot manipulator as:

$$u = -L^{-1}(x)\left(u_{eq} + u_s\right), \tag{7}$$

$$u_{eq} = N(x) - \ddot{x}_r + \lambda_2 |\varepsilon_2|^{\phi_2} \mathrm{sgn}(\varepsilon_2) + \lambda_1 |\varepsilon_1|^{\phi_1} \mathrm{sgn}(\varepsilon_1), \tag{8}$$

and

$$\dot{u}_s = (\bar{\nabla} + \kappa)\mathrm{sgn}(s). \tag{9}$$

Where, the initial value of the switching control law is $\dot{u}_s(0) = 0$, $\bar{\nabla}$ is the boundary value which is determined in Eq. (5), and κ is a positive constant. The control system is presented in Theorem 1.

Theorem 1: Consider the robot system in Eq. (3). If the suggested control input is designed for the system (3) as Eqs. (7)–(9), then the stability of the system and the tracking error convergence are secured.

Proof: Adopting the control input from Eqs. (7)–(9) to Eq. (6) yields:

$$\begin{aligned}
s &= N(x) - L(x)\left(L^{-1}(x)\left(\begin{array}{c} N(x) - \ddot{x}_r + \lambda_2|\varepsilon_2|^{\phi_2}\mathrm{sgn}(\varepsilon_2) \\ + \lambda_1|\varepsilon_1|^{\phi_1}\mathrm{sgn}(\varepsilon_1) + u_s \end{array} \right) \right) \\
&\quad + \Delta(x) - \ddot{x}_r + \lambda_2|\varepsilon_2|^{\phi_2}\mathrm{sgn}(\varepsilon_2) + \lambda_1|\varepsilon_1|^{\phi_1}\mathrm{sgn}(\varepsilon_1) \\
&= \Delta(x) - u_s
\end{aligned} \tag{10}$$

Taking time derivative of Eq. (10) gives:

$$\dot{s} = \dot{\Delta}(x) - (\bar{\nabla} + \kappa)\mathrm{sgn}(s) \tag{11}$$

To confirm the correctness of Theorem 1, the following Lyapunov function is considered:

$$V = 1/2s^T s \tag{12}$$

Using Eq. (11), the time derivative of the Lyapunov function is derived as follows:

$$\begin{aligned}
\dot{V} &= s^T\left(\dot{\Delta}(x) - (\bar{\nabla} + \kappa)\mathrm{sgn}(s) \right) \\
&= \left(\dot{\Delta}(x)s - \bar{\nabla}|s| \right) - \kappa|s| < -\kappa|s|
\end{aligned} \tag{13}$$

Consequently, according to the Lyapunov criteria, the stability and the tracking error of the system are guaranteed under the control input commands (7)–(9).

Generally, it is not trivial to precisely determine the uncertainty upper boundaries and attain an exact robot function in the equivalent control term. In order to deal with these challenges, a robust control scheme will be developed for industrial robots based on an adaptive neural full-order sliding mode control (ANN-FOSM) algorithm. Here, an ARBFNN will be used to approximate an unknown robot function, while an adaptive law will be used to estimate the uncertainty upper boundaries and estimated error of the NN. In our work, ARBFNN is used to approximate the dynamic robot model as follows:

$$N^*(x) = N(x) \tag{14}$$

Define $\hat{N}^*(x)$ as an approximated function of $N^*(x)$, $\hat{N}^*(x)$ can be described by an integral NN, as follows

$$\hat{G}^*(x) = \int_0^t \hat{\varphi}^T \Psi(x) dt \tag{15}$$

Here, $\hat{\varphi}$ is the adaptable parameter vector.
The optimal parameter φ^* can be described, as follows:

$$\varphi_P^* = \arg\min\left\{ \sup_{x \in \Theta_x} \left| N^*(x) - \hat{N}^*(x, \hat{\varphi}) \right| \right\} \tag{16}$$

Accordingly, RBFNN (15) can exactly approximate the arbitrary value of $N^*(x)$ which is given by the following Lemma.

Lemma 1: For any given real continuous function $N^*(X)$ on the compact set $\Theta_X \in R^n$ and arbitrary positive coefficient $e > 0$, there is a neural approximator existence $\hat{N}^*(X)$ which possesses a similar form as Eq. (15), such that

$$\sup_{X \in \Theta_x} \left| N^*(X) - \hat{N}^*(X, \hat{\varphi}) \right| < e \tag{17}$$

Therefore, the robot dynamic model can be described as

$$\dot{x}_2 = \int_0^t \varphi^{*T} \Psi(x) dt + L(x)u + \prod(x) \tag{18}$$

where $\prod(x) = \Delta(x) + e$ is the lumper uncertainty, including disturbances, dynamic uncertainties, and NN approximation error. In this step, the lumper uncertainty and it's the time derivative are assumed to be limited by unknown positive constants, $|\prod(x)| \leq \sigma$ and $|\dot{\prod}(x)| \leq \delta$.

To achieve the prescribed control performance, the control torque system is constructed for robot manipulator, as follows:

$$u = -L^{-1}(x)\left(u_{eq} + u_s\right), \tag{19}$$

$$u_{eq} = \int_0^t \hat{\varphi}^T \Psi(x) dt - \ddot{x}_r + \lambda_2 |\varepsilon_2|^{\phi_2} \mathrm{sgn}(\varepsilon_2) + \lambda_1 |\varepsilon_1|^{\phi_1} \mathrm{sgn}(\varepsilon_1), \tag{20}$$

and

$$\dot{u}_s = \left(\hat{\delta} + \kappa\right) \text{sgn}(s). \tag{21}$$

Here, the adaptive updating rules are given as

$$\dot{\hat{\delta}} = \theta^{-1}|s| \text{ and } \dot{\hat{\varphi}} = \beta^{-1}s\Psi(x) \tag{22}$$

where $\hat{\nabla}$ is the estimated value of ∇, the adaptive gains are θ, β.

The control design procedure for robot has been stated in Theorem 2 below.

Theorem 2: For the system (3), if the control input signal is constructed as (19)–(21) with its parameter updating rules designed as (22), then the stability of the system is guaranteed along with the tracking error variables reach to zero.

Proof: Define the adaptive estimation error and NN weight approximation error, respectively, as follows

$$\tilde{\delta} = \hat{\delta} - \delta \text{ and } \tilde{\varphi} = \varphi^* - \hat{\varphi} \tag{23}$$

With Eq. (18), the sliding mode surface (6) is rewritten as follows:

$$s = \int_0^t \varphi^{*T}\Psi(x)dt + L(x)u + \Delta(x) - \ddot{x}_r + \lambda_2|\varepsilon_2|^{\phi_2}\text{sgn}(\varepsilon_2) + \lambda_1|\varepsilon_1|^{\phi_1}\text{sgn}(\varepsilon_1) \tag{24}$$

Adopting the control input from Eqs. (19)–(20) to Eq. (24) yields:

$$s = \int_0^t \hat{\varphi}^T\Psi(x)dt + \Pi(x) - u_s \tag{25}$$

Taking time derivative of Eq. (25) yields:

$$\dot{s} = \tilde{\varphi}^T\Psi(x) + \dot{\Pi}(x) - \dot{u}_s \tag{26}$$

The positive-definite Lyapunov functional is selected as

$$V = 1/2s^Ts + 1/2\theta\tilde{\delta}^T\tilde{\delta} + 1/2\beta\tilde{\varphi}^T\tilde{\varphi} \tag{27}$$

With the result of Eq. (26), the time derivative of Eq. (27) is derived as

$$\begin{aligned} \dot{V} &= s^T\dot{s} + \theta\tilde{\delta}^T\dot{\tilde{\delta}} - \beta\tilde{\varphi}^T\dot{\hat{\varphi}} \\ &= s^T\left(\tilde{\varphi}^T\Psi(x) + \dot{\Pi}(x) - \left(\hat{\delta} + \kappa\right)\text{sgn}(s)\right) + \theta\left(\hat{\delta} - \delta\right)\dot{\hat{\delta}} - \beta\tilde{\varphi}^T\dot{\hat{\varphi}} \end{aligned} \tag{28}$$

Applying the updating laws (22) to (28) yields

$$
\begin{aligned}
\dot{V} &= s^T\left(\tilde{\varphi}^T\Psi(x) + \dot{\Pi}(x) - \left(\left(\hat{\delta}+\kappa\right)\mathrm{sgn}(s)\right)\right) + \left(\hat{\delta}-\delta\right)|s| - \tilde{\varphi}^T s\Psi(x) \\
&= (\dot{\Pi}(x)s - \delta|s|) - \kappa|s| \le -\kappa|s|
\end{aligned}
\tag{29}
$$

It is seen that $\dot{V}$ is a negative definite. Consequently, the system will be guaranteed, and the tracking error variables will converge to zero. Therefore, Theorem 2 is confirmed.

4 Numerical Simulation Results and Discussion

To evaluate the useful properties of the ANN-FOSMC, the ANN-FOSMC is adopted to a path tracking control for the first three joints of a PUMA560 robot manipulator, and its tracking performance is compared with the ANN-SMC [17]. The dynamic model found in a 3-DOF robot manipulator was described by Armstrong et al. [24]. In our work, only the first three joints of a robot were investigated. The simulations were employed to compare the controllers in expressions of their pathway tracking accuracy and the removal of the chattering in their control signals.

To ascertain the robustness of two controllers, we evaluated the system performance in two work stages, where disturbances and uncertainties are modelled as follows:

$$
F_r(\dot{q}) + \tau_d(t) = \begin{bmatrix} 1.1\dot{q}_1 + 1.2\sin(3q_1) + 1.9\sin(\dot{q}_1) \\ 1.65\dot{q}_2 + 2.14\sin(2q_2) + 2.03\sin(\dot{q}_2) \\ -3.01\dot{q}_3 + 1.3\sin(2q_3) + 1.76\sin(\dot{q}_3) \end{bmatrix}
\tag{30}
$$

The first Phase: Assume that robot was operated under normal work condition from time 0 s to 10 s.

The second Phase: Assume that robot was operated under work condition, but there is an external perturbation impacting the first joint between 10 s and 40 s. This external perturbation has a value defined as $(25\sin(q_1q_2) + 3.7\cos(\dot{q}_1q_2) + 10.2\cos(\dot{q}_1\dot{q}_2))$.

The third Phase: Assume that robot was operated under work condition, but there is a partial loss (60%) of control input effectiveness at the second joint between 20 s and 40 s, and a perturbation impacting the third joint between 20 s and 40 s. This perturbation has a value defined as $(29\sin(q_1q_2) + 3.6\cos(\dot{q}_1q_2) + 7.7\cos(\dot{q}_1\dot{q}_2))$.

The prescribed joint pathways for the position tracking are

$$
q_r = \left[\cos\left(\tfrac{t}{5\pi}\right) - 1, \quad \sin\left(\tfrac{t}{5\pi} + \tfrac{\pi}{2}\right), \quad \sin\left(\tfrac{t}{5\pi} + \tfrac{\pi}{2}\right) - 1\right]^T
\tag{31}
$$

The RBFNN structural design consists of 7 nodes, the initial weight matrix of the RBFNN is selected as 0, the width and center of the Gaussian function are respectively set as $d = 0.2$, $\eta_l = 0.5$, and η is selected in range $(-1.5 \div 1.5)$. The matrix used in an adaptive law of RBFNN is selected as $\beta = 15I_7$, and the RBFNN input is selected as $\xi = [\varepsilon_1 \quad \varepsilon_2 \quad q_r \quad \dot{q}_r \quad \ddot{q}_r \,]$.

The ANN-SMC control input is set as

$$\tau = \hat{\varphi}^T \Psi(x) + K_D s + K_V \text{sgn}(s) \tag{32}$$

Here, K_D, K_V, α are positive diagonal matrices, $s = \dot{\varepsilon} + \alpha\varepsilon$ is a linear sliding function. The design parameter for the ANN-SMC and the ANN-FOSMC is given in Table 1.

Table 1. The control parameter selection for the varying controllers

Control system	Control parameters	Parameter value
ANN-SMC	$K_D; K_V$	$diag(55, 55, 55); diag(10, 10, 10)$
	α	$diag(10, 10, 10)$
ANN-FOSMC	$\phi_1; \phi_2; \gamma_1; \gamma_2$	$9/23; 9/16; 2; 2$
	$\varpi_1; \varpi_2; \lambda_1; \lambda_2\kappa, \theta$	$0.1; 0.1; 15; 60.02; 0.05$

Table 2. The averaged tracking errors under the varying controllers

Error control system	ε_1^{av}	ε_2^{av}	ε_3^{av}
ANN-SMC	0.0213	0.2541	0.0075
ANN-FOSMC	0.0039	0.0039	0.0037

The averaged path tracking errors are computed by the following expression

$$\varepsilon_i^{av} = \sqrt{1/n \sum_{k=1}^{n} \left(\|\varepsilon_i\|^2 \right)} \quad i = 1, \ 2, \ 3, \text{ where } n \text{ is the number of simulation step.}$$

The trajectory tracking performances at each of the first three joints with two controllers, are illustrated in Figs. 2 and 3. In Phase 1 (from 0 s to 10 s), both control systems give similar good path tracking performance. In Phase 2 (from time greater than 10 s) and in Phase 3 (from time greater than 20 s), it is clear that the NN-SMC provides the poor path tracking performance, where robot working becomes unstable when a large disturbance or uncertainty is used. From Table 2 and Fig. 3, it is notable that the selected sliding surface has a significant role in improving fast transient response and robustness against uncertainty and disturbances. Consequently. the proposed control system gives better path tracking performance compared to the ANN-SMC, because of the role of the FOSM manifolds, an adaptive compensator, and an important benefit of the proposed system.

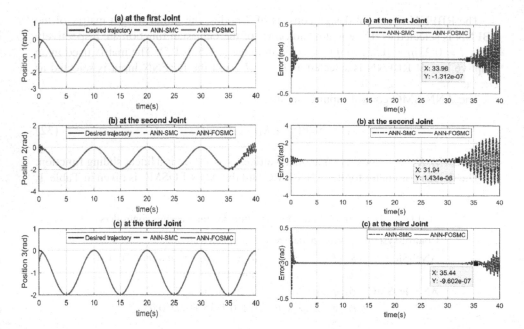

Fig. 2. Trajectory tracking positions: (a) at the first Joint, (b) at the second Joint 2, and (c) at the third Joint 3.

Fig. 3. Trajectory tracking errors: (a) at the first Joint, (b) at the second Joint 2, and (c) at the third Joint 3.

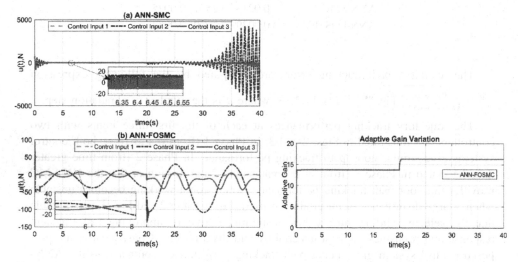

Fig. 4. Control input signals: (a) ANN-SMC and (b) ANN-FOSMC.

Fig. 5. Time history of adaptive gain.

The control signals for both control manners, including ANN-SMC, and the ANN-FOSMC are illustrated in Fig. 4. In Fig. 4a, the ANN-SMC shows a discontinuous control signal with serious chattering. On the contrary, the ANN-FOSMC shows a continuous control signal for the robot without the loss of its effectiveness, as illustrated in Fig. 4b.

The adjustment of the adaptive parameter is shown in Fig. 5. This adaptive gain is estimated along with the change of the effects of perturbations and uncertainties, and they will achieve a constant value while the tracking error variables approach to zero.

From the simulated performance, it is concluded that the proposed controller gives the best performance compared to the ANN-SMC in provisions of tracking precision, chattering deletion, and small steady state error.

5 Conclusion

In our work, the ANN-FOSMC is developed for industrial robots. The ANN-FOSMC has the following values: (1) inherits the benefits of the ARBFNN and FOSMC, including non-singularity, fast transient response, approximation capability, low steady-state errors, and high trajectory tracking accuracy; (2) Attains smooth with removal of chattering phenomenon; (3) does not require an exact dynamic model for the control system by adopting an ARBFNN to approximate an unknown robot function; (4) compared to ANN-SMC, the ANN-FOSMC provides better tracking performance and stronger resistance against disturbances and uncertainties; (5) stability and tracking error convergence of the robot system has been confirmed fully by Lyapunov criterion.

Acknowledgement. This research was supported by Basic Science Research Program through the National Research Foundation of Korea (NRF) funded by the Ministry of Education (NRF-2016R1D1A3B03930496).

References

1. Edwards, C., Spurgeon, S.: Sliding Mode Control: Theory and Applications. CRC Press, Boca Raton (1998)
2. Utkin, V.I.: Sliding Modes in Control and Optimization. Springer, Heidelberg (2013)
3. Tang, Y.: Terminal sliding mode control for rigid robots. Automatica 34(1), 51–56 (1998)
4. Wu, Y., Yu, X., Man, Z.: Terminal sliding mode control design for uncertain dynamic systems. Syst. Control Lett. 34(5), 281–287 (1998)
5. Mobayen, S.: Fast terminal sliding mode controller design for nonlinear second-order systems with time-varying uncertainties. Complexity 21(2), 239–244 (2015)
6. Madani, T., Daachi, B., Djouani, K.: Modular-controller-design-based fast terminal sliding mode for articulated exoskeleton systems. IEEE Trans. Control Syst. Technol. 25(3), 1133–1140 (2017)
7. Eshghi, S., Varatharajoo, R.: Nonsingular terminal sliding mode control technique for attitude tracking problem of a small satellite with combined energy and attitude control system (CEACS). Aerospace Sci. Technol. 76, 14–26 (2018)
8. Safa, A., Abdolmalaki, R.Y., Shafiee, S., Sadeghi, B.: Adaptive nonsingular terminal sliding mode controller for micro/nanopositioning systems driven by linear piezoelectric ceramic motors. ISA Trans. 77, 122–132 (2018)

9. Tuan, V.A., Kang, H.J.: A new finite time control solution for robotic manipulators based on nonsingular fast terminal sliding variables and the adaptive super-twisting scheme. J. Comput. Nonlinear Dyn. **14**(3), 031002 (2019)
10. Yang, L., Yang, J.: Nonsingular fast terminal sliding-mode control for nonlinear dynamical systems. Int. J. Robust Nonlinear Control **21**(16), 1865–1879 (2011)
11. Chen, G., Jin, B., Chen, Y.: Nonsingular fast terminal sliding mode posture control for six-legged walking robots with redundant actuation. Mechatronics **50**, 1–15 (2018)
12. Vo, A.T., Kang, H.: A chattering-free, adaptive, robust tracking control scheme for nonlinear systems with uncertain dynamics. IEEE Access (2019) https://doi.org/10.1109/access.2019.2891763
13. Vo, A.T., Kang, H.: An adaptive terminal sliding mode control for robot manipulators with non-singular terminal sliding surface variables. IEEE Access (2018). https://doi.org/10.1109/access.2018.2886222
14. Feng, Y., Zhou, M., Zheng, X., Han, F., Yu, X.: Full-order terminal sliding-mode control of MIMO systems with unmatched uncertainties. J. Franklin Inst. **355**(2), 653–674 (2018)
15. Feng, Y., Han, F., Yu, X.: Chattering free full-order sliding-mode control. Automatica **50**(4), 1310–1314 (2014)
16. Utkin, V.: Discussion aspects of high-order sliding mode control. IEEE Trans. Autom. Control **61**(3), 829–833 (2016)
17. Liu, J., Wang, X.: Advanced Sliding Mode Control for Mechanical Systems: Design, Analysis and MATLAB Simulation. Springer, Heidelberg (2012). https://doi.org/10.1007/978-3-642-20907-9
18. Vo, A.T., Kang, H.J., Nguyen, V.C.: An output feedback tracking control based on neural sliding mode and high order sliding mode observer. In: 2017 10th International Conference on Human System Interactions (HSI), pp. 161–165. IEEE (2017)
19. Sun, R., Wang, J., Zhang, D., Shao, X.: Neural network-based sliding mode control for atmospheric-actuated spacecraft formation using switching strategy. Adv. Space Res. **61**(3), 914–926 (2018)
20. Vo, A.T., Kang, H.-J., Le, T.D.: An adaptive fuzzy terminal sliding mode control methodology for uncertain nonlinear second-order systems. In: Huang, De-Shuang, Bevilacqua, Vitoantonio, Premaratne, Prashan, Gupta, Phalguni (eds.) ICIC 2018. LNCS, vol. 10954, pp. 123–135. Springer, Cham (2018). https://doi.org/10.1007/978-3-319-95930-6_13
21. Shen, Q., Jiang, B., Cocquempot, V.: Adaptive fuzzy observer-based active fault-tolerant dynamic surface control for a class of nonlinear systems with actuator faults. IEEE Trans. Fuzzy Syst. **22**(2), 338–349 (2014)
22. Spong, M.W., Vidyasagar, M.: Robot Dynamics and Control. Wiley, New York (1989)
23. Islam, S., Liu, X.P.: Robust sliding mode control for robot manipulators. IEEE Trans. Ind. Electron. **58**(6), 2444–2453 (2011)
24. Armstrong, B., Khatib, O., Burdick, J.: The explicit dynamic model and inertial parameters of the PUMA 560 arm. In: Proceedings of 1986 IEEE International Conference on Robotics and Automation, vol. 3, pp. 510–518. IEEE (1986)

Continuous PID Sliding Mode Control Based on Neural Third Order Sliding Mode Observer for Robotic Manipulators

Van-Cuong Nguyen[1], Anh-Tuan Vo[1], and Hee-Jun Kang[2]($\boxtimes$)

[1] Graduate School of Electrical Engineering, University of Ulsan,
Ulsan 44610, South Korea
[2] School of Electrical Engineering, University of Ulsan,
Ulsan 44610, South Korea
hjkang@ulsan.ac.kr

Abstract. This paper proposes a continuous PID sliding mode control strategy based on a neural third-order sliding mode observer for robotic manipulators by using only position measurement. A neural third-order sliding mode observer based on radial basis function neural network is first proposed to estimate both the velocities and the dynamic uncertainties and faults. In this observer, the radial basis function neural networks are used to estimate the parameters of the observer, therefore, the requirement of prior knowledge of the dynamic uncertainties and faults is eliminated. The obtained velocities and lumped uncertainties and fault information are then employed to design the continuous PID sliding mode controller based on the super-twisting algorithm. Consequently, this controller provides finite-time convergence, high accuracy, chattering reduction, and robustness against the dynamic uncertainties and faults without the need of velocity measurement and the prior knowledge of the lumped dynamic uncertainties and faults. The global stability and finite-time convergence of the controller are guaranteed in theory by using Lyapunov function. The effectiveness of the proposed method is verified by computer simulation for a PUMA560 robot.

Keywords: Neural third-order sliding mode observer ·
Continuous PID sliding mode control · Super-twisting algorithm ·
Radial basis function neural network

1 Introduction

Robot manipulators are high nonlinear systems – which have very complex dynamics with coupling terms. It is arduous to get its exact dynamics or even impossible in practice because of the frictions, disturbances, and payload changes – well-known as dynamic uncertainties. In rare case, unavoidable faults happen when the robot working. To deal with these uncertainties and faults, various control methods have been proposed in literature, such as PID control [1, 2], adaptive control [3], fuzzy control [4, 5], neural network control [6, 7], sliding mode control (SMC) [8–10]. Compared with other controllers, SMC standouts with salient features such as easy design procedure,

© Springer Nature Switzerland AG 2019
D.-S. Huang et al. (Eds.): ICIC 2019, LNAI 11645, pp. 167–178, 2019.
https://doi.org/10.1007/978-3-030-26766-7_16

robustness against the effect of dynamic uncertainties and faults. Basically, SMC uses a proportional-derivative (PD) sliding function to deal with the overshoot problem and get a fast response, respectively. To archive higher accuracy, an integral part has been added into the sliding function which is called proportional-integral-derivative SMC (PID-SMC) [11]. However, both conventional SMC and PID-SMC still include the disadvantages, such as: (1) chattering phenomenon; (2) do not provide the finite-time convergence; (3) require the prior knowledge of the dynamic uncertainties and faults, and (4) the design procedure need the velocity information.

In order to eliminate the chattering problem, this paper employs super-twisting algorithm (STA) to provide continuous control signal instead of the discontinuous one. On the other hand, the finite-time convergence of STA has been successfully proved in literature [12, 22]. Unfortunately, its design procedure still requires the velocity measurement information and the prior knowledge of the lumped uncertainties and faults which do not exist in realization. In order to obtain the dynamic uncertainties and faults information, a lot of estimation methods have been developed, such as time delay estimation [13, 14], neural network observer [15], third-order sliding mode observer (TOSMO) [16, 17]. Among them, TOSMO is the remarkable estimation method because of its ability to estimate not only the lumped dynamic uncertainties and faults but also the velocity information without the need of a lowpass filter.

Although the requirement of the prior knowledge of uncertainties and faults is eliminated in controller design procedure, however, it is still needed in the process of designing of TOSMO. Generally, the observer's parameters are selected based on experience and the upper bound of the lumped dynamic uncertainties and faults. In literature, various methods have been developed to estimate system parameters, such as adaptive law [18], fuzzy logic [19], radial basis function neural network (RBFN) [20]. In this paper, RBFN is used to approximate the parameters of TOSMO because of its fast learning ability, high approximation accuracy, and simple structure [17].

This paper proposes a neural TOSMO which can estimate both the velocities and the dynamic uncertainties and faults with high accuracy fast response without the need of lowpass filter. Additionally, the assumption of prior knowledge of the lumped uncertainties and faults are eliminated. Based on the obtained information of the observer, a PID-SMC control method based on super-twisting algorithm is designed to deal with the robot dynamic uncertainties and faults. The proposed control strategy provides high position tracking accuracy with less chattering, highly robust against the effects of the uncertainties and faults with only position measurement. The global stability and finite-time convergence are guaranteed by Lyapunov theory.

The structure of this paper is as follows. The dynamic equation of a n-link robotic manipulator is first described in Sect. 2. Then, the process of designing of TOSMO and neural TOSMO for robotic manipulators is presented in Sect. 3. Section 4 introduces the controller design procedure. In Sect. 5, the simulation results of the proposed control method are shown for a PUMA560 robotic manipulator. Finally, conclusions are presented in Sect. 6.

2 Problem Statement

The dynamic equation of a serial n-link robotic manipulator with faults is given as

$$\ddot{\theta} = M^{-1}(\theta)\left[\tau - C\left(\theta,\dot{\theta}\right)\dot{\theta} - G(\theta) - F\left(\theta,\dot{\theta}\right) - \tau_d\right] + \gamma\left(t - T_f\right)\Psi(t) \qquad (1)$$

where $\theta, \dot{\theta}, \ddot{\theta} \in \mathbb{R}^n$ denote position, velocity, and acceleration of robot joints, respectively. $\tau \in \mathbb{R}^n$, $M(\theta) \in \mathbb{R}^{n \times n}$, $C\left(\theta,\dot{\theta}\right) \in \mathbb{R}^n$, $G(\theta) \in \mathbb{R}^n$ represent the control input torque, inertia matrix, Coriolis and centripetal forces, gravitational force term, respectively. $F\left(\theta,\dot{\theta}\right) \in \mathbb{R}^n$, $\tau_d \in \mathbb{R}^n$ denote the friction vector and disturbance vector, respectively, these two elements constitute the dynamic uncertainties. $\Psi(t) \in \mathbb{R}^n$ indicates unknown faults with the time of occurrence T_f.

$\gamma\left(t - T_f\right) = diag\{\gamma_1\left(t - T_f\right), \gamma_2\left(t - T_f\right), \ldots, \gamma_n\left(t - T_f\right)\}$ is the time profile of the faults

$$\gamma_1\left(t - T_f\right) = \begin{cases} 0 & \text{if } t \leq T_f \\ 1 - e^{-\varsigma_i\left(t-T_f\right)} & \text{if } t \geq T_f \end{cases} \qquad (2)$$

where $\varsigma_i > 0$ is the unknown faults evolution rate. The small and large values of ς_i characterize incipient faults and abrupt faults, respectively.

For simpler in design procedure, the robot dynamics of (1) can be rewritten in state space as

$$\begin{aligned} \dot{x}_1 &= x_2 \\ \dot{x}_2 &= \Upsilon(x) + M^{-1}(x_1)u(t) + \Xi(x,t) \end{aligned} \qquad (3)$$

where $x_1 = \theta$, $x_2 = \dot{\theta}, x = [x_1 \quad x_2]^T$, $u(t) = \tau$, $\Upsilon(x) = M^{-1}(\theta)\left[-C\left(\theta,\dot{\theta}\right)\dot{\theta} - G(\theta)\right]$ and $\Xi(x,t) = M^{-1}(\theta)\left[-F\left(\theta,\dot{\theta}\right) - \tau_d\right] + \gamma(t - T_f)\Psi(t)$ indicates the lumped dynamic uncertainties and faults.

The purpose of this paper is to design a continuous PID-SMC based on neural TOSMO which can handle the effect of lumped dynamic uncertainties and faults with only available position measurement. This controller is designed based on the following assumption.

Assumption 1: The lumped dynamic uncertainty and fault $\Xi(x,t)$ of the system and its time derivative are bounded as

$$|\Xi(x,t)| \leq \bar{\Xi} \qquad (4)$$

$$|\dot{\Xi}(x,t)| \leq \bar{\Lambda} \qquad (5)$$

where $\bar{\Xi}$ and $\bar{\Lambda}$ are positive constants.

3   Observer Design

3.1   Third-Order Sliding Mode Observer

Based on robot dynamics (3), the third-order sliding mode observer is designed as [17]

$$
\begin{aligned}
\dot{\hat{x}}_1 &= \mu_1 |x_1 - \hat{x}_1|^{2/3} sign(x_1 - \hat{x}_1) + \hat{x}_2 \\
\dot{\hat{x}}_2 &= \Upsilon(\hat{x}) + M^{-1}(\hat{x}_1) u(t) + \mu_2 |x_1 - \hat{x}_1|^{1/3} sign(x_1 - \hat{x}_1) + \hat{z} \\
\dot{\hat{z}} &= \mu_3 sign(x_1 - \hat{x}_1)
\end{aligned}
\tag{6}
$$

where $\hat{x}$ is the estimation of x and μ_i denotes the sliding mode gains.

The estimation error can be obtained by subtracting (6) from (3)

$$
\begin{aligned}
\dot{\tilde{x}}_1 &= -\mu_1 |\tilde{x}_1|^{2/3} sign(\tilde{x}_1) + \tilde{x}_2 \\
\dot{\tilde{x}}_2 &= -\mu_2 |\tilde{x}_1|^{1/3} sign(\tilde{x}_1) + \Xi(x,t) + d(\tilde{x},t) - \hat{z} \\
\dot{\tilde{z}} &= \mu_3 sign(\tilde{x}_1)
\end{aligned}
\tag{7}
$$

where $\tilde{x} = x - \hat{x}$ and $d(\tilde{x},t) = \Upsilon(x) + M^{-1}(x_1)u(t) - (\Upsilon(\hat{x}) + M^{-1}(\hat{x}_1)u(t))$.

Defining $\tilde{x}_3 = -\hat{z} + \Xi_o(x,t)$, where $\Xi_o(x,t) = \Xi(x,t) + d(\tilde{x},t)$. Equation (7) becomes

$$
\begin{aligned}
\dot{\tilde{x}}_1 &= -\mu_1 |\tilde{x}_1|^{2/3} sign(\tilde{x}_1) + \tilde{x}_2 \\
\dot{\tilde{x}}_2 &= -\mu_2 |\tilde{x}_1|^{1/3} sign(\tilde{x}_1) + \tilde{x}_3 \\
\dot{\tilde{x}}_3 &= -\mu_3 sign(\tilde{x}_1) + \dot{\Xi}_o(x,t)
\end{aligned}
\tag{8}
$$

By choosing the candidate Lyapunov function L_0 and using the same proving procedure as in [21], we can prove that the system (8) is stable and the estimation errors $\tilde{x}_1, \tilde{x}_2$, and $\tilde{x}_3$ converge to zero in finite time with convergence time as

$$
T_0(\tilde{x}_0) \le \frac{L_0^{1/5}(\tilde{x}_0)}{\frac{1}{5}\chi}
\tag{9}
$$

where $\tilde{x}_0$ is the initial condition of (8) and $0 \le \chi \le 2.8 \times 10^{-4}$.

After the convergence process, the estimated states will achieve the real states $(\hat{x}_1 = x_1, \hat{x}_2 = x_2)$ and $d(\tilde{x},t) = 0$, the third term of system (8) becomes

$$
\dot{\tilde{x}}_3 = -\mu_3 sign(\tilde{x}_1) + \dot{\Xi}(x,t) \equiv 0
\tag{10}
$$

The lumped dynamic uncertainties and the faults can be reconstructed as

$$
\hat{\Xi}(x,t) = \int \mu_3 sign(\tilde{x}_1)
\tag{11}
$$

We can see that the estimated dynamic uncertainties and faults in (11) can be reconstructed directly without the need of lowpass filter. Besides, an integral element is included, therefore, the chattering of the estimated function is partially eliminated.

Although the TOSMO provides high estimation accuracy and finite time convergence, its parameters in still selected based on experience and the upper bound of the lumped dynamic uncertainties and faults. In the next Section, the neural TOSMO will be designed to surpass this drawback of TOSMO.

3.2   Neural Third-Order Sliding Mode Observer

Based on robot dynamics (3), the neural TOSMO is designed as

$$
\begin{aligned}
\dot{\hat{x}}_1 &= \hat{\mu}_1 |\tilde{x}_1|^{2/3} sign(\tilde{x}_1) + \hat{x}_2 \\
\dot{\hat{x}}_2 &= \Upsilon(\hat{x}) + M^{-1}(\hat{x}_1)u(t) + \hat{\mu}_2 |\tilde{x}_1|^{1/3} sign(\tilde{x}_1) + \hat{z} \\
\dot{\hat{z}} &= \hat{\mu}_3 sign(\tilde{x}_1)
\end{aligned}
\tag{12}
$$

where $\hat{\mu}_i$ is the estimation of μ_i

$$
\hat{\mu}_i = \hat{V}_i^T \Phi_i(X_k), \; i = 1, 2, 3
\tag{13}
$$

where $X_k = [e_1 \; e_2]^T$ is the neural network input with $e_1 = x_1 - \hat{x}_1$ and $e_2 = \dot{\hat{x}}_1 - \hat{x}_2$.
The RBFN is used as an activation function

$$
\Phi_i(X_k) = \exp\left(\frac{\|X_k - c_{ij}\|^2}{\sigma_{ij}^2} \right)
\tag{14}
$$

where c_{ij} denotes the center vector, σ_{ij} denotes the spread factor, $j = 1, 2, \ldots, N$ denotes the number of nodes in the hidden layer.
The weigh update law is as

$$
\dot{V}_i = \eta_i \Phi_i(X_k) \|X_k\|
\tag{15}
$$

where η_i denote the learning rate.
By the same procedure with Sect. 3.1, we can reconstruct the lumped dynamic uncertainties and faults as

$$
\hat{\Xi}(x, t) = \int \hat{\mu}_3 sign(\tilde{x}_1)
\tag{16}
$$

Consequently, the requirement of prior knowledge of the lumped dynamic uncertainties and faults has been removed. The obtained velocities and faults information are used to design controller in Sect. 4. The effectiveness of the proposed observer will be shown in simulation results.

4 Controller Design

We define the tracking error and velocity error as

$$\hat{e} = \hat{x}_1 - x_d \tag{17}$$

$$\dot{\hat{e}} = \hat{x}_2 - \dot{x}_d \tag{18}$$

where x_d and $\dot{x}_d$ describe the desired trajectory and velocity, respectively.

The PID sliding function is selected as [11]

$$\hat{s} = K_p \hat{e} + K_i \int \hat{e} + K_d \dot{\hat{e}} \tag{19}$$

where $K_i, K_p,$ and K_d denote the proportional, integral, and derivative gains, respectively.

Taking the first-time derivative of sliding function, we obtain

$$
\begin{aligned}
\dot{\hat{s}} &= K_p \dot{\hat{e}} + K_i \hat{e} + K_d \ddot{\hat{e}} \\
&= K_p \left(\dot{\hat{x}}_1 - \dot{x}_d \right) + K_i \hat{e} + K_d \left(\dot{\hat{x}}_2 - \ddot{x}_d \right) \\
&= K_p \left(\dot{\hat{x}}_1 - \dot{x}_d \right) + K_i \hat{e} + K_d \left(\begin{array}{c} \Upsilon(\hat{x}) + M^{-1}(\hat{x}_1) u(t) \\ + \hat{\mu}_2 |\tilde{x}_1|^{1/3} sign(\tilde{x}_1) + \hat{\mu}_3 \int sign(\tilde{x}_1) - \ddot{x}_d \end{array} \right)
\end{aligned}
\tag{20}
$$

The control law is designed as follows

$$u = -\frac{1}{K_d} M(x_1) \left(u_{eq} + u_{smc} \right) \tag{21}$$

where the equivalent control law u_{eq} is designed as

$$u_{eq} = K_p \left(\dot{\hat{x}}_1 - \dot{x}_d \right) + K_i \hat{e} + K_d \left(\Upsilon(\hat{x}) + \hat{\mu}_2 |\tilde{x}_1|^{1/3} sign(\tilde{x}_1) + \hat{\mu}_3 \int sign(\tilde{x}_1) - \ddot{x}_d \right) \tag{22}$$

and the switching control law u_{smc} is designed based on super-twisting algorithm as

$$
\begin{aligned}
u_{smc} &= \rho_1 |\hat{s}|^{1/2} \operatorname{sgn}(\hat{s}) + \vartheta \\
\dot{\vartheta} &= \rho_2 \operatorname{sgn}(\hat{s})
\end{aligned}
\tag{23}
$$

Substituting the control laws (21–23) into (20), we obtain

$$
\begin{aligned}
\dot{\hat{s}} &= \rho_1 |\hat{s}|^{1/2} \operatorname{sgn}(\hat{s}) + \vartheta \\
\dot{\vartheta} &= \rho_2 \operatorname{sgn}(\hat{s}) + \dot{\delta}
\end{aligned}
\tag{24}
$$

Since the estimation of the lumped dynamic uncertainties and faults $\hat{\mu}_3 \int sign(\tilde{x}_1)$ is included in the equivalent control signal, the stability and finite time convergence of the system (24) will be guaranteed in [22] if the sliding gains are chosen as

$$\rho_1 > 0$$
$$\rho_2 > \left(3 + 2\frac{\bar{\delta}}{\rho_1^2}\right)\bar{\delta} \tag{25}$$

where $\delta = \hat{\Xi} - \Xi$ denotes the uncertainties and faults estimation error and $|\dot{\delta}| \leq \bar{\delta}$.

The stability and finite time convergence of system (24) have been successfully proved by using Lyapunov theory in [22].

5 Simulation Results

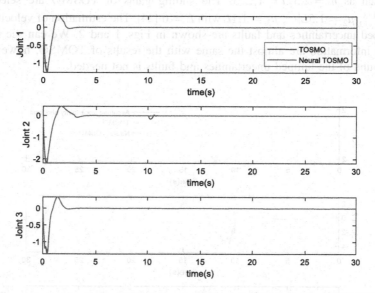

Fig. 1. Comparison of velocity estimation error between TOSMO and neural TOSMO among three joints.

In this section, the effectiveness of control method is proved by computer simulation for a PUMA560 robot with the last three links are locked. The explicit dynamic model and parameter values of PUMA560 robot are given in [23].

The friction $F\left(\theta,\dot{\theta}\right) \in \mathbb{R}^n$ and disturbance $\tau_d \in \mathbb{R}^n$ vectors are assumed as

$$F\left(\theta,\dot{\theta}\right) = \begin{bmatrix} 0.8\dot{\theta}_1 + 1.2\cos(3\theta_1) \\ 1.3\dot{\theta}_2 + 0.5 sin(5\theta_2) \\ 0.7\dot{\theta}_3 + 0.3\sin(3.5\theta_3) \end{bmatrix} \quad \text{and} \quad \tau_d = \begin{bmatrix} 1.2\sin\left(0.95\dot{\theta}_1\right) \\ 0.9\cos\left(1.7\dot{\theta}_2\right) \\ 0.5\sin\left(3.2\dot{\theta}_3\right) \end{bmatrix} \qquad (26)$$

The fault occurs at $T_f = 10s$ and is assumed as

$$\Psi = \begin{bmatrix} 2\cos(0.8t) & 3\sin(t) & 1.2\sin(t) \end{bmatrix}^T \qquad (27)$$

In this simulation, three RBFNs with 20 neurons in each hidden layer are used to approximate three parameters of the proposed observer, respectively. The learning rates are chosen as $\eta_i = 2.5$, $i = 1, 2, 3$. The sliding gains of TOSMO are selected as $\mu_1 = 3L^{1/3}$, $\mu_2 = 1.5L^{1/2}$, $\mu_3 = 1.1L$ with $L = 6$ [24]. The estimation of velocities and the lumped uncertainties and faults are shown in Figs. 1 and 2. We can see that the obtained information are almost the same with the results of TOMSO, however, the upper bound of the lumped uncertainties and faults is not needed.

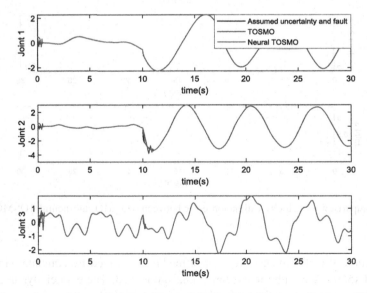

Fig. 2. Comparison of lumped uncertainty and fault estimation error between TOSMO and neural TOSMO among three joints.

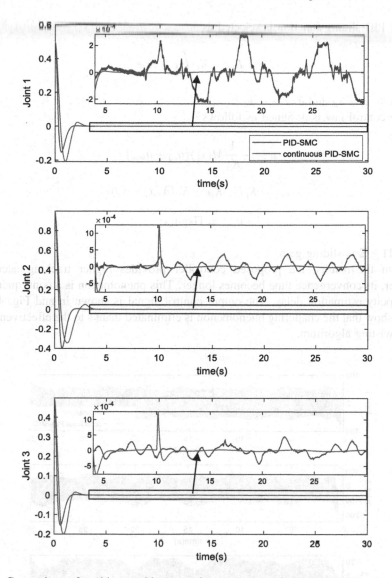

Fig. 3. Comparison of position tracking error between conventional PID-SMC and proposed controller among three joints.

In order to verify the effectiveness of the proposed controller, the parameters of proposed controller are selected as $K_i = 7, K_p = 3, K_d = 0.5, \rho_1 = 1.5, \rho_2 = 1.1$. Figure 3 shows the comparison of position tracking error between the proposed controller and the conventional PID-SMC which is designed as follows:

The PID sliding function is selected as

$$s = K_p e + K_i \int e + K_d \dot{e} \tag{28}$$

where $e = x_1 - x_d$ and $\dot{e} = x_2 - \dot{x}_d$

The control law is designed as follows

$$u = -\frac{1}{K_d} M(x_1)(u_{eq} + u_{smc}) \tag{29}$$

$$u_{eq} = K_p \dot{e} + K_i e + K_d(\Upsilon(x) - \ddot{x}_d) \tag{30}$$

$$u_{smc} = \Pi \text{sgn}(s) \tag{31}$$

where $\Pi \geq \bar{\Xi}$ is sliding gain.

From the results, the proposed controller provides higher tracking accuracy, however, its convergence time becomes longer. This phenomenon is a consequence of the velocity estimation delay. The output control signal is shown in and Fig. 4. The results show that the chattering phenomenon is eliminated thanks to the effectiveness of super-twisting algorithm.

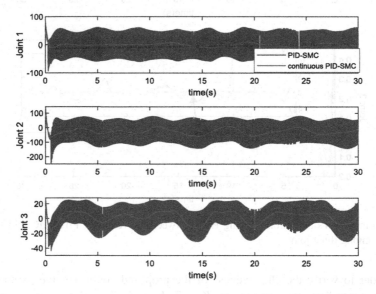

Fig. 4. Comparison of control signal between conventional PID-SMC and proposed controller among three joints.

6 Conclusions

This paper proposes a continuous PID-SMC strategy based neural TOSMO for uncertainty robotic manipulator without velocity measurement. The neural TOSMO provides high estimation accuracy without the requirement of the prior knowledge of lumped dynamic uncertainties and the faults. The tracking error accuracy is improved by using the proposed control algorithm. The global stability and finite-time convergence of system is guaranteed. The simulation results on a 3-DOF PUMA560 robot verify the effectiveness of the proposed strategy.

Acknowledgement. This research was supported by Basic Science Research Program through the National Research Foundation of Korea (NRF) funded by the Ministry of Education (NRF2016R1D1A3B03930496).

References

1. Song, Y., Huang, X., Wen, C.: Robust adaptive fault-tolerant PID control of MIMO nonlinear systems with unknown control direction. IEEE Trans. Ind. Electron. **64**(6), 4876–4884 (2017)
2. Alibeji, N., Sharma, N.: A PID-type robust input delay compensation method for uncertain Euler-Lagrange systems. IEEE Trans. Control Syst. Technol. **25**(6), 2235–2242 (2017)
3. Slotine, J.-J.E., Li, W.: On the adaptive control of robot manipulators. Int. J. Robot. Res. **6**(3), 49–59 (1987)
4. Wang, H.O., Tanaka, K., Griffin, M.F.: An approach to fuzzy control of nonlinear systems: stability and design issues. IEEE Trans. Fuzzy Syst. **4**(1), 14–23 (1996)
5. Tong, S., Wang, T., Li, Y.: Fuzzy adaptive actuator failure compensation control of uncertain stochastic nonlinear systems with unmodeled dynamics. IEEE Trans. Fuzzy Syst. **22**(3), 563–574 (2014)
6. Park, J., Sandberg, I.W.: Universal approximation using radial-basis-function networks. Neural Comput. **3**(2), 246–257 (1991)
7. Song, Y,, Guo, J.: Neuro-adaptive fault-tolerant tracking control of Lagrange systems pursuing targets with unknown trajectory. IEEE Trans. Ind. Electron. **64**(5), 3913–3920 (2017)
8. Utkin, V.I.: Sliding Modes in Control and Optimization. Springer, Heidelberg (2013)
9. Islam, S., Liu, X.P.: Robust sliding mode control for robot manipulators. IEEE Trans. Ind. Electron. **58**(6), 2444–2453 (2011)
10. Guo, Y., Woo, P.-Y.: An adaptive fuzzy sliding mode controller for robotic manipulators. IEEE Trans. Syst. Man Cybern.-Part A: Syst. Hum. **33**(2), 149–159 (2003)
11. Eker, I.: Sliding mode control with PID sliding surface and experimental application to an electromechanical plant. ISA Trans. **45**(1), 109–118 (2006)
12. Moreno, J.A., Osorio, M.: Strict Lyapunov functions for the super-twisting algorithm. IEEE Trans. Autom. Control **57**(4), 1035–1040 (2012)
13. Elmali, H., Olgac, N.: Implementation of sliding mode control with perturbation estimation (SMCPE). IEEE Trans. Control Syst. Technol. **4**(1), 79–85 (1996)
14. Van, M., Ge, S.S., Ren, H.: Finite time fault tolerant control for robot manipulators using time delay estimation and continuous nonsingular fast terminal sliding mode control. IEEE Trans. Cybern. **47**(7), 1681–1693 (2017)

15. Van, M., Kang, H.-J.: Robust fault-tolerant control for uncertain robot manipulators based on adaptive quasi-continuous high-order sliding mode and neural network. Proc. Inst. Mech. Eng. Part C: J. Mech. Eng. Sci. **229**(8), 1425–1446 (2015)

16. Van, M., Kang, H.-J., Suh, Y.-S., Shin, K.-S.: Output feedback tracking control of uncertain robot manipulators via higher-order sliding-mode observer and fuzzy compensator. J. Mech. Sci. Technol. **27**(8), 2487–2496 (2013)

17. Chalanga, A., Kamal, S., Fridman, L.M., Bandyopadhyay, B., Moreno, J.A.: Implementation of super-twisting control: super-twisting and higher order sliding-mode observer-based approaches. IEEE Trans. Ind. Electron. **63**(6), 3677–3685 (2016)

18. Bahrami, M., Naraghi, M., Zareinejad, M.: Adaptive super-twisting observer for fault reconstruction in electro-hydraulic systems. ISA Trans. **76**, 235–245 (2018)

19. Le, T.D., Kang, H.-J.: A fuzzy adaptive sliding mode controller for tracking control of robotic manipulators. J. Inst. Control Robot. Syst. **18**(6), 555–561 (2012)

20. Hoang, D.-T., Kang, H.-J.: Fuzzy neural sliding mode control for robot manipulator. In: Huang, D.-S., Han, Kyungsook, Hussain, Abir (eds.) ICIC 2016. LNCS (LNAI), vol. 9773, pp. 541–550. Springer, Cham (2016). https://doi.org/10.1007/978-3-319-42297-8_50

21. Ortiz-Ricardez, F.A., Sánchez, T., Moreno, J.A.: Smooth Lyapunov function and gain design for a second order differentiator. In: 2015 IEEE 54th Annual Conference on Decision and Control (CDC), pp. 5402–5407 (2015)

22. Moreno, J.A., Osorio, M.: A Lyapunov approach to second-order sliding mode controllers and observers. In: 2008 47th IEEE Conference on Decision and Control, pp. 2856–2861 (2008)

23. Armstrong, B., Khatib, O., Burdick, J.: The explicit dynamic model and inertial parameters of the PUMA 560 arm. In: Proceedings of 1986 IEEE International Conference on Robotics and Automation, vol. 3, pp. 510–518 (1986)

24. Levant, A.: Higher-order sliding modes, differentiation and output-feedback control. Int. J. Control **76**(9–10), 924–941 (2003)

Intelligent Bus Shunting Control System Based on Face Recognition

Xiaohang Li, Binggang Xiao$^{(\boxtimes)}$, and Fan Yang

College of Information Engineering, China Jiliang University,
Hangzhou 310018, Zhejiang, China
bgxiao@cjlu.edu.cn

Abstract. This paper introduces the bus distribution control system based on face recognition. It adopts face recognition technology to realize early payment when waiting in the bus station. The terminal judges the current waiting number by the feedback of face recognition to control the departure frequency of the bus. At the same time, it avoids the difficult situation of taking a taxi with the regular departure system. The face recognition payment technology before getting on the bus allows passengers to enter the car directly without using the bus card to pay. It could speed up the speed of getting on and off, and alleviate the crowded traffic during the peak period. The system includes face recognition payment by using the station's touchscreen or mobile phone APP; ultrasonic ranging prevents the passengers from waiting too close to the exit of the station to generate security hazards; infrared detection detects the number of people getting off; the terminal system processes the station data, etc. The application of these modules could solve the problems of slow departure, slow boarding, and unreasonable charges. The convenient and efficient operation mode will attract a large number of travellers. Compared with traditional buses, it has the advantages of price and high efficiency. It could replace the tradition bus after development and improvement.

Keywords: Shunt control · Infrared detection · Ultrasound ranging · Face recognition payment

1 Introduction

In the face of increasingly complex traffic conditions, more and more people tend to choose public transport. After decades of twists and turns, unmanned driving has finally made a big breakthrough in recent years. In 2017, a Unmanned bus named "Alphaba Smart Driving Bus" in Shenzhen officially went on the road, and followed by the first batch of unmanned buses in the United States also began to test the next year [1]. Unmanned buses can now realize the road environment through industrial computer, vehicle controller and CAN network, and can accurately respond to other road users and emergencies in real time. Under automatic driving, it has realized the functions of pedestrians and vehicle detection, deceleration avoidance, emergency stop, obstacle bypass, lane change, automatic stop by station, etc. [2]. Faced with a series of problems such as traffic congestion, high operating costs, waste of resources, delays in

© Springer Nature Switzerland AG 2019
D.-S. Huang et al. (Eds.): ICIC 2019, LNAI 11645, pp. 179–188, 2019.
https://doi.org/10.1007/978-3-030-26766-7_17

scheduling measures, untimely information feedback, and low service levels, unmanned public transportation is still a long-term business. To effectively solve urban public transportation problems, it is an effective way to operate intelligently and optimize passengers' travel experience. The application of face recognition and clean energy technologies can effectively solve the problems of environmental pollution, waste of resources and low travel efficiency caused by traditional public transportation [3] (Fig. 1).

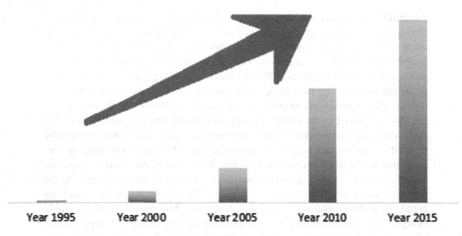

Fig. 1. The development of the smart bus

In the intelligent public transport system in this paper, the intelligent system is used to replace the operation of the human brain, and the sensor can be used to replace the human administrator. It can also realize automation, reduce labor costs through intelligent control, increase the safety factor of public rides, and solve the problems in current public transport system such as slow rides, long waiting times, and slow charges [4].

An innovation of the smart bus system is the use of face recognition payment. When using the mobile phone APP to complete the payment process in advance, it not only adapts to the current social situation that people go outside without cash, but also saves time on board. Face recognition is an emerging industry in recent years, there are many studies about face recognition in the world. The well-known ones include google, facebook, stanford, carnegie, etc. Many companies in China have emerged over the years, such as the face recognition of Hikvision. It has been able to be very precise, and all major companies have integrated Deep Learning algorithms on face recognition, and front-end depth intelligence provides powerful support for face big data applications. Face recognition has been widely used in recent years, including authentication systems, attendance systems, travel entertainment, access control, etc., but few face recognition is applied to traffic. Face recognition inbound payment is a major innovation in the smart unmanned public transportation systems [5].

In the face recognition technology, data training is performed by convolutional neural networks (NNS) [6]. The convolutional NNS is a multi-layer neural network. Through a series of methods, the image recognition problem with huge data volume is reduct dimensionality continuously, until it can be trained. In this project, face recognition needs to be recognized on the mobile APP by relatively static face. This relative error would be very small. When the camera faces the mobile crowd when people getting on the bus, it is necessary to identify different faces. So we should train it by NNS.

2   System Design

2.1   Overall Design

2.1.1   Three Major Parts

The smart bus system described in this paper can be divided into three parts: terminal, station, bus. At the station, the mobile APP or the fixed touchscreen of the station is used to select the route and face recognition payment. The terminal receives the station information to monitor the stations and buses in real time whether there is any waiting and waiting number for each route, and controls the departure of the bus (Fig. 2).

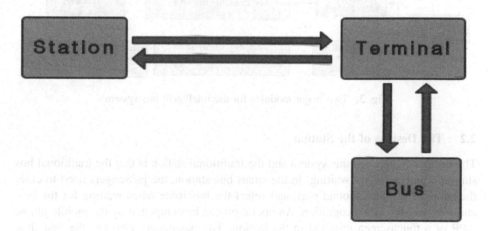

Fig. 2. Smart bus system

2.1.2   Two Major Modules

The intelligent bus system described in this paper is mainly divided into two major modules - the passenger module and the control module. Passengers select routes and make face recognition payments on the mobile app or on the touchscreen in the station before boarding, and then wait for the bus in the station. When the bus enters the station, a camera for face recognition is installed at the front door of the bus, and it is detected that the paid passenger will enter the compartment directly, and the unpaid passenger is recognized to automatically deduct the fee and enter the vehicle. This avoids the passengers who do not complete the payment while waiting in the station. When the destination platform is reached, the infrared detection of the rear door can detect the number of people getting off the vehicle and monitor the number of people in the station and the bus in real time.

In the control module, the departure is mainly carried out in two cases: the first is that the terminal monitors the number of waiting persons in the station in real time through the face recognition information, and when the number of waiting persons is greater than the threshold, the vehicle is immediately started. The second is a regular start-up similar to the traditional bus, so long-term waiting is avoided when the number of waiting people is small (Fig. 3).

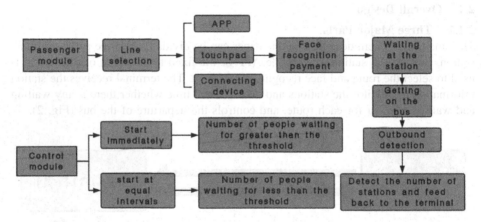

Fig. 3. Two major modules for the intelligent bus system

2.2   The Designs of the Station

The difference between this system and the traditional station is that the traditional bus station is only used for waiting. In the smart bus station, the passengers need to enter the station from the inbound port, and select the bus route when waiting for the bus, and then pay for face recognition. An operation can be completed by the mobile phone APP or a touchscreen installed in the station. The passengers who use the first time need to complete the information registration, and then only need to scan the face to directly deduct the fee (Fig. 4).

When the car enters the station, the passengers leave the station from the "Upper entrance" to enter the bus. Among them, an ultrasonic distance measuring module is installed at the "Upper entrance" to prevent the passenger from being too close to the road when waiting in the station.

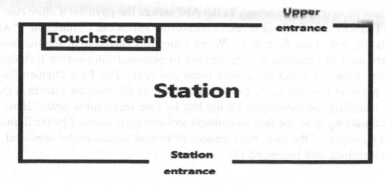

Fig. 4. The smart station

2.3 APP Module

The payment of this system is realized by the mobile phone APP or the touchscreen face recognition in the station, and the passenger needs to register personal information including face information when using for the first time. Then just select the route, enter the page and perform face recognition. Point your face at the phone camera until the interface shows "Deduction is successful". At this time, the face recognition data has been sent to the terminal through Bluetooth, so that the terminal can monitor the waiting situation in station in real time to regulate the departure (Fig. 5).

Fig. 5. Mobile phone App

The face recognition technology in the APP makes the payment a more convenient process. The APP interface has several buttons for "Bluetooth Connection", "Account Registration", and "Line Selection". When using this bus system for the first time, passengers need to complete the registration of personal information including face information, then just select the desired route and click "Pay Face Payment" to complete the payment process. In the process of getting on the bus, the camera at the door needs to recognize the passengers on the bus by face recognition again. Here, in the face of the walking flow, the face recognition technology is affected by the illumination and the movement of the face. Face images of natural scenes under non-ideal conditions are identified and processed [7].

2.4   Terminal Module

The terminal of the system belongs to the control and adjustment part, and it is necessary to continuously receive the information of the number of people in the station and the bus, and thereby control the frequency of the departure. The display screen of the terminal continuously displays the information of the people in the station and the bus. When a passenger enters the station and completes the face recognition payment, the terminal displays the information of the waiting person. When the number of waiting people increases, the number of waiting people displayed by the terminal increases, until the number reaches the threshold, and the terminal controls departure immediately. At the same time, the terminal the station display the departure information. The operator could see the number of people entering and leaving the station through the terminal. In addition to the terminal automatically controlling the departure, the operator can also manually control the departure (Fig. 6).

Fig. 6.   The display of terminal

2.5 Smart Floor

The floor inside the bus is based on pressure sensing technology, which induces the distribution of people and avoids excessive concentration of people. The pressure floor transmits the flow distribution data to the terminal. Through the data analysis, the terminal points to the area with less flow of people by the shining direction arrow on the floor, guiding the passenger to move to the empty area (Fig. 7).

Fig. 7. Pressure sensing floor

2.6 Equipments on the Bus

Intelligent hardware devices are also placed on the bus to match other modules. After the car enters the station, the entrance of the unmanned bus should be exactly aligned with the exit of the station. Passengers enter the bus through the entrance, and a camera for face recognition is placed at the entrance of the bus to detect the passengers on board. Whether the face recognition payment has been completed at the station itself, the paid passengers directly enter the bus, and the passengers who have not completed the payment at the station will be automatically charged by the background terminal.

An infrared detection module is installed at the exit of the bus to detect the number of people getting off at each station, and the data is transmitted to the terminal, so that the terminal can monitor the passenger data in the car in real time (Fig. 8).

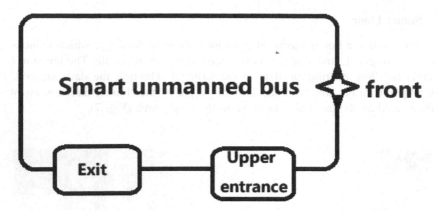

Fig. 8. Smart unmanned bus

3 Hardware Equipment

This equipment includes infrared detection, ultrasonic ranging, Bluetooth transmission, face recognition, APP module, pressure sensor, using STM32F103RCT6 single-chip microcomputer as the processor, the core CPU frequency can reach 72 MHz, fully consider the two contradictions of cost and function [8].

The infrared detection is controlled by the single-chip microcomputer to control the infrared emission tube and the infrared receiving tube [9]. In the system, the infrared detection is used as detection for the exit of bus, and the ultrasonic ranging is used as the safety warning device of the exit of station port, which considers the beam divergence of the ultrasonic sensor is large, it could adapt to the regional characteristics of the dangerous alarm range, and infrared has the characteristics of small beam divergence, only when the passenger gets off the vehicle and passes the lower door to detect the flow of people, fully adapt to the system needs.

In the smart floor part, the voltage dividing circuit can be pressed according to whether the varistor attached to the floor is pressed or not, and the voltage comparator outputs high and low levels [10]. The processor then counts the condition of each floor based on the level of the output, and then transmits it to another processor (station) using the NRF24L01 wireless transmission module. A monolithic wireless transceiver chip manufactured by NORDIC that operates in the 2.4 GHz to 2.5 GHz ISM band. The output power channel selection and protocol settings can be set via the SPI interface. Can be connected to a variety of microcontroller chips, and complete wireless data transfer. After the other processor receives the data, the data is analyzed to obtain the approximate distribution of personnel in the cabin, and finally displayed on the OLED [11].

4 Experimental Testing and Analysis

The experiment uses the back-end server that the self-made terminal is intended to be a reality, and displays the waiting number information at any time through the display screen. The ultrasonic device is proposed as an ultrasonic danger warning system at the exIt of the station, and the infrared detection device is intended to be used for infrared detection at the lower door. The "Smart Bus App" was built and connected to the terminal via Bluetooth instead of the 4G network, and the face registration and face recognition payment information was sent to the terminal. At present, face recognition has been implemented and information is transmitted to the terminal, and the terminal displays the real-time number information through the display screen and determines whether the threshold is reached to control the departure. Infrared detection should control the speed of the flow of people, which will affect the recognition results when passing through the infrared device continuously. Face recognition is affected by illumination and facial morphological changes, and it is necessary to continuously strengthen the training of data through neural networks (Fig. 9).

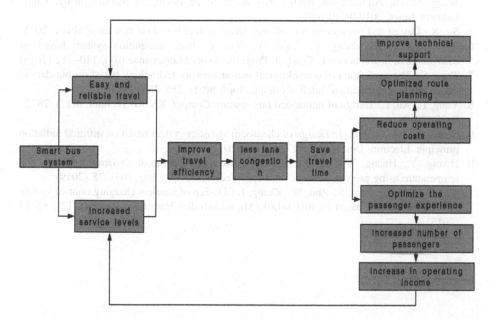

Fig. 9. Smart bus service architecture diagram

The smart bus system can facilitate travel and improve service levels, thereby improving people's travel efficiency. Since the bus body has a large body and has a large impact on traffic, it can effectively alleviate traffic congestion and save passengers' travel time by standard starting and driving. Under such an unmanned public transport system, not only does the operator have the cost-saving benefits, but also has a better travel experience for passengers and citizens.

References

1. Feiran, Z.: The application and development of artificial intelligence in life—taking unmanned bus and automatic picking robot as an example. China New Commun. **20**(11), 130–131 (2018)
2. Fan, D., He, X., Wei, H.: Research on key technologies and application of smart public transportation. Smart City (17) (2018)
3. Liu, L., Duan, W.: Research and application of smart urban transportation network system [J/OL]. Comput. Technol. Dev. (07), 1–4 (2019). http://kns.Cnki.net/kcms/detail/61.1450. TP.20190321.0942.074.html. Accessed 13 Apr 2019
4. Member of the Central Science and Technology Committee of Jiu San Society, and Dean of Beijing Huaxia Institute of Industrial Network Intelligent Technology Research Institute Wang Xiwen. Artificial intelligence has come to be intelligent manufacturing. China Industry News, 2019-04-10 (001)
5. Su, X.: Design and implementation of attendance system based on face recognition (2015)
6. Fan, A., Bao, G., Zheng, Y., Yang, Y., Yu, Y.: Face recognition system based on convolutional neural network. Comput. Program. Skills Maintenance (01), 110–112 (2019)
7. Tang, S.: The application of convolutional neural network technology based on big data in classroom face recognition. Intell. Comput. Appl. **9**(02), 255–256+259 (2019)
8. Yang, L., Xu, L.: Design of unmanned bus system. Comput. Knowl. Technol. **8**(12), 2872–2873 (2012)
9. Shang, J., Xu, Z., Song, J.: Design of classroom statistics system based on infrared radiation principle. Electron. Des. Eng. **25**(17), 90–92 (2017)
10. Huang, Y., Huang, W., Yu, Y.: Simultaneous measurement of dynamic pressure and temperature using pressure sensors. Electron. Technol. Softw. Eng. (03), 75 (2019)
11. Sun, G., Wu, R., Cang, S., Zhu, W., Zhang, L.: Design of wireless charging control system for electric vehicle based on nRF24L01. Microcontroller Embedded Syst. **18**(12), 81–85 (2018)

Tilt-Scrolling: A Comparative Study of Scrolling Techniques for Mobile Devices

Chuanyi Liu[1]([⊠]) [iD], Chang Liu[1], Hao Mao[2], and Wei Su[1]

[1] School of Information Science and Engineering, Lanzhou University,
Lanzhou, China
liuchuanyi96@hotmail.com, {chliu2017,suwei}@lzu.edu.cn
[2] Fiberhome Telecommunication Technologies Co., LTD., Wuhan, China
hmao@fiberhome.com

Abstract. Scrolling is a frequent operation on a mobile screen. The current scrolling method for a mobile device is both time-consuming and fatigue-prone, especially for one-handed interaction mode. In this study, we systematically evaluated six tilt-based scrolling techniques together with the base line through a repeated measure experiment participated by twenty-one subjects. The experimental results revealed that tilt-based scrolling techniques were more suitable for one-handed interaction with higher interaction performance and were less fatigue-prone than the base line. Tilt scrolling techniques with adaptive reference points outperformed those with the horizontal reference point in terms of scrolling speed and accuracy. The Step control function had higher interaction efficiency than the other two functions but was more error-prone. The results of our study have some implications for the design of scrolling techniques on mobile screens.

Keywords: Mobile devices · Tilt input · Scrolling

1 Introduction

Mobile devices are widely used in our daily life. We use mobile devices not only for communication but also for information processing. A mobile device has the following characteristics. First, it is portable. Second, a mobile device can be used in various scenarios in which we possibly interact with the device using one- or two-handed mode. Third, a mobile device typically possesses a smaller display vs. a desktop computer. We usually read information from a display. When we retrieve information from a computer, typically, not all the information can be displayed on one screen, then we have to scroll the window repeatedly. But repeated scrolling is both time-consuming and fatigue-prone. This issue is severer for a mobile device since its screen is much smaller than that of a desktop computer, thus scrolling happens more frequently.

The current scrolling technique on a mobile device is to scroll the screen by swipe and flick with our fingers. But repeated scrolling is fatigue-prone, especially when we interact with a mobile device unimanually, the repeated and continuous sliding of our thumbs makes them fatigue very soon. There are some studies aiming at scrolling techniques for mobile devices, but these researches either needed additional sensors

D.-S. Huang et al. (Eds.): ICIC 2019, LNAI 11645, pp. 189–200, 2019.
https://doi.org/10.1007/978-3-030-26766-7_18

(e.g., [1–6]) or were suitable for two-handed interaction (e.g., [7]) only. On the other hand, most mobile devices are integrated with sensors to sense tilt input, and tilt input has the potential on scrolling techniques. In this study, we systematically evaluated six tilt-based scrolling techniques together with the base line (the current scrolling technique) for mobile devices. The results of our work have some implications for design of scrolling techniques on mobile screens.

2 Related Work

Researchers have explored tilt-input for decades. These researches have spanned mobile phones, tablets, and wrist-worn watches. We survey the literature from general human performance on tilt-input control abilities, scrolling techniques for mobile devices, and other tilt-based techniques.

2.1 General Human Performance on Tilt Control

Wang et al. [8] utilized vision-based motion tracking system to explore user abilities on tilt control in terms of Fitts' law [9]. Sad and Poirier [10] proved that pointing and scrolling tasks conformed to Fitts' law, but they didn't report pointing throughput. Teather and Mackenzie [11] determined that tilt input can be modeled by Fitts' law through ISO9241-9 [12] 2D pointing task criterion. The authors compared velocity- and position-control for tilt-input interaction in the further study with the similar task [13]. They discovered that position-control outperformed its competitor with a higher task completion speed and a better pointing throughput.

Besides these researches of human performance on tilt control from a viewpoint of Fitts' law, there existed other studies exploring human performance on tilt control from different aspects. Guo and Paek [14] evaluated *ObjectPoint* and *AnglePoint*, two techniques of tilt input on smartwatches, and found out that *ObjectPoint* owned a better performance. Baglioni et al. [15] investigated human ability on controlling tilt input with a jerk tilt gesture. From three human wrist motion axes, Rahman et al. [16] systematically explored human ability of tilt input with a mobile phone.

2.2 Mobile Scrolling

Researchers have explored scrolling techniques for different mobile devices from different aspects. Many of these studies utilized additional sensors (e.g., [1–6]), and some of others were only suitable for two-handed interaction (e.g., [7]). Bartlett [2] utilized additional tilt sensors to a handheld device to scroll the display or control a picture sliding in the screen, but the study didn't include quantitative evaluation of the proposed technique. Oakley and O'Modhrain [5] proposed a one-handed scrolling technique with an iPaq expansion pack, and a quantitative evaluation was conducted. But in their experiment, the largest number of the candidate items was only fifteen and they didn't compare the proposed technique with the baseline. Miyaki and Rekimoto [4] used force sensitive resistor which was attached to the backside of a mobile phone to evaluate the effectiveness of a pressure-based scrolling control model. Spelmezan et al. [6] also employed additional pressure sensors attached to the left and the right sides of a mobile phone to perform large document scrolling tasks. They had proved that their proposed technique outperformed the traditional drag and flick touch gestures. Antoine et al. [1] proposed a novel auto-scroll technique for both desktop and mobile computers. The experimental results revealed that the auto-scroll technique improved over macOS and iOS systems baselines. But the technique relied on touch surfaces with fore-sensing capabilities, which are not available for all the commercial mobile devices. Bezel Swipe [7] introduced a set of gestures starting from the device bezel. Those gestures supported scrolling, pasting, etc., and did not interfere with other predefined gestures (e.g., zooming, panning, and tapping). However, those gestures could be performed two-handed only, thus were not suitable for all application scenarios of a mobile device. Sandnes [17] investigated user performance with left-to-right and right-to-left scrolling in interfaces on constrained mobile devices, and suggested that bidirectional scrolling functionality was appropriate. Dou and Sundar [18] conducted a user investigation and revealed that swiping technique positively affected user behavioral intentions to use the website.

Being different from the above studies, our work has the following characters. First, we utilize only the integrated sensors to sense tilt input of a mobile device. And devices with these sensors are common (if not all) in the market. Second, we have explored both one- and two-handed interaction methods. Third, we have compared different control functions and found out those being suitable for tilt-based scrolling techniques. Fourth, we have evaluated different scrolling techniques beside the baseline, and the results have more practical implications for some App designs for mobile devices.

2.3 Other Interaction Techniques with Tilt Input

Tilt input have been widely investigated in text entry [19–24] for different mobile devices. For example, Yeo et al. [24] proposed a tilt-based gesture keyboard which supported single-handed usage for various mobile devices. Some touch-free interaction methods were proposed for smartwatches [14, 25, 26]. For example, MyoTilt [26] combined a tilt operation and electromyography on smartwatches: tilt set the cursor's motion direction and fore (sensing from the electromyography) pulled the cursor to the target. Other implementation using tilt input also included mobile game [27–29], document [30] and menu [31] navigation, and 3D interaction [32], etc.

3 The Tilt-Scrolling Techniques

Algorithm 1 shows how the tilt-based scrolling techniques work. Combining two different reference points with three control functions, we explored six tilt-based scrolling techniques (i.e., Horizontal Reference Point (*HRP*) + Linear control function = HLinear, *HRP* +Step control function = HStep, *HRP* + Log control function = HLog, Adaptive Reference Point (*ARP*) + Linear control function = ALinear, *ARP* + Step control function = AStep, and *ARP* + Log control function = ALog).

A reference point was a tilt-input magnitude angle (*magnitudeA*) from which a delta angle (*deltaA*) was calculated. We investigated two types of reference points:

horizontal and adaptive. The value of *magnitudeA* was calculated from $0°$ when we employed a horizontal point, i.e., when the reference type (*referT*) was horizontal, in order to scroll the screen, the subjects started to tilt a mobile device from its screen was horizontal and upward. An adaptive reference point was a tilt magnitude angle of a mobile device when a subject took the device and read its screen comfortably, and it was not the same between subjects.

Algorithm 1 Scroll the List

```
 1:  deltaA=magnitudeA
 2:  if referT==Adaptive then deltaA=deltaA-referA
 3:  sPixel=a*function(deltaA)
 4:  if |deltaA|≥10° && No Touched Finger then
 5:     if deltaA>0° then
 6:        UpScroll(sPixel, tDuration)
 7:     elseif deltaA<0° then
 8:        DownScroll(sPixel, tDuration)
 9:     end if
10: end if
```

We employed three control functions: linear, step, and logarithm. As shown in Algorithm 1, we calculated a pixel number(*sPixel*) from *a* function(deltaA)*, where *function* was one of the three control functions, and *a* was an optimizing constant (different across the control functions). We set a neutral range, in which scrolling was suspended, when the absolute value of *deltaA* was less than $10°$. When *deltaA* was not in a neutral range and no fingers were touching on the screen, we scrolled the screen. The scrolling was upward or downward bidirectional according to *deltaA* was greater or less than $0°$, when we tilted the device inward or outward, respectively. We then smoothly scrolled the screen by distance pixels (*sPixel*) over duration milliseconds (*tDuration*).

4 Experiment

4.1 Subjects and Apparatus

Twenty-one volunteers served as subjects in our experiment. All were recruited from the local university, undergraduate or graduate students, right-handed, aged 18–33 years old. They all had some tilt usage experience.

The experiment was performed on three mobile phones:

- Huawei P10 with a 5.1" touch screen (1920 × 1080 pixels resolution),
- Huawei P20 with a 5.8" touch screen (2244 × 1080 pixels resolution),
- Huawei Mate 20 Pro with a 6.39" touch screen (3120 × 1440 pixels resolution).

All ran Android OS on Hisilioon CPUs, each had a built-in 3-axis accelerometer. The three mobile phones had different processing capacities and memories. Unfortunately, we could not find three mobile phones with the same components except their screen sizes. But these differences did not introduce significant bias of task completion time, since the experimental program conducted minute computation (only recorded a trial time and correct or not when an item was tapped) during time recording.

The experimental program was developed in Java with the Android SDK. The accelerometer was utilized to detect tilt magnitude of the mobile phone, sampling at a rate of 50 samples per second.

4.2   Procedure, Task, and Design

Prior to the formal experiment, we instructed the subjects to read texts from a mobile screen (holding the device in one hand and keeping in a comfortable posture) for 3 min. During the process, the experimental software recorded a magnitude tilt angle once per 10 s. The median of all the recorded angles was chosen to be the adaptive reference point for each subject. In the experiment, the subjects were required to perform a sequence of target selection from a list containing 150 items. The subjects started with a 5-minute exercise to familiarize themselves with the task. In the formal experiment, they were instructed to complete a trial as quickly and accurately as possible. Between two experimental conditions, the subjects were allowed to have a rest within 20 min; and a longer break was mandatory between two experimental blocks. It took each subject approximately one hour to complete the whole experiment. A questionnaire was given to the subjects to collect their subjective comments on the seven scrolling techniques for both one- and two-handed interaction modes. The questionnaire included an open comment and three subjective measures: usability, fatigue, and personal preference. The subjects were instructed to rate each measure with 1–7 rating levels (1 standing for the worst and 7 the best).

Twenty targets were pseudo-randomly arrayed in the list in a full-screen view. The subjects were instructed to scroll the list from top to bottom with different scrolling techniques, and select the targets by tapping on them in the same order. Each target was filled with yellow background, and the background turned into green if the target was selected correctly; while its background turned into red accompanying with an audio alarm if there was another target had been missed in front of this one, and a missing error was recorded. A target selection would not proceeded until the prior target had been selected correctly. A trial completion time was recorded from last target was tapped to the current one.

The experimental factors were: Block (B, 3 levels), Interaction (I, 2 levels, i.e., one- and two-handed), Control Mode (CM, 7 levels: 1 base line + 2 Reference Points (RP) $\times$ 3 Functions (F)), and Device (D, 3 levels). The base line is the current common scrolling method on a touch screen, i.e., scrolling by swipe and flick. We conducted a full factorial within-subjects design experiment. A 7×7 Latin square was utilized to counterbalance order effects of levels of CM. The level orders of the other factors were randomized between subjects. Totally, there were: 21 subjects $\times$ 3 Bs $\times$ 2 Is $\times$ 7 CMs $\times$ 3 Ds $\times$ 20 selections = 52920 correct selections in the whole experiment.

4.3   Hypotheses and Dependent Measures

Based on insights into different control modes, we would verify the following primary hypotheses in the experiment.

H1: Tilt-based scrolling were more suitable for one-handed interaction.
H2: The adaptive reference points performed better than the horizontal.
H3: The base line was more error-prone in terms of missing selection (caused by over scrolling).

Therefore, primary dependent measures were the total time to complete a trial and the error rates of target missing selection.

5   Results

We first performed a 3 *Bs* × 2 *Is* × 7 *CMs* × 3 *Ds* repeated measures ANOVA on mean data of trial completion times and error rates of target missing selection. Then we eliminated the data of the base line and conducted a 3 *Bs* × 2 *Is* × 2 *RPs* × 3*Fs* × 3 *Ds* repeated measures ANOVA on the rest data to obtain further insights of the tilt-based scrolling techniques. The results of the two times of ANOVA are analyzed in the following subsections.

5.1   Time

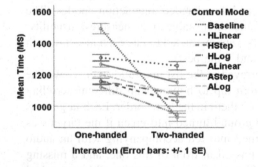
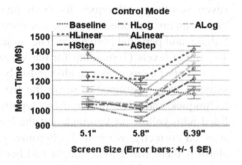

Fig. 1. Trial time across levels of *I* and *CM*.   **Fig. 2.** Trial time across levels of *D* and *CM*.

Results with the Base Line. There was a significant effect of block on mean trial completion time ($F_{2,40}$ = 121.5, p < 0.0001, η_p^2 = 0.859). This indicated that the subjects gained significant learning effects on the controlling the scrolling techniques across the blocks. Interaction (see Fig. 1) also had a significant effect on mean trial time ($F_{1,20}$ = 125.3, p < 0.0001, η_p^2 = 0.862). Figure 1 shows that one-handed interaction (1241 MS) took more time than two-handed (1066 MS). From Fig. 1, we also observed a significant interaction effect between *I* and *CM* ($F_{6,120}$ = 141.8, p < 0.0001, η_p^2 = 0.876). For two-handed interaction, the base line performed faster than all the tilt-based scrolling; for one-handed interaction, whereas all the tilt-based performed faster than the base line: **H1** was confirmed in terms of trial time. Figures 1 and 2 show that there was a significant effect of *CM* on trial time ($F_{6,120} = 65$, $p < 0.0001$, $\eta_p^2 = 0.765$), and all the *CM*s performed differently across levels of both *I* and *D*. Figure 2 also reveals a significant interaction effect between *D* and *CM* ($F_{12,240} = 91.5$, $p < 0.0001$, $\eta_p^2 = 0.821$), and this indicated that device screen sizes affected differently across different control modes (i.e., scrolling techniques). We observed from Fig. 2 that all the scrolling techniques except the base line performed the best on the 5.8" screen. The base line performed better and better in terms of trial time with the increase of screen sizes. There was a significant effect of *D* (screen size) on trial time ($F_{2,40} = 280$, $p < 0.0001$, $\eta_p^2 = 0.933$), see Fig. 3. We can see from Fig. 3 that generally the 5.8" screen outperformed the other two screens in terms of trial time, whereas the base line performed better on larger screens. This revealed that neither small nor oversized screen was good to tilt-based scrolling techniques.

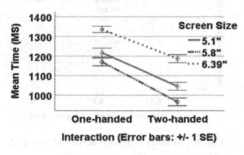

Fig. 3. Trial time across levels of *I* and *D*.   **Fig. 4.** Trial time across levels of *F* and *RP*.

Results Without the Base Line. There was a significant effect of *RP* on trial time ($F_{1,20} = 30.4$, $p < 0.0001$, $\eta_p^2 = 0.603$). We observed that the adaptive reference point outperformed the horizontal across levels of *F* (except the function Log, see Fig. 4), *I* (see Fig. 5), and *D* (see Fig. 6): **H2** was confirmed in terms of trial time. Function also significantly affected trial time ($F_{2,40} = 222.6$, $p < 0.0001$, $\eta_p^2 = 0.918$), see Fig. 4. Step function performed better in terms of trail time than Linear ($F_{1,20} = 530.4$, $p < 0.0001$, $\eta_p^2 = 0.964$) and Log ($F_{1,20} = 46.4$, $p < 0.0001$, $\eta_p^2 = 0.699$).

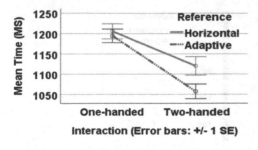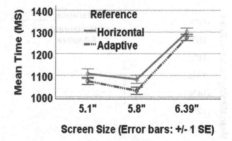

Fig. 5. Trial time across levels of *I* and *RP*.   **Fig. 6.** Trial time across levels of *D* and *RP*.

5.2   Missing Selection

Missing Selection with the Base Line. There was no significant effect of block ($F_{2,40}$ = 2.17, p > 0.1) or interaction ($F_{1,20}$ = 0.051, N.S) on error rates of missing selection. Control Mode significantly affected missing selection ($F_{6,120}$ = 9.49, p < 0.0001, η_p^2 = 0.322). We observed that the base line was more error-prone in terms of missing selection than all the tilt-based scrolling techniques across levels of I (Fig. 7) and D (Fig. 8): **H1** was confirmed in terms of over scrolling (i.e., missing selection), and **H3** was confirmed. However, error rates of missing selection were minute (<0.6%) under all experimental conditions.

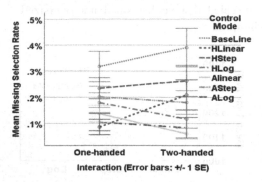

Fig. 7. Missing selection across I and CM.   Fig. 8. Missing selection across D and CM.

Missing Selection Without the Base Line. There was a significant effect of RP on error rates of missing selection ($F_{1,20}$ = 8.28, p < 0.01, η_p^2 = 0.293). We observed that the adaptive reference point outperformed the horizontal across levels of I (Fig. 9), F (Fig. 10), and D (Fig. 11): **H2** was confirmed in terms of over scrolling. There was also a significant effect of F on missing selection ($F_{2,40}$ = 9.19, p < 0.005, η_p^2 = 0.315), see Fig. 10. The error rates of Step function were significantly higher than that of Linear ($F_{1,20}$ = 13.3, p < 0.005, η_p^2 = 0.399) and Log ($F_{1,20}$ = 9.23, p < 0.01, η_p^2 = 0.316). But the error rates of the tilt-based scrolling techniques were minute (<0.25%) under all the experimental conditions.

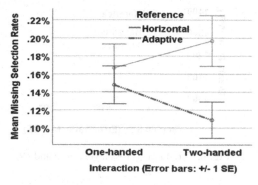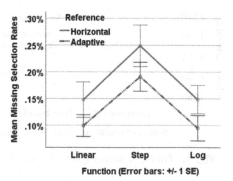

Fig. 9. Missing selection across I and RP.   Fig. 10. Missing selection across F and RP.

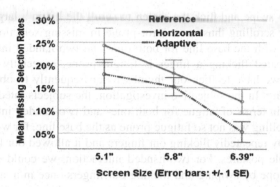

Fig. 11. Missing *selection* across *D* and *RP*.

5.3 Subjective Comments

For two-handed interaction, the subjects rated the base line the most highly in terms of usability (6.3, SD = 1.6) but the most slightly in fatigue (4.3, SD = 2.1), see Fig. 12. Among the tilt-based scrolling techniques, AStep was rated the most highly across the three subjective measures for two-handed interaction. As for one-handed interaction, the tilt-based scrolling techniques obtained higher levels than the base line across all the three subjective measures. And AStep achieved the highest levels across all the measures: usability (6.7, SD = 0.8), fatigue (5.9, SD = 1.6), and personal preference (6.7, SD = 0.6). For both one- and two-handed interaction modes, the three tilt scrolling techniques with adaptive reference points gained higher ratings than those with the horizontal reference point.

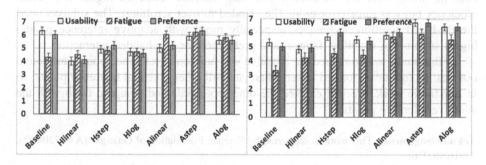

Fig. 12. Subjective comments: (left) two-handed, and (right) one-handed.

6   Discussion

For two-handed interaction, the base line outperformed the tilt-based scrolling techniques in both quantitative (speed only) and qualitative investigations. This might be partially caused by that we scroll a mobile screen by swipe and flick daily, so the subjects were familiar with the operation of the base line. In the experiment, the subjects tended to swipe and flick the screen to scroll the list in a large scale, but this easily led to over scrolling thus more error-prone in missing selection. We observed higher error rates with the base line for both one- and two-handed interaction modes. Furthermore, repeated flicking a finger is fatigue-prone, especially for one-handed interaction when we have to flick our thumbs more frequently to obtain a rapid and continuous scrolling. In the subjective investigation, the subjects rated the base line at the lowest levels in terms of fatigue for both one- and two-handed interaction modes.

Tilt-based scrolling was not so fatigue-prone as the base line, for we did not have to scroll the screen by repeatedly flicking our fingers and it allowed our fingers always to stay in comfortable postures. For two-handed interaction, we could scroll the screen once in a large scope by swiping and flicking our fingers once in a large scale, so the base line obtained higher interaction speed. But we could not swipe and flick our fingers in such a same large scope for one-handed interaction as two-handed, thus interaction efficiency of the base line for one-handed interaction was not as high as for two-handed.

Tilt scrolling techniques with adaptive reference points outperformed those with the horizontal in terms of speed and accuracy, for the former allowed the subjects to tilt and observe the screen more comfortably.

7 Conclusion and Future Work

We systematically evaluated six tilt-based scrolling techniques together with the base line. The experimental results revealed that tilt scrolling techniques were more suitable for one-handed interaction with better interaction performance and were less fatigue-prone than the base line. Techniques with adaptive reference points performed better than the those with the horizontal. The Step function achieved higher interaction efficiency but was more error-prone (i.e., tended to over scrolling). The subjects rated the tilt-based scrolling techniques (especially those with adaptive reference points) highly in terms of usability, fatigue, and preference for one-handed interaction.

In this study, we compared the adaptive and horizontal reference points and found out that the former performed better. There are still other potential reference points, so in our future work, we will explore the other potentials and compare them with the adaptive reference points.

Acknowledgement. This work is supported by Science Foundation of Guangxi (AA17204096, AD16380076).

References

1. Antoine, A., Malacria, S., Casiez, G.: ForceEdge: controlling autoscroll on both desktop and mobile computers using the force. In: Proceedings of the 2017 CHI Conference on Human Factors in Computing Systems, pp. 3281–3292. ACM, Denver (2017)
2. Bartlett, J.F.: Rock 'n' scroll is here to stay. IEEE Comput. Graph. Appl. **20**(3), 40–45 (2000)

3. Hinckley, K., Pierce, J., Sinclair, M., Horvitz, E.: Sensing techniques for mobile interaction. In: Proceedings of the 13th Annual ACM Symposium on User Interface Software and Technology, pp. 91–100. ACM, San Diego (2000)
4. Miyaki, T., Rekimoto, J.: Graspzoom: zooming and scrolling control model for single-handed mobile interaction. In: Proceedings of MobileHCI 2009. ACM, Bonn (2009)
5. Oakley, I., O'Modhrain, S.: Tilt to scroll: evaluating a motion based vibrotactile mobile interface. In: Proceedings of the First Joint Eurohaptics Conference and Symposium on Haptic Interfaces for Virtual Environment and Teleoperator Systems, pp. 40–49. IEEE Computer Society (2005)
6. Spelmezan, D., Appert, C., Chapuis, O., Pietriga, E.: Side pressure for bidirectional navigation on small devices. In: Proceedings of MobileHCI 2013, pp. 11–20. ACM, Munich (2013)
7. Roth, V., Turner, T.: Bezel swipe: conflict-free scrolling and multiple selection on mobile touch screen devices. In: Proceedings of the 27th International Conference on Human Factors in Computing Systems, pp. 1523–1526. ACM, Boston (2009)
8. Wang, J., Zhai, S., Canny, J.: Camera phone based motion sensing: interaction techniques, applications and performance study. In: Proceedings of the 19th Annual ACM Symposium on User Interface Software and Technology, pp. 101–110. ACM, Montreux (2006)
9. Fitts, P.M.: The information capacity of the human motor system in controlling the amplitude of movement. J. Exp. Psychol. 121(3), 381–391 (1954)
10. Sad, H.H., Poirier, F.: Evaluation and modeling of user performance for pointing and scrolling tasks on handheld devices using tilt sensor. In: Proceedings of the 2009 Second International Conferences on Advances in Computer-Human Interactions, pp. 295–300. IEEE Computer Society (2009)
11. MacKenzie, I.S., Teather, R.J.: Fittstilt: the application of Fitts' law to tilt-based interaction. In: Proceedings of the 7th Nordic Conference on Human-Computer Interaction: Making Sense Through Design, pp. 568–577. ACM, Copenhagen (2012)
12. Natapov, D., Castellucci, S.J., MacKenzie, I.S.: ISO 9241-9 evaluation of video game controllers. In: Proceedings of the 2009 Conference on Graphics Interface, pp. 223–230 (2009)
13. Teather, R.J., MacKenzie, I.S.: Position vs. velocity control for tilt-based interaction. In: Proceedings of Graphics Interface Conference 2014, pp. 51–58. Canadian Information Processing Society, Toronto (2014)
14. Guo, A., Paek, T.: Exploring tilt for no-touch, wrist-only interactions on smartwatches. In: Proceedings of MobileHCI 2016, pp. 17–28. ACM, Florence (2016)
15. Baglioni, M., Lecolinet, E., Guiard, Y.: Jerktilts: using accelerometers for eight-choice selection on mobile devices. In: Proceedings of the 13th International Conference on Multimodal Interfaces, pp. 121–128. ACM, Alicante (2011)
16. Rahman, M., Gustafson, S., Irani, P., Subramanian, S.: Tilt techniques: investigating the dexterity of wrist-based input. In: Proceedings of the 27th International Conference on Human Factors in Computing Systems, pp. 1943–1952. ACM, Boston (2009)
17. Sandnes, F.E.: Directional bias in scrolling tasks: a study of users' scrolling behaviour using a mobile text-entry strategy. Behav. IT 27(5), 387–393 (2008)
18. Dou, X., Sundar, S.S.: Power of the swipe: why mobile websites should add horizontal swiping to tapping, clicking, and scrolling interaction techniques. Int. J. Hum.-Comput. Interact. 32(4), 352–362 (2016)
19. Partridge, K., Chatterjee, S., Sazawal, V., Borriello, G., Want, R.: Tilttype: accelerometer-supported text entry for very small devices. In: Proceedings of the 15th Annual ACM Symposium on User Interface Software and Technology, pp. 201–204. ACM, Paris (2002)

20. Sazawal, V., Want, R., Borriello, G.: The unigesture approach one-handed text entry for small devices. In: Paternò, F. (ed.) Mobile HCI 2002. LNCS, vol. 2411, pp. 256–270. Springer, Heidelberg (2002). https://doi.org/10.1007/3-540-45756-9_20

21. Wigdor, D., Balakrishnan, R.: Tilttext: using tilt for text input to mobile phones. In: Proceedings of UIST 2003, pp. 81–90 (2003). ACM

22. Jones, E., Alexander, J., Andreou, A., Irani, P., Subramanian, S.: Gestext: accelerometer-based gestural text-entry systems. In: Proceedings of the 28th International Conference on Human Factors in Computing Systems, pp. 2173–2182. ACM, Atlanta (2010)

23. Dunlop, M.D., Roper, M., Imperatore, G.: Text entry tap accuracy and exploration of tilt controlled layered interaction on smartwatches. In: Proceedings of MobileHCI 2017, pp. 23:11–23:21. ACM, Vienna (2017)

24. Yeo, H.-S., Phang, X.-S., Castellucci, S.J., Kristensson, P.O., Quigley, A.: Investigating tilt-based gesture keyboard entry for single-handed text entry on large devices. In: Proceedings of the 2017 CHI Conference on Human Factors in Computing Systems, pp. 4194–4202. ACM, Denver (2017)

25. Sun, K., Wang, Y., Yu, C., Yan, Y., Wen, H., Shi, Y.: Float: one-handed and touch-free target selection on smartwatches. In: Proceedings of the 2017 CHI Conference on Human Factors in Computing Systems, pp. 692–704. ACM, Denver (2017)

26. Kurosawa, H., Sakamoto, D., Ono, T.: Myotilt: a target selection method for smartwatches using the tilting operation and electromyography. In: Baillie, L., Oliver, N. (eds.) Proceedings of the 20th International Conference on Human-Computer Interaction with Mobile Devices and Services, pp. 43:1–43:11. ACM, Barcelona (2018)

27. Browne, K., Anand, C.: An empirical evaluation of user interfaces for a mobile video game. Entertainment Comput. 3(1), 1–10 (2012)

28. Medryk, S., MacKenzie, I.S.: A comparison of accelerometer and touch-based input for mobile gaming. In: Proceedings of the International Conference on Multimedia and Human Computer Interaction, Toronto, Ontario, Canada, pp. 1171–1178 (2013)

29. Zaman, L., MacKenzie, I.S.: Evaluation of nano-stick, foam buttons, and other input methods for gameplay on touchscreen phones. In: International Conference on Multimedia and Human-Computer Interaction-MHCI, pp. 1–8. Citeseer (2013)

30. Eslambolchilar, P., Murray-Smith, R.: Tilt-based automatic zooming and scaling in mobile devices – a state-space implementation. In: Brewster, S., Dunlop, M. (eds.) Mobile HCI 2004. LNCS, vol. 3160, pp. 120–131. Springer, Heidelberg (2004). https://doi.org/10.1007/978-3-540-28637-0_11

31. Rekimoto, J.: Tilting operations for small screen interfaces. In: Proceedings of the 9th Annual ACM Symposium on User Interface Software and Technology, pp. 167–168. ACM, Seattle (1996)

32. Pietroszek, K., Wallace, J.R., Lank, E.: Tiltcasting: 3D interaction on large displays using a mobile device. In: Proceedings of the 28th Annual ACM Symposium on User Interface Software & Technology, pp. 57–62. ACM, Daegu (2015)

Optimizing Self-organizing Lists-on-Lists Using Pursuit-Oriented Enhanced Object Partitioning

O. Ekaba Bisong and B. John Oommen[(✉)]

School of Computer Science, Carleton University, Ottawa K1S 5B6, Canada
ekaba.bisong@carleton.ca, oommen@scs.carleton.ca

Abstract. Central to the field of Computer Science is the issue of storing, maintaining and retrieving data, and the "list" structure, maintained as a Singly-Linked-List (SLL), leads to one such Adaptive Data Structure (ADS). Recently, researchers have proposed the concept of hierarchical Singly-Linked-Lists on Singly-Linked-Lists (SLLs-on-SLLs), where the primitive elements of the primary list are, in and of themselves, sub- lists. The question of knowing which elements should be in the respective sub-lists is far from trivial, and is achieved using Learning Automata (LA). This paper demonstrates how we can incorporate one such LA, namely the Pursuit-Enhanced Object Migration Automaton (PEOMA) to improve the performance of the ADS operating in Non-stationary Environments (NSEs). The ADS is designed as a hierarchical SLL-on- SLL. The hierarchical concept employs a sub-context data structure to group together the elements that have a probabilistic dependence in the "unknown" distribution of the Environment. In this paper, we propose that the PEOMA reinforcement scheme can be powerful in learning the probabilistic distribution of the Environment to capture dependent elements within the sub-groups. The PEOMA improves on its predecessor algorithm by incorporating the Pursuit technique, which increases the likelihood of selecting superior actions with higher reward estimates by "pursuing" the currently known "best" action. The research shows that the PEOMA-enhanced SLLs-on-SLLs provide results that are an order of magnitude superior to the "de-facto" MTF and TR schemes used in such Environments with so-called "locality of reference". Also, the results surpass the performances of EOMA-enhanced hierarchical SLLs-on-SLLs schemes in NSEs including when the data-structure has some knowledge of the change in the dependence distribution of the Environment.

Keywords: Learning Automata (LA) · Object Migration Automaton (OMA) · Adaptive Data Structure (ADS) · Lists on Lists · The Pursuit Concept

1 Introduction

Self-organization allows a data structure to re-order its constituent elements in response to queries from the underlying query system, that serves as an Environment. The probability distribution of the query accesses is unknown to the data structure. The goal of this re-organization, among others, is to minimize the asymptotic cost or access-time

© Springer Nature Switzerland AG 2019
D.-S. Huang et al. (Eds.): ICIC 2019, LNAI 11645, pp. 201–212, 2019.
https://doi.org/10.1007/978-3-030-26766-7_19

of record retrieval. In the design of adaptive lists, it is assumed that the Environment will not request a record absent from the list, and that each record is retrieved at least once.

A list data structure contains a sequence of elements, R = {i: 1 ≤ i ≤ J}, and the average access-cost of the list is given as:

$$C(\Pi) = \Sigma_{1 \le i \le J} s_i \pi_i, \tag{1}$$

where,

- s_i is the unknown probability of accessing the element 'i', as defined by the probability distribution of the Environment, and
- π_i is the position of element with index i in the list.

The probability distribution of the query accesses from the Environment must be non-uniform. Otherwise, the list re-ordering will have no intrinsic worth as the average access-cost would remain unchanged [7]. In a Non-stationary Environment (NSE), the Environment is time-varying and exhibits a dependency property called "Locality of Reference" where the elements are probabilistically dependent on one another. In this work, we use the Markovian Switching Environment (MSE) and the Periodic Switching Environment (PSE) as qualifying examples of NSEs in the real-world.

The "de-facto" schemes for list re-organization in such Environments are the Move-To-Front (MTF) and the Transposition (TR) rules proposed by McCabe in [8]. The MTF rule updates the list by moving the accessed element to the head of the list, while the TR rule moves the accessed element one-step towards the list head. It has been shown that the MTF rule is characterized by a quicker convergence rate as it rapidly responds to query requests from the Environment. The TR rule, on the other hand, is more conservative in its update approach, and leads to more stable asymptotic costs. The preferability of these schemes is because they are shown to empirically out-perform other deterministic and probabilistic approaches such as the FC, TR, MRI(0) and TS(0) [2], and the time and space complexities involved in implementing other schemes render most of them impractical for real-world settings.

The hierarchical data-"sub" structure employed in this work combines the MTF and TR rules in its design as pioneered by Amer and Oommen in [1] to mitigate the rapid response of the MTF rule to queries from the Environment, as well as to improve the convergence speed of the TR rule. This idea of combining the MTF and the TR rules is, indeed, not entirely novel in designing "Adaptive" schemes in NSEs. Other researchers have also merged both these rules for precisely the same reasons. This has led to a number of composite MTF and TR strategies such as the Move-ahead-k, MHD(k) [12], where a queried element is moved k positions to the head of the list, the POS(k), where the queried element is moved to the kth position of the list if j > k, and otherwise it uses the TR rule so long the element is in positions 2 to k. Another mechanism is the SWITCH(k) rule which is similar to the POS(k) except that it applies the MTF rule when the element is in positions 2 to k. However, these schemes are not as robust as the MTF and the TR rules in NSEs possessing "locality of reference".

A key advantage of the MTF and TR hierarchical formulation is the "en masse" promotion of list elements within the same sub-list, where the elements within a sub-list context are assumed to have a probabilistic dependence. However, if such a dependence is not accurately inferred, it is far from the case that the elements within the sub-lists are probabilistically dependent. Consequently, in the absence of a learning mechanism, the hierarchical schemes end-up having an inferior performance to the MTF and TR rules in NSEs.

The Object Migration Automaton (OMA) family of reinforcement schemes from the theory of Learning Automata (LA) have been employed to capture the probabilistic dependence distribution in the Environment to modify its clustering of elements in the sub-list context. The pioneering work of Amer and Oommen in [1] utilized the OMA algorithm to capture the probabilistic dependence of the queries coming from the Environment. This resulted in the MTF-MTF- OMA, MTF-TR-OMA, TR-MTF-OMA and the TR-TR-OMA, where the third component in the triple is the reinforcement learning scheme.

However, a critical drawback of the OMA is the observed "deadlock"[1] scenario that occurs when an accessed element is swapped from one sublist to another and then back to the original sublist, thus preventing the LA from capturing the optimal dependence grouping. The Enhanced Object Migration Automaton (EOMA) introduced in [6] mitigated this "deadlock" effect by introducing conditions to avoid pointless swaps between the sublists, as well as modifying the LA's convergence criteria to include a few of the internal states.

The work by Bisong and Oommen in [4] further lowered the access cost of hierarchical SLLs-on-SLLs formulations in NSEs by augmenting the data- structure with the EOMA reinforcement algorithm. The results showed superior performances to the "de-facto" MTF and TR updates and the OMA-augmented hierarchical schemes. This work introduces the Pursuit concept from the theory of LA [14, 15] to filter the quality of query pairs from the Environment by eliminating divergent queries. The Pursuit concept is incorporated into the EOMA to result in the Pursuit EOMA (PEOMA), which has been shown to be approximately forty times faster than the vanilla OMA [13]. In this paper, the PEOMA has been used to further augment the speed and accuracy of the SLLs-on-SLLs formulations in NSEs, resulting in the MTF-MTF-PEOMA, MTF-TR-PEOMA, TR-MTF-PEOMA and the TR-TR-PEOMA adaptive data structures.

1.1   Contributions of This Paper

The novel and previously-unreported contributions of this paper include:

- Designing and implementing the PEOMA-enhanced SLLs-on-SLLs;
- Demonstrating the superiority of the PEOMA-augmented hierarchical schemes to the MTF and TR rules;
- Demonstrating the superiority of the PEOMA-augmented hierarchical schemes to the OMA-augmented schemes that pioneered the SLL-on-SLL approach;

---

[1] Although this is referred to as a "deadlock" in the literature, it could probably, be better termed as a "livelock".

- Demonstrating the superiority of the PEOMA-augmented hierarchical schemes to the EOMA-augmented hierarchical schemes;
- Showing that as the periodicity T increases in the PSE, the asymptotic cost of the PEOMA-augmented hierarchical schemes are further minimized;
- Showing that the "Periodic"/"UnPeriodic" versions of the PEOMA-augmented hierarchical schemes yield even more superior performances in PSEs.

1.2   Outline of This Paper

Section 1 provides an introduction to hierarchical SLLs-on-SLLs as the state-of- the-art scheme for minimizing retrieval costs of lists in NSEs, and introduces the case for the Pursuit-OMA augmented hierarchical schemes. Section 2 surveys the theory of LA^2 as the framework for the PEOMA algorithm. It then addresses the OMA and its drawbacks, the case for the "Enhanced"-OMA, and the Pursuit concept as it improves the EOMA partitioning algorithms. Section 3 explains the PEOMA reinforcement algorithm as it augments the SLLs-on-SLLs. Section 4 presents the results and discussions, and Sect. 5 concludes the paper.

2   Theoretical Background

2.1   Learning Automata

LA theory is a formalism for choosing the optimal action from a finite-hypothesis space of legal actions in stochastic (non-stationary) Environments. The operation of the LA can be modeled as a feedback loop between the Environment and the Automaton. The Automaton is the key mechanism for choosing the set of actions or policies based on the responses from the Environment so as to maximize the utility function. This mechanism is termed a "Learning Automaton" [4, 9, 13].

The Object Migration Automaton (OMA) improved on the Tsetlin/Krinsky strategies for partitioning objects into groups to solve the Equi-Partitioning Problem (EPP), which is a special case of the Object Partitioning Problem (OPP). The OMA rendered various real-life application domains^3 tractable by offering an improved computational cost in comparison to prior approaches like the Basic Adaptive Method (BAM) and other Hill Climbing solutions [10, 11].

2.2   OMA and Its Drawbacks

In the OMA, the number of actions represents the number of groups, or partitions, R, where each action contains a certain number of states, N. The OMA partitions the object set W into groups R by moving the abstract objects O around the action-states of

---

^{2} Due to space limitations, the background material is only briefly surveyed here. The seminal work by [9] and the theses by Shirvani [13] and the first author of this paper [3] contain exhaustive details of the theory and applications of LA.

^{3} A comprehensive list of these application domains is given in [13].

the automaton. This strategy is unlike the Tsetlin and Krinsky methods in that it moves the entire set of objects towards an action. So, for example, if object O_i is in action α_k, then O_i belongs to the kth group.

The state-action setup of the OMA is similar to Tsetlin automaton where an action α_k has N states $\{\varphi_{k1}, \varphi_{k2}, ..., \varphi_{kN}\}$ where φ_{k1} is the innermost state and φ_{kN} is the boundary state[4]. The automaton receives queries in the form of tuples, $<A_i, A_j>$. Hence, given a set of query elements, we formalize the transitions of the OMA for the abstract objects $<O_i, O_j>$ on reward and on penalty as follows:

On Reward: If the query objects $<O_i, O_j>$ are in the same class (or group, or action), the automaton is rewarded, and the objects are moved towards the innermost state of the action, φ_{k1}.

On Penalty: If the query objects $<O_i, O_j>$ are **not** in the same class, the automaton is penalized, and the objects are moved towards the boundary states φ_{Nk} of their respective classes. The movement on penalty depends on the states that the objects pairs O_i, O_j are in:

1. **If both objects, $<O_i, O_j>$ are not at boundary states:** Move O_i and O_j one step towards the boundary state of their respective actions. Formally, if $O_i \neq \varphi_{mN}$ and $O_j \neq \varphi_{kN}$, move O_i towards φ_{mN} and O_j towards φ_{kN}.
2. **If one object, O_i or O_j is at a boundary state:** In this case, either $O_i = \varphi_{mN}$ or $O_j = \varphi_{kN}$, but not both. Assuming $O_i = \varphi_{mN}$, move O_j a step towards its boundary state φ_{kN}. And if $O_j = \varphi_{kN}$, move O_i a step towards its boundary state φ_{mN}.
3. **If both objects $<O_i, O_j>$ are on boundary states:** In this case, both $O_i = \varphi_{mN}$ and $O_j = \varphi_{kN}$. Hence, there are two choices for a move and the program can execute either of them. When $O_i = \varphi_{mN}$, move O_j to the boundary state of O_i. Note that this will result in an unbalanced class for φ_{mN}. To remedy this, take the closest object to φ_{mN} that is not O_i or O_j and move it to the boundary state φ_{kN}. Also, when $O_j = \varphi_{kN}$, move O_i to the boundary state of O_j, and then take the closest object to φ_{kN} that is not O_i or O_j and move it to the boundary state φ_{mN}.

The OMA update scheme, however, results in a "deadlock" situation. The deadlock phenomenon occurs when there is a query pair $<O_i, O_j>$ in a stream of query pairs belonging to different actions, αm and αk. If one object is in the boundary state of its action, and the other is not, the query pairs are prevented from converging to their optimal groups, and can lead to an "infinite" loop.

The "Enhanced" OMA (EOMA) upgrades the OMA algorithm to mitigate the "deadlock" phenomenon which prevents the algorithm from converging to the objects' optimal partitioning. The deadlock condition is actually exacerbated when the algorithm is interacting with a near-optimal Environment (e.g., when the probability of obtaining well-paired objects is close to unity) by considerably slowing down the convergence rate even if the problem complexity is small. To mitigate this, if there

---

[4] The schematic of the OMA and the associated formal algorithms are given in [3, 5, 10, 11, 13], and omitted here in the interest of space. Those descriptions, figures and algorithms should be considered as the formal and accurate ones.

exists an object in the boundary state of the group containing O_j, the EOMA swaps O_i with the object in this boundary state. Otherwise, the update is identical to the OMA.

The EOMA also redefines the concept of the convergence condition so as to reduce the algorithm's vulnerability to divergent queries. This modification designates the two-innermost states as the "final" states, as opposed to just the innermost state in the vanilla OMA. A marginally superior solution specifies a parameter Z, to designate the Z innermost states of each action to be the convergence condition. More detailed analyses on the EOMA is found in [6, 13].

2.3   The EOMA vs. the OMA in SLLs-on-SLLs

From a naive perspective, it is easy to reckon that the effect of such a modification will be trivial. However, this is far from being true. For the first part, because of the deadlock scenario that the OMA, unfortunately, possesses, the accessed element can be swapped from one sublist to another and then back to the original sublist. Such an adverse occurrence has a further consequence because an unintended sublist could also be moved towards the front of the overall list structure. By mitigating such a deadlock scenario, the EOMA prohibits such "false alarm" swaps of elements between sublists, and also futile swaps between the various sublists themselves.

The second major issue that the EOMA takes care of involves the determination of the convergence criteria. Rather than wait "endlessly" for the accessed element to move into the most internal state of the OMA, the EOMA declares that the sublist has converged when the elements are within a few of the most internal states. This means that the EOMA is reasonably certain about the identity of the elements that should constitute a sublist, and does not permit a divergence from these sublists once it has declared that they have converged.

By making sure that both of these are in place, one can obtain a superior "en masse" reorganization of the sublist groupings that allow us to minimize the query access costs. We shall presently show that the Pursuit concept further improves on the EOMA to minimize query accesses in SLLs by filtering divergent query pairs from the Environment.

2.4   Improvements to the EOMA with the "Pursuit" Concept

Pursuit schemes employ Maximum Likelihood (ML) or Bayesian estimators to make use of estimates of the reward probabilities to update the actions. MLEs do this by keeping an accumulator that records the number of times a chosen action receives a corresponding reward. Additionally, for every iteration, the estimated reward vector probability is employed in order to update the action probability vector rather than merely use the feedback from the Environment. This strategy minimizes the likelihood of selecting actions with lower reward estimates and increases the probability of choosing actions with higher reward estimates. The Pursuit technique increases the likelihood of selecting actions with higher reward estimates by "pursuing" the currently known "best" action [13–15].

The pursuit concept has also been incorporated into the EOMA to enhance it to yield the Pursuit EOMA (PEOMA) [13]. It updates the joint query probabilities by using the ranked estimates of the reward probabilities to asymptotically choose or "pursue" better actions. In collecting estimates of query pairs to filter divergent pairs, it has a consequent effect of rendering the Environment noise-free tending to create a deadlock scenario. However, the EOMA caters for this inconvenience in its enhanced swap update schemes when a query element is at a boundary state (Fig. 1).

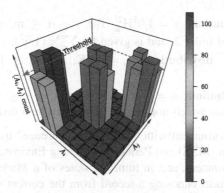

Fig. 1. This figure displays the magnitude of the elements in the Pursuit matrix showing the joint probability distribution of query pairs with a defined cut-off threshold.

The divergent pairs are filtered by collecting the probability estimates of the query accesses in a square matrix where each entry is the estimate of the joint probability access for the pair. Pairs less than a predefined threshold are considered divergent. For a more detailed discussion on the pursuit concept as it enhances the EOMA, the reader is referred to [13].

3   PEOMA-Augmented Hierarchical SLLs-on-SLLs

The hierarchical SLLs-on-SLLs involves the sub-structure formulation of dividing a list of size J into k sublists. The list outer and sub-contexts is then operated upon by either of the MTF or TR update schemes. This arrangement yields the MTF-MTF, MTF-TR, TR-MTF, and TR-TR hierarchical schemes. The PEOMA reinforcement scheme is attached to the hierarchical schemes to learn the dependent probability distribution of the query generator in the Environment. The learned information from the PEOMA is critical in updating the elements within a list's sub-context. The augmentation of hierarchical SLLs with the PEOMA reinforcement scheme gives rise to the MTF-MTF-PEOMA, MTF-TR-PEOMA, TR- MTF-PEOMA and the TR-TR-PEOMA.

The query generators we have used in the simulation of this design in NSEs are the Zipf, 80-20, Lotka, Exponential and Linear generators. For a given list of size J, divided into k sublists and each containing J/k elements, the probability distribution $\{si\}$ where $1 \leq i \leq m$ describes the query accesses for the elements in the subset k. In this way, the total probability mass for the accesses in each group is the same. The generators are:

1. **The Zipf distribution:** $s_i = 1/iH_m$, for $1 \leq i \leq m$, where H_m is the mth Harmonic number and defined as $H_m = \sum_{j=1}^{m} (1/j)$. The Zipf distribution is the most commonly-used one for modelling real-life access probabilities.

2. **The 80-20 distribution:** $s_i = 1/i^{(1-\theta)}H_m^{(1-\theta)}$, for $1 \leq i \leq m$ and $\theta = \log 0.80/ \log 0.20 \approx 0.1386$, where $H_m^{(1-\theta)}$ is the mth Harmonic number of order $(1 - \theta)$, and is given by $\sum_{j=1}^{m} (1/j^{(1-\theta)})$.

3. **The Lotka distribution:** $s_i = 1/i^2 H_m^2$, for $1 \leq i \leq m$, where H_m^2 is the mth harmonic number of order 2, and is given by $\sum_{j=1}^{m} (1/j^2)$.

4. **The Exponential distribution:** $s_i = 1/2^i K$, for $1 \leq i \leq m$, where $K = \sum_{j=1}^{m} (1/2^j)$.

5. **The Linear distribution:** $si = K(m - i+1)$, for $1 \leq i \leq m$, where K is determined as the constant which normalizes the $\{si\}$ to be a distribution.

The models of Environment with "locality of reference" used are the Markovian Switching Environment (MSE) and Periodic Switching Environment (PSE). In MSEs, the states of the Environment, are, in turn, the states of a Markov chain. If the probability of the Environment choosing a record from the current subset is 0.9, then the probability of switching to another subset is equally divided among the other subsets. In PSEs, the states of the Environment changes after every T queries from Qi to Qi + 1 mod k, where, each set of T consecutive queries belong to the same sub-context. For both models, the SLLs-on-SLLs data-structure can be made aware of the state change either directly or by observing the queries to the PEOMA when two successive queries are not in the same group.

4 Results and Discussions

We now comparatively discuss the simulations results obtained from the previous OMA-based and EOMA-based solutions, and our new solutions. The entire suite of results involved all the methods and for all the distributions listed above, and for lists for various sizes. The suite also included the comprehensive testing for various types of MSE and PSEs. Indeed, the results that we have obtained are extensive and involve all these "parameters" for numerous settings. These are given briefly in [5], but the more detailed set of results is included in the respective chapter of the thesis of the first author of this paper [3].

The experimental setup for the simulations involving the PEOMA-augmented hierarchical schemes in MSEs involved splitting a list of size 128 into k sublists with k $\in$ 2, 4, 8, 16, 32, 64. The degree of dependence of the MSE, α, was also set to 0.9 and the period in the PSE, T = 30. For all the results discussed in this section, the simulation involved an ensemble of 10 experiments, each evaluating 300,000 query accesses, and for the various aforementioned query generators. For conciseness sake, we present here only the pertinent results for k = 8.

From Table 1, observe that the performance of the PEOMA-augmented hierarchical schemes are an order of magnitude superior to the stand-alone MTF and TR rules and the EOMA-augmented schemes in MSEs when considering both their asymptotic and amortized costs. This observation is true for all the dependence generators under consideration. As an example, consider the Exponential distribution, which is one for which the MTF and TR stand-alone schemes are competitive due to its L-shaped logarithmic curve that assigns higher probabilities to a small subset of the elements in the query space. For this distribution, the EOMA-augmented hierarchical schemes found it difficult to compete with the MTF and TR rules, and was, for the most part, at-par to them.

However, with the PEOMA-augmented hierarchical schemes for the Exponential distribution, the results demonstrated that it was significantly superior to the stand-alone MTF and TR rules with asymptotic and amortized costs of the MTF-MTF-PEOMA, MTF-TR-PEOMA, TR-MTF-PEOMA and TR- TR-PEOMA being (2.45, 2.97, 2.88, 2.99) and (4.00, 4.23, 2.65, 4.77) respectively. In these cases, the MTF and TR had asymptotic and amortized costs of (8.72, 10.52) and (8.71, 10.93). Indeed, irrespective of the distribution shape that favours the MTF and TR rules, when the Pursuit concept to filter divergent queries was incorporated into the EOMA, the sublist had a superior convergence performance as it learned the Environment's dependence distribution.

For the Exponential distribution, the EOMA-enhanced hierarchical schemes which had the the outer-list context as the MTF, was superior to the stand-alone MTF and TR schemes. However, when the outer list context was the TR rule, the List-on-List performance was slightly-inferior to the MTF rule. However, in the PEOMA-enhanced versions, the schemes with TR as the outer list context were also an order of magnitude superior to the MTF and TR rules.

Table 2 showed the performance of the PEOMA-augmented hierarchical schemes in the PSEs. We observed that the schemes were superior to the MTF and TR rules for all distributions under consideration. This performance is significant given that for the

Table 1. Asymptotic (top) and Amortized (bottom) costs in MSEs with $\alpha = 0.9$ and k = 8.

Scheme	Zipf	80-20	Lotka	Exp.	Linear
MTF	43.35	43.76	39.30	8.72	43.60
TR	55.44	56.74	48.25	10.52	56.79
MTF-MTF-EOMA	19.14	19.23	18.70	12.34	19.31
MTF-TR-EOMA	27.80	27.77	27.17	16.89	28.04
TR-MTF-EOMA	18.84	18.99	18.37	12.87	18.96
TR-TR-EOMA	27.55	27.62	26.96	17.17	27.70
MTF-MTF-PEOMA	5.80	6.73	1.25	2.45	6.76
MTF-TR-PEOMA	5.35	6.21	1.25	2.97	6.77
TR-MTF-PEOMA	4.67	6.00	0.96	2.88	5.61
TR-TR-PEOMA	5.07	6.49	0.98	2.99	6.34
MTF	43.25	43.82	39.17	8.71	43.64
TR	55.85	56.96	48.66	10.93	57.26
MTF-MTF-EOMA	19.35	19.40	19.26	12.90	19.45
MTF-TR-EOMA	27.93	28.02	27.54	16.57	28.08
TR-MTF-EOMA	19.09	19.18	19.07	13.35	19.18
TR-TR-EOMA	27.72	27.80	27.25	17.10	27.87
MTF-MTF-PEOMA	6.97	7.77	2.32	4.00	7.79
MTF-TR-PEOMA	7.14	7.87	2.63	4.23	8.31
TR-MTF-PEOMA	5.95	7.13	2.04	4.65	6.71
TR-TR-PEOMA	6.84	8.05	2.29	4.77	7.91

Table 2. Asymptotic (top) and Amortized (bottom) costs in PSEs with $T = 30$ and k = 8.

Scheme	Zipf	80-20	Lotka	Exp.	Linear
MTF	49.64	50.24	44.52	8.46	50.08
TR	55.65	56.91	48.51	11.18	57.19
MTF-MTF-EOMA	14.63	14.70	14.12	8.59	14.72
MTF-TR-EOMA	25.82	25.90	25.32	13.88	25.92
TR-MTF-EOMA	14.39	14.49	13.76	8.92	14.50
TR-TR-EOMA	25.58	25.69	24.97	13.70	25.70
MTF-MTF-EOMA-P	7.16	7.24	6.66	6.14	7.26
MTF-MTF-EOMA-UP	7.69	7.78	7.19	8.90	7.79
MTF-MTF-PEOMA	11.80	11.29	10.57	4.10	10.30
MTF-TR-PEOMA	23.31	23.22	21.64	5.56	21.42
TR-MTF-PEOMA	12.00	11.04	10.21	4.32	10.04
TR-TR-PEOMA	20.94	20.11	19.28	5.56	21.17
MTF-MTF-PEOMA-P	7.13	7.21	6.53	6.10	7.21
MTF-MTF-PEOMA-UP	7.40	7.57	5.45	6.31	7.49
MTF	49.62	50.23	44.53	8.48	50.06
TR	56.09	57.28	48.91	11.58	57.60
MTF-MTF-EOMA	14.76	14.84	14.31	8.68	14.84
MTF-TR-EOMA	25.93	26.01	25.44	12.49	26.02
TR-MTF-EOMA	14.54	14.62	14.03	9.69	14.63
TR-TR-EOMA	25.71	25.80	25.12	13.11	25.80
MTF-MTF-EOMA-P	7.28	7.38	6.82	7.53	7.40
MTF-MTF-EOMA-UP	7.86	7.95	7.48	10.57	7.92
MTF-MTF-PEOMA	12.44	11.35	10.81	4.88	10.37
MTF-TR-PEOMA	23.84	23.78	21.64	5.90	21.63
TR-MTF-PEOMA	12.59	12.13	10.41	5.25	10.13
TR-TR-PEOMA	17.33	17.05	19.32	6.32	21.22
MTF-MTF-PEOMA-P	7.27	7.34	6.75	7.07	7.36
MTF-MTF-PEOMA-UP	7.44	7.68	5.51	6.90	7.59

EOMA-augmented variant, the MTF and TR rules were competitive in the Exponential distribution. But this was not the case when divergent queries were filtered with the Pursuit matrix. That being said, we still see that for a hierarchical scheme augmented with an OMA-based strategy, when the outer-list context is the MTF, the performance was superior to the analogous scheme having the outer-list context as the TR.

As an example to highlight this observation in the Exponential distribution, the asymptotic and amortized costs for the MTF-MTF-PEOMA, MTF- TR-PEOMA, TR-MTF-PEOMA and TR-TR-PEOMA were (4.10, 5.56, 4.32, 5.56) and (4.88, 5.90, 5.25, 6.32) respectively. The corresponding costs for the MTF and TR were (8.46, 11.18) and (8.48, 11.58) respectively.

Figure 2 provides a more holistic view of the results, across different sub-list partitions from k = 2 to k = 64, and for all the Environment distributions un- der consideration in the MSE by displaying their performances as the ratio of the PEOMA-augmented hierarchical schemes to the MTF. When the asymptotic cost ratio is greater than unity, the graph indicates that the MTF results were superior to the corresponding hierarchical PEOMA schemes, otherwise the hierarchical PEOMA scheme is superior to the MTF rule, which we see is true.

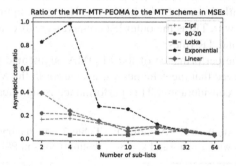

Ratio of the MTF-MTF-PEOMA to the MTF scheme in MSEs

Fig. 2. The asymptotic cost ratio of the MTF-MTF-PEOMA to the MTF scheme for different values of the sub-list partitions k ∈ {2, 4, 8, 10, 16, 32, 64} for MSE-dependent query Environments in which $\alpha = 0.9$.

In considering the effects of the Periodic (i.e. *"unperioidc"* and *"periodic"*) variations with the PEOMA-augmented SLLs-on-SLLs in the PSEs, we see that their performances were generally superior to their vanilla versions and, indeed, to the MTF and TR stand-alone rules. This result is further illustrated in Fig. 3, which shows the performances for Periodic variations of the MTF-MTF-PEOMA data-structure in the Zipf distribution for k = {k: 2, 4, 8, 10, 16, 32, 64}.

Figure 4 displays the performance of the PEOMA-augmented hierarchical schemes as the degree of dependence in the MSE goes from a noisy Environment with $\alpha = 0.1$ to a less-noisy Environment with $\alpha = 0.9$. Here, we see that the performance of the PEOMA-augmented hierarchical schemes became increasingly superior to the MTF and TR rules when $\alpha > 0.2$. This performance was a marked improvement to the EOMA-augmented

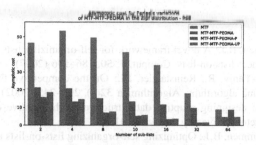

Fig. 3. Asymptotic cost of Periodic variations of MTF-MTF-PEOMA in the Zipf distribution. PSE with period T = 30 and k = {k: 2,4,8,10,16,32,64}.

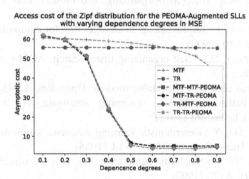

Fig. 4. Changes in the asymptotic cost of the stand-alone and hierarchical schemes with PEOMA in the MSE as the parameter α goes from 0.1 to 0.9.

hierarchical schemes whose performance was superior to the MTF when $\alpha > 0.6$. This further proves the robustness of the PEOMA in creating noise-free Environments by eliminating divergent query pairs while avoiding the "deadlock" scenario.

5 Conclusion

In this paper we have considered hierarchical Singly-Linked-Lists on Singly- Linked-Lists (SLLs-on-SLLs), where the primitive elements of the primary list are sub-lists. The question of knowing which elements should be in the respective sub-lists was achieved using Learning Automata (LA). This paper showed how we can incorporate one such LA, i.e., the PEOMA to improve the performance of the ADS operating in Non-stationary Environments. The results for the PEOMA-augmented hierarchical schemes showed an order of magnitude superior performance in the MSE Environments when compared to the stand- alone MTF and TR rules. In the PSE Environment, the PEOMA-augmented hierarchical schemes also boasted superior performances. When the knowledge of the query periods were incorporated into the PEOMA-augmented schemes, the results were further improved to minimize the asymptotic/ amortized costs.

212 O. E. Bisong and B. J. Oommen

References

1. Amer, A., Oommen, B.J.: A novel framework for self-organizing lists in environments with locality of reference: lists-on-lists. Comput. J. **50**, 186–196 (2007)
2. Bachrach, R., El-Yaniv, R., Reinstadtler, M.: On the competitive theory and practice of online list accessing algorithms. Algorithmica **32**(2), 201–245 (2002)
3. Bisong, E.O.: On designing adaptive data structures with adaptive data "sub"-structures. MCS thesis, Carleton University, Ottawa (2018)
4. Bisong, O.E., Oommen, B.J.: Optimizing self-organizing lists-on-lists using enhanced object partitioning. In: MacIntyre, J., Maglogiannis, I., Iliadis, L., Pimenidis, E. (eds.) AIAI 2019. IAICT, vol. 559, pp. 451–463. Springer, Cham (2019). https://doi.org/10.1007/978-3-030-19823-7_38
5. Bisong, E.O., Oommen, B.J.: Self-organizing lists-on-lists that invoke Pursuit-based deadlock-free object partitioning. Unabridged version of this paper
6. Gale, W., Das, S., Yu, C.T.: Improvements to an algorithm for equipartitioning. IEEE Trans. Comput. **39**(5), 706–710 (1990)
7. Hester, J.H., Hirschberg, D.S.: Self-organizing linear search. ACM Comput. Surv. (CSUR) **17**(3), 295–311 (1985)
8. McCabe, J.: On serial files with relocatable records. Oper. Res. **13**(4), 609–618 (1965)
9. Narendra, K.S., Thathachar, M.A.L.: Learning Automata: An Introduction. Courier Corporation, North Chelmsford (2012)
10. Oommen, B.J., Ma, D.C.Y.: Deterministic learning automata solutions to the equipartitioning problem. IEEE Trans. Comput. **37**(1), 2–14 (1988)
11. Oommen, B.J., Ma, D.C.Y.: Stochastic automata solutions to the object partioning problem. Comput. J. **35**, A105–A120 (1992)
12. Rivest, R.: On self-organizing sequential search heuristics. Commun. ACM **19**(2), 63–67 (1976)
13. Shirvani, A.: Novel solutions and applications of the object partitioning problem. Ph.D. thesis, Carleton University, Ottawa (2018)
14. Thathachar, M.A.L., Sastry, P.S.: A new approach to the design of reinforcement schemes for learning automata. IEEE Trans. Syst. Man Cybern. **SMC-15**(1), 168–175 (1985)
15. Zhang, X., Granmo, O.-C., Oommen, B.J., Jiao, L.: A formal proof of the e-optimality of absorbing continuous pursuit algorithms using the theory of regular functions. Appl. Intell. **41**, 974–985 (2014)

Research on Location Prediction of Moving Objects Based on Grey Markov Model

Yanxia Yang$^{(\boxtimes)}$ and Bowen Ke

Faculty of Information Engineering,
City College Wuhan University of Science and Technology,
Wuhan 430083, China
yxy_job@163.com, daisyeah@163.com

Abstract. In view of the poor fitting of the traditional GM (1,1) model in the long-term prediction and its failure in data series with high volatility, a grey Markov prediction model is established by combining grey theory with Markov theory. The location of moving objects is predicted by this model, and more accurate prediction results are obtained. Firstly, the GM (1,1) model is constructed by abstracting the problem into a single variable grey system. Then, at one-minute interval, some GPS trajectory data in GeoLife are extracted and substituted into the prediction model, and the model is tested according to the results. Finally, the Markov model is introduced into GM (1,1) model for further experiments. Experiments show that compared with the traditional GM (1,1) model, the grey Markov prediction model has less error in the location prediction of moving objects and improves the accuracy of the prediction.

Keywords: Position prediction · GM (1,1) model · Grey Markov model

1 Introduction

Location prediction is an important branch of trajectory data mining. Location prediction can solve a series of problems such as location-based information recommendation, urban road traffic optimization, and national population migration analysis.

The problem of moving object location prediction is abstracted into a single variable grey system [1] to construct Gray Model (GM 1,1) model [2]. The advantage of the model is that the amount of data needed is small and the short-term prediction accuracy is high. The disadvantage is that the fitting of long-term prediction and volatile data series is poor. On the contrary, Markov model can better predict the long-term and large random volatility data series [3]. This paper tries to introduce Markov model into grey model and construct grey Markov model [4]. It can reveal the general trend of the development and change of moving position data series by using GM (1,1) model, and determine the state rule by using Markov model, so as to enhance the accuracy of prediction.

© Springer Nature Switzerland AG 2019
D.-S. Huang et al. (Eds.): ICIC 2019, LNAI 11645, pp. 213–224, 2019.
https://doi.org/10.1007/978-3-030-26766-7_20

2 GM(1,1) Model

The grey system theory was put forward by Deng Julong, a famous Chinese scholar, in 1982 [5]. It is an innovative method to study the uncertainty of minority data and poor information. It takes the "small sample" and "poor information" uncertain system with "partial information known" as the research object, and extracts valuable information through the generation and development of "partial information known" to achieve the correct description and effective monitoring of the system's operation behavior and evolution law.

All grey series can be generated by some way to weaken its randomness and show its regularity. The common methods of data generation are cumulative generation, cumulative generation and weighted cumulative generation.

Set the original data as a list:

$$X^{(0)} = (x^{(0)}(1), x^{(0)}(2), \ldots, x^{(0)}(n))$$

Add it up once to get new data listed:

$$X^{(1)} = (x^{(1)}(1), x^{(1)}(2), \ldots, x^{(1)}(n)) \tag{1}$$

In the formula (1),

$$x^{(1)}(k) = \sum_{i=1}^{k} x^{(0)}(i), k = 1, 2, \ldots, n$$

Establishment of Linear Differential Equation for $X^{(1)}(t)$ Based on Grey Theory

$$\frac{dX^{(1)}(t)}{dt} + ax^{(1)} = u \tag{2}$$

In the formula (2), a and u are coefficients to be solved, which are called development coefficient and grey action quantity respectively.

This equation to initial conditions: when $t = t_0$, $x^{(1)}(t) = x^{(1)}(t_0)$

Solve the differential equation is $X^{(1)}(t) = (X^{(0)}(t_0) - \frac{u}{a})e^{-a(t-t_0)} + \frac{u}{a}$

Discrete values due to equidistant sampling $t_0 = 1$, so

$$X^{(1)}(t) = (X^{(0)}(1) - \frac{u}{a})e^{-a(t-1)} + \frac{u}{a} \tag{3}$$

In the formula (3), $X^{(1)}(1) = X^{(0)}(1)$ is the initial value.

Solve parameters a, u by least square method, expressed as

$$\begin{pmatrix} a \\ u \end{pmatrix} = (BB^T)^{-1}B^T Y \tag{4}$$

In the formula (4) Y and B are respectively

$$Y = (x^{(0)}(2), x^{(0)}(3), \ldots, x^{(0)}(n))^T$$

$$B = \begin{bmatrix} -\frac{1}{2}(x^{(1)}(2) + x^{(1)}(1) & -\frac{1}{2}(x^{(1)}(3) + x^{(1)}(2) & \cdots & -\frac{1}{2}(x^{(1)}(n) + x^{(1)}(n-1) \\ 1 & 1 & \cdots & 1 \end{bmatrix}$$

According to GM(1,1) model, the predicted value of one-time cumulative production is

$$\widehat{X}^{(1)}(t+1) = (x^{(0)}(1) - \frac{u}{a})e^{-at} + \frac{u}{a} \quad (t = 1, 2, \ldots, n) \tag{5}$$

If formula (5) is reduced, the predicted value of the original sequence is

$$\widehat{X}^{(0)}(t+1) = \widehat{X}^{(1)}(t+1) - \widehat{X}^{(0)}(t).$$

3 Application of GM(1,1) Model

3.1 Data Selection

Taking GeoLife user's GPS trajectory data as the research object, and taking one minute as the interval, a grey model is established based on a user's trajectory data of a certain day in 2008, and the accuracy of the model is evaluated.

Table 1. Some GPS trajectory data of a user on a certain date in 2008

Time	04:08:07	04:09:07	04:10:07	04:11:07	04:12:07	04:13:07
Longitude	116.286798	116.299569	116.299405	116.299625	116.300151	116.302055
Latitude	39.995777	39.984421	39.984499	39.984671	39.984789	39.984953
Time	04:14:07	04:15:07	04:16:07	04:17:07	04:18:07	04:19:07
Longitude	116.30426	116.306363	116.308579	116.309941	116.31007	116.309793
Latitude	39.985038	39.985127	39.985218	39.98584	39.987325	39.988901

Fig. 1. Moving trajectory diagram

Label the latitude and longitude on the map and make the moving path of the object from 04:08:07 to 04:19:07 during this period as shown in Fig. 1.

3.2   GM(1,1) Grey Model

The GM (1,1) model is established from the longitude and latitude data of user's moving trajectory in Table 1. The grey cumulative sequence of longitude $\widehat{x}_l^{(1)}$ is obtained as follows:

$$\widehat{x}_l^{(1)}(t+1) = 10198273.159152e^{0.0000114036t} - 10198156.872354$$

Sequence $\widehat{x}_l^{(1)}$ is reduced and $\widehat{x}_l^{(0)}$ is obtained:

$$\widehat{x}_l^{(0)}(t+1) = 116.296570e^{0.0000114036t}$$

Similarly, the grey cumulative sequence of latitudes is obtained as follows:

$$\widehat{x}_r^{(1)}(t+1) = 4598537.625423e^{0.000008694854t} - 4598497.629646$$

Sequence $\widehat{x}_r^{(1)}$ is reduced and $\widehat{x}_r^{(0)}$ is obtained:

$$\widehat{x}_r^{(0)}(t+1) = 39.983439e^{0.000008694854t}.$$

3.3   Forecast Results and Model Testing

The predicted longitude and latitude of the object in the next five minutes are shown in Table 2.

Table 2. Prediction value of GM(1,1) model for GPS moving trajectory data

Time	04:20:07	04:21:07	04:22:07	04:23:07	04:24:07
Original longitude	116.310258	116.310088	116.311312	116.313681	116.31579
Predicted longitude	116.312486	116.313812	116.315139	116.316465	116.317791
Original latitude	39.990325	39.990689	39.990969	39.991035	39.991191
Forecast latitude	39.987611	39.987959	39.988307	39.988654	39.989002

The comparison between the original data column and the GM (1,1) predicted data column is shown in Figs. 2 and 3.

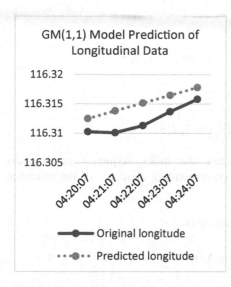

GM(1,1) Model Prediction of
Longitudinal Data

───◆─── Original longitude

···◆··· Predicted longitude

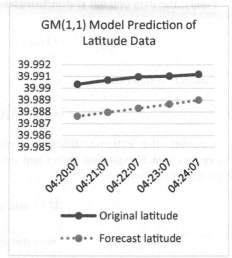

GM(1,1) Model Prediction of
Latitude Data

───◆─── Original latitude

···◆··· Forecast latitude

Fig. 2. Comparisons between actual data columns and predicted data columns

Fig. 3. Comparisons between actual data columns and predicted data columns

The path trajectory of predicted data and actual data on the map is shown in Fig. 4. The green marker in the map represents the actual geographical location, and the red marker represents the geographical location predicted by GM (1,1) model.

Fig. 4. Prediction data and actual data moving trajectory diagram (Color figure online)

There are three main methods to test the grey model: grey correlation degree, relative residual test and posterior error test. This paper adopts the first method. Grey relational degree analysis is a branch of grey system theory. Its basic idea is to judge whether a series of curves are closely related according to the similarity of their geometric shapes. The closer the curve is, the greater the correlation between the corresponding sequences, and vice versa.

Firstly, the data column is dimensionless:

$$X_i' = \frac{X_i}{x_i(1)} = (x_i'(1), x_i'(2), \ldots, x_i'(n)), i = 1, 2, 3, \ldots$$

Calculate difference:

$$\Delta_i(t) = |X_i'(t) - X_1'(t)|, i = 1, 2, 3, \ldots$$

Calculate the extreme difference between the index sequence (comparison sequence) of each evaluated object and the corresponding elements of the reference sequence one by one.

$$M = \max_i \max_t \Delta_i(t)$$

$$m = \min_i \min_t \Delta_i(t)$$

Calculate correlation coefficient:

$$\gamma_{1i}(t) = \frac{m + \xi M}{\Delta_i(t) + \xi M}, i = 2, 3, 4 \ldots \tag{7}$$

In the formula (7) ξ is called resolution coefficient. The smaller ξ is, the greater the resolution is. The value of ξ depends on the situation. Generally, the range of value is (0,1). At that time $p \leq 0.5643$ had the best resolution, and usually took $\xi = 0.5$.

Finally, the correlation degree is calculated.

$$\gamma_{1n} = \frac{1}{i} \sum_{k=1}^{n} \varepsilon_i(k), i = 1, 2, 3$$

The absolute residual sequence $e(t)$ and relative residual sequence $\varepsilon(t)$ of the predicted sequence $\hat{x}^{(0)}$ and the original sequence $X^{(0)}$ of the computational model are as follows:

$$e(t) = x^{(0)}(t) - \hat{x}^{(0)}(t), t = 1, 2, 3 \ldots$$

$$\varepsilon(t) = \left| \frac{x^{(0)}(t) - \hat{x}^{(0)}(t)}{x^{(0)}(t)} \right|, t = 1, 2, 3 \ldots$$

In order to test the reliability of grey model, a posterior error test is also needed. The variances of original data column $X^{(0)}$ and residual sequence $e(t)$ are S_1^2 and S_2^2:

$$S_1^2 = \frac{1}{n} \sum_{t=1}^{n} (x^{(0)}(t) - \bar{x})^2 \tag{8}$$

$$S_2^2 = \frac{1}{n} \sum_{t=1}^{n} (e(t) - \bar{e})^2 \tag{9}$$

Among (7) and (8),

$$\bar{x} = \frac{1}{n} \sum_{t=1}^{n} x^{(0)}(t), \bar{e} = \frac{1}{n} \sum_{t=1}^{n} e(t)$$

Calculate the posterior variance ratio:

$$C = \frac{S_2}{S_1} \tag{10}$$

The smaller index C is, the better it is. The smaller index C is, the larger index S_1 is and the smaller index S_2 is. S_1 large means that the variance of original data is large, that is, the degree of dispersion of original data is large, S_2 small means that the degree of dispersion of residual variance is small, and C small means that although the original data column is very discrete, the difference between the calculated value and the actual value of the model is not too discrete, and the credibility of the model is high. The results of model test are shown in Table 3.

Table 3. Precision test of GM(1,1) model

	Correlation degree γ	Relative residuals ε	Posterior variance ratio C
Longitude	0.66	0.0006	0.09
Latitude	0.66	0.0003	0.14

Because of the posterior difference ratio $C < 0.35$ and the correlation degree $\gamma > 0.6$, the model is reasonable, and because of the relative residual $\varepsilon < 0.01$, the model is believed to be highly consistent and credible.

4 Grey Markov Model and Its Application

Random processes with Markov property are called Markov processes. Markov property refers to the condition distribution of the state of a process after time t0 is independent of the state of the process before time t0 under the condition that the state of the process at time t0 is known. That is to say, the "future" state of the process has nothing to do with the "past" state.

4.1 Determining State Transition Probability Matrix

After dividing the object $E_i (i = 1,2,...,n)$ into equal states, the probability of transforming the p_{ij} representation data from state E_i to state Ej is defined, and the one-step state transition probability matrix is obtained as follows [6]:

$$R = \begin{pmatrix} p_{11} & \cdots & p_{1j} \\ \vdots & \ddots & \vdots \\ p_{j1} & \cdots & p_{jj} \end{pmatrix}$$

4.2   Dividing Prediction State

Because the values of the residual sequence $e(t)$ data columns are too small, they are all enlarged by 10,000 times. The state of the residual sequence $e(t) \in (0.40, 16.7)$ is divided into five states, as shown in Table 4.

Table 4. State division of residuals

State	Range
E_1	$(-\infty, 2.0)$
E_2	$[2.0, 6.0)$
E_3	$[6.0, 10.0)$
E_4	$[10.0, 14.0)$
E_5	$[14.0, +\infty)$

The state of longitude prediction values at each time point in the original data column can be expressed as shown in Table 5.

Table 5. The longitudinal state at each time point

Time	04:08:07	04:09:07	04:10:07	04:11:07	04:12:07	04:13:07
Residual	0	16.72	1.81	9.24	17.24	11.46
State	E_1	E_5	E_1	E_3	E_5	E_4
Time	04:14:07	04:15:07	04:16:07	04:17:07	04:18:07	04:19:07
Residual	2.68	5.08	13.98	14.33	2.36	13.66
State	E_2	E_2	E_4	E_5	E_2	E_4

The following is a detailed description of the method of finding matrix R. In the first row case, the first row of R is E_1 state. From Table 5, we can see that two points are in E_1 state. After one step of transfer, one point is in E_5 state and the other point is in E_3 state. Therefore, the first row is 0, 0, 1/2, 0, 1/2. By analogy, we can get that the matrix R is

$$R = \begin{pmatrix} 0 & 1/2 & 0 & 1/2 & 0 \\ 0 & 1/3 & 0 & 2/3 & 0 \\ 0 & 0 & 0 & 0 & 1 \\ 0 & 1/2 & 0 & 0 & 1/2 \\ 1/3 & 1/3 & 0 & 1/3 & 0 \end{pmatrix}$$

04:19:07 is in E_4 state, if its initial vector is $V_0 = (0,0,0,1,0)$, $V_0 = (0,0,0,1,0)$ will be obtained, so 04:20:07 will be in E_2 or E_5 state, so it may be assumed that it is in E_5 state.

$V_2 = V_0 \bullet R^2 = (1/6, 1/3, 0, 1/2, 0)$, it can be predicted that 04:21:07 is in E_4 state.
$V_3 = V_0 \bullet R^3 = (0, 4/9, 0, 11/36, 1/4)$, it can be predicted that 04:22:07 is in E_2 state.
$V_4 = V_0 \bullet R^4 = (1/12, 83/216, 0, 41/108, 11/72)$, it can be predicted that 04:23:07 is in E_2 state.
$V_5 = V_0 \bullet R^5 = (11/216, 133/324, 0, 113/324, 41/216)$, it can be predicted that 04:24:07 is in E_2 state.

As can be seen from Fig. 2, the longitude prediction value of the grey model is gradually larger, so the original prediction result is modified by Markov, as shown in Table 6.

Table 6. Longitudinal revised value of Grey Markov model

Time	Actual longitude	GM(1,1) predictive value	Predictive state	Markov Revise
04:20:07	116.310258	116.312486	E_5	116.311686
04:21:07	116.310088	116.313812	E_4	116.311412
04:22:07	116.311312	116.315139	E_2	116.314339
04:23:07	116.313681	116.316465	E_2	116.315665
04:24:07	116.31579	116.317791	E_2	116.316991

Similarly, as can be seen from Fig. 2, the latitude prediction value of the grey model is getting smaller and smaller, so the latitude prediction value is corrected by Markov correction as shown in Table 7.

Table 7. Latitude revised value of Grey Markov model

Time	Actual latitude	GM(1,1) predictive value	Predictive state	Markov revise
04:20:07	39.990325	39.987611	E_3	39.989211
04:21:07	39.990689	39.987959	E_2	39.988759
04:22:07	39.990969	39.988307	E_3	39.989907
04:23:07	39.991035	39.988654	E_3	39.990254
04:24:07	39.991191	39.989002	E_2	39.989802

The longitude and latitude values corrected by Markov are compared with the actual data values as shown in Figs. 5 and 6:

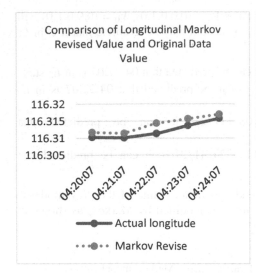

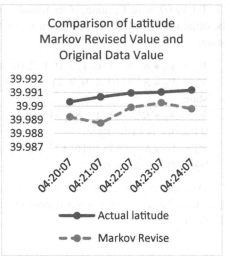

Fig. 5. Comparison of Markov corrected value and original data value

Fig. 6. Comparison of Markov corrected value and original data value

The grey model predicts longitude and latitude, and the revised longitude and latitude are compared with the actual path as shown in Fig. 6. The Brown markers in the graph indicate the location of the predicted data from GM (1,1) after Markov correction.

Fig. 7. Prediction, revision and comparison of actual moving paths

As shown in Fig. 7, the accuracy of GM (1,1) model is significantly improved after Markov correction, and the prediction results are closer to the actual values. Thus, the accuracy and reliability of grey Markov model are better than GM (1,1) model.

5   Comparison Experiment

The author chooses three models which are Linear Regression, Random Gradient Decline Linear Regression (SGD) and ARIMA to compare with the same type of research. Using the same trajectory data in Table 1 above, the data of the next 12 time points are predicted, and the predicted results are compared with the real data. The experimental results are shown in Figs. 8, 9 and 10. The blue line in the figure is the real data, and the orange line is the predicted result data. It is easy to see from the test results that the effects of these models are not very satisfactory.

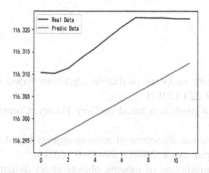

Fig. 8. Comparison chart of linear regression prediction results (Color figure online)

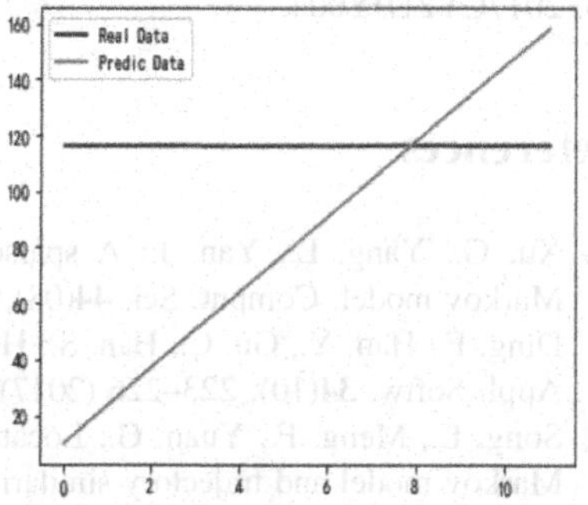

Fig. 9. Comparison chart of random gradient decline linear regression prediction results (Color figure online)

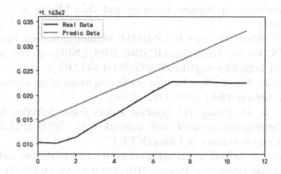

Fig. 10. ARIMA model prediction result comparison chart (Color figure online)

6   Conclusion

On the basis of studying grey model and Markov model, the changing trend of moving object position is analyzed emphatically, and the grey Markov location prediction algorithm is proposed. This algorithm not only applies the grey model to the research of location prediction, but also uses Markov model to modify the model. Compared with the classical GM (1,1) location prediction algorithm, this algorithm has better performance.

In summary, the application of grey Markov model in mobile location prediction has high accuracy, simple algorithm, strong portability and stronger practicability.

Acknowledgment. The work in this paper is partially supported by the National Innovation and Entrepreneurship Training Program for College Students in 2017 under Grant No. 201713235003 and in part supported by the Project of College-level Teaching and Research Fund of City College Wuhan University of Science and Technology in 2017 under Grant No. 2017CYZDJY004.

References

1. Xu, G., Yang, L., Yan, J.: A sparse trajectory endpoint prediction algorithm based on Markov model. Comput. Sci. **44**(08), 193–197,224 (2017)
2. Ding, F., Han, Y., Gu, C., Han, S.: Human gait prediction based on Grey Theory. Comput. Appl. Softw. **34**(10), 223–226 (2017)
3. Song, L., Meng, F., Yuan, G.: Location prediction algorithm of moving objects based on Markov model and trajectory similarity. Comput. Appl. **36**(01), 39–43+65 (2016)
4. Li, W.: Research on the method of location prediction of moving objects under dynamic relationship. China University of Mining and Technology (2017)
5. Li, L.: Research on Mobile Object Travel Peer Mining and Location Prediction. Yangzhou University (2014)
6. Yang, D.: Research on location prediction method based on hybrid multi-step Markov model. Northeast University (2014)
7. Morzy, M.: Mining frequent trajectories of moving objects for location prediction. In: 2018 International Workshop on Machine Learning and Data Mining in Pattern Recognition, pp. 667–680 (2018)
8. Huang, J., Wu, F., Meng, W., Yao, J.: LP-HMM: location preference-based hidden Markov model. In: Sun, S., Fu, M., Xu, L. (eds.) ICSINC 2018. LNEE, vol. 550, pp. 3–12. Springer, Singapore (2019). https://doi.org/10.1007/978-981-13-7123-3_1
9. Wang, H., Yang, Z., Shi, Y.: Next location prediction based on an adaboost-Markov model of mobile users. Sensors **19**(6), 1475–1493 (2019)
10. Fu, W., Li, X., Ji, H., Zhang, H.: Optimal access points selection based on mobility prediction in heterogeneous small cell networks. In: 2018 IEEE/CIC International Conference on Communications in China (ICCC)
11. Singh, J.: Tracking of moving object using centroid based prediction and boundary tracing scheme. Int. J. Image Graph. Sig. Process. (IJIGSP) **9**(8), 59–66 (2017)
12. Yuan, G., Sun, P., Zhao, J., Li, D., Wang, C.: A review of moving object trajectory clustering algorithms. Artif. Intell. Rev. **47**(1), 123–144 (2017)

Mildew Prediction Model of Warehousing Tobacco Based on Particle Swarm Optimization and BP Neural Network

Zhixia Ye[1,2], Lijun Yun[1,2(✉)], Huijun Li[3], Jingchao Zhang[1], and Yibo Wang[1]

[1] Information School, Yunnan Normal University,
Kunming 650500, Yunnan, China
yezx_leaf@aliyun.com, yunlj@163.com
[2] Yunnan Key Laboratory of Optoelectronic Information Technology,
Kunming 650500, Yunnan, China
[3] YCIH NO.2 Construction Co. Ltd., Kunming 650203, Yunnan, China

Abstract. In order to reduce the economic loss caused by mildew of warehousing tobacco, in this paper, particle swarm optimization algorithm is introduced into BP neural network mode. Particle swarm optimization is used to dynamically adjust the initial weights and thresholds of BP neural network. PSO-BP neural network prediction model is established to predict mildew of warehousing tobacco. Simulation experiment results show that the PSO-BP neural network model proposed in this paper is compared with the traditional BP neural network model. The prediction accuracy of warehousing tobacco mildew is higher. The effectiveness of the algorithm is verified.

Keywords: Particle swarm optimization · BP neural network · PSO-BP · Warehousing tobacco mildew

1 Introduction

Tobacco warehouse plays an important role in cigarette industry. The quality of tobacco leaves directly affects the quality of cigarette products. The huge economic losses caused by tobacco mildew caused by poor storage are the main problems facing the tobacco industry. Tobacco companies and major research institutes are working on how to reduce the damage caused by moldy leaves. At present, there are two main ways to prevent tobacco mildew. One is to completely prevent the mildew from being completely the moldy microorganisms of tobacco leaves by the physical, chemical or biological methods. The other is to use the Internet of Things technology to automatically monitor the storage environment of tobacco leaves to make early warning of mildew in advance. Environmental temperature, humidity and moisture content of

Key Project of Yunnan Applied Basic Research Program (grant No. 2018FA033)National Science Foundation of China (grant No. 61865015)

D.-S. Huang et al. (Eds.): ICIC 2019, LNAI 11645, pp. 225–234, 2019.
https://doi.org/10.1007/978-3-030-26766-7_21

tobacco leaves are the key factors causing tobacco mildew [1]. Zhang [2] selected the temperature and humidity of the storage environment and the water content of the tobacco leaf as the input layer parameters of the neural network, tobacco leaf mildew rate as output layer. A BP (Back-Propagation Network) neural network was established to predict tobacco mildew, realized the function of real-time prediction of tobacco leaf mold. The BP neural network model is simple, however, due to the initialization weights and thresholds of BP neural network are generated randomly, BP neural network is easy to fall into local minimum, and the global search ability is poor, resulting in low accuracy of prediction [3]. Particle Swarm Optimization (PSO) is a heuristic search technique with simple implementation, strong global search capability and superior performance, which can optimize artificial neural network algorithm [4]. In this paper, PSO algorithm is combined with BP neural network, the PSO-BP neural network model was constructed, to avoid BP neural network falling into local minimum, PSO algorithm is used to dynamically adjust the weights and thresholds of BP neural network. The model is used to predict tobacco leaf mildew to improve the accuracy of prediction.

2   PSO-BP Neural Network Model

PSO-BP neural network model combines PSO algorithm with BP neural network. Firstly, BP neural network model is constructed. Then the PSO algorithm is used to optimize the parameters of the BP neural network. A PSO-BP neural network prediction model is established.

2.1   PSO-BP Neural Network Structure

BP neural network is a widely used neural network model, which consists of input layer, hidden layer and output layer [5]. In this model, the input layer selects three main factors that affect the mildew rate of the tobacco, the ambient temperature, humidity and the water content of the tobacco leaf itself, the hidden layer uses only one layer, and the tobacco leaf mildew rate is used as the output. Therefore, the number of nodes in the input layer is 3, and the number of nodes in the output layer is 1. The structure of the PSO-BP neural network is shown in Fig. 1.

In the BP neural network prediction model, the number of nodes in the input layer, the number of hidden layers and the number of nodes in the hidden layer are the most important parameters affecting the network prediction ability [6]. In this model, the number of nodes in the input layer is 3, and the number of hidden layers is 1, therefore, the influence of the number of nodes in the hidden layer on the prediction accuracy is mainly analyzed. The number of nodes in the hidden layer is directly related to the number of nodes in the input and output layers and the setting of the expected error. Too many nodes will lead to a large amount of computation and easy over-fitting.

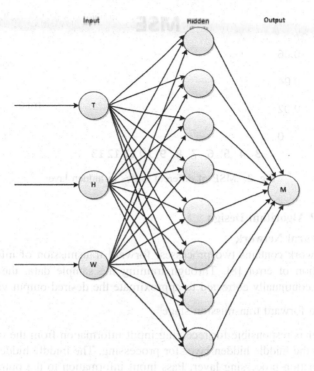

Fig. 1. Structure diagram of PSO-BP

The number of nodes is too small to achieve the desired results. Let the number of input layer nodes be n, the number of output layer nodes be m, and "a" be a constant between 1 to 10. Then the selection of the number of hidden layer nodes is generally calculated according to the following formula:

$$l = \sqrt{m+n} + a \tag{1}$$

According to the formula (1), the number of nodes in the hidden layer is 3–13, then the same tobacco data set is selected, and the same training times are performed, and the result of the mean square error (MSE) under different number nodes in the hidden layer is obtained as shown in Fig. 2.

It can be seen from the experimental results that when the number of hidden layer nodes is 8, the network performance is the best, so the number of hidden layer nodes is determined to be 8.

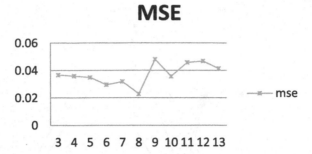

Fig. 2. MSE of different nodes in hidden layer

2.2 PSO-BP Algorithm Design

2.2.1 BP Neural Network

BP neural network contains two processes: forward transmission of information and back propagation of error [8]. Through training the sample data, the weights and thresholds are continually corrected to approximate the desired output value.

- Information forward transmission stage

The input layer is responsible for receiving input information from the outside world. And pass it to the middle hidden layer for processing. The middle hidden layer is the internal information processing layer. Pass input information to the output layer. The output layer outputs information processing results to the outside world. Let the number of input layer nodes be n, the number of output layer nodes be m. For node j, let the input value of each input node connected to it be xi, the weight between nodes is wij, the threshold is bj, and the net input value Sj of node j is:

$$S_j = \sum_{i=1}^{n} w_{ij}.x_i + b_j \tag{2}$$

Select the activation function f, the common activation function is sigmoid, tanh and ReLU, then the output yj of node j is:

$$y_j = f(S_j) \tag{3}$$

- Error back propagation stage

The actual output is compared with the expected output. And the error is calculated. The error is corrected by the output layer in the way of gradient descent, and the weight of each layer is transferred back layer by layer to the hidden layer and the input layer. Assuming that the expected output of the output layer is dj, and the actual output is yj, the mean square error is used to measure the error between the predicted value and the true value. The error function is as follows:

$$E = \frac{1}{2} \sum_{j=1}^{m} (d_j - y_j)^2 \qquad (4)$$

The goal of reverse delivery is to make E as small as possible. According to formula (2) and formula (3), the output value of each node is determined by the weight corresponding to the point connection line and the threshold of the layer. By adjusting the weight w and the threshold b, the error function E can be minimized. Finding the partial derivatives of w and b for the objective function E can obtain the update amount of w and b for the specified output node j, there are:

$$\Delta w(i,j) = -\eta \frac{\partial E(w,b)}{\partial w(i,j)} \qquad (5)$$

Where "η" is the learning rate parameter of the back propagation algorithm, and the negative sign means that the gradient falls in the weight space. Select the activation function as f, and define δij:

$$\delta_{ij} = (d_j - y_j).f'(S_j) \qquad (6)$$

Finding the partial derivative of w for the function E gives the update amount of the weight:

$$\Delta w(i,j) = \eta.\delta_{ij}.x_i \qquad (7)$$

Similarly, the partial derivative of b for function E is used to obtain the threshold update amount:

$$\Delta b_j = \eta.\delta_{ij} \qquad (8)$$

Information forward propagation and error back propagation are processes in which weights and thresholds of each layer are continuously adjusted until the network error is minimized or the maximum number of iteration.

2.2.2 PSO Particle Swarm Optimization Algorithm

PSO particle swarm optimization algorithm is a population-based search process. The system is initialized as a set of random solutions and the optimal solution is searched through iteration. Many experiences show that PSO particle swarm optimization algorithm is a very effective optimization tool. PSO particle swarm optimization algorithm is used to optimize the initial weight and threshold parameters of BP neural network, and the global optimal solution can be found in a small search space. The basic principle of the PSO particle swarm optimization algorithm [9, 10] is: In the D dimensional search space, there are n particles forming a population. The solution to each problem to be optimized is a particle in the search space. The position of each particle in the search space is represented as a D dimensional vector $xi = (xi1, xi2, xi3, \ldots, xiD)(i = 1, 2, 3, \ldots, n)$. Particle i flies in the search space at speed $vi = (vi1, vi2, vi3, \ldots, viD)$. Each particle has a fitness value-determined by the function being optimized). Through learning and adaptation of the environment, in the each iteration, each particle dynamically updates its speed and position based on individual flight experience (the optimal position Pbest discovered by the particle itself so far) and the group flight experience (the best position Gbest found by all particles in the whole population so far) The evolution equation of a particle swarm can be described as:

$$v_{ij}^{k+1} = w * v_{ij}^k + c1 * r1 * \left(p_{ij}^k - x_{ij}^k\right) + c2 * r2 * \left(p_{gj}^k - x_{ij}^k\right) \tag{9}$$

$$x_{ij}^{k+1} = x_{ij}^k + v_{ij}^{k+1} \tag{10}$$

Among them, c1 and c2 are learning factors, also known as acceleration weight, which is usually evaluated between (0, 2), c1 is used to adjust the step size of the particles flying to their best position, and c2 is used to adjust the step size of particles flying to their best position globally. w is the inertia factor, r1 and r2 are two independent random numbers between [0, 1] until the equilibrium or optimal state is reached.

2.2.3 PSO-BP Algorithm

The PSO-BP algorithm maps the weights and thresholds of nodes in the BP neural network to the particles in the PSO algorithm. The fitness value of each particle (the training error of BP neural network) is calculated by the fitness function of the particle. At the end of the iteration, the PSO particle swarm optimization algorithm outputs the global optimal particle as the initial weight and threshold of the BP neural network. The particle dimension D of the PSO particle swarm optimization algorithm and the number of nodes in input layer ni, hidden layer nh and output layer no of the BP neural network satisfy the following formula:

$$D = n_h + n_i * n_h + n_h * n_o + n_o \tag{11}$$

The specific algorithm flow is shown in Fig. 3:

① Initialize BP neural network parameters, including the number of nodes in input layer, hidden layer and output layer;
② Initialize parameters of the PSO particle swarm optimization algorithm, including population size, maximum number of iterations, and learning factors;
③ Calculate the fitness value of each particle;
④ Comparing the fitness of each particle, determining individual extreme points and global optimal extreme points;
⑤ Constantly update the position of the particles;
⑥ The optimal initialization weight and threshold of the BP neural network model are obtained when the condition is satisfied or the maximum iteration number is reached;
⑦ Train the BP neural network, test and output the prediction results.

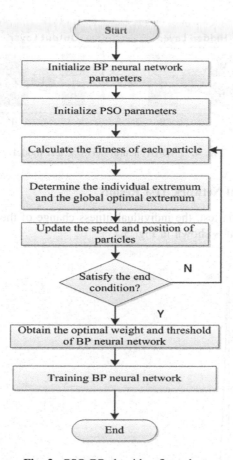

Fig. 3. PSO-BP algorithm flow chart

3 Simulation Experiment and Result Analysis

The tobacco leaf mold data in the literature [2] was used to simulate the proposed PSO-BP neural network model through Matlab 2017a software platform.

3.1 PSO-BP Neural Network Model Parameters

The structure of the PSO-BP neural network test model is shown in Fig. 4: In the PSO-BP neural network model, the number of input layer nodes is 3, the number of output layer nodes is 1, and the number of hidden layer nodes is 8. The intermediate layer neuron transfer function of BP neural network is tansig, the output layer neuron transfer function is purlin, the training function is traingdx, the target error is 0.0001, the learning rate is 0.1, and the maximum iteration number is set to 1000 times. The particle dimension of the PSO particle swarm optimization algorithm is 41, the maximum speed limit is 0.5, the maximum allowable iteration number is 300, and the learning factors c1 and c2 are both 2.

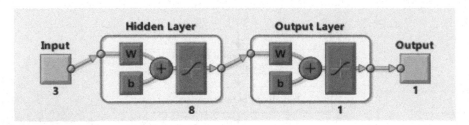

Fig. 4. PSO-BP neural network test model

3.2 PSO-BP Neural Network Test Results

Through the simulation test, the individual fitness change of the PSO particle swarm optimization algorithm is shown in Fig. 5.

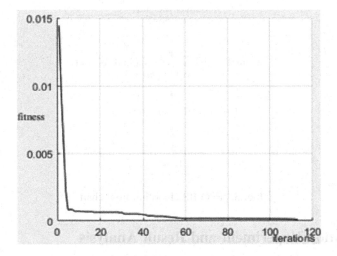

Fig. 5. Individual fitness of PSO

It can be seen from Fig. 5 that the fitness of the PSO particle swarm optimization algorithm changes significantly. After 80 iterations, the mean square error of the test sample that satisfies the condition is obtained, and the optimal initialization weight and threshold of BP neural network are obtained.

The predicted value of the optimized PSO-BP neural network model for the mildew rate of tobacco leaves and the comparison between the predicted value of the BP neural network model and the actual value of the mildew rate of tobacco leaves are shown in Fig. 6.

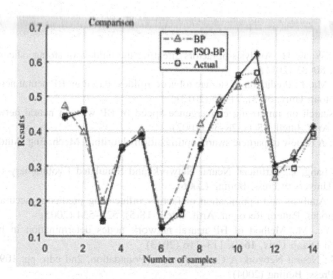

Fig. 6. Comparison of test results

It can be seen from Fig. 6 that both BP neural network and PSO-BP neural network can well predict the mildew rate of tobacco leaves. The PSO-BP neural network model optimized by PSO particle swarm optimization algorithm has higher prediction accuracy than BP neural network model.

4 Conclusion

In this paper, PSO particle swarm optimization algorithm is introduced into the BP neural network model. The PSO particle swarm optimization algorithm is used to optimize the initialization weight and threshold of the BP neural network model. A PSO-BP neural network prediction model was established. Simulation results show that both BP neural network and PSO-BP neural network can predict the mildew rate of tobacco leaves very well. The PSO-BP neural network model optimized by PSO particle swarm optimization algorithm has higher prediction accuracy than BP neural network model. The proposed model is applied to the environmental monitoring of storage tobacco leaves, which can give early warning of the mildew of tobacco leaves and reduce economic losses caused by mildew of tobacco leaves.

Acknowledgment. This work is funded by grants from the National Science Foundation of China (grant No. 61865015) and by Key Project of Applied Basic Research Program of Yunnan Province (grant No. 2018FA033).

234   Z. Ye et al.

References

1. Zhang, C., Wang, H.: Analysis of factors affecting the mildew of storage slices. Chin. Tob. Sci. **32**(3), 80–83 (2011)
2. Zhang, L., Ma, J.: Prediction of storage tobacco mildew based on BP neural network. J. East China Jiaotong Univ. **30**(3), 71–75 (2013)
3. Li, J.: Research on improving convergence speed of BP wavelet neural network. Pattern Recognit. Artif. Intell. **15**(1), 28–35 (2002)
4. Huang, L.: A review of particle swarm optimization algorithm. Mech. Eng. Autom. (5), 197–199
5. Yan, P., Zhang, C.: Artificial Neural Network and Simulated Evolutionary Computation. Tsinghua University Press, Beijing (2006)
6. Chen, G.: Analysis and optimization of factors influencing prediction accuracy of neural network model. Pattern Recognit. Artif. Intell. **18**(5), 528–534 (2005)
7. Jiao, B., Ye, M.: Method of BP neural network nodes determination in hidden layer. J. Shanghai Dianji Univ. **16**(3), 113–116 (2013)
8. Haykin, S.: Neural Network A Comprehensive Foundation, 2nd edn, pp. 109–123. China Machine Press, Beijing (2004)
9. Yang, W., Zhang, X.: Overview of particle swarm optimization algorithm. Gansu Sci. Technol. **28**(5), 88–92 (2012)
10. Zhang, Q., Meng, X.: Particle swarm optimization algorithm based on random dimension partitioning and learning. J. Zhejiang Univ. (Eng. Ed.) **52**(02), 367–378 (2018)

Exploring the Pan-Relationship Between Disease and Symptom Related to Coronary Heart Disease from Chinese Electronic Medical Records

Ronggen Yang[1], Yichen Wang[2,3], Bo Wang[2,3], and Lejun Gong[2(✉)]

[1] Faculty of Intelligent Science and Control Engineering,
Jinling Institute of Technology, Nanjing 211169, China
[2] Jiangsu Key Lab of Big Data Security and Intelligent Processing School
of Computer Science, Nanjing University of Posts and Telecommunications,
Nanjing 210046, China
glj98226@163.com
[3] School of Geography and Bioinformatics,
Nanjing University of Posts and Telecommunication, Nanjing, China

Abstract. Electronic Medical Records (EMRs) arises from the clinical treatment process, which reflects a patient's physical condition, the disease, the treatment process, etc. It has an important significance for the recovery of the patient. As the amount of electronic medical records continues to increase, people are actively looking for ways to get the valid data contained in electronic medical records. More effective methods are urgently needed for processing EMRs. In this paper, 108 real EMRs of coronary heart disease were analyzed, through named entity recognition and entity relation extraction. Pan-relationship between entities was defined in the process. Experimental results showed the analysis of pan-relation between coronary heart disease and symptom is helpful for treatment of coronary heart disease.

Keywords: Pan-relationship · Entity relation extraction · Linear relationship · Chi square test

1 Introduction

Coronary heart disease is a common and frequently-occurring disease, which refers to a narrowing of the coronary arteries, the blood vessels that supply oxygen and blood to the heart. It is also known as coronary artery disease. It is a major cause of illness and death. Some possible reasons of it could be unhealthy diet, overweight, obesity, high blood pressure, lack of exercise, excessive diet or some psychological reasons. It can also be affected by the genetics. The main symptoms of coronary heart disease are chest tightness, chest pain, angina, etc.

Electronic Medical Record (EMR), computer patient record, is the systematized collection of patient and population electronically-stored health information in a digital format [1]. These records can be shared across different health care settings. Coronary

© Springer Nature Switzerland AG 2019
D.-S. Huang et al. (Eds.): ICIC 2019, LNAI 11645, pp. 235–243, 2019.
https://doi.org/10.1007/978-3-030-26766-7_22

heart disease EMR is concerned in this work. There is enormous information about the symptom and treatment of coronary heart disease in the EMR. This attract the attention of many researchers. Large volumes of processing and utilization methods for EMR were reported in the recent years. Cai provides a deep learning [2] model for Chinese electronic medical records, which incorporating part of speech and self-matching attention for named entity recognition of EMR to solve the long entity recognition problem.

A hybrid approach [3] including bi-directional long term memory (BiLSTM) and conditional random fields (CRFs) was reported for named entity recognition in Chinese EMR. It is also can be used to monitor pneumococcal vaccination rates in a geriatric clinic [4]. Attention-based deep residual learning network can help for entity relation extraction in Chinese EMRs [5]. After all, EMR is a kind of data with text format, which can be processed by general natural language processing methods [6, 7] and machine learning technology [8, 9]. Researches about coronary heart disease EMR are also increased in recent years. Ding reclassify coronary heart disease Risk probability by Incorporating genetic markers, combined with insights from the Mayo EMRs and GEnomics Study [10]. Genotype-informed estimation of risk of coronary heart disease method was presented [11], which based on genome-wide association data linked to the EMR. A testing whether acute coronary heart disease and heart failure events was classified with EMR [12]. Some researchers alerted for patients warranting secondary prevention of coronary heart disease through 5-year evaluation of EMR [13].

In this paper, we focus on the Chinese EMR of coronary heart dis-ease, and extract the disease and symptom entities from the EMR of coronary heart disease. The pan-relationships between diseases and symptoms, diseases and diseases, symptoms and symptoms are extracted after the entity recognition. Finally, give the experimental results and discussions. More details are presented the following sections.

2 Materials and Methods

2.1 Data Source

The experimental data set is from the internal medicine of a hospital in Henan Province, China, which has 108 Chinese medical records. There are 43736 characters in the data set. Almost each record has five fields such as history of treatment, reasons for this visit, specific examination results, symptoms and medical history.

History of treatment described the symptoms of the patient in the past few years and measures taken against the symptoms. Some other information is also recorded, such previous diagnosis, whether they have been treated, how the treatment is processed, etc. Reasons for this visit field includes the physical condition and current symptoms of the patient, the frequency and severity of the symptoms. Most of them are from patient's statement. Specific examination results show detailed examination of the patient after the visit, and the doctor's diagnosis. Some specific treatment measures and methods can be found in this field. Symptoms field records the patient's symptoms and physical conditions includes diet, sleep, consciousness and defecation. Medical history records the patient's previous illness, including the name of the disease, the time and severity of the illness, the treatment. Not all the records have the alignment five fields, some lack history or medical history, which can not affect the processing and relation extraction to them.

International classification of diseases version 10(ICD-10) is the disease dictionary utilized in the work, which contains nearly 26,000 disease records, and the content is comprehensive and accurate, covering all diseases of the hospital, which is the most complete in China.

2.2   Methods

(1) Preprocessing of text

Preprocessing of original text is necessary before extraction relations. Stop words in the document appear frequently and have less significance. Results in the work [14] showed that elimination of stop words can improve the Mean average precision values of Gujarati information retrieval. So, we delete the stop words in the data set before name entity identification.

The ICD-10 disease bank also needs to be pre-processed, including changing or removing symbols, such as parentheses, brackets, etc., and sorting ICD-10 disease names for easy identification.

Each record in the dataset was annotated manually according to ICD-10 disease bank. The whole set is organized into samples to be used as train and test sets.

(2) Name entity recognition

We concerned entities in the work including symptoms and diseases. Dictionary based named entity recognition method was performed to the data set. Disease dictionary is from ICD-10 and symptom dictionary is downloaded by web crawler.

(3) Pan-relationship between entities from EMR

Relationships between disease, symptom, examination and treatment in EMR have the significance to treatment of new patients. These physical relationships reflect the patient's health situation and his medical treatment, as well as the doctor's professional knowledge. A common method in entities relation extraction is to calculate the Pearson correlation coefficient and the chi-square value.

Pan-relationship refers to the relationship that exists between entities of coronary heart disease EMR. This relationship can be used for text analysis of medical records. In this paper, relationships between disease and disease, symptoms and symptoms, disease and symptoms were concerned.

Pearson correlation coefficient is an important measure of the relationship between entities. The coefficient can show the correlation between two entities. Coefficient 0-0.3 indicates a low linear correlation; coefficient 0.3-0.6 shows a moderate linear correlation; coefficient 0.6-1.0 expresses a high linear correlation. This work focuses the relationships between diseases and diseases, symptoms and symptoms, and diseases and symptoms.

Chi-square test is to determine the independence of two variables by observing the deviation between the actual value and the theoretical value. It is often assumed that the two variables are independent, and observe the degree of deviation between the actual value and the theoretical value. If the deviation is small enough, we can determine the error is from random error, which is caused by uncertain factors or accidental occurrence. The two variables are indeed independent, and we accept the null hypothesis at this time. If the deviation is so large that such errors are unlikely to be accidental or inaccurate, we can think the two are actually relevant. That is to say, we reject the null hypothesis and accept the alternative hypothesis. In this paper, we calculate the relationship between disease entities and disease, symptom and symptom, and disease and symptom by calculating the chi-square value.

3   Results and Discussions

3.1   Result of Name Entity Recognition

Diseases and symptoms are identified from the EMR before the relationship extract. Dictionary based method performance on coronary heart disease EMR is showed in Table 1. As can be seen from the table, the performance of symptoms recognition is significantly better than the diseases generally. When the diseases and symptoms are identified together, the performance turned slightly less. However, to the top 10 disease and symptom entities, the performance is slightly improved. So, we give the emphasis on the top 10 disease and symptom entities, and relations between them.

Table 1.

	Precision	Recall	F-Measure
Overall disease	0.551	0.496	0.522
Disease TOP10	0.625	0.556	0.588
Disease TOP20	0.610	0.540	0.573
Overall symptom	0.724	0.628	0.673
Symptom TOP10	0.809	0.740	0.773
Symptom TOP20	0.771	0.702	0.734

3.2   Result of Relation Extraction

(1) Relations between disease and disease

Some diseases may depend each other, e.g., some diseases are developed from other diseases. Relation between diseases can reveal the law of health and diseases. Figure 1 shows the relationship between the top three diseases of frequency in the coronary heart disease EMR and diseases most relevant to it obtained through Pearson correlation calculation and chi-square test. It could be seen that if a patient with coronary heart disease has hypertension, then it is a high probability that he has high blood pressure and cerebral infarction from the result of chi-square test. Similarly, high blood pressure and cerebral infarction also have high correlation. A coronary heart disease patient with angina pectoris is likely to have both high blood pressure and chest pain. In other words, patients with coronary heart disease mainly have hypertension, cerebral infarction, chest pain, angina pectoris, and palpitation. Simultaneously, he may suffer from arterial sclerosis.

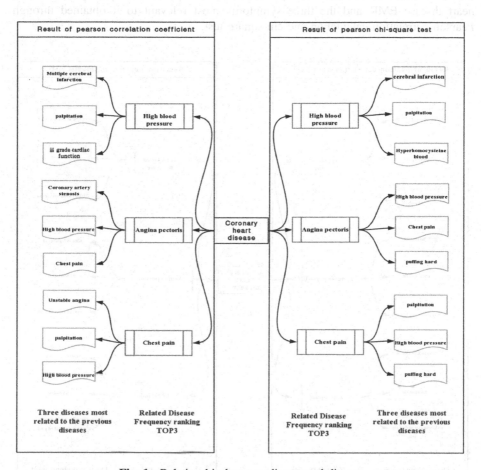

Fig. 1. Relationship between disease and disease

(2) Relations between disease and symptom

Symptoms are the direct manifestation of diseases. Symptoms imply some diseases and some diseases display some symptoms. So, symptoms can guide the treatment for diseases. Figure 2 shows the relations between the top three diseases of word frequency in the coronary heart disease EMR and three symptoms most relevant to it obtained through Pearson correlation calculation and chi-square test. It can be seen that a patient with high blood pressure will high probably show symptoms of chest pain, palpitation. The main symptoms of angina are chest pain, chest tightness, shortness of breath and fatigue, which is in complete agreement with our common sense.

(3) Relations between symptom and symptom

Similar symptoms can result in similar treatment event to different diseases. Figure 3 shows the relations between the top three symptoms of word frequency in coronary heart disease EMR and the three symptoms most relevant to it obtained through Pearson correlation calculation and chi-square test.

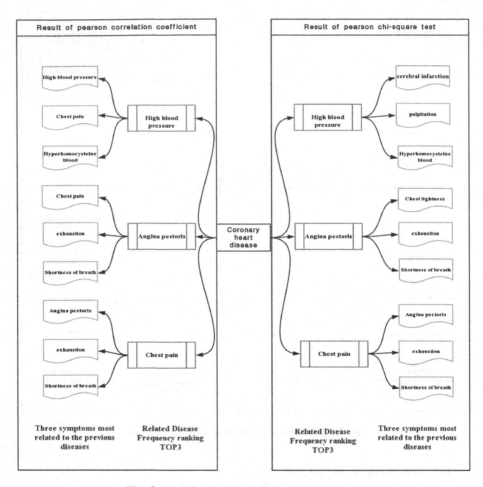

Fig. 2. Relations between disease and symptom

We identified five symptoms such as dizziness, nausea, heavy headache, chest tightness, and fatigue from the coronary heart disease EMR. Other diseases such as pharyngitis, high blood pressure and anemia can also cause dizziness and nausea. Anxiety disorders and depression can also cause dizziness and heavy-headedness at the same time. Myocardial Ischemia can cause dizziness and chest tightness at the same time. Migraine, vascular neuropathic headaches and colds can cause dizziness and fatigue.

Chest pain, heartache can occur with chest tightness accompanied by palpitations. In addition to coronary heart disease, patients with hypertension or rheumatic heart disease may feel palpitations and shortness of breath. Patients with viral myocarditis have a high probability of palpitations.

Chest tightness, palpitations, shortness of breath, fatigue, such four symptoms are very close. It was found that cardiac neurosis can cause chest tightness and shortness of breath. Patients with hypoglycemia and hypotension have a high probability of chest tightness, palpitations and fatigue.

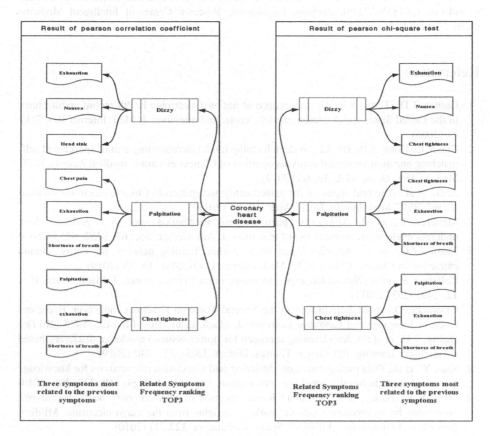

Fig. 3. The relationship between symptom and symptom

4 Conclusions

This work is about disease, symptom entity recognition and Exploring the pan-relationship between disease and symptom related to coronary heart disease from Chinese EMR. Based on the results of entity recognition, we extracted the relations between disease and disease, disease and symptom, symptom and symptom. Experiments showed that pan-relationship exists in the text of coronary heart disease. In other words, we can use these relations to analyze the similarity of the symptoms and diseases, which directly guide the treatment of diseases.

In addition, we can construct a network of the relationship between disease and disease, symptoms and symptoms, disease and symptoms. These networks of relations can be used to disease diagnosis and estimate the probability of different diagnosis for a patient.

Acknowledgements. This work is supported by the Natural Science Foundation of the Higher Education Institutions of Jiangsu Province in China (No. 16KJD520003), National Natural Science Foundation of China (61502243, 61502247, 61572263), China Postdoctoral Science Foundation (2018M632349), Zhejiang Engineering Research Center of Intelligent Medicine under 2016E10011.

References

1. Gunter, T.D., Terry, N.P.: The emergence of national electronic health record architectures in the United States and Australia: models, costs, and questions. J. Med. Internet Res. **7**(1), e3 (2005)
2. Cai, X.L., Dong, S.B., Hu, J.L.: A deep learning model incorporating part of speech and self-matching attention for named entity recognition of Chinese electronic medical records. BMC Med. Inform. Decis. Mak. **19**, 65 (2019)
3. Ji, B., et al.: A hybrid approach for named entity recognition in Chinese electronic medical record. BMC Med. Inform. Decis. Mak. **19**, 64 (2019)
4. Menezes, R., et al.: Monitoring pneumococcal vaccination rates in a geriatric clinic undergoing electronic medical record transition. J. Am. Geriatr. Soc. **67**, S252–S253 (2019)
5. Zhang, Z.C., et al.: Attention-based deep residual learning network for entity relation extraction in Chinese EMRs. BMC Med. Inform. Decis. Mak. **19**, 55 (2019)
6. Collobert, R., et al.: Natural language processing (almost) from scratch. J. Mach. Learn. Res. **12**, 2493–2537 (2011)
7. Glowacka, D., et al.: Introduction to the Special Topic on Grammar Induction, Representation of Language and Language Learning. J. Mach. Learn. Res. **12**, 1425–1428 (2011)
8. Huang, T.E., et al.: A deep learning approach for power system knowledge discovery based on multitask learning. IET Gener. Transm. Distrib. **13**(5), 733–740 (2019)
9. Sato, Y., et al.: Data mining based on clustering and association rule analysis for knowledge discovery in multiobjective topology optimization. Expert Syst. Appl. **119**, 247–261 (2019)
10. Ding, K.Y., Bailey, K.R., Kullo, I.: Reclassification of 10-y coronary heart disease risk probability by incorporating genetic markers; insights from the mayo electronic MEdical Records and GEnomics (eMERGE) Study. Circulation, **122**(21) (2010)
11. Ding, K.Y., Bailey, K.R., Kullo, I.J.: Genotype-informed estimation of risk of coronary heart disease based on genome-wide association data linked to the electronic medical record. BMC Cardiovasc. Disord. **11**, 66 (2011)
12. Kottke, T.E., Baechler, C.J.: Testing whether acute coronary heart disease and heart failure events can be classified with electronic medical record data. Circulation **125**(19), E804–E804 (2012)

13. Whitley, H.P., Fermo, J.D., Chumney, E.C.G.: 5-Year evaluation of electronic medical record flag alerts for patients warranting secondary prevention of coronary heart disease. Pharmacotherapy **26**(5), 682–688 (2006)
14. Joshi, H., et al.: To stop or not to stop - Experiments on stopword elimination for information retrieval of Gujarati text documents. In: 3rd Nirma University International Conference on Engineering (Nuicone 2012) (2012)

The Actuator and Sensor Fault Estimation Using Robust Observer Based Reconstruction for Mini Motion Package Electro-Hydraulic Actuator

Tan Van Nguyen and Cheolkuen Ha[✉]

Robotics and Mechatronics Lab, Ulsan University, Ulsan 44610, Korea
nvtan@hueic.edu.vn, cheolkeun@gmail.com

Abstract. In the period of industrialization and modernization of technology 4.0. Especially applying information technology and its applications in industry in general and for industrial equipment using the hydraulic systems, in particular, are very interested. Therefore, applying the control compensation algorithm to these devices for ensuring safety and accuracy is essential. In this paper, we study the estimation process of the actuator and sensor fault. The development process consists of the following steps. First, the mini motion package electro-hydraulic actuator is formulated with actuator and sensor faults. Second, unknown input observer (UIO) is constructed to estimate the actuator and sensor faults based on Lyapunov's stability condition and a linear matrix inequality (LMI) optimization algorithm in order to obtain the control signal error asymptotically stable. Finally, numerical simulations were run to show the effectiveness of the fault estimator.

Keywords: Unknown input observer · Fault-tolerant control · Fault diagnosis

1 Introduction

In the past decades, electro-hydraulic actuators (EAH) have been considered as the potential choices for modern industries ranging from manipulators to accurate machine tools because of their advantages such as high power-weight ratio, controllability, high accuracy, reliability, costs, etc. They can lift and hold heavy loads without breaking, and can move heavy objects at slow speeds without the need for gearing. With those benefits, a commercial product, known as mini motion packages (MMPs), is introduced by Kayaba industrial company, which is considered as one of the most feasible solutions for applications requiring short operating ranges and high output power. Many studies of the MMPs are carried out as [1–3]. However, in the practice, many faults or failures arise from some components of the EHA system such as failures in the electrical machine of the pump, the leakage faults of the pump and cylinder, aging and cable break faults of the sensor, and the others as polluted oil, and etc. Moreover, when working in a disturbed environment, system nonlinearities with large uncertainties become critical challenges in utilizing EHAs to obtain high precision positioning tracking control. To address this issue, the fault-tolerant control (FTC) technique for EHAs is proposed. One of the important issues in the FTC is fault estimation. There have been many researchers

© Springer Nature Switzerland AG 2019
D.-S. Huang et al. (Eds.): ICIC 2019, LNAI 11645, pp. 244–256, 2019.
https://doi.org/10.1007/978-3-030-26766-7_23

to study estimation of the actuator and sensor fault using various algorithms such as unknown input observer (UIO) as in [4, 5], using the augmented system performed by [6], using sliding mode observer as shown in [7–9], using the Kalman filter algorithm as shown in [10–12] and etc. Estimating faults is a difficult task and challenge for EHA. However, among various fault estimation techniques, more enhanced methods deal with simultaneous estimation of state and fault as shown in [2, 4, 5, 13, 14]. Therein, sensor fault based on UIO reconstruction for EHA system was proposed in [2].

In this paper, UIO is implemented based on the Lyapunov's stability condition and LMI optimization algorithm to reconstruct the actuator and sensor faults, in which the error dynamic for estimation is asymptotically stable. The fault decoupled process is performed by UIO reconstruction. Numerical simulation results show that from comparing the predefined actuator and sensor fault signal with the estimated signal our proposed method is efficient in estimating simultaneous faults of actuator and sensor in the MMP.

The important contributions of this paper are summarized as follows:

- A UIO reconstruction is constructed of the nonlinear continuous-time system for the stability of the system to obtain asymptotically stable of the estimated error dynamic system under Lyapunov's stability condition using LMI optimization algorithm.
- Proofs of LMI optimization algorithm implemented to obtain
- Numerical faults of actuator and sensor estimation show successfully performing to the MMP system.

2 EHA Model Formulation

The EHA system shown in Fig. 1, the dynamics of the piston position can be written as [2]:

$$m_p \ddot{x}_p + B_v \dot{x}_p + F_{sp} + F_{frc} + d = A_h P_h - A_r P_r \qquad (1)$$

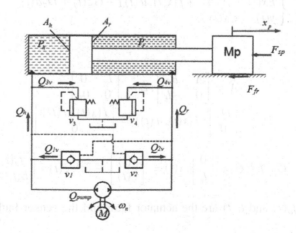

Fig. 1. Schematic of an electro-hydraulic actuator EHA system [2].

where m_p is the equivalent mass of the object M_p, x_p is the piston position, $\ddot{x}_p$ and $\dot{x}_p$ are the acceleration and piston velocity, respectively, P_h and P_r are the pressure in two chambers, respectively, A_h and A_r are the piston area in two chambers, respectively, F_{sp} is the external load force on the cylinder, d is the disturbance, B_v is the viscosity damping coefficient, and F_{frc} is the friction force.

Based on [2], the dynamic of the EHA system in state space is described as

$$\dot{x} = Ax + f(x, u) \tag{2}$$

where

$$\begin{cases} x_1 = x_p; \\ \dot{x}_1 = x_2 = \dot{x}_p; \\ \dot{x}_2 = x_3 = \ddot{x}_p \end{cases} \begin{cases} x = \begin{bmatrix} x_1^T & x_2^T & x_3^T \end{bmatrix}^T \\ f(x, u) = \begin{bmatrix} 0 & 0 & g^T(x, u) \end{bmatrix}^T \end{cases}; A = \begin{bmatrix} 0 & 1 & 0 \\ 0 & 0 & 1 \\ 0 & -\gamma_2 & -\gamma \end{bmatrix}$$

$g(x, u)$ can find in Sect. 2 of [2].

3 UIO-Based Reconstruction Approach for Nonlinear System

Considering the nonlinear system subject to an unknown input in the following form:

$$\begin{cases} \dot{x}(t) = Ax(t) + f(x(t), u(t)) + B_a \xi(t) + D_d d(t) \\ y(t) = Cx(t) + B_s \xi(t) \end{cases} \tag{3}$$

where $x(t) \in \mathbb{R}^n$ is the state vector, $y(t) \in \mathbb{R}^p$ is the outputs vector and $d(t) \in \mathbb{R}^{r_d}$ is the unknown input or disturbance vector, and $\xi(t) \in \mathbb{R}^f$ is the actuator and sensor fault. A, C, D_d, B_a, and B_s are known constant matrices with suitable dimensions. $f(x(t), u(t))$ is a nonlinear function vector $\forall x(t), x(t) \in \mathbb{R}^n, u(t) \in \mathbb{R}^m$.

The equation of (3) can be rewritten in the following form:

$$\begin{cases} \bar{E}\dot{\bar{x}}(t) = \bar{A}\bar{x}(t) + \bar{f}(x(t), u(t)) + \bar{G}\xi(t) + \bar{D}_d d(t) \\ y(t) = \bar{C}\bar{x}(t) \end{cases} \tag{4}$$

where

$$\bar{A} = \begin{bmatrix} A & B_f \\ 0 & -I_f \end{bmatrix}; \bar{E} = \begin{bmatrix} I_n & 0 \\ 0 & 0_f \end{bmatrix}$$

$$\bar{D}_d = \begin{bmatrix} D_d \\ 0 \end{bmatrix}; \bar{f}(x(t), u(t)) = \begin{bmatrix} f(x(t), u(t)) \\ 0 \end{bmatrix}$$

$$\bar{C} = \begin{bmatrix} C & I_s \end{bmatrix}; \bar{G} = \begin{bmatrix} 0 \\ I_f \end{bmatrix}; \bar{x}(t) = \begin{bmatrix} x(t) \\ \xi(t) \end{bmatrix} \in \mathbb{R}^{\bar{n}}; \xi(t) = \begin{bmatrix} f_a(t) \\ B_s f_s(t) \end{bmatrix}$$

with $\bar{n} = n + p$, $f_a(t)$ and $f_s(t)$ are the actuator fault and the sensor fault.

By using [2], a Lipschitz constraint of the nonlinear function vector $f(x(t), u(t))$ can be expressed as:

$$\|\Delta f(x(t), u(t))\| \leq \zeta \|x(t) - \hat{x}(t)\| \tag{5}$$

where

$$\|\Delta f(x(t), u(t))\| = \|f(x(t), u(t)) - f(\hat{x}(t), u(t))\|$$

Based on [2, 13], the UIO model can be constructed in the influences of unknown inputs in the system (4) as:

$$\begin{cases} \dot{\bar{z}}(t) = F\bar{z}(t) + Ly(t) + \Gamma\bar{f}(\hat{x}(t), u(t)) \\ \dot{\hat{x}}(t) = \bar{z}(t) + Hy(t) \\ \hat{y}(t) = \bar{C}\hat{\bar{x}}(t) \end{cases} \tag{6}$$

where $\hat{\bar{x}}(t) \in \mathbb{R}^{\tilde{n}}$, and $\hat{y}(t) \in \mathbb{R}^p$ are state vector estimation of $\bar{x}(t)$, and measurement output estimation vector, respectively. $\bar{z}(t) \in \mathbb{R}^{\tilde{n}}$ is the state vector of the observer. $F \in \mathbb{R}^{\tilde{n} \times \tilde{n}}$, $\Gamma \in \mathbb{R}^{\tilde{n} \times \tilde{n}}$, $H \in \mathbb{R}^{\tilde{n} \times p}$, $L \in \mathbb{R}^{\tilde{n} \times p}$, $L_1 \in \mathbb{R}^{\tilde{n} \times p}$, and $L_2 \in \mathbb{R}^{\tilde{n} \times p}$ are the observer matrices and these matrices should be designed according to the state estimation error vector.

The estimation error can be defined as:

$$\bar{e}(t) = \bar{x}(t) - \hat{\bar{x}}(t) \tag{7}$$

and

$$\begin{aligned} \dot{\bar{e}}(t) &= \dot{\bar{x}}(t) - \dot{\hat{\bar{x}}}(t) \\ &= (I_{\tilde{n}} - H\bar{C})\dot{\bar{x}}(t) - \dot{\bar{z}}(t) \\ &= \Gamma\dot{\bar{x}}(t) - \dot{\bar{z}}(t) \end{aligned} \tag{8}$$

where

$$\Gamma = I_{\tilde{n}} - H\bar{C}$$

In addition

$$\bar{e}_y(t) = \bar{C}\bar{e}(t) \tag{9}$$

Equation (4) can be rewritten as:

$$\Gamma\bar{E}\dot{\bar{x}}(t) = \Gamma\bar{A}\bar{x}(t) + \Gamma\bar{f}(x(t), u(t)) + \Gamma\bar{G}\xi(t) + \Gamma\bar{D}_d d(t) \tag{10}$$

and from (6), we obtain

$$\dot{\hat{x}}(t) = F\bar{z}(t) + Ly(t) + H\bar{C}\dot{\bar{x}}(t) + \Gamma\bar{f}(\hat{x}(t), u(t))$$
$$= F\hat{x}(t) - FHy(t) + Ly(t) + H\bar{C}\dot{\bar{x}}(t) + \Gamma\bar{f}(\hat{x}(t), u(t)) \qquad (11)$$
$$= F\hat{x}(t) - (FH - L_2)y(t) + L_1\bar{C}\bar{x}(t) + H\bar{C}\dot{\bar{x}}(t) + \Gamma\bar{f}(\hat{x}(t), u(t))$$

where $L = L_1 + L_2$
  Using (7)–(11), we have:

$$\dot{\bar{e}}(t) = F\bar{e}(t) + [I_{\bar{n}} - (\Gamma\bar{E} + H\bar{C})]\dot{\bar{x}}(t) + [(\Gamma\bar{A} - L_1\bar{C}) - F]\bar{x}(t)$$
$$+ (FH - L_2)y(t) + \Gamma\Delta\bar{f}_{(x,u)} + \Gamma\bar{G}\zeta(t) + \Gamma\bar{D}_d d(t) \qquad (12)$$

where $\Delta\bar{f}_{(x,u)} = \bar{f}(x(t), u(t)) - \bar{f}(\hat{x}(t), u(t))$
  From (12), the estimation error (13) can be represented as:

$$\dot{\bar{e}}(t) = F\bar{e}(t) + \Gamma\Delta\bar{f}_{(x,u)} + \Gamma\bar{D}_d d(t) \qquad (13)$$

If the following conditions are satisfied

$$\Gamma\bar{E} + H\bar{C} = I_{\bar{n}} \qquad (14)$$

$$\Gamma\bar{G} = 0 \qquad (15)$$

$$\Gamma\bar{A} - L_1\bar{C} = F \qquad (16)$$

$$FH = L_2 \qquad (17)$$

without loss of generality, the matrix Γ can be defined:

$$\Gamma = \begin{bmatrix} I_n & \sigma_1 \\ -C & \sigma_2 \end{bmatrix} \qquad (18)$$

where $\sigma_1 \in \mathbb{R}^{n \times f}$ and $\sigma_2 \in \mathbb{R}^{f \times f}$ are arbitrary matrices.
  By solving (14–17) and (18), we obtain:

$$\Gamma = \begin{bmatrix} I_n & 0 \\ -C_k & 0_f \end{bmatrix}; \sigma_1 = 0_{n \times f}; \sigma_2 = 0_f; H = \begin{bmatrix} 0 \\ I_f \end{bmatrix} \qquad (19)$$

Lemma [14]. *For the equation in the following form:*

$$\dot{\zeta}(t) = \Phi\zeta(t) + \Psi u(t) \qquad (20)$$

The eigenvalues of a given matrix $\Phi \in \mathbb{R}^{n \times n}$ belong to the circular region $D(\alpha, \rho)$ with the center $\alpha + j0$ and the radius λ if and only if there exists a symmetric positive definite matrix $P \in \mathbb{R}^{n \times n}$ such that the following condition holds:

$$\begin{bmatrix} -P & P(\Phi - \alpha I_n) \\ * & -\lambda^2 P \end{bmatrix} < 0 \tag{21}$$

In addition, from (5), we have:

$$\left\| \Delta \bar{f}_{(x,u)} \right\| \leq \bar{\zeta} \|\bar{e}(t)\| \tag{22}$$

where

$$\bar{\zeta} = \begin{bmatrix} \zeta I_n & 0 \\ 0 & 0_f \end{bmatrix} \in \mathbb{R}^{\bar{n}}; \bar{e}(t) = \bar{x}(t) - \hat{\bar{x}}(t).$$

The matrix Ω can be inferred from (22) as:

$$\Omega = \chi^T \begin{bmatrix} -\bar{\zeta}^T \bar{\zeta} & 0 & 0 \\ 0 & I_{\bar{n}} & 0 \\ 0 & 0 & 0 \end{bmatrix} \chi \leq 0 \tag{23}$$

where

$$\chi = \begin{bmatrix} \bar{e}^T(t) & \Delta \bar{f}_{(x,u)}^T & d^T(t) \end{bmatrix}^T.$$

Theorem. *For system (4), there exists a robust UIO in the form of (6) such that the output estimation error satisfies* $\|e_y(t)\| \leq \rho \|d(t)\|$ *and a prescribed circular region* $D(\alpha, r)$ *if there exists a positive-definite symmetric matrix* $P \in \mathbb{R}^{\bar{n} \times \bar{n}}$, *matrix* $Q \in \mathbb{R}^{\bar{n} \times p}$, *and positive scalars* ρ *and* ε *such that the following inequalities hold:*

$$\begin{bmatrix} (\Gamma \bar{A})^T P + P(\Gamma \bar{A}) - \bar{C}^T Q^T - Q\bar{C} + \varepsilon \bar{\zeta}^T \bar{\zeta} & P\Gamma & P(\Gamma \bar{D}_d) & \bar{C}^T \\ * & -\varepsilon I_{\bar{n}} & 0 & 0 \\ * & * & -\rho I_d & 0 \\ * & * & * & -\rho I_p \end{bmatrix} < 0 \tag{24}$$

and

$$\begin{bmatrix} -P & P\Gamma \bar{A} - Q\bar{C} - \alpha P \\ * & -r^2 P \end{bmatrix} < 0 \tag{25}$$

where $Q = PL_1$;
Then the state observer (6) is asymptotically stable.

Proof of (24): Consider a Lyapunov function as:

$$V(t) = \bar{e}^T(t)P\bar{e}(t) \tag{26}$$

Derivative the Eq. (26), we have:

$$
\begin{aligned}
\dot{V}(t) \\
&= \dot{\bar{e}}^T(t)P\bar{e}(t) + \bar{e}^T(t)P\dot{\bar{e}}(t) \\
&= \left[F\bar{e}(t) + \Gamma\Delta\bar{f}_{(x,u)} + \Gamma\bar{D}_d d(t)\right]^T P\bar{e}(t) + \bar{e}^T(t)P\left[F\bar{e}(t) + \Gamma\Delta\bar{f}_{(x,u)} + \Gamma\bar{D}_d d(t)\right] \\
&= \begin{bmatrix} \bar{e}(t) \\ \Delta\bar{f}_{(x,u)} \\ d(t) \end{bmatrix}^T \begin{bmatrix} F^T P + PF & P\Gamma & P(\Gamma\bar{D}_d) \\ * & 0 & 0 \\ ** & 0 & \end{bmatrix} \begin{bmatrix} \bar{e}(t) \\ \Delta\bar{f}_{(x,u)} \\ d(t) \end{bmatrix}
\end{aligned} \tag{27}
$$

Based on (23) and (26), we obtain:

$$V_1 = \dot{V}(t) - \varepsilon\Omega \leq 0 \tag{28}$$

Equation (25) leads to:

$$
\begin{aligned}
V_1 &\leq \chi^T \begin{bmatrix} F^T P + PF & P\Gamma & P(\Gamma\bar{D}_d) \\ * & 0 & 0 \\ ** & 0 & \end{bmatrix} \chi - \varepsilon\Omega \\
&= \chi^T \begin{bmatrix} F^T P + PF + \varepsilon\bar{\zeta}^T\bar{\zeta} & P\Gamma & P(\Gamma\bar{D}_d) \\ * & -\varepsilon I_{\bar{n}} & 0 \\ ** & 0 & \end{bmatrix} \chi
\end{aligned} \tag{29}
$$

Based on the measurement error condition $\|e_y(t)\| \leq \rho\|d(t)\|$ of the output, the matrix V_2 can be presented as:

$$
V_2 = \begin{bmatrix} \bar{e}(t) \\ \Delta\bar{f}_{(x,u)} \\ d(t) \end{bmatrix}^T \begin{bmatrix} \frac{1}{\rho}\bar{C}^T\bar{C} & 0 & 0 \\ 0 & 0 & 0 \\ 0 & 0 & -\rho I_d \end{bmatrix} \begin{bmatrix} \bar{e}(t) \\ \Delta\bar{f}_{(x,u)} \\ d(t) \end{bmatrix} \leq 0 \tag{30}
$$

From (29) and (30), the matrix V_{12} can be presented as:

$$V_{12} = V_1 + V_2 \leq 0 \tag{31}$$

Equation (31) is represented as:

$$
V_{12} = \chi^T \begin{bmatrix} F^T P + PF + \varepsilon \bar{\zeta}^T \bar{\zeta} & P\Gamma & P(\Gamma \bar{D}_d) \\ * & -\varepsilon I_{\tilde{n}} & 0 \\ ** & 0 \end{bmatrix} \chi + \chi^T \begin{bmatrix} \frac{1}{\rho} \bar{C}^T \bar{C} & 0 & 0 \\ 0 & 0 & 0 \\ 0 & 0 & -\rho I_d \end{bmatrix} \chi
$$

$$
= \chi^T \Pi \chi
$$

where

$$
\Pi = \begin{bmatrix} F^T P + PF + \varepsilon \bar{\zeta}^T \bar{\zeta} & P\Gamma & P(\Gamma \bar{D}_d) \\ * & -\varepsilon I_{\tilde{n}} & 0 \\ * & * & -\rho I_d \end{bmatrix} + \begin{bmatrix} \frac{1}{\rho} \bar{C}^T \bar{C} & 0 & 0 \\ * & 0 & 0 \\ * & * & 0 \end{bmatrix} \tag{33}
$$

Applying the Schur lemma [15] to (33) with $\Pi < 0$, then the inequality (24) is satisfied once $V_{12} < 0$ satisfies the condition $\| e_y(t) \| \le \rho \| d(t) \|$.

Proof of (22): Applying (13) to Lemma 2, then (25) is satisfied. $\qquad\square$

In summary, the design procedure for this fault estimator is implemented in the following steps:

- Step 1: Determine the matrices Q, P, and $L_1 = P^{-1}Q$ to solve the LMI defined by the matrix inequality (24, 25).
- Step 2: Calculate the gain matrices F, L_2 and L using (14).
- Step 3: Obtain the state and fault estimation as $\hat{x}(t) = [I_n \quad o_{n \times r}] \hat{\bar{x}}(t)$, and $f_s(t) = \bar{C}_s \hat{\bar{x}}(t)$, respectively, with $\bar{C}_s = [o_{r \times n} \quad I_r]$.

4 Simulation Results

Simulation results are applied for the EHA system using parameters in [2]. The nonlinear continuous time state space model of the EHA system is described as

$$
\dot{x} = Ax + f(x, u) \tag{34}
$$

Based on the parameters of the EHA system in [2], item A in (34) can be calculated as:

$$
A = \begin{bmatrix} 0 & 1 & 0 \\ 0 & 0 & 1 \\ 0 & -5.5457e + 02 & -1.4643e + 03 \end{bmatrix}
$$

Based on the design experience, the positive scalars ζ, r, α, ε, and μ were selected as $\zeta = 15$; $r = 0.1$, $\alpha = 0.2$, $\varepsilon = 0.05$, and $\mu = 0.1$. We can be solved for P, Q, and

F using the optimal algorithm LMI. If the solution is feasible, then the results can be obtained as:

$$\Gamma = \begin{bmatrix} 1 & 0 & 0 & 0 & 0 \\ 0 & 1 & 0 & 0 & 0 \\ 0 & 0 & 1 & 0 & 0 \\ -1 & 0 & 0 & 0 & 0 \\ 0 & -1 & 0 & 0 & 0 \end{bmatrix} ; H = \begin{bmatrix} 0 & 0 \\ 0 & 0 \\ 0 & 0 \\ 1 & 0 \\ 0 & 1 \end{bmatrix} ; Q = \begin{bmatrix} -1.6846e+00 & 7.7317e-04 \\ 7.7891e-04 & -1.7638e+00 \\ 1.0886e-08 & 9.6434e-06 \\ -1.6846e+00 & 7.7321e-04 \\ 7.7892e-04 & -1.7638e+00 \end{bmatrix} ;$$

$$L_1 = \begin{bmatrix} -4.5358e+02 & -9.0619e-01 \\ -4.7876e-01 & -3.2464e+02 \\ 1.3676e-02 & 1.0323e+02 \\ 4.5352e+02 & 9.0691e-01 \\ 4.7944e-01 & 3.2459e+02 \end{bmatrix} ; L_2 = \begin{bmatrix} 4.5358e+02 & 9.0619e-01 \\ 4.7861e-01 & 3.2464e+02 \\ -1.3676e-02 & -1.0323e+02 \\ -4.5353e+02 & -9.0691e-01 \\ -4.7929e-01 & -3.2459e+02 \end{bmatrix} ;$$

$$L = \begin{bmatrix} 6.5000e-03 & -3.3307e-16 \\ -1.5000e-04 & -1.1369e-13 \\ 5.2042e-18 & 2.8422e-14 \\ -6.5000e-03 & 3.3307e-16 \\ 1.5000e-04 & 1.1369e-13 \end{bmatrix} ;$$

$$P = \begin{bmatrix} 5.1953e-01 & -2.2902e-04 & 4.6933e-14 & 5.1953e-01 & -2.2902e-04 \\ -2.2902e-04 & 5.1901e-01 & -1.0426e-10 & -2.2902e-04 & 5.1897e-01 \\ 4.6933e-14 & -1.0426e-10 & 2.1145e-07 & 4.6941e-14 & 1.6400e-15 \\ 5.1953e-01 & -2.2902e-04 & 4.6941e-14 & 5.1953e-01 & -2.2902e-04 \\ -2.2902e-04 & 5.1897e-01 & 1.6400e-15 & -2.2902e-04 & 5.1897e-01 \end{bmatrix}$$

$$F = \begin{bmatrix} 4.5458e+02 & 9.0719e-01 & 3.2437e-07 & 4.5358e+02 & 9.0619e-01 \\ 4.7876e-01 & 3.2564e+02 & 5.2496e-04 & 4.7861e-01 & 3.2464e+02 \\ -1.3676e-02 & -1.0352e+02 & 2.3113e-01 & -1.3676e-02 & -1.0323e+02 \\ -4.5452e+02 & -9.0791e-01 & -3.2437e-07 & -4.5353e+02 & -9.0691e-01 \\ -4.7944e-01 & -3.2559e+02 & -5.2496e-04 & -4.7929e-01 & -3.2459e+02 \end{bmatrix}$$

The reference input is expressed as

$$y_r(t) = 1.5 \sin 0.85t + 1.5 \tag{35}$$

Assume that the actuator fault is given by equation as following

$$f_a(t) = \begin{cases} 0 & \text{if } t \le 15 \\ 30 - 2t & \text{if } 15 < t \le 15.5 \\ 2t - 32 & \text{if } 15.5 < t \le 16.5 \\ 34 - 2t & \text{if } 16.5 < t \le 17 \end{cases} \tag{36}$$

and the sensor fault is given by equation as following

$$f_s(t) = \begin{cases} 0 & \text{if } t \le 5 \\ \sin(6.35t) & \text{if } 5 < t \le 5.9369 \\ 0 & \text{if } 5.9369 < t \le 10 \\ 4t - 40 & \text{if } 10 < t \le 10.25 \\ 42 - 4t & \text{if } 10.25 < t \le 10.75 \\ 4t - 44 & \text{if } 10.75 < t \le 11.25 \\ 46 - 4t & \text{if } 11.25 < t \le 11.5 \\ 0 & \text{if } 1 \; t > 11.5 \end{cases} \tag{37}$$

4.1 With Disturbance d(t) = 0

The simulation result of the position response is shown in Fig. 2. These results show the response and estimation signal following the reference signal at the times without impact of actuator fault or sensor fault. This proves that the estimated error dynamic is asymptotically stable under UIO reconstruction of the proposed method.

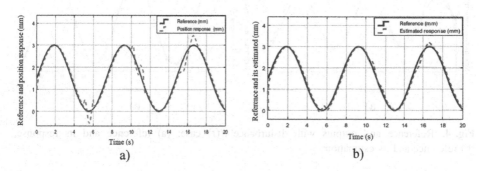

Fig. 2. Reference and outputs: (a) reference and its response, (b) reference and its estimation

The simulation results of the actuator and sensor faults, as well as their fault estimations, are shown in Fig. 3.

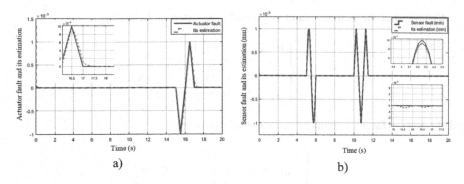

Fig. 3. Faults and their fault estimations: (a) actuator fault, (b) sensor fault

The result in Fig. 3(a) shows the actuator fault estimation following the predetermined actuator fault signal. A similar result in Fig. 3(b) also shows the sensor fault estimation according to a predetermined sensor fault signal. This proves that the actuator and sensor estimator works well.

4.2   With Disturbance d(t) = 0.000025 random(2, t)

The simulation result of the position response is shown in Fig. 4.

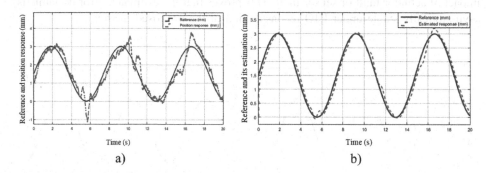

Fig. 4. Reference and outputs with disturbance $d(t)$ case: (a) reference and its response, (b) reference and its estimation

The simulation result of the actuator fault and its fault estimation is shown in Fig. 5(a) while Fig. 5(b) shows the sensor fault and its fault estimation for disturbance case.

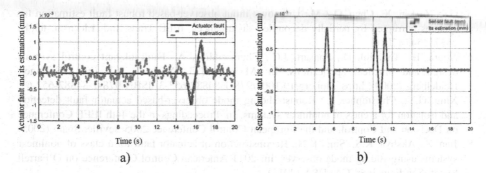

Fig. 5. Faults and their fault estimations with disturbance $d(t)$ case: (a) actuator fault, (b) sensor fault

We can see the difference between the position response under the influences of the disturbance in Fig. 4(a) compared with Fig. 2(a). There is a small difference between the sensor fault estimation in Figs. 5(a) and 3(b) under the impact of the disturbance d (t). However, the actuator fault estimation in Fig. 5(a) has a big difference in comparison to the actuator fault estimation in Fig. 3(a). This proves that there is a great effect of disturbance on actuators.

5 Conclusions

In this paper, a UIO reconstruction was applied to estimate the actuator fault and sensor fault. The result showed the estimated error dynamic is asymptotically stable under UIO reconstruction using optimization algorithm LMI and Lyapunov's stability condition. The results of the estimated actuator and sensor fault were achieved the desired result. The influences in the case without and with disturbance were also compared to show their effects in the system. The actuator and sensor fault estimation approach was successfully applied for the MMP system.

Acknowledgments. This work was supported by Korea Hydro & Nuclear Power company through the project "Nuclear Innovation Center for Haeoleum Alliance".

References

1. Ahn, K.K., Nam, D.N.C., Jin, M.: Adaptive back-stepping control of an electrohydraulic actuator. IEEE/ASME Trans. Mechatron. **19**, 987–995 (2014)
2. Tan, V.N., Cheolkeun, H.: Sensor fault-tolerant control design for mini motion package electro-hydraulic actuator MDPI. Processes **7**, 89 (2019)
3. Tri, M.N., Nam, C.N.D., Park, G.H., Ahn, K.K.: Trajectory control of an electro-hydraulic actuator using an iterative backstepping control scheme. Mechatronics **29**, 96–102 (2014)
4. Liu, X., Gao, Z.: Unknown input observers for fault diagnosis in Lipschitz nonlinear systems. In: Proceedings of 2015 IEEE International Conference on Mechatronics and Automation, Beijing, China (2015)

5. Gao, Z., Liu, X., Chen, Q.Z.M.: Unknown input observer-based robust fault estimation for systems corrupted by partially decoupled disturbances. IEEE Trans. Ind. Electron. **63**(4), 2537–2547 (2016)
6. Noura, H., Theilliol, D., Ponsart, J.C., Chamseddine, A.: Fault-tolerant control systems design and practical applications. In: Michael, J.G., Michael, A.J. (eds.) Springer, Heidelberg (2009). https://doi.org/10.1007/978-1-84882-653-3. ISBN 978-1-84882-652-6
7. Xing, G.Y., Christopher, E.: Robust sliding mode observer-based actuator fault detection and isolation for a class of nonlinear systems. In: Proceedings of the 44th IEEE Conference on Decision and Control, and the European Control Conference 2005 Seville, Spain (2005)
8. Jian, Z., Akshya, K.S., Sing, K.N.: Reconstruction of actuator fault for a class of nonlinear systems using sliding mode observer. In: 2011 American Control Conference on O'Farrell Street, San Francisco, CA, USA (2011)
9. Ngoc, P.N., Sung, K.H.: Sliding mode thau observer for actuator fault diagnosis of quadcopter UAVs. Appl. Sci. **8**, 1893 (2018). https://doi.org/10.3390/app8101893
10. Zhang, Q.: Adaptive kalman filter for actuator fault diagnosis. Automatica **93**, 333–342 (2018)
11. Xilin, Y., Michael, W., et al.: A UKF-based estimation strategy for actuator fault detection of UASs. In: 2013 International Conference on Unmanned Aircraft Systems (ICUAS), May 2013
12. Bahareh, P., Nader, M., Khashayar, K.: Sensor fault detection, isolation and identification using multiple model-based hybrid Kalman filter for gas turbine engines. IEEE Trans. Control Syst. Technol. **24**, 1184–1200 (2016)
13. Liu, X., Gao, Z., Zhang, A.: Robust fault tolerant control for discrete-time dynamic systems with applications to aero engineering systems. IEEE Access. **6**, 18832–18847 (2018)
14. Jia, Q., Li, H., Zhang, Y., Chen, X.: Robust observer-based sensor fault reconstruction for discrete-time systems via a descriptor system approach. Int. J. Control Autom. Syst. **13**, 274 (2015)
15. Boyd, S., Ghaoui, L.E., Feron, E., Balakrishnan, V.: Linear matrix inequalities in systems and control theory, SIAM, Philadelphia, PA, USA (1994). ISBN 0-89871-334-X

The Actuator Fault Estimation Using Robust Sliding Mode Observer for Linear System Applied for Mini Motion Package Electro-Hydraulic Actuator

Tan Van Nguyen and Cheolkuen Ha[✉]

Robotics and Mechatronics Lab, Ulsan University, Ulsan 44610, Korea
nvtan@hueic.edu.vn, cheolkeun@gmail.com

Abstract. In this paper presents a sliding mode observer (SMO) method to detect and isolate the actuator faults for a linear system. A proposed SMO method for the stability of the system has been derived and explained based on the Lyapunov's stability condition and linear matrix inequality (LMI) optimization algorithm. The constraint conditions of the switch gain for the proposed SMO to obtain the error dynamic system is asymptotically stable performed. In addition, the mathematical model is constructed for the electro-hydraulic actuator (EHA). The effectiveness of the actuator fault in the proposed SMO construction has been applied for numerical simulation of the mini motion package system (MMPs).

Keywords: Sliding mode observer · Fault diagnosis · Electro-hydraulic actuator (EHA)

1 Introduction

During the past two decades, there have been significant research activities in improving the new methodologies to apply for fault diagnosis as [1–3]. These approaches were classed into two categories: hardware residual method and analytical residual method (ARM). However, the ARM method has been applied by researchers because it is simple to use based on the analytical and mathematical model of the system to track the changes in the plant dynamics. Different fault effects were created by the residual vectors of output signals and its estimations so that fault diagnosis and isolation (FDI) can be obtained [4]. Some redundancy fault diagnosis methods were applied such that signal based FDI methods and model-based FDI methods. Here, model-based FDI methods have been considered in recent years [5, 6]. Model-based FDI methods use analytically computed outputs and compare these values with the sensor measurements to indicate the presences of faults in the system. These differences were called residuals. However, the model-based FDI methods were highly depended on the corresponding mathematical models, a major disadvantages of these methods is that an accurate mathematical model of the system is essential. which is often difficult to obtain in many practical situations. Moreover, the system was impacted by unknown input disturbances and system uncertainties which can cause misleading alarm.

© Springer Nature Switzerland AG 2019
D.-S. Huang et al. (Eds.): ICIC 2019, LNAI 11645, pp. 257–270, 2019.
https://doi.org/10.1007/978-3-030-26766-7_24

Therefore, these issues make the model-based fault diagnosis system ineffective. Hence, the robust fault diagnosis systems have been developed, and some of which consist such that unknown-input observer (UIO)-based fault diagnosis, adaptive observer (AO)-based fault diagnosis and sliding-mode observer (SMO)-based fault diagnosis. However, AO based fault diagnosis methods have the main disadvantage were that they were normally only suitable for the constant fault case [7]. UIO-based fault diagnosis has attracted considerable attention in the past and different UIOs have been developed. For example, in [8], fault diagnosis schemes based on UIO were proposed for linear systems with uncertainties. In [9, 10], UIO-based fault diagnosis methods were designed for Lipschitz nonlinear systems. In addition, SMO-based fault diagnosis has been widely studied in recent years because of the inherent robustness of sliding-mode algorithms to unknown input disturbances and modeling uncertainties. Considerable success has been achieved in many areas; for example, in [11] was applied for the aerospace field. In automotive vehicles was shown in [12]. Especially, its applications in the hydraulics systems were shown in [13], etc. Nonetheless, FDI functions can only indicate when and where a fault occurs in the system that can not provide the magnitude of the fault. Thus, in this paper, we introduce an actuator fault estimation approach based on SMO algorithm to determine the magnitude of the actuator fault. In the practice, some actuator faults or failures arise from some components of the MMP system such as failures in the electrical machine of the pump, the leakage faults of the pump and cylinder, aging, and the others as polluted oil, and etc. However, estimating the actuator fault is a difficult and complex work to apply for the MMP system.

In this paper, first, a mathematical model of the MMP system is constructed to implement the numerical simulation process. Second, an SMO construction is designed based on the Lyapunov's stability condition and LMI optimization algorithm to estimate the actuator faults, in which the error dynamic for estimation is asymptotically stable. The actuator fault decoupled process has been performed. Third, numerical simulation results show that from comparing the predefined actuator fault signal with the estimated signal our proposed method is efficient in estimating simultaneous faults of the actuator in the MMP.

The important contributions of this paper are summarized as follows:

- A mathematical model of the MMP system is developed from the system [14] to realize the numerical simulation procedure of the system.
- An SMO construction based on Lyapunov's stability condition and LMI optimization algorithm for the estimated error dynamic system is asymptotically stable.
- Actuator fault estimation is successfully applied to the MMP system.

2   EHA Model Formulation

The EHA system shown in Fig. 1, the dynamics of the piston position can be written as [14]:

$$m_p \ddot{x}_p + B_v \dot{x}_p + F_{sp} + F_{frc} + d = A_h P_h - A_r P_r \tag{1}$$

where m_p is the equivalent mass of the object M_p, x_p is the piston position, $\ddot{x}_p$ and $\dot{x}_p$ are the acceleration and piston velocity, respectively, P_h and P_r are the pressure in two chambers, respectively, A_h and A_r are the piston area in two chambers, respectively, F_{sp} is the external load force on the cylinder, d is the disturbance, B_v is the viscosity damping coefficient, and F_{frc} is the friction force.

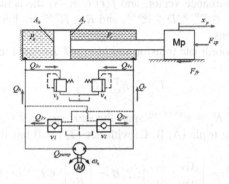

Fig. 1. Schematic of an electro-hydraulic actuator EHA system [14].

Equation (1) can be transferred under a form

$$m_p \ddot{x}_p + B_v \dot{x}_p + F_{sp} + F_{frc} + d = A_h P_L \tag{2}$$

where $P_L = (P_h - \sigma P_r)$; $A_r = \sigma A_h$.

The hydraulic continuity equations of the EHA system can be expressed as [14]:

$$\dot{P}_h = \Delta_1 (Q_h - Q_i - \dot{x}_p A_h) \text{ and } \dot{P}_r = \Delta_2 (Q_r + Q_i + \dot{x}_p A_r) \tag{3}$$

where $\Delta_1 = \beta_e / (V_{ch} + x_p A_h)$; $\Delta_2 = \beta_e / (V_{cr} - x_p A_r)$

β_e is the effective bulk modulus in each chamber, and V_{ch}, and V_{cr} are the initial control volumes of the two chambers.

The flow rates into the two chambers of the cylinder Q_h and Q_r are computed as [14]. Q_i is the internal leakage flow rate of the cylinder.

Equation (2) can be rewritten in the state space form as

$$\dot{x} = Ax + Bu + f(x,t) + Dd \tag{4}$$

where $x = \begin{bmatrix} x_1^T & x_2^T \end{bmatrix}^T = \begin{bmatrix} x_p^T & \dot{x}_p^T \end{bmatrix}^T$ is the state vector, and $f(x,t)$ is a nonlinear equation

$$A = \begin{bmatrix} 0 & 1 \\ -K_{sp}/m_p & -B_v/m_p \end{bmatrix}; B = \begin{bmatrix} 0 \\ A_1/m_p \end{bmatrix}; f(x,t) = \begin{bmatrix} 0 \\ -F_{frc}/m_p \end{bmatrix}; D = \begin{bmatrix} 0 \\ -1/m_p \end{bmatrix}; u = P_L$$

3 Robust Actuator Fault Estimation for Linear System

Considering the linear system subject to an unknown input in the following form:

$$\begin{cases} \dot{x}(t) = Ax(t) + Bu\,(t) + Rf_a(t) + Dd(t) \\ y\,(t) = Cx(t) \end{cases} \tag{5}$$

where $x(t) \in \mathbb{R}^n$ is the state vector, $y(t) \in \mathbb{R}^p$ is the outputs vector and $d(t) \in \mathbb{R}^{r_d}$ is the unknown input or disturbance vector, and $f(t) \in \mathbb{R}^f$ is the actuator and sensor fault. $A \in \mathbb{R}^{n \times n}, B \in \mathbb{R}^{n \times q}, C \in \mathbb{R}^{p \times n}, D \in \mathbb{R}^{n \times r}$, and $R \in \mathbb{R}^{n \times q}$ are known constant matrices with suitable dimensions.

Based on [15], a coordinate transformation $x \mapsto T_c x$ related to the invertible matrix

$$T_c = \begin{bmatrix} N_c^T & C^T \end{bmatrix}^T \tag{6}$$

where the columns of $N_c \in \mathbb{R}^{n \times (n-p)}$ span the null space of C. Using the change of coordinate $x \mapsto T_c x$, the triple (A, B, C) with $\det(T_c) \neq 0$ has the following form

$$T_c A T_c^{-1} = \begin{bmatrix} A_{11} & A_{12} \\ A_{21} & A_{22} \end{bmatrix}; T_c B = \begin{bmatrix} B_1 \\ B_2 \end{bmatrix}; C T_c^{-1} = \begin{bmatrix} 0 & I_p \end{bmatrix} \tag{7}$$

Assumption 1. The matrix pair (A, C) is detectable

It follows the assumption 1 that there exists a matrix $L \in \mathbb{R}^{n \times p}$ such that $A - LC$ is stable, and thus for any $Q > 0$, the Lyapunov equation

$$(A - LC)^T P + P(A - LC) = -Q \tag{8}$$

has a unique solution $P > 0$[15].

Assume that $P \in \mathbb{R}^{n \times n}, Q \in \mathbb{R}^{n \times n}$ are in the form following:

$$P = \begin{bmatrix} P_1 & P_2 \\ P_2^T & P_3 \end{bmatrix}, \quad Q = \begin{bmatrix} Q_1 & Q_2 \\ Q_2^T & Q_3 \end{bmatrix} \tag{9}$$

where $P_1 \in \mathbb{R}^{(n-p) \times (n-p)} > 0$, $P_1 \in \mathbb{R}^{p \times p} > 0$, $Q_1 \in \mathbb{R}^{(n-p) \times (n-p)} > 0$ and $Q_3 \in \mathbb{R}^{p \times p} > 0$ are submatrices follow from $P > 0$ and $Q > 0$.

Assumption 2. There exists an arbitrary matrix $F \in \mathbb{R}^{r \times p}$ such that

$$R^T P = FC \tag{10}$$

Assumption 3. The actuator fault vector $f_a(t)$ and disturbance vector $d(t)$ satisfies the following constraint:

$$|f_a| \leq \delta_a \text{ and } |d| \leq \delta_d \tag{11}$$

where δ_a and δ_d two known positive constants.

Suppose that the pair (R, D) has the following structure:

$$R = \begin{bmatrix} R_1^T & R_2^T \end{bmatrix}^T; \quad D = \begin{bmatrix} D_1^T & D_2^T \end{bmatrix}^T \tag{12}$$

where $D_1 \in \mathbb{R}^{(n-p) \times r}$, $D_2 \in \mathbb{R}^{p \times r}$ $R_1 \in \mathbb{R}^{(n-p) \times r}$, and $R_2 \in \mathbb{R}^{p \times r}$

Lemma 1 [15]. If P and Q have been partitioned as in (9), then

1. $D_1 + P_1^{-1} P_2 D_2 = 0$ and $R_1 + P_1^{-1} P_2 R_2 = 0$ if (10) is satisfied
2. The matrix $A_1 + P_1^{-1} P_2 A_3$ is stable if (8) is satisfied.

3.1 Sliding Mode Observer Design

Design of sliding mode observer performs based on a linear transformation construction of coordinates $z = Tx$ [15] so as to impose a specific structure on the fault distribution matrix R. The transformation matrix T has the following form:

$$T = \begin{bmatrix} I_{n-p} & P_1^{-1} P_2 \\ 0 & I_p \end{bmatrix} \tag{13}$$

Equation (25), can be transformed into the new coordinate z as:

$$\begin{cases} \dot{z}(t) = A_z x(t) + B_z u(t) + R_z f_a(t) + D_z d(t) \\ y(t) = C_z z(t) \end{cases} \tag{14}$$

where $A_z = TAT^{-1} = \begin{bmatrix} \bar{A}_{11} & \bar{A}_{12} \\ \bar{A}_{21} & \bar{A}_{22} \end{bmatrix}$; $B_z = TB = \begin{bmatrix} \bar{B}_1 \\ \bar{B}_2 \end{bmatrix}$; $C_z = CT^{-1} = \begin{bmatrix} 0 & I_p \end{bmatrix}$

$$R_z = \begin{bmatrix} R_1 + P_1^{-1} P_2 R_2 \\ R_2 \end{bmatrix} = \begin{bmatrix} 0 \\ R_2 \end{bmatrix}; \text{ and } D_z = \begin{bmatrix} D_1 + P_1^{-1} P_2 D_2 \\ D_2 \end{bmatrix} = \begin{bmatrix} 0 \\ D_2 \end{bmatrix};$$

System (15), can be rewritten as

$$\begin{cases} \dot{z}_1(t) = \bar{A}_{11} z_1(t) + \bar{A}_{12} z_2(t) + \bar{B}_1 u(t) \\ \dot{z}_2(t) = \bar{A}_{21} z_1(t) + \bar{A}_{22} z_2(t) + \bar{B}_2 u(t) + R_2 f_a(t) + D_2 d(t) \\ y(t) = z_2(t) \end{cases} \tag{15}$$

where $z = \begin{bmatrix} z_1^T & z_2^T \end{bmatrix}^T$ with the column $z_1 \in \mathbb{R}^{(n-p)}$ and $z_2 \in \mathbb{R}^p$.

Based on [16], the sliding mode observer is designed as:

$$
\begin{cases}
\dot{\hat{z}}_1(t) = \bar{A}_{11}\hat{z}_1(t) + \bar{A}_{12}y(t) + \bar{B}_1 u(t) \\
\dot{\hat{z}}_2(t) = \bar{A}_{21}\hat{z}_1(t) + \bar{A}_{22}\hat{z}_2(t) + \bar{B}_2 u(t) + (\bar{A}_{22} - A_0)(y(t) - \hat{y}(t)) + v \\
\hat{y}(t) = \hat{z}_2(t)
\end{cases} \tag{16}
$$

where $\hat{z}_1(t)$ and $\hat{z}_2(t)$ are the estimates of $z_1(t)$ and $z_2(t)$, respectively; $\hat{y}(t)$ denotes the estimate of $y(t)$; $A_0 \in \mathbb{R}^{p \times p}$ is a stable design matrix and the discontinuous vector v is computed by

$$
v = \begin{cases}
0 & \text{if } y(t) - \hat{y}(t) = 0 \\
k \dfrac{P_0(y(t) - \hat{y}(t))}{\|P_0[y(t) - \hat{y}(t)]\|} & \text{if } y(t) - \hat{y}(t) \neq 0
\end{cases} \tag{17}
$$

where P_0 is a symmetric positive definite matrix and the positive constant $k = |R_2|\rho + \eta$. With ρ and η are the positive constants. If the state estimation error is defined by $e_1(t) = z_1(t) - \hat{z}_1(t)$ and $e_2(t) = z_2(t) - \hat{z}_2(t)$. Then state error dynamic system can be described as

$$
\dot{e}_1(t) = \bar{A}_{11}e_1(t) \tag{18}
$$

$$
\dot{e}_2(t) = \bar{A}_{21}e_1(t) + \bar{A}_0 e_2(t) + R_2 f_a(t) + D_2 d(t) - v \tag{19}
$$

Theorem 1. For system (5) with Assumptions 1–3. If there exist matrices $P_1 = P_1^T > 0$, P_2, $P_0 = P_0^T > 0$, A_0, and the condition $\|e(t)\| \leq \gamma \|d(t)\|$ with a positive constant γ, such that

$$
P_1 R_1 + P_2 R_2 = 0 \tag{20}
$$

$$
\begin{bmatrix}
\bar{A}_{11}^T P_1 + P_1 \bar{A}_{11} + I_{n-p} & \bar{A}_{21}^T P_0 & 0 \\
* & A_0^T P_0 + P_0 A_0 + I_p & P_0 D_2 \\
* & * & -\gamma I_{n-p}
\end{bmatrix} < 0 \tag{21}
$$

Then the observer error dynamic is asymptotically stable.

Proof of (20).
Based on Lemma 1 that if Assumption 2 is satisfied, then

$$
R_1 + P_1^{-1} P_2 R_2 = 0 \tag{22}
$$

From (20) can be inferred from (22).

Proof of (21).

Consider a Lyapunov function as

$$V = e^T(t)P_z e^T(t) \tag{23}$$

where $P_z = T^{-T}PT^{-1}$ and $e(t) = \begin{bmatrix} e_1^T(t) & e_2^T(t) \end{bmatrix}$.

In the new coordinate, P_z has the following quadratic as

$$P_z = \begin{bmatrix} P_1 & 0 \\ 0 & P_0 \end{bmatrix} \text{ with } P_0 = -P_2^T P_1^{-T} P_2 + P_3 \tag{24}$$

The time derivative (23), we have

$$
\begin{aligned}
\dot{V} &= \dot{e}^T(t)P_z e(t) + e^T(t)P_z \dot{e}(t) \\
&= \dot{e}_1^T(t)P_1 e_1(t) + e_1^T(t)P_1 \dot{e}_1(t) + \dot{e}_2^T(t)P_0 e_2(t) + e_2^T(t)P_0 \dot{e}_2(t) \\
&= \dot{V}_1 + \dot{V}_2
\end{aligned}
\tag{25}
$$

where

$$\dot{V}_1 = \dot{e}_1^T(t)P_1 e_1(t) + e_1^T(t)P_1 \dot{e}_1(t) = e_1^T(t)\bar{A}_{11}^T P_1 + P_1 \bar{A}_{11} e_1(t) \tag{26}$$

and

$$
\begin{aligned}
\dot{V}_2 &= \dot{e}_2^T(t)P_0 e_2(t) + e_2^T(t)P_0 \dot{e}_2(t) \\
&\leq e_2^T(t)\left[\bar{A}_0^T P_0 + P_0 \bar{A}_0\right]e_2(t) + e_1^T(t)\bar{A}_{21}^T P_0 e_2(t) + e_2^T(t)P_0 \bar{A}_{21} e_1(t) \\
&\quad + e_2^T(t)P_0 D_2 d(t) + d^T(t)D_2^T P_0 \dot{e}_2(t)
\end{aligned}
\tag{27}
$$

To obtain the asymptotical stability of the state estimation errors $e_1(t)$, and $e_2(t)$ then the Eq. (23) needs to satisfy the following condition:

$$\dot{V}_1 \leq 0; \quad \dot{V}_2 \leq 0$$

Based on the initial condition $\|e(t)\| \leq \gamma\|d(t)\|$, the matrix J can be described as:

$$J = \begin{bmatrix} e_1(t) \\ e_2(t) \\ d(t) \end{bmatrix}^T \begin{bmatrix} I_{n-p} & 0 & 0 \\ 0 & I_p & 0 \\ 0 & 0 & -\gamma I_{n-p} \end{bmatrix} \begin{bmatrix} e_1(t) \\ e_2(t) \\ d(t) \end{bmatrix} \leq 0 \tag{28}$$

From (25)–(28), we have

$$
\Pi = \dot{V} + J
$$

$$
= \begin{bmatrix} e_1(t) \\ e_2(t) \\ d(t) \end{bmatrix}^T \begin{bmatrix} \bar{A}_{11}^T P_1 + P_1 \bar{A}_{11} + I_{n-p} & \bar{A}_{21}^T P_0 & 0 \\ * & A_0^T P_0 + P_0 A_0 + I_p & P_0 D_2 \\ ** & & -\gamma I_{n-p} \end{bmatrix} \begin{bmatrix} e_1(t) \\ e_2(t) \\ d(t) \end{bmatrix} \quad (29)
$$

With Eq. (29), then (21) is satisfied.
Proof is complete.
Equation (20) need to transform into a linear matrix inequality to solve by Matlab toolbox. This transformation is performed into the issue of finding the minimum of a positive scalar μ satisfying the following inequality constraint:

$$
\begin{bmatrix} \mu I_{n-p} & P_1 R_1 + P_2 R_2 \\ * & \mu I_r \end{bmatrix} > 0 \quad (30)
$$

By solving the LMI (21) and (30), we obtain P_1, P_2, P_0 and Q. from this, we can calculate the observer gain $A_0 = P_0^{-1} Q$ by substituting $C = C_z T$, $R = T^{-1} R_z$ and $P = T^T P_z T$.

For the error dynamic system (18), and (19), the sliding mode surface is defined as

$$
S = \{(e_1, e_2) | e_2 = 0\} \quad (31)
$$

Theorem 2. Under the Assumptions 1–3, and the observer (17). Then the error system (18) and (19) can be driven to the sliding surface (31) in finite time if gain k is chosen to satisfy

$$
k = \|R_2\|\delta_a + \|D_2\|\delta_d + \zeta_1 \geq \|R_2\|\delta_a + \|D_2\|\delta_d + \|\bar{A}_{21}\|\varepsilon + \zeta_2 \quad (32)
$$

where $\zeta_1 \geq \|\bar{A}_{21}\|\varepsilon + \zeta_2$, ε is the upper bound of $\|e\|$ and ζ_2 is a positive scalar.

Proof of (32).
Consider a Lyapunov function as

$$
V_s = e_2^T(t) P_0 e_2(t) \quad (33)
$$

The derivative (33), one has

$$
\begin{aligned}
\dot{V}_s &= \dot{e}_2^T(t) P_0 e_2(t) + e_2^T(t) P_0 \dot{e}_2(t) \\
&= e_1^T(t) \bar{A}_{21}^T P_0 e_2(t) + e_2^T(t) \bar{A}_0^T P_0 e_2(t) + f_a^T(t) R_2^T P_0 e_2(t) + d^T(t) D_2^T P_0 e_2(t) \\
&\quad - v^T P_0 e_2(t) + e_2^T(t) P_0 \bar{A}_{21} e_1(t) + e_2^T(t) P_0 \bar{A}_0 e_2(t) + e_2^T(t) P_0 R_2 f_a(t) \\
&\quad + e_2^T(t) P_0 D_2 d(t) - e_2^T(t) P_0 v
\end{aligned} \quad (34)
$$

Based on the inequation $2X^T Y \leq X^T X + Y^T Y$ is true for any X, Y, and since the matrix A_0 is a stable matrix by design. Therefore $A_0^T P_0 + P_0 A_0 < 0$, from these problems, (34) is obtained as

$$
\begin{aligned}
\dot{V}_s &\leq 2e_2^T(t) P_0 \bar{A}_{21} e_1(t) + 2e_2^T(t) P_0 R_2 f_a(t) + 2e_2^T(t) P_0 D_2 d(t) - 2e_2^T(t) P_0 v \\
&\leq 2\|P_0 e_2(t)\| (\|\bar{A}_{21}\| \varepsilon + \|R_2\| \delta_a + \|D_2\| \delta_d - \zeta_1)
\end{aligned}
\tag{35}
$$

If the condition (35) holds, then

$$
\dot{V}_s \leq -2\zeta_2 \|P_0 e_2(t)\| \leq -2\zeta_2 \sqrt{\lambda_{\min}(P_0)} V_s^{0.5}
\tag{36}
$$

where $\lambda_{\min}(P_0)$ is the smallest eigenvalue of P_0. This shows that the reachability condition is satisfied. As a consequence, an ideal sliding motion will take place on the surface S in finite time [15].   □

3.2 Actuator Fault Estimation

The actuator fault estimation based on the proposed observer in the form of (17) is to estimate actuator faults using the so-called equivalent output injection [16]. Assuming that a sliding motion has been obtained, then $e_2 = 0$ and $\dot{e}_2 = 0$. Equation (20) is presented as

$$
0 = \bar{A}_{21} e_1(t) + R_2 f_a(t) + D_2 d(t) - v_{eq}
\tag{37}
$$

where v_{eq} is the named equivalent output error injection signal which is required to maintain the motion on the sliding surface [16].

The discontinuous component in (18) can be approximated by the continuous approximation as

$$
v = k \frac{P_0 (y(t) - \hat{y}(t))}{\|P_0[y(t) - \hat{y}(t)]\| + \xi}
\tag{38}
$$

where ξ is a small positive scaler to reduce the chattering effect, with this approximation, the error dynamics can not slide on the surface S perfectly, but within a small boundary layer around it [16].

Based on [16], the actuator fault estimation is defined as

$$
\hat{f}_a = R_2^+ v_{eq}.
\tag{39}
$$

where $R_2^+ = (R_2^T R_2)^{-1} R_2^T$.

From Eqs. (37) and (39) can be represented as

$$f_a(t) - \hat{f}_a(t) = -R_2^+ \bar{A}_{21} e_1(t) - R_2^+ D_2 d(t) \tag{40}$$

By considering the norm of (40), we obtain

$$\begin{aligned}
\left\| f_a(t) - \hat{f}_a(t) \right\| &= \left\| R_2^+ \bar{A}_{21} e_1(t) + R_2^+ D_2 d(t) \right\| \\
&\leq \sigma_{max} \left(R_2^+ \bar{A}_{21} \right) \|e(t)\| + \sigma_{max} \left(R_2^+ D_2 \right) \|d(t)\|
\end{aligned} \tag{41}$$

With using the initial condition, $\|e(t)\| \leq \gamma \|d(t)\|$, we have

$$\left\| f_a(t) - \hat{f}_a(t) \right\| \leq \left[\gamma \sigma_{max} \left(R_2^+ \bar{A}_{21} \right) + \sigma_{max} \left(R_2^+ D_2 \right) \right] \|d(t)\| = \varsigma \|d(t)\| \tag{42}$$

where $\varsigma = \gamma \sigma_{max} \left(R_2^+ \bar{A}_{21} \right) + \sigma_{max} \left(R_2^+ D_2 \right)$.

Therefore, for a rather small $\varsigma \|d(t)\|$, then the actuator can be approximated as

$$\hat{f}_a = k \frac{R_2^+ P_0 [y(t) - \hat{y}(t)]}{\|P_0 [y(t) - \hat{y}(t)]\| + \xi} \tag{43}$$

4 Simulation Results

Simulation results are applied for the EHA system using parameters in [14]. The nonlinear continuous time state space model of the EHA system is described as

$$\dot{x}(t) = Ax(t) + Bu(t) + f(x(t), u(t)) + Dd(t) \tag{44}$$

However, to apply for the linear system, the friction force is ignored and therefore, the component $f(x(t), u(t))$ in (44) is equal to zeros. With the parameters of the EHA system in [14]; we obtain

$$A = \begin{bmatrix} 0 & 1 \\ -5.5457e+02 & -1.4643e+03 \end{bmatrix}; B = \begin{bmatrix} 0 \\ 1.8571e-04 \end{bmatrix}; R = \begin{bmatrix} 0.1 \\ 0.002 \end{bmatrix}$$

The matrix T_c is chosen by

$$T_c = \begin{bmatrix} -7.0711e-01 & -7.0711e-01 & 7.0711e-01 \\ 1 & 0 & 1 \\ 0 & 1 & 0 \end{bmatrix}$$

Based on the design experience, the positive scalar $\gamma = 0.002$, disturbance vector $D = \begin{bmatrix} 1e-7 & 1.8571e-2 \end{bmatrix}^T$. We can be solved for P_1, P_2, P_0, Q, A_0, and F by using the optimal algorithm LMI if the solution is feasible, then the results be obtained as:

$P_1 = 1.0460e\text{-}07$;

$P_0 = \begin{bmatrix} 1.0003e+00 & -3.3249e\text{-}09 \\ -3.3249e\text{-}09 & 2.1264e\text{-}05 \end{bmatrix}$;

$Q = \begin{bmatrix} -9.9985e\text{-}01 & 0 \\ 1.5161e\text{-}04 & -1.2257e+00 \end{bmatrix}$;

$P_2 = \begin{bmatrix} 3.9827e\text{-}13 & 7.3963e\text{-}08 \end{bmatrix}$;

$A_0 = \begin{bmatrix} -9.9957e\text{-}01 & -1.9160e\text{-}04 \\ 7.1295e+00 & -5.7642e+04 \end{bmatrix}$;

$F = \begin{bmatrix} 9.9967e\text{-}08 & 3.9490e\text{-}07 \end{bmatrix}$

The reference input is given as

$$y_r(t) = \sin \pi t \tag{45}$$

4.1 For the Fault-Free Case and Without Disturbance

The simulation result of the position and velocity response is shown in Fig. 2.

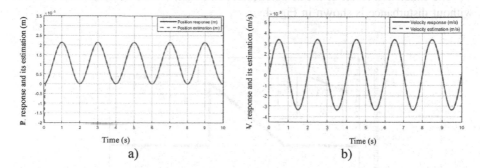

Fig. 2. Responses and their estimations: (a) position and (b) velocity

The results in Fig. 2 show the estimation signal following the response signal. This proves that the estimated error dynamic is asymptotically stable under SMO construction of the proposed method.

4.2 For Actuator Fault Case and Without Disturbance

Assume that the actuator fault is given by equation as following

$$f_a(t) = \begin{cases} 0 & \text{if } t \le 3 \\ (1/600)(t-3) & \text{if } 3 < t \le 6 \\ 0 & \text{if } t \ge 6 \end{cases} \tag{46}$$

The simulation result of the position and velocity response and their estimations for the actuator fault (46) and without disturbance is shown in Fig. 3.

The results in Fig. 3 show the influence of the response signals under the impact of actuator fault.

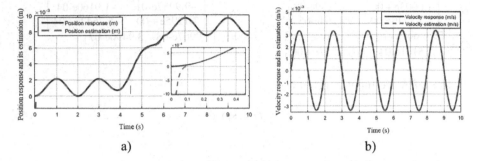

a)   b)

Fig. 3. Responses and their estimations: (a) position and (b) velocity

The simulation result of the actuator fault estimation for the actuator fault (46) and without disturbance is shown in Fig. 4.

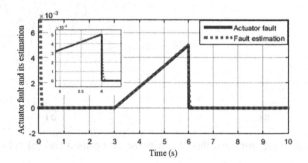

Fig. 4. Actuator fault and its estimation

4.3 For Actuator Fault Case and with Disturbance

Assuming that the disturbance is given as

$$d(t) = 0.0025 \, random(1, t) \tag{47}$$

and the actuator fault is given in (46). The simulation result of the position and velocity response and their estimation for the actuator fault (46) and with disturbance (47) is shown in Fig. 5.

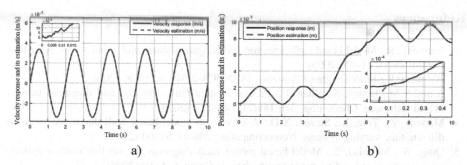

a)   b)

Fig. 5. Responses and their estimations with disturbance $d(t)$: (a) position (b) velocity

The simulation result of the actuator fault estimation for the actuator fault (46) and with disturbance (47) is shown in Fig. 6.

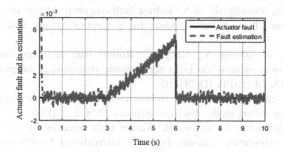

Fig. 6. Actuator fault and its estimation in disturbance case

We can see the difference between the position response under the influences of the disturbance in Fig. 5(a) compared with Figs. 3(a) and 5(b) compared with Fig. 3(b) under the impact of the disturbance $d(t)$. We can see the magnitude of the actuator fault and disturbance estimation in Fig. 6.

5 Conclusions

In this paper, an SMO construction was applied to estimate the actuator fault. The result showed the estimated error dynamic is asymptotically stable under SMO construction based on LMI optimization algorithm and Lyapunov's stability condition. The results of the estimated actuator were achieved the desired result. The influences in the case without and with disturbance were also compared to show their effects in the system. The actuator fault estimation approach was successfully applied to the MMP system model.

Acknowledgments. This work was supported by Korea Hydro & Nuclear Power company through the project "Nuclear Innovation Center for Haeoleum Alliance".

References

1. Isermann, R.: Model-based fault detection and diagnosis-status and applications. Ann. Rev. Control **29**(1), 71–85 (2005)
2. Isermann, R.: Fault-Diagnosis Systems. ISBN 978-3-540-30368-8, (2006)
3. Amin, T.M., Imtiaz, S., Khan, F.: Process system fault detection and diagnosis using a hybrid technique. Chem. Eng. Sci. **189**(2), 191–211 (2018)
4. Maiying, Z., Ting, X., Steven, X.D.: A survey on model-based fault diagnosis for linear discrete time-varying systems. Neurocomputing **306**(6), 51–60 (2018)
5. Qing, W., Mehrdad, S.: Model-based robust fault diagnosis for satellite control systems using learning and sliding mode approaches. J. Comput. **4**, 10 (2009)
6. Bokor, J.: Fault detection and isolation in nonlinear systems. IFAC Proc. Vol. **42**(8), 1–11 (2009)
7. Xiaodong, Z.: Sensor bias fault detection and isolation in a class of nonlinear uncertain systems using adaptive estimation. IEEE Trans. Autom. Control **56**, 5 (2011)
8. Aiguo, W., Guangren, D.: Robust fault detection in linear systems based on full-order state observers. J. Control Theor. Appl. **5**(4), 325–330 (2007)
9. Ziyabari, S.H.S., Shoorehdeli, M.A.: Robust fault diagnosis scheme in a class of nonlinear system based on UIO and fuzzy residual. Int. J. Control Autom. Syst. **15**(3), 1145–1154 (2017)
10. Jia, Q., Chen, W., Zhang, Y., Jiang, Y.: Robust fault reconstruction in discrete-time Lipschitz nonlinear systems via Euler-approximate proportional integral observers. Math. Prob. Eng. **2015**, 14 (2015). ID 741702
11. Xiaodong, C., Jinquan, H., Feng, l.: Robust in-flight sensor fault diagnostics for aircraft engine based on sliding mode observers. Sensors **17**, 835 (2017)
12. Bouibed, K., Aitouche, A., Bayart, M.: Sensor fault detection by sliding mode observer applied to an autonomous vehicle. In: 2009 International Conference on Advances in Computational Tools for Engineering Applications, Zouk Mosbeh, pp. 621–626 (2009)
13. Zhao, B., Li, C., Liu, D., Li, Y.: Decentralized sliding mode observer based dual closed-loop fault tolerant control for reconfigurable manipulator against actuator failure. PLoS ONE **10**, 7 (2015)
14. Tan, V.N., Cheolkeun, H.: Sensor fault-tolerant control design for mini motion package electro-hydraulic actuator. MDPI Process. **7**, 89 (2019)
15. Shtessel, Y., Edwards, C., Fridman, L., Levant, A.: Sliding Mode Control and Observation (2015). ISBN 978-0-8176-4893-0
16. Yuri, S., Christopher, E., Leonid, F., Arie, L.: Sliding Mode Control and Observation. Springer, Heidelberg (2013)
17. Edwards, C., Spurgeon, S.: Sliding Mode Control: Theory and Applications. Taylor and Francis, Abingdon (1998)

A Fault Tolerant Strategy Based on Model Predictive Control for Full Bidirectional Switches Indirect Matrix Converter

Van-Tien Le[1] and Hong-Hee Lee[2(⊠)]

[1] Graduate School of Electrical Engineering, University of Ulsan,
Ulsan, South Korea
tien94.levan@gmail.com
[2] School of Electrical Engineering, University of Ulsan, Ulsan, South Korea
hhlee@mail.ulsan.ac.kr

Abstract. This paper proposes an open-switch fault tolerant strategy based on the model predictive control for a full bidirectional switches indirect matrix converter (FBS-IMC). Compared to the conventional IMC, the FBS-IMC can provide healthy current path when the open-switch fault is occurred. To keep the continuous operation, the fault tolerant strategy is developed by means of reversing the DC link voltage polarity regardless of the faulty switch location in the rectifier or inverter stages. Therefore, the proposed fault tolerant strategy can maintain the same input and output performances during the faulty condition as the normal condition. The simulation results are given to verify the effectiveness of the proposed strategy.

Keywords: Full bidirectional switches indirect matrix converter ·
Fault tolerant · Predictive control for indirect matrix converter

1 Introduction

The matrix converter (MC) is a direct AC/AC converter which has several advantages such as sinusoidal input and output current waveforms, bidirectional power flow, controllable input power factor and no bulky intermediate energy storage. The MCs are classified into direct matrix converter (DMC) and indirect matrix converter (IMC). Compared to the DMC, the IMC requires simpler commutation and clamp circuit [1]. Figure 1 shows a topology of the IMC, which consists of a separated rectifier stage and an inverter stage. Due to the absence of intermediate energy storage on the DC-link, IMCs have longer life time and more compact design compared to the back-to-back converters with equivalent power ratings. With these advantages, IMCs have become attractive to many applications such as in wind power generation [2], induction motor drives [3] and power supply [4].

The applications in [2–4] often require a high reliability and robustness against the power switch failures. Therefore, the control method with fault tolerant capability is necessary to improve the system reliability. In recent years, many publications have focused on the open-switch fault diagnosis and fault tolerance methods for MCs

© Springer Nature Switzerland AG 2019
D.-S. Huang et al. (Eds.): ICIC 2019, LNAI 11645, pp. 271–282, 2019.
https://doi.org/10.1007/978-3-030-26766-7_25

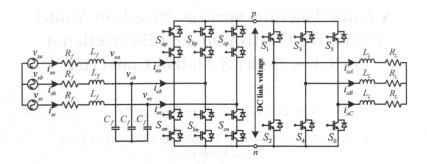

Fig. 1. Topology of IMC.

[5–10]. For example, in [5], the fault tolerant control strategies for MC are proposed based on the concept of four-leg MC. In [6], the optimal duty cycles of the most appropriate switching states are obtained based on the Karush-Kuhn-Tucker conditions to drive the MC as close as possible to its desired normal operation. In [7], the fault tolerance method based on the model predictive control (MPC) is proposed by the optimal switching state among the remaining healthy switching states to drive the MC during each control period. In [8], a fault tolerant switching strategy is developed for MCs based on the optimum Alesina-Venturini Modulation without additional devices. Even though these methods [5–8] can maintain the continuous operation of the MCs, the system performance is degraded under open-switch fault conditions.

Most of the aforementioned researches [5–8] are focused on the fault tolerance control method for the DMCs. The fault tolerance control method for IMCs has not been sufficiently studied until now. In [9], a back-up phase is added into the inverter stage to continuously operate the IMC when an open-switch fault occurs in the inverter stage. In [10], the full bidirectional switches indirect matrix converter (FBS-IMC) is presented by adding six additional switches into the inverter stage as backup switch to improve the fault-tolerant capacity of the IMC. However, the fault tolerant strategy in [10] has only treated the open-switch fault in the inverter stage of the FBS-IMC.

In order to fulfill the fault tolerant capacity of the FBS-IMC, this paper proposes the fault tolerant strategy based on MPC to maintain the continuous operation of the system irrespective of the open-switch fault location in the rectifier or inverter stages. The proposed fault tolerant strategy is implemented simply by reversing the DC link voltage polarity when the open-switch fault is detected. Compared to the previous methods, the proposed method can maintain the good performances in both the input and output sides during the faulty condition as the normal condition. The proposed fault tolerant strategy was verified by simulation results.

2 Model Predictive Control for IMC

The MPC has been applied to achieve fast dynamic response and simple implementation for IMC [16]. The main idea of this method is using the system model to predict the future behavior of the control variables. Based on the system model, the cost function is defined as the absolute error between the control variable references and their predicted value. After that, the switching state which minimizes the cost function is selected to control the converter in the next sampling period. Our control objectives are how to track the output current reference and achieve unity input power factor for IMC. The input power factor is indirectly control by means of the instantaneous input reactive power. Therefore, the modeling of the IMC, input filter and the load are firstly required to implement the MPC.

2.1   Modeling of IMC System

In the conventional topology of IMC, the rectifier stage is directly connected to the inverter stage through a fictitious DC-link, as shown in Fig. 1. Therefore, the DC-link voltage v_{DC} is obtained as a function of the state of the rectifier switches and the input phase voltage v_e as follows:

$$v_{DC} = [S_{ap} - S_{an} \quad S_{bp} - S_{bn} \quad S_{cp} - S_{cn}]v_e, \tag{1}$$

where S_{xy} represents the state of each six rectifier switches ($x \in \{a, b, c\}$ and $y \in \{p, n\}$). The state is equal to 1 when the switch is turned on and 0 as the switch is turned off. The input current i_e is a function of the DC-link current i_{DC} and the state of the rectifier switches:

$$i_e = [S_{ap} - S_{an} \quad S_{bp} - S_{bn} \quad S_{cp} - S_{cn}]^T i_{DC}, \tag{2}$$

where the superscription T is the transposition of matrix. For the mathematical model of the inverter stage, the DC-link current i_{DC} is a function of the load current i_o and the state of the inverter switches:

$$i_{DC} = [S_1 - S_2 \quad S_3 - S_4 \quad S_5 - S_6]i_o. \tag{3}$$

where S_i represents the state of each six inverter switches ($i = 1, 2, \dots 6$). The state is equal to 1 when the switch is turned on and 0 when the switch is turned off. The load voltage v_o is a function of the state of the inverter switches and the DC-link voltage v_{DC}:

$$v_o = [S_1 - S_2 \quad S_3 - S_4 \quad S_5 - S_6]^T v_{DC}. \tag{4}$$

2.2   Modeling of Input Filter and RL Load

Mathematical models of the input filter and RL load are used to predict the future values of the input currents and load currents, which are needed to evaluate the cost function. To attenuate the high-order harmonics caused by the high switching frequency, the second-order LC filter is used in the input side of the rectifier stage. The state-space model of input filter is expressed as follows:

$$\begin{bmatrix} \frac{di_s}{dt} \\ \frac{dv_e}{dt} \end{bmatrix} = A \begin{bmatrix} i_s \\ v_e \end{bmatrix} + B \begin{bmatrix} i_e \\ v_s \end{bmatrix}, \tag{5}$$

where the matrices A and B are calculated as follows:

$$A = \begin{bmatrix} -1/L_f & -R_f/L_f \\ 0 & 1/C_f \end{bmatrix}, \ B = \begin{bmatrix} 1/L_f & 0 \\ 0 & -1/C_f \end{bmatrix}.$$

where v_s and i_s are the source voltage and source current, respectively, and R_f, L_f, and C_f are the resistance, inductance and capacitance of the input filter, respectively.

The dynamic model of the RL load is expressed as follows:

$$L_L \frac{di_o}{dt} = v_o - R_L i_o. \tag{6}$$

Considering sampling period T_s, the dynamic models (5) and (6) are discretized as:

$$\begin{bmatrix} i_s^{k+1} \\ v_e^{k+1} \end{bmatrix} = C \begin{bmatrix} i_s^k \\ v_e^k \end{bmatrix} + D \begin{bmatrix} i_e^k \\ v_s^k \end{bmatrix}.$$

$$i_o^{k+1} = \left(1 - \frac{R_L T_s}{L_L}\right) i_o^k + \frac{T_s}{L_L} v_o^k \tag{7}$$

where C and D are calculated as follows:

$$C = e^{AT_s} \quad D = A^{-1}(C - I)B$$

By using (7), the source current and load current in the next sampling time $(k+1)^{th}$ are predicted at every sampling time k^{th}.

2.3   The Cost Function Design

The goals of the control method are how to track the output current reference and achieve unity input power factor for IMC. Therefore, the error between the predicted load currents and their references are expressed in the following cost function

$$\Delta i_o^{k+1} = \left| i_{o\alpha}^* - i_{o\alpha}^{k+1} \right| + \left| i_{o\beta}^* - i_{o\beta}^{k+1} \right|, \tag{8}$$

where $i_{o\alpha}^{k+1}$ and $i_{o\beta}^{k+1}$ are the load currents in $\alpha\beta$ coordinates at $(k+1)^{th}$ sampling time, and $i_{o\alpha}^*$ and $i_{o\beta}^*$ are the load current references in $\alpha\beta$ coordinates, respectively. On the other hand, the instantaneous input reactive power is regulated to indirectly control the input power factor. The error between the predicted value of the instantaneous input reactive power and its reference is expressed as:

$$\Delta q^{k+1} = \left| q^*_{k+1} - (v^{k+1}_{s\alpha} i^{k+1}_{s\beta} - v^{k+1}_{s\beta} i^{k+1}_{s\alpha}) \right|, \tag{9}$$

where the $v^{k+1}_{s\alpha}, v^{k+1}_{s\beta}, i^{k+1}_{s\alpha}$ and $i^{k+1}_{s\beta}$ denote the source voltages and source currents in $\alpha\beta$ coordinates at $(k+1)^{th}$ sampling time, respectively, and q^*_{k+1} is the instantaneous input reactive power reference. To achieve the unity input power factor, the instantaneous input reactive power reference q^*_{k+1} is set to zero.

In order to avoid the short circuit in the inverter stage, the DC-link voltage must be positive at all the time. To satisfy this condition, the following equation is defined:

$$h^{k+1} = \begin{cases} 0, & v^{k+1}_{DC} > 0 \\ M, & v^{k+1}_{DC} \leq 0 \end{cases}, \tag{10}$$

where the value of the positive number M is big enough in the cost function to maintain the positive DC-link voltage at all the time. The DC-link voltage v^{k+1}_{DC} is estimated by using (1) and (7). It is worth noting that the given term (10) is not needed in the proposed method in Sect. 4 because the negative DC-link voltage is allowed in the proposed fault tolerance method.

Finally, three above conditions are merged into the single cost function as follows:

$$g^{k+1} = \Delta q^{k+1} + \lambda.\Delta i^{k+1}_o + h^{k+1}, \tag{11}$$

where λ is weighting factor, which is tuned empirically. As shown in Tables 1 and 2, there are 9 and 8 valid switching states for the rectifier and inverter stages, respectively. Therefore, at each sampling time, 72 combination valid switching states of IMC are used to calculate the cost function (11), and the switching state N that minimizes the cost function is applied in the next sampling time.

$$g[N] = \min\{g[1], g[2], g[3], \ldots, g[72]\}. \tag{12}$$

Table 1. The valid switching states of the rectifier stage

No.	S_{ap}	S_{bp}	S_{cp}	S_{an}	S_{bn}	S_{cn}	v_{DC}
1	1	0	0	0	1	0	v_{ab}
2	1	0	0	0	0	1	v_{ac}
3	0	1	0	1	0	0	v_{ba}
4	0	1	0	0	0	1	v_{bc}
5	0	0	1	1	0	0	v_{ca}
6	0	0	1	0	1	0	v_{cb}
7	1	0	0	1	0	0	0
8	0	1	0	0	1	0	0
9	0	0	1	0	0	1	0

Table 2. The valid switching states of the inverter stage

No.	S_1	S_3	S_5	S_2	S_4	S_6	v_{AB}	v_{BC}	v_{CA}
1	1	1	0	0	0	1	0	v_{DC}	$-v_{DC}$
2	1	0	0	0	1	1	v_{DC}	0	$-v_{DC}$
3	1	0	1	0	1	0	v_{DC}	$-v_{DC}$	0
4	0	0	1	1	1	0	0	$-v_{DC}$	v_{DC}
5	0	1	1	1	0	0	$-v_{DC}$	0	v_{DC}
6	0	1	0	1	0	1	$-v_{DC}$	v_{DC}	0
7	1	1	1	0	0	0	0	0	0
8	0	0	0	1	1	1	0	0	0

3   Fault Tolerant Topology for Indirect Matrix Converter

Figure 2 shows the topology of the FBS-IMC. This topology is first presented in [10] for fault tolerant strategy when the open-switch fault occurs in the inverter stage. Comparing with the conventional IMC in Fig. 1, the FBS-IMC has the same structure of the rectifier stage, whereas the inverter stage of the FBS-IMC consists of six bidirectional switches that are arranged as the same structure as the inverter stage of conventional IMC. When an open-switch fault occurs, this topology can provide the alternative healthy current path to replace the open faulty current path. The fault tolerant strategy for the FBS-IMC is presented in Sect. 4.

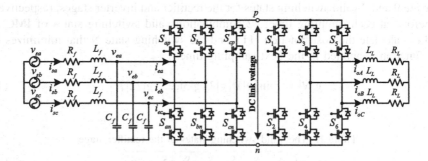

Fig. 2. The topology of FBS-IMC.

4   The Fault Tolerant Strategy for FBS-IMC

When the open-switch fault occurs, the faulty switch must be rapidly detected and isolated from the power circuit. After that, the fault tolerant strategy is applied to keep the continuous operation of the FBS-IMC and avoid any damages caused by overvoltage and overcurrent. To achieve these purposes, we propose the open-switch fault tolerant strategy based on MPC for faulty switch either in the rectifier or inverter stages. The principle of the fault tolerant strategy is the same for both the rectifier and inverter stages of FBS-IMC. When a specific bidirectional faulty switch is detected, the other healthy bidirectional switch in the faulty leg is triggered immediately to give an alternative current path to the faulty current path. And, the driving signals of the healthy switches are adjusted according to the state of the detected faulty switch. For more specific, the fault tolerant strategy is analyzed in two cases:

Case 1. The driving signal of the faulty switch is 0.
Noting that the power switch is turned on when its driving signal is 1, and turned off when its driving signal is 0. Therefore, the operation of FBS-IMC is not affected by the open-switch fault when the driving signal of the faulty switch is 0 regardless of the faulty switch location in the rectifier or inverter stages. In this case, the fault tolerant strategy keeps the same driving signals for all power switches as the normal condition.

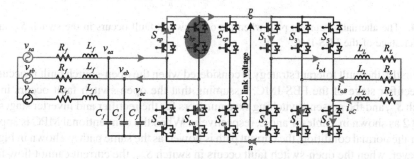

Fig. 3. Current path when the open-switch fault occurs in the switch S_{bp}.

For example, assuming that the open-switch fault occurs in the switch S_{bp}, and the present switching state numbers for the rectifier and inverter stages are 1 and 2 as shown in Tables 1 and 2, respectively. It is clearly that the operation of the converter is not affected by the faulty switch, and the current flows consistently between the source and the load as shown in Fig. 3. The driving signals of all switches are kept the same of that in the conventional MPC under normal condition:

$$\begin{cases} G^f_{S_{ap}} = G_{S_{ap}} \\ G^f_{S_{an}} = G_{S_{an}}, \\ G^f_{S_{bp}} = G_{S_{bp}} \end{cases} \begin{cases} G^f_{S_{bn}} = G_{S_{bn}} \\ G^f_{S_{cp}} = G_{S_{cp}}, \\ G^f_{S_{cn}} = G_{S_{cn}} \end{cases} \begin{cases} G^f_{S_1} = G_{S_1} \\ G^f_{S_2} = G_{S_2}, and \\ G^f_{S_3} = G_{S_3} \end{cases} \begin{cases} G^f_{S_4} = G_{S_4} \\ G^f_{S_5} = G_{S_5}, \\ G^f_{S_6} = G_{S_6} \end{cases}$$ (13)

where G is the driving signal when the conventional MPC is applied under normal condition and G^f is the driving signal when the fault tolerant strategy is activated.

Case 2. The driving signal of the faulty switch is 1.
In this case, the faulty switch is always opened even though its driving signal is 1. Therefore, there is no current path that connects the source and the load if the conventional MPC is applied. To overcome this problem, the proposed fault tolerant strategy controls all the healthy switches by means of exchanging the driving signal between the upper and lower switches in the same leg. Therefore, the alternative

current path is replaced for the open faulty current path by reversing the polarity of the DC-link voltage.

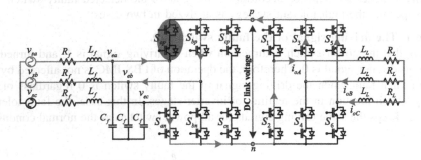

Fig. 4. The alternative current path when the open-switch fault occurs in the switch S_{ap} in the rectifier stage of the FBS-IMC.

Firstly, the fault tolerant strategy is considered when the open-switch fault is occurs in the rectifier stage of the FBS-IMC. Assuming that the open-switch fault occurs in the switch S_{ap} and the present switching state numbers for the rectifier and inverter stages are 1 and 2 as shown in Tables 1 and 2, respectively. When the conventional MPC is applied under the normal condition, the current path is flowed as the same path as shown in Fig. 3. However, when the open-switch fault occurs in switch S_{ap}, the current cannot flow from the source to the load through the current path in Fig. 3 because the connection between input phase a and DC-link polarity p is cut out. After the proposed fault tolerant strategy is activated, the faulty switch S_{ap} is quickly isolated by turning off its driving signal. And, the current path from the input phase a to the load phase A through the DC-link polarity p is replaced by the alternative current path through the DC-link polarity n by turning on the switches S_{an} and S_2 instead of S_{ap} and S_1, respectively. The currents back to the source through the switches S_3, S_5 and S_{bp} instead of S_4, S_6 and S_{bn}, respectively. Figure 4 shows the healthy current path in the proposed fault tolerant strategy. Compared to the conventional MPC, the proposed fault tolerant strategy can keep the continuous operation of the FBS-IMC without faulty effect on the input and output performances. In general, the driving signals of all switches are modified as follows:

$$
\begin{cases} G^f_{S_{ap}} = 0 \\ G^f_{S_{an}} = G_{S_{ap}}, \\ G^f_{S_{bp}} = G_{S_{bn}} \end{cases}
\begin{cases} G^f_{S_{bn}} = G_{S_{bp}} \\ G^f_{S_{cp}} = G_{S_{cn}}, \\ G^f_{S_{cn}} = G_{S_{cp}} \end{cases}
\begin{cases} G^f_{S_1} = G_{S_2} \\ G^f_{S_2} = G_{S_1}, and \\ G^f_{S_3} = G_{S_4} \end{cases}
\begin{cases} G^f_{S_4} = G_{S_3} \\ G^f_{S_5} = G_{S_6} \\ G^f_{S_6} = G_{S_5} \end{cases}. \qquad (14)
$$

Secondly, the fault tolerant strategy is considered when the open-switch fault is occurs in the inverter stage of the FBS-IMC. Assuming that the open-switch fault occurs in the switch S_1 and the present switching state numbers for the rectifier and inverter stages are 1 and 2 as shown in Tables 1 and 2, respectively. Figure 5 shows the healthy current path when the proposed fault tolerant strategy is activated. In this case, the current flows from the input phase a to the load phase A through the switches S_{an}

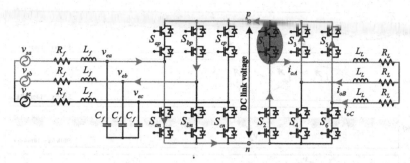

Fig. 5. The alternative current path when the open-switch fault occurs in the switch S_1 in the inverter stage of the FBS-IMC.

and S_2, respectively, and the currents back to the source through S_3, S_5 and S_{bp}. Therefore, the connection between the source and the load is maintained even if the open-switch fault is occurred in the inverter stage of the FBS-IMC. The driving signals of all switches in the proposed fault tolerant strategy are modified as follows:

$$\begin{cases} G_{S_{ap}}^f = G_{S_{an}} \\ G_{S_{an}}^f = G_{S_{ap}}, \\ G_{S_{bp}}^f = G_{S_{bn}} \end{cases} \begin{cases} G_{S_{bn}}^f = G_{S_{bp}} \\ G_{S_{cp}}^f = G_{S_{cn}}, \\ G_{S_{cn}}^f = G_{S_{cp}} \end{cases} \begin{cases} G_{S_1}^f = 0 \\ G_{S_2}^f = G_{S_1}, and \\ G_{S_3}^f = G_{S_4} \end{cases} \begin{cases} G_{S_4}^f = G_{S_3} \\ G_{S_5}^f = G_{S_6} \\ G_{S_6}^f = G_{S_5} \end{cases} \quad (15)$$

5   Simulation Results

In order to verify the effectiveness of the proposed fault tolerant strategy, the simulation is carried out by using PSIM software. The simulation parameters are listed in Table 3. The load current reference is set to be 10A/50 Hz for all simulation tests.

Table 3. Simulation parameters

Source voltage (V_{RMS})	220 V
Input frequency	60 Hz
Input filter parameters	$R_f = 5\,\Omega; L_f = 130\,\mu H; C_f = 40\,\mu F$
Load parameters	$R_L = 10\,\Omega; L_L = 35\,mH$
Output frequency	50 Hz
Sampling period	25 μs

Figure 6 shows the performance of the conventional MPC for the FBS-IMC under the normal condition. As shown in Fig. 6(a) and (b), the source and load currents are balanced sinusoidal. The positive DC-link voltage is maintained at all the time, as shown in Fig. 6(c). Figure 6(d) shows that the source current phase a is in phase with the source voltage phase a, which means that the unity input power factor is achieved.

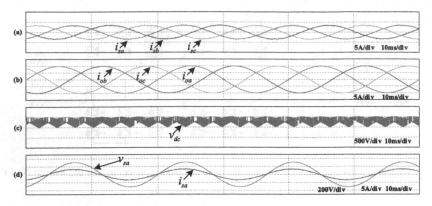

Fig. 6. Simulation results of the conventional MPC for the FBS-IMC under the normal condition: (a) The source currents, (b) the load currents, (c) the DC-link voltage, (d) the source voltage and current phase *a*.

Figure 7 shows the performance of the FBS-IMC when the open-switch fault occurs in the switch S_{ap} in the rectifier stage. When the conventional MPC is applied in the first half of Fig. 7, the source and load currents are severely deteriorated and the overvoltage is occurred in DC-link voltage, which may damage the power circuit. In contrast, when the proposed fault tolerant strategy is activated in the latter half of Fig. 7, the source and load currents are balanced sinusoidal and the overvoltage is the DC-link voltage is avoided, as shown in Fig. 7(a) to (c). We can see that the polarity of the DC-link voltage is reversed at some periods due to exchanging the driving signals between the upper and lower switches. The latter half of Fig. 7(d) shows that the source current phase *a* is in phase with the source voltage phase *a*, which means that the unity input power factor is achieved when the proposed fault tolerant strategy is activated.

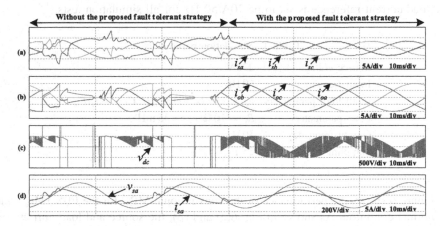

Fig. 7. Simulation results of the performance of the FBS-IMC for the conventional MPC and the proposed fault tolerant strategy when the open-switch fault occurs in the switch S_{ap} in the rectifier state: (a) The source currents, (b) the load currents, (c) the DC-link voltage, and (d) the source voltage and current phase *a*.

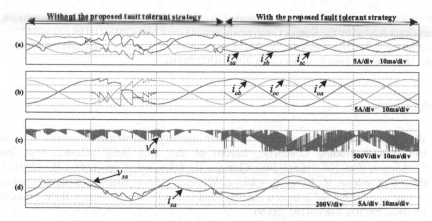

Fig. 8. Simulation results of the performance of the FBS-IMC for the conventional MPC and the proposed fault tolerant strategy when the open-switch fault occurs in the switch S_1 in the inverter state: (a) The source currents, (b) the load currents, (c) the DC-link voltage, and (d) the source voltage and current phase a.

Figure 8 shows the performance of the FBS-IMC when the open-switch fault occurs in the switch S_1 in the inverter stage. When the conventional MPC is applied, the performance of the source and load currents are severely deteriorated, as shown in the first half of Fig. 8. However, when the proposed fault tolerant strategy is activated in the latter half of Fig. 8, the source and load currents are kept balanced sinusoidal. And similarly with the case when the faulty switch is located in the rectifier stage, the proposed fault tolerant strategy reverses the polarity of the DC-link voltage at some periods to keep continuous operation of the FBS-IMC. The latter half of Fig. 8(d) shows that the source current phase a is in phase with the source voltage phase a, which means that the unity input power factor is achieved when the proposed fault tolerant strategy is activated. From Figs. 6, 7 and 8, it is clear that the proposed fault tolerant strategy can maintain the same input and output performances during the faulty condition as the normal condition regardless of the faulty switch location in the rectifier or inverter stages.

6 Conclusion

This paper proposed an effective open-switch fault tolerant strategy based on MPC for the FBS-IMC. By reversing the DC-link voltage polarity, the proposed fault tolerant strategy provides the alternative healthy current path to keep the continuous operation of the converter regardless of the faulty switch location in the rectifier or inverter stages. Moreover, compared to the previous methods, the proposed strategy can achieve the good performance in both the input and output side during the faulty condition as the normal condition. The effectiveness of the proposed strategy was verified by simulation results.

Acknowledgments. This work was supported in part by the National Research Foundation of Korea Grant funded by the Korean Government under Grant NRF-2018R1D1A1A09081779 and in part by the Korea Institute of Energy Technology Evaluation and Planning and the Ministry of Trade, Industry and Energy under Grant 20194030202310.

References

1. Wei, L., Lipo, T.A.: A novel matrix converter topology with simple commutation. In: 36th IAS Annual Meeting Conference Record of the 2001 IEEE Industry Applications Conference, (Cat. No. 01CH37248), Chicago, IL, USA, vol. 3, pp. 1749–1754 (2001)
2. Pena, R., Cardenas, R., Reyes, E., Clare, J., Wheeler, P.: A topology for multiple generation system with doubly fed induction machines and indirect matrix converter. IEEE Trans. Ind. Electron. **56**(10), 4181–4193 (2009)
3. Lee, H.H., Nguyen, M., Chun, T.: A study on rotor FOC method using matrix converter fed induction motor with common-mode voltage reduction. In: 2007 7th International Conference on Power Electronics, Daegu, pp. 159–163 (2007)
4. Mohamad, A.S.: Modeling of a steady-state VSCF aircraft electrical generator based on a matrix converter with high number of input phases. In: 2015 IEEE Student Conference on Research and Development (Scored), Kuala Lumpur, pp. 500–505 (2015)
5. Kwak, S.: Four-leg-based fault-tolerant matrix converter schemes based on switching function and space vector methods. IEEE Trans. Ind. Electron. **59**(1), 235–243 (2012)
6. Dasika, J.D., Saeedifard, M.: A fault-tolerant strategy to control the matrix converter under an open-switch failure. IEEE Trans. Ind. Electron. **62**(2), 680–691 (2015)
7. Dasika, J.D., Saeedifard, M.: An on-line fault detection and a post-fault strategy to improve the reliability of matrix converters. In: 2013 Twenty-Eighth Annual IEEE Applied Power Electronics Conference and Exposition (APEC), Long Beach, CA, pp. 1185–1191 (2013)
8. Cruz, S.M.A., Mendes, A.M.S., Cardoso, A.J.M.: A new fault diagnosis method and a fault-tolerant switching strategy for matrix converters operating with optimum alesina-venturini modulation. IEEE Trans. Ind. Electron. **59**(1), 269–280 (2012)
9. Tran, Q.H., Chun, T.W., Lee, H.H.: Fault tolerant strategy for inverter stage in indirect matrix converter. In: IECON 2013 - 39th Annual Conference of the IEEE Industrial Electronics Society, Vienna, pp. 4844–4849 (2013)
10. Han, N., Zhou, B., Qin, X., Zhou, X.: Fault-tolerant modulation strategy for inverter stage of full bidirectional switches two stage matrix converter. In: IECON 2016 - 42nd Annual Conference of the IEEE Industrial Electronics Society, Florence, pp. 6255–6260 (2016)
11. Correa, P., Rodriguez, J., Rivera, M., Espinoza, J.R., Kolar, J.W.: Predictive control of an indirect matrix converter. IEEE Trans. Ind. Electron. **56**(6), 1847–1853 (2009)

Rotary Machine Fault Diagnosis Using Scalogram Image and Convolutional Neural Network with Batch Normalization

Duy Tang Hoang[1] (iD) and Hee Jun Kang[2(✉)] (iD)

[1] Graduate School of Electrical Engineering, University of Ulsan,
Ulsan, South Korea
hoang.duy.tang@gmail.com
[2] School of Electrical Engineering, University of Ulsan,
Ulsan, South Korea
hjkang@ulsan.ac.kr

Abstract. Deep learning or deep neural network is a type of machine learning which exploits deep structures of neural networks with many layers of nonlinear data processing units. Deep neural networks have the ability to automatically extracting abstract data presentation for raw data such as time series signals, texts, and images. Deep learning has been extensively applied in vibration signal-based fault diagnosis for rotary machine since it can learn features from the raw signal of a rotary machine without requiring hand-crafted feature extraction. Batch normalization is a technique proposed to accelerate the training process and convergence of deep neural networks. In this paper, we propose a method for diagnosing faults in a rotary machine based on vibration signal using convolutional neural network with batch normalization technique, which is an advantaged variant of traditional convolutional neural network. The effectiveness of the proposed method is then verified with experiments on bearings and gearboxes fault data.

Keywords: Convolutional neural network · Batch normalization ·
Rotary machine fault diagnosis · Bearing fault diagnosis ·
Gearbox fault diagnosis

1 Introduction

Rotary machines are widely used in a numerous number of applications in industry. Any fault or malfunction will cause inevitable effects on the operating performance of the machine, the system, and the quality of the final products. Therefore, early fault detection of rotary machines is a critical task in industrial environments. Intelligent based fault diagnosis using vibration signal is an extensively applied method. In this approach, the fault diagnosis is considered as a pattern classification task which employed machine learning (ML) algorithms.

© Springer Nature Switzerland AG 2019
D.-S. Huang et al. (Eds.): ICIC 2019, LNAI 11645, pp. 283–293, 2019.
https://doi.org/10.1007/978-3-030-26766-7_26

Fig. 1. Traditional intelligent fault diagnosis

A typical intelligent fault diagnosis consists of four steps as shown in Fig. 1. In this procedure, feature extraction is the most important step which has a profound influence on the accuracy of fault detection. To carry out this step, it requires signal processing techniques, human labor, and expert knowledge. To the best knowledge of the authors, there doesn't exist any standard procedure of extracting feature. As a result, for each fault diagnosis task, a new feature extractor must be redesigned manually.

Deep learning (DL) or deep neural network (DNN) is a type of ML algorithm which exploits neural networks which many layers (deep structure) for learning multiple data abstraction from input data. The most important advantage of DNN over traditional ML algorithms is its ability of automatically learning feature from the raw input data without the requirement of hand-crafted feature extraction. DNN can effectively overcome the problem of insufficient adaptability of traditional feature extraction methods based on human labor, expert knowledge. Many DNN based applications have been proposed such as in computer vision [1], natural language processing [2], signal processing [3], medical image processing [4], and machine health monitoring [5].

Nowadays, there are numerous types of DNNs. However, all those DNNs can be considered being extensions of the four basic algorithms: convolutional neural network (CNN) [6], autoencoder [7], restricted Boltzmann machine [8], and recurrent neural networks [9]. Among those types of basic algorithms, CNN is the most popular method applied in machine health monitoring. Review some CNN papers in bearing fault diagnosis. L. Eren proposed a 1-D CNN model to handle the raw vibration signals [10]. An ensemble technique using DS evidence fusion for deep convolutional network was proposed by Li et al. [11]. Wen et al. converted vibration signals into 64×64 gray pixel image and fed into a CNN to diagnose fault of bearing [12].

In monitoring heath for machines, vibration signal-based approaches are applied extensively in industry. Vibration signals are easily measured from rotary machines by using accelerometers. Measured vibration signals reflecting machine health status [13]. However, they also contain noise which makes the fault detection become harder. The role of feature extraction step is to extract the most discriminate features from the raw signals. There are a huge number of feature extraction methods have been proposed in time domain, frequency domain, and time-frequency domain [14].

Normally, training DNNs is difficult because of their complex structure with many layers and a huge number of trainable parameters (weights and biases). Batch normalization (BN) a technique proposed for increase training performance of neural network, especially in the case of the neural networks with many layers (DNN). In CNN, BN is usually placed next to the convolutional layer and fully-connected layer to reduce the shift of internal covariance and accelerate the training process [15].

In this paper, we propose a method for machine health monitoring using CNN with BN. First, the vibration signals measured from machines are represented in 2-D form scalogram images by a transformation based on wavelet transform. Next, the obtained image dataset is used to training the CNN model. The proposed diagnosis method is verified its effectiveness with two types of popular component in the rotary machine: bearing and gearbox. The comparison between CNN with and without BN technique is also conducted to demonstrated the strength of the BN in training deep neural networks.

The remainder of the paper is organized as follows: Sect. 2 describes the proposed DNN, Sect. 3 describes the proposed fault diagnosis method. Experiments are shown in Sect. 4. Section 5 concludes the paper.

2 Convolutional Neural Network with Batch Normalization

A normal convolutional neural network often consists of four types of layers as shown in Fig. 2. A convolutional layer (CL) convolves the input data with the kernels of that layer. The convolved results then are fed into the nonlinear activate function to generate the output features of the CL. Sub-sampling layers (SL) are added after CLs. SLs have the role of down-sampling the input features which reduces the spatial sizes of the input features. The sub-sampling layers can be installed with max operation or average operation. The fully-connected layer (FL) of CNN has the same structure as the traditional multilayer perception (MLP). The structure of the CNN can be considered to consist of two parts. The first part composed of CLs, SLs, and FLs, plays the role of extracting features from input data. The second part is the classifier, which of exploits Softmax classifier to classify the extracted features.

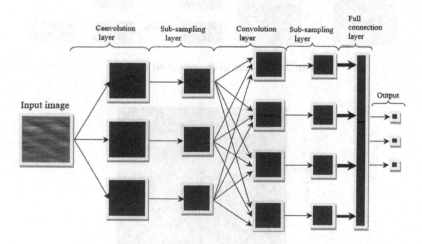

Fig. 2. Convolutional neural network

Batch normalization [16] is a technique proposed to reduce the shift of internal covariance and accelerate the training process of DNNs. The BN is usually added right after the CLs and FLs. Given the p-dimension input to a BN layer $x = \begin{pmatrix} x^1 & \cdots & x^p \end{pmatrix}$, the transformation of the BN layer is described as follows:

$$\hat{x}^{(i)} = \frac{x^{(i)} - E[x^{(i)}]}{Var[x^{(i)}]}$$

$$y^{(i)} = \gamma^{(i)} \hat{x}^{(i)} + \beta^{(i)} \qquad (1)$$

In the above equation, the scale parameters are denoted by γ^i and the shift parameters are denoted by β^i, and y^i denote the outputs of neuron. Firstly, the feature in each dimensionality of the input of that layer are standardized independently. Secondly, these features are shifted and normalized by the parameters β^i and γ^i. In short, we can say that the goal of γ^i and β^i is to restore the representation power of the network.

3 Proposed Fault Diagnosis Method

The flow chart of the fault diagnosis method is shown in Fig. 3. First, vibration signals are measured from rotary machines. Ordinarily, the vibration signals are in 1-D form. In other to used vibration signals with CCN, we will transform the vibration signals into 2-D form. The wavelet transform is exploited to do the transformation. The corresponding 2-D form of a vibration signal is called a scalogram image as shown in Fig. 4.

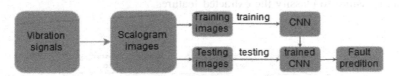

Fig. 3. Diagram of the proposed method

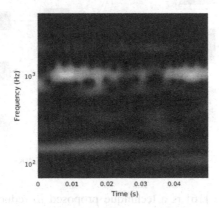

Fig. 4. Scalogram of the vibration signal

By this transformation, the fault information in the vibration signals are represented under the features in the scalogram image. The vibration-signal-based fault diagnosis task becomes an image classification task. We will solve this task by using a CNN. We conduct the training process of CNN by the obtained scalogram image dataset. In other to ensure the generalization of the CNN, the original scalogram image dataset is randomly split into two separate datasets: the training and testing dataset.

4   Experiment

In this section we are going to verity the effectiveness of the proposed fault diagnosis method via experiments with actual rotary machines. Normally, in rotary machines, bearings and gearboxes are the most often have defects component. We will apply the proposed method to detect faults in two separate experiments: bearing testbed and gearbox testbed.

4.1   Bearing Testbed

In this section, we apply the proposed method in bearing fault diagnosis. The bearing fault data are supplied by the Bearing data center of Case Western Reverse University [17]. The bearing testbed is shown in Fig. 5.

Fig. 5. Bearing testbed

The testbed consisted of a 2-hp motor (left), a torque transducer/encoder (center), a dynamometer (right), and control electronics. The test bearing supports the motor shaft. Single point faults were introduced to the test bearings using electro-discharge machining with three types of faults: fault at ball, fault at inner race, and fault at outer race with different fault diameters.

Vibration data was measured using accelerometers and digitalized at sampling frequency 12000 Hz. Ten types of bearing conditions are shown in Table 1. The bearing vibration signals are shown in Fig. 6. The scalogram images are shown in Fig. 7. After

288   D. T. Hoang and H. J. Kang

the pre-process, totally we obtain 2400 scalogram images. The image dataset is split into training dataset and testing dataset by stratified sampling method.

Table 1. Bearing faults and labels

No.	Bearing condition	Fault size (mils)	Label
1	No fault		0
2	Inner race fault	7	1
3	Inner race fault	14	2
4	Inner race fault	21	3
5	Ball fault	7	4
6	Ball fault	14	5
7	Ball fault	21	6
8	Outer race fault	7	7
9	Outer race fault	14	8
10	Outer race fault	21	9

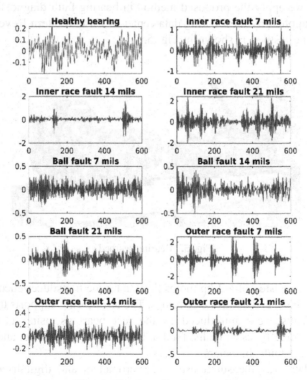

Fig. 6. Bearing vibration signals

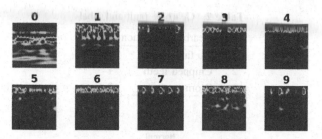

Fig. 7. Scalogram images of vibration signals

4.2   Gearbox Testbed

In this section, we verify the proposed method with gearbox fault diagnosis. The gearbox fault data was supplied by Amir Kabir University of Technology [18]. The gearbox testbed is shown in Fig. 8. The vibration signals were measured by an accelerometer mounted on gearbox frame digitized at sampling frequency 10 kHz.

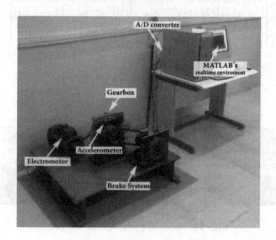

Fig. 8. Gearbox testbed

The gearbox vibration signals are shown in Fig. 9. By applying the same pre-processing as in the previous experiment with bearing vibration signal, we obtain totally 1500 scalogram images of gearbox fault signals (Fig. 10). This image set is split into training dataset (1050 images) and testing dataset (450 images) (Table 2).

Table 2. Gearbox fault and label

No.	Gearbox condition	Label
1	No fault	0
2	Chipped teeth	1
3	Worn teeth	2

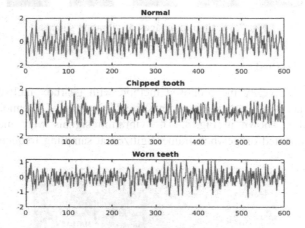

Fig. 9. Gearbox vibration signals

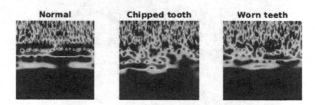

Fig. 10. Gearbox scalogram images

4.3 Experimental Result

The CNN models have configurations as shown in Table 3. For both experiments, we use two CNN models with the structures are almost the same. The only difference is that the sizes of the output layers. In the case of CNN models for bearing fault diagnosis, the output layer has the size of 10×1. In the case of the gearbox experiment, the respective output layer has the size of 3×1. All the activate function is Rectified Linear Unit (ReLU) function.

Table 3. CNN configuration

Layer	Kernel size	Kernel number	Input feature size	Output feature size
CL1	3 × 3	8	3 × 224 × 224	8 × 224 × 224
SL1	2 × 2	8	8 × 224 × 224	8 × 112 × 112
CL2	3 × 3	16	8 × 112 × 112	16 × 112 × 112
SL2	2 × 2	16	16 × 112 × 112	16 × 56 × 56
CL3	3 × 3	32	16 × 56 × 56	32 × 56 × 56
SL3	2 × 2	32	32 × 56 × 56	32 × 28 × 28
CL4	3 × 3	64	32 × 28 × 28	64 × 28 × 28
SL4	2 × 2	64	64 × 28 × 28	64 × 14 × 14
SL5	2 × 2	64	64 × 14 × 14	64 × 7 × 7
FL1			64 × 7 × 7	1024 × 1
FL2			1024 × 1	256 × 1
FL3			256 × 1	32 × 1
Softmax			32 × 1	10 × 1 (or 3 × 1)

The CNN models are trained with the training datasets. When the training process converges, the testing datasets are used to evaluate the performance of the fault diagnosis models. We train the CNN models with both methods: with and without BN. The training process is shown in Fig. 11.

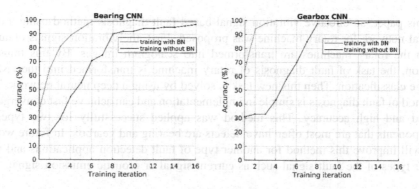

Fig. 11. Training process via convergence rate

We can see that, in both experiments with bearing and gearbox, the BN technique help the training process more quickly and accurately.

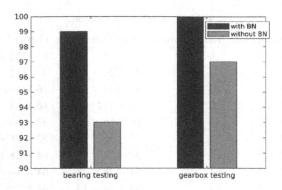

Fig. 12. Testing results via accuracy

Figure 12 shows the results when testing trained CNN models with the testing datasets. We can observe that the CNN models which trained with BN technique have higher accuracy. By using the BN technique, in the case of bearing test, the accuracy is 99.99%, and in the case of gearbox test, the accuracy is 100%. Without using the BN technique, in the case of bearing test, the accuracy is 93%, and in the case of gearbox test, the accuracy is 97%.

5   Conclusion

In this paper, an effective vibration signal-based fault diagnosis method using a deep neural network for rotary machine was proposed. First, the vibration signals obtained from the rotary machines are transformed into scalogram images. By this transformation, the task of fault diagnosis of rotary machine is transformed into the task of image classification. Then this new task is solved by using a deep neural network. This method of fault diagnosis is simple in implementation and can achieve fast convergence speed and high accuracy. This method was applied successfully for two types of components that are most often have defects are bearing and gearbox. In future works, we will improve this method for another type of fault detection applications and with other types of the fault signal such as current signal or acoustic emission signal.

Acknowledgment. This research was supported by Basic Science Research Program through the National Research Foundation of Korea (NRF) funded by the Ministry of Education (NRF-2016R1D1A3B03930496).

References

1. Szegedy, C., Ioffe, S., Vanhoucke, V., Alemi, A.A.: Inception-v4, Inception-ResNet and the impact of residual connections on learning. In: AAAI 2017, pp. 4278–4284 (2017)
2. Yan, R.: 'Chitty-chitty-chat bot': deep learning for conversational AI. In: IJCAI International Joint Conference on Artificial Intelligence (2018)

3. Yu, D., Deng, L., Jang, I., Kudumakis, P., Sandler, M., Kang, K.: Deep learning and its applications to signal and information processing. IEEE Signal Process. Mag. **28**, 145–154 (2011)
4. Xu, Y., et al.: Deep learning of feature representation with multiple instance learning for medical image analysis. In: 2014 IEEE International Conference on Acoustics, Speech and Signal Processing (ICASSP), pp. 1626–1630 (2014)
5. Zhao, R., Yan, R., Chen, Z., Mao, K., Wang, P., Gao, R.X.: Deep learning and its applications to machine health monitoring. Mech. Syst. Signal Process. **115**, 213–237 (2019)
6. Karpathy, A.: LeNet-5, convolutional neural networks (2015). http://yann.lecun.com/exdb/lenet
7. Rumelhart, D.E., Hinton, G.E., Williams, R.J., et al.: Learning representations by back-propagating errors. Cogn. Model. **5**(3), 1 (1988)
8. Smolensky, P.: Information processing in dynamical systems: Foundations of harmony theory (1986)
9. Medsker, L.R., Jain, L.C.: Recurrent Neural Networks: Design and Applications (2001)
10. Eren, L.: Bearing fault detection by one-dimensional convolutional neural networks. Math. Probl. Eng. **2017**, 9 p. (2017). Article ID 8617315. https://doi.org/10.1155/2017/8617315
11. Li, S., Liu, G., Tang, X., Lu, J., Hu, J.: An ensemble deep convolutional neural network model with improved DS evidence fusion for bearing fault diagnosis. Sensors **17**(8), 1729 (2017)
12. Wen, L., Li, X., Gao, L., Zhang, Y.: A new convolutional neural network-based data-driven fault diagnosis method. IEEE Trans. Ind. Electron. **65**, 5990–5998 (2018)
13. Kharche, P.P., Kshirsagar, S.V.: Review of fault detection in rolling element bearing. Int. J. Innov. Res. Adv. Eng. **1**(5), 169–174 (2014)
14. Van, M., Kang, H.-J.: Bearing-fault diagnosis using non-local means algorithm and empirical mode decomposition-based feature extraction and two-stage feature selection. IET Sci. Meas. Technol. **9**(6), 671–680 (2015)
15. Zhang, W., Peng, G., Li, C., Chen, Y., Zhang, Z.: A new deep learning model for fault diagnosis with good anti-noise and domain adaptation ability on raw vibration signals. Sensors **17**(2), 425 (2017)
16. Ioffe, S., Szegedy, C.: Batch normalization: accelerating deep network training by reducing internal covariate shift. In: International Conference on Machine Learning, pp. 448–456 (2015)
17. Loparo, K.A.: Bearing data center. Case Western Reserve University (2013)
18. Zamanian, A.H., Ohadi, A.: Gear fault diagnosis based on Gaussian correlation of vibrations signals and wavelet coefficients. Appl. Soft Comput. J. **11**, 4807–4819 (2011)

Real Implementation of an Active Fault Tolerant Control Based on Super Twisting Technique for a Robot Manipulator

Quang Dan Le[1] and Hee-Jun Kang[2]([⊠])

[1] Graduate School of Electrical Engineering, University of Ulsan,
Ulsan, South Korea
ledanmt@gmail.com
[2] School of Electrical Engineering, University of Ulsan, Ulsan, South Korea
hjkang@ulsan.ac.kr

Abstract. In this paper, real implementation of an active fault-tolerant control for a robot manipulator based on the combination of an external linear observer and the super-twisting algorithm is proposed. This active fault-tolerant scheme uses an external linear observer to identify faults. Then, the fault information is used to compensate the uncertainties/disturbance and faults with the super twisting controller. Finally, the effectiveness of proposed control is verified by simulation and implementation for a 3-DOF robot manipulator. The results were illustrated that the proposed control can tolerate the relatively bigger faults due to the design of the observer and then show the better performances than the conventional super-twisting controller does.

Keywords: Fault tolerant control · External linear observer ·
Super twisting sliding mode control

1 Introduction

In these days, fault-tolerant control has paid a lot of attention to many researchers. There are wide industrial demands [1–3]. Generally, the fault tolerant control can be categorized into two kinds (1) Passive fault-tolerant control (PFTC) [4–6] and (2) Active fault-tolerant control (AFTC) [7–9]. PFTC is based on the tolerant ability dealing with uncertainties/disturbances of control when faults occur. This method cannot handle the faults with high magnitude. Unlike the PFTC, the AFTC uses the fault estimation process to estimate faults, after that, by using the faults information, the control system was changed into the controller with a fault observer. Then the AFTC can solve the fault with high magnitude occurred during operation of the system. In addition, the operator system can monitor and identify faults in the system.

In the field of robot manipulator control, to tolerate the faults, it can be solved by the hardware or the software. By addition hardware, robot increase the price and need space to install. Therefore, the solution with software has paid a lot of attention to many researchers by costly. With the ability dealing with uncertainties/disturbances, sliding mode controller was known as the PFTC [10–12]. This control technique is based on a

© Springer Nature Switzerland AG 2019
D.-S. Huang et al. (Eds.): ICIC 2019, LNAI 11645, pp. 294–305, 2019.
https://doi.org/10.1007/978-3-030-26766-7_27

discontinuous control action and sliding surface. The most disadvantage of this technique is chattering phenomenal which comes from the neglected fast dynamics and the switching frequency of control signal. However, in fact, this controller is robustness, easily implement in practical system, easily be stabilized with a proper choice of the sliding surface. To reduce the chatting phenomenon, in [13], Levant was introduced the high-order sliding mode concept. Then, the terminal sliding mode [14, 15], twisting controller [16, 17] and the supper-twisting controller [10, 18] are kind of second-order sliding mode algorithm. Among of them, the super-twisting algorithm is preferable over the classical the sliding mode due to the ability reducing the chatting phenomenon and robustness.

In AFTC scheme, the performance depends on the accuracy of both fault estimation and the quality of controller. By using the faults estimation process, the operator can easy monitor, identify faults and the system can deal with the high magnitude faults. It can guarantee that robot manipulator has acceptable performance when faults occur. To estimate faults, several methods were proposed such as parameter estimation [19], parity relations [20], and state observers [21]. Similarly to the sliding mode controller, high order sliding model observer (HSMO) has paid a lot of attention to many researchers [9, 22, 23]. However, the most disadvantage of this technique is a proof of stability and the knowledge about bound of uncertainties/disturbances to choose the parameter of observer. To solve the proof of stability problem, Lavent and co-authors have several publishes about this issue [24–26]. Second issue of this method is how to choose the parameter of HSMO. It was proposed by Kumari and Chalangain [27]. However, this technique requires the knowledge about bound of uncertainties/disturbances. Therefore, in [18, 28], the authors used the adaptive technique to solve this problem. Comparison with HSMO in practical, the external linear observer (ELO) is commonly used in real system. ELO based on the linear observer for non-linear feedback control was proposed by Khalil and Laurent Praly in [29]. In [30], ELO is proposed to apply in cases system. By using ELO, we no need to exactly know about the bound of uncertainties/disturbances. In addition, the chattering does not exit due to it did not use discontinuous term in observer. Therefore, in this paper, the ELO is used to compensate faults and uncertainties/disturbances with the supper-twisting controller. By using this combination, the fault-tolerant control can reduce chattering phenomenal and achieve the acceptable performance when faults occur during operation the system.

In this paper, the fault-tolerant control for robot manipulator based on the combination of the external linear observer and supper-twisting method is proposed. By using the ELO we can easily monitor the faults and no need the knowledge about bound of uncertainties/disturbances and faults. The information about faults were compensated in the supper-twisting controller. By using this combination, the accuracy of robot is increased and shown acceptable performance when faults occur. The simulation and experimental results were shown in 3-DOF robot manipulator to illustrate the effectiveness of the proposed fault-tolerant control.

The rest of this paper follows as. Section 2, the dynamic of robot manipulator is described. In Sect. 3, the proposed fault-tolerant control is shown. The simulation results are shown for 3-DOF robot manipulator to validate the effectiveness of proposed control in Sect. 4. Section 5, experimental results on 3-DOF FARA-AT2 robot manipulator is presented. Finally, some conclude are given in Sect. 6.

2   Dynamics Model of Robot Manipulator

Dynamics of a n-degree of freedom robot manipulator was defined as

$$M(q)\ddot{q} + C(q,\dot{q}) + G(q) + F_f + \gamma(t - T_f)\xi(t) = \tau \tag{1}$$

where $\ddot{q}, \dot{q}, q \in \Re^n$ are the vectors of joint acceleration, velocity and position, respectively. $M(q) \in \Re^{n \times n}$, $C(q,\dot{q}) \in \Re^n$ and $G(q) \in \Re^n$ represent the inertia matrix, the centripetal and Coriolis matrix, and the gravitation force, respectively. $F_f \in \Re^n$ is friction term. $\tau \in \Re^n$ is the torque at joints. $\xi(t) \in \Re^n$ is a vector composed of actuator faults and component faults. $\gamma(t - T_f) \in \Re^n$ presents the time profile of the faults. T_f is the time of occurrence of the faults.

In practical, we do not exactly know about the dynamic model of robot so that Eq. (1) can be written as

$$\begin{aligned}(M(q) + \Delta M(q))\ddot{q} &+ (C(q,\dot{q}) + \Delta C(q,\dot{q}))\dot{q} \\ &+ (G(q) + \Delta G(q)) + (F_f(\dot{q}) + \Delta F_f(\dot{q})) + \delta + \gamma(t - T_f)\xi(q,\dot{q},t) = \tau\end{aligned} \tag{2}$$

where $\Delta M, \Delta C, \Delta G$ and ΔF are dynamic uncertainties and δ is external disturbance. Equation (2) can be simply rewritten as

$$M(q)\ddot{q} + C(q,\dot{q})\dot{q} + G(q) + F_f(\dot{q}) + \psi(q,\dot{q},t) = \tau \tag{3}$$

where $\psi(q,\dot{q},t) = \Delta M + \Delta C + \Delta G + \Delta F + \delta + \gamma(t - T_f)\xi(q,\dot{q},t)$ which includes uncertainties, external disturbances and actuator faults.

To estimate the faults and uncertainties/disturbances, the external linear observer is used in this paper. An active fault-tolerant control based on the combination of an external linear observer and supper-twitting is proposed.

3   Proposed Fault-Tolerant Control

3.1   Faults Estimation Based on an External Linear Observer

In this section, the faults estimation based on an external linear observer is designed.
In the position control, the Eq. (3) can be rewritten as

$$\ddot{q} = M^{-1}(q)(\tau - H(q,\dot{q})) - M^{-1}(q)\psi(q,\dot{q},t) \tag{4}$$

where $H(q,\dot{q}) = C(q,\dot{q}) + G(q) + F_f(\dot{q})$.
The dynamics model in (4) can be rewritten in state space as

$$\begin{cases} \dot{x}_1 = x_2 \\ \dot{x}_2 = f(x_1,x_2,u) + \phi(x_1,x_2,t) \end{cases} \tag{5}$$

where $\quad x_1 = q \in \Re^n, \quad x_2 = \dot{q} \in \Re^n, \quad u = \tau, \quad f(x_1, x_2, u) = M^{-1}(q)(\tau - H(q, \dot{q})),$
$\phi(x_1, x_2, t) = -M^{-1}(q)\psi(q, \dot{q}, t)$ An external linear state observer [17] was given as

$$\begin{cases} \dot{\hat{x}}_1 = \hat{x}_2 + \frac{\alpha_1}{\varepsilon}(x_1 - \hat{x}_1) \\ \dot{\hat{x}}_2 = f(x_1, \hat{x}_2, u) + \frac{\alpha_2}{\varepsilon}(x_1 - \hat{x}_1) + \hat{\phi} \\ \dot{\hat{\phi}} = \frac{\alpha_3}{\varepsilon}(x_1 - \hat{x}_1) \end{cases} \tag{6}$$

where $\hat{x}_1, \hat{x}_2$ and $\hat{\phi}$ are estimate of x_1, x_2 and ϕ, respectively. α_1, α_2 and α_3 are positive constant. $0 < \varepsilon < 1, |\dot{f}| \leq L.$

Theorem 1: Considering the system (5) with observer (9) and satisfy $0 < \varepsilon < 1$ and $|\dot{f}| \leq L$ then $\hat{x}_1(t) \to x_1(t)$, $\hat{x}_2(t) \to x_2(t)$ and $\hat{\phi}(q, \dot{q}, t) \to \phi(q, \dot{q}, t)$ as $t \to \infty.$

Proof: We define observer error as $\tilde{e} = [\tilde{e}_1, \tilde{e}_2, \tilde{e}_3]^T$ where $\tilde{e}_1 = (x_1 - \hat{x}_1)/\varepsilon^2$, $\tilde{e}_2 = (x_2 - \hat{x}_2)/\varepsilon$ and $\tilde{e}_3 = f - \hat{\phi}.$
We define the Lyapunov function as

$$V = \varepsilon \tilde{e}^T P \tilde{e} \tag{7}$$

where matrix P satisfying the Lyapunov equation as

$$A^T P + PA = -Q \tag{8}$$

in which A and Q are positive matrix.
The time derivative V in (7), we have

$$\begin{aligned} \dot{V} &= \varepsilon \dot{\tilde{e}}^T P \tilde{e} + \varepsilon \tilde{e}^T P \dot{\tilde{e}} \\ &= (A\tilde{e} + \varepsilon B\dot{f})^T P \tilde{e} + \tilde{e}^T P(A\tilde{e} + \varepsilon B\dot{f}) \\ &= \tilde{e}^T A^T P \tilde{e} + \varepsilon (B\dot{f})^T P \tilde{e} + \tilde{e}^T PA\tilde{e} + \varepsilon \tilde{e}^T PB\dot{f} \\ &= \tilde{e}^T (A^T P + PA)\tilde{e} + 2\varepsilon \tilde{e}^T PB\dot{f} \leq -\tilde{e}^T Q\tilde{e} + 2\varepsilon \|PB\| \cdot \|\tilde{e}\| \cdot |\dot{f}| \end{aligned} \tag{9}$$

where B is positive matrix.
From (9), we have

$$\dot{V} \leq -\lambda_{\min}(Q)\|\tilde{e}\|^2 + 2\varepsilon L\|PB\|\|\tilde{e}\| \tag{10}$$

where $\lambda_{\min}(Q)$ is the minimum eigenvalue of Q. L is positive constant satisfy $|\dot{f}| \leq L.$
To guaranties stable of the system, $\dot{V} \leq 0$ we get observer error convergence conclusion as:

$$\|\tilde{e}\| \leq \frac{2\varepsilon L\|PB\|}{\lambda_{\min}(Q)} \tag{11}$$

3.2   The Proposed Fault-Tolerant Controller

In this section, the fault-tolerant controller based on the combination of an external linear observer, fuzzy logic and super-twitting is proposed.

Sliding surface is designed as

$$s = \dot{e} + ce \qquad (12)$$

where $e = q_d - q$ and $\dot{e} = \dot{q}_d - \dot{q} \in \Re^n$ are position error and velocity error. $c = diag(c) \in \Re^{n \times n}$ is constant.

The fault-tolerant controller is proposed as

$$\tau = \tau_0 + \tau_{smc} + \tau_{ob} \qquad (13)$$

where $\tau_0 = M(q)(\ddot{q}_d + c\dot{e}) + H(q,\dot{q})$, $\tau_{smc} = M(q)\left(K_1|s|^{\frac{1}{2}}\text{sgn}(s) - v\right)$, $\dot{v} = -K_2 sign(s)$, $\tau_{ob} = -M(q)\hat{\phi}$ which K_1 and $K_2 \in \Re^{n \times n}$ is diagonal positive matrices.

Theorem 2: Consider the system (3) with the proposed controller (16), then the position error approach to zero at time infinity.

Proof: From (4) and (16) the system closed as

$$\begin{aligned}
\ddot{q} &= \ddot{q}_d + c\dot{e} + K_1|s|^{\frac{1}{2}}sign(s) - v - \hat{\phi} - M^{-1}(q)\psi \\
&= \ddot{q}_d + c\dot{e} + K_1|s|^{\frac{1}{2}}sign(s) - v - \hat{\phi} + \phi \\
&= \ddot{q}_d + c\dot{e} + K_1|s|^{\frac{1}{2}}sign(s) - v
\end{aligned} \qquad (14)$$

where $\dot{v} = -K_2 sign(s)$

Differencing (15) with time given as

$$\dot{s} = \ddot{e} + c\dot{e} \qquad (15)$$

Substituting (18) into (17) we have

$$\begin{cases} \dot{s} = -K_1|s|^{\frac{1}{2}}sign(s) + v \\ \dot{v} = -K_2 sign(s) \end{cases} \qquad (16)$$

Lyapunov function

$$V = \zeta^T P \zeta \qquad (17)$$

where $\zeta = \left[|s|^{\frac{1}{2}}sign(s), v\right]^T$, P is symmetric and positive matrix and

$$\dot{\zeta} = \frac{1}{2}\frac{1}{|s|^{1/2}} \begin{bmatrix} -k_1 & 1 \\ -2k_2 & 0 \end{bmatrix} \begin{bmatrix} |s|^{\frac{1}{2}}sign(s) \\ v \end{bmatrix} = \frac{1}{2}\frac{1}{|s|^{1/2}}A\zeta \qquad (18)$$

Differencing (20) with time given as

$$\dot{V} = \zeta^T P \zeta + \zeta^T P \dot{\zeta} \tag{19}$$

Substituting (21) into (22) we have

$$
\begin{aligned}
\dot{V} &= \left(\frac{1}{2} \frac{1}{|s|^{1/2}} A\zeta \right)^T P\zeta + \zeta^T P \left(\frac{1}{2} \frac{1}{|s|^{1/2}} A\zeta \right) \\
&= \frac{1}{2} \frac{1}{|s|^{1/2}} \zeta^T \left(A^T P + AP \right) \zeta \\
&= -\frac{1}{2} \frac{1}{|s|^{1/2}} \zeta^T Q \zeta \leq 0
\end{aligned}
$$

where Q is symmetric and positive-definite matrix

$$A^T P + AP = -Q \tag{20}$$

Therefore, the position error converges to zero at time infinity.

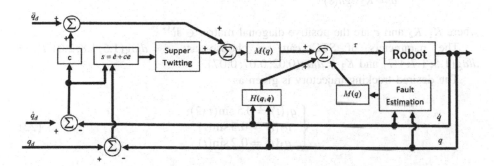

Fig. 1. Block diagram of the proposed fault-tolerant control for robot manipulator

4 Simulation Results

In this section, a simulation results of 3-DOF robot manipulator were shown. The robot manipulator was built on Solidwork then exported to Mathlab SimMechainics (Fig. 2). To comparison, the conventional supper-twisting controller was considered in addition to the proposed AFTC with super-twisting method. In this simulation, the scenario is assumed that fault only occurs at joint 2 at 5th second.

Fig. 2. 3-DOF robot manipulator in SimMechanics

The parameters of ELO are suitably chosen to be $\alpha_1 = 6, \alpha_2 = 25, \alpha_3 = 3$ and $\varepsilon = 0.01$. The parameters of proposed AFTC were suitably chosen as: $c = diag(1.5; 1.8; 2.0)$, $K_1 = diag(1.5; 1.7; 1.7)$ and $K_2 = diag(0.02; 0.07; 0.07)$. Then, the conventional Super-Twisting controller (STC) gives as

$$\tau = M(q)(\ddot{q}_d + c\dot{e}) + H(q, \dot{q}) + M(q)(K_1|s|^{\frac{1}{2}}sign(s) - \upsilon)$$
$$\dot{\upsilon} = K_2 sign(s) \tag{21}$$

where K_1, K_2 and c are the positive diagonal matrix $\in \Re^{3 \times 3}$.

The parameters of STC can be choose as: $c = diag(1.5; 1.8; 2.0), K_1 = diag(1.5; 1.7; 1.7)$ and $K_2 = diag(0.02; 0.07; 0.07)$.

The desired tracking trajectory is given as

$$\begin{cases} q_1(t) = 0.5\sin(t/2) \\ q_2(t) = 0.3\sin(t) \\ q_3(t) = 0.2\sin(t) \end{cases} \tag{22}$$

The faults function are assumed that occur at 5th second with attitude as

$$\begin{cases} \xi_1 = 0 \\ \xi_2 = -70\cos(t - 5) \\ \xi_3 = 0 \end{cases} \tag{23}$$

In Fig. 3, the fault estimation results were shown. It is easy to see that the estimation well estimates the fault and uncertainties/disturbances. In Fig. 4, the tracking trajectory error is presented. When fault occurs at joint 2, it highly affected to the joint 3 due to these joints have affected by gravity. Before fault occurs, the conventional super-twisting controller and the proposed AFTC have similar quality. However, after fault occurs, the convention STC has large error due to the ability dealing with high magnitude fault of STC is limited. In Fig. 4b and c, the STC needs nearly 3 s to achieve the accuracy similar with the proposed AFTC. From that results, we can see that with small uncertainties/disturbances the conventional STC (or PFTC with STC) has well performance. However, in case high magnitude faults, the proposed AFTC with STC shown better performance.

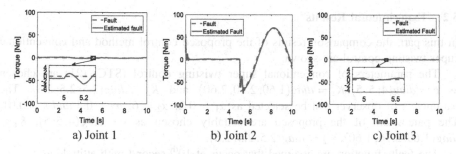

a) Joint 1   b) Joint 2   c) Joint 3

Fig. 3. Fault estimation results

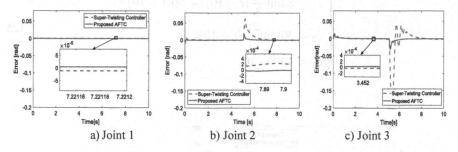

a) Joint 1   b) Joint 2   c) Joint 3

Fig. 4. The tracking trajectory error results

5 Experimental Results

5.1 Hardware Setup

In this part, experimental setup of hardware is presented. In ordinary case, the FARA-AT2 robot manipulator is shown in Fig. 5. The robot has 6 joints but the joints 4, 5 and 6 are blocked. The 3-DOF FARA-AT2 robot used CSMP series motor to drive the actuated joints. The gear box at joints is 120:1, 120:1, 100:1, respectively. The encoder at each joint produce 8192 pulses per revolution in quadratic mode. The controller runs on Labview FPGA NI-PXI-8110 and NI-PXI-7842R PXI card. The frequency of control loop was set up at 500 Hz. The controller NI-PXI-8110 runs window operating system.

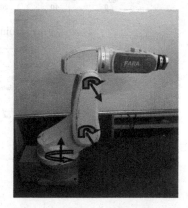

Fig. 5. Real-system 3-DOF robot manipulator

The desired tracking trajectory is given as

$$q_i(t) = \frac{\pi}{6}\sin(\pi t)\,(i = 1, 2, 3) \tag{24}$$

5.2   Experimental Results

In this part, the comparison results of the proposed control method and conventional super-twisting control are shown.

The parameters of conventional super twisting control (STC) (21) are chosen as $c = diag(4; 5; 5), K_1 = diag(1.60; 2.70; 2.60)$ and $K_2 = diag(2.5; 2.5; 2.5)$. The parameters of observer can be selected as $\alpha_1 = 0.2$, $\alpha_2 = 6$, $\alpha_3 = 0.3$ and $\varepsilon = 0.01$. The parameters of the proposed are suitably chosen as $c = diag(4; 5; 5)$, $K_1 = diag(1.60, 2.70, 1.60)$, $K_2 = diag(2.5, 2.5, 2.5)$.

The faults function are assumed that occur at 10th second with attitude as

$$\begin{cases} \xi_1 = 8\sin(\frac{2\pi}{3}(t - 10)) + 50(t - 10) \\ \xi_2 = 30\sin(\frac{2\pi}{3}(t - 10)) \\ \xi_3 = 40(t - 10) \end{cases} \tag{25}$$

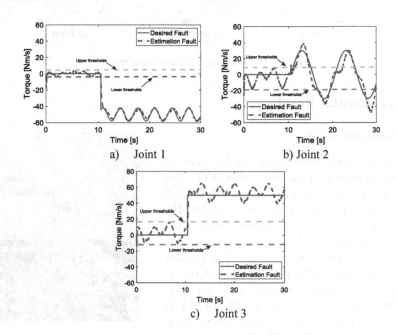

a)   Joint 1

b) Joint 2

c)   Joint 3

Fig. 6. Estimate fault results

In Fig. 6, the estimation fault results were shown. Due to the low accuracy of identify the dynamic parameter of robot manipulator, the upper thresholds and lower thresholds were used to detect faults. In Fig. 7, the error at each joint was presented. We can easily see that before and after faults occur the accuracy of the proposal AFTC better than the conventional super-twisting controller because the high magnitude of uncertainties/disturbances and faults in practical system reduce the effectiveness of conventional STC. By using the ELO to compensate the uncertainties/disturbances and faults with STC, the system has high accuracy and tolerating faults with acceptable performance in practical system.

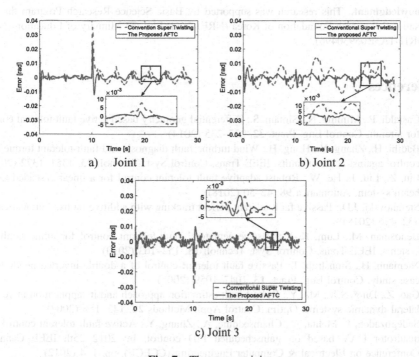

a)   Joint 1   b) Joint 2

c) Joint 3

Fig. 7. The error at joint

6 Conclusion

In this paper, the real implementation of active fault-tolerant control based on the combination of an external linear observer and super-twisting controller for a robot manipulator was proposed. This proposed improved the ability dealing with high magnitude of the uncertainties/disturbances and faults. By using external observer to estimate faults, the system no needs to identify faults step and easily apply in real system. From the simulation results we can see that the conventional super twisting controller have similar accuracy with the proposed AFTC in case free fault. However, in the experimental results, the proposed AFTC has better accuracy than conventional super twisting controller (STC). It can be explained that in simulation, the STC well deals with small magnitude of uncertainties/disturbances. However, in practical, the STC was reduced the ability dealing with high magnitude of uncertainties/disturbances. Therefore, in practical, an external linear observer is usefulness and effectiveness in case free fault and faults. This issue is strongly confirmed by simulation and experimental results in case faults. After faults occur, the STC needs time to achieve the same accuracy with the proposed AFTC in simulation and in practical, the proposed AFTC shown better performance than STC. Therefore, the proposed AFTC with super twisting control can apply in real system and has acceptable performance when faults occur.

Acknowledgment. This research was supported by Basic Science Research Program through the National Research Foundation of Korea (NRF) funded by the Ministry of Education (NRF-2016R1D1A3B03930496).

References

1. Castaldi, P., Mimmo, N., Simani, S.: Differential geometry based active fault tolerant control for aircraft. Control Eng. Pract. **32**, 227–235 (2014)
2. Badihi, H., Zhang, Y., Hong, H.: Wind turbine fault diagnosis and fault-tolerant torque load control against actuator faults. IEEE Trans. Control Syst. Technol. **23**, 1351–1372 (2015)
3. Liu, Z., Liu, J., He, W.: Robust adaptive fault tolerant control for a linear cascaded ODE-beam system. Automatica **98**, 42–50 (2018)
4. Stefanovski, J.D.: Passive fault tolerant perfect tracking with additive faults. Automatica **87**, 432–436 (2018)
5. Benosman, M., Lum, K.-Y.: Passive actuators' fault-tolerant control for affine nonlinear systems. IEEE Trans. Control Syst. Technol. **18**, 152–163 (2010)
6. Niemann, H., Stoustrup, J.: Passive fault tolerant control of a double inverted pendulum-a case study. Control Eng. Pract. **13**, 1047–1059 (2005)
7. Gao, Z., Ding, S.X., Ma, Y.: Robust fault estimation approach and its application in vehicle lateral dynamic systems. Optim. Control Appl. Methods **28**, 143–156 (2007)
8. Sadeghzadeh, I., Mehta, A., Chamseddine, A., Zhang, Y.: Active fault tolerant control of a quadrotor UAV based on gainscheduled PID control. In: 2012 25th IEEE Canadian Conference on Electrical & Computer Engineering (CCECE), pp. 1–4 (2012)
9. Van, M., Ge, S.S., Ren, H.: Robust fault-tolerant control for a class of second-order nonlinear systems using an adaptive third-order sliding mode control. IEEE Trans. Syst. Man Cybern. Syst. **47**, 221–228 (2017)
10. Bahrami, M., Naraghi, M., Zareinejad, M.: Adaptive super-twisting observer for fault reconstruction in electro-hydraulic systems. ISA Trans. **76**, 235–245 (2018)
11. Edwards, C., Spurgeon, S.K., Patton, R.J.: Sliding mode observers for fault detection and isolation. Automatica **36**, 541–553 (2000)
12. Wen, S., Chen, M.Z.Q., Zeng, Z., Huang, T., Li, C.: Adaptive neural-fuzzy sliding-mode fault-tolerant control for uncertain nonlinear systems. IEEE Trans. Syst. Man. Cybern. Syst. **47**, 2268–2278 (2017)
13. Levant, A.: Sliding order and sliding accuracy in sliding mode control. Int. J. Control **58**, 1247–1263 (1993)
14. Xu, S.S.-D., Liu, Y.-K.: Study of Takagi-Sugeno fuzzy-based terminal-sliding mode fault-tolerant control. IET Control Theory Appl. **8**, 667–674 (2014)

15. Han, Z., Zhang, K., Yang, T., Zhang, M.: Spacecraft fault-tolerant control using adaptive non-singular fast terminal sliding mode. IET Control Theory Appl. **10**, 1991–1999 (2016)
16. Emel'Yanov, S.V., Korovin, S.K., Levantovskii, L.V: Higher-order sliding modes in binary control systems. In: Soviet Physics Doklady, p. 291 (1986)
17. Martínez-Fuentes, C.A., Moreno, J.A., Fridman, L.: Anti-chattering strategy using twisting controller. IFAC-PapersOnLine **51**, 384–389 (2018)
18. Luo, D., Xiong, X., Jin, S., Kamal, S.: Adaptive gains of dual level to super-twisting algorithm for sliding mode design. IET Control Theory Appl. **12**, 2347–2356 (2018)
19. Jiang, T., Khorasani, K., Tafazoli, S.: Parameter estimation-based fault detection, isolation and recovery for nonlinear satellite models. IEEE Trans. Control Syst. Technol. **16**, 799–808 (2008)
20. Chan, C.W., Hua, S., Hong-Yue, Z.: Application of fully decoupled parity equation in fault detection and identification of DC motors. IEEE Trans. Ind. Electron. **53**, 1277–1284 (2006)
21. Fridman, L., Shtessel, Y., Edwards, C., Yan, X.-G.: Higher-order sliding-mode observer for state estimation and input reconstruction in nonlinear systems. Int. J. Robust Nonlinear Control IFAC-Affiliated J. **18**, 399–412 (2008)
22. Dávila, J., Salazar, S.: Robust control of an uncertain UAV via high-order sliding mode compensation. IFAC-PapersOnLine **50**, 11553–11558 (2017)
23. Kommuri, S.K., Lee, S.B., Veluvolu, K.C.: Robust sensors-fault-tolerance with sliding mode estimation and control for PMSM drives. IEEE/ASME Trans. Mechatron. **23**, 17–28 (2018)
24. Cruz-Zavala, E., Moreno, J.A.: Homogeneous high order sliding mode design: a Lyapunov approach. Automatica **80**, 232–238 (2017)
25. Cruz-Zavala, E., Moreno, J.A.: Levant's arbitrary order exact differentiator: a Lyapunov approach. IEEE Trans. Autom. Control **64**, 3034–3039 (2018)
26. Moreno, J.A., Osorio, M.: Strict Lyapunov functions for the super-twisting algorithm. IEEE Trans. Autom. Control **57**, 1035–1040 (2012)
27. Kumari, K., Chalanga, A., Bandyopadhyay, B.: Implementation of super-twisting control on higher order perturbed integrator system using higher order sliding mode observer. IFAC-PapersOnLine **49**, 873–878 (2016)
28. Xiong, X., Kamal, S., Jin, S.: Adaptive gains to super-twisting technique for sliding mode design. arXiv preprint arXiv:1805.07761 (2018)
29. Khalil, H.K.: High-gain observers in nonlinear feedback control. In: 2008 International Conference on Control, Automation and Systems, pp. xlvii–lvii (2008)
30. Khalil, H.K.: Cascade high-gain observers in output feedback control. Automatica **80**, 110–118 (2017)

Credibility Assessment of Complex Simulation Models Using Cloud Models to Represent and Aggregate Diverse Evaluation Results

Xiaojun Yang[✉] ⓘ, Zhongfu Xu, Rongmao He, and Fangxia Xue

Luoyang Electronic Equipment Test Center of China,
Luoyang 471003, He'nan, China
yangxiaojun2007@gmail.com, xuzf6904@aliyun.com

Abstract. Comprehensive and objective credibility assessments of complex simulation models are crucial to the successful application of models and simulation results in various critical evaluation and decision making problems. However, the credibility assessment of a complex simulation model usually encounters many challenges, involves the measurements and evaluations of hundreds of qualitative and quantitative indicators, and requires the integration of heterogeneous results. Therefore, cloud models which can describe both fuzziness and randomness are adopted to represent and aggregate diverse evaluation results of various qualitative and quantitative indicators. Then, crisp values, interval numbers, statistical data and linguistic terms can all be represented and aggregated by normal cloud models. The main advantages of our methods are that diverse evaluation results of various indicators can be represented and aggregated, and uncertainties associated with these results of leaf indicators can be preserved and propagated into the final assessment result. A missile simulation model credibility assessment example is presented to illustrate the proposed methods.

Keywords: Credibility assessment · Cloud model · Simulation ·
Linguistic term · Weighted average cloud · Knowledge representation

1 Introduction

Simulation enables a very economical and efficient way to observe and study real world systems or phenomena, especially when these systems do not yet exist or it is too difficult or expensive to experiment directly with them. The widespread use of simulation models in various areas has provoked more and more concerns about their credibility.

As modeling and simulation become more and more elaborate and complicated, credibility assessment of complex simulation models is encountering many challenges, involves the measurements and evaluations of hundreds of qualitative and quantitative indicators, mandates Subject Matter Expert (SME) evaluations, and requires the integration of heterogeneous measurements and evaluations [1]. Integration of both qualitative and quantitative methods to assess validity and credibility of a simulation model may lead to a more robust and reliable result [2]. Liao *et al.* [3] also pointed out that a

© Springer Nature Switzerland AG 2019
D.-S. Huang et al. (Eds.): ICIC 2019, LNAI 11645, pp. 306–317, 2019.
https://doi.org/10.1007/978-3-030-26766-7_28

single characteristic is usually not enough to obtain a reliable assessment but multiple characteristics need to be aggregated to a comprehensive combination. They used time series of density and velocity, pedestrian flow, spatiotemporal profiles, pedestrian trajectories and model stability to assess pedestrian models. Olsen *et al.* [4] calculated model credibility by aggregating the assessment results of structure, behavior, and data.

Validations and evaluations of diverse characteristics of a complex simulation model make the credibility assessment become a difficult multiple criteria decision making problem that involves both tangible and intangible indicators. What can make this challenging is that the evaluation results of leaf indicators can be crisp values, interval numbers, statistical data, or even linguistic terms, when various qualitative and quantitative methods are used.

Balci [5] suggested that the credibility score of a leaf indicator can be represented by a single real value between 0 and 100, an interval within the range of 0 and 100, or a nominal score predefined by a single real value or an interval. Then, the scores were all transformed into interval numbers on a scale between 0 and 100. Finally, low value, mid value and high value of the interval scores were aggregated respectively in a bottom-up fashion throughout the indicator hierarchy. Therefore, the final assessment contained the high, average, and low scores of all indicators.

Considering it is difficult for SMEs to determine exact numerical information for qualitative indicators, Azadeh *et al.* [6] incorporated linguistic variables, triangular fuzzy numbers and crisp numbers into the evaluation process of ranking different maintenance policies. Farther more, to deal with both intrapersonal uncertainty and interpersonal uncertainty, diverse inputs including numbers, intervals, type-1 fuzzy sets, and words modeled by interval type-2 fuzzy sets, were all used in the evaluation process of a weapon system [7].

Fuzziness can be modeled by type-1 fuzzy sets or type-2 fuzzy sets which can describe people's intangible concepts. However, fuzzy sets cannot effectively deal with many other uncertainties, especially the randomness of objective phenomena and subjective opinions in both natural and social domains. Randomness and fuzziness are the two most important uncertainties inherent in human cognition and natural laws. The cloud model brought forward by Li *et al.* [8] is a knowledge representation model which can describe both fuzziness and randomness of uncertain concepts simultaneously. Therefore, cloud models which are developed based on fuzzy sets and probability theory can be used to describe many phenomena and concepts with both fuzziness and randomness in various practical applications.

In this paper, we use cloud models to represent and aggregate diverse evaluation results of various qualitative and quantitative indicators with both fuzziness and randomness. We have shown that, diverse results involving crisp values, interval numbers, statistical data and linguistic terms can all be represented and aggregated by normal cloud models in the credibility assessment of a complex simulation model. Consequently, diverse evaluation results of various indicators can be represented and aggregated, and uncertainties associated with these results of leaf indicators can be preserved and propagated into the final assessment result. Therefore, comprehensive and objective credibility assessments of complex simulation models will encourage applying these models to support various critical evaluation and decision making problems.

The rest of the paper is organized as follows. In Sect. 2, the concept and operation rules of normal cloud models are introduced. Section 3 describes the proposed representation methods of crisp values, interval numbers, statistical data and linguistic terms using normal cloud models, and also presents the weighted average cloud aggregation algorithm. In Sect. 4, an example of credibility assessment of a missile simulation model is used to illustrate our proposed methods. Finally, Sect. 5 draws our conclusions and suggestions for future research.

2 Cloud Models

Based on fuzzy sets and probability theory, Li *et al.* [9] defined the cloud model to express uncertainties of both fuzziness and randomness simultaneously. As a transformation model representing uncertainties between a qualitative concept and its quantitative representation, cloud models can describe many phenomena with uncertainties in both natural and social sciences.

Many kinds of cloud models were proposed, including normal, triangular, trapezoidal, half and combined cloud models etc. Thus far, the normal cloud model based on normal distribution and Gaussian membership function is the most commonly used one. It has been applied in many fields successfully [10], such as decision making, image processing, data mining, uncertainty reasoning, time series prediction and knowledge representation. In this paper, we focus on the normal cloud model.

A normal cloud model is characterized by three parameters fully, namely, Expectation (*Ex*), Entropy (*En*), and Hyper-entropy (*He*). *Ex* is the mathematical expectation of the cloud drops belonging to a concept in the universe. *En* represents the uncertainty measurement of a qualitative concept. *He* is the entropy of entropy *En*. Therefore, a qualitative concept *T* can be expressed by a normal cloud model with three parameters, i.e. *T(Ex, En, He)*.

The basic concept of the normal cloud model and the forward normal cloud generator algorithm can be found in References [11, 12]. The operation rules of normal cloud models are summarized in Reference [10].

3 Representation and Aggregation of Diverse Evaluation Results Using Cloud Models

The evaluation results of different indicators or from different SMEs may have different representation forms. They may be expressed by crisp values, interval numbers, statistical data, or even linguistic terms. To process these heterogeneous results, the representation and aggregation methods for diverse data based on normal cloud models are proposed.

3.1 Representation of Crisp Values

There is no uncertainty for a crisp value, i.e. both entropy and hyper-entropy are zero. Thus, the representation of a crisp value v by a normal cloud model is defined as

$$v \rightarrow T(v, 0, 0) \tag{1}$$

It is obvious that, all the cloud drops generated from $T(v, 0, 0)$ are the same certain value v with fixed certainty degree 1.

3.2 Representation of Interval Numbers

The representation of an interval number $I = [I^L, I^U]$ by a normal cloud model is defined as

$$I \rightarrow T\left(\frac{I^L + I^U}{2}, \frac{I^U - I^L}{6}, 0\right) \tag{2}$$

As aforementioned, Ex is the mathematical expectation of a normal cloud model. Thus, the median of the interval is calculated as the normal cloud model parameter Ex. Similar to the "3σ rule" of normal distribution, about 99.7% cloud drops are within the interval $[Ex - 3 \times En, Ex + 3 \times En]$. Therefore, 1/6 length of the interval is calculated as the normal cloud model parameter En. The uncertainty of entropy En cannot be inferred from only one interval. Thus, the normal cloud model parameter He of an interval number should be equal to 0.

3.3 Representation of Statistical Data

If statistical data obey the normal distribution or approximate normal distribution, the mean value μ and standard deviation σ of the data are first calculated. Then the statistical data can be represented by a normal cloud model as

$$S_{\sim N(\mu, \sigma)} \rightarrow T(\mu, \sigma, 0) \tag{3}$$

As similar to that in Reference [13], the transformation is a straightforward parameter mapping in accordance with the definition of the normal cloud model.

Due to the universality of normal distribution, many phenomena in practice can be regarded as (approximate) normal distribution, such as miss distances of a missile.

As another option, the algorithm of the *Backward Normal Cloud Generator* [14] can be used as a substitute for the above parameter mapping method to transform statistical data into a normal cloud model.

3.4 Representation of Linguistic Terms

Balci [5] defined a nominal score set involving fifteen nominal terms. However, the experiments performed by psychologist George Miller [15] found out that people could deal with information simultaneously involving only a few grades, at most seven plus or minus two. With more, they become confused and cannot handle the information. Thus, we define nine linguistic terms with different grades herein using normal cloud models as listed in Table 1. Figure 1 shows these normal cloud models of the nine linguistic terms.

Table 1. Linguistic terms represented by normal cloud models for score evaluations.

Linguistic terms	Normal cloud models
none (n)	(0.00, 1.00, 0.20)
very small (vs)	(1.97, 4.57, 0.22)
small (s)	(12.46, 10.62, 0.34)
slightly small (ss)	(30.83, 10.30, 0.33)
medium (med)	(50.58, 11.09, 0.56)
slightly large (sl)	(71.05, 10.27, 0.41)
large (l)	(87.43, 12.42, 0.41)
very large (vl)	(97.75, 10.54, 0.27)
maximum (m)	(100.00, 1.00, 0.20)

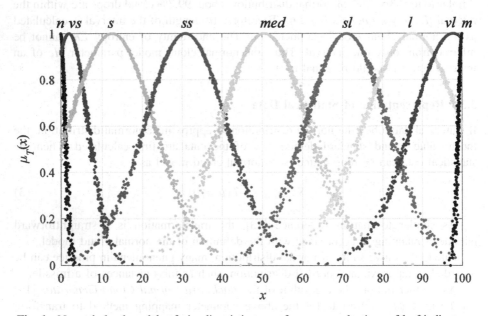

Fig. 1. Normal cloud models of nine linguistic terms for score evaluations of leaf indicators.

Yang *et al.* [13] encoded 32 words into individual normal cloud models using fuzzy statistics and membership function fitting based on a dataset collected from 175 individuals on a scale of 0–10. What is slightly different is that the universe of discourse for the credibility assessment of leaf indicators is on a scale of 0–100 as suggested by Balci [5]. Therefore, in Table 1, five linguistic terms *very small*, *small*, *medium*, *large* and *very large* are duplicated from a straight scale transformation (×10) of the corresponding encoded words in Reference [13]. Other two linguistic terms *slightly small* and *slightly large* correspond to the two encoded words *some* and *good amount* in Reference [13] respectively. Moreover, two linguistic terms *none* and *maximum* are newly-defined herein to represent the two extremes.

An SME can select a linguistic term from the predefined linguistic term set for each leaf indicator.

3.5 Aggregation Algorithms

There are two types of aggregations. One is the aggregation of all SMEs' scores for the same indicator when group decision making is performed. The other is the aggregation of all child indicators to obtain the credibility score of their parent indicator or to obtain the overall assessment result of a simulation model.

Heterogeneous evaluation results expressed by crisp values, interval numbers, statistical data, or linguistic terms are first transformed into normal cloud models using the representation methods described in Sect. 3.1–3.4. Then, the aggregation result is calculated by an aggregation algorithm for normal cloud models.

There are two kinds of aggregation algorithms for normal cloud models [16]. One is the synthetic cloud aggregation algorithm, and the other is the weighted average cloud aggregation algorithm. The latter calculates the weighted average cloud of all clouds and can be considered as the result of group decision or comprehensive assessment. Therefore, the weighted average cloud aggregation algorithm is used in this paper to aggregate evaluation results.

The weighted average cloud $T_{wa}(Ex_{wa}, En_{wa}, He_{wa})$ of n normal cloud models $T_i(Ex_i, En_i, He_i)$, $i = 1, 2, \ldots, n$, is calculated as

$$T_{wa} = \sum_{i=1}^{n} w_i T_i \Big/ \sum_{i=1}^{n} w_i \tag{4}$$

where w_i is the weight of T_i and can also be a linguistic term predefined by a normal cloud model constructed in Reference [13] as listed in Table 2. The addition, multiplication and division of normal cloud models are presented in Reference [10].

Table 2. Linguistic terms represented by normal cloud models for weight assignments.

Linguistic terms	Normal cloud models
unimportant	(0.38, 0.97, 0.06)
more or less unimportant	(2.89, 0.72, 0.09)
moderately unimportant	(3.32, 0.61, 0.08)
more or less important	(6.44, 0.60, 0.07)
moderately important	(6.86, 0.71, 0.09)
very important	(9.17, 0.93, 0.11)

Alternatively, the improved analytic hierarchy process [10] can also be used to determine the weights.

4   An Illustrative Example

In this section, we present an illustrative example for the credibility assessment of a missile simulation model to demonstrate the applicability of our methodology.

Missile performance characterization factors include both static data and dynamic data. Examples of static data are terminal miss distance, kill probability, damage area *etc*. Dynamic data varies during the flight of the missile, involving height, velocity, thrust, weight, roll position, behavioral feature *etc*. Suppose there are four indicators for assessing the credibility of a missile simulation model as shown in Fig. 2. The first indicator is the flight height of the missile from launch to landing which is recorded by time series data. The second indicator is the terminal miss distance which usually obeys a statistical distribution. The third indicator is the damage area of the missile to a given target. The fourth indicator is the behavioral feature which is determined by a series of guidance and control rules.

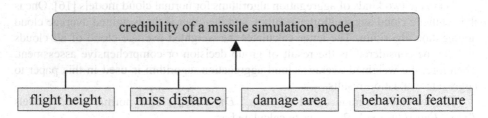

Fig. 2. Four indicators for credibility assessment of a missile simulation model.

As the output data of the four indicators are very different, distinct quantitative measure and evaluation method are used for each indicator.

4.1   Evaluation of Flight Height

The simulated and observed flight height data under identical inputs are listed in Table 3. Each time series consists of 20 data points.

Table 3. Simulated and observed flight height time series.

Time	Flight height		Time	Flight height	
	Simulated	Observed		Simulated	Observed
t_1	0	0	t_{11}	602	581
t_2	233	208	t_{12}	584	567
t_3	461	422	t_{13}	562	543
t_4	559	525	t_{14}	486	507

(continued)

Table 3. (*continued*)

Time	Flight height		Time	Flight height	
	Simulated	Observed		Simulated	Observed
t_5	615	593	t_{15}	358	380
t_6	635	615	t_{16}	223	209
t_7	637	627	t_{17}	199	195
t_8	633	618	t_{18}	195	185
t_9	628	610	t_{19}	162	153
t_{10}	608	592	t_{20}	25	0

The Theil's inequality coefficient (TIC) [17, 18] presented in Eq. (5) is used as the distance measure to quantify the discrepancy between simulated and observed data. The calculated result of TIC is 0.0216.

$$TIC = \frac{\sqrt{\frac{1}{n}\sum_{i=1}^{n}(s_i - o_i)^2}}{\sqrt{\frac{1}{n}\sum_{i=1}^{n}s_i^2} + \sqrt{\frac{1}{n}\sum_{i=1}^{n}o_i^2}} \tag{5}$$

Then, the mapping function [19] presented in Eq. (6) with parameters $x_{min} = 0.01$, $x_{max} = 0.05$ and $k = 1.5$, is used to transform the TIC measure into a credibility score. Therefore, the evaluation result of the first indicator *flight height* is 81.30. It is a crisp value, and can be represented as a normal cloud model $T_1(81.30, 0, 0)$ as defined in Sect. 3.1.

$$f(x) = \begin{cases} 100 & x \leq x_{min} \\ 100 \cdot \left(\frac{x_{max}-x}{x_{max}-x_{min}}\right)^k & x_{min} < x < x_{max} \\ 0 & x \geq x_{max} \end{cases} \tag{6}$$

4.2   Evaluation of Miss Distance

The simulated and observed miss distance data are listed in Table 4. There are 6 observed miss distances, and 15 miss distances produced by the simulation model under identical inputs.

Table 4. Simulated and observed miss distance data.

Source	Miss distances						
Simulated	6.70	6.70	7.02	7.06	6.64	7.90	7.41	6.14
	5.87	7.01	7.72	7.52	5.09	5.65	6.21	
Observed	5.85	6.28	4.80	7.03	6.50	6.41		

The calculated mean value and standard deviation of the observed data are 6.1450 and 0.7608 respectively. Thus, the corresponding normal cloud model is $T_O(6.1450, 0.7608, 0)$ as described in Sect. 3.3. Mean value and standard deviation of the simulated data are 6.7093 and 0.7984, and corresponding normal cloud model is $T_S(6.7093, 0.7984, 0)$. Then, the absolute value of error measure of two expectations is calculated as $\Delta Ex = abs(6.7093–6.1450) = 0.5643$, and the absolute value of error measure of two entropies is $\Delta En = abs(0.7984 - 0.7608) = 0.0376$.

Analogously, the mapping function presented in Eq. (6) with parameters $x_{min} = 0.0$, $x_{max} = 10.0$ and $k = 2.0$ is used to transform ΔEx into a credibility score as $f(\Delta Ex) = 89.03$. The entropy of the credibility score is calculated approximately as $0.5 \times (f(\Delta Ex–\Delta En) – f(\Delta Ex + \Delta En)) = 0.71$.

Therefore, the evaluation result of the second indicator *miss distance* can be represented as a normal cloud model $T_2(89.03, 0.71, 0)$.

4.3 Evaluation of Damage Area

Suppose observed damage areas of the missile to a given target are within the interval [408, 627] without any distribution assumption. The damage area to this target produced by the missile simulation model is a fixed value 600. Thus, the minimum and maximum relative errors are 0 and 0.4706 respectively.

Then, the minimum and maximum relative error measures are transformed into an interval credibility score as [55.73, 100] using the mapping function presented in Eq. (6) with parameters $x_{min} = 0.05$, $x_{max} = 1.0$ and $k = 1.0$. Therefore, the evaluation result of the third indicator *damage area* is an interval number [55.73, 100], and can be represented as a normal cloud model $T_3(77.87, 7.38, 0)$ according to Sect. 3.2.

4.4 Evaluation of Behavioral Feature

As there are a large number of different behavioral features under various situations, it is hard to calculate a quantitative discrepancy between the outputs of the simulation model and the actual system. Thus, the credibility scores of the fourth indicator are assigned by SMEs directly based on examination and using their expertise, knowledge and experience. The score can be expressed by a crisp value, an interval number, or a predefined linguistic term, according to preferences of different SMEs.

Five SMEs engage in assessing the behavioral feature of the missile simulation model. SME_1 and SME_2 are two technical experts with expertise in the development of missile simulation models. SME_3 is a technical expert specializing in missile simulation model designs. SME_4 and SME_5 are two operational experts with knowledge and experience about the operations of this missile system. The five SMEs are assigned with different weights considering their different degrees of knowledge and experience. The scores determined by the five SMEs and weights of SMEs are listed in Table 5.

Table 5. Scores determined by five SMEs and their weights.

SMEs	Scores by SMEs	Weights of SMEs
SME_1	85	very important
SME_2	large	moderately important
SME_3	[80, 86]	moderately important
SME_4	slightly large	more or less important
SME_5	large	very important

The scores and weights expressed by crisp values, interval numbers, or linguistic terms presented in Table 5 are transformed into corresponding normal cloud models using the methods described in Sect. 3.1, 3.2, 3.4 and 3.5. Then, the weighted average cloud aggregation algorithm defined by Eq. (4) is used to aggregate these results. The calculated evaluation result of the fourth indicator *behavioral feature* is T_4(83.32, 6.78, 0.67).

4.5   Overall Assessment

The credibility scores of the four indicators are summarized in Table 6.

Table 6. Credibility scores of four indicators.

Indicators	Credibility Scores
flight height	T_1(81.30, 0.00, 0.00)
miss distance	T_2(89.03, 0.71, 0.00)
damage area	T_3(77.87, 7.38, 0.00)
behavioral feature	T_4(83.32, 6.78, 0.67)

Inspection of the results in Table 6 brings some suggestions as follows. The indicator *miss distance* gains the highest score with very small entropy and zero hyperentropy, indicating that the simulated miss distances are more convincible and more stable. In this case, it can be supposed that the modeling and simulation of functional mechanism and distribution characteristic of *miss distance* should be reasonable. On the contrary, the indicator *damage area* gains the lowest score, and the entropy is considerable large. This indicates that the simulated *damage area* is in lower credibility level. This might be caused by the use of a fixed value with no uncertainty to model a characteristic with many uncertainties in nature. Moreover, the assessment result of *behavioral feature* indicate that subjective methods may produce more uncertainties than objective methods.

In practice, several simulation models usually need to be compared with each other to choose the best one. Thus, it is necessary to obtain the overall assessment result of a simulation model by aggregating all indicators in a bottom-up fashion.

Suppose the weights of the four indicators are *moderately important*, *very important*, *very important* and *more or less important*, respectively, considering their different degrees of influence and importance with respect to the intended use of the missile simulation model. Then, the weighted average cloud aggregation algorithm defined by Eq. (4) is used to aggregate the scores of four indicators. The calculated overall assessment result is $T(82.96, 6.50, 0.73)$. It indicates that the overall credibility of this missile simulation model ranks between *slightly large* and *large*, much closer to *large*.

The final overall assessment result represented by a normal cloud model can reflect the degree of uncertainty quantificationally by values of parameters En and He of T. These uncertainties stems from both subjective and objective uncertainties during evaluating the leaf indicators and assigning weights of indicators. The utilization of normal cloud models brings many advantages as diverse inputs can be aggregated, and uncertainties associated with these inputs can be preserved and propagated into the final evaluation, etc.

The overall credibility assessment is also very important to tendency analysis, management and decision making, ranking and selection, prediction and optimization *etc.*, in different life cycle development stages of a simulation model or when making an optimal decision from within several optional simulation models.

5 Conclusion

Complex simulation models involve various characteristics and lots of qualitative and quantitative indicators for credibility assessment. Thus, the representation and processing of heterogeneous data bring many challenges for the credibility assessment. This paper has shown that crisp values, interval numbers, statistical data and linguistic terms can all be represented and aggregated using normal cloud models.

Distinguishing features and advantages of our approach are summarized as follows.

(1) A wide range of heterogeneous data, from numbers to linguistic terms, can be represented and computed in a simple way.

(2) Without having to convert linguistic terms into (interval) numbers, no information is lost.

(3) Both subjective and objective uncertainties associated with evaluation scores and weights of leaf indicators propagated through the entire calculation process, and the final overall assessment result not only contains magnitude and grade information but also preserves uncertainty information.

(4) Group decision making technique is used for the evaluation of qualitative indicators to overcome the knowledge limitation of a single SME.

Although we just illustrated our methods through a quite simple example, the proposed methods are intended to be applicable to many similar domains, such as multi-criteria evaluation, ranking, selection, optimization, prediction, and decision making under uncertainty in a group scenario and in diverse areas. Future research will be concerned with more practical applications.

Acknowledgment. This work was supported by the Equipment Pre-Research Project of the 'Thirteenth Five-Year-Plan' of China under Grant 6140001010506.

References

1. Balci, O., Adams, R.J., Myers, D.S., Nance, R.E.: A collaborative evaluation environment for credibility assessment of modeling and simulation applications. In: Proceedings of the 2002 Winter Simulation Conference, pp. 214–220 (2002)
2. Yang, Y.N., Kumaraswamy, M.M., Pam, H.J., Mahesh, G.: Integrated qualitative and quantitative methodology to assess validity and credibility of models for bridge maintenance management system development. J. Manag. Eng. 27(3), 149–158 (2011)
3. Liao, W.C., Zhang, J., Zheng, X.P., Zhao, Y.: A generalized validation procedure for pedestrian models. Simul. Model. Pract. Theory 77, 20–31 (2017)
4. Olsen, M.M., Raunak, M., Setteducati, M.: Enabling quantified validation for model credibility. In: Proceedings of the 50th Computer Simulation Conference, pp. 1–10 (2018)
5. Balci, O.: A methodology for certification of modeling and simulation applications. ACM Trans. Model. Comput. Simul. 11(4), 352–377 (2001)
6. Azadeh, A., Abdolhossein Zadeh, S.: An integrated fuzzy analytic hierarchy process and fuzzy multiple-criteria decision-making simulation approach for maintenance policy selection. Simulation 92(1), 3–18 (2016)
7. Wu, D.R., Mendel, J.M.: Computing with words for hierarchical decision making applied to evaluating a weapon system. IEEE Trans. Fuzzy Syst. 18(3), 441–460 (2010)
8. Li, D.Y., Meng, H.J., Shi, X.M.: Membership clouds and membership cloud generators. Comput. Res. Dev. 42(8), 32–41 (1995)
9. Li, D.Y., Han, J.W., Shi, X.M., Chan, M.C.: Knowledge representation and discovery based on linguistic atoms. Knowl.-Based Syst. 10(7), 431–440 (1998)
10. Yang, X.J., Yan, L.L., Zeng, L.: How to handle uncertainties in AHP: the cloud Delphi hierarchical analysis. Inf. Sci. 222, 384–404 (2013)
11. Li, D.Y., Liu, C.Y., Gan, W.Y.: A new cognitive model: cloud model. Int. J. Intell. Syst. 24, 357–375 (2009)
12. Li, D.Y., Du, Y.: Artificial Intelligence with Uncertainty. Chapman & Hall/CRC Press, Boca Raton (2007)
13. Yang, X.J., Yan, L.L., Peng, H., Gao, X.D.: Encoding words into cloud models from interval-valued data via fuzzy statistics and membership function fitting. Knowl.-Based Syst. 55, 114–124 (2014)
14. Liu, C.Y., Feng, M., Dai, X.J., et al.: A new algorithm of backward cloud. J. Syst. Simul. 16 (11), 2417–2420 (2004)
15. Miller, G.A.: The magical number seven, plus or minus two: some limits on our capacity for processing information. Psychol. Rev. 63, 81–97 (1956)
16. Yang, X.J., Zeng, L., Zhang, R.: Cloud Delphi method. Int. J. Uncertain. Fuzziness Knowl.-Based Syst. 20(1), 77–97 (2012)
17. Kheir, N.A., Holmes, W.M.: On validating simulation models of missile systems. Simulation 30(4), 117–128 (1978)
18. Yang, X.J., Xu, Z.F., Ouyang, H.B., Zhang, X.: Experimental comparison of some classical distance measures for time series data in simulation model validation. In: Proceedings of the 2019 IEEE 8th Data Driven Control and Learning Systems Conference (2019, accepted)
19. Yang, X.J., Xu, Z.F., Ouyang, H.B., Wang, L.H.: Credibility assessment of simulation models using flexible mapping functions. In: Proceedings of the 2019 IEEE 8th Joint International Information Technology and Artificial Intelligence Conference (2019, accepted)

Dual Sparse Collaborative Matrix Factorization Method Based on Gaussian Kernel Function for Predicting LncRNA-Disease Associations

Ming-Ming Gao[1], Zhen Cui[1], Ying-Lian Gao[2], Feng Li[1],
and Jin-Xing Liu[1,3](✉)

[1] School of Information Science and Engineering, Qufu Normal University,
Rizhao, China
gaommzjr@126.com, cuizhensdws@126.com,
lifeng_10_28@163.com, sdcavell@126.com
[2] Library of Qufu Normal University, Qufu Normal University, Rizhao, China
yinliangao@126.com
[3] Co-Innovation Center for Information Supply and Assurance Technology,
Anhui University, Hefei, China

Abstract. With the development of science and technology, there are increasing evidences that lncRNAs are associated with diseases, so it is particularly important to find some novel lncRNA-disease associations (LDAs). Finding some novel lncRNA-disease associations is benefit to us in the treatment and prevention of diseases, but this process requires a lot of energy and time, so there is an urgent need to discover some new methods. In this paper, a dual sparse collaborative matrix factorization method based on gaussian kernel function (DSCMF) is proposed to predict novel LDAs. DSCMF is based on the traditional collaborative matrix factorization method. The lncRNA network similarity matrix is combined with the lncRNA expression similarity matrix, and the disease network similarity matrix is combined with the disease expression similarity matrix. Therefore, the Gaussian interaction profile kernel is added. To increase the sparsity, the $L_{2,1}$-norm is also added. Finally, the AUC value is obtained by ten-fold cross-validation method as the evaluation indicator of the performance of this method. The simulation experiment is used to predict some novel associations. The experimental results prove that this method will be of great help to future research.

Keywords: lncRNA-disease associations · Gaussian interaction profile kernel ·
$L_{2,1}$-norm

1 Introduction

In recent years, long non-coding RNAs (lncRNAs) have received extensive attention from many experts and scholars. A lncRNA is a non-coding RNA greater than 200 nucleotides in length and is not involved in encoding protein functions [1]. A large number of experiments have shown that lncRNAs play an important role in epigenetic

© Springer Nature Switzerland AG 2019
D.-S. Huang et al. (Eds.): ICIC 2019, LNAI 11645, pp. 318–326, 2019.
https://doi.org/10.1007/978-3-030-26766-7_29

regulation, cell cycle control and cell differentiation regulation [2–4]. There are increasing evidences that lncRNAs are closely linked to many human diseases, such as common cardiovascular diseases [5, 6], diabetes [7], Alzheimer's [8] and some cancers. Although many databases have been provided for research by some experts and scholars, since lncRNAs have just been recognized, only a small part of the known lncRNA-disease associations (LDAs) have been discovered. Therefore, it is urgent to find some novel LDAs. Most importantly, it is necessary to find a way to identify LDAs correctly and quickly.

At present, many methods have been proposed in the aspect of LDAs prediction. These methods have helped more or less for predictions. For example, Sun et al. proposed a new computational model that used random walk and restart methods on the lncRNA functional similarity network [9]. A lncRNA-lncRNA functional similarity network was constructed, and the relationship between similar phenotypic diseases and functionally similar lncRNAs was used to predict novel associations. Finally, it was found through experiments that this method is indeed feasible. Ding et al. proposed a way to combine the gene-disease association network with the lncRNA-disease association network into a lncRNA-disease-gene tripartite graph for prediction [10]. The advantage of this method is that it can better describe the heterogeneity of coding-noncoding genes-disease associations than other methods.

In this paper, an improved matrix factorization model is proposed to predict LDAs. This method mainly uses the collaborative matrix factorization, and then joins the Gaussian interaction profile kernel, that is, the lncRNA network similarity matrix and disease network similarity matrix are added. At the same time, dual sparseness constraints are added, that is, the $L_{2,1}$-norm is added to prevent over-fitting. Since there may be some missing associations in the course of the experiment, the accuracy of our predictions will be reduced, so we also added the weight K nearest known neighbors (WKNKN) pre-processing process. The cross-validation method is used to obtain the AUC value of this method. At the end of the paper, the simulation experiment is carried out. The results show that our method is indeed superior to other methods.

Materials such as the dataset used in the method can be seen in the second part of the article. The third part introduces the method and its formula and specific algorithm in detail. Experiments and conclusions are in the fourth part of the paper. The paper concludes with a description of our future work plans.

2 Materials

2.1 Human LncRNA-Disease Associations

The gold standard dataset used in this paper has 178 diseases, 115 lncRNAs [11] and 540 lncRNA-disease associations, as listed in Table 1. Y is an adjacency matrix. When its element value is 1, it means that this lncRNA $l(i)$ is associated with disease $d(j)$. Otherwise, it means that the lncRNA is not associated with disease.

Table 1. LncRNAs, diseases, and associations in Gold Standard Dataset

Datasets	LncRNAs	Diseases	Associations
Gold Standard Dataset	115	178	540

2.2 LncRNA Expression Similarity

The matrix $\mathbf{S}^l$ can be used to represent the lncRNA expression similarity matrix, and the expression similarity between lncRNA l_i and lncRNA l_j can be expressed in the form of $\mathbf{S}^l(l_i, l_j)$.

2.3 Disease Semantic Similarity

The matrix $\mathbf{S}^d$ can be used to represent the lncRNA expression similarity matrix, and the expression similarity between lncRNA d_i and lncRNA d_j can be expressed in the form of $\mathbf{S}^d(d_i, d_j)$.

3 Method

3.1 Gaussian Interaction Profile Kernel Similarity

Regardless of whether the disease is associated with the lncRNA in the lncRNA-disease network, it is likely to have a similar association with the new disease. The Gaussian interaction profile (GIP) kernel similarity used in this method is based on this assumption [12]. The GIP kernel similarity can be used in this method to represent the network topological structure of lncRNA-disease associations. The expressions for the topological structure of lncRNA l_i, l_j and disease d_i, d_j are as follow:

$$GIP_{\text{lncRNA}}(l_i, l_j) = \exp(-\gamma \|\mathbf{Y}(l_i) - \mathbf{Y}(l_j)\|^2), \tag{1}$$

$$GIP_{disease}(d_i, d_j) = \exp(-\gamma \|\mathbf{Y}(d_i) - \mathbf{Y}(d_j)\|^2). \tag{2}$$

The parameters of the adjustable kernel bandwidth represented by γ in the above two formulas. The lncRNA expression similarity matrix and the disease semantic similarity matrix are combined with the network similarity matrix, respectively. The resulting formulas are as follow:

$$\mathbf{K}_l = \alpha \mathbf{S}^l + (1 - \alpha)\mathbf{GIP}_l, \tag{3}$$

$$\mathbf{K}_d = \alpha \mathbf{S}^d + (1 - \alpha)\mathbf{GIP}_d. \tag{4}$$

In the above two formulas, $\alpha \in [0, 1]$, and it is a parameter that can be adjusted. Where $\mathbf{K}_l$ is the linear combination of the lncRNA expression similarity matrix and the lncRNA network similarity matrix, and $\mathbf{K}_d$ is the linear combination of the disease semantic similarity matrix and the disease network similarity matrix.

3.2 DSCMF

Collaborative filtering is introduced in the traditional CMF method [13], which can accurately predict some novel LDAs. The objective function of the traditional CMF is as follows:

$$
\min_{\mathbf{A},\mathbf{B}} = ||\mathbf{Y} - \mathbf{AB}^T||_F^2 + \lambda_h(||\mathbf{A}||_F^2 + ||\mathbf{B}||_F^2) \\
+ \lambda_l||\mathbf{S}^l - \mathbf{AA}^T||_F^2 + \lambda_d||\mathbf{S}^d - \mathbf{BB}^T||_F^2, \tag{5}
$$

where $|| \cdot ||_F$ is Frobenius norm. λ_h, λ_l and λ_d are positive parameters.

Traditional collaborative matrix factorization method can predict LDAs efficiently, but it ignores the network information of lncRNA and disease. Therefore, the lncRNA network similarity matrix and the disease network similarity matrix are added in this paper. That is, $\mathbf{S}^l$ in the original formula is replaced with $\mathbf{K}_l$, and $\mathbf{S}^d$ in the original formula is replaced with $\mathbf{K}_d$. The improved formula is as follows:

$$
\min_{\mathbf{A},\mathbf{B}} = ||\mathbf{Y} - \mathbf{AB}^T||_F^2 + \lambda_h(||\mathbf{A}||_F^2 + ||\mathbf{B}||_F^2) \\
+ \lambda_l||\mathbf{K}_l - \mathbf{AA}^T||_F^2 + \lambda_d||\mathbf{K}_d - \mathbf{BB}^T||_F^2. \tag{6}
$$

At the same time, to increase the sparsity, the method in this paper adds $L_{2,1}$-norm to matrix $\mathbf{A}$ and $\mathbf{B}$ respectively. The final objective function can be written as:

$$
\min_{\mathbf{A},\mathbf{B}} = ||\mathbf{Y} - \mathbf{AB}^T||_F^2 + \lambda_h(||\mathbf{A}||_F^2 + ||\mathbf{B}||_F^2) + \lambda_h||\mathbf{A}||_{2,1} \\
+ \lambda_h||\mathbf{B}||_{2,1} + \lambda_l||\mathbf{K}_l - \mathbf{AA}^T||_F^2 + \lambda_d||\mathbf{K}_d - \mathbf{BB}^T||_F^2. \tag{7}
$$

The matrices $\mathbf{A}$ and $\mathbf{B}$ of this formula are two latent feature matrices produced by the decomposition of the matrix $\mathbf{Y}$. The first term is to construct an approximate model, the purpose is to find the matrix $\mathbf{A}$ and $\mathbf{B}$. The second part is to add the Tikhonov regularization terms to prevent overfitting. The third part is to add the $L_{2,1}$-norm to matrix $\mathbf{A}$. The fourth part is to add the $L_{2,1}$-norm to matrix $\mathbf{B}$. The last two parts are the collaborative regularization terms of lncRNA expression similarity matrix and disease semantic similarity matrix.

3.3 Optimization and Algorithm of DSCMF Method

In this paper, we use the least squares method to update $\mathbf{A}$ and $\mathbf{B}$ to optimize the new method of this paper. In the first step, the values of $\mathbf{A}$ and $\mathbf{B}$ need to be initialized, so the singular value decomposition (SVD) method is used in this paper. The initial formula is:

$$[\mathbf{U}, \mathbf{S}, \mathbf{V}] = SVD(\mathbf{Y}, k), \mathbf{A} = \mathbf{US}_k^{1/2}, \mathbf{B} = \mathbf{VS}_k^{1/2}, \tag{8}$$

where $\mathbf{S}_k$ represents a diagonal matrix that contains the k largest singular values. Next, the objective function of the proposed method is replaced by H. Then we can obtain partial derivatives for $\mathbf{A}$ and $\mathbf{B}$, respectively, and the values of their partial derivatives are zero. Finally, updating is stopped once $\mathbf{A}$ and $\mathbf{B}$ converge. The iteration formula is as follows:

$$\mathbf{A} = (\mathbf{YB} + \lambda_l \mathbf{K}_l \mathbf{A})(\mathbf{B}^T \mathbf{B} + \lambda_h \mathbf{I}_k + \lambda_l \mathbf{AA}^T + \lambda_h \mathbf{D}_1 \mathbf{I}_K)^{-1}, \tag{9}$$

$$\mathbf{B} = (\mathbf{Y}^T \mathbf{A} + \lambda_d \mathbf{K}_d \mathbf{B})(\mathbf{A}^T \mathbf{A} + \lambda_h \mathbf{I}_k + \lambda_d \mathbf{B}^T \mathbf{B} + \lambda_h \mathbf{D}_2 \mathbf{I}_k)^{-1}, \tag{10}$$

where λ_h, λ_l and λ_d are a combination of the best parameters automatically selected from $\lambda_h \in \{2^{-2}, 2^{-1}, 2^0, 2^1\}$ and $\lambda_l/\lambda_d \in \{0, 10^{-4}, 10^{-3}, 10^{-2}, 10^{-1}\}$. $\mathbf{D}_1$, $\mathbf{D}_2$ are two diagonal matrices, where the values of the j-th diagonal element are denoted as $d_{jj}^A = 1/2\|(\mathbf{A})^j\|_2$, $d_{jj}^B = 1/2\|(\mathbf{B})^j\|_2$, respectively.

Through the detailed description of the above process, the algorithm of the DSCMF method can be organized as follows:

Algorithm 1: DSCMF

Input: lncRNA-Disease matrix $\mathbf{Y} \in R^{n \times m}$, lncRNA expression similarity matrix $\mathbf{S}^l$, disease semantic similarity matrix $\mathbf{S}^d$

Parameters: $K, P, k, \lambda_h, \lambda_l, \lambda_d$

Output: prediction score matrix $\hat{\mathbf{Y}}$

Pre-processing: $\mathbf{S}_l \rightarrow \mathbf{K}_l, \mathbf{S}^d \rightarrow \mathbf{K}_d$, $\mathbf{Y} = WKNKN(\mathbf{Y}, \mathbf{K}_l, \mathbf{K}_d, K, P)$

Initialization: $[\mathbf{U}, \mathbf{S}, \mathbf{V}] = SVD(\mathbf{Y}, k)$, $\mathbf{A} = \mathbf{US}_k^{1/2}, \mathbf{B} = \mathbf{VS}_k^{1/2}$

Repeat

    Update $\mathbf{A}$ using Eq.(12)

    Update $\mathbf{B}$ using Eq.(13)

Until convergence

$\hat{\mathbf{Y}} = \mathbf{AB}$

Return $\mathbf{Y}$

---

4 Results and Discussions

4.1 Cross Validation

In terms of experiments, cross-validation is used to conduct experiments. And compared with the previously proposed LRLSLDA [14], ncPred [15] and TPGLDA [10] methods. The experiment process mainly uses the ten-fold cross-validation method, and repeats 30 experiments for each method to ensure the stability and correctness of the experimental results. It should be noted that before the experiment of running the DSCMF method, we add a pre-processing process to prevent some unknown associations.

At the end of the final experiment, a corresponding AUC value [16] will be generated. This AUC value is an evaluation indicator used to evaluate the quality of our method. The area under the ROC curve is the AUC value. The area under the ROC curve is a number not greater than 1, that is, the AUC value is a number between 0 and 1.

4.2   Comparison with Other Methods

The final experimental results are listed in Table 2. Table 2 details the AUC values of several methods compared to the DSCMF method. The reason why the AUC value is chosen as the evaluation indicator of the method is because the AUC is not sensitive to the distribution of the skew class. The dataset used in our experiments is unbalanced, which will lead to more negative impacts, so the AUC value is more reasonable as an evaluation indicator.

Table 2. AUC results of cross validation experiments

Methods	Gold Standard Dataset
LRLSLDA	0.6625(0.0089)
ncPred	0.7566(0.0218)
TPGLDA	0.7586(0.0306)
DSCMF	**0.8523(0.0049)**

In Table 2, the method and result with the highest AUC value are bolded, and the values in parentheses indicate the standard deviation. The experimental results show that the method with the lowest AUC value is LRLSLDA. The DSCMF method proposed in this paper has the highest AUC value. The AUC values of the other two methods are 0.7566 and 0.7586, respectively, which are lower than the DSCMF method. The value of DSCMF is 18.98% higher than LRLSLDA, and 9.37% higher than TPGLDA.

The above results fully show that the DSCMF method is better than the previous methods, which is more conducive to the prediction of LDAs.

4.3   Sensitivity Analysis from WKNKN

To deal with some missing unknown associations, a WKNKN pre-processing process is added. Among them, the set parameters are different, which will result in different AUC values, so the choice of parameters is very important. It includes the choice of two parameters, one is the K value representing the nearest known neighbor, and the other is the attenuation parameter P. According to previous experience, when setting K to 5 and P to 0.7, AUC tends to be stable.

4.4 Case Study

In this section, simulation experiment is performed to predict some novel LDAs. For the predicted results, one common diseases is selected for research: prostate cancer. The experimental procedure is as follows: For one of the diseases, the predicted score matrix obtained is sorted from high to low. Then several lncRNAs with the highest scores are selected for analysis and verified by the databases LncRNADisease and Lnc2cancer.

Prostate cancer is an epithelial malignant tumor that occurs in the prostate and has an important relationship with genetic factors. For more detailed information on prostate cancer, please visit the https://www.omim.org/entry/176807 website. A total of 13 lncRNAs associated with prostate cancer have been identified in the gold standard dataset. The top 20 lncRNAs are selected from the prediction matrix, of which 12 known associations are successfully predicted, and 8 potential associations are predicted, as listed in Table 3. The 15th, 16th, and 18th lncRNAs in the table are confirmed in the LncRNADisease database to be associated with prostate cancer. Their PMIDs are 26975529 [17], 19767753 [18] and 23660942 [19], respectively. The XIST in the table is confirmed to be associated with prostate cancer in the database Lnc2cancer, and its PMID is 29212233 [20]. The PTENP1 in the table is confirmed to be associated with prostate cancer in both the LncRNADisease and Lnc2cancer databases. Their PMIDs are 24373479 [21] and 20577206 [22] respectively. Detailed introduction to the relationship between various lncRNAs in prostate cancer.

Table 3. Predicted LncRNAs for prostate cancer

Rank	lncRNA	Evidence	Rank	lncRNA	Evidence
1	MALAT1	known	11	HOTTIP	known
2	MEG3	known	12	DANCR	known
3	H19	known	13	XIST	Lnc2cancer
4	HOTAIR	known	14	PTENP1	LncRNADisease; Lnc2cancer
5	GAS5	known	15	TUG1	LncRNADisease
6	PVT1	known	16	IGF2-AS	LncRNADisease
7	UCA1	known	17	ZFAS1	Unconfirmed
8	HULC	known	18	CDKN2B-AS1	LncRNADisease
9	KCNQ1OT1	known	19	CCAT1	Unconfirmed
10	NEAT1	known	20	SNHG16	Unconfirmed

It turns out that lncRNAs have important associations with many diseases in humans, so finding novel lncRNA-disease associations is particularly important for people to prevent and treat some major diseases. So, if you find some new ways to predict LDAs, this will be of great help to our research. In the DSCMF method proposed in this paper, Gaussian interaction profile kernel is introduced in the traditional collaborative matrix factorization, and the network similarity matrices are combined with the lncRNA expression similarity matrix and the disease semantic similarity matrix. Also, the $L_{2,1}$-norm is added, which increases the sparsity. The results prove that the DSCMF method is more suitable for predicting LDAs than some existing methods.

5 Conclusion

A ten-fold cross-validation method is used in the experimental part of this paper. And WKNKN pre-processing method is also used in the paper to solve those unknown interactions, so the accuracy of prediction is improved to the greatest extent.

In the future work, we will make up for the shortcomings of the old methods and continue to look for some new ways to predict lncRNA-disease associations. At the same time, there will be more new datasets to help us conduct research. At the end of the paper, I hope that the DSCMF method can be helpful for predicting lncRNA-disease associations, and we will be more committed to this research and contribute to human society.

Acknowledgment. This work was supported in part by the grants of the National Science Foundation of China, No. 61872220, and 61572284.

References

1. Ponting, C.P., Oliver, P.L., Reik, W.: Evolution and functions of long noncoding RNAs. Cell **136**, 629–641 (2009)
2. Wang, C., et al.: LncRNA structural characteristics in epigenetic regulation. Int. J. Mol. Sci. **18**, 2659 (2017)
3. Wapinski, O., Chang, H.Y.: Long noncoding RNAs and human disease. Trends Cell Biol. **21**, 354–361 (2011)
4. Alvarez-Dominguez, J.R., Hu, W., Lodish, H.F.: Regulation of eukaryotic cell differentiation by long non-coding RNAs. In: Khalil, A., Coller, J. (eds.) Molecular Biology of Long Noncoding RNAs, pp. 15–67. Springer, New York (2013). https://doi.org/10.1007/978-1-4614-8621-3_2
5. Micheletti, R., et al.: The long noncoding RNA Wisper controls cardiac fibrosis and remodeling. Sci. Transl. Med. **9**, eaai9118 (2017)
6. Zhu, Y.P., Hedrick, C.C., Gaddis, D.E.: Hematopoietic stem cells gone rogue. Science **355**, 798–799 (2017)
7. Bai, X., et al.: Long noncoding RNA LINC01619 regulates microRNA-27a/Forkhead box protein O1 and endoplasmic reticulum stress-mediated podocyte injury in diabetic nephropathy. Antioxid. Redox Signal. **29**, 355–376 (2018)
8. Luo, Q., Chen, Y.: Long noncoding RNAs and Alzheimer's disease. Clin. Interv. Aging **11**, 867 (2016)
9. Sun, J., et al.: Inferring novel lncRNA–disease associations based on a random walk model of a lncRNA functional similarity network. Mol. BioSyst. **10**, 2074–2081 (2014)

10. Ding, L., Wang, M., Sun, D., Li, A.: TPGLDA: Novel prediction of associations between lncRNAs and diseases via lncRNA-disease-gene tripartite graph. Sci. Rep. **8**, 1065 (2018)
11. Parkinson, H., et al.: ArrayExpress—a public database of microarray experiments and gene expression profiles. Nucleic Acids Res. **35**, D747–D750 (2006)
12. van Laarhoven, T., Nabuurs, S.B., Marchiori, E.: Gaussian interaction profile kernels for predicting drug–target interaction. Bioinformatics **27**, 3036–3043 (2011)
13. Shen, Z., Zhang, Y.-H., Han, K., Nandi, A.K., Honig, B., Huang, D.-S.: miRNA-disease association prediction with collaborative matrix factorization. Complexity **2017** (2017)
14. Chen, X., Yan, G.-Y.: Novel human lncRNA–disease association inference based on lncRNA expression profiles. Bioinformatics **29**, 2617–2624 (2013)
15. Alaimo, S., Giugno, R., Pulvirenti, A.: ncPred: ncRNA-disease association prediction through tripartite network-based inference. Front. Bioeng. Biotechnol. **2**, 71 (2014)
16. Huang, J., Ling, C.X.: Using AUC and accuracy in evaluating learning algorithms. IEEE Trans. Knowl. Data Eng. **17**, 299–310 (2005)
17. Du, Z., et al.: Integrative analyses reveal a long noncoding RNA-mediated sponge regulatory network in prostate cancer. Nat. Commun. **7**, 10982 (2016)
18. Eeles, R.A., et al.: Identification of seven new prostate cancer susceptibility loci through a genome-wide association study. Nat. Genet. **41**, 1116 (2009)
19. Cheetham, S., Gruhl, F., Mattick, J., Dinger, M.: Long noncoding RNAs and the genetics of cancer. Br. J. Cancer **108**, 2419 (2013)
20. Du, Y., et al.: LncRNA XIST acts as a tumor suppressor in prostate cancer through sponging miR-23a to modulate RKIP expression. Oncotarget **8**, 94358 (2017)
21. Martens-Uzunova, E.S., Böttcher, R., Croce, C.M., Jenster, G., Visakorpi, T., Calin, G.A.: Long noncoding RNA in prostate, bladder, and kidney cancer. Eur. Urol. **65**, 1140–1151 (2014)
22. Poliseno, L., Salmena, L., Zhang, J., Carver, B., Haveman, W.J., Pandolfi, P.P.: A coding-independent function of gene and pseudogene mRNAs regulates tumour biology. Nature **465**, 1033 (2010)

Applying Support Vector Machine, C5.0, and CHAID to the Detection of Financial Statements Frauds

Der-Jang Chi[1], Chien-Chou Chu[2(✉)], and Duke Chen[3]

[1] Department of Accounting, Chinese Culture University,
No. 55, Hwa-Kang Road, Yang-Ming-Shan, Taipei City 11114, Taiwan
[2] Graduate Institute of International Business Administration,
Chinese Culture University, No. 55, Hwa-Kang Road, Yang-Ming-Shan,
Taipei City 11114, Taiwan
julian.cpa@gmail.com
[3] Ernst & Young Global Limited, 9F, No. 333, Sec.1 Keelung Rd,
Taipei 11012, Taiwan

Abstract. This paper applies support vector machine (SVM), decision tree C5.0, and CHAID to the detection of financial reporting frauds by establishing an effective detection model. The research data covering 2007-2016 is sourced from the Taiwan Economic Journal (TEJ). The sample consists of 28 companies engaged in financial statement frauds and 84 companies not involved in such frauds (at a ratio of 1 to 3), as listed on the Taiwan Stock Exchange and the Taipei Exchange during the research period. This paper selects key variables with SVM and C5.0 before establishing the model with CHAID and SVM. Both financial and non-financial variables are used to enhance the accuracy of the detection model for financial reporting frauds. The research suggests that the C5.0-SVM model yields the highest accuracy rate of 83.15%, followed by SVM-SVM (81.91%), the C5.0-CHAID model (80.93%), and the SVM-CHAID model (77.16%).

Keywords: Financial statements fraud · Support vector machine (SVM) · C5.0 · CHAID

1 Introduction

Financial reports are an important reference for investors, which is why listed companies must disclose financial reports and operating results to investors. However, if financial statements are manipulated and fabricated by insiders or the company assets are embezzled and misused, investors incur great losses, the capital market loses confidence, and the national economy suffers as a result. Financial statements are the main basis for decision making by stakeholders and other accounting information users, as well as the presentation of the operating results and financial status. However, there are increasing incidences of financial reporting frauds over recent years [1–3].

Infamous financial frauds in the US include Enron in 2001, K-Mart in 2002, WorldCom in 2003, AIG in 2005, and IBM in 2008. In Taiwan, the year 2004 saw a

© Springer Nature Switzerland AG 2019
D.-S. Huang et al. (Eds.): ICIC 2019, LNAI 11645, pp. 327–336, 2019.
https://doi.org/10.1007/978-3-030-26766-7_30

series of corporate frauds, e.g. Summit Technology, Procomp Informatics, ABIT Computer, and Infodisc Technology. In 2016, the XPEC Entertainment fraud also caused significant damage to investors and the capital market. All these cases demonstrate the importance of the prevention of financial reporting frauds.

The U.K., Germany, and Japan all have robust legal systems regarding corporate governance. In the U.S., the bitter lesson from the Enron scandal prompted Congress to pass the Sarbanes-Oxley Act in 2002 on the transparency of corporate financial information. This act also imposes a heavier burden of responsibility and penalty for corporate management, financial supervisors, and external auditors, with the purpose of enhancing the oversight on corporate operations and the accuracy of financial disclosure. The Sarbanes-Oxley Act specifies the responsibility of senior management on financial statements and severe punitive measures on fraud in financial reporting. Taiwan's Audit Standards Bulletin No. 43 on the Auditing Consideration for Frauds in Financial Statement aims to empower auditors to effectively identify errors in financial reporting, as well as any material misrepresentation in financial statements due to fraudulent activities. The standards also stipulate the auditing measures required to prevent earnings manipulation by management.

Financial reporting fraud is an intentional and illegal action, as it directly causes material misleading in financial statements or financial disclosure [2–5]. The Statement on Auditing Standard (SAS No. 99, 2002) defines financial reporting fraud as a purposeful or reckless action, either in misrepresentation or omission, which results in significantly misleading financial statements. The Securities and Exchange Commission (SEC) stipulates that companies must "provide a comprehensive overview of the company's business and financial condition, including audited financial statements" [2].

Humpherys et al. indicated that the prevention cost for financial reporting frauds are at least in the billions of dollars each year in the U.S. [6]. In fact, the prevention costs for financial statement frauds are high in all countries with well-developed capital and financial markets. It is important to prevent and detect frauds in financial reporting in a timely manner, and make amends in the fastest manner possible. An innovative approach to the detection of financial reporting frauds is data mining, which achieves robustness and accuracy [1–3, 5, 7–9].

To ensure rigorous and credible research results, this paper screens out important variables by running the data with the support vector machine (SVM) and decision tree C5.0, which is followed by model construction with the decision tree CHAID and SVM for the detection of financial reporting frauds.

2   Materials and Methods

2.1   Data Source and Samples

The research data is sourced from the Taiwan Economic Journal (TEJ) regarding the companies accused by the Securities and Futures Investors Protection Center in 2007–2016 for false information in financial reporting, false statements in the prospectus, share price manipulation, and insider trading, by quoting Article 20, the first paragraph of Article 20, Article 32, Article 155, and the first paragraph of Article 157 of the

Securities and Exchange Act. By referring to the prosecutions and verdicts published by the Securities and Futures Bureau on major securities crimes, this paper selects a total of 28 companies engaged in financial statement frauds and 84 companies not involved in financial statement frauds, at a ratio of 1 to 3, from the pool of the companies listed on the Taiwan Stock Exchange or the Taipei Exchange. In summary, this paper uses a total of 112 companies (28 FFS companies and 84 non-FFS companies), 25 research variables, and 10 years of data (2007–2016), with a total of 38,720 data.

2.2 Variables

The dependent variables are measured as a dummy variable based on whether the company concerned was involved in financial reporting frauds. It is 1 if yes and 0 if no. The selection of research variables is based on the literature review and the frequently used measurements in practice for financial statement frauds. A total of 25 variables are selected, including 21 financial variables (X01–X21) and four non-financial variables (X22–X25). The variables and equations are summarized in Table 1.

Table 1. Variables description

No.	Variable description/definition or formula (The year before the year of financial statements fraud)
X01	Debt-to-equity ratio: Total liabilities/Total equity
X02	Debt ratio: Total liabilities/Total assets
X03	Total assets turnover: Net Sales/Average total assets
X04	Gross profit/Net sales
X05	Operating expense/Net sales
X06	Net income/Net sales
X07	Current ratio: Current assets/Current liabilities
X08	Quick ratio: Quick assets/Current liabilities
X09	Inventory turnover: Cost of goods sold/Average inventory
X10	Operating cash flow ratio: Operating cash flow/Current liabilities
X11	Accounts receivable turnover: Net sales/Average accounts receivable
X12	Sales revenue growth rate: Δ Sales revenue/Sales revenue prior year
X13	Return on assets (ROA): [Net income + interest expense $\times$ (1 − tax rate)]/Average total assets
X14	Return on equity (ROE): Net income/Average equity
X15	Inventory/Total assets
X16	Net income/Total assets
X17	Gross profit/Total assets
X18	Cash/Total assets
X19	Current liabilities/Total assets
X20	Operating cash flow/Total sales
X21	Interest expense/Total liabilities

(continued)

Table 1. (*continued*)

No.	Variable description/definition or formula (The year before the year of financial statements fraud)
X22	The ratio of pledged stocks held by directors and supervisors: Number of pledged stocks held by directors and supervisors/Number of stocks held by directors and supervisors
X22	Stockholding ratio of directors and supervisors: Number of stocks held by directors and supervisors/Total number of common stock outstanding
X23	Audited by BIG4 (the big four CPA firms) or not: 1 for companies audited by BIG4, otherwise is 0
X24	Change CFO during 3 years or not: 1 is for change; 0 is for non-change
X25	Change CPA firm (CPAs) or not: 1 is for change; 0 is for non-change

2.3   Statistical Methods

This paper applies the following statistical methods: support vector machine (SVM), C5.0, and CHAID. Literature suggests that SVM and C5.0 are suitable for selecting important variables [1–3]; while CHAID and SVM are appropriate for detecting fraud [1–3, 8].

2.3.1   Support Vector Machine (SVM)

SVM is a statistical technique for classification developed by Boser et al. [10]. Assuming a training data set (x_i, y_i), x_i is a random sector, and y_i is the target variable and comes in $(1, -1)$. SVM is about data classification with hyperplane classifiers, as expressed in Eq. (1):

$$w^t X_i + \tau = 0 \tag{1}$$

where w and τ denote the parameters for hyperplane classifiers. The following condition can be met for linear classifications:

$$\begin{cases} w^t X_i + \tau \geq 1, \text{ if } y_i = 1 \\ w^t X_i + \tau \leq -1, \text{ if } y_i = -1 \end{cases} \equiv y_i((w^t X_i) + \tau) \geq 1 \tag{2}$$

In general data sets, while there may be multiple planes for classification, there is only one maximal margin classifier that allows for the maximum frontier, most concentrated distribution of data points, and minimum testing errors. The frontier can be expressed with Eq. (3):

$$\text{maximize: } \frac{2}{w^t W} \equiv \text{minimize: } \frac{w^t w}{2} \tag{3}$$

If linear classifications are not possible, SVM will be transformed in a non-linear manner into feature spaces with high dimension. The training errors ϵ_i will be added, in order that the optimization question can be rewritten into the minimization of the regularized risk function to derive the new values for w and τ, as expressed in Eq. (4):

$$\text{minimize} : \frac{\mathbf{w}^t\mathbf{w}}{2} + C\left(\sum_i \epsilon_i\right) \tag{4}$$

$$\text{subject to} : y_i = ((\mathbf{w}^t k(x_i)) + \tau) \geq 1 - \epsilon_i, \epsilon_i \geq 0$$

where C denotes the coefficient of correction, which measures the alternative relation between empirical risks and structural risks. The higher the value, the greater the empirical risks. The lower the value, the greater the structural risks. K (x_j, x_i) is the kernel function that transforms data to the feature space of a high dimension. In general, the radial basis function can resolve most classification problems. The function is expressed, as follows:

$$k(x_i, x_j) = \exp\left(-\gamma\|x_i - x_j\|^2\right) \tag{5}$$

2.3.2   C5.0

A decision tree is an algorithm for predictive modeling in artificial intelligence and classification modeling. Decision trees consist of roots, nodes, and leaves, and data is classified according to conditional rules into a tree structure. The C5.0 algorithm is the optimized version of C4.5, as developed by Quinlan [11], which conducts a machine-learning algorithm for classification. The core is the ID3 algorithm. The C5.0 algorithm is highly efficient and accurate. It applies the concept of entropy in order to identify the best data attribute by estimating information gain. Information gain can be defined as the information before testing less the information after testing. Therefore, the greater the information gain, the better and the smaller the entropy.

An important aspect of the C5.0 algorithm is the pruning rules, which conduct error-based pruning (EBP) to trim the decision tree in an appropriate manner in order to boost classification accuracy. The C5.0 algorithm has a lot of strengths; for example, it is more robust in dealing with missing data and a large number of input columns; it usually does not require long training sessions for estimates; nor does it require powerful enhancement techniques for better classifying accuracy.

2.3.3   CHAID

CHAID (chi-squared automatic interaction detector), as developed by Kass [12], is a quick and efficient multi-tree statistical algorithm for data exploration. It can establish partitions and describe data according to the desired classifications. The partition rules can be determined by chi-square testing, and the p value can be used to determine whether the partitioning should continue. CHAID prevents data overfitting and allows the decision tree to stop partitioning. The Chi-square measures the deviation between the actual value and the theoretic value of the sample. The larger the deviation, the larger the Chi-square value. The Chi-square value is close to one when the difference is small.

Data characteristics are measured according to the p value. The closer the relation between eigen variables and nominal variables, the smaller the p value. The eigen variable will be then chosen as the optimal nominal variable. The CHAID decision tree is constructed according to the abovementioned principles, until the entire sample has been classified.

2.4   Research Procedure

This paper collects data, screens the variables in terms of importance with support vector machine (SVM) and decision tree C5.0, and then establishes models with decision tree CHAID and SVM for the detection of financial reporting frauds. This is concluded with analysis of the prediction results. The research procedures are illustrated in Fig. 1.

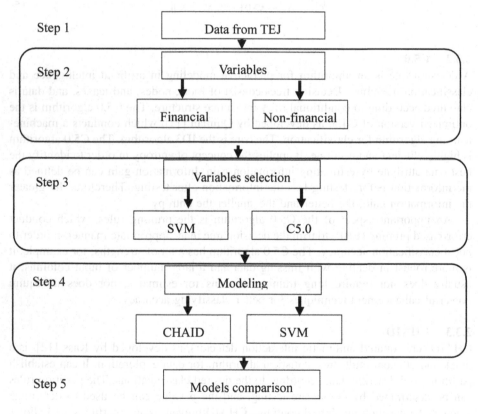

Fig. 1. Research procedure

3   Results and Analysis

This paper uses the SPSS Clementine by first applying SVM and C5.0 to select important variables, and then, for modeling by CHAID and SVM. Ten-fold cross-validation is used as a robust method to gain detection accuracy [1–3, 8]. The results are, as follows.

3.1   Variables Selection by SVM

A total of 11 variables are selected by SVM (importance value $\geq$ 0.05), as shown in Table 2. The order of importance of the variables is: X22: The ratio of pledged stocks held by directors and supervisors; X15: Inventory/Total assets; X04: Gross profit/Net sales; X11: Accounts receivable turnover; X16: Net income/Total assets; X13: Return on assets (ROA); X20: Operating cash flow/Total sales; X21: Interest expense/Total liabilities; X01: Debt-to-equity ratio; X03: Total assets turnover; and X08: Quick ratio.

Table 2. Variables selection by SVM

Variable	Importance
X22 The ratio of pledged stocks held by directors and supervisors	0.10
X15 Inventory/Total assets	0.07
X04 Gross profit/Net sales	0.06
X11 Accounts receivable turnover	0.06
X16 Net income/Total assets	0.05
X13 Return on assets (ROA)	0.05
X20 Operating cash flow/Total sales	0.05
X21 Interest expense/Total liabilities	0.05
X01 Debt-to-equity ratio	0.05
X03 Total assets turnover	0.05
X08 Quick ratio	0.05

3.2   Modeling by SVM

This paper uses the SPSS Clementine for model building. The variables selected with SVM are used for model construction with CHAID and SVM for FFS and non-FFS classifications. The training group and the testing group are trained and tested at a ratio of eight to two, i.e. 80% of the sampled data for training and model construction and 20% of the data for testing and accuracy rate calculations.

The accuracy of SVM models by ten-fold cross-validation is: the SVM-SVM model has higher detection accuracy (81.91%) than the SVM-CHAID model (77.16%), as shown in Table 3.

Table 3. Detection accuracy of SVM models by ten-fold cross-validation

No.	SVM-CHAID	SVM-SVM
1	78.67%	83.33%
2	75.31%	81.71%
3	80.00%	84.00%
4	76.54%	80.52%
5	77.01%	80.90%
6	77.78%	78.57%
7	75.58%	83.95%
8	79.55%	80.72%
9	75.58%	82.72%
10	75.61%	82.67%
Average	77.16%	81.91%

3.3 Variables Selection by C5.0

A total of 7 variables are selected by C5.0 (importance value ≥ 0.05), as shown in Table 4. The order of importance of the variables is: X10: Operating cash flow ratio; X02: Debt ratio; X25: Change CPA firm (CPAs) or not; X18: Cash/Total assets; X19: Current liabilities/Total assets; X04; and X11: Accounts receivable turnover.

Table 4. Variables selection by C5.0

Variable	Importance
X10 Operating cash flow ratio	0.30
X02 Debt ratio	0.17
X25 Change CPA firm (CPAs) or not	0.12
X18 Cash/Total assets	0.11
X19 Current liabilities/Total assets	0.10
X04 Gross profit/Net sales	0.07
X11 Accounts receivable turnover	0.07

3.4 Modeling by C5.0

This paper uses the SPSS Clementine to establish models with CHAID and SVM by using the variables selected by the C5.0 algorithm for FFS and non-FFS classifications. According to the results of ten-fold cross-validation, the C5.0-SVM model has higher detection accuracy (83.15%) than the C5.0-CHAID model (80.93%), as shown in Table 5.

Table 5. Detection accuracy of C5.0 models by ten-fold cross-validation

No.	C5.0-CHAID	C5.0-SVM
1	80.68%	83.52%
2	81.18%	82.05%
3	82.35%	82.35%
4	79.76%	83.33%
5	80.00%	82.93%
6	80.46%	82.72%
7	82.93%	84.21%
8	79.22%	82.05%
9	82.05%	81.61%
10	80.68%	86.75%
Average	80.93%	83.15%

The empirical results suggest that, among the four financial statements fraud detection models, the C5.0-SVM model yields the highest accuracy at 83.15%, followed by the SVM-SVM model (81.91%), the C5.0-CHAID model (80.93%), and the SVM-CHAID model (77.16%).

4   Conclusions

As capital markets around the globe continue to develop, there is an increasing number of large, privately owned listed companies. However, financial statement frauds occurring one after another have caused great damage to investors and stakeholders, as well as tremendous costs to the economy and society. The Enron scandal was a wake-up call to the U.S. and the rest of the world, and led to the creation of the Sarbanes-Oxley Act (2002). It is essential to prevent and detect financial statement frauds by establishing effective detection models.

While most literature on financial reporting frauds used traditional regression methods, data mining has become widely used as a more robust method in academic research over recent years. This paper selects important variables by using SVM and C5.0, and then, constructs the detection models by applying CHAID and SVM for financial reporting frauds. Meanwhile, in order to enhance research rigor, this paper chooses 25 variables (21 financials and 4 non-financials) frequently seen in literature and in practice for the measurement of financial statement frauds. The ten-fold cross-validation, which is a research technique well regarded for its robustness, is used to evaluate detection accuracy. The research suggests that the C5.0-SVM model shows the highest accuracy at 83.15%, followed by the SVM-SVM model (81.91%), the C5.0-CHAID model (80.93%), and the SVM-CHAID model (77.16%).

This paper hopes to contribute to academic research and practical aspects by providing a framework for CPAs and auditors in the detection of financial statement frauds. It is hoped that the research findings serve as a template for credit ratings agencies, securities analysts for decision making, and academic studies on financial reporting frauds.

References

1. Yeh, C.C., Chi, D.J., Lin, T.Y., Chiu, S.H.: A hybrid detecting fraudulent financial statements model using rough set theory and support vector machines. Cybern. Syst. **47**, 261–276 (2016)
2. Chen, S.: Detection of fraudulent financial statements using the hybrid data mining approach. SpringerPlus **5**, 89 (2016). https://doi.org/10.1186/s40064-016-1707-6
3. Jan, C.L.: An effective financial statements fraud detection model for the sustainable development of financial markets: evidence from Taiwan. Sustainability **10**(2), 513 (2018). https://doi.org/10.3390/su10020513
4. Beasley, M.: An empirical analysis of the relation between the board of director composition and financial statement fraud. Account Rev. **71**(4), 443–466 (1996)
5. Ravisankar, P., Ravi, V., Rao, G.R., Bose, I.: Detection of financial statement fraud and feature selection using data mining techniques. Decis. Support Syst. **50**, 491–500 (2011)
6. Humpherys, S.L., Moffitt, K.C., Burns, M.B., Burgoon, J.K., Felix, W.F.: Identification of fraudulent financial statements using linguistic credibility analysis. Decis. Support Syst. **50**, 585–594 (2011)
7. Salehi, M., Fard, F.Z.: Data mining approach to prediction of going concern using classification and regression tree (CART). Glob. J. Manag. Bus. Res. Account. Audit. **13**, 25–29 (2013)
8. Chen, S., Goo, Y.J., Shen, Z.D.: A hybrid approach of stepwise regression, logistic regression, support vector machine, and decision tree for forecasting fraudulent financial statements. Sci. World J. (2014). https://doi.org/10.1155/2014/968712
9. Kotsiantis, S., Koumanakos, E., Tzelepis, D., Tampakas, V.: Forecasting fraudulent financial statements using data miming. World Enfor. Soc. **12**, 283–288 (2014)
10. Boser, B.E., Guyon, I.M., Vapnik, V.N.: A training algorithm for optimal margin classifiers. In: Haussler D. (ed.) Proceedings of the Annual Conference on Computational Learning Theory, pp. 144–152. ACM Press, Pittsburgh (1992)
11. Quinlan, J.R.: C4.5–Programs for Machine Learning. Morgan Kaufmann, San Francisco (1993)
12. Kass, G.V.: An exploratory technique for investigating large quantities of categorical data. Appl. Statist. **29**(2), 119–127 (1980)

A New Manifold-Based Feature Extraction Method

Zhongbao Liu[✉]

School of Information Engineering & Art Design,
Zhejiang University of Water Resources and Electric Power,
Hangzhou 310018, China
liu_zhongbao@hotmail.com

Abstract. Many traditional feature extraction methods takes the global or the local characteristics of training samples into consideration during the process of feature extraction. How to fully utilize the global or the local characteristics to improve the feature extraction efficiencies is worthy of research. In view of this, a new Manifold-based Feature Extraction Method (MFEM) is proposed. MFEM takes both the advantage of Linear Discriminant Analysis (LDA) in keeping the global characteristics and the advantage of Locality Preserving Projections (LPP) in keeping the local characteristics into consideration. In MFEM, Within-Class Scatter based on Manifold (WCSM) and Between-Class Scatter based on Manifold (BCSM) are introduced and the optimal projection can be obtained based on the Fisher criterion. Compared with LDA and LPP, MFEM considers the global information and local structure and improves the feature extraction efficiency.

Keywords: Feature extraction · Fisher criterion · Global characteristics · Local structure

1 Introduction

Linear Discriminant Analysis (LDA) [1] is popular in practice, in which the non-singularity problem has greatly influent its improvement of efficiencies. In view of this, many effective improvements are made by scientists: Regularized Discriminant Analysis (RDA) proposed by Friedman [2] efficiently solves the above problem; 2D-LDA is proposed to directly extract the features based on Fisher criterion [3]; Orthogonal LDA (OLDA) tries to diagonalize the scatter matrix so as to obtain the discriminant vectors [4]; Direct LDA (DLDA) [5] carries no discriminative information by modifying the simultaneous diagonalization procedure. Besides, the commonly-used improvement approach include Pseudo-inverse LDA (PLDA) [6], Two-stage LDA [7], Penalized Discriminant Analysis (PDA) [8], Enhanced Fisher Linear Discriminant Model (EFM) [9]. In recent years, as to the under-sampled problems, we proposed Scalarized LDA (SLDA) [10], and Matrix Exponential LDA (MELDA) [11].

The general strategy of above approach is to solve the singularity problem firstly and then uses the Fisher criterion to obtain the optimal projections. LDA only takes the sample global information into consideration but always neglects the local structure.

© Springer Nature Switzerland AG 2019
D.-S. Huang et al. (Eds.): ICIC 2019, LNAI 11645, pp. 337–346, 2019.
https://doi.org/10.1007/978-3-030-26766-7_31

On the other hand, many popular manifold learning approach such as Locality Preserving Projection (LPP) [12], only focus on the local structure.

Therefore, all the sample information including the global characteristics and local structure is taken into consideration, and propose a new Manifold-based Feature Extraction Method (MFEM). MFEM inherits the advantage of Fisher criterion and manifold learning and effectively improve the feature extraction efficiency.

2   Background Knowledge

2.1   LDA

Given a dataset matrix $X = [x_1, x_2, ..., x_N] = [x_1, x_2, ..., x_c]$ where $x_i(i = 1, ..., N)$, N and c are respectively the training size and the class size. N_i denoting the number of sample in the i th class.

In LDA, two scatters named between-class scatter S_B and within-class scatter S_W are defined as:

$$S_B = \sum_{i=1}^{c} \frac{N_i}{N} (\bar{x}_i - \bar{x})(\bar{x}_i - \bar{x})^{\mathrm{T}} \qquad (1)$$

$$S_W = \sum_{i=1}^{c} \sum_{j=1}^{N_i} \frac{1}{N} (x_{ij} - \bar{x}_i)(x_{ij} - \bar{x}_i)^{\mathrm{T}} \qquad (2)$$

where $\bar{x}_i = \frac{1}{N_i} X_i e_i$ with $e_i = [1, ..., 1]^T \in R^{N_i}$ is the centroid of class i and $\bar{x} = \frac{1}{N} X e$ with $e_i = [1, ..., 1]^T \in R^N$ is the global centroid.

The optimal function of LDA is:

$$J(W_{opt}) = \max_{W} \frac{W^{\mathrm{T}} S_B W}{W^{\mathrm{T}} S_W W} \qquad (3)$$

The Eq. (3) is equivalent to:

$$\max_{W} W^{\mathrm{T}} S_B W \qquad (4)$$

and

$$\min_{W} W^{\mathrm{T}} S_W W \qquad (5)$$

where W is the optimal projection.

The projection W can be obtained by calculating the eigenvectors.

It can be seen from the above analysis, LDA tries to preserve the global characteristics invariant before and after feature extraction. Its efficiency can not be improved because it neglects the local structure of each class.

2.2 LPP

The optimal problem of LPP is:

$$\min_{W} \sum_{i,j} (W^T x_i - W^T x_j)^2 S_{ij} \tag{6}$$

$$s.t \sum_{i} W^T x_i D_{ii} x_i^T W = 1 \tag{7}$$

where W is the optimal projection, S_{ij} is the weight function which reflects the similarity of samples, $D_{ii} = \sum_j S_{ij}$.

The above optimization problem can be transformed as follows based on the linear algebra theory:

$$\min_{W} W^T XLX^T W \tag{8}$$

$$s.t. \ W^T XDX^T W = 1 \tag{9}$$

where $L = D - S$.

The optional projection matrix is obtained by computing all the nonzero eigenvectors of $XLX^T W = \lambda XDX^T W$.

In conclusion, LPP tries to preserve the local characteristic but does not take the global characteristics into consideration, especially, when encountering noise, the feature extraction efficiency of LPP is greatly influenced.

3 MFEM

Feature extraction is a classical preprocessed approach in dealing with the high-dimensional samples. Though they are widely-used in practice, the feature extraction efficiency is limited due to neglecting the global characteristics and local structure. In order to take all the characteristics of the training samples, a new Manifold-based Feature Extraction Method (MFEM) is proposed. In MFEM, Within-Class Scatter based on Manifold (WCSM) and Between-Class Scatter based on Manifold (BCSM) are introduced and the optimal projection can be obtained based on the Fisher criterion.

3.1 Between-Class Scatter Based on Manifold

Inspired by manifold learning, we firstly construct the adjacency graph $G_D = \{X, D\}$ where G_D donates a graph with different classes, X and D donate the dataset and the weight of different classes respectively. The different-class weight function of two random samples x_i and x_j can be defined:

$$D_{ij} = \begin{cases} \exp(-d/\|x_i - x_j\|^2), & l_i \neq l_j \\ 0, & l_i = l_j \end{cases} \tag{10}$$

where l_i ($i = 1, 2, \ldots, N$) donates the class label and d is a constant.

The different-class weight function verifies that if the samples x_i and x_j belong to different classes, the weight of them is large; or else, the weight is zero.

In order to preserve the local characteristics of different classes, the samples x_i and x_j belonged to different classes will be far away after feature extraction. The optimization problem can be described as follows.

$$\max_{W} \sum_{i,j} (y_i - y_j)^2 D_{ij} \tag{11}$$

Where $y_i = W^T x_i$, W donates the projection matrix and $x_i \in X$.

$\sum_{i,j} (y_i - y_j)^2 D_{ij}$ is reformulated to the following equations based on the algebraic transformation.

$$\begin{aligned} &\frac{1}{2} \sum_{i,j} (y_i - y_j)^2 D_{ij} \\ &= \frac{1}{2} \sum_{i,j} (W^T x_i - W^T x_j)^2 D_{ij} \\ &= \sum_{i,j} (W^T x_i D_{ii} x_i^T W - W^T x_i D_{ij} x_j^T W) \\ &= W^T X D' X^T W - W^T X D X^T W \\ &= W^T X (D' - D) X^T W \\ &= W^T S_D W \end{aligned} \tag{12}$$

where $S_D = X(D' - D)X^T$, D' is a diagonal matrix and $D' = \sum_j D_{ij}$.

By taking (12) to (11), (11) is reformulated to

$$\max_{W} W^T S_D W \tag{13}$$

Based on the above analysis, we can see Eqs. (4) and (13) reflect the global characteristics of different classes and local structure of each class respectively. In order to fully utilize all the above information, we can obtain the following optimization expression based on (4) and (13).

$$\begin{aligned} &\max_{W} \alpha W^T S_B W + (1 - \alpha) W^T S_D W \\ &= \max_{W} W^T [\alpha S_B + (1 - \alpha) S_D] W \\ &= \max_{W} W^T M_B W \end{aligned} \tag{14}$$

where $M_B = \alpha S_B + (1 - \alpha) S_D$ and α is a parameter balancing S_B and S_D. M_B is called Between-Class Scatter based on Manifold (BCSM).

3.2   Within-Class Scatter Based on Manifold

Similarity with BCSM, the same-class weight function of two random samples x_i and x_j is defined:

$$S_{ij} = \begin{cases} \exp(-\|x_i - x_j\|^2 / s), & l_i = l_j \\ 0, & l_i \neq l_j \end{cases} \qquad (15)$$

where l_i $(i = 1, 2, .., N)$ donates the class label and s is a constant.

The same-class weight function verifies that if the samples x_i and x_j with the same class label, the weight of them is large; or else, the weight is zero.

In order to keep the neighborhood close, it can be described as:

$$\min_{W} \sum_{i,j} (y_i - y_j)^2 S_{ij} \qquad (16)$$

where $y_i = W^T x_i$, W donates the projection matrix and $x_i \in X$.

$\sum_{i,j} (y_i - y_j)^2 S_{ij}$ is reformulated to the following equations based on the algebraic transformation.

$$\begin{aligned}
&\tfrac{1}{2}\sum_{i,j} (y_i - y_j)^2 S_{ij} \\
&= \tfrac{1}{2}\sum_{i,j} (W^T x_i - W^T x_j)^2 S_{ij} \\
&= \sum_{i,j} (W^T x_i S_{ii} x_i^T W - W^T x_i S_{ij} x_j^T W) \qquad (17) \\
&= W^T X S' X^T W - W^T X S X^T W \\
&= W^T X (S' - S) X^T W \\
&= W^T S_S W
\end{aligned}$$

where $S_S = X(S' - S)X^T$, S' is a diagonal matrix and $S' = \sum_j S_{ij}$.

By taking (17) to (16), (16) is reformulated to

$$\max_{W} W^T S_S W \qquad (18)$$

The following optimization expression based on (5) and (18).

$$\begin{aligned}
&\max_{W} \beta W^T S_W W + (1 - \beta) W^T S_S W \\
&= \max_{W} W^T [\beta S_W + (1 - \beta) S_S] W \qquad (19) \\
&= \max_{W} W^T M_W W
\end{aligned}$$

where $M_W = \beta S_W + (1 - \beta) S_S$ and β is a parameter balancing S_W and S_S. M_W is called Within-Class Scatter based on Manifold (WCSM).

3.3   The Optimization Problem

Inspired by the Fisher criterion, the above optimization problem can be described as follows.

$$J = \max_{W} \frac{M_B}{M_W} = \max_{W} \frac{W^{\mathrm{T}}(\alpha S_B + (1-\alpha)S_D)W}{W^{\mathrm{T}}(\beta S_W + (1-\beta)S_S)W} \tag{20}$$

The solution of the maximization of (20) is given by computing all the nonzero eigenvectors of $M_B W = \lambda M_W W$.

From the optimization expression of MFEM, it can be seen MFEM not only takes the global characteristics into consideration, but also preserves the local structure. MFEM inherits the advantages of LDA and LPP and improves the feature extraction efficiency to some extent. When $\alpha = \beta = 1$ or $d = s = \infty$, MFEM is equivalent to LDA; When $\alpha = \beta = 0$, $d = \infty$ and $s < \infty$, MFEM is equivalent to LPP.

In practice, M_W maybe singular and the optimal projection can not be obtained by the above approach. For the sake of convenience, the singular value perturbation by adding a little positive number to the diagonal of M_W is introduced to solve the singular problem.

3.4   Optimization Algorithm

Input: the original dataset X and the reduced dimension d
Output: the corresponding lower dimensional dataset $Y = [y_1, y_2, ..., y_d]$
Step1: Construct the adjacency graph $G_D = \{X, D\}$ and $G_S = \{X, S\}$ where $X = \{x_1, x_2, ..., x_N\}$ donates the original dataset, D and S respectively donate the weights of different classes and the same class. We put an edge between x_i and x_j if they are in different classes in the G_D, or else, put an edge between them in the G_s.
Step2: Compute the different-class weights and the same-class weights. If different-class samples x_i and x_j are connected, utilize Eq. (10) to compute the different-class weights; else utilize Eq. (15) to compute the same-class weights.
Step3: Compute S_W, S_B, M_W and M_B.
Step4: Solving the singular problem of M_W. The singular value perturbation is introduced to solve the singular problem. Let M_W transform to M'_W after perturbation.
Step5: Compute the optimal projection W. The solution of the optimal projection W is given by computing all the nonzero eigenvectors of $M_W^{-1} M_B W = \lambda W$ or $M'^{-1}_W M_B W = \lambda W$. The nonzero eigenvectors corresponding to the biggest d eigen-values are combine to form the optimal projection $W = [w_1, ..., w_d]$.
Step6: As to a certain sample $x_i \in X$, the corresponding lower dimensional sample can be obtained by $y_i = W^{\mathrm{T}} x_i$.

4 Experimental Analysis

4.1 UCI Two-Dimensional Visualization

Wine dataset, including 178 samples with 3 classes, in UCI machine learning repository is used in the experiment. Set the reduced dimension is two, successively run PCA, LPP, LDA, MFEM on the wine dataset, and we can obtain the experimental results, shown in Table 1.

Table 1. Recognition rates of PCA, LPP, LDA, MFEM on the face datasets

Data sets	k	PCA	LPP	LDA	MFEM
ORL	3	0.711(28)	0.789(28)	0.814(30)	**0.875(20)**
	4	0.808(28)	0.867(30)	0.875(30)	**0.954(18)**
	5	0.845(22)	0.890(24)	0.905(30)	**0.950(21)**
	6	0.863(22)	0.906(30)	0.950(30)	**0.963(25)**
	7	0.892(20)	0.917(22)	0.925(26)	**0.958(20)**
	8	0.873(20)	0.925(30)	0.938(26)	**0.963(23)**
Yale	4	0.619(12)	0.733(14)	0.667(14)	**0.733(12)**
	5	0.667(14)	0.763(14)	0.767(14)	**0.767(13)**
	6	0.653(12)	0.770(14)	0.747(10)	**0.787(14)**
	7	0.750(12)	0.833(12)	0.833(14)	**0.900(14)**
	8	0.800(10)	**0.899(14)**	0.822(14)	0.867(14)

It can be seen from Fig. 1, some different-class samples are overlapped after feature extraction by PCA. LPP, LDA and MFEM can mainly fulfill the feature extraction task but the efficiencies are different. After feature extraction by LPP, some samples lying near the three-class boundary are overlapped. Therefore, compared with LDA and MFEM, the efficiency of LPP is lowest. The efficiencies of LDA and MFEM are both high, but in the respect of distribution, MFEM shows much more perfect than LDA. This is because MFEM tries to preserve the original distribution by taking both the global and local characteristics, while LDA only focus on the global characteristics based on the Fisher criterion so as to make the different-class samples far and the same-class samples close. Although the within-class scatter reflects the closeness of the same-class samples, yet it does not take the relationship of adjacent samples before and after feature exaction.

4.2 Experiments on Face Datasets

Experiment datasets include ORL face dataset and Yale face dataset. We will discuss the relationship between the sizes of training samples and recognition rates as well as the relationship between the reduced dimensions and recognition rates.

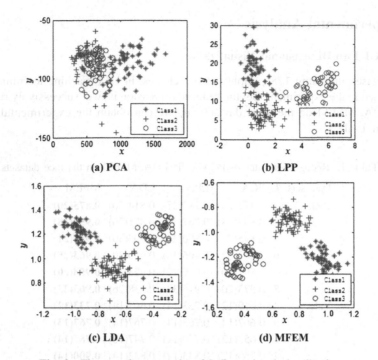

(a) PCA   (b) LPP

(c) LDA   (d) MFEM

Fig. 1. The experiment results of 2-dimensional visualization

Relationship Between the Size of Training Samples and the Recognition Rate. The training dataset consists of the first k images of each subject, and the remainders are used for test. The values of k in ORL and Yale dataset are selected from 3, 4, 5, 6, 7 and from 3, 5, 7, 9 respectively. The comparative experimental results are show in Table 1. In order to overcome the small size problem in LDA, we utilize PCA + LDA instead for LDA in the experiment.

It can be seen from Table 1, compared with PCA, LPP, LDA, MFEM performs best on the ORL dataset and except $k = 8$, the efficiency of MFEM is highest on the Yale dataset.

Relationship Between the Reduced Dimensions and Recognition Rates. The training dataset consists of the first 5 images of each subject, and the remainders are used for test. We can obtain the experiment results shown in Fig. 2.

It can be seen from Fig. 2, as the reduced dimension becomes higher, the corresponding recognition rate mainly has an upward tendency. Compared with PCA, LPP, LDA, MFEM performs best.

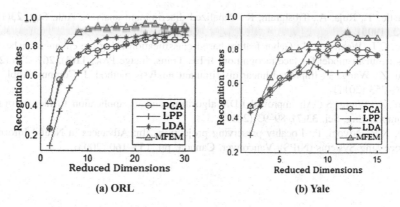

Fig. 2. Relationship between reduced dimensions and recognition rates

5   Conclusions

Researches on current feature extraction approaches can be reduced to two ways, one pays more attention on the global structure, and the other originates from local structure and tries to make the relationship between samples before and after feature extraction be invariant. In view of shortages of classical feature extraction approaches, MFEM is proposed. MFEM considers all the information and improves the feature extraction efficiency. Experiments on some standard datasets verify the effectiveness of MFEM. In practice, linear inseparability is a quite common problem and how to solve it is attracting more and more researchers' interest. MFEM proposed in this paper is suitable to the linear separability situation, how to expand MFEM to linear inseparability is our next work.

References

1. Martinez, A.M., Kak, A.C.: PCA versus LDA. IEEE Trans. Pattern Anal. Mach. Intell. **23**(2), 228–233 (2001)
2. Friedman, H.: Regularized discriminant analysis. J. Am. Stat. Assoc. **84**(405), 165–175 (1989)
3. Li, M., Yuan, B.: 2D-LDA: a novel statistical linear discriminant analysis for image matrix. Pattern Recogn. Lett. **26**(5), 527–532 (2005)
4. Ye, J.P., Xiong, T.: Computational and theoretical analysis of null space and orthogonal linear discriminant analysis. J. Mach. Learn. Res. **7**, 1183–1204 (2006)
5. Yu, H., Yang, J.: A direct LDA algorithm for high-dimensional data with application to face recognition. Pattern Recogn. **34**(11), 2067–2070 (2001)
6. Ji, S.W., Ye, J.P.: Generalized linear discriminant analysis: a unified framework and efficient model selection. IEEE Trans. Neural Netw. **19**(10), 1768–1782 (2008)
7. Belhumeur, P.N., Hespanha, J.P., Kriegman, D.J.: Eiegnfaces vs. Fisherfaces: recognition using class specific linear projection. IEEE Trans. Pattern Anal. Mach. Intell. **19**(7), 711–720 (1997)

8. Hastle, T., Ruja, A., Tibshirani, R.: Penalized discriminant analysis. Ann. Stat. **23**(1), 73–102 (1995)
9. Liu, C.J., Wechsler, H.: Gabor feature based classification using the enhanced fisher linear discriminant model for face recognition. IEEE Trans. Image Proc. **11**(4), 267–276 (2002)
10. Liu, Z., Wang, S.: Improved linear discriminant analysis method. J. Comput. Appl. **31**(1), 250–253 (2011)
11. Liu, Z., Wang, S.: An improved LDA algorithm and its application to face recognition. Comput. Eng. Sci. **33**(7), 89–93 (2011)
12. He, X.F., Niyogi, P.: Locality preserving projections. In: Advances in Neural Information Processing Systems (NIPS), Vancouver, Canada, pp. 153–160 (2003)

lncRNA-LSTM: Prediction of Plant Long Non-coding RNAs Using Long Short-Term Memory Based on p-nts Encoding

Jun Meng[1], Zheng Chang[1], Peng Zhang[1], Wenhao Shi[1],
and Yushi Luan[2(✉)]

[1] School of Computer Science and Technology,
Dalian University of Technology, Dalian 116024, Liaoning, China
[2] School of Bioengineering, Dalian University of Technology,
Dalian 116024, Liaoning, China
luanyush@dlut.edu.cn

Abstract. Long non-coding RNA (lncRNA) plays an important role in regulating biological activities. Traditional feature engineering methods for lncRNA prediction rely on prior experience and require manual feature extraction from some related datasets. Besides, the structure of plant lncRNA is complex. It is difficult to extract features with good discrimination. This paper proposes a method based on long short-term memory networks (LSTM) for lncRNA recognition called lncRNA-LSTM. K-means clustering is used to solve the problem of unbalanced sample size at first, p-nts coding is performed according to the characteristics of RNA sequences, and it is input into a recurrent neural network including embedded layer, LSTM layer and full connection layer. lncRNA-LSTM is more effective than support vector machine, Naive Bayes and other model with feature fusing of open reading frame, second structure and k-mers. Using the same *Zea mays* dataset, lncRNA-LSTM achieves 96.2% accuracy which is 0.053, 0.173, 0.211 and 0.162 higher than that of CPC2, CNCI, PLEK and LncADeep, the precision and recall are much more effective and robust.

Keywords: Deep learning · Recurrent Neural Network · LSTM · lncRNA classification

1 Introduction

RNA is a carrier of genetic information presented in biological cells and some viruses and viroids, regulates the life activities of organisms. In recent years, related research on non-coding RNA (ncRNA) recognition has become a hot spot of concern. Among these ncRNAs, an RNA that is longer than 200 nt and unable to encode protein is of particular interest, called long non-coding RNA (lncRNA) [1, 2]. It has been found that lncRNA plays an important role in regulating the life activities of organisms [3, 4], and various traditional experimental methods to identify lncRNA are time consuming and expensive. Besides, because of low levels of expression and limited cross-species conservation, the recognition of lncRNA is affected to varying degrees [5].

© Springer Nature Switzerland AG 2019
D.-S. Huang et al. (Eds.): ICIC 2019, LNAI 11645, pp. 347–357, 2019.
https://doi.org/10.1007/978-3-030-26766-7_32

In the early years, the researchers conducted a large number of experiments on humans and animals, and some robust lncRNA recognition software were proposed [6]. For the prediction of lncRNA by bioinformatics methods, many studies use machine learning algorithms to build predictive models. Physicochemical feature, sequence and structural features are used as input vectors to construct a classifier model for identifying lncRNA. Studies have shown that the classifier is obtained by extracting features such as open reading frames (ORF), codon frequency preference, and similarity of known proteins as input, and training linear regression, support vector machine (SVM) or other models have a good classification effect. Most of the prediction software derived in recent years adopts the above characteristics. Among them, CPC [7], new version CPC2 [8] and CPAT [9] use sequence features to distinguish between encoded and non-coding RNA; CNCI [10] uses ATN score matrix and sequence structure; PLEK [11] uses k-mer and sliding window to analyze transcripts, and selects k-mers frequency as its characteristic. And there is very few deep learning based approaches such as LncRNAnet which incorporates recurrent neural networks (RNNs) and convolutional neural networks (CNNs) to learn features successfully [12] and LncADeep integrates intrinsic and homology features into a deep belief network and constructs models targeting both full- and partial-length transcripts [13].

However, traditional method requires manual feature engineering or exhaustive searching, and these features are determined based on prior knowledge [14]. Each feature is a dimension. If the number of features is too small, it may not be accurately classified, which may cause under-fitting; too many features may lead to over-emphasis on a certain feature in the classification process and then classification error, that is, over-fitting. Furthermore, most methods only work well on animal datasets. Compared with humans and animals, plant lncRNA is quite different from them [15]. Because the research is still in its infancy, we could not predict plant lncRNA directly by using software based on animal training sets. The software for plant lncRNA recognition is still very rare.

To further improve the accuracy of plant lncRNA prediction, this paper first specifically encodes RNA data, and chooses the deep learning method to identify plant lncRNA. Recurrent neural network (RNN) has a good performance for the processing of sequence data. In this paper, a special RNN, long short term memory (LSTM) [16] is used to construct the lncRNA classification model.

The main contributions of this paper are as follows:

(1) Consider the characteristics of RNA sequence data, p-nts encoding of RNA sequence was proposed to adapt it to the input of the recurrent neural network.
(2) Multi-layer neural networks are designed to build lncRNA classification models, which is called lncRNA-LSTM.

2   Materials and Methods

2.1   Datasets

Deep learning model requires a large number of data. lncRNA data of *Zea mays* are relatively large, and it's genetic annotation information is relatively abundant. In the paper, lncRNA positive dataset is downloaded from the Green Non-Coding Database (GreeNC) [17] (http://greenc.sciencedesigners.com/wiki/Main_Page). There are 18,110 validated *Zea mays* lncRNAs, randomly select 18,000 of them as positive dataset. Negative dataset is 57654 mRNA sequences downloaded from the RefSeq database. In order to solve the imbalanced data problem between positive and negative samples, remove noise and redundant data. To avoid over-fitting, we need to undersampling the mRNA. First, removing the sequences with higher similarity, which belong to the same gene, 37,851 mRNAs remaining at this time. Next use K-means which is a good unsupervised clustering method, to cluster *Zea mays* mRNA dataset and select representation samples of the same number as positive dataset. The specific steps of clustering are shown below.

First, extract the k-mers characteristics of RNA sequences. A k-mer consists of k (Here, in order to reduce experimental time, we only set k to 1, 2) bases. So there is $4 + 4 * 4 = 20$ dimension features. We use a sliding window of length of k to perform a sliding match along the RNA sequence with a step size of 1, and then use c_i to indicate the number of matches ($i = 1, 2, ..., 20$), and assign a coefficient w_k to each k-mer so that each type of frequency has the same effect on the clustering effect, the specific formula is as follows:

$$f_i = w_k \frac{c_i}{s_k}, \ k = 1,2, i = 1,2, ..., 20 \tag{1}$$

$$s_k = L - k + 1, \ k = 1,2 \tag{2}$$

$$w_k = \frac{1}{4^{3-k}}, \ k = 1,2 \tag{3}$$

In the paper, consider the time factor, we set the center point of K-means to 200. After clustering, calculate the number of samples in each category as x_i ($i = 1, 2, ..., 200$), then each category was formed. Randomly select O_i ($i = 1, 2, ..., 200$) samples as training samples, the formula is as follows:

$$O_i = \left\lceil \frac{x_i}{37851} \times 18000 \right\rceil \tag{4}$$

$$N_{mRNA} = 18000 \times \left(\left\lceil \frac{O_1}{37851} \right\rceil + \left\lceil \frac{O_2}{37851} \right\rceil + ... + \left\lceil \frac{O_{200}}{37851} \right\rceil \right) = 18000 \tag{5}$$

The final positive and negative datasets are shown in Table 1.

Table 1. Datasets information (80% for training data, 20% for test data)

Data	Quantity	Database
lncRNA	18000	GreeNC
mRNA	18000	RefSeq

2.2   p-nts Encoding

In natural language processing, an important step is word segmentation, which is the premise of text classification or text summarization. After segmentation, each sentence can be represented by a sequence of numbers. For our dataset, RNA is a sequence composed of nucleotide permutations, therefore, analogous to the process of natural language processing, each RNA sequence is imagined as a sentence, and then the corresponding word segmentation is processed according to the biological character-istics of the RNA sequence.

RNA is a chain-like molecule formed by ribonucleotide condensation via phos-pholipid bonds [18]. A ribonucleotide molecule consists of phosphoric acid, ribose and base. The bases of RNA are A-adenine, U-uracil, C-cytosine, G-guanine. Three adjacent nucleotides are programmed to represent a certain amino acid during protein synthesis. Therefore, we regard the continuous k nucleotides in the RNA sequence as the "words" in the "sentence" of RNA, so as to encode the RNA sequence.

A p-nts has p nucleotides, and each nucleotide can be A, C, U, G. In this paper, analog k-mer feature, the value of p is 2, 3, 4, 5, 6, so there are 16, 64, 256, 1024, and 4096 codes respectively. Using a sliding window of length of p to match the above p-nts, considering time and memory factors, the sliding window was matched along the RNA sequence with a step size of p nucleotides [19]. Figure 1 shows the 3-nts coding.

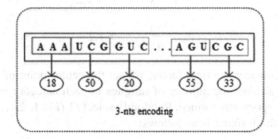

Fig. 1. A p-nts encoding example when p is set to 3.

2.3   RNN and LSTM

RNN is a neural network that is particularly suitable for processing sequential data [12]. The basic RNN includes input layer, hidden layer and output layer. The input of the hidden layer not only comes from the input layer at the current moment, but also includes the output of the hidden layer at the previous moment. At present, the most popular RNN network is LSTM. It's main component is memory cell, which consists of four elements: input gate, self-loop connection, forgetting gate and output gate. The input gate is used to control whether the input signal changes the state of the memory unit, whether it is activated or blocked. The output gate can control the influence of the state of memory unit on other neurons. Finally, the forgetting gate is used to adjust the self-cyclic connection of the memory unit, so that the memory unit can remember or forget the state of the previous time points as needed.

The calculation process of the memory unit of each layer of LSTM at time t is as follows [20]:

(1) The state of the input gate:

$$i_t = sigmoid(W_i x_t + U_i h_{t-1} + b_i) \qquad (6)$$

(2) The state of the forgotten gate:

$$f_t = sigmoid(W_f x_t + U_f h_{t-1} + b_f) \qquad (7)$$

(3) Update the state of the memory unit at time t:

$$C_t = i_t * \tanh(W_c x_t + U_c h_{t-1} + b_c + f_t * C_{t-1}) \qquad (8)$$

(4) Calculate the state of the output gate and the output of the LSTM layer:

$$o_t = sigmoid(W_o x_t + U_o h_{t-1} + V_o C_t + b_o) \qquad (9)$$

$$h_t = o_t * \tanh(C_t) \qquad (10)$$

where W and U are weight matrices, b represents bias, x_t represents the current input and h_{t-1} represents the input of the previous cell.

2.4 lncRNA-LSTM

Figure 2 shows the overview of proposed lncRNA-LSTM method. The datasets are firstly preprocessed, then the RNA is p-nts encoded, and the structure of the classification model is followed. In this method, the first layer of the model is the word embedding layer, which can transform the encoded index into a dense vector of fixed size. The second layer is the bidirectional LSTM layer, units is set to 64, to solve the problem of overfitting, the dropout mechanism is set to 0.4 here. Finally, there is a full connection layer, sigmoid is selected as the activation function, binary cross entropy is chosen as the loss function, and Adam is used as the optimization method.

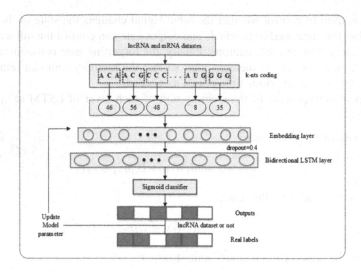

Fig. 2. The overview of lncRNA-LSTM. There are three steps: (1) data preprocessing; (2) p-nts coding; (3) build a classification model based on LSTM.

The source code for the experiments was written in python 3.6.7 programming language using Keras 2.2.4 with Tensorflow 1.8.0-gpu backend.

2.5 Evaluation Indicator

This paper chooses binary cross entropy as the loss function, it is very suitable for the two-class problem, and uses the accuracy rate (ACC), precision rate (P), recall rate (R), and F1 score ($F1\_score$) to evaluate the trained classification model. $F1\_score$ is the harmonic mean of precision rate and recall rate, which is equivalent to the comprehensive evaluation index of precision rate and recall rate. These indicators are defined as follows:

$$ACC = \frac{TP + TN}{TN + FP + TP + FN} \tag{11}$$

$$P = \frac{TP}{TP + FP} \tag{12}$$

$$R = \frac{TP}{TP + FN} \tag{13}$$

$$F1\_score = \frac{2TP}{2TP + FP + FN} \tag{14}$$

where TP refers to predicting a positive class as a positive class, FN refers to a positive class as a negative class, FP refers to a negative class as a positive class, and TN refers to a negative class as a negative class.

3 Results and Discussion

3.1 Experiment Results with Different p-nts Coding

For p-nts coding, when p is assigned different values, the results obtained by constructing LSTM classification model are shown in Fig. 3.

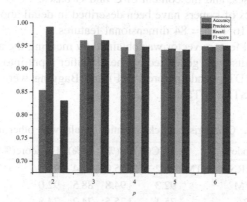

Fig. 3. Classification results based on p-nts coding. Red represents accuracy, blue represents precision, green represents recall, and purple represents F1 score from left to right respectively. (Color figure online)

Figure 3 shows the classification model constructed by 3-nts coding has the highest overall accuracy of 96.2%, which is largely due to every three adjacent nucleotides in the RNA sequence determine a codon, which in turn determines the amino acid. When p is 2, the precision rate shows the highest accuracy nearly 100%, but the overall accuracy is only 85.4%. When p is 4, 5, 6, the accuracy is 94.7%, 94.2% and 95.0%, respectively. The highest recall rates and F1 score values are based on the 3-nts coding. It can be seen that when p is given a reasonable value, the proposed lncRNA-LSTM method can effectively identify lncRNA. In conclusion, when p is 3, the classification performance is the best.

3.2 Performance Comparison with Feature Engineering Methods

In our preliminary work, we carried out a large number of experiments based on feature engineering. In these experiments, we found a method with high performance by comparing the experimental results. ORF, secondary structure (SS) and k-mers are integrated as input vectors, and the classifiers obtained by inputting these features into SVM and other classification models for training have better classification performance.

ORF has potential translation ability, and the ORF coverage rate of mRNA is significantly higher than that of lncRNA [21]. For ORF, integrity, ORF coverage and normalized ORF value are extracted. Integrity is defined as a Boolean variable bool, 0 means there is no integrity ORF, 1 means there is integrity ORF. Coverage represents the ratio of the length of all normalized ORFs to the length of RNA sequences, and normalized ORF is the ratio of the number n of ORF in sequences to the length L of RNA sequences.

The more stable the secondary structure (the more free energy is released), the stronger the ability of RNA to be encoded. At the same time, the stability of the secondary structure is related to the number of paired bases in the RNA sequence and the content of nucleotide C and G [22]. For SS, extract the minimum free energy, the number of paired bases, and the content of C and G bases.

The characteristics of k-mers have been described in detail above. Where k is 1, 2, and 3, there are $4 + 16 + 64 = 84$ dimensional features.

A 90-dimensional feature vector was obtained by merging the above three kinds of features, and the results were generated by the classifier input into SVM, Naive Bayes (NB), decision tree (DT), random forest (RF) and Bagging were compared with the results of lncRNA-LSTM as Table 2.

Table 2. Comparison of classification results with other models.

Model	ACC(%)	P(%)	R(%)	F1_score(%)
lncRNA-LSTM	96.2	95.0	97.4	96.2
SVM	92.3	94.8	88.5	92.0
NB	75.1	75.5	74.2	74.8
DT	93.2	93.3	93.0	93.1
RF	95.0	96.3	93.5	94.9
Bagging	91.9	98.2	85.1	91.2

It can be seen that lncRNA-LSTM has the highest overall accuracy of 96.2%, RF, DT and SVM also shows good classification performance, with accuracy of 95.0%, 93.2% and 92.3%, respectively. In terms of precision, Bagging is the best, and the proposed model ranks third, only 0.32 lower. The lncRNA-LSTM has the best performance in terms of recall and F1 score, reaching 97.4% and 96.2% respectively. These results indicate that the proposed method can effectively predict lncRNA.

3.3   Performance Comparison with Other Existing Tools

In order to further prove the superiority of lncRNA-LSTM, we compared it with other four advanced predictors, CPC2, CNCI, PLEK and LncADeep. CPC2 uses hierarchical feature selection procedure to identify effective features with recursive feature elimination method, and trains a SVM model using these features, training species includes human, *Arabidopsis thaliana*, et al. CNCI is a classifier to differentiate protein-coding and noncoding transcripts by profiling the intrinsic composition of the sequence, training species includes human and plant. PLEK uses a computational pipeline based on an improved k-mer scheme and a SVM algorithm to distinguish lncRNAs from mRNAs, training species includes human and maize. LncADeep is novel lncRNA identification and functional annotation tool which based on deep learning algorithms, training species includes human and mouse. The results of several classification methods on the same *Zea mays* dataset are shown in Table 3. It can be seen that the overall accuracy of lncRNA-LSTM is 96.2%, which is 0.053, 0.173, 0.211 and 0.162 higher than that of CPC2, CNCI, PLEK and LncADeep. That may because CNCI and PLEK use SVM as their classification model which can't extract the intrinsic features of sequence. Although LncADeep uses deep learning method, but the training data is human's lncRNA, so result in the lower accuracy. Precision of lncRNA-LSTM is 95.0%, which is 0.019, 0.045, 0.266 and 0.040 higher than that of CPC2, CNCI, PLEK and LncADeep. For recall only two are over 90% and the proposed method is best. F1 score of lncRNA-LSTM reached 96.2%, which is 0.054, 0.209, 0.173 and 0.193 more than CPC2, CNCI, PLEK and LncADeep. These results show that the lncRNA-LSTM method exhibits powerful performance for identify lncRNA.

Table 3. Comparison of classification results with other methods.

Method	ACC(%)	P(%)	R(%)	F1_score(%)
lncRNA-LSTM	96.2	95.0	97.4	96.2
CPC2	90.9	93.1	88.6	90.8
CNCI	78.9	90.5	64.5	75.3
PLEK	75.1	68.4	93.3	78.9
LncADeep	80.0	91.0	66.6	76.9

By comparing the accuracy and recall rate of lncRNA-LSTM, CPC2, CNCI, PLEK and LncADeep, it can be found from Fig. 4 that the recall rate of PLEK is 0.249 higher than the accuracy, which indicates that PLEK tends to predict mRNA as lncRNA. The recall rates of CNCI and LncADeep are about greater than 0.25, indicating that they tend to predict lncRNA as the negative sets. The difference between the accuracy and recall rate of CPC2 is 0.045, which is 93.1% and 88.6%, respectively. The robustness is general. For lncRNA-LSTM, its accuracy and recall rate are approximately equal and both above 95.0%, indicating that lncRNA-LSTM prediction method is more robust than CPC2, CNCI and other predictors.

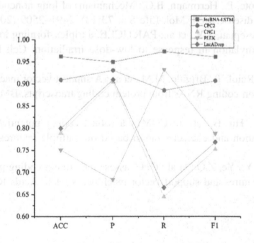

Fig. 4. Comparison of classification results with other methods.

4 Conclusions

In order to overcome the shortcomings of traditional feature engineering needs to manually extract features based on previous experience, such as under-fitting and over-fitting, further improve the accuracy of identifying lncRNA, this paper proposes the lncRNA-LSTM method, which constructs the lncRNA classification model based on the deep learning method. Firstly, the positive and negative sample datasets are clustered with k-mers feature to achieve the quantity balance. Then p-nts coding is performed on lncRNA and mRNA, and we find that the classification performance is best when p is 3. Finally, input into the designed recurrent neural network classification model with embedded layer and LSTM layer, and the final classifier is obtained after 5 rounds of training. Comparing with the existing CPC2, CNCI and other methods of lncRNA on the *Zea mays* dataset, the lncRNA-LSTM method is strong and robust and shows powerful ability for identifying lncRNA. But the proposed method has a time consuming problem, we will find a faster optimization method to reduce the training time in the future. At present, the number of plant lncRNAs is relatively small, our future work is to test the effectiveness of the proposed method by using more species of plants. Moreover, most biological processes regulated by lncRNA are still unclear, our future work is to identify the interaction between lncRNA and miRNA and predict the function of lncRNA.

Acknowledgment. The current study was supported by the National Natural Science Foundation of China (Nos. 61872055 and 31872116).

References

1. Falazzo, A.F., Lee, E.S.: Non-coding RNA: what is functional and what is junk? Front. Genet. **6**, 2 (2015)
2. Aryal, B., Rotllan, N., Fernándezhernando, C.: Noncoding RNAs and atherosclerosis. Curr. Atherosclerosis Rep. **16**(5), 1–11 (2014)
3. Schmitz, S.U., Grote, P., Herrmann, B.G.: Mechanisms of long noncoding RNA function in development and disease. Cell. Mol. Life Sci. **73**(13), 2491–2509 (2016)
4. O'Leary, V.B., Ovsepian, S.V., et al.: PARTICLE, a triplex-forming long ncRNA, regulates locus-specific methylation in response to low-dose irradiation. Cell Rep. **11**(3), 474–485 (2015)
5. Schneider, H.W., Raiol, T., Brigido, M.M., et al.: A support vector machine based method to distinguish long non-coding RNAs from protein coding transcripts. BMC Genom. **18**(1), 804 (2017)
6. Long, H., Xu, Z., Hu, B., et al.: COME: a robust coding potential calculation tool for lncRNA identification and characterization based on multiple features. Nucleic Acids Res. **45**(1), e2 (2017)
7. Kong, L., Zhang, Y., Ye, Z.Q., et al.: CPC: assess the protein-coding potential of transcripts using sequence features and support vector machine. Nucleic Acids Res. **36**, W345–W349 (2007)

8. Kang, Y.J., Yang, D., Kong, C.L., et al.: CPC2: a fast and accurate coding potential calculator based on sequence intrinsic features. Nucleic Acids Res. **45**(W1), W12–W16 (2017)

9. Wang, L.G., Hyun, J.P., Surendra, D., et al.: CPAT: coding-potential assessment tool using an alignment-free logistic regression model. Nucleic Acids Res. **41**(6), e74 (2013)

10. Sun, L., Luo, H.T., Bu, D.C., et al.: Utilizing sequence intrinsic composition to classify protein-coding and long non-coding transcripts. Nucleic Acids Res. **41**(17), e166 (2013)

11. Li, A.M., Zhang, J.Y., Zhou, Z.Y.: PLEK: a tool for predicting long non-coding RNAs and messenger RNAs based on an improved k-mer scheme. BMC Bioinform. **15**, 311 (2014)

12. Baek, J., Lee, B., Kwon, S., et al.: LncRNAnet: long non-coding RNA identification using deep learning. Bioinformatics **34**(22), 3889–3897 (2018)

13. Yang, C., Yang, L.S., Zhou, M., et al.: LncADeep: an ab initio lncRNA identification and functional annotation tool based on deep learning. Bioinformatics **34**(22), 3825–3834 (2018)

14. Pan, X.Y., Shen, H.B.: RNA-protein binding motifs mining with a new hybrid deep learning based cross-domain knowledge integration approach. BMC Bioinform. **18**, 136 (2017)

15. Bai, Y., Dai, X., Harrison, A.P., et al.: RNA regulatory networks in animals and plants: a long noncoding RNA perspective. Brief. Funct. Genomics **14**(2), 91–101 (2015)

16. Liu, G., Guo, J.B.: Bidirectional LSTM with attention mechanism and convolutional layer for text classification. Neurocomputing **337**, 325–338 (2019)

17. Andreu, P.G., Antonio, H.P., Irantzu, A.L., et al.: GREENC: a Wiki-based database of plant lncRNAs. Nucleic Acids Res. **44**(D1), D1161–D1166 (2016)

18. Li, X., Yang, L., Chen, L.-L.: The biogenesis, functions, and challenges of circular RNAs. Mol. Cell **71**(3), 428–442 (2018)

19. Ehsaneddin, A., Mohammad, R.K., et al.: Continuous distributed representation of biological sequences for deep proteomics and genomics. PLoS ONE **10**(11), e0141287 (2015)

20. Hochreite, S., Schmidhuber, J.: Long short-term memory. Neural Comput. **9**(8), 1735–1780 (1997)

21. Dinger, M.E., Pang, K.C., Mercer, T.R., et al.: Differentiating protein-coding and noncoding RNA: challenges and ambiguities. PLoS Comput. Biol. **4**(11), e1000176 (2008)

22. Ronny, L., Stephan, H.B., Christian, H.S., et al.: ViennaRNA package 2.0. Algorithms Mol. Biol. **6**, 26 (2011)

Prediction of Plant lncRNA-Protein Interactions Using Sequence Information Based on Deep Learning

Haoran Zhou[1], Yushi Luan[2], Jael Sanyanda Wekesa[1],
and Jun Meng[1(✉)]

[1] School of Computer Science and Technology,
Dalian University of Technology, Dalian 116024, Liaoning, China
mengjun@dlut.edu.cn
[2] School of Bioengineering, Dalian University of Technology,
Dalian 116024, Liaoning, China

Abstract. Plant long non-coding RNA (lncRNA) plays an important role in many biological processes, mainly through its interaction with RNA binding protein (RBP). To understand the function of lncRNA, a basic step is to determine which proteins are interacted with lncRNA. Therefore, RBP can be predicted by computational approaches. However, the main challenge is that it is difficult to find interaction patterns or primitives. In this study, we propose a method based on sequences to predict plant lncRNA-protein interaction, namely PLRPI uses k-mer frequency feature for RNA and protein, stacked denoising autoencoder and gradient boosting decision tree to learn the hidden interaction between plant lncRNAs and proteins sequences. The experimental results show that PLRPI achieves good performance on the test datasets ATH948 and ZEA22133 based on lncRNA-protein interaction of *Arabidopsis thaliana* and *Zea mays*. Our method gets an accuracy of 90.4% on ATH948 and 82.6% on ZEA22133. PLRPI is also superior to other methods in some public RNA-protein interaction datasets. The result shows PLRPI has strong generalization ability and high robustness. It is an effective model for predicting plant lncRNA-protein interactions.

Keywords: lncRNA-protein · k-mer · Stacked denoising autoencoder · Gradient boosting decision tree

1 Introduction

Long non-coding RNA (lncRNA) [1] is a kind of RNA molecule with specific functions in eukaryotes. Its length is generally more than 200 nt. Basically, they have no ability to encode proteins, which are large in number and are presented in the nucleus or cytoplasm. It has been found that lncRNA can participate in various levels of gene expression regulation by interacting with proteins such as chromatin-modified complexes and transcription factors. lncRNA also plays a regulatory role in many important biological processes. Their interactions are closely related to the most basic life activities of organisms [2–5]. Many key cellular processes such as signal transduction,

© Springer Nature Switzerland AG 2019
D.-S. Huang et al. (Eds.): ICIC 2019, LNAI 11645, pp. 358–368, 2019.
https://doi.org/10.1007/978-3-030-26766-7_33

chromosome replication, material transport, mitosis, transcription and translation, are closely related to the interaction between RNA and protein [6–8]. Although there is no doubt about the role of lncRNA in the regulation of gene expression, only a few functions and mechanisms of lncRNA have been studied. Since the regulatory role of lncRNA mostly requires the coordination of protein molecules, it is necessary to identify the interactions of lncRNA and protein molecules.

Research on plant lncRNA is still in its infancy compared with animals. To date, nearly 10,000 lncRNAs have been found in several plants such as *Arabidopsis thaliana,* wheat, corn, soybeans, and rice, accounting for 1% of total lncRNAs. They play an important role in guiding reproductive development, growth, stress response, chromosome modification, and protein interactions.

The interaction of lncRNA with protein is ubiquitous. At present, there are few structural data of protein complexes obtained by conventional methods such as X-ray diffraction, nuclear magnetic resonance, electron microscopy and neutron diffraction. This is mainly because the experimental methods have disadvantages like high cost, long time-consuming and complicated measurement process. With the development of high-throughput sequencing technology, people can quickly obtain a large amount of transcriptome and proteomic information, including a large number of potential RPI needs analysis. However, traditional experimental methods can only be studied on specific protein, RNA or protein-RNA complexes, which is far from technically sufficient. Therefore, machine learning is widely used in bioinformatics to extract features from samples and analyze them.

Traditional machine learning models require manual feature extraction, which may not be able to pinpoint hidden relationships in raw data. Deep learning provides a powerful solution to this problem. It consists of multi-layer neural network model architecture [9–11] that automatically extracts high-level abstractions from the data. At the same time, in the fields of image recognition [12], speech recognition, signal recognition [13], deep learning shows better performance than other commonly used machine learning methods. It has also been well applied in the field of bioinformatics [14, 15]. For example, deep learning has been successfully applied to predict RNA splicing patterns [16]. Compared with other sequence-based methods, deep learning can automatically learn the sequence characteristics of RNA and protein, discover the specific correlation between these sequences [17, 18], and reduce the influence of noise in the original data by learning the real hidden advanced features. In addition, some methods based on deep learning artificially introduce noise to reduce over-fitting, which can enhance the generalization ability and robustness of the model.

This study presents a new model, PLRPI, for predicting plant lncRNA-protein interactions based on sequence information. For a particular plant protein and lncRNA pair, PLRPI can predict whether there are interactions between them. In the experiment, we first extracted the 4-mer features of lncRNA and the 3-mer features of proteins [19]. 20 amino acids of proteins were divided into 7 groups according to their physico-chemical properties [20]. They are embedded into matrices and features are extracted using stacked denoising autoencoder. Then the extracted features of lncRNAs and proteins are contacted and added into the softmax layer, which is compared with the data labels for supervised learning, the advanced features are obtained and fine-tuned. The gradient boosting descent tree classifier is used for ensemble classification, and the final result is obtained. We evaluated the performance of PLRPI on plant datasets and other RNA-protein datasets from previous studies for comparison with other advanced methods. The results show that PLRPI not only has high prediction accuracy, but also has good generalization ability and high robustness. It can effectively predict the interaction between plant lncRNAs and proteins.

2   Materials and Methods

2.1   Datasets

To test the performance of PLRPI, we created the datasets ATH948 and ZEA22133 based on *Arabidopsis thaliana* and *Zea mays*. Firstly, we downloaded *Arabidopsis thaliana* and *Zea mays* lncRNA-protein datasets from Ming Chen's bioinformatics group (http://bis.zju.edu.cn/PlncRNADB/index.php?p=network&spe=Zea%20mays). In order to reduce the bias of sequence homology, the redundant sequences with sequence similarity greater than 90% for both protein and lncRNA sequences were excluded by using CD-HIT [21]. For constructing non-interaction pairs, the same number of negative pairs were generated through randomly pairing proteins with lncRNAs and further removing the existing positive pairs [19]. After redundancy removal, ATH948 dataset, including 948 interactive pairs and 948 non-interactive pairs, was obtained consisting of 35 protein chains and 109 lncRNA chains. Similarly, ZEA22133 dataset, including 22133 interactive pairs and 22133 non-interactive pairs, was obtained consisting of 42 protein chains and 1704 lncRNA chains. It should be pointed out that compared with other datasets, it is more difficult to extract features from plant lncRNA-protein interaction datasets. This is due to the poor homology of plant lncRNA and the fact that a larger number of interactions require only a smaller number of lncRNAs and proteins. It may increase the noise which is more evident in ZEA22133. The details are shown in Table 1.

Table 1. Experimental datasets.

Dataset	lncRNA	Protein	Interaction pair	Non-interaction pair
ATH948	109	35	948	948
ZEA22133	1704	42	22133	22133
RPI2241	842	2043	2241	2241
RPI369	332	338	369	369
RPI488	25	247	243	245
RPI1807	1078	1807	1807	1436

To test the robustness of PLRPI, we also collected other RNA-protein datasets from previous studies, such as RPI1807 [22], RPI369 [19], RPI2241 [19] and RPI488 [23], which were all extracted based on structure-based experimental complexes. RPI1807, RPI369 and RPI2241 datasets are RNA-protein interactions from many species, including human, animals and plants. Only RPI488 dataset is lncRNA-protein interaction.

2.2 Methods

We first extracted 4-mer features of lncRNAs and 3-mer features of proteins, and then put them into stacked denoising autoencoder models, respectively. The results are fine-tuned using label information from RNA-protein pairs. After high-level features were fine-tuned and they were classified using gradient boosting decision tree to get the output. The detailed process is shown in Fig. 1.

The datasets and python code supporting the findings of this study are available at https://github.com/zhr818789/PLRPI. The source code for the experiments was written in python 3.5.2 using Keras 2.2.2 with Tensorflow 1.10.0 backend.

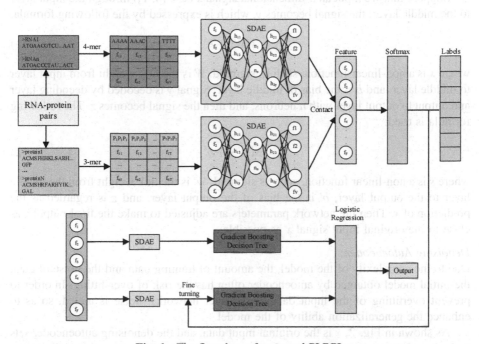

Fig. 1. The flowchart of proposed PLRPI.

Sequence Information Processing

In order to obtain the raw features of autoencoder, we extracted simple sequence component composition features from both RNAs and proteins. For RNA sequences, 4-mer frequency features of RNA sequences (A, C, G, T) are extracted, we got $4 \times 4 \times 4 \times 4 = 256$ dimensional features. Each feature value is the normalized frequency of 4-mer nucleotides in RNA sequences, which is AAAA...CATC...TTTT. For protein sequences, analysis by existing studies indicates that RNA-binding residues are prone to amino acids with certain properties. According to the physicochemical properties of amino acids and the effects of interactions, the 20 amino acids are divided into 7 categories. They include: {Val, Gly, Ala}, {Phe, Pro, Leu, Ile}, {Ser, Tyr, Met, Thr}, {His, Asn, Tpr, Gln}, {Arg, Lys}, {Glu, Asp} and {Cys}. We divided the protein sequences into 7 groups according to the rules above. Since the conjoint triad (3-mer) of protein is composed by 3 amino acids, we extracted the 3-mer features of protein trimer and got $7 \times 7 \times 7 = 343$ dimensional features.

Stacked Denoising Autoencoder (SDAE)

Autoencoder (AE)

Autoencoder belongs to unsupervised learning and does not need to label training samples. It is composed of two parts. The first part is an encoding network consisting of input layer and middle layer which is used to compress the signal. The second part is a decoding network consisting of middle layer and output layer which is used to restore the compressed signal.

Suppose that we input an *n*-dimensional signal x ($x < [0, 1]$) through the input layer to the middle layer, the signal becomes y, which is expressed by the following formula:

$$y = s(Wx + b) \tag{1}$$

where s is a non-linear function, such as sigmoid. W is the link weight from input layer to middle layer, and b is the bias of middle layer. Signal y is decoded by decoding layer and output to output layer with n neurons, and then the signal becomes z. The following formula is used:

$$z = s(W'y + b') \tag{2}$$

where s is a non-linear function, such as sigmoid. W' is the link weight from the middle layer to the output layer, b' is the bias of the output layer, and z is regarded as the prediction of x. Then the network parameters are adjusted to make the final output z as close to the original input signal x as possible.

Denoising Autoencoder

Due to the complexity of the model, the amount of training data and the noise of data, the initial model obtained by autoencoder often has the risk of over-fitting. In order to prevent overfitting of the input data (input layer network), noise is added, so as to enhance the generalization ability of the model.

As shown in Fig. 2, x is the original input data, and the denoising autoencoder sets the value of the input layer node to 0 with a certain probability, so as to get the model input $x^{\hat{}}$ with noise. This is similar to dropout, except that dropout sets the neurons in the hidden layer to 0. By calculating y and z with the corrupted data x' and iterating errors with z and the original x, the network learns the corrupted data.

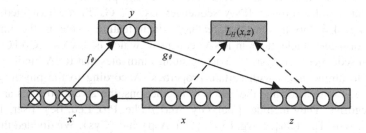

Fig. 2. The flowchart of denoising autoencoder.

Through the comparison with non-corrupted data training, the weight noise of corrupted data is relatively small. This is because the input noise is accidentally removed, and the corrupted data alleviates the generation gap between training data and test data to a certain extent. Because part of the data is removed, the corrupted data is close to the test data to a certain extent.

Stacked Denoising Autoencoder (SDAE)
The idea of SDAE is to stack multiple DAEs together to form a deep architecture [24]. Noise is added to the input when training the model. A SDAE with two hidden layers is shown in Fig. 3.

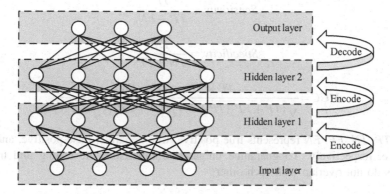

Fig. 3. A SDAE with two hidden layers.

Each encoding layer carries out unsupervised training separately. The training objective is to minimize the error between input (input is the hidden output of the previous layer) and reconstruction results. The output of layer K is obtained through forward propagation, and then layer $K + 1$ is trained with the output of layer K as the input.

Once SDAE training is completed, its high-level features are used as input of traditional supervised algorithms. A layer of logistic regression layer can be added at the top level, and then the network can be fine-tuned with labeled data.

Gradient Boosting Decision Tree (GBDT)
GBDT is one of the best algorithms to fit the real distribution in traditional machine learning algorithms. Its effect is good and it is used for classification and regression.

GBDT uses multiple iterations, and each iteration produces a weak classifier. Each classifier is trained on the basis of the residual of the previous one. The requirement for weak classifiers is usually simple enough with low variance and high deviation, because the training process is to improve the accuracy of the final classifier by reducing the deviation. The weak classifier will generally choose CART (classification and regression tree). Because of the above high deviation and simple requirement, the depth of each classification regression tree will not be very deep. The final total classifier is the sum of weighted weak classifiers obtained from each round of training (that is the additive model).

Evaluation Criteria

In this study, we classify protein and lncRNA pairs to be interacting or not. We follow the widely used evaluation measure by means of the classification accuracy, precision, sensitivity, specificity and MCC defined respectively as follows:

$$Accuracy = \frac{TP + TN}{TP + TN + FP + FN} \tag{3}$$

$$Precision = \frac{TP}{TP + FP} \tag{4}$$

$$Sensitivity = \frac{TP}{TP + FN} \tag{5}$$

$$Specificity = \frac{TN}{TN + FP} \tag{6}$$

$$MCC = \frac{TP \times TN - TP \times FN}{\sqrt{(TP + FP)(TP + FN)(TN + FP)(TN + FN)}} \tag{7}$$

where *TP, TN, FP, FN* represents true positive, true negative, false positive, and false negative, respectively. To guarantee unbiased comparison, the testing and training datasets do not overlap with each other.

3   Results and Discussion

3.1   Results

In this study, PLRPI method is tested on ATH948 and ZEA22133 datasets which are the interactions between lncRNA and protein. The test results are shown in Table 2.

Table 2. Performance of proposed method on our constructed datasets (%).

Dataset	Accuracy	Precision	Sensitivity	Specificity	MCC
ATH948	90.4	92.8	87.6	93.2	81.1
ZEA22133	82.6	99.9	67.5	99.6	69.6

Through the experimental results, we find that our method not only has high accuracy, but also has excellent sensitivity and precision. This indicates that PLRPI has a strong ability to recognize negative samples, and the proportion of actual positive set samples in the predicted positive set is large. Although deep learning models generally require enough data as support, the larger amount of data, does not yield higher accuracy. The data of ZEA22133 is more, however, its accuracy is not as good as that of ATH948.

3.2   Comparing with Other Methods

We compared PLRPI with other sequence-based methods IPMiner [23], RPISeq [19] and lncPro [25] on our datasets. In study [19], the authors proposed RPISeq-RF and RPISeq-SVM for predicting RNA-protein interaction, and RPISeq-RF performed better than RPISeq-SVM on most datasets. Accordingly, here we only compared PLRPI with RPISeq-RF. As shown in Table 3, on data ATH488 and ZEA22133, PLRPI achieved the best performance. On dataset ATH488 it increased the accuracy with 10% over IPMiner. Compared with other methods, it obtained the best performance in other indexes with a little advantage over IPMiner, RPISeq-RF and lncPro. On dataset ZEA22133, PLRPI achieved a prediction accuracy of 82.6% with an increase of about 20% over other methods. It achieved a precision of 99.9% and a specificity of 99.6% with an increase of about 50% over other methods. This shows that our model performs well in plant lncRNA-protein interactions datasets, and can effectively extract advanced features and make predictions.

Table 3. Performance compared with other methods on our constructed datasets (%).

Dataset	Method	Accuracy	Precision	Sensitivity	Specificity	MCC
ATH948	PLRPI	**90.4**	**92.8**	**87.6**	**93.2**	**81.1**
	IPMiner	88.2	89.2	86.9	89.5	76.5
	RPISeq-RF	75.6	76.2	75.2	73.0	79.4
	lncPro	75.4	76.9	75.4	74.7	71.5
ZEA22133	PLRPI	**82.6**	**99.9**	**67.5**	**99.6**	**69.6**
	IPMiner	68.7	69.6	66.5	70.9	37.5
	RPISeq-RF	65.4	64.1	62.5	70.3	35.9
	lncPro	60.3	61.3	60.8	69.6	30.9

PLRPI outperforms other models on ATH948 and ZEA22133 datasets is because it uses GBDT as a classifier. For GBDT, trees are not a multi-training average relationship. They are interrelated, hierarchical, and the variance must be relatively large. However, because its learning ability is relatively strong, its deviation is very small, and the more trees there are, the stronger the learning ability and the smaller the deviation. Thus, as long as the number of trees for learning is enough, the predicted mean will be infinitely close to the target.

3.3   Testing the Robustness of PLRPI

To test the robustness of PLRPI, we also compared it with other sequence-based methods on other published ncRNA-protein and RNA-protein datasets. On dataset RPI2241 and RPI369, the proposed method achieved higher performance than the other methods. This shows that our method has strong robustness (Table 4).

Table 4. Performance comparison with different methods on other public datasets (%).

Dataset	Method	Accuracy	Precision	Sensitivity	Specificity	MCC
RPI2241	PLRPI	**70.7**	**72.9**	**65.9**	**75.5**	**41.7**
	IPMiner	64.8	65.7	62.0	67.6	29.7
	RPISeq-RF	64.6	66.3	65.2	63.0	29.3
	lncPro	65.4	66.9	65.9	64.0	31.0
RPI369	PLRPI	**74.5**	**73.3**	**77.2**	71.8	**49.2**
	IPMiner	72.3	72.4	72.3	**72.3**	44.7
	RPISeq-RF	70.4	70.7	70.5	70.2	40.9
	lncPro	70.4	71.3	70.8	69.6	40.9
RPI488	PLRPI	89.0	**93.9**	83.3	**94.6**	78.5
	IPMiner	**89.1**	93.5	**84.0**	94.4	**78.8**
RPI1807	PLRPI	97.2	97.2	**98.2**	96.5	94.3
	IPMiner	**97.4**	**97.3**	98.1	96.5	**94.8**
	RPISeq-RF	97.3	96.0	96.8	**98.4**	94.6
	lncPro	96.9	95.5	96.5	98.1	93.8

On dataset RPI488 and RPI1807, PLRPI has not achieved the best performance but its indicators are almost the same as other methods. The reason is that the datasets are mixed with samples of different organisms, and our model is better at dealing with the plant lncRNA with poor homology, that is, our single species dataset.

PLRPI achieves good results on public datasets, mainly because of the use of stacked denoising autoencoder. When the amount of training data is small, if we use the traditional autoencoder to build the learning network, after passing the first few layers, the error is extremely small. In addition, the training becomes invalid, and the learning speed is slow. SDAE first performs unsupervised pre-training on each single hidden layer of the denoising autoencoder, then stacks them, and finally performs overall fine-tuning training to avoid the above problems and obtain better results.

In the process of training, the early stop method is used, which means that training is stopped when the performance of the model begins to decline on the verification set, thus avoiding the problem of over-fitting caused by continued training. PLRPI stops training when the generalization loss exceeds the threshold, which reduces the impact of over-fitting and save time. To further reduce the impact of over-fitting, we set dropout to 0.5 [26], which is a common setting.

It can be found that PLRPI is not strict with the requirement of data quantity. From hundreds to tens of thousands of sequences, the performance is excellent, but if the number of interaction between lncRNA and protein is large and the number of their respective sequences is relatively small (which is common in plant data), other general models do not perform well, and our model still maintains a good performance. This proves that PLRPI can adapt well to the data of plant lncRNA-protein interaction and obtain higher performance.

4 Conclusion

In this study, we propose a computational method PLRPI based on stacked denoising autoencoder and gradient boosting decision tree to predict plant lncRNA-protein interactions. It achieved a better performance on our constructed lncRNA-protein datasets ATH948 and ZEA22133. The comprehensive experimental results of other previously published datasets also show the effectiveness of PLRPI. In dataset ZEA22133, it improves the performance of the model by about 20% compared with other existing sequence-based methods. The results show that stacked denoising autoencoder extracts discriminant high-level features, which is very important for building deep learning model. The high-level features are the features automatically learned from multiple layers of neural network. PLRPI has shown good performance in plant lncRNA-protein, which is better than other advanced methods. In future work, we will apply different methods for sequence information of lncRNA and protein such as OPT, PSSM, One-hot, and adjust the network structure according to different datasets. We hope that we can use this model to construct network for plant lncRNAs and proteins, which can be used to infer the functions of plant lncRNAs.

Acknowledgment. The current study was supported by the National Natural Science Foundation of China (Nos. 61872055 and 31872116).

References

1. Okazaki, Y., Furuno, M., Kasukawa, T., Adachi, J., Bono, H., Kondo, S., et al.: Analysis of the mouse transcriptome based on functional annotation of 60,770 full-length cDNAs. Nature **420**(6915), 563–573 (2002)
2. Chen, Y., Varani, G.: Protein families and RNA recognition. FEBS J. **272**(9), 2088–2097 (2005)
3. Cooper, T.A., Wan, L., Dreyfuss, G.: RNA and disease. Cell **136**(4), 777–793 (2012)
4. Lukong, K.E., Chang, K.W., Khandjian, E.W., Richard, S.: RNA-binding proteins in human genetic disease. Trends Genet. **24**(8), 416–425 (2008)
5. Chen, X., Sun, Y.Z., Guan, N.N., et al.: Computational models for lncRNA function prediction and functional similarity calculation. Brief. Funct. Genomics **18**(1), 58–82 (2018)
6. Lunde, B.M., Moore, C., Varani, G.: RNA-binding proteins: modular design for efficient function. Nat. Rev. Mol. Cell Biol. **8**(6), 479–490 (2007)
7. Zhang, L., Zhang, C., Gao, R., Yang, R., Song, Q.: Prediction of aptamer-protein interacting pairs using an ensemble classifier in combination with various protein sequence attributes. BMC Bioinform. **17**(1), 225–238 (2016)
8. Gawronski, A.R., Uhl, M., Zhang, Y., et al.: MechRNA: prediction of lncRNA mechanisms from RNA-RNA and RNA-protein interactions. Bioinformatics **34**(18), 3101–3110 (2018)
9. Bengio, Y., Courville, A., Vincent, P.: Representation learning: a review and new perspectives. IEEE Trans. Pattern Anal. Mach. Intell. **35**(8), 1798–1828 (2012)
10. Hinton, G.E., Salakhutdinov, R.R.: Reducing the dimensionality of data with neural networks. Science **313**, 504–507 (2006)
11. Lecun, Y., Bengio, Y., Hinton, G.: Deep learning. Nature **521**(7553), 436–444 (2015)
12. Litjens, G., Kooi, T., Bejnordi, B.E., et al.: A survey on deep learning in medical image analysis. Med. Image Anal. **42**(9), 60–88 (2017)

13. Deng, L., Yu, D.: Deep learning: methods and applications. Found. Trends® Sig. Process. **7**(3), 197–387 (2014)
14. Alipanahi, B., Delong, A., Weirauch, M.T., Frey, B.J.: Predicting the sequence specificities of DNA- and RNA-binding proteins by deep learning. Nat. Biotechnol. **33**(8), 831–838 (2015)
15. Zhou, J., Troyanskaya, O.G.: Predicting effects of noncoding variants with deep learning–based sequence model. Nat. Methods **12**(10), 931–934 (2015)
16. Leung, M.K.K., Xiong, H.Y., Lee, L.J., Frey, B.J.: Deep learning of the tissue-regulated splicing code. Bioinformatics **30**(12), i121–i129 (2014)
17. Ray, D., Kazan, H., Cook, K.B., Weirauch, M.T., et al.: A compendium of RNA-binding motifs for decoding gene regulation. Nature **499**(7457), 172–177 (2013)
18. Cook, K.B., Hughes, T.R., Morris, Q.D.: High-throughput characterization of protein–RNA interactions. Brief. Funct. Genomics **14**(1), 74–89 (2015)
19. Muppirala, U.K., Honavar, V.G., Dobbs, D.: Predicting RNA-protein interactions using only sequence information. BMC Bioinform. **12**(1), 489–500 (2011)
20. Pan, X.Y., Zhang, Y.N., Shen, H.B.: Large-scale prediction of human protein-protein interactions from amino acid sequence based on latent topic features. J. Proteome Res. **9**(10), 4992–5001 (2010)
21. Huang, Y., Niu, B., Gao, Y., Fu, L., Li, W.: CD-HIT Suite: a web server for clustering and comparing biological sequences. Bioinformatics **26**(5), 680–682 (2010)
22. Suresh, V., Liu, L., Adjeroh, D., Zhou, X.: RPI-Pred: predicting ncRNA-protein interaction using sequence and structural information. Nucleic Acids Res. **43**(3), 1370–1379 (2015)
23. Pan, X., Fan, Y.X., Yan, J., Shen, H.B.: IPMiner: hidden ncRNA-protein interaction sequential pattern mining with stacked autoencoder for accurate computational prediction. BMC Genom. **17**(1), 582–596 (2016)
24. Vincent, P., Larochelle, H., Lajoie, I., Bengio, Y., Manzagol, P.A.: Stacked denoising autoencoders: learning useful representations in a deep network with a local denoising criterion. J. Mach. Learn. Res. **11**(12), 3371–3408 (2010)
25. Lu, Q., Ren, S., Lu, M., Zhang, Y., Zhu, D., Zhang, X., et al.: Computational prediction of associations between long non-coding RNAs and proteins. BMC Genom. **14**(1), 651–661 (2013)
26. Dahl, G.E., Sainath, T.N., Hinton, G.E.: Improving deep neural networks for LVCSR using rectified linear units and dropout. In: International Conference on Acoustics, Speech and Signal Processing, pp. 8609–8613. IEEE, Vancouver (2013)

Predicting of Drug-Disease Associations via Sparse Auto-Encoder-Based Rotation Forest

Han-Jing Jiang[1,2], Zhu-Hong You[1,2(✉)], Kai Zheng[3],
and Zhan-Heng Chen[1,2]

[1] The Xinjiang Technical Institute of Physics and Chemistry,
Chinese Academy of Sciences, Urumqi 830011, China
zhuhongyou@ms.xjb.ac.cn
[2] University of Chinese Academy of Sciences, Beijing 100049, China
[3] School of Computer Science and Technology,
China University of Mining and Technology, Xuzhou, China

Abstract. Computational drug repositioning, designed to identify new indications for existing drugs, significantly reducing the cost and time involved in drug development. Confirming the association between drugs and disease is a critical process in drug development. At present, the most advanced method is to apply the recommendation system or matrix decomposition method to predict the similarity between drugs and diseases. In this paper, the association between drugs and diseases is integrated and a novel computational method based on sparse auto-encoder combined with rotation forest (SAEROF) is proposed to predict new drug indications. First, we constructed a drug-disease similarity based on drug-disease association and integrated it into sparse auto-encoder to obtain the final drug-disease similarity. Then, we adopt a rotation forest algorithm to predicted scores for unknown drug–disease pairs. Cross validation and independent test results show that this model is better than the existing model and has reliable prediction performance. In addition, the case study of two diseases further proves the practical value of this method, and the results obtained can be found in CTD database.

Keywords: Drug-disease associations · Rotation forest · Sparse auto-encoder

1 Introduction

In the past few decades, development of new drugs has been a time-consuming and costly process. Government spending on new drugs has been rising for decades, but the number of drugs approved by the US food and drug administration is falling. The key issue for drug development is the need to identify and avoid potential drug disease associations. Drug development is divided into three phases: the discovery phase, the clinical phase, and the clinical development phase. Drug candidates for drug retargeting have passed the second phase of drug development, thus speeding up the drug development process. So, the reposition of drugs is a hot topic for researchers. In addition, successful drug repositioning cases such as sildenafil, thalidomide, and raloxifene have been a boon for patients.

© Springer Nature Switzerland AG 2019
D.-S. Huang et al. (Eds.): ICIC 2019, LNAI 11645, pp. 369–380, 2019.
https://doi.org/10.1007/978-3-030-26766-7_34

For instance, *Chandrasekaran et al.* opened up a new way to study drug-disease associations by adding additional information to disease targets [1]. Chen *et al.* constructed a three-layer heterogeneity network with the addition of drug-mirna associations and mirna-disease associations [2]. Yang *et al.* constructed a network that includes a drug-target-path-gene disease. The other is to predict new drug-disease links through known drug-disease associations, drug characteristics, and disease characteristics [3]. Liang *et al.* proposed an information integration method based on Laplace regular sparse subspace learning (LRSSL) [4]. Zhang *et al.* proposed an expression method based on network topology for drugs and diseases. This approach describes the association between known drugs and disease as two networks, based on the similarities between the two networks and the linear domain [5]. Wu *et al.* proposed to divide the heterogeneous data into three layers, and calculate the association between the drugs and the disease at each layer, and let the similarity between the drugs and the disease be measured through the similar points of the two layers [6]. Lu *et al.* proposed a drug that can be found in known diseases or related to known drugs. Their method is a semi-supervised global learning method called DR2DI [7].

In addition, deep learning strategy has attracted much attention because they can effectively process high-throughput biological data, which makes it possible to construct interactive models of complex biological relationships with deep learning. Deng *et al.* proposed a feature learning method for speech emotion recognition based on sparse auto-encoder [8]. Xu *et al.* proposed a method for pathological image detection of breast cancer tissue based on a multi-layer sparse auto-encoder [9]. Based on previous research, A new computational drug repositioning method is proposed, which applies sparse auto-encoder after integrating various biological data. In this study, drug - disease similarity was constructed by combining drug similarity, and disease similarity. Then, we propose a drug reposition model based on sparse auto-encoder combined with rotating forest (SAEROF) to predict the latent indications of existing drugs. In our model, innovations are made in two aspects based on similarities in drug-related and disease-related characteristics. As shown in Fig. 1. Firstly, since previous studies have adjusted the weak similarity between drug and disease, we calculated the gaussian interaction profile kernel of the disease and drug, respectively, and fused the gaussian interaction profile kernel obtained with the similarity. Second, sparse auto-encoder is applied to reduce dimension of features, retaining effective features.

Furthermore, experimental results show that SAEROF has good performance in two datasets under the general criteria, which can effective on discovering new drug indications. In the case study, two drugs were selected for prediction in the CTD dataset [10]. The CTD database is constantly being updated to include the newly validated drug-disease relationship and to provide a basis for our validation. Most of the new drug-disease associations that rank high are strongly supported by public databases, so our effectiveness is further confirmed.

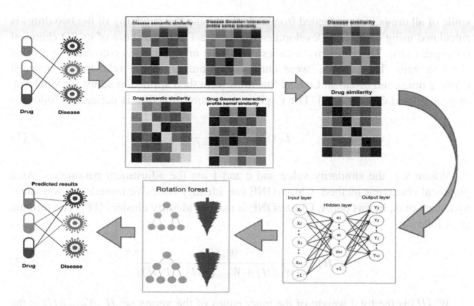

Fig. 1. Flowchart of GIPSAE model to predict potential drug-disease associations.

2 Materials and Methods

2.1 Dataset

In the experiment, we validate our model using the gold standard data set provided by Gottlieb et al. and the Cdataset provided by Luo *et al.* [11, 12]. The gold standard dataset included 593 drugs, 313 diseases and 1933 confirmed drug-disease associations. Here, the gold standard dataset is referred to as Fdataset for short. Cdataset contains 663 drugs, 409 diseases and 2532 proven associations between them. Drug-Bank is a database that contains a large amount of drug result information and extracts drug information [13]. Disease information is collected from the human phenotype defined by the online Mendelian genetic database (OMIM) [14], which provides information on human diseases. The number of diseases, drugs, and drug-disease associations is shown in Table 1.

Table 1. Drug, diseases and interactions in each dataset

Datasets	Drugs	Diseases	Interactions
Cdataset	663	409	2532
Fdataset	593	313	1933

2.2 Drug Chemical Structures Similarity

The structure similarity of drugs is calculated from the chemical structure of drugs. SMILES is a line symbol that describes the chemical structure [15], the Canonical smile of all drugs is downloaded from DrugBank, and the similarity of the two drugs is calculated using Chemical Development Kit as the Tanimoto score of its 2D chemical fingerprint [16]. The similarity with less predicted information is converted to a value close to zero. Then, Drugs were clustered according to the relationship between existing drugs and diseases. Logical functions were then applied to adjust for similarity in gene-related diseases [17]. The Logistic regression function is defined as follows:

$$L(x) = \frac{1}{1 + e^{(cx+f)}} \tag{1}$$

Where x is the similarity value and c and f are the adjustment parameters. As a graphical clustering method, ClusterONE can identify cohesive modules in formaldehyde networks. In this paper, ClusterONE is used to identify clusters [18]. The formula is as follows:

$$f(H) = \frac{W_{in}(H)}{(W_{in}(H) + W_{bound}(H) + P(H))} \tag{2}$$

$W_{in}(H)$ is the total weight of the inner edges of the vertex set H, $W_{bound}(H)$ is the total weight of the edges connecting the set to the remainder of the group, and $P(H)$ is the penalty term. The resulting similarity in drug structure is referred to as DE [12].

2.3 Disease Semantic Similarity

Disease semantic similarity is calculated by MimMiner [19], which measures disease similarity by calculating the similarity between MeSH terms [20]. The known drug-disease association information is considered, and the similarity is adjusted by analyzing the correlation between the drug and the disease [21]. Last, the disease was clustered by ClusterONE, and the comprehensive disease similarity DS was obtained [12].

2.4 Gaussian Interaction Profile Kernel Similarity

The Gaussian interaction profile of the disease is calculated based on the assumption that similar diseases (e.g., different subtypes of lung cancer) can often bind to the same drug molecule, and vice versa [22]. The definition binary vector $V(d(i))$ represents the interaction profiles with disease $d(i)$, and $V(d(j))$ represents the interaction profiles with disease $d(j)$ [23]. We apply $V(d(i))$ to represent the row vector of the adjacency matrix [24]. For the Gaussian interaction profile between disease $d(i)$ and disease $d(j)$, our calculation formula is as follows [25]:

$$GD(d(i), d(j)) = \exp\left(-\theta_d \|V(d(i)) - V(d(j))\|^2\right) \tag{3}$$

Where parameter ∂_d was implemented to tune the kernel band width which was calculated via normalizing original parameter as ∂_d' follows:

$$\partial_d = \partial_d' \Big/ \left[\sum_{i=1}^{md} \|V(d(i))\|^2 \right] \tag{4}$$

Similarly, the definition binary vector $V(e(i))$ represents the interaction profiles with disease $e(i)$, and $V(e(j))$ represents the interaction profiles with the drug $e(j)$. We apply the binary vector $V(e(i))$ to represent the row vector of the adjacency matrix [24]. The following formula apply to calculate the Gaussian interaction profile kernel similarity between the drug $e(i)$ and the drug $e(j)$ in $GE(e(i), e(j))$ [26]:

$$GE(e(i), e(j)) = \exp\left(-\theta_e \|V(e(i)) - V(e(j))\|^2\right) \tag{5}$$

$$\partial_e = \partial_e' \Big/ \left[\sum_{i=1}^{nd} \|V(e(i))\|^2 \right] \tag{6}$$

2.5 Multi-source Feature Fusion

In this part, we ultimately used descriptors that integrated multiple data sources including disease similarity, drug similarity and the Gaussian interaction profile kernel to predict the drug-disease association [27]. The method has the advantage that it can reflect the characteristics of disease and drug from different perspectives [24], which is conducive to deeply dig out exploration of the potential relationship between drug and disease and the improvement of the prediction performance of the model [28].

For the similarity of diseases, we construct disease semantic similarity DS and disease Gaussian interaction profile kernel similarity GD [29]. The disease similarity matrix $SIM_{dis}(d(i), d(j))$ between disease $d(i)$ and $d(j)$ can be obtained by integrating the above disease similarities [22]. The formula is as follows:

$$SIM_{dis}(d(i), d(j)) = \begin{cases} GD(d(i), d(j)) & \text{if } d(i) \text{ and } d(j) \text{ has Gaussian} \\ & \text{interaction profile} \\ & \text{kernel similarity} \\ DS(d(i), d(j)) & \text{otherwise} \end{cases} \tag{7}$$

If $d(i)$ and $d(j)$ have the Gaussian interaction profile kernel [23], then the Gaussian interaction profile kernel is applied; otherwise, the disease semantic similarity is used to fill in the kernel, thus forming a new disease similarity matrix [30]. For the similarity of drug, we combined drug semantic similarity DE and drug Gaussian interaction profile kernel similarity GE to form drug similarity matrix SIM_{drug} [31]. The drug similarity matrix $SIM_{drug}(e(i), e(j))$ formula for drug $e(i)$ and drug $e(j)$ is as follows:

$$SIM_{drug}(e(i), e(j)) = \begin{cases} GE(e(i), e(j)) \ if \ e(i) \ and \ e(j) \ has \ Gaussian \\ \qquad\qquad\qquad\qquad\quad interaction \ profile \\ \qquad\qquad\qquad\qquad\quad kernel \ similarity \\ \quad DE(e(i), e(j)) \qquad\qquad otherwise \end{cases} \quad (8)$$

If the Gaussian interaction profile kernel exists in $e(i)$ and $e(j)$ [32], then the Gaussian interaction profile kernel is used; otherwise, the similarity of drug semantic is filled in [29].

2.6 Sparse Auto-Encoder

Auto-encoder is an unsupervised neural network model that learns the implicit characteristics of input data. The basic structure of the auto-encoder is shown in Fig. 2 [33]. At the same time, the original input data can be reconstructed with the new features learned [34]. By learning the hidden layer, the data of the output layer can be reconstructed. The downside is that it simply copies the input layer to the hidden layer, does not effectively extract meaningful features to restore the input data. Therefore, a sparse auto-encoder is introduced, which has the advantage of introducing sparse auto-encoder to excite sparse penalty terms and learning relatively sparse features. Compared with auto-encoder, it has more advantages in performance and is more valuable in practical applications.

Sparse auto-encoder (SAE) is a symmetric three-layer neural network with the feature matrix as the input matrix of the SAE. Select the sigmoid function $\sigma(x) = 1/(1 + e^{-x})$ as the activation function for the network. For the input feature matrix X, the feature expression to be learned and obtained is [34]:

$$h = \sigma(Wx(i) + b) \quad (9)$$

The hidden layer makes the output close to the input, and the output expression of the hidden layer is [35]:

$$\sigma(Wh + b) \quad (10)$$

An effective feature can be learned by adding a sparse penalty to the objective function of the autoencoder. The sparse penalty term acts on the hidden layer [34], so that the average value of the node output of the hidden layer is as zero as possible. Suppose that $a_z(x)$ denotes the activation of hidden unit z. The average activation amount of hidden unit j is defined as:

$$\widehat{\rho}_z = \frac{1}{n} \sum_{i=1}^{n} [a_z(x(i))] \quad (11)$$

We added a sparse term to the objective function. If ρ_a and ρ have significant deviations, the objective function will be penalized. The penalty term is expressed as [34]:

$$P_{penalty} = \sum_{J=1}^{S_2} KL(\rho||\widehat{\rho}_z) \tag{12}$$

S_2 is the number of neurons in the hidden layer. KL is the Kullback-Leibler divergence. KL is defined as follows:

$$KL(\rho||\widehat{\rho}_z) = \rho log \frac{\rho}{\widehat{\rho}_z} + (1-\rho)log \frac{1-\rho}{1-\widehat{\rho}_z} \tag{13}$$

After sparse penalty terms are added, the cost function becomes as follows:

$$J_{sparse}(W,b) = J(W,b) + \beta \sum_{z=1}^{S_2} KL(\rho||\widehat{\rho}_z) \tag{14}$$

Initialize the parametric parameter vector and then use the gradient descent method to find the optimal neural network parameters. The formula for the gradient descent method is as follows:

$$W_{ij}(l) = W_{ij}(l) - \alpha \frac{\partial}{\partial W_{ij}(l)} J_{sparse}(W,b) \tag{15}$$

$$b_i(l) = b_i(l) - \alpha \frac{\partial}{\partial b_i(l)} \tag{16}$$

Where α is stands for learning rate. Through the forward transfer of all training examples, the average activation amount is calculated to obtain the sparse error, and then the parameters are updated by the backpropagation algorithm. Then, the sparse auto-encoder can learn a valid sparse feature representation.

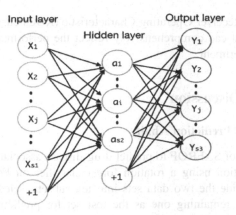

Fig. 2. The structure of an auto-encoder.

2.7 Rotation Forest

Rotating Forest is a novel proposed ensemble system that simultaneously increases the accuracy and ensemble diversity of a single classifier [36]. The basic idea of rotation a forest is as follows: First, the input feature set is randomly divided into K subsets, and the divided subsets may or may not intersect. To improve diversity, we select disjoint subsets. Principal component analysis (PCA) is then performed on all feature subsets. In the original literature, principal component analysis (PCA), nonparametric discriminant analysis (NDA) and random projection (RP) were used as transformation methods. PCA-based rotation forests performed best. Saving all the principal components saves the variability information in the data [37]. Finally, K-axis rotation is performed to encourage individual accuracy and diversity. The advantage of rotation forest is the rotation matrix constructed by linear transformation subset [38].

2.8 Evaluation Criteria

To have a comprehensive assessment of the performance of SAEROF, we follow the widely used evaluation indicators and strategies. The 10-fold cross-validation was applied to evaluate the performance of SAEROF. We follow common evaluation criteria to evaluate the model, including precision (Prec.), Recall, F1-score and accuracy (Acc.). Their calculation formulas are defined as follows:

$$Prec. = \frac{TP}{TP + FP} \tag{17}$$

$$Recall. = \frac{TP}{TP + FN} \tag{18}$$

$$F1 - score = \frac{2PR}{P + R} \tag{19}$$

$$Acc. = \frac{TP + TN}{TP + TN + FP + FN} \tag{20}$$

In addition, the Receiver Operating Characteristic (ROC) curve and the area under the curve (AUC) that can comprehensively reflect the performance of the model are also used in the experiment.

3 Results and Discussion

3.1 Assessment of Prediction Ability

To assess the ability of SAEROF to predict drug-disease associations, we performed a ten-fold cross-validation using a rotation forest classifier on Fdataset and Cdataset, respectively. We divide the two data sets into ten subsets, select nine subsets as the training set, and the remaining one as the test set for prediction. Finally, different subsets are selected in turn as test sets. We collected the results of these ten experiments and used the mean and standard deviation as the final experimental results.

As can be seen from Tables 2 and 3, the average value of the 10-fold cross-validation results in Fdataset is 81.33% precision, 84.05% f1-score, 87.00% recall and 83.49% accuracy, respectively. The standard deviations are 1.88%, 1.52%, 2.18% and 1.58% respectively. The average values of the ten-fold cross-validation results in Cdataset are: precision is 78.64%, f1-score is 81.64%, recall is 84.06%, accuracy is 80.70%. The standard deviations are 1.49%, 1.71%, 2.83% and 1.64% respectively. As can be seen from Fig. 3, the AUC obtained after the 10-fold cross validation on Fdataset and Cdataset were 0.9157 and 0.9306, respectively.

Table 2. Lists the results of the ten-fold cross-validation obtained by SAEROF on Fdataset

Test set	Acc.(%)	Pre.(%)	Recall.(%)	F1-score.(%)
1	81.44	77.73	88.14	82.61
2	82.21	81.09	84.02	82.53
3	82.73	79.26	88.66	83.70
4	82.12	78.44	88.60	83.21
5	80.83	76.92	88.08	82.13
6	80.83	77.42	87.05	81.95
7	81.09	79.70	83.42	81.52
8	81.35	76.65	90.16	82.86
9	79.02	76.42	83.94	80.00
10	84.46	81.82	88.60	85.07
Average	81.61 ± 1.35	78.55 ± 1.78	87.07 ± 2.27	82.56 ± 1.28

Table 3. List the results of the ten-fold cross-validation obtained by SAEROF on Cdataset

Test set	Acc.(%)	Pre.(%)	Recall.(%)	F1-score.(%)
1	84.06	82.64	86.22	84.39
2	85.43	82.37	90.16	86.09
3	81.42	78.49	86.56	82.33
4	84.39	81.29	89.33	85.12
5	83.20	83.07	83.40	83.23
6	83.40	82.13	85.38	83.72
7	80.63	79.47	82.61	81.01
8	83.20	81.58	85.77	83.62
9	82.41	81.78	83.40	82.58
10	85.38	84.29	86.96	85.60
Average	83.35 ± 1.49	81.71 ± 1.60	85.98 ± 2.35	83.77 ± 1.50

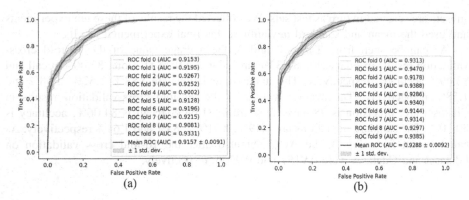

Fig. 3. (a) and (b) show the ROC curves yielded by GIPSAE using 10-fold cross validation on the Fdataset and Cdataset, respectively.

4 Conclusion

In this paper, we proposed a predictive model based on rotation forest and sparse autoencoder to predict drug-disease associations. SAEROF is a new computational method for identifying new associations between drugs and diseases. In SAEROF, the synthetical similarity consists of four portions: the semantic similarity of the disease, the similarity of the drug chemical structures, drug and disease Gaussian interaction profile kernel similarity. In order to test the property of our approach in identifying novel indications of existing drugs, a comprehensive experiment including cross validation and case study was conducted. In future studies, SAEROF performance could be further enhanced by improving the algorithm of deep learning in combination with more valid drug and disease similarity.

References

1. Chandrasekaran, S.N., Koutsoukas, A., Huan, J.: Investigating multiview and multitask learning frameworks for predicting drug-disease associations. In: ACM International Conference on Bioinformatics (2016)
2. Chen, H., Peng, W., Zhang, Z.: miRDDCR: a miRNA-based method to comprehensively infer drug-disease causal relationships. Sci. Rep. **7**(1), 15921 (2017)
3. Yang, J., et al.: Drug-disease association and drug-repositioning predictions in complex diseases using causal inference-probabilistic matrix factorization. J. Chem. Inf. Model. **54** (9), 2562–2569 (2014)
4. Liang, X., et al.: LRSSL: predict and interpret drug-disease associations based on data integration using sparse subspace learning. Bioinformatics **33**(8), 1187–1196 (2017). (p. btw770)
5. Zhang, W., et al.: Predicting drug-disease associations and their therapeutic function based on the drug-disease association bipartite network. Methods **145**, 51–59 (2018). (S1046202318300045)

6. Wu, G., Liu, J., Wang, C.: Predicting drug-disease interactions by semi-supervised graph cut algorithm and three-layer data integration. BMC Med. Genomics 10(Suppl 5), 79 (2017)

7. Lu, L., Yu, H.: DR2DI: a powerful computational tool for predicting novel drug-disease associations. J. Comput. Aided Mol. Des. 32(5), 633–642 (2018)

8. Deng, J., et al.: Sparse autoencoder-based feature transfer learning for speech emotion recognition. In: Affective Computing & Intelligent Interaction (2013)

9. Xu, J., et al.: Stacked sparse autoencoder (SSAE) for nuclei detection on breast cancer histopathology images. IEEE Trans. Med. Imaging 35(1), 119 (2016)

10. Mattingly, C.J., et al.: The comparative toxicogenomics database (CTD): a resource for comparative toxicological studies. J. Exp. Zool. Part A Ecol. Genet. Physiol. 111(6), 793–795 (2003)

11. Gottlieb, A., et al.: PREDICT: a method for inferring novel drug indications with application to personalized medicine. Mol. Syst. Biol. 7(1), 496 (2014)

12. Luo, H., et al.: Drug repositioning based on comprehensive similarity measures and bi-random walk algorithm. Bioinformatics 32(17), 2664 (2016)

13. Wishart, D.S., et al.: DrugBank 5.0: A major update to the DrugBank database for 2018. Nucleic Acids Res. 46, D1074–D1082 (2017). (Database issue)

14. Ada, H., et al.: Online mendelian inheritance in man (OMIM), a knowledgebase of human genes and genetic disorders. Nucleic Acids Res. 33(1), 514–517 (2005)

15. Weininger, D.: SMILES, a chemical language and information system. 1. Introduction to methodology and encoding rules. J. Chem. Inf. Comput. Sci. 28(1), 31–36 (1988)

16. Steinbeck, C., et al.: The chemistry development kit (CDK): an open-source Java library for chemo- and bioinformatics. Cheminform 34(21), 493–500 (2003)

17. Vanunu, O., et al.: Associating genes and protein complexes with disease via network propagation. PLoS Comput. Biol. 6(1), e1000641 (2010)

18. Nepusz, T., Yu, H., Paccanaro, A.J.N.M.: Detecting overlapping protein complexes in protein-protein interaction networks. Nat. Methods 9(5), 471–472 (2012)

19. Van Driel, M.A., et al.: A text-mining analysis of the human phenome. Eur. J. Hum. Genet. 14(5), 535–542 (2006)

20. Lipscomb, C.E.: Medical subject headings (MeSH). Bull. Med. Libr. Assoc. 88(3), 265–266 (2000)

21. You, Z.H., Li, X., Chan, K.C.: An improved sequence-based prediction protocol for protein-protein interactions using amino acids substitution matrix and rotation forest ensemble classifiers. Neurocomputing 228, 277–282 (2017)

22. Chen, X., et al.: GIMDA: graphlet interaction-based MiRNA-disease association prediction. J. Cell Mol. Med. 22(3), 1548–1561 (2017)

23. Chen, X., et al.: WBSMDA: within and between score for MiRNA-disease association prediction. Sci. Rep. 6, 21106 (2016)

24. Zhao, H., et al.: Prediction of microRNA-disease associations based on distance correlation set. BMC Bioinform. 19(1), 141 (2018)

25. Manchanda, S., Anand, A.: Representation learning of drug and disease terms for drug repositioning. In: IEEE International Conference on Cybernetics (2017)

26. Zeng, X., et al.: Predict the relationship between gene and large yellow croaker's economic traits. Molecules 22(11), 1978 (2017)

27. Chen, X., Cheng, J.-Y., Yin, J.: Predicting microRNA-disease associations using bipartite local models and hubness-aware regression. RNA Biol. 15(9), 1192–1205 (2018)

28. You, Z.H., et al.: Large-scale protein-protein interactions detection by integrating big biosensing data with computational model. Biomed. Res. Int. 2014(2), 598129 (2014)

29. Chen, X., et al.: Novel human miRNA-disease association inference based on random forest. Mol. Therapy-Nucleic Acids 13, 568–579 (2018)

30. Chen, X., et al.: MDHGI: matrix decomposition and heterogeneous graph inference for miRNA-disease association prediction. PLoS Comput. Biol. **14**(8), e1006418 (2018)
31. Chen, X., et al.: DRMDA: deep representations-based miRNA-disease association prediction. J. Cell. Mol. Med. **22**(1), 472–485 (2017)
32. Peng, L.-H., et al.: HNMDA: heterogeneous network-based miRNA–disease association prediction. Mol. Genet. Genomics **293**, 983–995 (2018)
33. Wang, L., et al.: Combining high speed ELM learning with a deep convolutional neural network feature encoding for predicting protein-RNA interactions. IEEE/ACM Trans. Comput. Biol. Bioinform. **PP**(99), 1 (2018)
34. Sun, W., et al.: A sparse auto-encoder-based deep neural network approach for induction motor faults classification. Measurement **89**, 171–178 (2016). (ISFA)
35. Te, J., et al.: Partial discharge pattern recognition algorithm based on sparse self-coding and extreme learning machine. In: 2018 2nd IEEE Conference on Energy Internet and Energy System Integration (EI2). IEEE (2018)
36. Wang, L., et al.: RFDT: a rotation forest-based predictor for predicting drug-target interactions using drug structure and protein sequence information. Curr. Protein Pept. Sci. **19**, 445–454 (2018)
37. Wang, L., et al.: An improved efficient rotation forest algorithm to predict the interactions among proteins. Soft. Comput. **22**(17), 1–9 (2018)
38. Rodríguez, J.J., Kuncheva, L.I., Alonso, C.J.: Rotation forest: a new classifier ensemble method. IEEE Trans. Pattern Anal. Mach. Intell. **28**(10), 1619–1630 (2006)

A Novel Compression Algorithm for Hardware-Oriented Gradient Boosting Decision Tree Classification Model

Xiao Wang[ID], Yafei Li[ID], Yinqi Li[ID], and Kuizhi Mei[(✉)][ID]

Xi'an Jiaotong University, No. 28, West Xianning Road, Xi'an, Shaanxi, China
{xyalwong, lnl314, yinqi98}@stu.xjtu.edu.cn,
meikuizhi@mail.xjtu.edu.cn

Abstract. Gradient boosting decision tree is a widely used machine learning algorithm. As the big data era comes, this algorithm has been applied to the multimedia fields. However, this algorithm suffers a lot when it comes to the mobility of multimedia application. In this paper, we propose a compression algorithm, GBDT Compression, to enhance the storage adaptability of a gradient boosting decision tree classification model. GBDT Compression introduces a new rule to pruning, addressing and encoding for a well trained gradient boosting decision tree model. By conduct GBDT Compression on small data sets, we show that this algorithm considerably reduces the space cost of original model up to dozens of times. Furthermore, as data increases and the original model grows, the compression rate also enhances. Which is meaningful for mobile multimedia application with limited memory.

Keywords: Gradient boosting decision tree · Classification model · Compression

1 Introduction

Machine learning and artificial intelligence are becoming more and more important in many areas. Among machine learning methods used in practice, Gradient Boosting Decision Tree (GBDT) is one technique that shines in many applications. Proposed by Friedman in 1999 [1], GBDT has been shown to give state-of-the-art results on many regression and classification problems theoretically. Meanwhile, GBDT has also found an increasingly wide utilization in many practical areas including computer vision [2], voice recognition [3], and text [4] etc.

As more and more effective algorithms proposed, it is necessary for GBDT to acquire a higher performance on both time and storage. Therefore, algorithms such as XGBoost [5] and FastBDT [6] and techniques such as GPU acceleration [7] appear. These approaches work on speed issue of GBDT. However, the problem of storage of GBDT attracts rather less attention, which is crucial for a system with limited memory. To the best of our knowledge, current researches about the hardware application of GBDT [8] focus more on the method to apply GBDT to hardware rather than adjust it to fit hardware.

© Springer Nature Switzerland AG 2019
D.-S. Huang et al. (Eds.): ICIC 2019, LNAI 11645, pp. 381–391, 2019.
https://doi.org/10.1007/978-3-030-26766-7_35

Similarly to GBDT, neural network also suffers a lot from time and storage issues with data growing larger. For storage problem, many researchers have done a lot of research on the compression and acceleration of neural network models. These methods mainly include the use of pruning [9, 10], quantification [11, 12], and encoding hardware-friendly [13]. These methods have a great inspiration for our GBDT Compression work.

In this paper, we propose a novel pipeline framework, "GBDT Compression", to modify the prediction model that runs on mobile multimedia system with limited memory. This pipeline framework aims at the application of GBDT classification algorithm. And it contains three stages. First, we acquire a more compact model by pruning. Then, we build the tree structure by a new addressing mode. Finally, we reduce the data redundancy by combined coding. Experiments show that the three-step pipeline works well on models trained on datasets in all sizes.

We organize the rest of the paper as follows. In Sect. 2, we discuss the workflow and storage method of a GBDT classification model. In Sect. 3 we illustrate the detail of GBDT Compression. In Sect. 4 we demonstrate the method to execute our method. Finally, in Sect. 5 we show the experimental performance of our method.

2   GBDT Classification Model

In this section, we will review the concept of GBDT classification model [1]. Besides, we will also analyze current storage methods for GBDT classification model.

2.1   GBDT Classification Model

A GBDT classification model consists of several forests which are made up of additive regression trees. Each forest represents a class. In each class, the first tree is a weak classifier trained on original data, and latter trees are classifiers trained on the residual of the last classifier. In each tree, parent-node splits the variable space, and leaf-node predicts the probability belonging to a specific class.

For a 4-class GBDT model as Fig. 1, we use *Node* to describe the information of each node.

$$
\begin{aligned}
p &= \max_{k \in K} \sum_{t=1}^{T_k} f_t(x), \\
s.t.\ f(x) &= score_{q(x)}, \\
q &: x \to N.
\end{aligned}
\tag{1}
$$

Here we use *layer* to represent the depth information of the node; *split* tells if there is bifurcation for the node; *attri* tells which attribute of the test data to consider for the node; *thres* tells which branch to choose for a parent-node by comparing with the test data; *score* represents the prediction score of a node. For a given test data, we will first traverse all the trees in each class by mapping the test data to the corresponding leaves. Then we sum up all the scores of these leaf-nodes. Finally we will output the class with the largest score as the final prediction. Table 1 shows an example of *Node*.

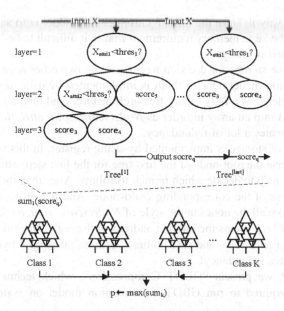

Fig. 1. Conduction of a GBDT classification model

Table 1. An example of 4-class GBDT Model.

		layer	score	split	attri	thres
Class k	Tree 0	1	0.297445	1	3	3.029999
		2	2.229729	1	0	0.144999
		3	0.386776	0		
		3	−0.092233	0		1.110000
		2	−3.103068	0		
	Tree 1	1	−0.476744	1	0	2.049999
		...	...	...	...	...
Class K + 1	Tree 0					
	Tree t					

In this case, we will conduct a classification assignment with (2).

Where p represents the predicted class; K represents the amount of classes in the model; T_k represents the amount of trees in k-th class; $f(x)$ represents the predicted results of each tree; q represents the structure of a tree, which reflects the mapping relationship between test data x and the corresponding leaf-node index N.

$$Node(layer, split, attri, thres, score) \qquad (2)$$

2.2 Model Storage

The storage of GBDT classification model is an urgent problem for hardware system especially mobile multimedia system with limited memory.

The simplest way to store the GBDT classification model is to save the model as itself. However, the large memory requirement makes it difficult to be incorporated into system with limited memory.

Referring to the storage for decision tree, there are two other approaches to store a GBDT classification model. One of them is implemented by using array. In this way, each tree is completed as a fully binary tree architecture. And then all the nodes of the tree will be stored into an array in order as *Node(score, split, attri, thres)*. It is easy to run, but it also creates a lot of redundancy.

Another type of storage is implemented by using register. In this way, the test data will directly traverse the root-node of the first tree for the first step. Then the *split, attri, thres* information help to decide which branch to follow. After that, the register will tell the indirect address of the corresponding child-node. And it will iterate until the end. So, we need to store all the nodes in the style of *Node(score, split, attri, thres, register)*, in which "register" represents the indirect address of the node. Nevertheless, each node needs two indirect addresses, and each indirect address takes up 8 byte space. So this method also creates redundancy.

In this paper, we present "GBDT Compression", which technically reduce the storage that is required to run GBDT classification model on system with limited memory.

3   Compression Algorithm

In this section, we propose a compression algorithm for GBDT classification model, "GBDT Compression". This algorithm reduces model storage resources by preprocessing, pruning, readdressing, and coding without changing original precision.

3.1   Preprocessing

After extracting parameters from original GBDT classification model as (1), we should adjust precision of parameters without changing the original model accuracy. From (1) we can tell that the structure q is related to the parameter *layer, split, attri, thres* and *score*. Among these parameters, values of *layer, split* and *attri* are integers while values of *thres* and *score* are floats with many bits, which also introduces redundancy. As we discussed in Sect. 2, *thres* is used to decide which child-node to choose by comparing with the test data. So let the decimal digits of *thres* equal the decimal digits of the viable (including test data and training data) will keep original precision. Note that the decimal digits of *thres* can be reduced if the final precision is guaranteed. On the other hand, *score* is used to predict the final score by summing up the score corresponding to the test data. As *score* ranges from negative integers to positive integers, the decimal place of *score* has little influence on the final result. So we choose to approximate *score* as integer here. And the practice in Sect. 4 proves that it is viable.

3.2   Pruning

In this part we introduce a simple way to prune the model.

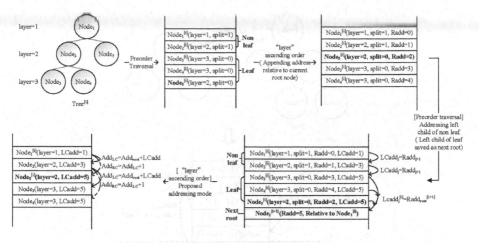

Fig. 2. An example of re-addressing

After rounding the parameter *thres* and *score*, we will prune each tree of the model from bottom to top by deleting nodes with four same parameters (except for *layer*) of (1). And then let their parent-node be the new leaf-node with their parameters.

3.3 Re-addressing

Having been pruned, we propose new addressing approaches in this part for both serial and concurrent computation.

Addressing Mode for Serial Computation For each tree in serial computation, we first preorder traverse it. Then we re-arrange the original tree structure in a new structure, in which the parameter *layer* is in ascending order (See Fig. 2). For nodes belonging to the same layer, the left node is just ahead of the right node. For each tree arranged in this new manner, the root-node is treated as reference, and the address of other nodes relative to root is *Radd*. After that, we return to preorder traversal to record every left child node *Radd* to *LCadd* of its parent node. For leaf-node, its *LCadd* is saved as the *Radd* of next root relative to current root. Eventually, we save each tree in *layer* ascending order, with which the left child node can be addressed by $Add_{root}+ LCadd$, the right child node can be addressed by $Add_{root}+ LCadd + 1$. Add_{root} is the true address of current root node. Note we save all data of each class respectively in an array.

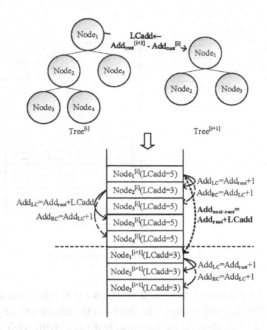

Fig. 3. One class of parallel addressing

Addressing Mode for Concurrent Computation. For concurrent computation, we can replace the add of the root-node with the relative address of next root-node (See Fig. 3). Note that the address of each root-node's left-child-node can be calculated by *add(root-node)* + *1*. In this way, we can easily transform the serial storage of each class into a parallel way. Therefore we can traverse all the trees before conducting the whole first tree.

Comparing to the addressing method using register referred in Sect. 2, the re-addressing method only stores the current node address relative to its root node. So the storage will be saved considerably. Meanwhile, the concurrent computation method is also time-saving.

3.4   Coding

In this part we demonstrate a combinational coding manner.

After readdressing, there are 6 parameters for each node in the model. Here We use parameter *score, split, attri, thres* and *add* of *Node* to code it. For parameter *split*, there are two values to choose, 1 or 0. So we code it with 1 bit binary code. For parameter *attri* and *add*, as they are both integers, the number of binary bits of them is determined by (3) after coding. For parameter *thres* modified in Part A, we will first left-shift its decimal point to acquire an integer, and record the amount of shifted places as m, which will be used in Part D. Then we complete it as a positive integer. Finally, we use (3) to determine the number of binary bits that *thres* needs after coding. For parameter *score* also modified in Part A, we will first calculate the complement of it, and then we acquire the corresponding number with binary bits satisfying (3) after encoding.

$$ceil(log_2 max(y))$$ (3)

To reduce the redundancy, we combine the code of all parameters of *Node* as one final code. For parent-node, we combine parameter *split*, *attri*, *thres* and *add* in turn. While for leaf-node, we combine parameter split, score and add in turn. Here we need to add zeroes between parameter *split* and *score*, so the code length of leaf-node meets the code length for parent-node. One example of the combination rule is shown in Fig. 4.

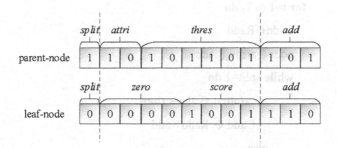

Fig. 4. An example of combination coding

The previous storage methods store the parameters of a node respectively, which enlarges the storage requirement to some extent. While our method combines all the parameters into one code, so it can be adopted by a limited memory system.

4 Conduction of Compression Model

In this part we will describe the procedure to run the whole compression model.

Algorithm 1: Conduction of Compression Model

Input: Test data x

Output: Predicted class Class

x←round(complete(x·10^m))

for k=1 to K **do**

    Radd, Cadd, Tscore$_k$←0

    for t=1 to T$_k$ **do**

        Cadd←Radd

        node←array$_k$(Cadd)

        while split=1 **do**

            if x(attri) ≤ thres **then**

                Cadd ← Radd+ add

            else

                Cadd ← Radd + add +1

            end if

            node←array$_k$(Cadd)

        end while

        Tscore$_k$ ← Tscore$_k$ + score

        Radd ←Radd + add

    end for

end for

Class←Index of max(T score)

Recall that the compression model is made up of K classes, respectively described as an array *array*. In each array, we store T_k trees (See Fig. 2). Nodes in a tree are coded as Fig. 4 with four parameters. Among these parameters, the *thres* has been left-shifted m places. So we need to process the test data similarly to keep the model running accurately. When the test data starts traversing nodes, we use *Radd* to indicate the absolute address of each root-node, and *Cadd* to indicate the current absolute address of each node. Simultaneously, we use *Tscore* to record the total scores of each class, and *Class* to indicate the predicted class. The pseudo code is shown in Algorithm 1.

Intuitively, before inputting a test data into the model, we will first conduct m left-shift operations and a complement.

When the data traverses the first root-node of the k-th class, we have $t = 1$; $Cadd = Radd = 0$; $Tscore_k = 0$. We derive the parameter *split* here by bitwise operations including Boolean AND and left-shift.

If *split* = 1, then the current node is a parent-node. Calculate the parameter *attri*, *thres* and *add*, and compare the *attri*-th feature of the test data x with *thres*. If *data* (*attri*) $\leq$ *thres*, then access the left-child-node of it, whose absolute address satisfys $Cadd = Radd + add$. Otherwise, access the right-child-node of it, whose absolute address satisfys $Cadd = Radd + add + 1$.

If *split* = 0, then the current node is a leaf-node. Calculate the parameter *score* and *add* here. Let $Tscore_k = Tscore_k + score$ to derive the scores of current class. Next traverse the root-node of the $(t + 1)$-th tree, whose absolute address satisfys $Cadd = LCadd + 1 = Radd + add + 1$.

Traverse all T_k trees in the k-th class in the same manner, then we will yield the final $Tscore_k$. Traverse all K classes in the same manner, then we will yield K *Tscore*. Choose the class with the largest *Tscore* as the predicted class.

5   Experiment

In this section, we evaluate the performance of our compression algorithm using models that are trained by four different datasets from UCI Machine Learning Repository. As there has been little research about GBDT compression so far, we only provide results of our algorithm in this paper.

As a preliminary work, we choose HUMAN ACTIVITY [14], BREAST CANCER, GLASS and IRIS [15] as our datasets to train the GBDT classification model. Both the amount of features and the amount of examples of these datasets range from large to small, so the experiment result will be more representative. The detail of these datasets is shown in Table 2. In which, "Features" excludes id for IRIS and BREAST CANCER, and subject for HUMAN ACTIVITY.

Table 2. Experimental datasets

Dataset	Examples	Features	Classes
HUMAN ACTIVITY	10300	561	6
BREAST CANCER	569	30	2
GLASS	214	9	7
IRIS	150	4	3

To show the effect of our proposed method more clearly, we store each model in two ways when it has been trained. One is the "register" method referred in Sect. 2. The other is our "GBDT Compression" method. Then we re-run the model separately with same test data. Finally calculate the compress rate of compressed models by comparing with the "register" models. The result is shown in Table 3. Notion, in those tables, Original indicates the original model, and Proposed indicates the compressed model using proposed method, and symbol "35*5" indicates model consist of 35 trees with 5 layers.

Table 3. Overall compression results (serial)

Model		Top-1 error	Parameters	Compress rate
HUMAN ACTIVITY 35*5	Original	7.33%	102.1 KB	3.8×
	Proposed	7.33%	27.0 KB	
BREAST CANCER 30*5	Original	8.70%	16.2 KB	5.2×
	Proposed	8.70%	3.1 KB	
GLASS 16*5	Original	22.45%	23.1 KB	9.2×
	Proposed	22.45%	2.5 KB	
IRIS 12*5	Original	0%	6.8 KB	17.9×
	Proposed	0%	0.38 KB	

From Table 3 we can tell that the compress rate grows larger as the scale of dataset reduces with Top-1 Error stays the same. One main reason is that most data of the first three datasets possess many decimal digits. While to keep the original precision, we need to memorize as much information. to, our has to be decimal. Which means that thres has to have many decimal digits. So the scale of parameters does not decrease apparently. Another point that we could conclude from Table 3 is that the small model, IRIS model, owns a large compress rate, which shows that the "GBDT Compression" could achieve good results when the complexity of data is not that high.

6 Conclusion

In this paper, we demonstrate "GBDT Compression", an algorithm that allows the GBDT classification model work on memory-limited system. GBDT Compression mainly includes a three stage pipeline: pruning, re-addressing and coding. By pruning, we decrease the model scale; By re-addressing, we change the storage architecture; By encoding, we reduce the storage for each parameter. We then apply this method to some datasets. Although the results are vary, we could conclude that our compression method works well in small datasets, and with data enlarging, its compression rate will increase accordingly. In other words, our method is meaningful for mobile multimedia application with limited storage spaces processing large amount of data.

References

1. Friedman, J.H.: Greedy function approximation: a gradient boosting machine. Ann. Stat. **29** (5), 1189–1232 (2001)
2. Tian, D., He, G., Wu, J., Chen, H., Jiang, Y.: An accurate eye pupil localization approach based on adaptive gradient boosting decision tree. In: Visual Communications and Image Processing, pp. 1–4 (2016)
3. Moon, J., Kim, S.: An approach on a combination of higher-order statistics and higher-order differential energy operator for detecting pathological voice with machine learning. In: 2018 International Conference on Information and Communication Technology Convergence (ICTC), pp. 46–51, October 2018
4. Prasad, A.G., Sanjana, S., Bhat, S.M., Harish, B.S.: Sentiment analysis for sarcasm detection on streaming short text data. In: 2017 2nd International Conference on Knowledge Engineering and Applications (ICKEA), pp. 1–5, October 2017
5. Chen, T., Guestrin, C.: XGBoost: a scalable tree boosting system. In: ACM SIGKDD International Conference on Knowledge Discovery and Data Mining, pp. 785–794 (2016)
6. Keck, T.: FastBDT: a speed-optimized and cache-friendly implementation of stochastic gradient-boosted decision trees for multivariate classification (2016)
7. Zhang, H., Si, S., Hsieh, C.J.: GPU-acceleration for large-scale tree boosting (2017)
8. Wang, Y., Feng, D., Li, D., Chen, X., Zhao, Y., Niu, X.: A mobile recommendation system based on logistic regression and gradient boosting decision trees. In: International Joint Conference on Neural Networks, pp. 1896–1902 (2016)
9. Han, S., Mao, H., Dally, W.J.: Deep compression: compressing deep neural networks with pruning, trained quantization and Huffman coding. Fiber **56**(4), 3–7 (2015)
10. Liu, Z., Li, J., Shen, Z., Huang, G., Yan, S., Zhang, C.: Learning efficient convolutional networks through network slimming, pp. 2755–2763 (2017)
11. Iandola, F.N., Han, S., Moskewicz, M.W., Ashraf, K., Dally, W.J., Keutzer, K.: SqueezeNet: AlexNet-level accuracy with 50x fewer parameters and <0.5 Mb model size (2016)
12. Ullrich, K., Meeds, E., Welling, M.: Soft weight-sharing for neural network compression (2016)
13. Gysel, P.: Ristretto: hardware-oriented approximation of convolutional neural networks (2016)
14. Oneto, L., Parra, X., Anguita, D., Ghio, A., Reyes-Ortiz, J.L.: A public domain dataset for human activity recognition using smartphones. In: 21st European Symposium on Artificial Neural Networks (2013)
15. Fisher, R.A.: The use of multiple measurements in taxonomic problems. Ann. Hum. Genet. **7** (2), 179–188 (2012)

MISSIM: Improved miRNA-Disease Association Prediction Model Based on Chaos Game Representation and Broad Learning System

Kai Zheng[1], Zhu-Hong You[2($\boxtimes$)], Lei Wang[2], Yi-Ran Li[1], Yan-Bin Wang[2], and Han-Jing Jiang[1]

[1] School of Computer Science and Technology,
China University of Mining and Technology, Xuzhou, China
[2] Xinjiang Technical Institute of Physics and Chemistry,
Chinese Academy of Sciences, Ürümqi, China
zhuhongyou@ms.xjb.ac.cn

Abstract. MicroRNAs (miRNAs) play critical roles in the development and progression of various diseases. However, traditional experimental approaches are difficult to detect potential human miRNA-disease associations from the vast amount of biological data. Therefore, computational techniques could be of significant value. In this work, we proposed a miRNA sequence similarity calculation model (MISSIM) to large-scale predict miRNA-disease associations by combined Chaos Game Representation (CGR) with Broad Learning System (BLS). In the five-cross-validation experiment, MISSIM achieved ACC of 0.8424 on the HMDD.

Keywords: miRNAs · Disease · Sequence information · Chaos Game Representation · Broad Learning System

1 Introduction

miRNA plays an important regulatory character in some processes of physiology such as cell differentiation, proliferation, and apoptosis [1]. In the past few years, many computational models were built [2–11]. They are mainly divided into network-based and machine learning-based methods [12–25]. In this study, we build a miRNA sequence similarity calculation model (MISSIM) based on machine learning which combines multiple sources including miRNA sequence information, disease semantic information, miRNA functional information, and confirmed miRNA-disease association information [26–35]. It shows that MISSIM is reliable for predicting potential disease-specific miRNA.

2 Results

Table 1 lists the yielded averages of accuracy, recall, precision and f1-score.

© Springer Nature Switzerland AG 2019
D.-S. Huang et al. (Eds.): ICIC 2019, LNAI 11645, pp. 392–398, 2019.
https://doi.org/10.1007/978-3-030-26766-7_36

Table 1. The comparison results of MISSIM and AUCs based on 5-fold cross validation

Testing set	Accuracy	Recall	Precision	F1-score
Average	84.24%	86.24%	83.18%	84.57%

3   Materials and Methods

3.1   Data Set

Our model is confirmed by the HMDD (Human microRNA Disease Database) dataset in this experiment [36]. After filtering, miRNA-disease association pairs were chosen as positive set in the experiment which built by 1057 miRNAs and 850 diseases. We selected the same number of miRNA-disease pairs from all possible miRNA-disease pairs except positive samples as a negative group.

3.2   miRNA Functional Similarity

In this method, we downloaded it and constructed a matrix *FS* where an entity $FS(m(a), m(b))$ is degree of comparability between miRNA$m(a)$ *and* $m(b)$[37].

3.3   Disease Semantic Similarity

Disease Semantic Similarity Model 1. The semantic value is described as follows:

$$\begin{cases} D_d(t) = 1 & \text{if } t = D \\ D_d(t) = max\{\Delta * D_d(t') | t' \in \text{children of } t\} & \text{if } t \neq D \end{cases} \quad (1)$$

Δ is the semantic decay factor. Besides, we described the semantic value $DV(D)$ as follows:

$$DV(D) = \sum_{t \in T_d} D_d(t) \quad (2)$$

*Sim*1 defined as follows:

$$Sim1(d(i), d(j)) = \frac{\sum_{t \in T_{d(i)} \cap T_{d(j)}} \left(D_{d(i)}(t) + D_{d(j)}(t)\right)}{DV(d(i)) + DV(d(j))} \quad (3)$$

3.4   Gaussian Interaction Profile Kernel Similarity

Gaussian Interaction Profile Kernel Similarity for Diseases. We described $KD(d(a), d(b))$ between $d(a)$ and $d(b)$ as follow:

$$KD(d(a), d(b)) = exp(-\gamma_d * \|IP(d(a)) - IP(d(b))\|^2) \quad (4)$$

γ_d is designed as follows:

$$\gamma_d = \frac{1}{\frac{1}{nd} \sum_{i=1}^{nd} \|P(d(i))\|^2} \quad (5)$$

Gaussian Interaction Profile Kernel Similarity for miRNAs. We computed the Gaussian profile kernel similarity between miRNAs in the same way:

$$KM(m(a), m(b)) = exp(-\gamma_m * \|IP(m(a)) - IP(m(b))\|^2) \quad (6)$$

$$\gamma_m = \frac{1}{\frac{1}{nm} \sum_{i=1}^{nm} \|IP(m(i))\|^2} \quad (7)$$

3.5 Integrated Similarity

Integrated Similarity for Diseases. It can be described as follows:

$$SD(d(a), d(b)) = \begin{cases} \frac{Sim1(d(a),d(b)) + Sim2(d(a),d(b))}{2} & if\ d(a), d(b)\ in\ Sim1\ and\ Sim2 \\ KD(d(a), d(b)) & others \end{cases}$$

$$(12)$$

Integrated Similarity for miRNAs. We calculated the similarity between miRNA $m(a)$ and $m(b)$ as follows:

$$SM(m(a), m(b)) = \begin{cases} FS(m(a), m(b)) & if\ m(a), m(b)\ in\ FS \\ KM(m(a), m(b)) & others \end{cases} \quad (8)$$

3.6 Sequence Similarity for MiRNAs

Define the length of a miRNA sequence to be n_G. The coordinates of each nucleotide can be expressed as CGR_i. g_i is the nucleotide coefficient, where A = (0, 0); C = (0, 1); G = (1, 1); U = (1, 0). The CGR position can be described as follow [38]:

$$CGR_i = CGR_{i-1} + \theta * (CGR_{i-1} - g_i) \quad (9)$$

Where parameter θ is the decay factor. In the meantime, set this parameter to 0.5 and define $i = 1 \ldots n_G$ and $CGR_0 = (0.5, 0.5)$.

3.7   Broad Learning System

Many methods have been proposed to solve classification problems [39–56]. Broad Learning System (BLS) based on the traditional RVFLNN was built to reduce the training time and update the system dynamically [57]. The core of broad learning is the incremental learning algorithms which can fast remodeling in broad expansion without a retraining process. Using the latest calculation result and the newly added data, only a small amount of calculation is needed to obtain the updated weight.

4   Conclusion

In this study, we have built a MIRNA sequence SIMilarity calculation model (MIS-SIM) based on machine learning, which fused multiple sources including miRNA sequence information, disease semantic information, MIRNA functional information and confirmed miRNA-disease related information. Results show that MISSIM is reliable for predicting potential disease-specific miRNA.

References

1. Ambros, V.: The functions of animal microRNAs. Nature **431**(7006), 350 (2004)
2. An, J.Y., et al.: Identification of self-interacting proteins by exploring evolutionary information embedded in PSI-BLAST-constructed position specific scoring matrix. Onco-target **7**(50), 82440–82449 (2016)
3. Bao, W., You, Z.-H., Huang, D.-S.: CIPPN: computational identification of protein pupylation sites by using neural network. Oncotarget **8**(65), 108867 (2017)
4. An, J.Y., et al.: Improving protein–protein interactions prediction accuracy using protein evolutionary information and relevance vector machine model. Protein Sci. **25**(10), 1825–1833 (2016)
5. Chan, K.C., You, Z.-H.: Large-scale prediction of drug-target interactions from deep representations. In: 2016 International Joint Conference on Neural Networks (IJCNN). IEEE (2016)
6. An, J.Y., et al.: Using the relevance vector machine model combined with local phase quantization to predict protein-protein interactions from protein sequences. Biomed. Res. Int. **2016**(6868), 1–9 (2016)
7. Chen, X., et al.: DRMDA: deep representations-based miRNA–disease association prediction. J. Cell Mol. Med. **22**(1), 472–485 (2018)
8. An, J.Y., et al.: RVMAB: using the relevance vector machine model combined with average blocks to predict the interactions of proteins from protein sequences. Int. J. Mol. Sci. **17**(5), 757 (2016)
9. Chen, W., et al.: Environment-map-free robot navigation based on wireless sensor networks. In: 2007 International Conference on Information Acquisition. ICIA 2007. IEEE (2007)
10. An, J.Y., et al.: Robust and accurate prediction of protein self-interactions from amino acids sequence using evolutionary information. Mol. BioSyst. **12**(12), 3702 (2016)
11. Chen, W., et al.: Design and implementation of wireless sensor network for robot navigation. Int. J. Inf. Acquis. **4**(01), 77–89 (2007)

12. You, Z., et al.: A localization algorithm nin wireless sensor networks using a mobile beacon node. In: 2007 International Conference on Information Acquisition. ICIA 2007. IEEE (2007)
13. Huang, Y.-A., et al.: EPMDA: an expression-profile based computational model for microRNA-disease association prediction. Oncotarget **8**(50), 87033 (2017)
14. You, Z., Lei, Y., Ji, Z., Zhu, Z.: A novel approach to modelling protein-protein interaction networks. In: Tan, Y., Shi, Y., Ji, Z. (eds.) ICSI 2012. LNCS, vol. 7332, pp. 49–57. Springer, Heidelberg (2012). https://doi.org/10.1007/978-3-642-31020-1_7
15. Ji, Z., et al.: Predicting dynamic deformation of retaining structure by LSSVR-based time series method. Neurocomputing **137**, 165–172 (2014)
16. You, Z., Ming, Z., Niu, B., Deng, S., Zhu, Z.: A SVM-based system for predicting protein-protein interactions using a novel representation of protein sequences. In: Huang, D.-S., Bevilacqua, V., Figueroa, J.C., Premaratne, P. (eds.) ICIC 2013. LNCS, vol. 7995, pp. 629–637. Springer, Heidelberg (2013). https://doi.org/10.1007/978-3-642-39479-9_73
17. Huang, Y.-A., et al.: Construction of reliable protein–protein interaction networks using weighted sparse representation based classifier with pseudo substitution matrix representation features. Neurocomputing **218**, 131–138 (2016)
18. You, Z., Wang, S., Gui, J., Zhang, S.: A novel hybrid method of gene selection and its application on tumor classification. In: Huang, D.-S., Wunsch, D.C., Levine, D.S., Jo, K.-H. (eds.) ICIC 2008. LNCS (LNAI), vol. 5227, pp. 1055–1068. Springer, Heidelberg (2008). https://doi.org/10.1007/978-3-540-85984-0_127
19. Huang, Z.-A., et al.: PBHMDA: path-based human microbe-disease association prediction. Front. Microbiol. **8**, 233 (2017)
20. Lei, Y.-K., et al.: Increasing reliability of protein interactome by fast manifold embedding. Pattern Recogn. Lett. **34**(4), 372–379 (2013)
21. Chen, X., et al.: IRWRLDA: improved random walk with restart for lncRNA-disease association prediction. Oncotarget **7**(36), 57919–57931 (2016)
22. Li, S., et al.: Distributed winner-take-all in dynamic networks. IEEE Trans. Autom. Control **62**(2), 577–589 (2017)
23. Luo, X., et al.: An efficient second-order approach to factorize sparse matrices in recommender systems. IEEE Trans. Industr. Inf. **11**(4), 946–956 (2015)
24. Wang, Y.-B., et al.: Improving prediction of self-interacting proteins using stacked sparse auto-encoder with PSSM profiles. Int. J. Biol. Sci. **14**(8), 983–991 (2018)
25. You, Z.-H., et al.: Highly efficient framework for predicting interactions between proteins. IEEE Trans. Cybern. **47**(3), 731–743 (2017)
26. Li, J.-Q., et al.: PSPEL: in silico prediction of self-interacting proteins from amino acids sequences using ensemble learning. IEEE/ACM Trans. Comput. Biol. Bioinform. (TCBB) **14**(5), 1165–1172 (2017)
27. Li, Z.-W., et al.: Highly accurate prediction of protein-protein interactions via incorporating evolutionary information and physicochemical characteristics. Int. J. Mol. Sci. **17**(9), 1396 (2016)
28. Luo, X., et al.: Incorporation of efficient second-order solvers into latent factor models for accurate prediction of missing QoS data. IEEE Trans. Cybern. **48**(4), 1216–1228 (2018)
29. Qu, J., et al.: In silico prediction of small molecule-miRNA associations based on HeteSim algorithm. Mol. Therapy-Nucleic Acids (2018)
30. Song, X.-Y., et al.: An ensemble classifier with random projection for predicting protein-protein interactions using sequence and evolutionary information. Appl. Sci. **8**(1), 89 (2018)
31. Wang, L., et al.: Combining high speed ELM learning with a deep convolutional neural network feature encoding for predicting protein-RNA interactions. IEEE/ACM Trans. Comput. Biol. Bioinform. (2018)

32. Wang, Y., et al.: PCVMZM: using the probabilistic classification vector machines model combined with a zernike moments descriptor to predict protein-protein interactions from protein sequences. Int. J. Mol. Sci. **18**(5), 1029 (2017)

33. Wen, Y.-T., et al.: Prediction of protein-protein interactions by label propagation with protein evolutionary and chemical information derived from heterogeneous network. J. Theor. Biol. **430**, 9–20 (2017)

34. Yu, H.-J., You, Z.-H.: Comparison of DNA truncated barcodes and full-barcodes for species identification. In: Huang, D.-S., Zhang, X., Reyes García, C.A., Zhang, L. (eds.) ICIC 2010. LNCS (LNAI), vol. 6216, pp. 108–114. Springer, Heidelberg (2010). https://doi.org/10.1007/978-3-642-14932-0_14

35. Zhu, H.-J., et al.: DroidDet: effective and robust detection of android malware using static analysis along with rotation forest model. Neurocomputing **272**, 638–646 (2018)

36. Li, Y., et al.: HMDD v2. 0: a database for experimentally supported human microRNA and disease associations. Nucleic Acids Res. **42**(D1), D1070–D1074 (2013)

37. Wang, D., et al.: Inferring the human microRNA functional similarity and functional network based on microRNA-associated diseases. Bioinformatics **26**(13), 1644–1650 (2010)

38. Jeffrey, H.J.: Chaos game representation of gene structure. Nucleic Acids Res. **18**(8), 2163–2170 (1990)

39. Sun, X., et al.: Modeling of signaling crosstalk-mediated drug resistance and its implications on drug combination. Oncotarget **7**(39), 63995 (2016)

40. Wang, Y.B., et al.: Predicting protein-protein interactions from protein sequences by a stacked sparse autoencoder deep neural network. Mol. BioSyst. **13**(7), 1336–1344 (2017)

41. Wang, Y.-B., et al.: Prediction of protein self-interactions using stacked long short-term memory from protein sequences information. BMC Syst. Biol. **12**(8), 129 (2018)

42. You, Z.-H., et al.: A MapReduce based parallel SVM for large-scale predicting protein-protein interactions. Neurocomputing **145**, 37–43 (2014)

43. Zhang, S., Wu, X., You, Z.: Jaccard distance based weighted sparse representation for coarse-to-fine plant species recognition. PLoS ONE **12**(6), e0178317 (2017)

44. Zhu, L., You, Z.-H., Huang, D.-S.: Increasing the reliability of protein–protein interaction networks via non-convex semantic embedding. Neurocomputing **121**, 99–107 (2013)

45. Chen, X., et al.: Long non-coding RNAs and complex diseases: from experimental results to computational models. Brief. Bioinform. **18**(4), 558 (2016)

46. Gao, Z.G., et al.: Ens-PPI: a novel ensemble classifier for predicting the interactions of proteins using autocovariance transformation from PSSM. Biomed. Res. Int. **2016**(4), 1–8 (2016)

47. Li, S., et al.: Inverse-free extreme learning machine with optimal information updating. IEEE Trans. Cybern. **46**(5), 1229 (2016)

48. You, Z.-H., et al.: Large-scale protein-protein interactions detection by integrating big biosensing data with computational model. Biomed. Res. Int. **2014**, 1–9 (2014)

49. You, Z.-H., et al.: A semi-supervised learning approach to predict synthetic genetic interactions by combining functional and topological properties of functional gene network. BMC Bioinform. **11**(1), 343 (2010)

50. Chen, X., et al.: MicroRNAs and complex diseases: from experimental results to computational models. Brief. Bioinform. **20**, 515–539 (2017)

51. Huang, Y.-A., et al.: Improved protein-protein interactions prediction via weighted sparse representation model combining continuous wavelet descriptor and PseAA composition. BMC Syst. Biol. **10**(4), 120 (2016)

52. Li, L.-P., et al.: PCLPred: a bioinformatics method for predicting protein-protein interactions by combining relevance vector machine model with low-rank matrix approximation. Int. J. Mol. Sci. **19**(4), 1029 (2018)

53. Luo, X., et al.: An incremental-and-static-combined scheme for matrix-factorization-based collaborative filtering. IEEE Trans. Autom. Sci. Eng. **13**(1), 333–343 (2016)
54. Wang, L., et al.: Using two-dimensional principal component analysis and rotation forest for prediction of protein-protein interactions. Sci. Rep. **8**(1), 12874 (2018)
55. You, Z.-H., et al.: A novel method to predict protein-protein interactions based on the information of protein sequence. In: 2012 IEEE International Conference on Control System, Computing and Engineering (ICCSCE). IEEE (2012)
56. Zhang, S., et al.: Fusion of superpixel, expectation maximization and PHOG for recognizing cucumber diseases. Comput. Electron. Agric. **140**, 338–347 (2017)
57. Chen, C.P., Liu, Z.: Broad learning system: an effective and efficient incremental learning system without the need for deep architecture. IEEE Trans. Neural Netw. Learn. Syst. **29**(1), 10–24 (2018)

Univariate Thiele Type Continued Fractions Rational Interpolation with Parameters

Le Zou[1,2,3], Liang-Tu Song[2,3], Xiao-Feng Wang[1(✉)],
Qian-Jing Huang[4], Yan-Ping Chen[1], Chao Tang[1], and Chen Zhang[1]

[1] Anhui Provincial Engineering Laboratory of Big Data Technology Application
for Urban Infrastructure, Department of Computer Science and Technology,
Hefei University, Hefei 230601, China
xfwang@hfuu.edu.cn

[2] Institute of Intelligent Machines, Hefei Institutes of Physical Science,
Chinese Academy of Sciences, P. O. Box 1130, Hefei 230031, China

[3] University of Science and Technology of China, Hefei 230027, China

[4] Department of Environmental Engineering, Hefei University, Hefei 230601,
Anhui, China

Abstract. Thiele-type continued fractions interpolation may be the classical
rational interpolation and plays critical role in image interpolation and numerical
analysis. Different from the traditional method, a new Thiele type continued
fractions rational interpolation method with parameters was presented to address
the interpolation problem efficiently. Firstly, in order to gain neat expressions in
terms of inverse differences, we chose the multiplicity of the points strategically.
Secondly, we constructed a univariate Thiele type continued fractions rational
interpolation with parameters, which can satisfy the interpolation condition. We
also discussed the interpolation algorithm, interpolation theorem. Numerical
examples were given to show that the presented method achieves state-of-the-art
performance.

Keywords: Unattainable point · Thiele continued fractions · Parameters ·
Virtual point · Geometry design · Inverse difference

1 Introduction

Interpolation problem plays a vital role in numerical analysis and it is a classical topic
in approximation theory [1]. The interpolation problem has been widely studied, many
scholars had been intensively investigated on this subject and obtained many interesting results over the last decades. It is well known that polynomial interpolation and
rational interpolation are the classical interpolation method. Both of them have many
practical applications, such as numerical approximation [1], image processing, extensive applications in the aspects of arc structuring [1, 2]. Zhan [3] proposed a rational
interpolation fitting algorithm, which kept endpoints through the parameterization
scheme of some middle points selected. The fitting functions possess property of
keeping end-points, regularity and flexibility. Zhang and Duan [4] reported some new
kinds of weighted bivariate blending rational spline interpolation. The value of the

© Springer Nature Switzerland AG 2019
D.-S. Huang et al. (Eds.): ICIC 2019, LNAI 11645, pp. 399–410, 2019.
https://doi.org/10.1007/978-3-030-26766-7_37

interpolation function at any point in the interpolating region can be modified under unaltered interpolating data by selecting suitable parameters and different coefficients. The interpolation geometric surfaces can be modified for the given data when needed in practical design. One of the authors [5] proposed the approximation problem of associated continued fractions and its Viscovatov algorithm to gain associated continued fractions through Taylor series. Zhao [6] presented the block-based Thiele-like blending rational interpolation, which is a flexible blending rational interpolation. For given interpolation data, many kinds of block based Thiele-like blending rational interpolation could be constructed based on different partitioning methods. One of the authors [7, 8] presented several different formats of bivariate interpolation and worked out some general interpolation functions.

At present, the Thiele-type continued fractions rational interpolation is the main research object of rational interpolation. The Thiele-type continued fractions rational interpolation has a better performance in numerical approximation, but it is unique for the given interpolation data. It may meet unattainable points, nonexistent inverse differences in classical Thiele-type continued fractions rational interpolations. To prevent the aforementioned problems of the Thiele-type continued fractions rational interpolation problem. This paper constructs a new kind of Thiele-type continued fractions rational interpolation by introducing one or more parameters. The new function can alter the curve without changing the given data, so as to meet the actual needs. Meanwhile, compared with the interpolation method in [1], our method gives a better performance. Compared with common interpolation method and the interpolation method studied by Tang and Zou [7, 8], it is unnecessary to select suitable parameters of function. It is hoped that the aforementioned problems will be resolved with our proposed method in this paper.

The organization of the paper is as follows. We construct unique Thiele-type continued fractions rational interpolation with parameters and present its algorithm and interpolation theorem in Sect. 2. We also present the Thiele-type continued fractions rational interpolation with single parameter and double parameters. As an application of the proposed method, numerical examples are shown to demonstrate the effectiveness of the method in Sect. 3.

2 Univariate Thiele Continued Fractions Interpolation with Parameters

It is well known that if a Thiele-like continued fractions rational interpolation function exists and then it is unique in the traditional method, however, it is inconvenient for practical application. As well known, many interpolation problem can't be solved by Thiele type continued fractions rational interpolation. So we construct a new kind of Thiele type continued fractions rational interpolation.

2.1 Thiele Type Continued Fractions Rational Interpolation with a Single Parameter

Let us consider the following interpolation problem. Given a function $y = f(x)$ and point sets $\{(x_0, y_0), (x_1, y_1), \ldots (x_n, y_n)\}$ on interval $[a, b]$, one can drive the following classical Thiele-type continued fractions rational interpolation representation [1]

$$R_n(x) = b_0 + \frac{x - x_0}{b_1} + \frac{x - x_1}{b_2} + \cdots + \frac{x - x_{n-1}}{b_n}, \tag{1}$$

Where $b_i (i = 1, \cdots, n)$ are the inverse differences of $f(x)$.

We construct a new Thiele-type continued fractions rational interpolation with a single parameter in this subsection. In order to obtain neat expressions in terms of inverse differences, we choose the multiplicity of the points strategically. Taking an arbitrary point of the original points $(x_k, y_k)(k = 0, 1, \cdots, n)$ as a double virtual point, the multiple numbers of other points remain unaltered. We can drive a new Thiele type continued fractions rational interpolation by introducing a parameter $\lambda (\lambda \neq 0)$.

Let

$$y_i^0 = y_i, \ i = 0, 1, \ldots, n, \tag{2}$$

When $j = 1, \cdots, k+1$, for $i = j, j+1, \cdots, n$,

$$y_i^j = \frac{x_i - x_{j-1}}{y_i^{j-1} - y_{j-1}^{j-1}}. \tag{3}$$

For $i = k+1, k+2, \cdots, n$,

$$z_i^{k+1} = \frac{x_i - x_k}{y_i^{k+1} - 1/\lambda}. \tag{4}$$

When $j = k+2, k+3, \cdots, n$, for $i = j, j+1, \cdots, n$,

$$z_i^j = \frac{x_i - x_{j-1}}{z_i^{j-1} - z_{j-1}^{j-1}}. \tag{5}$$

One can construct Thiele type continued fractions rational interpolation with a single parameter

$$R_n^{(0)}(x) = c_0 + \frac{x - x_0}{c_1} + \cdots + \frac{x - x_{k-1}}{c_k} + \frac{x - x_k}{c_{k+1}^0} + \frac{x - x_k}{c_{k+1}} + \frac{x - x_{k+1}}{c_{k+2}} + \cdots + \frac{x - x_{n-1}}{c_n}, \tag{6}$$

Where

$$c_i = \begin{cases} y_i^i, & i = 0, 1, \cdots, k \\ z_i^i, & i = k+1, k+2, \cdots, n \end{cases}, \ c_{k+1}^0 = \frac{1}{\lambda}. \tag{7}$$

One can construct an inverse difference table with a single parameter for the formula (6).

Theorem 1. Suppose the given diverse interpolation data $\{(x_0, y_0), (x_1, y_1), \cdots, (x_n, y_n)\}$, then

$$R_n^{(0)}(x_i) = f(x_i) = y_i, i = 0, 1, \cdots, n. \tag{8}$$

Proof. Suppose $0 \le i \le k$,

As you know, Eq. (6) changes into the classical Thiele-type continued fractions rational interpolation, obviously we have

$$R_n^{(0)}(x_i) = f(x_i) = y_i, i = 0, 1, \cdots, k.$$

For $i = k+1$,

$$
\begin{aligned}
R_n^{(0)}(x_i) &= c_0 + \frac{x_{k+1} - x_0}{c_1} + \cdots + \frac{x_{k+1} - x_{k-1}}{c_k} + \frac{x_{k+1} - x_k}{c_{k+1}^0} + \frac{x_{k+1} - x_k}{c_{k+1}} \\
&= y_0^0 + \frac{x_{k+1} - x_0}{y_1^1} + \cdots + \frac{x_{k+1} - x_{k-1}}{y_k^k} + \frac{x_{k+1} - x_k}{1/\lambda} + \frac{x_{k+1} - x_k}{z_{k+1}^{k+1}} \\
&= y_0^0 + \frac{x_{k+1} - x_0}{y_1^1} + \cdots + \frac{x_{k+1} - x_{k-1}}{y_k^k} + \frac{x_{k+1} - x_k}{1/\lambda} + \frac{x_{k+1} - x_k}{\frac{x_{k+1} - x_k}{y_{k+1}^{k+1} - 1/\lambda}} \\
&= y_0^0 + \frac{x_{k+1} - x_0}{y_1^1} + \cdots + \frac{x_{k+1} - x_{k-1}}{y_k^k} + \frac{x_{k+1} - x_k}{y_{k+1}^{k+1}} \\
&= \cdots = y_{k+1}^0 = y_{k+1}
\end{aligned}
$$

If $n \ge i \ge k+2$, using formula (5) and the previous proof process, we can drive

$$
\begin{aligned}
R_n^{(0)}(x_i) &= c_0 + \frac{x_i - x_0}{c_1} + \cdots + \frac{x_i - x_{k-1}}{c_k} + \frac{x_i - x_k}{c_{k+1}^0} + \frac{x_i - x_k}{c_{k+1}} + \frac{x_i - x_{k+1}}{c_{k+2}} + \cdots + \frac{x_i - x_{i-1}}{c_i} \\
&= y_0^0 + \frac{x_i - x_0}{y_1^1} + \cdots + \frac{x_i - x_{k-1}}{y_k^k} + \frac{x_i - x_k}{1/\lambda} + \frac{x_i - x_k}{z_{k+1}^{k+1}} + \frac{x_i - x_{k+1}}{z_{k+2}^{k+2}} + \cdots + \frac{x_i - x_{i-1}}{z_i^i} \\
&= y_0^0 + \frac{x_i - x_0}{y_1^1} + \cdots + \frac{x_i - x_{k-1}}{y_k^k} + \frac{x_i - x_k}{1/\lambda} + \frac{x_i - x_k}{z_{k+1}^{k+1}} + \frac{x_i - x_{k+1}}{z_{k+2}^{k+2}} + \cdots + \frac{x_i - x_{i-2}}{z_i^{i-1}} + \frac{x_i - x_{i-1}}{\frac{x_i - x_{i-1}}{z_i^{i-1} - z_{i-1}^{i-1}}} \\
&= y_0^0 + \frac{x_i - x_0}{y_1^1} + \cdots + \frac{x_i - x_{k-1}}{y_k^k} + \frac{x_i - x_k}{1/\lambda} + \frac{x_i - x_k}{z_{k+1}^{k+1}} + \frac{x_i - x_{k+1}}{z_{k+2}^{k+2}} + \cdots + \frac{x_i - x_{i-2}}{z_i^{i-1}} \\
&= \cdots = \\
&= y_0^0 + \frac{x_i - x_0}{y_1^1} + \cdots + \frac{x_i - x_{k-1}}{y_k^k} + \frac{x_i - x_k}{1/\lambda} + \frac{x_i - x_k}{z_i^{k+1}} \\
&= \cdots = y_i^0 = y_i,
\end{aligned}
$$

So we have

$$R_n^{(0)}(x_i) = f(x_i) = y_i, \ i = 0, 1, \cdots, n.$$

The proof is completed.

2.2 Univariate Thiele Type Continued Fractions Interpolation with Multiple Parameters

Without loss of generality, we just take the Thiele-type continued fractions rational interpolation with two parameters into consideration. There are two categories, one is that taking an arbitrary node as a treble virtual point, the other is that taking two arbitrary nodes in original points as the virtual double points. One can study the Thiele type continued fractions rational interpolation with more parameters similarly.

2.2.1 Thiele Type Continued Fractions Rational Interpolation with Single Virtual Treble Point and Two Parameters

Considering to taking an arbitrary point in original points $(x_k, y_k)(k = 0, 1, \cdots, n)$ as a treble virtual point, and other points remain unchanged. One can construct the osculatory rational interpolation with parameters $\alpha, \beta(\alpha, \beta \neq 0)$ as the following function

$$R_n^{(1)}(x) = c_0 + \frac{x - x_0}{c_1} + \cdots + \frac{x - x_{k-1}}{c_k} + \frac{x - x_k}{c_{k+1}^0} + \frac{x - x_k}{c_{k+1}^1} + \frac{x - x_k}{c_{k+1}} + \frac{x - x_{k+1}}{c_{k+2}} + \cdots + \frac{x - x_{n-1}}{c_n} \quad (9)$$

Algorithm 1
Step 1: Initialization.

$$y_i^0 = y_i, \ i = 0, 1, \cdots, n, \tag{10}$$

Step 2: for $j = 1, \cdots, k+1, \ i = j, j+1, \cdots, n$,

$$y_i^j = \frac{x_i - x_{j-1}}{y_i^{j-1} - y_{j-1}^{j-1}}, \tag{11}$$

Step 3: for $i = k+1, k+2, \cdots, n$,

$$z_i^k = \frac{x_i - x_k}{y_i^{k+1} - 1/\alpha}, \tag{12}$$

Step 4: for $i = k+1, k+2, \cdots, n$,

$$z_i^{k+1} = \frac{x_i - x_k}{z_i^k - 1/\beta}, \tag{13}$$

Step 5: when $j = k+2, k+3, \cdots, n$, for $i = j, j+1, \cdots, n$,

$$z_i^j = \frac{x_i - x_{j-1}}{z_i^{j-1} - z_{j-1}^{j-1}}, \tag{14}$$

Step 6: we can construct the osculatory rational interpolation with parameters α, β

$$R_n^{(1)}(x) = c_0 + \frac{x - x_0}{c_1} + \cdots + \frac{x - x_{k-1}}{c_k} + \frac{x - x_k}{c_{k+1}^0} + \frac{x - x_k}{c_{k+1}^1} + \frac{x - x_k}{c_{k+1}} + \frac{x - x_{k+1}}{c_{k+2}} + \cdots + \frac{x - x_{n-1}}{c_n}, \tag{15}$$

Where

$$c_i = \begin{cases} y_i^i, & i = 0, 1, \cdots, k \\ z_i^i, & i = k+1, k+2, \cdots, n \end{cases}, \quad c_{k+1}^0 = \frac{1}{\alpha}, \quad c_{k+1}^1 = \frac{1}{\beta}. \tag{16}$$

Theorem 2. For the given diverse interpolation data $\{(x_0, y_0), (x_1, y_1), \cdots (x_n, y_n)\}$, the rational interpolation driven by (15)–(16) satisfies

$$R_n^{(1)}(x_i) = f(x_i) = y_i, \; i = 0, 1, \cdots, n. \tag{17}$$

Proof: if $0 \le i \le k$,

Equation (15) changes into the classical Thiele-type continued fractions rational interpolation, obviously we have

$$R_n^{(0)}(x_i) = f(x_i) = y_i, \; i = 0, 1, \cdots, k.$$

If $i = k+1$,

$$R_n^{(1)}(x_{k+1}) = c_0 + \frac{x_{k+1} - x_0}{c_1} + \cdots + \frac{x_{k+1} - x_{k-1}}{c_k} + \frac{x_{k+1} - x_k}{c_{k+1}^0} + \frac{x_{k+1} - x_k}{c_{k+1}^1} + \frac{x_{k+1} - x_k}{c_{k+1}}$$

$$= y_0^0 + \frac{x_{k+1} - x_0}{y_1^1} + \cdots + \frac{x_{k+1} - x_{k-1}}{y_k^k} + \frac{x_{k+1} - x_k}{1/\alpha} + \frac{x_{k+1} - x_k}{1/\beta} + \frac{x_{k+1} - x_k}{z_{k+1}^{k+1}}$$

$$= y_0^0 + \frac{x_{k+1} - x_0}{y_1^1} + \cdots + \frac{x_{k+1} - x_{k-1}}{y_k^k} + \frac{x_{k+1} - x_k}{1/\alpha} + \frac{x_{k+1} - x_k}{1/\beta} + \frac{x_{k+1} - x_k}{\frac{x_{k+1} - x_k}{z_{k+1}^{k+1} - 1/\beta}}$$

$$= y_0^0 + \frac{x_{k+1} - x_0}{y_1^1} + \cdots + \frac{x_{k+1} - x_{k-1}}{y_k^k} + \frac{x_{k+1} - x_k}{1/\alpha} + \frac{x_{k+1} - x_k}{\frac{x_{k+1} - x_k}{y_{k+1}^{k+1} - 1/\alpha}}$$

$$= y_0^0 + \frac{x_{k+1} - x_0}{y_1^1} + \cdots + \frac{x_{k+1} - x_{k-1}}{y_k^k} + \frac{x_{k+1} - x_k}{y_{k+1}^{k+1}}$$

$$= \cdots = y_{k+1}^0 = y_{k+1}$$

If $n \ge i \ge k+2$, we can get the following results by using Eqs. (12)–(14)

$$R_n^{(1)}(x_i) = c_0 + \frac{x_i - x_0}{c_1} + \cdots + \frac{x_i - x_{k-1}}{c_k} + \frac{x_i - x_k}{c_{k+1}^0} + \frac{x_i - x_k}{c_{k+1}^1} + \frac{x_i - x_k}{c_{k+1}} + \frac{x_i - x_{k+1}}{c_{k+2}} + \cdots + \frac{x_i - x_{i-1}}{c_i}$$

$$= y_0^0 + \frac{x_i - x_0}{y_1^1} + \cdots + \frac{x_i - x_{k-1}}{y_k^k} + \frac{x_i - x_k}{1/\alpha} + \frac{x_i - x_k}{\frac{1}{\beta}} + \frac{x_i - x_k}{z_{k+1}^{k+1}} + \frac{x_i - x_{k+1}}{z_{k+2}^{k+2}} + \cdots + \frac{x_i - x_{i-1}}{z_i^i}$$

$$= \cdots =$$

$$= y_0^0 + \frac{x_i - x_0}{y_1^1} + \cdots + \frac{x_i - x_{k-1}}{y_k^k} + \frac{x_i - x_k}{1/\alpha} + \frac{x_i - x_k}{1/\beta} + \frac{x_i - x_k}{z_i^{k+1}}$$

$$= \cdots = y_i^0 = y_i$$

So we have $R_n^{(1)}(x_i) = f(x_i) = y_i$, $i = 0, 1, \cdots, n$.
The proof is completed.

2.2.2 Thiele Type Continued Fractions Rational Interpolation with Two Virtual Double Points and Two Parameters

Considering to taking arbitrary two points in original points $(x_k, y_k), (x_s, y_s)(s > k,$ $s, k = 0, 1, \cdots, n)$ as two double virtual points, and other points remain unchanged. We can construct the osculatory rational interpolation with two parameters $\phi, \delta(\phi, \delta \neq 0)$

$$R_n^{(2)}(x) = c_0 + \frac{x - x_0}{c_1} + \cdots + \frac{x - x_{k-1}}{c_k} + \frac{x - x_k}{c_{k+1}^0} + \frac{x - x_k}{c_{k+1}} + \frac{x - x_{k+1}}{c_{k+2}}$$
$$+ \cdots + \frac{x - x_{s-1}}{c_s} + \frac{x - x_s}{c_{s+1}^0} + \frac{x - x_s}{c_{s+1}} + \frac{x - x_{s+1}}{c_{s+2}} + \cdots + \frac{x - x_{n-1}}{c_n}. \tag{18}$$

Algorithm 2

Step 1: Initialization.

$$y_i^0 = y_i, \quad i = 0, 1, \cdots, n, \tag{19}$$

Step 2: if $j = 1, \cdots, k+1$, for $i = j, j+1, \cdots, n$,

$$y_i^j = \frac{x_i - x_{j-1}}{y_i^{j-1} - y_{j-1}^{j-1}}, \tag{20}$$

Step 3: for $i = k+1, k+2, \cdots, n$,

$$z_i^{k+1} = \frac{x_i - x_k}{y_i^{k+1} - 1/\delta}, \tag{21}$$

Step 4: if $j = k+2, k+3, \cdots, s+1$, for $i = j, j+1, \cdots, n$,

$$z_i^j = \frac{x_i - x_{j-1}}{z_i^{j-1} - z_{j-1}^{j-1}}, \tag{22}$$

Step 5: for $i = s+1, s+2, \cdots, n$,

$$w_i^{s+1} = \frac{x_i - x_s}{z_i^{s+1} - 1/\phi},\tag{23}$$

Step 6: if $j = s+2, s+3, \cdots, n$, for $i = j, j+1, \cdots, n$,

$$w_i^j = \frac{x_i - x_{j-1}}{w_i^{j-1} - w_{j-1}^{j-1}},\tag{24}$$

Step 7: One can construct the osculatory rational interpolation with two parameters $\phi, \delta (\phi, \delta \neq 0)$

$$R_n^{(2)}(x) = c_0 + \frac{x - x_0}{c_1} + \cdots + \frac{x - x_{k-1}}{c_k} + \frac{x - x_k}{c_{k+1}^0} + \frac{x - x_k}{c_{k+1}} + \frac{x - x_{k+1}}{c_{k+2}}$$
$$+ \cdots + \frac{x - x_{s-1}}{c_s} + \frac{x - x_s}{c_{s+1}^0} + \frac{x - x_s}{c_{s+1}} + \frac{x - x_{s+1}}{c_{s+2}} + \cdots + \frac{x - x_{n-1}}{c_n},\tag{25}$$

Where

$$c_i = \begin{cases} y_i^i, & i = 0, 1, \cdots, k \\ z_i^i, & i = k+1, j+2, \cdots, s \\ w_i^i, & i = s+1, s+2, \cdots, n \end{cases}, \quad c_{k+1}^0 = \frac{1}{\delta}, c_{s+1}^0 = \frac{1}{\phi},\tag{26}$$

Theorem 3. For the given diverse interpolation data $\{(x_0, y_0), (x_1, y_1), \cdots (x_n, y_n)\}$, the rational interpolation driven by (18) satisfies

$$R_n^{(2)}(x_i) = f(x_i) = y_i, \ i = 0, 1, \cdots, n.\tag{27}$$

The proof is similar to Theorem 1.

It can be seen from the interpolation algorithm and interpolation theorem that the new kind of Thiele type continued fractions is not unique, and it satisfy the given interpolation condition. As you known, it may meet unattainable points, nonexistent inverse differences in classical Thiele-type continued fractions rational interpolations. The new kind of interpolation can solve the above interpolation problem. We will give experimental explanation in numerical examples.

3   Numerical Examples

In this section, we present some examples to show how the algorithms are implemented and how flexible our method is. The first example is given to show the interpolation with nonexistent inverse differences in classical Thiele-type continued fractions rational interpolations. The second example is given to show the interpolation with unattainable points in classical Thiele-type continued fractions rational interpolations.

Example 1 [1]. Given the interpolation data of given points $\{(0,0),(1,1),(2,0.5),$ $(3,3),(4,-0.5),(5,5)\}$.

Through calculating the inverse differences, one can find one of the inverse differences does not exist. We couldn't get the classical Thiele-type continued fractions rational interpolation function. If we exchange the two points $(3,3),(4,-0.5)$, as the method in the paper [9], we can construct the Thiele-type continued fractions rational interpolation function,

$$R_1(x) = 0 + \frac{x}{1} + \frac{x-1}{1/3} + \frac{x-2}{-3} + \frac{x-4}{-1/3} + \frac{x-3}{3}$$

It is easy to be found that $R_1(x)$ does not satisfy the interpolation condition. By using the classical Thiele continued fractions rational interpolation and the method in [9], one could not get the satisfied interpolation function.

By using the method in this paper, we regard the point $(0,0)$ as a double virtual point, by introducing parameter $c(c \neq 0)$ we can get

$$R_2(x) = \frac{x}{c} + \frac{x}{\frac{-1}{c-1}} + \frac{x-1}{\frac{-(c-1)(c-4)}{c+2}} + \frac{x-2}{-\frac{1}{2}(c-1)-\frac{1}{6}} + \frac{x-3}{\frac{18c(c-1)}{(c+2)(c+8)}} + \frac{x-4}{\frac{-3}{8}(c-1)-\frac{1}{24}}$$

It is easy to verify

$$R_2(x_i) = f_i, \ (i = 0, 1, \cdots, 5)$$

So, we can see that the proposed method can solve the above interpolation problem. The method is very easy to problem and calculate. The result is meet to interpolation conditions. Meanwhile, the function $R_2(x)$ can change into many interpolation functions by choosing the parameter $c(c \neq 0)$. So the new Thiele-type continued fractions rational interpolation has a better performance in numerical approximation, it is not unique for the given interpolation data.

Example 2. Given interpolation data as following (Table 1),

Table 1. Interpolation data

i	0	1	2
x_i	2	1	0
f_i	1	0	0

If we use the classical Thiele type continued fractions rational interpolation, we can get

$$r(x) = 1 + \frac{x-2}{1} + \frac{x-1}{-1} = 0.$$

Because $r(0) = r(2) = 0 \neq 1$, we can get that the point $(2, 1)$ is an unattainable point. So we cannot get the satisfied interpolation function by using classical Thiele-type continued fractions rational function. But, if we add the multiple numbers of point $(2, 1)$, construct the osculating interpolation which has first-order derivative in point $(2, 1)$. By introducing parameter $\lambda(\lambda \neq 0)$, then we can drive the following Thiele type osculatory rational interpolation

$$R_3(x) = 1 + \frac{x-2}{\lambda} + \frac{x-2}{\frac{1}{\lambda-1}} + \frac{x-1}{3 - \frac{2}{\lambda} - \lambda}$$

It is easy to verify

$$R_3(x_i) = f_i, \quad (i = 0, 1, 2).$$

Therefore, the method can solve this kind of special interpolation problem very well, and it is easy to construct and calculate. In addition, the function $R_3(x)$ can change into many interpolation function. For example, we can choose as follows

If we choose $\lambda = -0.8$, then we can get

$$R_3(x) = 1 + \frac{x-2}{-\frac{4}{5}} + \frac{x-2}{\frac{10x}{63} - \frac{5}{7}}$$

If we choose $\lambda = 80$, then we can get

$$R_3(x) = 1 - \frac{x-2}{-80} + \frac{x-2}{\frac{40x}{3081} - \frac{1}{39}}$$

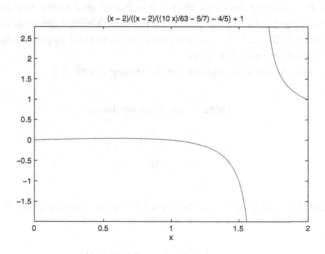

$$(x - 2)/((x - 2)/((10\ x)/63 - 5/7) - 4/5) + 1$$

Fig. 1. Picture of $R_3(x)$ with $\lambda = -0.8$

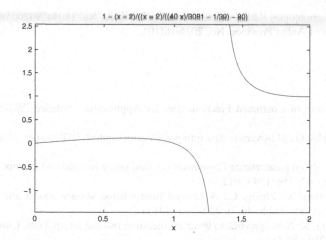

$$1 - (x - 2)/((x - 2)/((40 x)/8081 - 1/39) - 80)$$

Fig. 2. Picture of $R_3(x)$ with $\lambda = 80$

As you seen from the figure, all two function satisfy the interpolation condition. We can choose other parameters $\lambda(\lambda \neq 0)$, the function $R_3(x)$ can change into other function (Figs. 1 and 2).

4 Conclusions and Future Work

In this study, we had proposed a kind of univariate Thiele-type continued fractions rational interpolation with parameters, by choosing the multiplicity of the interpolation point strategically. We discuss the algorithms, interpolation theorems. The new kinds of Thiele-type continued fractions rational interpolation are easing to be used, which extend the theoretical research of the rational interpolation. By selecting suitable parameters, the value of the interpolation function at any point in the interpolating region can be altered under unchanged interpolating data, so it can be used to design a curve. However, it is still a relatively complicated problem on how to select suitable parameters and then construct a suitable interpolation for actual geometric design requirement. But it is a very interesting and practical problem, we will study it in future. We can generalize the proposed method to the multivariate case using similar methods. We will continue to study the multivariate interpolation with parameters in future. We conclude this paper by pointing out that it is uncomplicated to generalize the Thiele-type continued fractions rational interpolation method with parameters to vector-valued case or matrix-valued case [1, 2, 9].

Acknowledgements. The authors would like to express their thanks to the referees for their valuable suggestions. This work was supported by the grant of the National Natural Science Foundation of China, Nos. 61672204, 61806068, the grant of Anhui Provincial Natural Science Foundation, Nos. 1508085QF116, 1908085MF184, the grant of the key Scientific Research Foundation of Education Department of Anhui Province, Nos. KJ2018A0555, KJ2018A0556, the grant of Major Science and Technology Project of Anhui Province, No. 17030901026, the

410   L. Zou et al.

grant of Key Technologies R&D Program of Anhui Province, No. 1804a09020058, the grant of Teaching Team of Anhui Province, No. 2016jxtd101.

References

1. Tan, J.: Theory of Continued Fractions and Its Applications. Science Publishers, Beijing (2007)
2. Cuyt, A., Celis, O.: Multivariate data fitting with error control. BIT Numer. Math. **59**(1), 35–55 (2018)
3. Zhan, T.: Study on parameterized continued fraction fitting method and its application. Math. Pract. Theory **42**, 156–159 (2012)
4. Zhang, Y., Bao, X., Zhang, C.: A rational interpolation surface model and visualization constraint. Sci. Sin. Math. **44**(7), 729–740 (2014)
5. Zou, L., Tang, S.: New approach to bivariate blending rational interpolants. Chin. Q. J. Math. **26**(2), 280–284 (2011)
6. Zhao, Q., Tan, J.: Block-based Thiele-like blending rational interpolation. J. Comput. Appl. Math. **195**, 312–325 (2006)
7. Zou, L., Tang, S.: General structure of block-based interpolational function. Comm. Math. Res. **28**(3), 193–208 (2012)
8. Zou, L., Tang, S.: A new approach to general interpolation formulae for bivariate interpolation. Abstr. Appl. Anal. **2014**, 1–11 (2014)
9. Zhu, X., Zhu, G.: A study of the existence of vector valued rational interpolation. J. Inf. Comput. Sci. **2**, 631–640 (2005)

Data Science Approaches for the Analysis of Animal Behaviours

Natasa Kleanthous[1]($\boxtimes$), Abir Hussain[1]($\boxtimes$), Alex Mason[2,3,4],
and Jennifer Sneddon[5]

[1] Department of Computer Science, Liverpool John Moores University,
Liverpool, UK
N.K.Orphanidou@2015.ljmu.ac.uk, A.Hussain@ljmu.ac.uk
[2] Animalia AS, Norwegian Meat and Poultry Research Institute, Oslo, Norway
alex.mason@animalia.no
[3] Faculty of Science and Technology, Norwegian University of Life Sciences,
1432 Ås, Norway
[4] Department of Built Environment, Liverpool John Moores University,
Liverpool, UK
[5] Department of Natural Sciences and Psychology,
Liverpool John Moores University, Liverpool, UK
J.C.Sneddon@ljmu.ac.uk

Abstract. Animals' welfare can be categorized and predicted by observing their daily routine and behaviour and consequently, conclusions about their health status and their ecology can be made. Yet, observing the animals' activity is time-consuming and labour intensive, therefore there is a need to develop automated behavioural monitoring systems for more efficient and effective computerised agriculture. In this study, accelerometer and gyroscope measurements were collected from seven Hebridean ewes located in Cheshire, UK. Once the activities of the animals were labelled as grazing, resting, walking, browsing, scratching, and standing, data analysis was conducted. The performance of the random forest was evaluated as we have previously suggested that this algorithm can provide advantages and has been proven to adequately classify the behaviours of the animals. When using features from both the accelerometer and gyroscope, the algorithm obtained the best results having accuracy and kappa value of 96.43%, and 95.02%, respectively. However, using data from the accelerometer exclusively, only decreased the accuracy by 0.40% and kappa value by 0.56%. Therefore, in future work, we will consider only the use of accelerometer sensor and we will test the performance of the same features and random forest for real-time activity recognition using larger cohort of animals as well as mixed breed animals.

Keywords: Machine learning · Random forest · Animal activity recognition

1 Introduction

Animals' welfare can be categorized and predicted by observing their daily routine and behaviour and consequently, conclusions about their health status and their ecology can be made [1]. The activity of the animal can be an indicator of its health status and

© Springer Nature Switzerland AG 2019
D.-S. Huang et al. (Eds.): ICIC 2019, LNAI 11645, pp. 411–422, 2019.
https://doi.org/10.1007/978-3-030-26766-7_38

usually indicates if the animal is stressed or diseased [2]. Additionally, lameness can be recognised based on the animals' activity [3]. Progress in the field of computer science and electronics engineering has resulted in a growth of interest towards the development of systems for more effective and automated agriculture [4, 5]. Through the use of smart systems, cognitive labour will replace intense manual labour by improving the decision-making process of the farmers and stakeholders [6]. Such systems will open up opportunities for a more flexible and cost-efficient way to monitor animal activity and identify any anomalies in their behaviour.

The monitoring of animals involved only humans for many years. This method of animal observation is challenging and labour intensive, especially with large number of free-ranging sheep, since they graze on hills and large grasslands. Hence, information regarding the location and activity of the animals can be collected with the utilisation of an intelligent system [6]. The possession of such data can advance decisions concerning animal health, distribution control and effective use of land [7], which can prevent erosion and contamination of soil [4]. Additionally, the health status of the animals will be known on a 24/7 basis without the need of intense human observation, which is not as effective as an automated monitoring system. Furthermore, the system has the additional advantage for giving early warning of any behavioural abnormalities.

Devices embedded with machine learning (ML) algorithms are considered to be useful tools in remote monitoring and diagnosing of the welfare of animals, as they can predict and identify animal diseases, food consumption and overall activity. Information about the position and daily routine of the animal can be combined and used to qualify patterns of pasture use and animal distribution for targeted animal behaviour [8].

In our previous work, we have tested various ML techniques in order to identify sheep and goat behaviour such as grazing, standing, lying, scratching or biting, and walking. The previous data was sampled at 100 Hz and the behaviours were segmented into 30 s windows [9]. In this current work, new data was collected from seven Hebridean ewes to categorize six behaviours; grazing, resting, walking, standing, scratching, browsing. The performance of the random forest was evaluated as we have previously suggested that this algorithm can provide advantages and has been proven to adequately classify the behaviours of the animals. Another aim of this study was to apply random forest with lower sampling frequency and smaller window size (i.e. 10 Hz and 10 s windows), and evaluate the resultant performance. This is predicated on previous findings which indicated that lower frequency rates improve memory and demand less power of the device [10]. It was also previously demonstrated that 16 Hz sample rate and 7-s windows can offer reduced energy requirements and can be used for real-time sheep activity monitoring [11].

2   Previous Work

Machine learning is very valuable and commonly used tool for the classification of animal behaviour including domestic and wild animals. Various data types are used so far such as images [12, 13], GPS coordinates [14–16], tilt coordinates [17], video [8], accelerometer, gyroscope, and magnetometer measurements [9, 10, 18–26]. However, because of their size, low power consumption and ability to provide information In regard to human and animal gait patterns, accelerometers are the most frequently used sensors.

For the classification of walking, grazing, and resting of cattle, Global positioning system (GPS) data was used with linear discriminant analysis and the algorithm correctly classified 71% of all observations [14]. Another study used collars which contained accelerometer and GPS sensors on cattle to classify foraging, ruminating, traveling, resting and 'other active behaviors' of cattle. Threshold-based decision trees method was used and the algorithm classified correctly 90.5% of all the data points [27]. Classification tree analysis [15], and k-means [16] was also used on GPS data to classify cattle behaviour.

Horse activity data was gathered from accelerometer, gyroscope, and magnetometer sensors, and was classified using an embedded multilayer perceptron [28]. The algorithm obtained 81% behaviour recognition in a real scenario. Accelerometer data was also gathered and was used with threshold-based statistical analysis for standing and feeding activities of cows, achieving 93.33%, 95.45%, and 95.56% sensitivity, precision, and specificity, respectively [29]. Previous research concerned with goat and sheep behaviour is presented in Table 1.

Table 1. Previous work concerned with the classification of sheep and goat behaviour

Ref	Animals	Class	Signal	Window	Method	Accuracy
[17]	Sheep	2	tilt	30-s	Linear Discriminant Analysis	94.4%
[30]	Sheep	5	acc	–	Multilayer perceptron	76.2%
[31]	Sheep	5	acc	5.12-s	Quadratic Discriminant Analysis	89.7%
[32]	Sheep	5	acc	10-s	Decision Trees	91.3%
[33]	Sheep	5	acc	5.3-s	Linear Discriminant Analysis	85.7%
[19]	Sheep, Goats	5	acc, gyr, magn	1-s	Deep Neural Network	94.0%
[34]	Sheep	3	acc	60-s	Discriminant Analysis	93.0%
[22]	Sheep	5	acc	5.3-s	Statistical on-board classifier	82.4%
[35]	Sheep	3	acc	10-s	Quadratic Discriminant Analysis	97.67%
[36]	Sheep	3	acc	30-s	Discriminant Analysis	89.7%
[37]	Goats	6	acc	2-s	Naïve Bayes	94.0%
[38]	Sheep	3	acc, gyr	7-s	Random Forest	92.0%
[11]	Sheep	3	acc, gyr	7-s	Random Forest	95.0%
[9]	Sheep, Goats	5	acc, gyr magn	30-s	Random Forest	96.47%
[39]	Sheep	5	acc	6.4-s	Linear support vector machine	88.4%

It is shown that a lot of research exists on animal behaviour classification as it is a very important aspect of agriculture community, behavioural studies, conservation and ecology. Many challenges exist, including degree of accuracy in animal monitoring and performance but also hardware implementation. Research is still needed to identify the optimal method for categorising animal activity and at the same time improving battery life of the device as well as memory because this could be used for real time animal monitoring.

3   Methods and Equipment

To evaluate the performance of the ML technique implemented on previous research [9] using new data, behavioural data was gathered from seven Hebridean ewes from a farm located in Cheshire, Shotwick. In the following sections, the description of the required steps is presented. The protocol of the experiment was approved by the Senior Research Officer and LSSU Manager of Liverpool John Moores university (approval AH_NKO/2018-13).

3.1   Animals and Location

Seven Hebridean ewes were located in a paddock in Shotwick OS 53.2391152, −3.0028081, Cheshire with an approximate area size and perimeter of 1500 m^2 and 250 m, respectively. The behaviour of the ewes (average age 11 years) was observed and recorded using IMU measurements daily from 4th July until 18th July 2018 during different times of a day. The animals were free to use the whole area of the paddock and had access to grass and water. The ewes were used in previous research [20] and were used to human interaction, so their behaviours were not affected by human presence during the video recordings.

3.2   Data Acquisition

During the experimental period, one sheep at a time was fitted with a smartphone device (Samsung Galaxy S5) on the top and side part of the harness. The orientation of the smartphone device was not fixed because we wanted to test whether we could achieve high accuracy similarly with previous work [9] without sensor orientation dependency. The smartphone device was set with a data logger application (HyperIMU [40]) which saved the x, y, and z coordinates of accelerometer and gyroscope data at a sample rate of 10 Hz as a csv file for offline data processing. During the data gathering, the animal carrying the smartphone was video recorded using the smartphone camera (Samsung Galaxy S5) and a human observer was at present all time. This information was due to be used as ground truth for activity labelling. In total, 35 h of recordings were obtained but only 30 h were selected for data analysis because they included behaviours with insufficient information.

3.3   Data Annotation and Preparation

For the annotation of the animals' behaviour, the ELAN_5.2 annotation tool was used. The csv files were synchronised with the video recordings to provide an accurate mark of each behaviour. Behaviours with insufficient information were discarded (biting, fighting, running, shaking). Table 2 describes the selected animal activities and the duration of each activity.

Table 2. Description and duration of the animals' activities

Activity	Description	Time (seconds)	Time in %
Grazing	Grazing while walking and standing	23150.10	20.95%
Browsing	The animal was reaching upwards for leaves on trees or bushes	593.80	0.54%
Resting	The animal was inactive in a lying posture or ruinating	34453.30	31.18%
Scratching	Scratching with leg movement or pushing against trees or bushes	168.90	0.15%
Standing	Standing idle or ruminating	36640.40	33.15%
Walking	Walking forward, backward or sideward	15506.00	14.03%
Total		110512.50	100%

After annotating the data, RStudio open-source IDE for R programming language was used for pre-processing and analysis. The data was tested for missing values and six behaviours were visualised to provide a better understanding of each activity.

3.4   Feature Extraction

For the analysis of the data and to test the performance of the random forest algorithm, firstly, the desired features were extracted from 10 s data blocks resulting in a total of 98755 windows. Time domain features such as the mean, standard deviation, and root mean square are the most usual features to be used for activity recognition according to [41]. Additional important features for activity recognition based on accelerometer measurements are integrals and squared integrals [42]. The features were prechosen and identified from our previous work [9] and are described in Table 3. We used 11 features from time and frequency domain resulting in 66 newly created features. Once the features were extracted, the data is split with 70% and 30% for training and testing, respectively.

Table 3. Feature formulas and feature sets

Feature	Formula [9]	Set A	Set B	Set C	Set D
Mean	$\frac{1}{n}\sum_{i=1}^{n} x_i$	acc gyr	acc gyr	acc	gyr
Standard Deviation	$\sqrt{\frac{\sum_{i=1}^{n}(x_i-\bar{x})^2}{n-1}}$	acc gyr	ax az	acc	gyr
Root Mean Square (RMS) velocity	$\sqrt{\frac{1}{n}\sum_{i=1}^{n} diffinv(x_i)^2}$	acc gyr	az ay	acc	gyr
Root Mean Square (RMS)	$\sqrt{\frac{1}{n}\sum_{i=1}^{n} x_i^2}$	acc gyr	acc	acc	gyr
Energy	$\sum_{i=1}^{n} x_i^2$	acc gyr	acc	acc	gyr
Sum of Changes	$\sum_{i=1}^{n} diff(x_i)$	acc gyr	–	acc	gyr
Mean Absolute Change	$\frac{1}{n}\sum_{i=1}^{n} |diff(x_i)|$	acc gyr	–	acc	gyr
Absolute Integrals	$\int_{t=0}^{T}|x|dt + \int_{t=0}^{T}|y|dt + \int_{t=0}^{T}|z|dt$	acc gyr	acc	acc	gyr
Squared Integrals	$\left(\int_{t=0}^{T}|x|dt\right)^2 + \left(\int_{t=0}^{T}|y|dt\right)^2 + \left(\int_{t=0}^{T}|z|dt\right)^2$	acc gyr	acc	acc	gyr
Madogram	$\gamma_p(t) = \frac{1}{2}\mathrm{E}|X_u - X_{i+u}|^p, p = 1$	acc gyr	gx	acc	gyr
Peak Frequency	$arg\left(\frac{Fs}{n} max_{i=0}^{n-1} P(i)\right)$	acc gyr	az ay	acc	gyr

acc: x,y,z coordinates of the accelerometer, gyr: x,y,z coordinates of the gyroscope
ax: x coordinate of the accelerometer, ay: y coordinate of the accelerometer
az: z coordinate of the accelerometer, gx: x coordinate of the gyroscope

3.5 Feature Selection

The algorithm was tested using four feature sets, Set A, Set B, Set C, and Set D, as described in Table 3. Firstly, we used all features from both accelerometer and gyroscope sensors resulting in 66 new features (Set A). The next step was to use only the features from the accelerometer as a new set (Set C), and only features from gyroscope sensor as Set D. Also, we wanted to isolate the 25 most important features from the whole set. For this method, we used Recursive feature elimination with backward selection based on random forest. At first, the significance level was selected (significance level = 0.05), and the model was fitted using the whole set of predictors. Then, the predictor with the highest p value (p > significance level) was removed and the model was fitted again. The process was repeated until all the remaining predictors had a p-value lower than the selected significance level. Once we have identified the most important features, the top 25 features were selected based on their importance ranking.

3.6   Random Forest and Performance Evaluation Metrics

Based on our previous findings [9], we used Random Forest (RF) [43] to train and test our model. For the multiclass classification problem, we used the One-vs-All technique [44] which is constructed from binary classification. Each time, a single class is treated as positive against the rest which are the negative ones. Sensitivity, specificity, precision, Kappa value, and accuracy were used to evaluate the performance of the ML algorithm as illustrated in Table 4.

Table 4. Performance evaluation metrics

Metric	Description	Formula
Sensitivity	The correctly Identified the positive class	$Se = TP/(TP + FN)$
Specificity	The correctly Identified negative classes	$Sp = TN/(TN + FP)$
Precision	The positive predictive value	$Pr = TP/(TP + FP)$
Accuracy	The degree of overall correctness	$Ac = (TP + TN)/(TP + TN + FP + FN)$
Kappa value	The inter-rater agreement measurement	$K = (Po - Pe)/(1 - Pe)$

4   Results

The performance of the proposed algorithm is shown in Table 5. The accuracy and kappa values for all four feature sets (Set A, Set B, Set C, Set D; refer to Table 3) yielded results higher than 91.69% and 88.40%, respectively. The best results obtained with the feature set A which included all features from the accelerometer and gyroscope coordinates having accuracy 96.43% and kappa value of 95.02%. The highest sensitivity was noted in walking behaviour followed by resting. The walking behaviour was at 100% which is obtained using all feature sets. Specificity was higher than 93.76% for all activities with walking having the highest at 100%. The lowest accuracy and kappa value were realized using only features extracted from gyroscope sensor with accuracy and kappa value of 91.69% and 88.40%, respectively. Using the feature set D, browsing activity obtained a very low sensitivity of 36.72%, which could suggest that gyroscope coordinates cannot distinguish well the browsing movements of the animal. Yet, browsing activity obtained lowest sensitivity than the rest of the activities from all feature sets as well. It should be noted that performance of the algorithm was slightly decreased when using only features from the accelerometer sensor having accuracy and kappa value of 96.03% and 94.46%, respectively.

Table 5. Random forest performance using four different feature sets

| | Animal activity | | | | | | | |
	Walking	Browsing	Grazing	Resting	Scratching	Standing	Accuracy	Kappa value
set A*								
Sens	100.00%	78.91%	94.90%	97.29%	90.91%	95.48%	96.43%	95.02%
Spec	100.00%	99.99%	98.21%	99.34%	100.00%	97.46%		
Prec	100.00%	98.06%	93.28%	98.73%	100.00%	94.81%		
set B*								
Sens	100.00%	75.00%	94.66%	96.63%	81.82%	93.88%	95.60%	93.88%
Spec	100.00%	99.98%	97.33%	99.18%	100.00%	97.44%		
Prec	100.00%	94.12%	90.31%	98.42%	100.00%	94.69%		
set C*								
Sens	100.00%	79.69%	94.64%	97.14%	90.91%	94.57%	96.03%	94.46%
Spec	100.00%	99.98%	97.85%	99.18%	100.00%	97.45%		
Prec	100.00%	95.33%	92.03%	98.43%	100.00%	94.75%		
set D*								
Sens	100.00%	36.72%	87.10%	93.29%	72.73%	90.77%	91.69%	88.40%
Spec	100.00%	100.00%	96.79%	97.61%	100.00%	93.76%		
Prec	100.00%	100.00%	87.68%	95.38%	100.00%	87.62%		

* Sens = sensitivity, Spec = specificity, Prec = precision

5 Discussion and Conclusion

Monitoring the activity of livestock and wildlife animals is of great importance because of the highly valuable information we can gain through such knowledge. For example, the animals' use of pasture can play an important role in the prevention of soil erosion and contamination. Additionally, the daily activity of the animals could be an indicator of their health status and could help the decision-making process of farm managers. The monitoring of the animals was always a humans' responsibility which is undoubtedly labour intensive and time consuming. For this reason, automatic classification of animal activity is considered to be a solution to this problem. Also, machine learning can play an important role in a more intense observation of the animals because the machine could identify information which could be more difficult to be noticed through human observation alone.

In this work, data from seven Hebridean ewes located in Shotwick, Cheshire was collected in order to be used for animal activity classification identifying six behaviours; walking, grazing, resting, browsing, scratching, and standing. Gyroscope and accelerometer measurements were obtained from a smartphone device placed on the animal's harness on the top or side using the HyperIMU application. Similarly with previous studies, we have selected random forest for the classification of the activities as it demonstrated able to detect the activities with high accuracy [9, 38].

Similarities and differences with our previous study [9]:

(a) In our previous work we tested several machine learning algorithms in order to identify the one which could yield the best results. Therefore, for the data collected in this current work we used random forest since it demonstrated the best performance and we wanted to evaluate it with the new data.

(b) In this work, we used only sheep (Hebridean ewes), however, the data used in the previous work consisted of sheep and goats as well. With the outcome, we could be confident that this technique is powerful in classifying sheep and goat behaviour regardless the breed and we will test the findings with other breeds as well in the future.

(c) Similarly, to previous work, the recording device was placed on the animals without fixed orientation. Hence, we have verified that the features used are independent of the sensors' orientation. This result is in agreement with previous work [37].

(d) Using only accelerometer measurements did not compromise the performance of the algorithm [37]. This conclusion could be used in future work when implementing a solution for real time monitoring of the animals without conceding the battery life of the device.

(e) The sampling frequency used in this work was at 10 Hz in contrast with our previous work which was at 100 Hz. The reason for this was because it was noted that higher frequency rates have a negative effect of memory and power of the device [10]. It was also previously demonstrated that 16 Hz sample rate and 7-s windows can offer reduced energy needs and can be used for real-time sheep activity monitoring [11].

To conclude, random forest yielded very high results in classification of six behaviours of the ewes. Additionally, the kappa value demonstrates that the results are reasonable and not due to chance. When using features from both accelerometer and gyroscope, the algorithm obtained the best results having accuracy and kappa value of 96.43%, and 95.02%, respectively. However, using only features extracted from accelerometer, only decreased the accuracy by 0.40% and kappa value with 0.56%. Therefore, in future work, we will consider only the use of accelerometer sensor and we will test the performance of the same features and random forest for real-time activity recognition using larger number of animals as well as mixed breed animals.

6 Future Work

In future work, we aim to develop a novel multifunctional system, which will combine accelerometer sensor signals and GPS coordinates to monitor in real time animal behaviour, provide information about the animal health status and grazing activities, and also act as a virtual fence to keep the animals away from restricted areas and control the distribution of them on the grassland.

Acknowledgements. This study is funded by The Douglas Bomford Trust and John Moores University.

References

1. McLennan, K.M., et al.: Technical note: validation of an automatic recording system to assess behavioural activity level in sheep (Ovis aries). Small Rumin. Res. **127**, 92–96 (2015)
2. Gougoulis, D.A., Kyriazakis, I., Fthenakis, G.C.: Diagnostic significance of behaviour changes of sheep: a selected review. Small Rumin. Res. **92**, 52–56 (2010)
3. Barwick, J., Lamb, D., Dobos, R., Schneider, D., Welch, M., Trotter, M.: Predicting lameness in sheep activity using tri-axial acceleration signals. Animals **8**, 12 (2018)
4. Rutter, S.M.: Advanced livestock management solutions. In: Ferguson, D.M., Lee, C., Fisher, A. (eds.) Advances in Sheep Welfare, pp. 245–261. Woodhead Publishing, Sawston (2017)
5. Frost, A., Schofield, C., Beaulah, S., Mottram, T., Lines, J., Wathes, C.: A review of livestock monitoring and the needs for integrated systems. Comput. Electron. Agric. **17**, 139–159 (1997)
6. Anderson, D.M., Estell, R.E., Holechek, J.L., Ivey, S., Smith, G.B.: Virtual herding for flexible livestock management - a review. Rangeland J. **36**, 205–221 (2014)
7. Norton, B.E., Barnes, M., Teague, R.: Grazing management can improve livestock distribution: increasing accessible forage and effective grazing capacity. Rangelands **35**, 45–51 (2013)
8. Shepard, E.L.C., et al.: Identification of animal movement patterns using tri-axial accelerometry. Endanger. Species Res. **10**, 47–60 (2010)
9. Kleanthous, N., et al.: Machine learning techniques for classification of livestock behavior. In: Cheng, L., Leung, A.C.S., Ozawa, S. (eds.) ICONIP 2018. LNCS, vol. 11304, pp. 304–315. Springer, Cham (2018). https://doi.org/10.1007/978-3-030-04212-7_26
10. Hounslow, J.L., et al.: Assessing the effects of sampling frequency on behavioural classification of accelerometer data. J. Exp. Mar. Bio. Ecol. **512**, 22–30 (2019)
11. Walton, E., et al.: Evaluation of sampling frequency, window size and sensor position for classification of sheep behaviour. R. Soc. Open Sci. **5**, 171442 (2018)
12. Krahnstoever, N., Rittscher, J., Tu, P., Chean, K., Tomlinson, T.: Activity recognition using visual tracking and RFID. In: 2005 Seventh IEEE Workshops on Application of Computer Vision. WACV/MOTIONS 2005, vol. 1, pp. 494–500 (2005)
13. Cangar, Ö., et al.: Automatic real-time monitoring of locomotion and posture behaviour of pregnant cows prior to calving using online image analysis. Comput. Electron. Agric. **64**, 53–60 (2008)
14. Schlecht, E., Hülsebusch, C., Mahler, F., Becker, K.: The use of differentially corrected global positioning system to monitor activities of cattle at pasture. Appl. Anim. Behav. Sci. **85**, 185–202 (2004)
15. Ungar, E.D., Henkin, Z., Gutman, M., Dolev, A., Genizi, A., Ganskopp, D.: Inference of animal activity from GPS collar data on free-ranging cattle. Rangeland Ecol. Manag. **58**, 256–266 (2005)
16. Schwager, M., Anderson, D.M., Butler, Z., Rus, D.: Robust classification of animal tracking data. Comput. Electron. Agric. **56**, 46–59 (2007)
17. Umstätter, C., Waterhouse, A., Holland, J.P.: An automated sensor-based method of simple behavioural classification of sheep in extensive systems. Comput. Electron. Agric. **64**, 19–26 (2008)
18. Brewster, L.R., et al.: Development and application of a machine learning algorithm for classification of elasmobranch behaviour from accelerometry data. Mar. Biol. **165**, 62 (2018)

19. Kamminga, J.W., Bisby, H.C., Le, D.V., Meratnia, N., Havinga, P.J.M.: Generic online animal activity recognition on collar tags. In: Proceedings of the 2017 ACM International Joint Conference on Pervasive and Ubiquitous Computing and Proceedings of the 2017 ACM International Symposium on Wearable Computers on – UbiComp 2017, pp. 597–606. ACM, New York (2017)
20. Mason, A., Sneddon, J.: Automated monitoring of foraging behaviour in free ranging sheep grazing a biodiverse pasture. In: 2013 Seventh International Conference on Sensing Technology (ICST), pp. 46–51 (2013)
21. Radeski, M., Ilieski, V.: Gait and posture discrimination in sheep using a tri-axial accelerometer. Animal 11, 1249–1257 (2017)
22. Le Roux, S., et al.: An overview of automatic behaviour classification for animal-borne sensor applications in South Africa (2017)
23. Werner, J., et al.: Evaluation and application potential of an accelerometer-based collar device for measuring grazing behavior of dairy cows. In: Animal, pp. 1–10 (2019)
24. Moreau, M., Siebert, S., Buerkert, A., Schlecht, E.: Use of a tri-axial accelerometer for automated recording and classification of goats' grazing behaviour. Appl. Anim. Behav. Sci. 119, 158–170 (2009)
25. Hu, W., Jia, C.: A bootstrapping approach to entity linkage on the Semantic Web. Web Semant. Sci. Serv. Agents World Wide Web 34, 1–12 (2015)
26. Ladds, M.A., et al.: Super machine learning: improving accuracy and reducing variance of behaviour classification from accelerometry. Animal Biotelemetry 5, 8 (2017)
27. González, L.A.A., Bishop-Hurley, G.J.J., Handcock, R.N.N., Crossman, C.: Behavioral classification of data from collars containing motion sensors in grazing cattle. Comput. Electron. Agric. 110, 91–102 (2015)
28. Gutierrez-Galan, D., et al.: Embedded neural network for real-time animal behavior classification. Neurocomputing 272, 17–26 (2018)
29. Arcidiacono, C., Porto, S.M.C.C., Mancino, M., Cascone, G.: Development of a threshold-based classifier for real-time recognition of cow feeding and standing behavioural activities from accelerometer data. Comput. Electron. Agric. 134, 124–134 (2017)
30. Nadimi, E.S., Jørgensen, R.N., Blanes-Vidal, V., Christensen, S.: Monitoring and classifying animal behavior using ZigBee-based mobile ad hoc wireless sensor networks and artificial neural networks. Comput. Electron. Agric. 82, 44–54 (2012)
31. Marais, J., et al.: Automatic classification of sheep behaviour using 3-axis accelerometer data (2014)
32. Alvarenga, F.A.P., Borges, I., Palkovič, L., Rodina, J., Oddy, V.H., Dobos, R.C.: Using a three-axis accelerometer to identify and classify sheep behaviour at pasture. Appl. Anim. Behav. Sci. 181, 91–99 (2016)
33. le Roux, S.P., Marias, J., Wolhuter, R., Niesler, T.: Animal-borne behaviour classification for sheep (Dohne Merino) and Rhinoceros (Ceratotherium simum and Diceros bicornis). Animal Biotelemetry 5, 25 (2017)
34. Giovanetti, V., et al.: Automatic classification system for grazing, ruminating and resting behaviour of dairy sheep using a tri-axial accelerometer. Livest. Sci. 196, 42–48 (2017)
35. Barwick, J., Lamb, D.W., Dobos, R., Welch, M., Trotter, M.: Categorising sheep activity using a tri-axial accelerometer. Comput. Electron. Agric. 145, 289–297 (2018)
36. Decandia, M., et al.: The effect of different time epoch settings on the classification of sheep behaviour using tri-axial accelerometry. Comput. Electron. Agric. 154, 112–119 (2018)
37. Kamminga, J.W., Le, D.V., Meijers, J.P., Bisby, H., Meratnia, N., Havinga, P.J.M.: Robust sensor-orientation-independent feature selection for animal activity recognition on collar tags. In: Proceedings of ACM Interactive, Mobile, Wearable Ubiquitous Technology, vol. 2, pp. 1–27 (2018)

38. Mansbridge, N., et al.: Feature selection and comparison of machine learning algorithms in classification of grazing and rumination behaviour in sheep. Sensors (Switzerland) **18**, 1–16 (2018)
39. le Roux, S.P., Wolhuter, R., Niesler, T.: Energy-aware feature and model selection for onboard behavior classification in low-power animal borne sensor applications. IEEE Sens. J. **19**, 2722–2734 (2019)
40. HyperIMU – ianovir.com. https://ianovir.com/works/mobile/hyperimu/
41. Yang, X., Dinh, A., Chen, L.: Implementation of a wearerable real-time system for physical activity recognition based on Naive Bayes classifier. In: 2010 International Conference on Bioinformatics and Biomedical Technology, pp. 101–105. IEEE (2010)
42. Bouten, C.V., Westerterp, K.R., Verduin, M., Janssen, J.D.: Assessment of energy expenditure for physical activity using a triaxial accelerometer. Med. Sci. Sports Exerc. **26**, 1516–1523 (1994)
43. Breiman, L.: Random forests. Mach. Learn. **45**, 5–32 (2001)
44. Rifkin, R., Klautau, A.: In defense of one-vs-all classification. J. Mach. Learn. Res. **5**, 101–141 (2004)

An Algorithm of Bidirectional RNN for Offline Handwritten Chinese Text Recognition

Xinfeng Zhang and Kunpeng Yan$^{(\boxtimes)}$

Beijing University of Technology, Beijing 100124, China
ykp_2016@bjut.edu.com

Abstract. Handwritten Chinese text recognition characters is a challenging problem as it involves a imbalanced training data, and the samples are very different even in same character. In this paper, we propose a novel algorithm based on the bidirectional Recurrent Neural Network (BiRNN) to recognize the characters in the text regions. We solve the problems with pre-processing and improved CNN network. In addition, we utilize RNN to analyze the correlation between characters. Compared with previous works, the algorithm has three distinctive properties: (1) It can predict characters by context analyzing from forward and backward. (2) It solve the problem of sample imbalance effectively. (3) The convergence rate of training has increased. Moreover, the proposed algorithm has achieved good results in recognition.

Keywords: OCR · Deep learning · RNN

1 Introduction

Compared with the general visual elements in pictures or videos, text can be recorded more conveniently. But a lot of information is recorded in the picture. So it is very important to convert the information in pictures into words. The purpose of Optical Character Recognition (OCR) is to make the computer recognize the text in the images and transfer it into the representation that the machine can store. It is a hot issue in computer vision. Currently, recognition methods for printed characters proposed by researchers have good performances [1]. However, handwriting character recognition (HCR) is more difficult than printed character recognition [2]. Since the HCR has great diversity and uncertainty [3, 4]. Each character may have a different size, font, color, brightness, contrast and so on. Text areas may also be deformed, incomplete and blurred.

Before text recognition, some pre-processing methods for segment characters are used in most Chinese text recognition algorithms as usual [5, 6, 7]. Therefore, the accuracy of segmentation has a great influence on the recognition performance [8, 9]. In this paper, we made an experiment in which the characters are not segmented by Connectionist Temporal Classification (CTC). In the method, the raw data is input directly into the Recurrent Neural Network (RNN).

The major problem of OCR is the limited number of characters' classes. The main risk of using CTC to train network parameters is that if the probability is not evenly distributed over too many character classes, it may cause over-fitting. We overcome the

D.-S. Huang et al. (Eds.): ICIC 2019, LNAI 11645, pp. 423–431, 2019.
https://doi.org/10.1007/978-3-030-26766-7_39

over-fitting problem by pre-processing, and improved accuracy rate through context analysis using RNN. The results given in Sects. 2 and 3 show that the effectiveness of pre-processing and Bidirectional-LSTM (Bi-LSTM) achieves the best performance of the data.

2   Data Pre-processing

As mentioned in the previous chapter, there are two challenges in HCR. One of them is that different people have very different handwriting styles, that is to say, the intra-class distance is very large. Compared with some other handwritten characters databases, the characters in our database are written by students, which are clear, legible and without altering. The another challenge is that there are too many classes of Chinese characters, the risk of sample imbalance is likely to increase dramatically. In order to improve the accuracy of the results in processing the data with our database, we expand the data set for which picture contain characters appear a little by increasing noise (see Fig. 1).

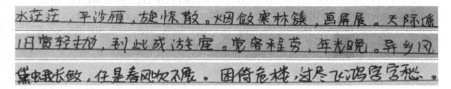

Fig. 1. Data sample

Firstly, we count the distribution of characters in the data set and calculate the frequency and width of each character. The maximum, minimum and average values are calculated respectively (see Figs. 1 and 2).

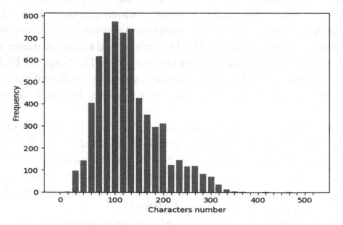

Fig. 2. Frequency of characters

As can be seen from the figure, there are many characters that appear less than 10 times. But the performance of some characters in recognition would be very poor, which appear less than 10 times in database.

Through the analysis of the data set, we expanded the data to 2000 training samples. Our method is to expand the data of which characters appeared less than 30 times. However, direct replication of data may lead to over-fitting of training results [11, 12]. So we do the following three ways to process the target image. The first way is to add salt and pepper noise, the second is to add Gauss noise, and the last is to do gamma transform on the original image. The parameters of Gaussian noise are Mean 0.05 variance 0.1. The parameter of salt and pepper noise is set to SNR = 0.8 and of gamma transform to gamma = 0.7 (see Figs. 1 and 3).

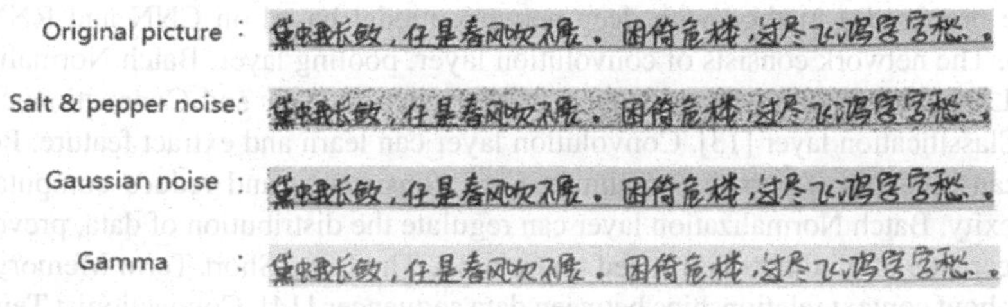

Original picture :

Salt & pepper noise:

Gaussian noise :

Gamma :

Fig. 3. Enhanced sample

In order to train data for RNN network, we normalized the data after analysis. We analyzed the length-width ratio of the data. After calculation, the average length-width ratio of the data is 8.32. Images are normalized according to the length-width ratio (see Figs. 1 and 4).

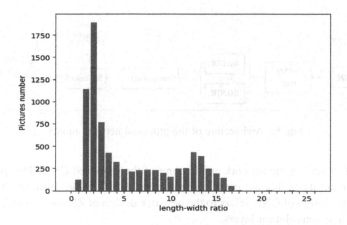

Fig. 4. Length-width ratio of picture

3   Recognition with Recurrent Neural Network

In order to solve the problem that handwritten characters are difficult to identify. We used the improved CNN to extract the deep features of the image, and predict characters by context analysis through BiRNN. The proposed a method as following. Given the training set D and a training instance (x, z) in which x represents the original pictures of handwriting written by students and z is the corresponding label sequence, our method aims at minimizing the Loss Function $L(D)$ with correctly labelling all the training samples in D:

$$L(D) = -ln \prod_{(x,z) \in D} p(z|x) = - \sum_{(x,z) \in D} ln\{p(z|x)\} \qquad (1)$$

We proposed a method with deep network model based on CNN and RNN (see Fig. 1). The network consists of convolution layer, pooling layer, Batch Normalization layer, Long Short-Term Memory layer, full connection layer and Connectionist Temporal Classification layer [13]. Convolution layer can learn and extract feature. Pooling layer can transform features to maintain local consistency and reduce computational complexity. Batch Normalization layer can regulate the distribution of data, prevent the disappearance of gradients and speed up training. The Long Short-Term Memory layer learns about context relationships between data sequences [14], Connectionist Temporal Classification Layer can translate the output of Long Short-Term Memory Layer and calculate the network loss value in order to optimize by gradient descent method (Fig. 5).

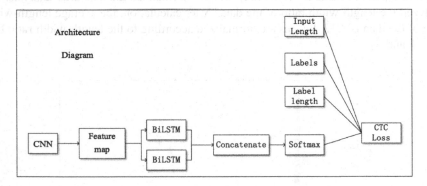

Fig. 5. Architecture of the proposed network model

Figure 6 describes the network architecture of the proposed CNN. This part outputs 64×32 feature maps that are used to characterize the character from the handwritten text data. The design of CNN refers to the network design of Zhuoyao et al. [15], which consists of five convolution layers.

For the purpose of automatically extracting high-level abstract features from the image, we adjust its structure to six-layer convolution. At the same time, we adjust the feature dimension conversion of the last layer, which is to facilitate the calculation of LSTM layer.

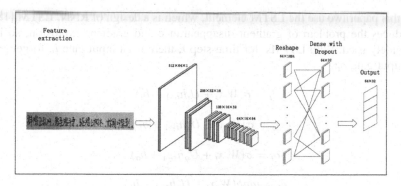

Fig. 6. Architecture of the proposed Convolutional Neural Network

Context information plays an important role in text recognition [16, 17]. We design a Bi-bidirectional RNN network to record the context information of text sequence. The 64×32 vector extract from CNN is used as the input of RNN network (see Fig. 1).

One shortcoming of conventional RNN is that they can only use previous context. In text recognition, where whole text are record at once, it is very import to exploit future context. BiRNN do this by processing the data in both directions with two separate hidden layers, which are then fed forwards to the same output layer. As shown in Fig. 7. BiRNN computes the forward hidden sequence h_f, the backward hidden sequence h_b, and the output sequence y by iterating the forward layer from time-step $t = 1$ to T, the backward layer from time-step $t = T$ to 1 and updating the output sequence y. The W_* is the input-to-hidden weight matrix, and b_* is the bias vector.

$$h_f = W_{xh_f}x_t + W_{h_f}h_{f(t-1)} + b_{h_f} \tag{2}$$

$$h_b = W_{xh_b}x_t + W_{h_b}h_{b(t-1)} + b_{h_b} \tag{3}$$

$$y_t = W_{h_f}h_{ft} + W_{h_b}h_{bt} + b_y \tag{4}$$

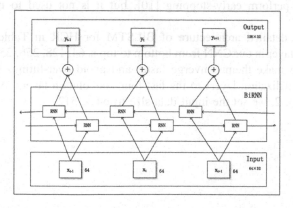

Fig. 7. Architecture of the Recurrent Neural Network

In this paper, we use the LSTM element, which is a design of RNN. LSTM [18, 19, 20] reduces the problem of gradient disappearance and gradient explosion, so it has been widely used. With LSTMs, for time-step t, there is an input gate i_t, forget gate f_t, and output gate o_t:

$$i_t = \sigma(W_i x_t + U_i h_{t-1} + b_i) \tag{5}$$

$$f_t = \sigma\big(W_f x_t + U_f h_{t-1} + b_f\big) \tag{6}$$

$$o_t = \sigma(W_o x_t + U_o h_{t-1} + b_o) \tag{7}$$

$$\tilde{c}_t = tanh(W_c x_t + U_c h_{t-1} + b_c) \tag{8}$$

$$c_t = f_t \odot c_{t-1} + i_i \odot \tilde{c}_t \tag{9}$$

$$h_t = o_t \odot \tanh(c_t) \tag{10}$$

The W_* is the input-to-hidden weight matrix, U_* is the state-to-state recurrent weight matrix, and b_* is the bias vector. The operation $\odot$ denotes the element-wise vector product, the σ uses the logistic sigmoid function. The hidden state of LSTM is the concatenation of (c_t, h_t) and c_t stores the long-term memory. As shown in Eq. (6), the forget gate and input gate are used to control the updates of c_t. As shown in Eq. (7), the output gate is used to control the updates of h_t.

4   Experimental Results and Analysis

In our experiments, training data samples are written by students, and each sample is a labeled image. All images in the database are long text images. The database has 10,000 pictures, including more than 100,000 characters and 2,382 character types. We divide the data into training set, verification set and test set according to the ratio of 8:1:1. That means there are 8000 training samples, 1000 validation samples and 1000 testing samples. The validation data is used to measure the convergence of the parameters and perform early-stopping [10], but it is not used to modify the free parameters.

We list our detailed architecture of BiLSTM for HCR in Table 1. The kernel number of each layer in our CNN from bottom to top is 64, 128, 256, 256, 512, 512 and 512. In order to make them converge faster and avoid over-fitting. We have added some batch normalization layers to the last four convolution layers. According to the analysis in Chap. 2, we set the input data size to 64 × 512.

Table 1.

Type	Configurations
CTC	
Full Connection	#hidden units:x
Bidirectional-LSTM	#hidden units:256
Bidirectional-LSTM	#hidden units:256
Map-to-Sequence	
Convolution	#map:512, k:2 × 2. s:1, p:0
MaxPooling	Window:1 × 2, s:2
Convolution	#map:512, k:3 × 3. s:1, p:1
Convolution	#map:512, k:3 × 3. s:1, p:1
MaxPooling	Window:1 × 2, s:2
Convolution	#map:256, k:3 × 3. s:1, p:1
Convolution	#map:256, k:3 × 3. s:1, p:1
MaxPooling	Window:2 × 2, s:2
Convolution	#map:128, k:3 × 3. s:1, p:1
MaxPooling	Window:2 × 2, s:2
Convolution	#maps:64, k:3 × 3, s:1, P:1
Input	Batchsize × 64 × 512 image

We use TensorFlow Deep Learning Framework to build our network and trained our model with four Geforce Titan-X 12G GPUs, in addition, our platform has 64G memory and Xeon E5 CPU. We recorded the training time and compared it with other models.

Table 2 shows the experimental data set. In this paper, the combinations of improved CNN and different RNN units are compared respectively. At the same time, the data are segmented equally, which is used in the training of only CNN network and the contrast experiment of base line. Its convergence time is the shortest, and the average testing time of single sheet is also the shortest. But the accuracy rate is not ideal, and the comprehensive word accuracy rate is only 50.25%. This is due to the loss of part of the text when data is segmented at an equal interval, which makes the accuracy rate drop sharply. Because of the need for recording the context when using RNN network, each iteration needs to wait for the current round of training to load into memory [21, 13]. Memory and bandwidth have a great impact on training speed. However, the accuracy rate of the experiment has been improved conspicuously by context semantics analysis and CTC-Loss function. When using double-layer RNN structure, the accuracy and recall rate have been improved by nearly 5%.

Table 2. Comparison with state-of-the-art methods based on recall rate, accuracy rate, train time and test speed on our dataset.

Methods	Recall	Acc	Train speed	Test speed
CNN (base line)	44.5	50.25	**5.1**	**0.8**
LSTM (base line)	66.25	79.32	9.9	1.5
CNN + LSTM (Dewi et al. 2016 [9])	69.55	77.63	11.3	1.5
GRU (Zhang et al. 2016 [8])	71.91	79.72	16.3	1.5
CNN + BiLSTM (our model)	**73.79**	**83.00**	15.5	1.6

Compared with Dewi et al. [8] and Zhang et al. [9], we apply a two-layer bidirectional neural network. It is possible to analyze the correlation between characters by forward and backward information. Zhang et al. used the full convolution network (FCN) to extract features. In addition, they used GRU element to accelerate the training time. However, the computational complexity of FCN is relatively high. As shown in Table 2, it spent longer training time than our method without obvious accuracy rate improvement. Furthermore, the improved CNN is utilized for improving the effect of feature extraction and it shows a better performance comparing to others with the 73.79% recall rate and 83.00% accuracy for our dataset.

5   Conclusion and Future Work

Aiming at the problem of handwriting Chinese text recognition written by student, we proposed a method using Bi-LSTM neural network. The method can be used for Chinese text recognition which written carefully and neatly. Firstly, according to the student's characteristics of handwriting, we normalized to the section of data. Then, we use CNN network to extract image features and adopt LSTM network to record the context diagram. We utilize the CTC-Loss function to accelerate convergence in our data model. The results validate the effectiveness of the proposed method. The experimental results show that the proposed algorithm has a competitive performance for Chinese handwritten images recognition based on our database.

In this paper, the text is processed all arranged horizontally. In the following work, we will consider the recognition ability of the extended algorithm for vertical or oblique characters [22]. In addition, due to the shortage of data, long text and short text are processed simultaneously. In the future, we will consider different processing methods for different lengths of images, that is, long text and short text are processed separately.

There is no doubt that RNN is an important future research topic. Considering its advantages in dealing with sequence problems, our another important future work is taken even more advantage of pre-trained convolutional filters for the specific purpose on tongue image sequence recognition.

Acknowledgment. Our work is supported by the national key research and development program (No. 2017YFC1703300) of China.

References

1. Luo, X.-L., Huang, S.-F.: Image character recognition technology and its application research based on the integration of multiple classifiers. Software **36**(3), 98–102 (2015)
2. Quan, Z., Lin, J.: Text-independent writer identification method based on Chinese handwriting of small samples. J. East China Univ. Sci. Technol. **44**(6), 882–886 (2018)
3. Wang, T., Wu, D.J., Coates, A., Ng, A.Y.: End-to-end text recognition with convolutional neural networks. In: ICPR, pp. 6–7 (2012)
4. Liu, W., Li, J.: A method for off-line handwritten Chinese character recognition based on hidden Markov model. CCSSTA **11**, 774–777 (2009)
5. Liu, C.-L., Marukawa, K.: Pseudo two-dimensional shape normalization methods for handwritten Chinese character recognition. Pattern Recogn. **32**(12), 2242–2255 (2005)
6. Yan, X., Wang, L.: Handwritten Chinese character recognition system based on neural network convolution depth. Comput. Eng. Appl. **53**(10), 246–250 (2017)
7. Lee, S.-G., Sung, Y., Kim, Y.-G., Cha, E.-Y.: Variations of AlexNet and GoogLeNet to improve Korean character recognition performance. J. Inf. Process. Syst. **14**(1), 205–217 (2018)
8. Nguyen, C.K., Nguyen, C.T., Masaki, N.: Tens of thousands of Nom character recognition by deep convolution neural networks. In: Historical Document Imaging and Processing, pp. 37–41 (2017)
9. He, K., Zhang, X., Ren, S., Sun, J.: Deep residual learning for image recognition. In: CVPR, pp. 770–778 (2016)
10. Srivastava, N., Hinton, G., Krizhevsky, A., Sutskever, I., Salakhutdinov, R.: Dropout: a simple way to prevent neural networks from overfitting. J. Mach. Learn. Res. **15**(1), 1929–1958 (2014)
11. Cho, K., Courville, A., Bengio, Y.: Describing multimedia content using attention-based encoder-decoder networks. Multimedia **17**(11), 1875–1886 (2015)
12. Kabade, A.L., Sangam, V.G.: Canny edge detection algorithm. IJARECE **5**(5), 12–16 (2016)
13. Garla, V.N., Brandt, C.: Ontology-guided feature engineering for clinical text classification. J. Biomed. Inform. **45**(6), 992–998 (2012)
14. Jordan, M.I., Mitchell, T.M.: Machine learning: trends, perspectives, and prospects. Science **6245**(349), 255–260 (2015)
15. Graves, A., Jaitly, N.: Towards end-to-end speech recognition with recurrent neural networks. In: ICML-14, pp. 1764–1772 (2014)
16. Sanchez, J., Romero, V., Toselli, A., Vidal, E.: ICFHR2014 competition on handwritten text recognition on transcriptorium datasets. In: ICFHR, pp. 181–186 (2014)
17. Wang, Q.-F., Yin, F., Liu, C.-L.: Handwritten Chinese text recognition by integrating multiple contexts. Pattern Anal. Mach. Intell. **32**(8), 1469–1481 (2012)
18. Deng, L., Yu, D.: Deep learning: methods and applications. Found. Trends Sig. Process. **7**(3), 197–287 (2014)
19. He, K., Zhang, X., Ren, S., et al.: Delving deep into rectifiers: surpassing human-level performance on image-net classification. In: ICCV, pp. 1026–1034 (2015)
20. Kalayeh, M.M., Gong, B., Shah, M.: Improving facial attribute prediction using semantic segmentation. In: CVPR, pp. 6942–6950 (2017)
21. Yang, L., Yu, X., Wang, J., Liu, Y.: Research on recurrent neural network. Comput. Appl. **38**(22), 16–26 (2018)
22. Liu, C.-L., Yin, F., Wang, D.-H., Wang, Q.-F.: CASIA online and offline Chinese handwriting databases. In: ICDAR, pp. 37–41 (2011)

A Novel Concise Representation of Frequent Subtrees Based on Density

Weidong Tian[✉], Chuang Guo, Yang Xie, Hongjuan Zhou,
and Zhongqiu Zhao

School of Computer Science and Information Engineering,
Hefei University of Technology, Hefei 230601, Anhui, China
wdtian@hfut.edu.cn

Abstract. Frequent subtree mining has wide applications in many fields. However the number of the frequent subtree is often too large because of the extensive redundancy in frequent subtree set, which makes it difficult to be used in practice. In this paper, density of frequent subtree in the lattice induced by frequent subtree set is introduced, then a novel concise representation of frequent subtree called FTCB is proposed, and the corresponding mining algorithm FTCBminer is proposed too. Experimental results show that FTCB keeps more information than MFT and reduces the size of frequent subtree set more efficiently than CFT.

Keywords: Frequent subtree · Concise representation · Density · Dynamic density

1 Introduction

Tree-structured data is one of the common data type in many fields, such as XML document mining [1], biological information [2], and Natural Language Processing [3]. By mining frequent subtrees in tree-structured data, some important structural features can be obtained and be applied to various fields [4]. For example, Hao [5] explores the information dissemination mode of social networks on the real dataset of weibo based on the idea of frequent subtrees. There are many algorithms for mining frequent subtrees, such as TreeMiner [6], FREQT [7], EvoMiner [8], and FRESTM [9].

One of the key problem in frequent tree pattern mining is that the number of frequent trees is often too large, which brings difficulties for practical application [10]. Therefore, many concise representations of frequent subtrees have been proposed. The Maximal Frequent Tree (MFT) [11] and the Closed Frequent Tree(CFT) [14] are two of the classic ones. MFT retains the positive boundary between frequent subtrees and infrequent subtrees, but cannot retain subtrees' support information. Algorithms for mining MFT include PathJoin [11], CMTreeMiner [12], MFTM [13], etc. CFT retains frequent subtrees which supports are lower than any of their subtrees. In practice, the size of CFT maybe large too, especially while support threshold is quite low. There are many algorithms for mining CFT, such as DRYADE [14] and FBMiner [15]. Wang [16] proposed a concise representation model based on stratification. According to the minimum support and error value, the frequent subtrees are divided into n different

D.-S. Huang et al. (Eds.): ICIC 2019, LNAI 11645, pp. 432–442, 2019.
https://doi.org/10.1007/978-3-030-26766-7_40

layers, and the maximal frequent subtrees are mined for each layer, which is the concise representation of the original frequent subtrees. However the model is sometime too rigid to conducive to the concentration of dataset. For example, for tree T_2 and its subtree T_1, while the minimum support is 10, the error range is 10, and the support are 21 and 19 respectively, then they belong to different layers and cannot be merged, but in fact their support are nearly equal.

Therefore, we define the density of the frequent subtrees which can be applied to merge those seemingly same frequent subtrees. Based on the density, we propose a concise representation of frequent subtrees and a mining algorithm too. While reducing the number of subtrees, the proposed concise representation of frequent subtrees can still ensure that the average error is not too high.

2 Tree and Subtree Pattern

2.1 Tree

Tree is a unidirectional acyclic connected graph, denoted as $T(r, E, V)$, where r is the only root node, V is the set of vertices and E is the set of edges. Each edge can be represented by (x, y), where $x, y \in V$.

For each node $x, y \in V$, if there has a path from x to y, we defined this as $x \preccurlyeq_i y$, where i represents the number of edges in the path, x is the ancestor node of y; when $i = 1$, x is the parent node of y, denoted as (x, y). If multiple nodes have the same parent node, they are called sibling nodes.

The label of each node is taken from a set of labels $L = \{l_1, l_2 \dots\}$, there exists a function $l: V \rightarrow L$, which maps node n_i to label $l(n_i) = l_j \in L$. An ordered tree is a rooted tree in which the children of each node are ordered. Different orders represent different trees.

$|T|$ represents the number of nodes in the tree. A tree is called k-tree when $|T| = k$.

2.2 Subtree Pattern

There are two types of subtree: embedded subtree and induced subtree.

Definition 1 (Induced Subtree): *Given a tree T_1 (r_1, V_1, E_1), tree T_2 (r_2, V_2, E_2) is called an induced subtree of T_1, denoted as $T_2 <_i T_1$, if ① there exists a function g: $V_2 \longrightarrow V_1$, which maps node $x \in V_2$ to node $g(x) = y \in V_1$; ② For any $(x, y) \in E_2$, $(g(x), g(y)) \in E_1$ is satisfied; ③ For any $x \in V_2$, $l(x) = l(g(x))$ is satisfied.*

Definition 2 (Embedded Subtree): *Given a tree T_1 (r_1, V_1, E_1), tree T_2 (r_2, V_2, E_2) is called an embedded subtree of T_1, denoted as $T_2 <_e T_1$, if ① there exists a function g: $V_2 \longrightarrow V_1$, which maps node $x \in V_2$ to node $g(x) = y \in V_1$; ② For any $(x, y) \in E_2$, $g(x) \preccurlyeq_i g(y)$ is satisfied; ③ For any $\in V_2$, $l(x) = l(g(x))$ is satisfied.*

As shown in Fig. 1, T_2 is an embedded subtree of T_1, and T_3 is an induced subtree of T_1. The embedded subtree and the induced subtree are collectively referred to as subtree, denoted as $T_2 < T_1$ and $T_3 < T_1$.

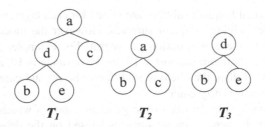

Fig. 1. Embedded subtree and induced subtree

Let sup_num(T) be the number of the T's super-tree included in tree dataset D, the support of tree T in D, denoted as Supp(T), is defined as Eq. (1).

$$\text{Supp}(T) = \text{sup\_num}(T)/|D| \tag{1}$$

Definition 3 (Frequent Subtree): *In the dataset D of trees, the tree T is a frequent subtree when the support of T is not lower than the specified minimum support. The specified minimum support was denoted as min_sup, min_sup $\in$ (0, 1].*

For a tree dataset, frequent subtree reveals the hidden structural features, so frequent subtree is called subtree pattern, or tree pattern.

3   Concise Representation of Frequent Subtrees

In order to effectively reduce the subtree set, the model selects the most representative trees from the frequent subtree set to form a set of constrict base. Based on the constrict base, the frequent subtree set is divided into several different fuzzy equivalence classes, and each class is represented by a corresponding constrict base. The difference in the support in each fuzzy equivalence class is not large, which ensures that the support error caused by the reduction is not high.

3.1   Concise Representation Model

Let $\theta \in [0, 1]$ be the tolerance threshold and FT be the collection of all frequent subtrees in a tree dataset D.

Definition 4 (Neighborhood): *Let $T_f \in FT$ be a frequent tree. The neighborhood of T_f is defined as follows:*

$$\text{REP}(T_f, \theta) = \left\{ T_i | \text{Supp}(T_i) - \text{Supp}(T_f) \leq \theta \wedge T_i < T_f, T_i \in FT \right\}. \tag{2}$$

Definition 5 (Frequent Tree Constrict Base, FTCB): *Let R be a subset of FT. For an arbitrary tree $T \in FT\backslash R$, if there exists $T_f \in R$ and $T \in REP$ (T_f, θ), then R is the frequent tree constrict base of FT, denoted as FTCB.*

Definition 6 (Blurred Frequent Represent Tree, BFRT): *Let $T_f \in$ FTCB be a frequent tree. The fuzzy equivalence class of T_f is defined as follows:*

$$\text{BFRT}(T_f, \theta) = \{T | T = T_f \cup \text{REP}(T_f, \theta)\} \tag{3}$$

The tree except the generator base T_f in each BFRT is within the neighborhood of T_f, and the support difference from T_f is less than θ. The size of the FTCB is equal to the number of BFRT.

Theorem 1. *FTCB is a concise representation of the whole set of subtrees FT with tolerance θ.*

Proof: (1) Let an arbitrary tree $T_f \in$ FTCB be a generator base of BRFT(T_f, θ), based on Definition 4 and 6, Since All trees in BFRT(T_f, θ) are subtrees of T_f, all frequent subtrees in BRFT(T_f, θ) can be generated by pruning on T_f. Based on Definition 5 and 6, since FT can be divided into several fuzzy equivalence classes by FTCB, and each fuzzy equivalence class BFRT can be represented by a generator base in FTCB, all frequent subtrees in FT can be restored by FTCB. (2) Let $T_f \in$ FTCB be a generator base of BRFT(T_f, θ). For an arbitrary tree $T \in$ BFRT(T_f, θ), based on Definition 4, since Supp(T)-Supp(T_f) θ, the support error of T after being restored remains between 0 and θ. Therefore, FTCB is a concise representation of the whole set of subtrees FT with tolerance θ.

Proposition 1: *When $\theta = 1$, the maximum frequent tree set MFT is FTCB.*

Proof: $\forall T \in$ FT\MFT, based on the definition of the maximum frequent tree, there must exists $T_f \in MFT$, satisfy T is a subtree of T_f. Also, since $0 <$ Supp$(T_f) \leq$ Supp$(T) \leq 1$, Supp(T) − Supp$(T_f) \leq \theta = 1$ is satisfied. Therefore, MFT is FTCB when $\theta = 1$.

Proposition 2: *When $\theta > 0$, the maximum frequent tree set MFT is a subset of the FTCB.*

Proof: $\forall T \in MFT$, assuming $T \notin$ FTCB, there must exists $T_f \in$ FTCB, satisfy $T \in$ REP(T_f, θ), based on Definitions 4 and 6, T is a subtree of T_f, which contradicts that $T \in MFT$, and the assumption is not true.

Proposition 3: *When $\theta = 0$, the closed frequent tree set CFT is FTCB.*

Proof: For any $T_f \in CFT$, based on the definition of the closed frequent tree, its subtrees' support are equal to its support, so the condition of $\theta = 0$ is always satisfied. Therefore, CFT is FTCB when $\theta = 0$.

3.2   Evaluation Criteria

Definition 7 (Average Error Rate): *In tree database D, for $\forall T \in$ FT, $\exists T_f \in$ FTCB, and $T \in$ REP (T_f, θ). The average error rate is defined as follows:*

$$\sigma = \frac{\sum_{T \in FT} (\text{Supp}(T) - \text{Supp}(T_f))}{|FT|} \tag{4}$$

For the neighborhood of the same constrict base T_f, the error values of the support of all frequent subtrees are kept within a specific range θ. Both the constrict base and tolerance threshold have an impact on the average error rate.

Definition 8 (Compression Ratio): *In database D, The compression ratio is defined as follows*:

$$p = |FTCB|/|FT| \tag{5}$$

The compression ratio reflects the degree of reduction, the lower the compression ratio, the smaller the reduced data set.

4   Mining and Restoring

4.1   Mine FTCB

In order to get a FTCB with low average error rate, we introduce the concept of dynamic density.

Definition 9 (Density, ρ): *Let $T \in FT$ be a frequent tree. The density of T is defined as follows:*

$$\rho(T, \theta) = |T \cup REP(T, \theta)|. \tag{6}$$

For a frequent tree T, the density value indicates the number of subtrees in its neighborhood. The greater the density value, the denser the data distribution around T, and the more frequent subtrees it can represent.

Definition 10 (Dynamic Density, mov_ρ): *Let $T \in FT$ be a frequent tree. In the dynamic execution of the algorithm FTCBminer, the density of T is defined as dynamic density, denoted as mov_$\rho(T, \theta)$.*

Proposition 4: *mov_$\rho(T, \theta)$ will constantly change with the operation of the algorithm. The initial value of mov_$\rho(T, \theta)$ is equal to $\rho(T, \theta)$.*

Definition 11 (θ-Neighborhood): *Let $T_0 \in FT$ be a frequent tree. The collection of all frequent trees whose neighborhood has an intersection with the neighborhood of T_0 is defined as the θ-Neighborhood of T_0:*

$$\alpha(T_0, \theta) = \left\{ T_f | T \in BFRT(T_0, \theta) \wedge T \in REP\left(T_f, \theta\right), T_f \in FT \right\} \tag{7}$$

For $\forall T \in FT$, there are three kinds of information related to T: dynamic density information mov_$\rho(T, \theta)$, the trees whose neighborhood covers T and the trees in the neighborhood of T. All information about the FT is stored in an array of nodes.

In general, for loss compression of frequent subtrees, the smaller the set obtained after reduction, the more information that is lost, and the larger the support error value. Therefore, it is important to select the most representative frequent subtrees so that each tree can represent more number of frequent subtrees. The specific strategy is as follows:

(1) Select T_{max} from FT to satisfy the conditions: (a) $markbit[T_{max}] = 0$, indicating that T_{max} has not been processed; (b) $T_{max} = \text{argmax}_{T \in FT}(\text{mov\_}\rho(T, \theta))$. The greater the dynamic density of T_{max}, the more subtrees it can represent. The purpose is to obtain the most representative frequent subtrees in the final FTCB.

(2) Set $markbit[T_{max}] = 1$, indicating that T_{max} has been processed, and add it to FTCB.

(3) For any unprocessed frequent subtree $T \in$ TF, if $T_{max} \in$ REP(T,θ), modify its dynamic density: $\text{mov\_}\rho(T, \theta) = \text{mov\_}\rho(T, \theta) - 1$.

(4) For any unprocessed frequent subtree $T \in$ REP(T_{max}, θ), if there exists an unprocessed tree $T_f \in \alpha(T_{max}, \theta)$, whose neighborhood contains T, then modify the dynamic density of T_f: $\text{mov\_}\rho(T_f, \theta) = \text{mov\_}\rho(T_f, \theta) - 1$.

The value of the decrease in the dynamic density of T_f is equal to the number of unprocessed trees in the set REP(T_{max}, θ) $\cap$ REP(T_f, θ).

(5) For any unprocessed frequent subtree $T \in$ REP(T_{max}, θ), set $markbit[T] = 1$, indicating that T has been processed.

(6) Repeat steps (1)–(5) until there is no unprocessed tree in FT.

The following is the mining algorithm FTCBminer.

Algorithm 1. FTCBminer

Input: FT: the collection of all frequent subtrees; θ: tolerance threshold.
Output: FTCB: frequent tree constrict base.
FTCB $\leftarrow \varnothing$;
$done \leftarrow 0$;
while $done \leqslant$ |FT|
   $T_{max} \leftarrow$ Find(FT, $markbit$)
   ++$done$;
   $markbit[T_{max}] \leftarrow 1$;
   FTCB $\leftarrow T_{max}$
   for $\forall T \in$ FT
      if $markbit[T] == 0$ **and** $T_{max} \in$ REP(T, θ) **then**
         $\text{mov\_}\rho(T, \theta) \leftarrow \text{mov\_}\rho(T, \theta)$ -1
   for $\forall T \in$ BFRT(T_{max}, θ)
      if $markbit[T] == 0$ **then**
         $markbit[T] \leftarrow 1$
         for $\forall T_f \in \alpha(T_{max}, \theta)$
            if $markbit[T_f] == 0$ **and** $T \in$ REP(T_f, θ) **then**
               $\text{mov\_}\rho(T_f, \theta) \leftarrow \text{mov\_}\rho(T_f, \theta)$ -1
         end for
      end for
end while

The $markbit$ is used to save the processing status of all subtrees.

4.2   Restoring Original Frequent Subtree Set

Based on the mined FTCB, the support of all frequent subtrees is obtained through restoration.

Definition 12 (Direct Subtree): *Given a k-tree T, T_b is called a direct subtree of T, if:* ① T_b *is subtree of T;* ② $|T_b| = k - 1$. *All the direct subtrees of T is denoted as Dch(T).*

Definition 13 (k-Frequent Subtrees): *All the frequent subtrees whose number of nodes is k in the FT is defined as k-frequent subtrees, denoted as FT(k).*

Definition 14 (k-Reduction Frequent Subtrees): *All the frequent subtrees whose number of nodes is k in the FTCB is defined as k-reduction frequent subtrees, denoted as FTCB (k).*

For the generation of the direct subtrees of tree T, there are the following situations:

When $|T| = 2$, only the root node and the leaf node, the two nodes can be seen as direct subtrees of T (Fig. 2).

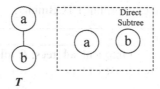

Fig. 2. Direct Subtree of 2- tree

When $|T| \geq 3$, there are two cases:

(1) The root node of T has only one child node. For any node x in T, after pruning x, the subtree formed by the ancestor-descendant relationship of the remaining nodes is retained, and $|T|$ direct subtrees can be generated. As shown in Fig. 3.

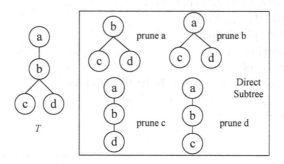

Fig. 3. Direct subtree of k-tree

(2) The root node of T has two or more child nodes. For any non-root node x in T, after pruning x, the direct subtree formed by the ancestor-descendant relationship

of the remaining nodes is retained, and $|T| - 1$ direct subtrees can be generated. As shown in Fig. 4.

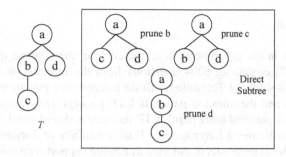

Fig. 4. Direct subtree of k-tree

Definition 15 (Restore Supertree): *Let $T \in FT\backslash FTCB$, all the frequent subtrees whose neighborhood contains T in FTCB is defined as the restoring supertrees of T, denoted as $RT(T)$.*

Proposition 5: *For an arbitrary frequent subtree $T \in FT\backslash FTCB$, $|RT(T)| \geq 1$.*

Definition 16 (Maximum Restore Supertree): *Let $T \in FT\backslash FTCB$ be a frequent subtree. The tree with the largest support in $RT(T)$ is defined as the maximum restoring supertree of T, denoted as $max(T)$.*

Proposition 6: *For an arbitrary frequent subtree $T \in FT\backslash FTCB$. if $Supp(T) = Supp$ $(max(T))$, the support error of the restored subtree is minimal.*

Proof: For an arbitrary subtree $T_i \in FTCB$ whose neighborhood contains T, Supp $(T_i) \leq Supp(max(T)) \leq Supp(T)$ is satisfied. Since the support_error $= Supp(T) - Supp(T_i)$, when restoring, the support error is minimal if $Supp(T) = Supp(max(T))$.

The restoring algorithm RFT is as follows:

Algorithm 2. RFT

Input: FTCB: Frequent Tree Constrict Base.
Output: FT: the collection of all frequent subtrees.
FT←FTCB
for k **from** k_max **to** 2
  for $\forall T \in$ FT(k)
    Dch(T) ← Creat_dch(T)
    for $\forall T_c \in$ Dch(T)
      if $T_c \notin$ FTCB(k-1) **then**
        Supp(T_c) ← Supp(max(T_c))
        FT(k-1) ← T_c
    end for
  end for
end for

The function Create_dch(T) produces a direct subtree set of T, and restores its support when the element in Dch(T) is not in FTCB.

5   Experiment

The tree set used in the experiments is extracted from the Chinese question classification dataset [17] containing 6294 questions from the Information Retrieval Laboratory of Harbin Institute of Technology. Basic frequent tree pattern mining algorithm is TreeMiner [6] and the question parser is LTP package [18]. TreeMiner is downloaded from Zaki's personal homepage[1]. LTP package is downloaded from the website of the Information Retrieval Laboratory of Harbin Institute of Technology[2].

We conduct the experiments in different minimum support thresholds. The first one is the minimum support threshold 0.005 and 10847 frequent subtrees extraced. And the other three minimum support thresholds are 0.004, 0.0035 and 0.0025 respectively.

5.1   FTCB vs. CFT

In this section, we conduct some experiments to show the effects of the representation of FTCB.

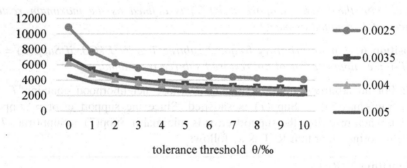

Fig. 5. FTCB vs. CFT

In Fig. 5, curves with different color represent FTCB with different *min_sup*.

As can be seen in Fig. 5, when tolerance threshold $\theta = 0$, FTCB is just the CFT. While the θ increases, the size of FTCB decreases.

5.2   Compression Ratio and Average Error Rate

In this section, we collect the compression ratio and average error rate of FTCB from the experiments.

---

[1] http://www.cs.rpi.edu/~zaki/www-new/pmwiki.php/Software/Software#toc16.

[2] http://www.ltp-cloud.com/.

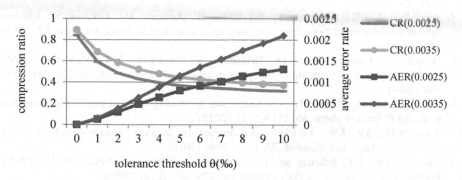

Fig. 6. Compression ratio (CR) and average error rate (AER)

As can be seen from Fig. 6, the value when $\theta = 0$ reflects the compression ratio of the closed frequent subtree set at different minimum support levels. The compression ratio and the average error rate increase with the increase of θ. When θ increases to one percent, the compression ratio reaches 0.31(0.37). At this time, the number of closed subtree set is 2.62(2.38) times that of FTCB, and the average error rate has never exceeded 0.0015(0.0025) during the change process.

6   Conclusion

In this paper, a novel density-based concise representation of frequent subtree pattern, FTCB, is proposed. FTCB can ensure excellent reduction effect and low recovery error rate. FTCBminer, an efficient FTCB mining algorithm, is designed too. By using the conditional independence of random variables, the redundant patterns caused by random noise may be effectively eliminated. The principles and concrete method about this approach will be the next research objectives.

References

1. Wang, S., Hong, Y., Yang, J.: XML document classification using closed frequent subtree. In: Bao, Z., et al. (eds.) WAIM 2012. LNCS, vol. 7419, pp. 350–359. Springer, Heidelberg (2012). https://doi.org/10.1007/978-3-642-33050-6_34
2. Milo, N., Zakov, S., Katzenelson, E., Bachmat, E., Dinitz, Y., Ziv-Ukelson, M.: Unrooted unordered homeomorphic subtree alignment of RNA trees. Algorithms Mol. Biol. 8, 13 (2013)
3. Nguyen, D.P.T., Matsuo, Y., Ishizuka, M.: Relation extraction from Wikipedia using subtree mining. In: National Conference on Artificial Intelligence, pp. 1414–1420 (2007)
4. Jimenez, A.D., Berzal, F., Cubero, J.: Frequent tree pattern mining: a survey. J. Intell. Data Anal. 14, 603–622 (2010)
5. Hao, Z., Huang, C., Cai, R., Wen, W., Huang, Y., Chen, B.: User interest related information diffusion pattern mining in microblog. Pattern Recog. Artif. Intell. 29, 924–935 (2016)
6. Zaki, M.J.: Efficiently mining frequent trees in a forest: algorithms and applications. IEEE Trans. Knowl. Data Eng. 17, 1021–1035 (2005)

7. Asai, T., Abe, K., Kawasoe, S., Sakamoto, H., Arimura, H., Arikawa, S.: Efficient substructure discovery from large semi-structured data. IEICE Trans. Inf. Syst. **87**, 2754–2763 (2004)
8. Deepak, A., Fernández-Baca, D., Tirthapura, S., Sanderson, M.J., McMahon, M.M.: EvoMiner: frequent subtree mining in phylogenetic databases. Knowl. Inf. Syst. **41**, 559–590 (2014)
9. Zhang, S., Du, Z., Wang, J.T.: New techniques for mining frequent patterns in unordered trees. IEEE Trans. Cybern. **45**, 1113–1125 (2015)
10. Tian, W.D., Xu, J.W.: Concise representation of frequent itemset based on fuzzy equivalence. Appl. Res. Comput. **33**, 1936–1940 (2016)
11. Xiao, Y., Yao, J.-F.: Efficient data mining for maximal frequent subtrees. In: Third IEEE International Conference on Data Mining, pp. 379–386. IEEE (2003)
12. Chi, Y., Yang, Y., Xia, Y., Muntz, R.R.: CMTreeMiner: mining both closed and maximal frequent subtrees. In: Dai, H., Srikant, R., Zhang, C. (eds.) PAKDD 2004. LNCS (LNAI), vol. 3056, pp. 63–73. Springer, Heidelberg (2004). https://doi.org/10.1007/978-3-540-24775-3_9
13. Yang, P., Tan, Q.: Maximum frequent tree mining and its applications. Comput. Sci. **35**, 150–153 (2008)
14. Termier, A., Rousset, M.-C., Sebag, M.: DRYADE: a new approach for discovering closed frequent trees in heterogeneous tree databases. In: Fourth IEEE International Conference on Data Mining (ICDM 2004), pp. 543–546. IEEE (2004)
15. Feng, B., Xu, Y., Zhao, N., Xu, H.: A new method of mining frequent closed trees in data streams. In: 2010 Seventh International Conference on Fuzzy Systems and Knowledge Discovery, pp. 2245–2249. IEEE (2010)
16. Wang, T., Lu, Y.S.: Mining condensed frequent subtree base. J. SE Univ. **22**, 48–53 (2006)
17. Yang, S.C.: Research on Question Classification for Chinese Question Answering System. Nanjing University, Nanjing (2013)
18. Che, W., Li, Z., Liu, T.: LTP: a Chinese language technology platform. In: Proceedings of the 23rd International Conference on Computational Linguistics: Demonstrations, pp. 13–16. Association for Computational Linguistics (2010)

Ranking Research Institutions Based on the Combination of Individual and Network Features

Wenyan Wang[1], Guangyu Wang[1], Jun Zhang[2], Peng Chen[2], and Bing Wang[1,2]($\boxtimes$)

[1] School of Electrical and Information Engineering,
Anhui University of Technology, Maanshan 243002, People's Republic of China
wangbing@ustc.edu
[2] The Institute of Health Sciences, Anhui University, Hefei 230601,
Anhui, China

Abstract. Regarding the fierce competition between research institutions, institutional rankings are widely carried out. At present, there are many factors affecting the ranking of institutions, but most of them are aimed at the attributes of the institutions themselves, and the feature selection is relatively simple. Therefore, this paper proposes a state-of-the-art method combining different types of features for predicting the influence of scientific research institutions. Based on the MAG dataset, this paper first calculates the institutional scores through the publication volume of the article, constructs an inter-institutional cooperation network, and calculates the importance characteristics of the institutions in the network. Then, considering the contribution of the faculty and staff to the organization, an individual characteristic based on the author's influence is constructed. Finally, a random forest algorithm is used to solve this prediction problem. As a result, this paper raises the ranking accuracy rate NDCG@20 to 0.865, which is superior to other methods. The experimental results show that this method has a good effect on the prediction of innovation capability.

Keywords: Institutional ranking · Feature combination · MAG · Random forest

1 Introduction

In recent years, the activities of ranking research institutes have flourished and developed, which is the product of the development of research institutes to a certain stage [1, 2]. However, the development of ranking research institutions is not perfect since most ranking methods are static and using only part of indicators. The KDD Cup 2016 would like to galvanize the community to address this very important problem through any publicly available datasets, like the Microsoft Academic Graph (MAG). The Microsoft Academic Graph is a heterogeneous graph containing scientific publication records, citation relationships between those publications, as well as authors, institutions, journals, and fields of study [3].

© Springer Nature Switzerland AG 2019
D.-S. Huang et al. (Eds.): ICIC 2019, LNAI 11645, pp. 443–454, 2019.
https://doi.org/10.1007/978-3-030-26766-7_41

For institutional rankings, good features is the key to reliable ranking results. Currently, there are about three major types of features were established. Firstly, based on institution attributes, Gupta et al. only use the feature of the affiliation score calculated from the amount of article received at a meeting to make predictions [6, 7]. Wilson et al. firstly classified the papers to determine whether the paper belonged to the full research papers or to all papers. The motivation for this is that some conferences listed by kdd cup 2016 receive only the full research papers [8–10]. In addition, the total number of papers of each institution is also used as a unique feature that replaces the affiliation score [9, 11]. Similarly, the inherent property of an institution, such as name and location of affiliations, state GDP, etc., can also be used as features of the institution. Last but not least, the relations between institutions were also taken into consideration to represent the importance of institutions [7, 10–12]. Secondly, paper features were established to demonstrate the paper's influence on the institution. Such as the number of papers, the trend of paper number and track record describing the publication history of an institution, and so on [4, 5, 8]. Lastly, the individual characteristics of the author had also been excavated. Such as, the feature of active degree measuring the institution active degree in one conference according to the number of active authors in the institution, the continuity evaluate feature representing how the research filed of an institution is insisting on were designed [5], and number of first, second authors and (author, paper) pair can also make a contribution to the assessment of institutional influence. What's more, the network of authors is also established to highlight the more influential authors [13–15]. In fact, in order to expand the limited availability of data sets, information about some related meetings can also be grouped together as a common feature [12, 13, 15].

In this work, although the scores of each institutions from 2011 to 2015 can be calculated directly from the dataset provided by the KDD Cup, it is not ideal to predict the institution score of 2016 by only using this single feature. On the issue of extreme lack of institutional information, it is urgent to construct more characteristics that describe the attributes of institutions. Therefore, in this paper, the individual characteristics based on the author and the network characteristics based on the institution were constructed from the micro and macro perspectives, and these features were treated as datasets and applied to random forest models to solve institutional ranking problems. As a result, our work raises the ranking accuracy NDCG@20 to 0.865, which outperformed current methods and the complementary information between the features had been proved.

2 Materials and Feature Generation

2.1 Dataset

Although the organizers of KDD Cup encourage the participants to use any publicly available information, they do provide us with Microsoft Academic Graph (MAG). The data can be downloaded from the website of http://aka.ms/academicgraph and the version "2016-02-05" is available. All the data is uniquely identified by the primary key, such as the conference ID, affiliation ID, etc. Take the year of 2015 as an example, some of the data in the MAG dataset shown in Table 1.

Table 1. Examples of MAG dataset.

Paper	Year	Author
76381EFA	2015	8084BB24
76381EFA	2015	75421677
76381EFA	2015	832818A2
716E3093	2015	10C312AC
8389DBD7	2015	112ACB8A

2.2 Feature Generation

2.2.1 Affiliation Score

According to the cooperation among authors, institutions and papers displayed in dataset, as shown in Table 1, and following the simple policy specified by organizers. The most direct and effective affiliation score can be calculated firstly since it has a fundamental effect on the predicting institutions future scores and it's the easiest to build in an existing data set. Figure 1 shows the method of calculating the affiliations score.

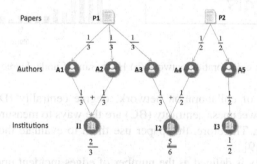

Fig. 1. Example of institutional score calculation method

2.2.2 Network Features

After statistics on the numbers of articles that belonging to one or more institutions published in the KDD conference from 2011 to 2015, as shown in Table 2, we found that articles written by multiple affiliations accounted for a large proportion. So we take the year of 2015 as an example, constructing the co-operative networks and the node degree distribution histogram, as shown in Fig. 2. In the network, each node (red circle) represents an affiliation, and they are connected by a straight line.

From the Fig. 2(a), we can see that some nodes occupy the central position in the network and have obvious importance. And in Fig. 2(b), it shows that the distribution of the node degree satisfies the power law distribution so that the graph is a scale-free network [16–18]. Based on these two factors, the collaboration networks from 2011 to 2015 were established.

Table 2. The Number of papers in KDD.

Year	Multiple affiliation	Single affiliation	Total
2011	71	79	150
2012	73	58	131
2013	67	58	125
2014	79	68	147
2015	100	60	160

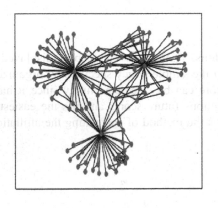 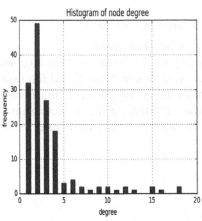

Fig. 2. (a) Affiliation collaboration network (b) Histogram of node degree (Color figure online)

In the analysis of collaboration network, degree centrality (DC), closeness centrality (CC) and betweenness centrality (BC) are the ways to measure the importance of nodes in a network. Therefore, this paper use them to evaluate the institutional innovation capability [19].

Degree centrality is defined as the number of edges incident upon a node. In order to make a comparison between different scale networks, the degree centrality of node i was calculated using (1). Closeness centrality reflects the degree of the node to the network center in the whole network structure. Using (2), the closeness centrality of node i was calculated. And betweenness centrality is defined as the more number of shortest paths through a node, the greater important of this node in the shortest path of all node pairs. Using (3), the betweenness centrality of node i was calculated.

$$DC(i) = \frac{k_i}{N - 1} \tag{1}$$

$$CC(i) = \frac{N - 1}{\sum_{j \neq i} d_{ij}} \tag{2}$$

$$BC(i) = \sum_{s \neq i \neq t} \frac{g_{st}^i}{g_{st}} \tag{3}$$

Where $k_i = \sum_{i=1}^{N} a_{ij}$, N is the number of all nodes, $N - 1$ is the maximum possible edges value of the node, and a_{ij} is an element with i row j column of the adjacency matrix, d_{ij} is the shortest path between node i and j. g_{st} is the total number of shortest path from node s to node t, and g_{st}^i is the number of shortest path through node i in the shortest path from node s to t [19, 20].

2.2.3 Individual Features of the Author

Inspired by the cooperation between affiliations, author collaboration network has also been established. In order to highlight the importance of the author, the number of cooperations between authors is used as the weight of the connection line.

In the author's collaboration network, this paper first assumes that the author's importance is only related to the authors in the two nearest layers of the network. For each author, the collaboration score can be expressed as

$$ACS(A_i) = \frac{N_1}{N_2} * \left(\sum_{j \in layer1} W_j + \sum_{j \in layer2} W_k \right) \tag{4}$$

Where $ACS(A_i)$ is the author collaboration score of $A_i N_1$ and N_2 are the number of authors included in the first and second layers, respectively. W_j, W_k are the weights corresponding to the author in the first and second layers.

The probability of an author publishing paper (PAPP) in the year $(t + 1)$ is computed by taking the ratio of total number of years the author has published papers in the past to total number years under consideration (t). The probability score thus calculated is a binomial probability for publishing paper in year $(t + 1)$.

$$P_{a,t+1} = \frac{\sum_{i=1}^{t} x_{a,i}, \, for \, x_{a,i} = 1}{t} \tag{5}$$

The Joint probability 1 (JP1) for publishing paper in year (t) and in the year $(t + 1)$ is computed by taking the ratio of the frequency of paper published in consecutive years to the sum of frequencies of publishing paper in year (t) and year $(t + 1)$ plus publishing paper in year (t) but not in $(t + 1)$.

$$p(x_{a,t+1} = 1, x_{a,t} = 1)$$

$$= \frac{\sum_{i=1}^{t} x_{a,i}, for \, x_{a,i} = 1 \ and \ x_{a,i} = 1}{(\sum_{i=1}^{t} x_{a,i}, for \, x_{a,i} = 1 \ and \ x_{a,i-1} = 1) + (\sum_{i=1}^{t} x_{a,i}, for \, x_{a,i} = 0 \ and \ x_{a,i-1} = 1)} \tag{6}$$

The Joint probability 2 (JP2) for not publishing paper during the year (t) but publishing paper in the year ($t + 1$) is computed by taking the ratio of the frequency of not publishing paper in the year (t) and publishing paper in the year ($t + 1$) to the sum of frequencies of publishing paper in year (t) and year ($t + 1$) plus not publishing paper in year (t) but in ($t + 1$).

$$P(x_{a,t+1} = 1, x_{a,t} = 0) = \frac{\sum_{i=1}^{t} x_{a,i} = 1, for\, x_{a,i} = 1 \text{ and } x_{a,i-1} = 0}{\left\{ (\sum_{i=1}^{t} x_{a,i}, for\, x_{a,i} = 1 \text{ and } x_{a,i-1} = 1) + (\sum_{i=1}^{t} x_{a,i}, for\, x_{a,i} = 1 \text{ and } x_{a,i-1} = 0) \right\}} \tag{7}$$

Where $x_{a,i} = 1$ if the author published a paper in that year, else $x_{a,i} = 0$, t represents a time range under consideration.

3   Individual Features of the Author

3.1   Data Normalization

In this work, each sample was represented by three type of features. However, these features contain different physical meanings or data ranges. So, in order to eliminating the impact of imbalanced information expression, all of features have to be normalized before applying to the model. Here, all values of each feature always fall within a fixed interval [0, 1] by

$$x^* = \frac{x - x_{min}}{x_{max} - x_{min}} \tag{8}$$

Where x_{max}, x_{min} represent the maximum, minimum of variables, respectively.

3.2   Random Forest Regression Model

Random forest uses the bootstrap resampling method to extract multiple samples from the original sample and building a model of decision tree for each bootstrap sample. This method was proved to have higher prediction accuracy, good tolerance to outliers and noise and it is not easy to over-fitting [20, 21].

The fitting process of a single decision tree in the random forest algorithm is as follows:

(a) Bagging sampling is used to form the training set that equal to the original sample number.

(b) The feature is chosen randomly when the internal node splits.

(c) Don't prune every tree in the forest, let it grow randomly [22].

Suppose the input space is divided into M units $R_1, R_2, \ldots, R_m$, and each unit R_m has a fixed output value c_m, the model of regression tree can be expressed as below:

$$f(x) = \sum_{m=1}^{M} c_m I(x \in R_m) \tag{9}$$

In this paper, through the grid search algorithm, the optimal number of trees and the ratio between the features required for each tree and the total features are found, 150 and 0.7 respectively, and there has not been a fitting phenomenon, so the pruning operation is not used in this work.

3.3 Evaluation Strategy

Normalized Discounted Cumulative Gain (NDCG) is one of the most popular ranking indicators. According to the requirements of the organizers, we only need to focus on the top 20 institutions. On the basis of this notion, NDCG@20 was adapted to measure the relevance and it can be computed as follows [23]

$$DCG@20 = \sum_{i=1}^{20} \frac{rel_i}{\log_2(1+i)}$$
$$NDCG@20 = \frac{DCG@20}{Ideal\,DCG@20} \tag{10}$$

Where i is the rank of an institution, and rel_i is this institution's relevance score.

4   Experimental Results and Discussion

4.1   Features and Samples Selection

For features, affiliation score are considered to be inherent features and they are combined with individual and network features respectively as authors' individual feature and institutional centrality feature. At the same time, the individual and network features are combined, called the fusion feature. Finally, three kinds of feature sets were constructed as shown in Table 3.

Table 3. Features in training and testing set.

Features				Time intervals
Fusion features	Institutional centrality feature	Network centrality	BC	2011–2015
			CC	2011–2015
			DC	2011–2015
	Individual feature	Affiliation score	AS	2011–2015
		Authors' individual	ACS	2011–2015
			JP1	2014–2015
			JP2	2014–2015
			PAPP	2014–2015

4.2   Results

Although this paper generates a variety of features, it is not clear whether they are really related to the influence of the organization. To verify this problem, the correlation coefficients between each feature and the corresponding institutional score (AS) for 2011-2015 were calculated and shown in the Table 4.

Table 4. The correlation coefficient between each feature and the corresponding institutional score

Year	BC	CC	DC	JP2	JP1	ACS	PAPP
2011	0.71	0.49	0.73			0.96	
2012	0.66	0.56	0.75			0.91	
2013	0.64	0.57	0.59			0.91	
2014	0.65	0.6	0.64	0.93	0.68	0.84	0.94
2015	0.75	0.47	0.74	0.94	0.76	0.91	0.95

From the data in Table 4, the correlation coefficient between the closeness centrality (CC) and the institutional scores in 2011-2015 are less than or equal to 0.6. The correlation coefficients of ACS were distributed around 0.9, which was significantly higher than the closeness centrality. In order to obtain the effect of each feature and its correlation coefficient on the forecasting influence of the institutions, this paper first uses the network features and fusion features including closeness centrality and not including closeness centrality to predict the institutional influence, respectively. The violin chart of 10 prediction results and average values are shown in Fig. 3(a). In addition, this paper attempts to make the effect of the betweenness centrality on the predictive mechanism's influence. The result is shown in Fig. 3(b). Where fusion feature-CC represents the prediction result after the closeness centrality is removed, fusion feature-BC represents the result after the closeness centrality is removed and betweenness centrality is also removed, and InCe in Fig. 3(b) represents the institution centrality feature. It can be seen from Fig. 3 that after removing the closeness central characteristics, the prediction accuracy and robustness of using only the institution centrality feature are improved, and the performance of using the fusion feature prediction is also improved. From Fig. 3(a), it can be seen that the average accuracy of the institution centrality and fusion feature is increased by about 3%, the height of the fusion feature-CC in the violin diagram is significantly lower, and the In Ce-CC is also shown the same performance in the violin diagram of the Fig. 3(b), which indicates the closeness centrality established in this paper does not apply to institutional impact prediction. However, from Fig. 3(b), it can be found that the removal of the betweenness centrality is the opposite of the removal of the closeness centrality. Although the stability of the fusion feature is increased, the prediction accuracy of the institution centrality and the fusion feature are reduced by 4% and 1% respectively.

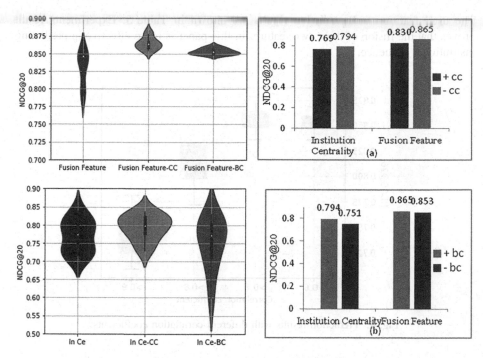

Fig. 3. Prediction results using different time periods.

In Table 4, after processing the features longitudinally with correlation coefficient values, this paper sorts all the generated features by the value of the feature correlation coefficients, and eliminates the features with correlation coefficients less than 0.6, 0.7, 0.8, and 0.9. The remaining features are used for prediction to observe the influence of each feature in the horizontal direction on the ranking prediction. The box plot of the predicted result is shown in Fig. 4. In the box plot, a line in the middle of the box indicates the median of the data. The upper and lower lines of the box indicate the upper quartile and the lower quartile of the data. A line above and below the box represents the maximal and minimum values, the circle represents the outlier. It can be seen from Fig. 3 and Table 4 that the smaller the feature correlation coefficient of the removal, the larger the prediction accuracy and the more stable. In other words, the more the feature amount used, the higher the prediction accuracy, the more robust. This further proves that there is information complementation between the features con-structed in this paper, and the fusion features have the best predictive performance. Therefore, this paper will use all the features except the closeness centrality to predict the influence, and the prediction accuracy is 0.865.

Finally, the central, individual and fusion features of the institutions are respec-tively applied to the random forest model, and the predicted results are shown in Fig. 5. In order to prove the validity of the confused features in this paper, we compare the prediction results with those who did the same work as described in the introduction.

The used features and prediction results are shown in Table 5. Experiments result proves that the fusion of these two features in this paper is more effective in predicting institutional influence.

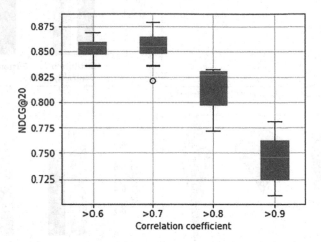

Fig. 4. Prediction results with different correlation coefficients.

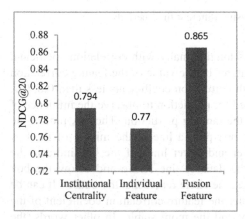 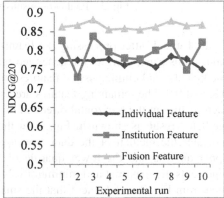

Fig. 5. Prediction results with three types of features.

Table 5. Comparison of results.

Method	Features	Prediction result
Paper 1 [11]	Network central features + Human knowledge	0.794
Paper 2 [14]	Individual features	0.77
This paper	Network central feature + Individual feature	0.865

5   Conclusion

In order to improve the prediction accuracy of scientific research institutions' innovative ability, a novel combined the individual and network attributes method had been proposed in this paper, which is based on the random forest algorithm. By analyzing the correlation coefficient between the constructed feature and the institutional score, As a result, this paper uses all features except closeness centrality, and the central, individual and fusion features of the institutions are respectively applied to the random forest model to realize the prediction of innovation ability of scientific research institutions. Experimental results demonstrate that the feature generate method proposed in this paper has a great effectiveness in both accuracy and stability of the institution influence prediction, and the information between the features is complementary.

Acknowledgement. This work is supported by the National Natural Science Foundation of China (Nos. 61472282, 61672035, and 61872004), Anhui Province Funds for Excellent Youth Scholars in Colleges (gxyqZD2016068), the fund of Co-Innovation Center for Information Supply & Assurance Technology in AHU (ADXXBZ201705), and Anhui Scientific Research Foundation for Returned Scholars.

References

1. Cuthbert, R.: University rankings, diversity, and the new landscape of higher education. Int. J. Lifelong Educ. **30**, 119–121 (2011)
2. Szentirmai, L., Radacs, L.: World university rankings qualify teaching and primarily research. In: IEEE International Conference on Emerging Elearning Technologies and Applications, pp. 369–374 (2013)
3. Sinha, A., et al.: An overview of microsoft academic service (MAS) and applications. In: International Conference on World Wide Web, pp. 243–246 (2015)
4. Mussard, M., James, A.P.: Engineering the global university rankings: gold standards, its limitations and implications. IEEE Access **PP**, 1 (2018)
5. Al-Juboori, A.F.M.A., Su, D.J., Ko, F.: University ranking and evaluation: trend and existing approaches. In: The International Conference on Next Generation Information Technology, pp. 137–142 (2011)
6. Gupta, A., Murty, M.N.: Finding influential institutions in bibliographic information networks (2016)
7. Orouskhani, Y., Tavabi, L.: Ranking research institutions based on related academic conferences. arXiv e-prints (2016)
8. Wilson, J., Mohan, R., Arif, M., Chaudhury, S., Lall, B.: Ranking academic institutions on potential paper acceptance in upcoming conferences (2016)
9. Sandulescu, V., Chiru, M.: Predicting the future relevance of research institutions - the winning solution of the KDD Cup 2016 (2016)
10. Zhang, J., Xu, B., Liu, J., Tolba, A., Al-Makhadmeh, Z., Xia, F.: PePSI: Personalized prediction of scholars' impact in heterogeneous temporal academic networks (2018)
11. Klimek, P.S., Jovanovic, A., Egloff, R., Schneider, R.J.S.: Successful fish go with the flow: citation impact prediction based on centrality measures for term–document networks. Scientometrics **107**(3), 1265-1282 (2016)

12. Xie, J.: Predicting institution-level paper acceptance at conferences: a time-series regression approach (2016)
13. Qian, Y., Dong, Y., Ma, Y., Jin, H., Li, J.: Feature engineering and ensemble modeling for paper acceptance rank prediction (2016)
14. Moed, H.: Bibliometric rankings of world universities (2006)
15. Bai, X., Zhang, F., Hou, J., Xia, F., Tolba, A., Elashkar, E.: Implicit multi-feature learning for dynamic time series prediction of the impact of institutions. IEEE Access **PP**, 1 (2017)
16. Crucitti, P., Latora, V., Marchiori, M., Rapisarda, A.: Error and attack tolerance of complex networks. Nature **340**, 378–382 (2000)
17. Holme, P., Edling, C.R., Liljeros, F.: Structure and time evolution of an Internet dating community. Soc. Netw. **26**, 155–174 (2004)
18. Barabási, A.L., Jeong, H., Néda, Z., Ravasz, E., Schubert, A., Vicsek, T.: Evolution of the social network of scientific collaborations. Physica A Stat. Mech. Appl. **311**, 590–614 (2002)
19. Ren, X., Lü, L.: Review of ranking nodes in complex networks. Chin. Sci. Bull. **59**, 1175 (2014)
20. Belgiu, M., Drăguţ, L.: Random forest in remote sensing: a review of applications and future directions. ISPRS J. Photogramm. Remote Sens. **114**, 24–31 (2016)
21. Désir, C., Bernard, S., Petitjean, C., Heutte, L.: One class random forests. Pattern Recogn. **46**, 3490–3506 (2013)
22. Zhou, Z.H.: Ensemble learning. In: Encyclopedia of Biometrics, pp. 270–273 (2009)
23. Wang, Y., Wang, L., Li, Y., He, D., Liu, T.Y., Chen, W.: A theoretical analysis of NDCG type ranking measures. J. Mach. Learn. Res. **30**, 25–54 (2013)

Chinese Temporal Expression Recognition Combining Rules with a Statistical Model

Mengmeng Huang[1], Jiazhu Xia[1], Xianyu Bao[2],
and Gongqing Wu[1(✉)]

[1] School of Computer Science and Information Engineering,
Hefei University of Technology, Hefei 230601, China
wugq@hfut.edu.cn
[2] Shenzhen Academy of Inspection and Quarantine, Shenzhen 518045, China

Abstract. Traditional rule-based methods for recognizing Chinese temporal expressions present a lower recall rate and they cannot recognize the event-type Chinese temporal expressions, thus, we propose a new Chinese temporal expression recognition method through combining rules with a statistical model. Firstly, we divide Chinese temporal expressions into seven categories and use basic time units as the smallest unit of recognition to simplify the complexity of rule-making. Then, we use regular rules to recognize Chinese temporal expressions and label the training data automatically. Meanwhile, we label the event-type temporal expressions that rule-based method cannot recognize. Lastly, we use the labeled training data to learn a Conditional Random Fields model for Chinese temporal expression recognition. Experimental results show that our proposed method significantly reduces the amount of annotation work and effectively improves the recognition performance. The F1 value reaches 88.73%, which is higher than the rule-based method by 6.13%.

Keywords: Chinese temporal expression · Regular expression · Basic time unit · Semantic role · Conditional random fields

1 Introduction

Internet has become an indispensable part of people's lives. People browse the news website for fresh information and use social media to share interesting things around them. Human network activities generate hundreds of millions of text messages. Facing massive amounts of information, using automatic event detection and analysis techniques to find useful information is particularly important. Time series is the basic element of event causality [1], and events can be understood by the time structure. Time information is one of the important elements of an event, which can be used for event detection and tracking. Therefore, automatic recognition of temporal expressions in texts has significant theoretical and practical value.

As an important task in the field of natural language processing and information retrieval, temporal expression recognition has attracted much attention in recent years. In 1995, the Message Understanding Conference first listed the temporal expression recognition task as a subtask for named entity recognition. In 2004, the National

© Springer Nature Switzerland AG 2019
D.-S. Huang et al. (Eds.): ICIC 2019, LNAI 11645, pp. 455–467, 2019.
https://doi.org/10.1007/978-3-030-26766-7_42

Institute of Standards and Technology (NIST) organized the first temporal expression recognition and normalization evaluation. ACE2005 (Automatic Content Extraction) and SemEval2007 (Semantic Evaluation) also included temporal expression recognition into their evaluation tasks. The TERQAS conference presented TimeML [2] in 2002 to provide a standard for event and time related research, and TimeML eventually became an ISO standard in 2007. On the basis of studying the event-oriented knowledge representation, Liu et al. [3] proposed an event topic model and listed the time element as one of basic elements of the model. Ge et al. [4] constructed a burst information network using keywords together with time information, and used graph clustering to detect events, which overcame the problem of cluster center offset. Du et al. [5] used time information and tag information for event sequence modeling, and used recurrent neural networks to predict events and achieved good results.

In summary, recognizing temporal expressions from texts automatically is an important task. But traditional methods have some drawbacks. Rule-based methods can achieve a high precision, but due to the variety of Chinese temporal expressions, it is difficult to develop a complete set of rules to recognize all types of Chinese temporal expressions, which leads to a low recall rate. Statistical models have good generalization capabilities, thus, statistics-based methods tend to achieve a high recall rate in recognizing temporal expressions. But the training data need to be labeled manually, which is time consuming. In view of the advantages and disadvantages of the above two methods, we propose a new Chinese temporal expression recognition method combining rules and statistics in this paper.

The contributions of this paper are below: (1) We divide Chinese temporal expressions into seven categories. Regular rules are formulated based on basic time units, which can reduce the complexity of rule-making. (2) Chinese temporal expressions with the regular rules are recognized to label the training data automatically, and then the labeled training data are used to train a CRF model. (3) Event-type Chinese temporal expressions are manually labeled, which they cannot be recognized by the rule-based recognition method. And meanwhile, the semantic role is introduced to construct feature vectors, which can significantly improve the recognition effect of event-type Chinese temporal expressions.

The structure of this paper is as follows: Sect. 2 introduces the related work in the field of Chinese temporal expression recognition. Section 3 introduces basic concepts involved in the paper. Section 4 first introduces the overall framework of the Chinese temporal expression recognition method combining rules and statistics, and then introduces the implementation details of these two modules. Section 5 describes the detail of experiments and analyzes experimental results. Section 6 finally summarizes this paper.

2 Related Work

Existing Chinese temporal expression recognition methods are mainly classified into two categories: rule-based methods and statistics-based methods. On one hand, since the format of Chinese temporal expressions is relatively fixed, rule-based methods have achieved good results in many occasions. More precisely, He et al. [6] proposed a method for recognizing temporal expressions based on dependency parsing and error-driven learning, which began with the time trigger word (namely, the syntactic head of dependency relation) and used Chinese dependency parsing to recognize temporal expressions. Subsequently, the transformation-based error-driven learning was used to improve the performance. Lin et al. [7] studied TIMEX2 [8] manual on temporal expression annotation, and designed a temporal expression labeling system based on regular rules. Wu et al. [9] proposed the concept of basic time unit. In the paper, rules were generated based on basic time unit firstly, which improved the recall rate. Then, the rules obtained by the error-driven method are pruned to reduce the "noise" in the training data, which can lead to a high precision.

On the other hand, statistics-based methods present a better generalization ability due to using statistical models and achieve a better recall rate due to the variety of temporal expressions. Conditional Random Fields (CRF) is one of the most used methods for recognizing temporal expressions. For instances, Zhu et al. [10] proposed a Chinese temporal expression recognition method based on Conditional Random Fields. Temporal expressions were divided into two types: time-type and event-type. The natural language processing tool was used for tokenization and feature extraction. Three vocabularies were summarized to label training data. The method achieved good recognition results in both types of temporal expressions. Liu et al. [11] introduced the semantic role as an artificial feature into temporal expression recognition, which can improve the recognition effect of temporal expressions. Wu et al. [12] considered that temporal expressions were consisted of time units and time affixes. First, the trained CRF model is used to recognize time units automatically, which can effectively utilize the generalization ability of statistical model. Then, a time prefix vocabulary and a time suffix vocabulary are summarized. Finally, time units were merged with time affixes into complete temporal expressions. Yan et al. [13] proposed a method based on the CRF model and the self-training semi-supervised learning, which achieved a good performance.

3 Preliminaries

3.1 Chinese Temporal Expressions

A temporal expression is a time series consisting of time units, which can be time points, time periods, or frequencies. Temporal expressions [10] are divided into time-type and event-type, in which time-type expressions contain explicit time words, such as "6月20日/June 20th", "10年前/10 years ago", and event-type expressions is specified by an event and does not contain explicit time words, such as "地震发生后/after the earthquake" and "李克强总理访问上海时/When Premier Li Keqiang visited Shanghai". Task 13 of SemEval 2010 [14] classified temporal expressions into four categories: DURATION, SET, TIME and DATE. On the basis of SemEval2010, Wu et al. [12] added three new categories based on the characteristics of Chinese temporal expressions. This paper combines the above research results and classifies Chinese temporal expressions into the following seven basic categories:

- **DATETIME**: indicates accurate time. e.g. 2008年5月12日/May 12, 2008, 14时28分/14:28
- **DURATION**: indicates time periods. e.g. 两个星期/two weeks
- **SET**: indicates frequencies. e.g. 每两天/every two days
- **LUNAR**: indicates festivals. e.g. 春节/Spring Festival, 圣诞节/Christmas
- **FUZZY**: indicates fuzzy time. e.g. 数十年/tens of years, 目前/currently
- **RELATIVE-TIME**: indicates relative time. e.g. 明天/tomorrow, 去年/last year
- **EVENT-TIME**: indicates event-type time. e.g. 地震发生后/after the earthquake, 火灾发生时/when the fire broke out

3.2 Basic Time Unit

Temporal expressions are generally composed of smaller units with independent semantics. Wu et al. [9] refer to such minimum units as "basic time unit". For example, the temporal expression. "2008年8月8日晚上8时/8:00 evening on August 8, 2008" is consist of basic time units "2008年/2008", "8月/August", "8日/8th", "晚上/evening" and "8时/8 o'clock", each basic time unit is a complete semantic unit.

Many rule-based temporal expression recognition methods make rules for entire temporal expressions. Due to the flexibility of temporal expressions, it is difficult to write a set of rules to recognize all forms of temporal expressions, and the rules tend to be very redundant. Although the form of temporal expressions is variable, basic time units that make up temporal expressions are relatively fixed. We can first recognize the basic time units, and then combine these time units into a complete temporal expression. The method can effectively reduce the complexity of the rule-making and achieve a high recall rate.

3.3 Conditional Random Fields

Conditional Random Fields (CRF) is a probability graph model. This model assumes that for a set of random inputs, the corresponding output is an undirected graph. For the first time, Lafferty et al. [15] applied linear Conditional Random Fields into labeling problems. Let P(Y|X) be a linear CRF, and the conditional probability distribution of the random variable Y under the condition that the random variable X takes x is as shown in Eqs. (1) and (2):

$$P(y|x) = \frac{1}{z(x)} \exp\left(\sum_{i,k} \lambda_k t_k(y_{i-1}, y_i, x, i) + \sum_{i,l} \mu_l s_l(y_i, x, i) \right) \tag{1}$$

$$z(x) = \sum_{y} \exp\left(\sum_{i,k} \lambda_k t_k(y_{i-1}, y_i, x, i) + \sum_{i,l} \mu_l s_l(y_i, x, i) \right) \tag{2}$$

where t_k and s_l are eigenfunctions, λ_k and μ_l are the corresponding weights, $z(x)$ is the normalization factor, and the summation is performed on all possible output sequences.

Temporal expression recognition can be transformed into a labeling problem, and the mentioned linear CRF above is used to automatically label temporal expressions. Given a conditional random field P(Y|X), for an input sequence x, CRF model performs the labeling of the sequence by calculating the output sequence y with the highest probability. CRF has no strong independent assumptions like the Hidden Markov Model [16], so it can adapt to a wide variety of contexts. In addition, it calculates the global optimal output sequence, overcoming the labeling bias problem of the maximum entropy model.

3.4 Semantic Role

Semantic role labeling (SRL) [17] is a kind of shallow semantic analysis technology, which is a simplification of semantic analysis. Semantic analysis is a method to formalize sentences based on the grammatical structure of sentences and the meaning of word. For example, "Li Ming eats apples" and "Apples are eaten by Li Ming", the result of formal analysis of semantic analysis is: eating (Li Ming, Apples). Semantic analysis can help researchers understand natural language. Unfortunately, after years of efforts, semantic analysis technology has not made breakthroughs.

Gildea et al. [18] proposed an empirical semantic role labeling method. Unlike semantic analysis, this method only labels the semantic roles played by the sentence components relative to the verb in a sentence, without detailed analysis of the entire sentence. Typical semantic roles include acting, receiving, time and place. The Language Technology Platform (LTP) [19] developed by the Harbin Institute of Social Computing and Information Retrieval Center provides excellent semantic role labeling. The core semantic roles of the LTP is A0-5. A0 usually represents the actuator of the action. A1 usually indicates the influence of actions. A2-5 has different semantic meanings depending on the predicate verbs. The remaining 15 semantic roles are additional semantic roles, such as LOC for location, TMP for time, and so on. For example, the labeled result of a temporal expression "接到报警后/when the police received the alarm, 消防人员火速赶到火灾现场/the firefighters rushed to the scene of the fire." is shown as Fig. 1. The verb in the sentence is "赶到/rushed to", and other components in the sentence are labeled with respect to the verb. Among them, "after receiving the alarm" is an event-type Chinese temporal expression, which is labeled as TMP successfully.

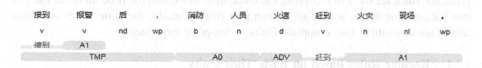

Fig. 1. An example of semantic role labeling

4 Our Method for Chinese Temporal Expression Recognition

4.1 System Architecture

This paper proposes a new Chinese temporal expression recognition method combining rules and statistics. As shown in Fig. 2, we first formulate the regular rules based on basic time units. Then, the rules are used to recognize Chinese temporal expressions automatically in the training data. At the same time, we manually label event-type Chinese temporal expressions. Finally, we extract features to construct feature vectors, and train a CRF model using the labeled training data to recognize Chinese temporal expressions. Details of the techniques are as follows.

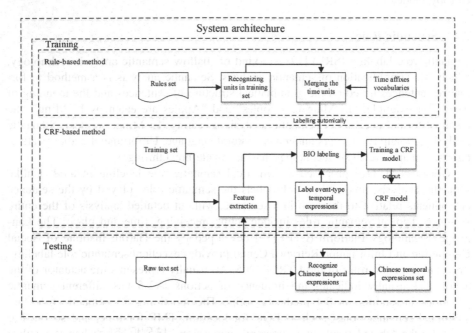

Fig. 2. System architecture

4.2   Chinese Temporal Expression Recognition Based on Rules

The method considers that Chinese temporal expressions consist of basic time units and time affixes. As shown in Fig. 2, first, regular rules are formulate for basic time units. Then, the rules are used to recognize the basic time units, and the time units are merged into longer time series according to merging rules. Finally, the time series and time affixes are assembled into complete Chinese temporal expressions.

4.2.1   Regular Rules Based on Basic Time Units

In order to facilitate the operation, treat the tokens of Chinese temporal expressions as basic time units. For example, "2008年/2008", "5月/May", "12日/12th", "下午/afternoon" and "14时/14 o'clock" are the tokens of the Chinese temporal expression "2008年5月12日下午14时/14 o'clock on the afternoon of May 12, 2008". Next, classify the time units into several categories and formulate regular rules for each category. For example, the basic time units "2008年/2008" and "08年/08 years" can be divided into a category, the corresponding regular rule is "{2, 4}/d |year".

In addition, this paper also treats the LUNAR and the RLATIVE-TIME Chinese temporal expressions in Sect. 3.1 as basic time units. They are limited in number and relatively fixed in format. We manually summarize two vocabularies and construct regular rules for them.

4.2.2   Merging Rules of Basic Time Units

Merging rules of basic time units aim to merge the units that are less than 3 Chinese characters apart. In view of the diversity of temporal expressions, it is difficult to formulate a complete set of rules to recognize all basic time units. In addition, many time units have different semantics in different contexts. In order to ensure the precision of recognition, we cannot treat this kind of tokens as time units. The above two factors probably lead to the problem that some basic time units cannot be correctly identified. Among the basic time units that make up a temporal expression, if a time unit is not successfully recognized, it will cause a single temporal expression to be incorrectly recognized as multiple ones. Through a large number of observations, we obtain that the distance between the time units of a temporal expression is generally less than 3 Chinese characters. The merging rule proposed in this paper can effectively solve the problem that basic time units cannot be correctly recognized.

4.2.3   Time Affixes

Chinese temporal expressions often contain some time affixes, which are divided into time prefixes and time suffixes. Commonly used time prefixes include "大约/about", "到/to" and "从/from" so on. Commonly used time suffixes include "左右/about", "之前/before" and "期间/during" so on. Time affixes are important components of Chinese temporal expressions. In many cases, the semantics of temporal expressions can be significantly changed. For example, "下午两点/2 pm" and "下午两点之前/Before 2 pm" have a large semantic difference, so it is necessary to recognize temporal expressions including time affixes completely. When time suffixes appear independently, they usually may not have temporal semantics. Therefore, it is possible to determine whether a time suffix is part of a certain time expression through the context. In view of the limited number of common time affixes, we manually summarize a time prefix vocabulary and a suffix vocabulary, and we combine with the basic time units according to the context information to obtain a complete temporal expression. Details of these two affix vocabularies are shown in Table 1.

Table 1. Details of affix vocabularies.

Type	Vocabulary
Prefix	约/about, 大约/about, 到/until, 从/from, 截至/up, 自/from, 到/to, 在/at, 直到/until, 至/to, 不到/less than, 大概/about, 近/near, 至/to
Suffix	左右/about, 许/about, 之后/after, 之前/before, 期间/during, 后/after, 中旬/middle third, 前/before, 内/within, 后/after, 多钟/more than, 为止/until, 前/before, 以来/since, 起/from

4.3   Chinese Temporal Expression Recognition Based on CRF

Firstly, we extract four kinds of artificial features (the token, part of speech, semantic role and vocabulary features) using LTP, and select features to construct feature vectors. Then, we automatically label the Chinese temporal expressions in the training data recognized using the rule-based method in Sect. 4.2, and meanwhile, we manually label event-type temporal expressions that cannot be recognized by the rule-based method. Lastly, use the training data to train a CRF model.

4.3.1   Feature Extraction and Selection

Language Technology Platform (LTP) is used for tokenization and feature extraction. Tokenization is typically the first step of text mining process, and consists in transforming a stream of characters into a stream of processing units called tokens (e.g., syllables, words, or phrases). For the characteristics of Chinese temporal expressions, this paper extracts four effective artificial features: the token, part of speech, semantic role and vocabulary feature. For each token, extract its part-of-speech and semantic role using LTP. LTP often labels part-of-speech and semantic roles of temporal expressions as /t and /TMP respectively. Therefore, they are effective features that can significantly improve the recognition effect of Chinese temporal expressions. Time affixes are part of Chinese temporal expressions, but they do not have any useful features. Therefore, through the analysis of a large number of texts, we summarize two vocabularies: a prefix vocabulary and a suffix vocabulary to label each token. Vocabularies here are the same as the vocabulary mentioned in Sect. 4.2.3.

The composition that a token acts as in Chinese depends on the context. Therefore, when judging whether a token is a part of a Chinese temporal expression, not only the features of the current token but also the adjacent tokens need to be considered. The selected features for constructing feature vectors are shown in Table 2. PS, SR, and VF are abbreviations for part of speech, semantic role, and vocabulary features, respectively.

Table 2. Feature selection.

Type	Features
Token	The current token The two tokens adjacent to the front The two tokens adjacent to the back
Part for Speech (PS)	The PS of the current token The PS of the two tokens adjacent to the front The PS of the two tokens adjacent to the back
Semantic Role (SR)	The SR of the current token The SR of the two tokens adjacent to the front The SR of the two tokens adjacent to the back
Vocabulary Feature (VF)	The VF of the current token The VF of the two tokens adjacent to the front The VF of the two tokens adjacent to the back

4.3.2   BIO Labeling

The Chinese temporal expression recognition method based on CRF can be transformed into a sequence labeling problem. We use the BIO sequence annotation method to label Chinese temporal expressions in the training data. BIO labeling has three labels (B, I, O), among them, B indicates that the token is at the beginning of a temporal expression, I states that the token is inside a temporal expression, and O means that the token is not a part of a temporal expression. For the sentence "2008年5月12日14时28分, 四川汶川发生里氏7.8级地震/At 14:28 on May 12, 2008, an earthquake measuring 7.8 on the Richter scale in Wenchuan, Sichuan", the result of the BIO labeling is: "2008年/B 5月/I12日/I 14时/I28 分/I, 四川/O汶川/O里氏/O7.8级/O地震/O". Features extracted in the previous section and the BIO tags together form the training data. Details of the training data is shown in Fig. 3. The first five columns represent features, namely token, part of speech, semantic role, and vocabulary features. The last column is the BIO label.

2	昨日	/t	/NNN	/NNN	/NNN	/B
3	七点	/t	/NNN	/NNN	/NNN	/I
4	左右	/t	/TMP	/SUF	/NNN	/I
5	,	/wp	/NNN	/NNN	/NNN	/O
6	东莞	/ns	/NNN	/NNN	/NNN	/O
7	一纸厂	/n	/NNN	/NNN	/NNN	/O
8	因	/p	/NNN	/NNN	/NNN	/O
9	雷电	/n	/NNN	/NNN	/NNN	/O
10	发生	/v	/V	/NNN	/NNN	/O
11	火灾	/n	/N	/NNN	/NNN	/O
12	。	/wp	/NNN	/NNN	/NNN	/O
13	接到	/v	/TMP	/NNN	/NNN	/B
14	报警	/v	/TMP	/NNN	/NNN	/I
15	后	/v	/TMP	/NNN	/NNN	/I
16	,	/wp	/NNN	/NNN	/NNN	/O
17	消防	/b	/A0	/NNN	/NNN	/O
18	人员	/n	/A0	/NNN	/NNN	/O
19	火速	/d	/ADV	/NNN	/NNN	/O
20	赶到	/v	/V	/NNN	/NNN	/O
21	火灾	/n	/A1	/NNN	/NNN	/O
22	现场	/n1	/A1	/NNN	/NNN	/O
23	。	/wp	/NNN	/NNN	/NNN	/O

Fig. 3. A fragment of a training data

5   Experiments

5.1   Experimental Corpus

In this paper, we use the Chinese Emergency Corpus (CEC) developed by the Semantic Intelligence Laboratory of Shanghai University as the experimental data set. The CEC includes 328 articles from five categories, including earthquake, fire, traffic accident, terrorist attack and food poisoning. Related texts are collected from the Internet. The CEC data is moderate in quantity and covers a wide range. Among them, the event-type Chinese temporal expressions account for about 15%. Details of the CEC data are shown in Table 3.

Table 3. Details of CEC.

Type	Documents	Sentences	Expressions
Earthquake	62	418	986
Fire	75	482	1047
Traffic accident	85	530	1295
Terrorist attack	45	377	771
Food poisoning	61	386	1095
Total	328	2193	5194

In this paper, the ten-fold cross-validation method is used to conduct experiments, and the corpus is divided into 10 parts. Among them, the proportion of the above five kinds of articles in each part is the same as the proportion in the whole corpus. For each experiment, 9 of them are used as the training data, and the remaining one is used as the testing set. The experiment is repeated by 10 times, and the average of 10 experimental results is taken as the final result of the experiment.

The CRF software used in the experiment is CRF++ −0.58 toolkit for the open source library, the feature selection rules in Sect. 4.3.1 are written into a template file, and the executable file can be called to train a CRF model and test the model on the testing set.

5.2 Performance Metrics

This paper uses precision rate (P), recall rate (R) and $F1$ value as performance metrics to evaluate experimental results. Let N_1 be the number of Chinese temporal expressions in the testing set. Let N_2 be the number of temporal expressions that are recognized in the testing set automatically. Three performance Metrics are defined as follows:

$$P = \frac{N_1 \cap N_2}{N_2} \quad R = \frac{N_1 \cap N_2}{N_1} \quad F1 = \frac{2 \times P \times R}{P + R}$$

5.3 Experimental Results and Analysis

Four sets of experiments have been carried out in this section. The first experiment is the traditional rule-based Chinese temporal expression recognition. The remaining three experiments all train CRF methods and the difference between them is the labeled training data. The second one only labels the event-type temporal expressions in the training data. The third one uses the regular rules to recognize the Chinese temporal expressions, and automatically labels the training data. In the last experiment, event-type temporal expressions are manually labeled on the basis of the third experiment's training data. Experimental results of the four sets of experiments are shown in Table 4. The precision rate of the first experiment is 91.05%, but the recall rate is only 75.18%. The recall rate of the second experiment is low, only 11.25%. Compared with the first experiment, the precision rate of the third is lower, but the recall rate increases by 7.23%, and the F1 value increases by 2.01%. The fourth experiment shows the recall rate increases by 8.85% compared with the second experiment, and the F1 value reaches 88.73%, which is 3.36% higher than that of the third one.

Table 4. Experimental results.

No.	Precision (%)	Recall (%)	F1(%)
1	91.05	75.18	82.36
2	82.17	11.25	19.79
3	88.56	82.41	85.37
4	86.34	91.26	88.73

The traditional rule-based method can achieve a high precision rate, but due to the diversity of Chinese temporal expressions, it is difficult to formulate a complete set of rules to recognize all types of Chinese temporal expressions. In particular, the method cannot recognize event-type Chinese temporal expressions. The above characteristics of Chinese temporal expressions lead to a lower recall rate in the first experiment, which only reached 75.18%. In the second experiment, the recall rate is hence lower in the case of only manually labeling the event-type Chinese temporal expressions, where the event-type Chinese temporal expressions account for only about 15%. The experimental results show that CRF-based method can effectively recognize event-type temporal expressions. The Chinese temporal expression recognition methods based on CRF use the natural language tool to extract artificial features, effectively utilizing the latest research results of natural language processing. In addition, because the statistical learning model has a good generalization ability, statistics-based methods tend to achieve a higher recall rate. In the third experiment, the recall rate is 7.23% higher than that of the first experiment. As compared with the third experiment, the recall rate of the last experiment increases by 8.85%, and the F1 value achieves 88.73%. And meanwhile, our proposed method outperforms traditional methods in Chinese temporal expression recognition.

6 Conclusions

This paper proposed a new Chinese temporal expression recognition method combining rules with a statistical model. More specifically, manually formulate regular rules based basic time units, and use the rules to label the training data, which can reduce the amount of labeling workload significantly. In addition, the method effectively utilizes the good generalization ability of the CRF model and the ability to recognize event-type Chinese temporal expressions, and significantly improves the recognition effect of Chinese temporal expressions. However, the extraction of artificial features depends on natural language processing tools. At the same time, due to the limited complexity of traditional statistical learning models, more and more bottlenecks are encountered in dealing with complex natural languages. In recent years, deep learning has made breakthroughs in the field of natural language processing. To reduce the dependence on natural language processing tools and improve recognition, our future work will use deep learning to recognize Chinese temporal expressions.

Acknowledgement. This work was supported by the National Key Research and Development Program of China [grant number 2018YFC1603601], the Program for Innovative Research Team in University of the Ministry of Education [grant number IRT17R32], and the National Natural Science Foundation of China [grant numbers [61673152, 91746209].

References

1. Zacks, J.M., Tversky, B.: Event structure in perception and conception. Psychol. Bull. **127** (1), 3 (2001)
2. Pustejovsky, J., Lee, K., Bunt, H., Romary, L.: ISO-TimeML: an international standard for semantic annotation. In: LREC, vol. 10, pp. 394–397 (2010)
3. Liu, Z., Huang, M., Zhou, W.: Researcher on event-oriented ontology model. Comput. Sci. **36**(11), 189–192 (2009)
4. Ge, T., Cui, L., Chang, B., Sui, Z., Zhou, M.: Event detection with burst information networks. In: Proceedings of COLING 2016, the 26th International Conference on Computational Linguistics: Technical Papers, pp. 3276–3286 (2016)
5. Du, N., Dai, H., Trivedi, R., Upadhyay, U., Gomez-Rodriguez, M., Song, L.: Recurrent marked temporal point processes: embedding event history to vector. In: Proceedings of the 22nd ACM SIGKDD International Conference on Knowledge Discovery and Data Mining, pp. 1555–1564 (2016)
6. He, R., Qin, B., Liu, T., Pan, Y., Li, S.: Recognizing the extent of Chinese time expressions based on the dependency parsing and error-driven learning. J. Chin. Inf. Process. **21**(5), 36–40 (2007)
7. Lin, J., Cai, D., Yuan, C.: Automatic timex2 tagging of Chinese temporal information. J. Tsinghua Univ. **48**(1), 117–120 (2008)
8. Ferro, L., Gerber, L., Mani, I., Sundheim, B., Wilson, G.: TIDES 2005 standard for the annotation of temporal expressions (2005)
9. Wu, T., Zhou, Y., Huang, X., Wu, L.: Chinese time expression recognition based on automatically generated basic-time-unit rules. J. Chin. Inf. Process. **24**(4), 3–11 (2010)
10. Zhu, S., Liu, Z., Fu, J., Zhu, F.: Chinese temporal phrase recognition based on conditional random fields. Comput. Eng. **37**(15), 164–167 (2011)
11. Liu, L., He, Z., Xing, X., Mao, X.: Chinese time expression recognition based on semantic role. Appl. Res. Comput. **28**(7), 2543–2545 (2011)
12. Wu, Q., Huang, D.: Temporal information extraction based on CRF and time thesaurus. J. Chin. Inf. Process. **28**(6), 169–174 (2014)
13. Yan, Z., Ji, D.: Exploration of Chinese temporal information extraction based on CRF and semi-surprised learning. Comput. Eng. Des. **36**(06), 1642–1646 (2015)
14. Verhagen, M., Sauri, R., Caselli, T., Pustejovsky, J.: SemEval-2010 task 13: TempEval-2. In: Proceedings of the 5th International Workshop on Semantic Evaluation, pp. 57–62 (2010)

15. Lafferty, J., McCallum, A., Pereira, F.C.N.: Conditional random fields: probabilistic models for segmenting and labeling sequence data. In: Proceedings of the 18th International Conference on Machine Learning, pp. 282–289. Morgan Kaufmann Publishers Inc. (2001)
16. Elliott, R.J., Aggoun, L., Moore, J.B.: Hidden Markov Models: Estimation and Control, vol. 29. Springer, New York (2008). https://doi.org/10.1007/978-0-387-84854-9
17. He, L., Lee, K., Lewis, M., Zettlemoyer, L.: Deep semantic role labeling: what works and what's next. In: Proceedings of the 55th Annual Meeting of the Association for Computational Linguistics, vol. 1, pp. 473–483 (2017)
18. Gildea, D., Jurafsky, D.: Automatic labeling of semantic roles. Comput. Linguist. **28**(3), 245–288 (2002)
19. Che, W., Li, Z., Liu, T.: LTP: a Chinese language technology platform. In: COLING 2010, 23rd International Conference on Computational Linguistics, Demonstrations, Beijing, China, vol. 23–27, pp. 13–16 (2010)

An Optimization Regression Model for Predicting Average Temperature of Core Dead Stock Column

Guangyu Wang[1], Bing Dai[2], Wenyan Wang[1], Hongming Long[2],
Jun Zhang[3], Peng Chen[3], and Bing Wang[1,2(✉)]

[1] School of Electrical and Information Engineering,
Anhui University of Technology, Maanshan 243002,
People's Republic of China
wangbing@ustc.edu
[2] Key Laboratory of Metallurgical Emission Reduction
and Resources Recycling, Anhui University of Technology,
Ministry of Education, Maanshan 243002, People's Republic of China
[3] The Institute of Health Sciences, Anhui University, Hefei 230601,
Anhui, China

Abstract. Hearth activity is one of the most important factors which affect the smooth progress of production and even the life of blast furnace. However, the calculation of hearth activity depends on the empirical model entirely, and the model parameter acquisition is difficult. To overcome this deficiency, this paper presents a novel method based on an improved multiple linear regression model to predict average temperature of core dead stock column for evaluating it. In the algorithm, the Pearson correlation analysis, metallurgical formulas and the Akaike Information Criterion based on least square method are used to establish a multiple linear regression model. The method makes the estimation of hearth activity out of the empirical formula. And it is easy for the evaluated model to obtain parameters. Meanwhile, experimental results show our proposed method can achieve 0.69% average relative error on the test data set and average relative error of 0.57% on the training data set. Moreover, the function of low average temperature of core dead stock column warning can be realized.

Keywords: Multiple linear regression model ·
Average temperature of core dead stock column · Data mining ·
Hearth activity · Blast furnace

1 Introduction

The long-term stability is important for equipment maintenance and production quality control in blast furnace production [1–3]. As one of the most critical parameters for evaluating the status of production, hearth activity, which can be characterized as the capability of slag iron penetrate the stock column in hearth, affects the smooth progress of production, and even the life of blast furnace [4]. While the hearth activity is being

© Springer Nature Switzerland AG 2019
D.-S. Huang et al. (Eds.): ICIC 2019, LNAI 11645, pp. 468–478, 2019.
https://doi.org/10.1007/978-3-030-26766-7_43

in a good situation, blast furnace will be stable and production will be smooth, otherwise, the quality and quantity of products will be fluctuated [12–16]. Therefore, the hearth activity monitoring has become a significant issue in the blast furnace production.

However, the calculation of hearth activity is difficult, which is mainly caused by two reasons [5–11]. One is that heart activity is influenced by multiple factors, such as theoretical combustion temperature, blast velocity, the CSR of coke, coke ratio, coal ratio, slag ration, central and edge gas flow, gas permeability, and so on. These factors are fluctuated in production, and hard to be measured accurately, which will lead to the difficulty in the calculation of hearth activity. On the other hand, the hearth is located in the interior of the blast furnace which is a closed system, and the change of its working status cannot be detected directly, which will cause a big delay for act in response.

Fortunately, there are some previous works had addressed this problem to estimate hearth activity in some indirect ways. Raipala et al. tried to evaluate hearth activity using an empirical formula of the core dead stock column temperature [12]. Chen developed a new index based on the flowing resistance coefficient of slag and hot metal to estimate hearth activity [13]. Dai et al. proposed an optimized on-line and off-line hearth activity quantitative calculation model based on the Chen's index [14]. Chen has established an estimation model based on the Kozeny-Carman equation [16]. Jin and Tao developed an active index to reflect hearth activity [15]. However, these methods are strongly experience-dependent and it is often difficult to establish the identifiability of their parameters.

Therefore, an improved multiple linear regression algorithm-based model (IMLRA) was proposed in this work to predict average core dead stock column temperature for hearth activity evaluation. As a highly complex system, there are many state parameters in the blast furnace production process can affect the hearth activity, and the coupling relationship among them is not clear. For example, 50 state parameters related with the production processing had been collected for inferring hearth activity in blast furnace. Therefore, therein a correlation analysis based Pearson correlation coefficient and physical relationship between the parameters are firstly used to figure out the inherent relationship between different state parameters, and remove the abundant information from the original data. Then a multiple linear regression model is established using the Akaike Information Criterion based on the least square method. IMLRA not only makes the acquisition of model parameters easy but also makes the estimation of hearth activity out of the empirical formula. The experimental results we achieved show that reliability of the established model, the accuracy of prediction and low temperature warning function can be realized of our proposed approach.

The framework for the rest of this paper is organized as follows. Section 2 introduces IMLRA method that includes data preparation, feature selection and regression modelling using the Akaike Information Criterion based on the least square method. A model established by multiple linear regression method will be evaluated in Sect. 3. Section 4 makes a conclusion about this research work.

2 Method

2.1 Data Preparation

Data Cleaning. The data used in this work was collected from the production processing of a blast furnace with 38 tuyeres and 4 iron notches, and its volume is 4,747 m^3. In this blast furnace, there is a PW type String Bell-less top with material distributor, cooling stave with thin-wall copper lining, closed-cycle system combined soft water, green Anglo-Pakistani sludge treatment system, ring seam gas cleaning system, and after pressure power generation system of TRT.

Because the core dead stock column temperature cannot be available directly, 50 types of technological or state parameters in productive process have been collected to delineate the status of blast furnace, such as air resistance, average top gas temperature, and so on, which are listed in the Table 1. These status parameters are collected one time an hour in a continuous time interval from October 19, 2017 to January 18, 2018. Generally, 24 data points can be acquired for each parameter within one day. However, some missing values have been found in our dataset for some reasons come from data gather techniques.

Table 1. Operation process parameters.

Parameters	Parameters
Air resistance	Theoretical combustion temperature
Average top gas temperature	Top gas riser temperature 4
Blast furnace bosh gas	Top gas CO
Blast furnace bosh gas index	Top pressure
Center coke ratio	Top gas CO2
Central charge	Top gas composition H2
Cross center temperature	Transmission gear box temperature
Cold air pressure	Tuyere wind speed
Cold air temperature	Utilization rate of CO
Differential pressure	W throat location temperature
Drum wind can	W cross position temperature
Gear box cooling water flow	Z medium position temperature
Hot air flow	Z position temperature
Hot air pressure	20.080 m furnace body
Lower pressure difference	Static pressure
Marginal mean temperature	20.080 m furnace body
Oxygen enrichment rate	Static pressure at position A
Permeability	20.080 m furnace body
Rich oxygen flow	Static pressure at position B
Rotating gear box temperature	20.080 m furnace body
Set the top pressure	Static pressure at position C
Set the air temperature	26.025 m furnace body
Set the oxygen-enriched	Static pressure

(continued)

Table 1. (*continued*)

Parameters	Parameters
Supply air temperature	26.025 m furnace body
Tilting gear box temperature	Static pressure at position A
Top gas N2	26.025 m furnace body
Top gas riser temperature 1	Static pressure at position B
Top gas riser temperature 2	26.025 m furnace body
Top gas riser temperature 3	Static pressure at position C

In this work, the status information in 2,011 points of time are successfully acquired, but there are still some missing values within the dataset should be preprocessed before it can be utilized into the modeling whereafter. In some situations, i.e., data collected in November 9, November 15, November 16, January 11, and January 18, there are too many missing values in one time point, thus all of the data within these days are removed from the dataset for data quality. If only a few values of some parameters are missing, such as the data collected from November 5 to 8, a simple average value filling strategy is adopted to fill the missing data using the two around neighbors. In addition, blast furnace bosh gas, theoretical combustion temperature and utilization rate of CO which is out of the vast majority of data, and can be seen as outliers and will be deleted from the original dataset. Therefore, a total number of 60 data were removed. Overall, 60 pieces of abnormal data were deleted and a total of 4 pieces of data were filled in, and the number of valid time points in the dataset is decreased from 2,011 to 1,955.

Core Dead Stock Column Temperature Calculation. Raipala *et al.* came up with a formula for the calculation of core dead stock column temperature in 2010 [12]. However, the equation is applied in large blast furnace inappropriately. Therefore, Dai et al. revised the calculation method of theoretical combustion temperature and slag mobility index based on the model proposed by the former. Then, a model of core dead stock column temperature calculation for evaluating hearth activity [14] becomes

$$DMT = (0.165 \times t_f \times V_{bosh})/D^3 + 2.445 \times (FR - 483) + 2.91 \times (\Delta_t - 107)$$
$$- 11.2 \times (\beta_{CO,C} - 27.2) + 28.09 \times (D_{pcoke} - 25.8) + 326 \tag{1}$$

where DMT is the temperature of core dead stock column, t_f is the theoretical combustion temperature, V_{bosh} is the blast furnace bosh gas, D is the hearth diameter, FR is the fuel ratio, Δ_t is the slag mobility index, $\beta_{CO,C}$ is the utilization rate of CO, and D_{pcoke} is the coke size of core dead stock column. Therein, D is 14.8m, FR changes in [500 kg, 530 kg], D_{pcoke} is [30 mm, 40 mm], and Δ_t is [−20 °C, 120 °C], V_{bosh}, t_f and $\beta_{CO,C}$ can be accessed in our collected data, but Δ_t, FR, and D_{pcoke} are varied during the production process of the blast furnace, which will result in the fluctuation of DMT. Therefore, the final value of DMT, the temperature of core dead stock column, is replaced using an average one to consider its variability, and

$$DMT_f = (0.165 \times t_f \times V_{bosh})/D^3 + 2.445 \times (515 - 483) + 2.91 \times (100 - 107)$$
$$- 11.2 \times (\beta_{CO,C} - 27.2) + 28.09 \times (35 - 25.8) + 326$$
(2)

Based on the above analysis, the DMT_f value of 1,955 valid time points can be calculated. After calculation, the DMT_f changes from 1,358.85 to 1,478.02 within the time interval of data acquisition, and the distribution can be found in Fig. 1.

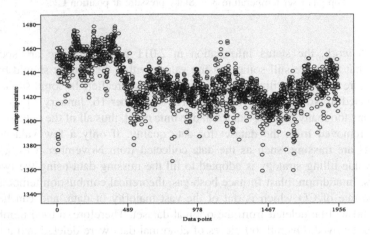

Fig. 1. Distribution of average temperature of core dead stock column.

2.2 Feature Selection

Degree of correlation between operation process parameters and target values is generally divided based on the absolute value of the Pearson correlation coefficient [17, 18]. It should be noted that data points of time series 1 to 1855 are regarded as training dataset for the analysis. According to the results of the analysis, the operation process parameters which are 26.025 m furnace body static pressure at position B, 26.025 m furnace body static pressure at position C, 20.080 m furnace body static pressure at position A, 20.080 m furnace body static pressure at position B, 26.025 m furnace body static pressure at position A, rich oxygen flow, cold air temperature, oxygen enrichment rate, lower pressure difference, theoretical combustion temperature and 20.080 m furnace body static pressure will be selected. It is worth noting that the absolute value of the correlation coefficient between these selected parameters and target values is greater than 0.6, which is considered to be strongly correlated with target values.

However, a review of the physical meaning of these 11 operation process parameters reveals that rich oxygen flow and oxygen enrichment rate depend on each other [19]. Therefore, rich oxygen flow and oxygen enrichment rate are just going to keep one. In this work, the parameter with a large absolute value of the coefficient with target values is regarded as a feature. Then, the oxygen enrichment rate is kept.

Based on the analysis above, a total of 10 operation process parameters which are shown in the Table 2 are regarded as features to establish multiple linear regression models.

Table 2. Features.

1. Cold air temperature	7. 20.080 m furnace body static pressure at position B
2. Lower pressure difference	
3. Oxygen enrichment rate	8. 26.025 m furnace body static pressure at position A
4. Theoretical combustion temperature	
5. 20.080 m furnace body static pressure	9. 26.025 m furnace body static pressure at position B
6. 20.080 m furnace body static pressure at position A	10. 26.025 m furnace body static pressure at position C

2.3 Multiple Linear Regression Models

Loss function of the simple multiple linear regression modeling is an empirical loss, which cannot control the complexity of the model. Therefore, in this work, the Akaike Information Criterion (AIC Criterion) will be taken as a loss function to establish a model with low model complexity and good generalization ability [20]. The Akaike Information Criterion (AIC) satisfy

$$AIC = n \log(2\pi\sigma_1^2) + n + 2(p+2) \tag{3}$$

where n is the number of features in the sample, σ_1^2 is unbiased estimation of random errors for a multiple linear regression model, and the calculation process of σ_1^2 can be found in [21]. p is the number of features contained in the established model. The model with the minimum AIC value is regarded as the optimal model.

Based on the analysis above, taking the data points of time series 1 to 1855 as training dataset, which are composed of 10 features, to write a computer program, a multiple linear regression model is

$$DMT_f = 790.5006 + 0.1252C_{at} + 1.3587O_x - 0.1611DP + 0.0836DP\_A \\ + 0.1349DP\_B + 0.4048DP\_b + 0.5430BP + 0.1957t_f \tag{4}$$

where DMT_f is the predicted value of the multiple linear regression model, C_{at} is the cold air temperature, the O_X is oxygen enrichment rate, DP is the 20.080mfurnace body static pressure, DP_A is the 20.080mfurnace body static pressure at position A, DP_B is the 20.080mfurnace body static pressure at position B,DP_b is the 26.025 m furnace body static pressure at position B, BP is the lower pressure difference, t_f is the theoretical combustion temperature.

3 Results and Discussion

3.1 Statistical Evaluation of the Multiple Linear Regression Model

For the established multiple linear regression model, the goodness of fit test and the significance test of model coefficients are required [22, 23]. A multiple linear regression model established in this work, the value of the R square is 0.647 which means that the established multiple linear regression model has a good fitting effect. Besides, all confidence values of the regression coefficient of conditional variables are 0, which indicates that conditional variables are significant and unrelated variables have not been introduced into the regression model. More importantly, from the coefficients of a multiple linear regression model, it can be seen that the contribution degree to the target value is ranked as oxygen enrichment rate, lower pressure difference, 26.025 m furnace body static pressure at position B, theoretical combustion temperature, 20.080mfurnace body static pressure, 20.080mfurnace body static pressure at position B, cold air temperature and 20.080mfurnace body static pressure at position A from high to low. Details are shown in the Table 3.

However, it is worth noting that the established model does not show the correlation between features that 26.025 m furnace body static pressure at position A and C and the average temperature of core dead stock column. It is different from the correlation distribution obtained by correlation analysis. The reason is that the AIC loss function balances the empirical risk and structural risk of multiple linear regression models, so as to grasp the main operational process parameters, then to establish simple multiple linear regression models. Moreover, it indicates that the other eight operation process parameters are more important than the two operation process parameters during the establishment of a multiple linear regression model.

Table 3. Statistical evaluation of the model.

Name of the variable	Value of significance test	Degree of significance (Significant√ or not ×)	Order of importance (Between 1 and 8)
C_{at}	0	√	7
t_f	0	√	4
O_X	0	√	1
BP	0	√	2
DP	0	√	5
DP_A	0	√	8
DP_B	0	√	6
DP_b	0	√	3

3.2 Model Performance on the Training and Test Dataset

Statistical model evaluation indicates that this model has good data fitting effect. But, the actual performance on the dataset needs exploration. It should be noted that data points of time series 1856 to 1955 are regarded as test dataset.

It is clear that target values have a close linear relationship with predicted values in the Fig. 2 and the Fig. 4, which proves that this multiple linear regression model is reasonable. Figure 3 shows that the maximum relative error is 3.33%, the minimum relative error is 0, the relative error of most samples is less than 1%, and the average relative error obtained by calculation is 0.57%, which proves that data points on the training dataset can be well fitted by this model. As is shown in the Fig. 5, the maximum test error is 1.60%, the minimum relative error is 0, the relative errors of most samples are less than 1%, and the average relative error obtained by calculation is 0.69%, which proves that this model has good effect on the test dataset. Therefore, conclusions can be made that this model has good performance on both the training set and the test set, which proves the validity of the model.

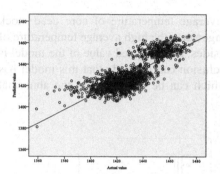

Fig. 2. Relationship between actual and predicted values on the training set.

Fig. 3. Target value error distribution on the training set

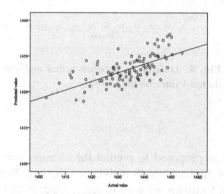

Fig. 4. Relationship between actual and predicted values on the test set.

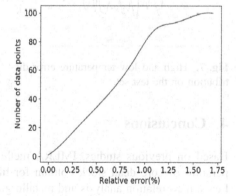

Fig. 5. Target value error distribution on the test set.

3.3   Low Average Temperature of Core Dead Stock Column Warning

The level of the average temperature of core dead stock column reflects the level of activity of the hearth. The low-temperature early warning capability can provide an important aid for artificially restoring the activity of the hearth and ensuring the stable antegrade of the blast furnace. Therefore, model's low temperature warning ability is important.

On the test dataset, the temperature from 1402 to 1427 °C is regarded as low temperature. And the temperature from 1428 to 1454 °C is regarded as high temperature. Figure 7 shows the low and high average temperature of core dead stock column relative errors. It is clear that the low temperature relative errors are smaller than the high temperature relative errors, which indicates that the low temperature prediction ability of this multiple linear regression model is stronger than the high temperature. Figure 8 shows the differences between the predicted values and the actual values. The differences are defined as predicted target values minus the actual target values. It is clear that target values predicted via our model is less likely than the actual average temperature.

Based on the analysis above, the low average temperature of core dead stock column prediction ability of the model is stronger than the high average temperature of core dead stock column prediction ability. Besides, the predicted value of the model is smaller than the actual value. Therefore, conclusion can be made that this model has good low temperature warning capability, which can be used to warn the abnormal activity of the hearth.

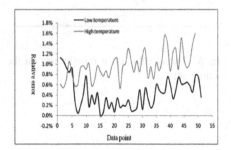 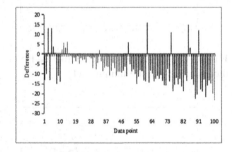

Fig. 7. High and low temperature error distribution on the test set.

Fig. 8. Differences between actual and predicted values on the test set.

4 Conclusions

Based on previous studies, IMLRA method is proposed to predict the average temperature of core dead stock column for blast furnace hearth activity estimation. The Pearson correlation analysis and metallurgical formulas are used to select features first. Based on the selected features, the Akaike Information Criterion which is regarded as a loss function is used to establish an optimal multiple linear regression model. Experimental results indicate that the proposed method can establish a simple model which has accessible parameters and hearth activity warning function, which has important significance in the field of blast furnace production and protection.

Acknowledgement. This work is supported by the National Natural Science Foundation of China (Nos. 61472282, 61672035, and 61872004), Key Laboratory of Metallurgical Emission Reduction & Resources Recycling in AHUT (KF 17-02), Anhui Province Funds for Excellent Youth Scholars in Colleges (gxyqZD2016068), Co-Innovation Center for Information Supply & Assurance Technology in AHU (ADXXBZ201705), and Anhui Scientific Research Foundation for Returness.

References

1. Zhou, D., Cheng, S., Zhang, R., Li, Y., Chen, T.: Uniformity and activity of blast furnace hearth by monitoring flame temperature of raceway zone. ISIJ Int. **57**, 1509–1516 (2017)
2. Luomala, M.J., Mattila, O.J., Härkki, J.J.: Physical modelling of hot metal flow in a blast furnace hearth. Scand. J. Metall. **30**, 225–231 (2010)
3. Shibata, K., Kimura, Y., Shimizu, M., Inaba, S.I.: Dynamics of dead-man coke and hot metal flow in a blast furnace hearth. Revue De Métallurgie **87**, 333–340 (2007)
4. Jiao, K., Zhang, J.L., Liu, Y.X., Li, S.F., Liu, F.: Analysis on the stamping coke dissolution of hot metal in the blast furnace hearth. Can. Metall. Q. **56**, 1–7 (2017)
5. Gao, C., Jian, L., Luo, S.: Modeling of the thermal state change of blast furnace hearth with support vector machines. IEEE Trans. Industr. Electron. **59**, 1134–1145 (2011)
6. Zhang, Y., Deshpande, R., Huang, D., Chaubal, P., Zhou, C.Q.: Numerical analysis of blast furnace hearth inner profile by using CFD and heat transfer model for different time periods. Int. J. Heat Mass Transf. **51**, 186–197 (2014)
7. Zolotykh, M.O., Dmitriev, A.N., Vitkina, G.Y.: The association of various approaches to the monitoring of lining condition in the blast furnace hearth. Defect Diffus. Forum **380**, 186–190 (2017)
8. Gomes, F.S.V., Coco, K.F., Salles, J.L.F.: Multistep forecasting models of the liquid level in a blast furnace hearth. IEEE Trans. Autom. Sci. Eng. **14**, 1286–1296 (2017)
9. Komiyama, K.M., Guo, B.Y., Zughbi, H., Zulli, P., Yu, A.B.: Improved CFD model to predict flow and temperature distributions in a blast furnace hearth. Metall. Mater. Trans. B **45**, 1895–1914 (2014)
10. Agrawal, A., Kor, S.C., Nandy, U., Choudhary, A.R., Tripathi, V.R.: Real-time blast furnace hearth liquid level monitoring system. Ironmaking Steelmaking **43**, 160128032747001 (2016)
11. Dmitriev, A.N., Zolotykh, M.O., Chen, K., Vitkina, G.Y.: The thermophysical bases of monitoring of the fireproof lining wear in the blast furnace hearth. Defect Diffus. Forum **370**, 113–119 (2017)
12. Raipala, K.: Deadman and hearth phenomena in the blast furnace. Scand. J. Metall. **29**, 39–46 (2010)
13. Chen, H., Wu, S.L., Yu, X.B.: New index of evaluating activity of blast furnace hearth. Iron Steel **42**, 12–15 (2007)
14. Dai, B., Liang, K., Wang, X.J., Xin, L.I., Guo, Y.W., Works, I.: Development and practice of quantitative calculation models of blast furnace hearth activity. China Metallurgy (2015)
15. Chen, C., Gang, A.N., Plant, I.: Activity index and improvement measure of BF hearth. Iron Steel **29**–33 (2018)
16. Jin, J.: Practice of active hearth condition during long-term production with High PCR. Bao Steel Technol. **13**–16 (2002)
17. Neto, A.M., Rittner, L., Leite, N., Zampieri, D.E.: Pearson's correlation coefficient for discarding redundant information in real time autonomous navigation system. In: IEEE International Conference on Control Applications, pp. 50–19 (2007)

18. Liu, C., Jin, R., Gong, E., Liu, Y., Yue, M.: Prediction for the performance of gas turbine units using multiple linear regression. Proc. CSEE **37**, 4731–4738 (2017)
19. Zhou, H.Y.: On oxygen enrichment blast in front of blower. Enterprise Sci. Technol. Dev. (2012)
20. Arnold, T.W.: Uninformative parameters and model selection using Akaike's information criterion. J. Wildlife Manag. **74**, 1175–1178 (2011)
21. Verhagen, S., Teunissen, P.J.G.: Least-squares estimation and Kalman filtering. In: Teunissen, P.J.G., Montenbruck, O. (eds.) Springer Handbook of Global Navigation Satellite Systems. SH, pp. 639–660. Springer, Cham (2017). https://doi.org/10.1007/978-3-319-42928-1_22
22. Mudgal, A., Baffaut, C., Anderson, S.H., Sadler, E.J., Thompson, A.L.: APEX model assessment of variable landscapes on runoff and dissolved herbicides. Trans. ASABE **53**, 1047–1058 (2010)
23. Weakley, A., Williams, J.A., Schmitteredgecombe, M., Cook, D.J.: Neuropsychological test selection for cognitive impairment classification: a machine learning approach. J. Clin. Exp. Neuropsychol. **37**, 899–916 (2015)

A Selection Method for Denoising Auto Encoder Features Using Cross Entropy

Jie Cai, Wei Huang, Sheng Yang[✉], Shulin Wang, and Jiawei Luo

College of Computer Science and Electronic Engineering,
Hunan University, Changsha, China
Yangsh0506@sina.com

Abstract. There are a lot of noise and redundant information in gene expression data, which will reduce the accuracy of the classification model. Denoising auto encoder can be used to reduce the dimension for high-dimensional gene expression data, and get high-level features with strong classification ability. In order to get a better classification model furtherly, a high-level feature selection method based on information cross-entropy is proposed. Firstly, denoising auto encoder is used to encode high-dimensional original data to get high-level features. Then, the high-level features with low cross entropy are selected to get the low-dimensional mapping of original data, which is used to generate optimized and simplified classification models. The high-level features obtained by the denoising auto encoder can improve the accuracy of the classification model, and the selection of high-level features can improve the generalization ability of the classification model. The classification accuracy of the new method under different Corruption Level values and selection rate are studied experimentally. Experimental results on several gene expression datasets show that the proposed method is effective. Compared with classical and excellent mRMR and SVM-RFE algorithms furtherly, the proposed method shows better accuracy.

Keywords: Feature selection · Denoising auto encoder · Cross entropy

1 Introduction

Deep learning [1] is a research hotspot in research and application fields. It can extract features with high recognition ability from complex raw data and completes a revolution in traditional artificial intelligence and machine learning. Deep learning solves the problem that artificial intelligence has not made progress for many years, and it is one of the most important advances in machine learning in recent years.

The auto encoder in deep learning is different from the traditional PCA, LDA and KPCA methods. After automatic encoding and fine tuning, it can extract different forms of high-level features, so it has strong feature extraction ability. In fact, this strong feature extraction ability makes deep learning achieve great success. Auto Encoder (AE) was first proposed by Rumelhart in 1986. In 2006, Hinton et al. [2] improved the learning algorithm of auto encoder. By comparing with the traditional feature extraction algorithm in image and text data, researchers proved the great superiority and effectiveness of auto encoder. Therefore, auto encoder has been taken as one main

© Springer Nature Switzerland AG 2019
D.-S. Huang et al. (Eds.): ICIC 2019, LNAI 11645, pp. 479–490, 2019.
https://doi.org/10.1007/978-3-030-26766-7_44

method for feature extraction, and many varieties are proposed constantly, including Sparse Auto Encoder (SAE) [3], Denoising Auto Encoder (DAE) [4], Convolutional Auto Encoder (CAE) [5] and etc. With the deepening of research and development, auto encoder has been applied to more and more fields, including image recognition, fault diagnosis, natural language processing and biological information processing, and achieved very good results.

Traditional machine learning methods have achieved very good results in the analysis of gene expression data and other bioinformatics data [6, 7]. As a new machine learning theory, deep learning method has been more and more applied to the analysis and research of these data. Gupta extracted the high-level features of gene expression data using denoising auto encoder, constructed a deep belief model, and then used K-Means and spectral clustering methods to achieve gene clustering [8]. Singh et al. used sparse auto encoder to extract high-level features from gene expression data, then used Individual Training Error Reduction index to select high-level features to simplify the network, and then used stacked sparse auto encoder network to classify gene expression data [9]. Usually, in the application of deep neural network, there will be over-fitting, that is to say, the error function of the model in the training sample set is small, and the classification accuracy rate is high; while in the test sample set, the error function is relatively large, and the classification accuracy rate is low. Eliminating noise or simplifying network structure can improve the generalization performance of the classification model. Antoniades used cross-entropy to select the input of the deep neural network to realize the network pruning and the feature selection process is completed by the auto encoder [10]. In fact, these feature selection methods are the same as Dropout strategy [11] in deep learning. The classifiers obtained by the above methods all show good accuracy and generalization ability. In the learning process of classification model, if there are too many learning parameters, it is easy for the model to be fitted, especially for the dataset with few training samples.

Deep learning models often meet the problem of over-fitting. Feature selection can prune and simplify the deep learning model and improve the generalization ability of the model. Aiming at the high-level features extracted from deep learning, a High-level Denoising Auto Encoding feature selection method based on Cross-Entropy (HDAECE) is proposed.

2   Information Cross Entropy Measure for DAE Feature

2.1   Denoising Auto Encoder

The auto encoder can get high-level features with more robustness by adding noise into the original input data [4], this extension of auto encoder is called denoising auto encoder. Its schematic is shown in Fig. 1, where F, F_g and F_{rcs} are original feature set, encoded feature set and reconstruction feature set respectively. The denoising auto encoder can avoid the problem of over-fitting of hidden layer nodes in general auto encoders, and build a learning model with better generalization performance. The denoising auto encoder and auto encoder have no difference in loss function ($L(F, F_{rcs})$ in Fig. 1) and training processing. The loss function may be a mean square error or a cross entropy function, as in Eqs. (1) and (2), respectively, where m is the size of the sample set and i is the sequence number of the feature.

$$L(F, F_{rcs}) = \sum_{i}^{m} (F_i - F_{rcs\,i})^2 \tag{1}$$

$$L(F, F_{rcs}) = -\sum_{i}^{m} [F_i \log F_{rcs\,i} + (1 - F_i) \log(1 - F_{rcs\,i})] \tag{2}$$

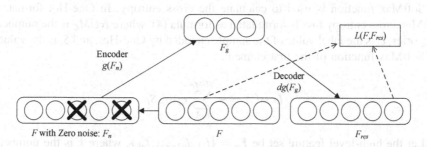

Fig. 1. Schematic of denoising auto encoder

Zero-noise and Gaussian noise are two common ways for adding noise in denoising auto encoder. In this study, zero-noise is used, and the noise is controlled by the noise ratio parameter, Corruption Level. Corruption Level is usually large (>0.2) for image data. However, for data with a small sample size such as gene expression spectrum, the addition of high noise may greatly reduce the regularity of data, and the following learning model may be invalid. The general Corruption Level is relatively small (<0.2) [8]. The size of the Corruption Level is based on the principle of not destroying the underlying patterns hidden in the data, otherwise the wrong classification model may be trained.

2.2 Information Cross Entropy

Cross entropy $H(P,Q)$ can be used as loss function in neural network, defined by formula (3), where P is the distribution of real label, and Q is the prediction label distribution of the trained model. The cross entropy loss can measure the difference between P and Q.

$$H(P, Q) = \sum_{x} (P(x) \log(1/Q(x))) \tag{3}$$

When the two random distributions are the same, their cross-entropy is zero. When the difference between the two random distributions increases, their cross entropy and relative entropy will also increase. Therefore, cross entropy can be considered for feature selection. Literature [12] maximizes the relative entropy of variants to select the optimal discriminant features for a single category. Firstly, the feature with the largest weight of the coding network is selected as the initial feature, and then the feature which reduces the cross-entropy reconstruction error of the coding network is continuously selected until all the features are evaluated. Literature [13] also points out that features with smaller reconstruction errors are more representative, and an image feature selection algorithm based on deep belief network is presented.

SoftMax function is used to calculate the cross entropy. In One-Hot format, the SoftMax cross entropy loss is formulated as formula (4), where NUM_C is the number of categories, c_i is the label value of the sample encoded by One-Hot, and S_i is the value of the SoftMax function of the first element.

$$Loss(.) = - \sum_{i=1}^{NUM_C} c_i \log S_i \qquad (4)$$

Let the high-level feature set be $F_g = \{f_{g1}, f_{g2}, \ldots, f_{gk}\}$, where k is the number of high-level features. The feature selection can be used for further compression processing to reduce the noise and redundancy of data and improve the computing speed [14, 15]. Given $f_{gi} \in F_g$ and the SoftMax classifier, the loss function of the SoftMax classifier is smaller if the classification ability is stronger. Therefore, the importance of a single high-level feature to classification can be measured by using the loss function (i.e. cross-entropy) as the criterion of feature selection evaluation. A criterion for selecting high-level features is denoted by formula (5), where m is the size of the sample set.

$$\underset{SoftMax}{Min} \sum_{1}^{m} Loss(f_{gi}) \qquad (5)$$

3 Denoising Auto Encoder Feature Selection Using Cross Entropy

Based on the above analysis, a selection algorithm for High-level Denoising Auto Encoder feature based on Cross Entropy (HDAECE) is proposed. The dataset is denoising auto encoded, and the denoising auto encoding layer is fine tuned by using SoftMax classifier; then the high-level features with small loss function value in the denoising auto encoding layer are selected; the new auto encoder network is generated by pruning the coding network, and the new dataset is obtained by calculation. This new dataset can be applied to train a general classification learning algorithm for evaluation and application. The specific steps of selecting high-level features are as follows.

The operation of HDAECE algorithm is mainly divided into two steps: denoising auto encoding and high-level feature selection. The denoising auto encoding includes two steps: auto encoding and fine tuning. If dataset D contains n original features and m samples, the number of auto encoding learning cycles is $Epoch_1$, the number of fine-tuning learning cycles is $Epoch_2$, and the network of one sample at a time is calculated as a basic unit of calculation, then the time complexity of denoising auto encoder is $O\ (m(Epoch_1 + Epoch_2))$. The selection of high-level features requires the number of high-level features ($|\ F_g\ |$) iterations of m samples. In fact, when calculating the cross entropy loss, the network has only one input, which is equivalent to only one iteration of the whole network. Therefore, the time complexity of high-level feature selection is $O\ (m)$. Therefore, the time complexity of the whole algorithm only needs to consider the time complexity of denoising auto encoder, which is $O\ (m\ (Epoch_1 + Epoch_2))$.

Method: HDAECE

Input : F_g : high-level feature set

   Num: the size of selected high-level feature set

Output : $\hat{F}_g$: high-level features with less cross entropy , $|\hat{F}_g|=Num$

Steps :

(1) $\hat{F}_g \leftarrow \phi$;

(2) For each high-level feature $f_{gi} \in F_g$:

   Take it as the only input of the SoftMax classifier that is used for fine tuning;

   Calculate the value of cross entropy loss $\sum_i^m Loss(f_{gi})$;

   End

(3) Sort $f_{gi} \in F_g$ according to $Loss(f_{gi})$ ascendingly;

(4) Select top Num high-level features, and add them into $\hat{F}_g$;

(5) Return $\hat{F}_g$.

---

4 Experiments

In order to test the classification performance of high-level feature subsets obtained by HDAECE algorithm, eight open datasets of cancer gene expression datasets are used to validate the proposed algorithm. They are Nervous System, 9 Tumors, Brain Tumor,

Breast Cancer, DLBCL, ALL Leukemia, Prostate and Gliomas. In order to facilitate comparison, the original data are also discretized. HDAECE algorithm is implemented in Python 3.6 and Tensorflow 1.09 programming environment. The hardware environment is CPU 3.7G, memory 32G and GPU 12G.

After the high-level feature selection is completed, the accuracy experiments are carried out using two classifiers: Support Vector Machine (SVM) and K-Nearest Neighbor (KNN1). The accuracy of each classifier is tested by 10-fold cross-validation under different Corruption Level and different number of selected feature. Furthermore, HDAECE algorithm is also compared with the classical mRMR and SVM-RFE methods.

Set the parameters of denoising auto encoder as follows:

$|F_g|/|F|$: 0.8
Bacth Size: 5
$Epoch_1$: 500
$Epoch_2$: 500
Learning Rate: 0.01

The test steps in detail are as following. A new auto encoding network is built to generate the new training and test sample set. The classification model is obtained by training SVM or KNN1 with the new training sample set, and the accuracy can be obtained by testing the new test sample set. The process is repeated 10 times, and the average accuracy is taken as the final accuracy.

4.1 Different Corruption Level

The Corruption Level will affect the classification performance of the high-level features. The accuracy of the HDAECE algorithm with Corruption Level of 0, 0.01, 0.02, 0.03, 0.04, 0.05, 0.1, 0.15, and 0.2 is examined. Note that when the Corruption Level is 0, the denoising auto encoder does not add any noise for training, meaning that the network is a normal auto encoder. Another important parameter, the number of the selected high-level features ($|\hat{F}_g|$), is a fixed value which is taken as $|\hat{F}_g|/|F_g| = 0.2$.

Figure 2 shows the classification accuracy under different Corruption Level values. For all datasets, most of the SVM and KNN1 classifiers have greater accuracies than 0.9. Among them, the Nervous System, Breast Cancer, ALL Leukemia, Prostate and Gliomas datasets achieve the optimal accuracies close to or equal to 1 on the KNN1 classifier, while the Breast Cancer, ALL Leukemia and Prostate datasets achieve the optimal accuracies close to or equal to 1 on the SVM classifier. It can be seen that the HDAECE algorithm performs well, mainly because the denoising auto encoder acquires high-level features with strong classification ability. Even in the case of general auto encoder (Corruption Level = 0), the HDAECE algorithm performs well.

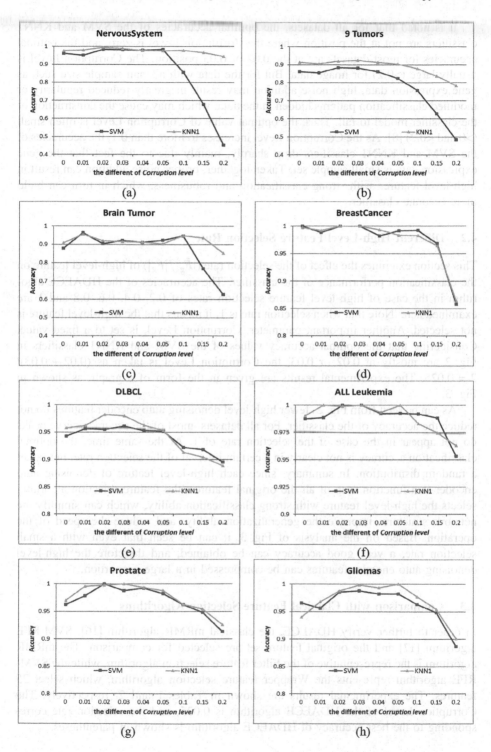

Fig. 2. Effect of corruption level on classification accuracy

It is noted that for all datasets, the optimal accuracies of the SVM and KNN1 classifiers are not at the position where the Corruption Level is zero, and the optimal accuracies for most datasets are at the 0.02 or 0.03 position. The Corruption Level is usually large (>0.2) for image data. But for the data with a small sample size such as gene expression data, high noise addition may result in greatly reduced regularity or modified classification patterns hidden in the data, which may cause the construction of the classifier model to fail. Thus, the optimal value of Corruption Level is often small (>0 and <0.2) [8]. As the Corruption Level increases to more than 0.1, the accuracies of the SVM and KNN1 classifiers drop sharply, mainly due to the fact that the gene expression data is small sample set. Taken together, proper noise addition can result in high-level features with strong classification and robustness, which in turn can build more accurate classifier.

4.2   Different High-Level Feature Selection Rate

This section examines the effect of the selection rate ($|\hat{F}_g|/|F_g|$) of high-level feature on the classification performance of the classifier. The accuracies of the HDAECE algorithm in the case of high-level feature selection rates of 0.2, 0.4, 0.6, 0.8 and 1 are examined here. Note that when selection rate is 1, it means that the high-level feature is not selected. Another important parameter, Corruption Level, is set to a fixed value. Considering that the optimal accuracy values of the SVM and KNN1 classifiers in Fig. 2 are mostly at 0.02 or 0.03, the Corruption Level is taken as $(0.02 + 0.03)/2 = 0.025$. The experimental results are given in the form of a graph, as shown in Fig. 3.

As can be seen from Fig. 3, fewer high-level denoising auto encoder features do not reduce the accuracy of the classifier. For all datasets, most of the accuracy curve peaks do not appear in the case of the selection rate of 1. At the same time, the optimal classification accuracy is not clearly in a certain interval of the selection rate, but rather a random distribution. In summary, since each high-level feature of denoising auto encoder is a function map of all the original features, the feature selection algorithm selects the high-level feature with strong classification ability, which can simplify the network structure, improve the generalization ability, and reduce the speed of the operation. Based on the analysis of Fig. 3, it can be seen that even with a small selection rate, a very good accuracy can be obtained, and therefore the high-level denoising auto encoder features can be compressed in a large proportion.

4.3   Comparison with Classical Feature Selection Algorithms

In order to further verify HDAECE, the classical mRMR algorithm [16], SVM-RFE algorithm [17] and the original feature set are selected for comparison. The mRMR algorithm is the representative of the Filter feature selection algorithm, while the SVM-RFE algorithm represents the Wrapper feature selection algorithm, which select 25 features. The experimental results are shown in Tables 1 and 2, respectively. The Corruption Level of the HDAECE algorithm is 0.025, and the selection rate corresponding to the best accuracy of HDAECE algorithm is shown in parentheses.

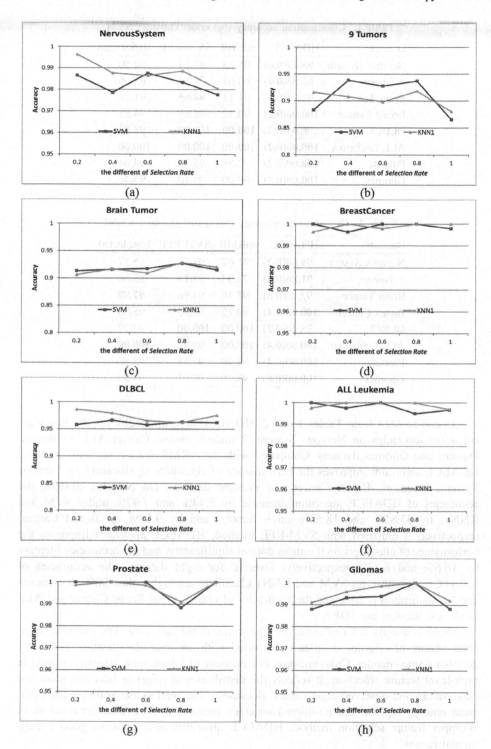

Fig. 3. Effect of selection rate on classification accuracy

Table 1. Classification accuracy (%) under classifier SVM

Datasets	HDAECE	mRMR	SVM-RFE	UnSelected
Nervous System	**98.77(0.6)**	95.33	97.74	94.50
9 Tumors	**93.84(0.4)**	60.01	77.24	65.06
Brain Tumor	92.73(0.8)	91.11	**94.65**	91.22
Breast Cancer	**100.00(0.2)**	90.72	98.86	94.85
DLBCL	96.64(0.4)	**100.00**	**100.00**	98.83
ALL Leukemia	**100.00(0.2)**	**100.00**	**100.00**	**100.00**
Prostate	**100.00(0.2)**	92.58	95.19	94.64
Gliomas	**100.00(0.8)**	98.00	99.10	95.40

Table 2. Classification accuracy (%) under classifier KNN1

Datasets	HDAECE	mRMR	SVM-RFE	Unselected
Nervous System	**99.64(0.2)**	91.67	97.52	85.17
9 Tumors	**91.86(0.8)**	73.33	83.13	85.50
Brain Tumor	92.86(0.8)	95.56	94.76	**97.89**
Breast Cancer	**100.00(0.4)**	90.72	98.92	95.98
DLBCL	98.64(0.2)	**100.00**	**100.00**	97.27
ALL Leukemia	**100.00(0.4)**	**100.00**	98.92	**100.00**
Prostate	**100.00(0.4)**	88.90	93.51	95.66
Gliomas	**100.00(0.8)**	98.10	98.67	97.20

As can be seen from Tables 1 and 2, HDAECE algorithm achieves the best classification accuracies on Nervous System, 9 tumors, Breast Cancer ALL Leukemia, Prostate and Gliomas datasets. Compared with the mRMR feature selection method, HDAECE algorithm improves the performance of classifiers significantly on Nervous System, 9 tumors, Breast Cancer and Prostate datasets. On Nervous System, the accuracies of HDAECE algorithm increase by 3.44% and 7.97% under SVM and KNN1, by 33.85% and 18.53% on 9 tumors, and by 9.28% on Breast Cancer, respectively. Compared with SVM-RFE method, HDAECE algorithm improves the performance of classifiers on 9 tumors dataset significantly, and the accuracies increase by 16.6% and 8.73% respectively. Overall, for eight datasets, the accuracies of HDAECE algorithm on SVM and KNN1 classifiers are higher than 90%. It achieves good classification results, and the accuracies of classifiers on Breast Cancer and ALL Leukemia datasets are 100%.

There is usually a lot of noise in high-dimensional raw data. The encoded high-level feature of original samples is extracted with accurate classification ability by denoising auto encoder. Cross entropy, as a measure of information diversity, is used in high-level feature selection. It reflects the distribution divergence between predicted class values and observed class values. Because of the need to use SoftMax classifier, cross entropy replaces the traditional accuracy measure as an evaluation measure of Wrapper feature selection method. HDAECE algorithm also achieves good experimental results.

5 Conclusions

In this paper, a high-level denoising auto encoder feature selection algorithm based on cross entropy (HDAECE) is proposed. Firstly, denoising auto encoder is used to transform the original data to obtain high-level features with strong characterization and robustness. Then cross entropy loss is used as evaluation criterion to select high-level features, and the classification model with good classification ability can be constructed. The rationality of the selection strategy is analyzed theoretically and verified on 8 gene expression datasets. The results show that the method of denoising auto encoder combined with feature selection achieves good classification accuracy. Compared with classical and excellent mRMR and SVM-RFE algorithms, HDAECE algorithm shows better accuracy. HDAECE can also be easily extended to stacked structures.

Acknowledgment. The authors would like to acknowledge the assistance provided by National Natural Science Foundation of China (Grant no. 61572180, no. 61472467 and no. 61672011).

References

1. Lecun, Y., Bengio, Y., Hinton, G.: Deep learning. Nature **521**(7553), 436 (2015)
2. Hinton, G.E., Salakhutdinov, R.R.: Reducing the dimensionality of data with neural networks. Science **313**(5786), 504–507 (2006)
3. Bengio, Y., Lamblin, P., Popovici, D., et al.: Greedy layer-wise training of deep networks. In: Advances in Neural Information Processing Systems, pp. 153–160 (2007)
4. Vincent, P., Larochelle, H., Bengio, Y., et al.: Extracting and composing robust features with denoising autoencoders. In: Proceedings of the 25th International Conference on Machine Learning, pp. 1096–1103. ACM (2008)
5. Masci, J., Meier, U., Cireşan, D., Schmidhuber, J.: Stacked convolutional auto-encoders for hierarchical feature extraction. In: Honkela, T., Duch, W., Girolami, M., Kaski, S. (eds.) ICANN 2011. LNCS, vol. 6791, pp. 52–59. Springer, Heidelberg (2011). https://doi.org/10.1007/978-3-642-21735-7_7
6. Wang, B., Huang, D.S., Jiang, C.J.: A new strategy for protein interface identification using manifold learning method. IEEE Trans. Nanobiosci. **13**(2), 118–123 (2014)
7. Wang, S.L., Li, X.L., Zhang, S.W., et al.: Tumor classification by combining PNN classifier ensemble with neighborhood rough set based gene reduction. Comput. Biol. Med. **40**(2), 179–189 (2010)
8. Gupta, A., Wang, H., Ganapathiraju, M.: Learning structure in gene expression data using deep architectures, with an application to gene clustering. In: IEEE International Conference on Bioinformatics and Biomedicine (BIBM), pp. 1328–1335. IEEE (2015)
9. Singh, V., Baranwal, N., Sevakula, R.K., et al.: Layerwise feature selection in stacked sparse auto-encoder for tumor type prediction. In: IEEE International Conference on Bioinformatics and Biomedicine (BIBM), pp. 1542–1548. IEEE (2016)
10. Antoniades, A., Took, C.C.: Speeding up feature selection: a deep-inspired network pruning algorithm. In: International Joint Conference on Neural Networks, pp. 360–366. IEEE (2016)
11. Srivastava, N., Hinton, G., Krizhevsky, A., et al.: Dropout: a simple way to prevent neural networks from overfitting. J. Mach. Learn. Res. **15**(1), 1929–1958 (2014)

12. Zhang, Y., Li, S., Wang, T., et al.: Divergence-based feature selection for separate classes. Neurocomputing **101**, 32–42 (2013)
13. Zou, Q., Ni, L., Zhang, T., et al.: Deep learning based feature selection for remote sensing scene classification. IEEE Geosci. Remote Sens. Lett. **12**(11), 2321–2325 (2015)
14. Taherkhani, A., Cosma, G., McGinnity, T.: Deep-FS: a feature selection algorithm for deep boltzmann machines. Neurocomputing **322**, 22–37 (2018)
15. Cai, J., Luo, J., Wang, S., et al.: Feature selection in machine learning: a new perspective. Neurocomputing **300**, 70–79 (2018)
16. Peng, H., Long, F., Ding, C.: Feature selection based on mutual information criteria of max-dependency, max-relevance, and min-redundancy. IEEE Trans. Pattern Anal. Mach. Intell. **27**(8), 1226–1238 (2005)
17. Guyon, I., Weston, J., Barnhill, S., et al.: Gene selection for cancer classification using support vector machines. Mach. Learn. **46**(1), 389–422 (2002)

Data Lineage Approach of Multi-version Documents Traceability in Complex Software Engineering

Fan Yang[1,2(✉)] ⓘ, Jun Liu[1], and Yi-wen Liang[2]

[1] City College of Wuhan University of Science and Technology,
Wuhan 430083, China
18341029@qq.com
[2] Computer School of Wuhan University, Wuhan 430072, China

Abstract. Due to the large amount of documents and versions in complex software engineering, it is very difficult to find and manage documents. In order to simplify the process of people finding and managing documents, this paper proposes a new documents organization mechanism, called DLROM, which provides dynamic and adaptive multi-version documents traceability management. DLROM introduces a new documents traceability approach based on the idea of Data Lineage, which draws on PROV Model to describe the software engineering documents and their relationship, establishes Lineage Relationship Model between documents, realizes the traceability of documents in the whole process of software development. DLROM automatically manages multi-version documents in traceability, avoiding editors manually maintaining document relationships. Finally, the paper proves that DLROM has the characteristics of low labor cost, comprehensive tracking dimension, appropriate tracking granularity and unified expression mechanism.

Keywords: Data lineage · Multi-version document · PROV model · Document traceability · Software engineering

1 Introduction

In a complex software engineering, a large number of documents are generated, such as requirement documents, design documents, code documents, test documents, etc., and the versions of these documents update constantly. We refer to these documents as Multi-Version Documents. For large-scale, long-period software projects, it is extremely difficult to maintain the consistency of document versions due to the large number of documents, complex communication, and frequent developers mobility [1]. To solve this problem, a documents traceability approach needs to be introduced, which is to trace the relation source between "requirement-design-code-test" in software engineering.

Foundation project: National Students' Innovation and Entrepreneurship Training Program (201813235005); School-level Key Research Project (2017CYZDKY006).

An overall document traceability method can guarantee the quality of software products, provide strong support for software development and maintenance, and reduce costs [2]. In recent years, a rich body of research on the document traceability method mainly divided into static and dynamic tracking technology.

The static tracking methods mainly include Tracking Graph, Tracking Matrix, Cross-reference, and Entity Relation models. Tracking Graph method [3] expresses the relationship between the requirements document, design document, source code and other information by means of graphs. Tracking Matrix (RTM) method [4] expresses the associations in the form of a matrix, marking the "Trace To" and "Trace From" relationships between the two versions of the document at the intersection of the matrix, enabling forward tracking and reverse track. Cross-reference method places a reference to one software product in another software product when there is a relationship between the two software products. Entity Relation method [5] uses E-R model to describe the tracking relationship, which can represent multiple tracking relationships, and manage the tracking relationship with a relational database. Static tracking requires manual establishment and maintenance of the tracking chain. The biggest challenge is to maintain the tracking chain. Every change in documents requires tracking chain changes. If not maintained in time, the tracking chain will deteriorate.

Dynamic tracking technology mainly includes: Information Retrieval (IR) [6], Heuristic Rules (HR) [7], System Running Real-time Analysis (SRRA) [8] and so on. The IR technology first calculates the similarity of the project document, then sorts the similarity based on the size, and finally sets a threshold to filter the relationship that satisfies the condition. The HR technology has established its own matching rules for the relationship between software requirements, use cases and software products, and uses natural language to deal with the logical relationship and context between software products. The SRRA records the execution relationship between the execution module and the software document by detecting the execution information generated when the software is executed, thereby establishing a tracking relationship between the execution module and the software document. Dynamic tracking technology solves the problem of static tracking, but these technologies have obvious limitations, mainly staying in academic research. There is no unified expression mechanism, and it is difficult to promote it in actual software engineering.

In summary, the existing document relationship management methods mainly classify by file name. Such methods are inefficient, lack a unified expression mechanism, and are difficult to maintain. The essential cause is that they cannot fully reveal the traceability relationship between documents. An effective method must consider four main issues, namely, labor cost issues, documents traceability dimension issues, documents traceability granularity issues, and standard issues of expression mechanisms.

We propose a new multi-version document traceability method called DLROM. To with the above four problems, DLROM uses the PROV model [9] developed by W3C as the standard expression mechanism. DLROM maps the people, documents, behaviors, associations and other elements involved in the software development process to the "entity", "activity" and "agent" in the PROV model. DLROM describes the origin information of the document in the software development life cycle, and establishes an automatic tracking scheme, which provides a better solution for the traceability problem of multi-version documents in complex engineering.

2 Backgrounds and Problem Statement

Data-Lineage [10] emerged in the 1990s as a description of the entities and processes that generate, transmit, or influence data. The W3C [11] organization defines it as "a source is a resource that describes the entities and processes that generate, transmit, or influence other resources. Lineage information is the basis for resource credibility assessment, giving trust, and allowing regeneration".

In the development of the Data Lineage model, Sudha et al. [12] proposed 7 W Model, which indicates that the Data-Lineage information consists of seven parts, what, who, where, when, which, how, and why. In the first IPAW meeting in May 2006, the researchers formed some unified opinions on the expression methods of Data Lineage, and drafted and proposed a new data model. Moreau et al. [13] proposed the OPM model by combing the consensus reached and describing it in a formal way. The OPM model defines a set of specific rules and objects to represent the context between data and specifies formal representations, which become the criterion for exchanging origin data in many fields.

In 2013, the W3C defined the first Data-Lineage standard PROV [14–17], which is a key milestone in the world's Data-Lineage specification, making the origin of data widely used and the exchange between WEB and other information management systems. As possible. The PROV model [15] is mainly composed of three parts, Entity, Activity, and Agent, as shown in Fig. 1.

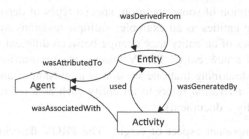

Fig. 1. PROV core framework

Entity. In PROV, objectively existing, numerical, conceptual, or other types of things are called *Entities*. For example, web pages, charts, or tools for spell checking. The origin of an entity is described by an origin record or by multiple other entities. For example, document D is an entity whose origin can point to an entity such as a chart in which document D is inserted, a data set used to create the chart, etc.; *Entities* can be described from different perspectives with different attributes, in this case representing different *Entities*. For example, the document D stored in "My Folders", the document D in the second edition, and the document D being written are 3 *Entities* of different origins.

Activity. *Activity* mainly describes how the *Entity* becomes the current state and how the properties of the *Entity* change, making it a new *Entity*. Usually, the Activity acts on an existing Entity and is described by actions, actions, and the like.

Agent. *Agent* assumes the responsibility of role in the *Activity*. An *Agent* can be software, an organization, or a person. When the *Agent* is responsible for *Activity*, PROV uses *wasAssociatedWith* to indicate the relationship between *Agent* and *Activity*. Multiple *Agents* can be related to the same *Activity*, and vice versa. *Entity* belongs to an Agent. Use *wasAttributedTo* to indicate the responsibility of the Agent for that Entity, or it may be jointly responsible with other Agents. In short, the Agent is responsible for the *Activity* that generates the *Entity*.

In addition to the above three core concepts, the PROV model also defines a variety of relationships to associate them. The following describes the relationships that will be used later.

Usage and Generation. *Activities* generate new *Entities*. For example, writing a document generates a new document, and modifying the document can also generate a new version. Activities also utilize entities. For example, modifying a misspelled update document takes advantage of the original version of the document and the updated list.

Derivation and Revision. *Derivation* and *Revision* refer to the relationship between entities. When the existence, content and nature of an entity originate from other entities, the former is derived from the latter. For example, a document contains content that is copied to other documents, and a chart comes from the data it describes. PROV allows for the description of some common, special types of derivatives.

Taking document entities as an example, multiple revisions may occur over time. One or more attributes of an entity may change between different versions. In PROV, each new version is a new entity. PROV allows one version to be associated with another version, by describing that one version is a revision of another version.

Another kind of *derivation* refers to an entity such as a reference that references another entity (usually a document).

Time. *Time* is an important aspect of origin. The PROV describes a time series of important events, including when the entity was generated or utilized, and when the activity started and ended. For example, the model can describe when a new version of the document was created (time of generation), or when the document was edited (the beginning and end of the editing activity).

The PROV model, as a computer-generated source of information representation framework, allows users to represent and exchange many pieces of information related to origin data in a variety of ways, such as XML [18]. In addition, PROV also proposes methods for acquiring, using, and verifying lineage information [19].

The process of software development is to generate the iterative development process of requirement document, summary design, detailed design, coding and testing in turn [20]. It can be recorded and described by Data Lineage technology, that is, Data Lineage can be used to record the lifecycle activities in software development, including document creators, time, notes, dependencies, modification information, etc., to establish a relationship between different versions of the documents.

3 Model and Design

Inspired by the idea of Data Lineage, we propose a Documents Lineage Relationship Organization Mechanism, called DLROM. We apply Data Lineage to software development in DLROM, to record the evolution information of software products (requirements documents, design documents, codes, etc.) throughout the life cycle. Then DLROM realizes the document traceability in software development. Based on the W3C's PROV model, this chapter extends the vocabulary of PROV and establishes the DLROM model of the software engineering documents traceability.

3.1 Data-Lineage Model of Multi-version Documents

According to the representation framework of the PROV model, this paper designs Documents Lineage model, as shown in Fig. 2. Record and maintain the evolution history of software products (requirement documents, design documents, code documents, test documents) and the relationship between software documents by extracting lineage information in the software development life cycle.

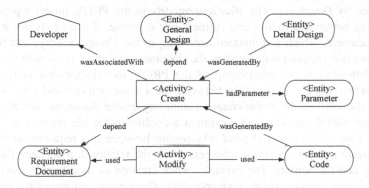

Fig. 2. Documents lineage model

In a software project, the developer's operation on the document mainly includes modifying the document, generating multiple versions of the document, and submitting a document according to the previous document in the logical relationship, thereby generating a dependency relationship. There are also some input of parameters for the operation of the document, such as the operator, the operation time, the description of the main content of the document, and so on. The PROV model describes the relationship between software products in software development, and mainly includes three types of information.

(1) *Entity* **information in software development**

This paper extends the *Entity* in the PROV model and applies it to the software development environment to specifically represent some entities in software development. *Require Document* represents the requirements document in software development, including software specifications, requirements changes, and so on. *General*

Design represents the overall design specification, which is dependent on the Require Document. *Detail Design* represents detailed design documentation, including database design, file system design, class design, etc., which is dependent on *General Design*. *Code* represents the implementation code of the system, which has dependencies on *General Design*. *Parameter* indicates the description information that the software developer fills in when submitting or modifying the document or code.

(2) *Activity* **information in software development**

This article extends the *Activity* in the PROV model to represent some of the operations in software development. *Create* represents a create operation, such as creating a document, code, and so on. *Modify* means modify operation, for example, modify the document, code, and so on.

(3) *Association* **information in software development**

Used in the PROV model indicates the relationship between the *Modify* activity and the *Entity*, such as the commit operation after modifying the requirements document, the commit operation after modifying the code, and so on. *WasAssociatedWith* in the PROV model represents the relationship between the *Activity* and the *Agent*. For example, if the software developer creates a requirement document, the *Create* activity is executed by *Developer*. The *WasGeneratedBy* in the PROV model represents the relationship between activities and entities. For example, if a user creates a detailed design document, which is generated according to the *Create* activity, and there is a *WasGeneratedBy* relationship between the document and the *Create* activity.

The following is the relationship of the PROV model extended in this article. *Depend* represents the relationship between the *Create* activity and the *Entity*. For example, the software developer creates a summary design document, which there is a relationship with the requirement document according to the life cycle of the software function. That is, there is a *Depend* relationship between this requirement document and the create activity. *HasParameter* represents the relationship between the Parameter entity and the activity. For example, when creating or modifying a document or code, the user inputs some corresponding description information, that is, a *HasParameter* relationship exists between the *Parameter* entity and the activity.

In the above, the PROV model is applied to the environment of software development by summarizing and summarizing the relationship between the whole life cycle of the document and the code, thus realizing the dynamic document traceability model in software development.

3.2 Expression of Documents Lineage Information

When a function changes in the process of software development, the software developer wants to know which design documents and code will be changed. The PROV model provides a rich vocabulary to abstract the origin process. For the software development process, Table 1 shows the *Entity* section.

The core entities in software development are various types of data objects. These data objects include Word documents, Cdm database design files, Asda class design files, code files, and so on.

Table 1. Data-Lineage entity in software development

Entity type	Description	Origin
Require Document	Documents generated during the requirements analysis phase	Expansion
General Design	Documents produced during the outline design phase	Expansion
Detail Design	Documents produced during the detailed design phase	Expansion
Parameter	Description information entered by the user when Create, Modify activity	Expansion
Code	Code generated during the project development phase	Expansion

In the PROV model, *Activity* refers to an action that acts on an entity and produces a new entity or changes the state of the entity. For the software development process, Table 2 shows *Activity* section.

Table 2. Data-Lineage activities in software development

Activity type	Description	Origin
Create	Create documents, code, etc.	Expansion
Modify	Modify the document, code, etc.	Expansion

Software development process activities include two interactions, *Create* and *Modify*. *Create* activity refers to the software developer to create a document or code. For example, the software developer creates a summary design document according to the requirements document. *Modify* activity refers to the software developer modifying the document or code. For example, for a summary design document, the software developer thinks that the previous design is unreasonable, modifies the summary design document, and then submits it to the document version management system, covering the previous version.

Table 3. Data-Lineage relation of software development

Relation type	Description	Origin
Used	Relationship between the Modify activity and the entity when modifying the document and code	PROV Model
WasAssociatedWith	Relationship between users and activities	PROV Model
WasGeneratedBy	When an activity acts on an entity, a new entity is generated, and the relationship between the activity and the entity	PROV Model
Depend	Create activity and entity relationship	Expansion
HadParameter	Relationship between activity and Parameter entity	Expansion

In PROV model, *Relation* refers to the relationship between the activity and the entity. For example, some content of a document is modified, then a new document is generated, so the modification activity has a *Used* relationship for the document entity. For the software development process, Table 3 shows *Relation* section.

The software development process mainly includes seven relationships, three of which are defined in the PROV model, and the other four are extended by this paper.

4 Implementation Scheme

The previous chapter proposed a Documents Lineage Model for the multi-version documents traceability. However, in a specific business environment, different process instances correspond to different DLROM instances. This chapter takes an *office system software development project* as example, constructs a corresponding DLROM instance, and designs a set of automatic document traceability Data-Lineage scheme. This scheme combines SVN version control tools to extract lineage information, completes the bidirectional trace of software engineering, automatically provides developers with traceability information between software products. DLROM implementation scheme consists of three parts, including the lineage information extraction part, the lineage information organization and storage part, and the lineage information reading, as shown in Fig. 3.

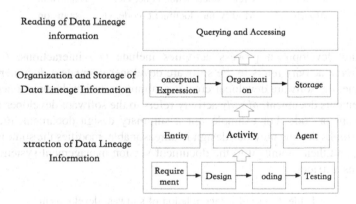

Fig. 3. DLROM instance overall framework

4.1 Extraction of Data Lineage Information

The process of software development is to sequentially generate the requirements document, summary design, detailed design, etc., and iterative development process of the code. As shown in Fig. 4.

Figure 4 depicts an example of a software development process. The software developer first created the requirement document "*Requirement_v1*" according to the user's requirements. Secondly developer generated the corresponding summary design

"*General_v1*" according to the requirements document. Thirdly, the developer modified the summary design and generated the summary design document "*General_v2*". Finally, the detailed design of the summary design is carried out, and the documents "*Detail_DB_v2*" and "*Detail_Class_v3*" are generated.

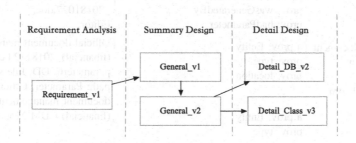

Fig. 4. Software development process

We constrain the developer to provide the document name and file type identifier on which the document depends when creating the document. For example, the requirement document is *RD,* the summary design is *GD*, the detailed design is *DD*, and the code is *CD*. In order to establish dependencies between documents, the annotation structure specified in this paper is as follow.

(Depended Full Name) - (Document Type Identifier) - Document Comment.

For example, "*General_v2.doc-DD-document management CMD model diagram*", where "*General_v2.doc*" represents the full name of the document it depends on, "*DD*" indicates that the type of the submitted document is a detailed design, and "*document management CMD model diagram*" indicates the main part of this document.

4.2 Organization and Storage of Documents Lineage Information

DLROM does not impose strict restrictions on the specific description granularity, but the expression ability can be only the combination and refinement of the lineage information under different granularity. As for which expression should be adopted, it depends on the specific application requirements. Users should be able to choose to record or view lineage information from different levels.

DLROM model corresponds to the actual process of software development. According to the original vocabulary and dynamic requirements tracking of the abstract expression of the PROV model, the above example scenario of software development can be described by the PROV model, which can describe the software development from the initial creation of the requirements document, and finally to the corresponding coding implementation. The lifecycle process of a software product, as well as the specific activities, activity performers, and associated information for that change.

The recording and storage of document origin information is in RDF format, which is one of the standard description languages of the PROV model provided by W3C. As shown in Table 4, the following code describes the lineage information of some documents.

Table 4. Storage of documents lineage information

Storage content	Model description	Remarks
% prefix part was ignored : op6	a prov: Activity; prov: opTime; prov: type; prov: depend prov: wasAssociatedWith prov: wasGeneratedBy prov: hadParameter	2018-12-06 T18:50:02; Create; Overall design specification _20181027.doc; Editor;
: Official document management (financial) _20181124.cdm	a prov: Entity prov: type prov: location	Official document management (financial)_20181124.cdm; parameter6; DD; File storage path; Parameter; Official document management
: parameter6	a prov: Entity prov: type prov: content	(financial) CDM Model;
: Yang	a prov: Agent prov: name	Developer

4.3   Querying of Documents Lineage Information

The Lineage information is queried using the SPARQL language provided by the W3C organization, which provides four query methods. The corresponding relationship with this solution is as follows.

(1) SELECT mode returns the associated subset of the specified document or agent.
(2) The CONSTRUCT method returns the corresponding software development RDF map information according to the given constraints.
(3) The ASK method uses a Boolean value to describe whether the corresponding information is matched.
(4) The DESCRIBE method describes the related information found in the form of an RDF graph.

Subsequent product query code for the requirements document is shown as follows.

```
SELECT ?documentName ?opTime ?documentType ?relationName
  FROM <DLROM_model.rdf>
  WHERE {
    ?document prov:type RD .
    ?document prov:name ?documentName .
    ?document prov:opTime ?opTime .
    ?document prov:type ?documentType .
    ?relation prov:depend ?document.
    ?relation prov:name ?relationName .
    ORDER BY [?opTime]
    }
```

4.4 Feasibility Experiment

According to the scheme proposed above, this chapter designs an automatic Multi-Version document traceability tool DRTT. In Lineage information collection stage, the DRTT obtains the evolution information, such as time, operator, operation, description, etc. by reading the log file of the SVN. In Lineage information storage and organization stage, the lineage log is stored as an RDF file by using RDF, and the organization is a directed acyclic graph, in other words, it is stored and organized according to the DLROM model. In the query and access phase, the qualified software product association diagram is queried using the SPARQL language.

According to statistics, most of the current document traceability methods use Tracking Matrix. So we compare Tracking Matrix method with DLROM method. In our experiment, the document traceability matrix indicates that there is a dependency between the software design document and the requirement document, but it does not indicate the type of relationship, and ignores the information of the dimensions of personnel and time. It is not possible to track the responsibility of the personnel and view the progress of the project. It is impossible to describe the ins and outs of the documents. However, the DLROM method does not have these defects.

5 Conclusion and Future Work

By analyzing the lineage information involved in multi-version document, this paper establishes DLROM model and expands the lineage vocabulary under the document version traceability application scenario. The model clearly describes the evolution process and the relationships of all documents in the software engineering. we designed the DLROM scheme for software development team management and project schedule management.

Of course, there is still a larger research space for DLROM of this paper. The future work we intend to do is as follows.

Firstly, we will create a more intelligent lineage information query system. This paper uses the SPARQL language provided by the W3C organization to query. Next, we are going to abandon the SPARQL language and create the intelligent query system based on human language.

Secondly, we will refine the lineage relationship of modules or classes in the code. The lineage relationship established in this paper is based on the entire code document. In the next step, we will refine the lineage relationship of modules or classes in the code.

References

1. Dekhtyar, A., Hayes, J.H.: Automating Requirements Traceability: Two Decades of Learning from KDD (2018)
2. Ali, N., Gueneuc, Y.G., Antoniol, G.: Trustrace: mining software repositories to improve the accuracy of requirement traceability links. IEEE Trans. Softw. Eng. **39**(5), 725–741 (2013)

3. Sangwan, R.S., Qiu, R.G., Jessen, D.: Using RFID tags for tracking patients, charts and medical equipment within an integrated health delivery network. In: Networking, Sensing & Control. IEEE (2005)
4. M-Net Demo, International Center for Software Engineering, University of Illionois at Chicago. www.iese.eecs.uic.edu
5. Saleiro, P., et al.: RELink: A Research Framework and Test Collection for Entity-Relationship Retrieval (2017)
6. Duarte, A.M.D., Duarte, D., Thiry, M.: TraceBoK: toward a software requirements traceability body of knowledge. In: Requirements Engineering Conference. IEEE (2016)
7. Spanoudakis, G.: Plausible and adaptive requirement traceability structures. In: International Conference on Software Engineering & Knowledge Engineering DBLP (2002)
8. Rempel, P., Mader, P.: Preventing defects: the impact of requirements traceability completeness on software quality. IEEE Trans. Softw. Eng. 43(8), 777–797 (2017)
9. Jing, N., Xianxue, M.: PROV model and its web application. Libr. Inf. Serv. 58(3), 13–19 (2014)
10. Niu, X., et al.: Provenance-aware query optimization. In: IEEE International Conference on Data Engineering (2017)
11. W3C Provenance Working Group. Provenance Working Group Wiki Main Page. http://www.w3.org/2011/prov/wiki/Main_Page. Accessed 21 June 2012
12. Simmhan, Y.L., et al.: A survey of data provenance in e-science. ACM Sigmod Rec. 34(3), 31–36 (2005)
13. Moreau, L., Freire, J., Futrelle, J., McGrath, R.E., Myers, J., Paulson, P.: The open provenance model: an overview. In: Freire, J., Koop, D., Moreau, L. (eds.) IPAW 2008. LNCS, vol. 5272, pp. 323–326. Springer, Heidelberg (2008). https://doi.org/10.1007/978-3-540-89965-5_31
14. Groth, P., Moreau, L.: An Overview of the PROV Family of Documents. https://www.w3.org/TR/prov-overview/. Accessed 30 Apr 2013
15. Moreau, L., Missier, P.: PROV-DM: The PROV Data Model. https://www.w3.org/TR/2013/REC-prov-dm-20130430/. Accessed 30 Apr 2013
16. Gil, Y., Miles, S.: PROV Model Primer. https://www.w3.org/TR/2013/NOTE-prov-primer-20130430/. Accessed 30 Apr 2013
17. Lebo, T., Sahoo, S., McGuinness, D.: PROV-O: The PROV Ontology. https://www.w3.org/TR/2013/REC-prov-o-20130430/. Accessed 30 Apr 2013
18. Moreau, L.: PROV-XML: The PROV XML Schema. https://www.w3.org/TR/2013/NOTE-prov-xml-20130430/. Accessed 30 Apr 2013
19. Davis, D.B., et al.: Data provenance for multi-agent models. In: IEEE International Conference on E-science (2017)
20. Baracaldo, N., et al.: Mitigating poisoning attacks on machine learning models: a data provenance based approach. In: The 10th ACM Workshop (2017)

Resource Efficiency Optimization for Big Data Mining Algorithm with Multi-MapReduce Collaboration Scenario

Zhou Fengli[✉] and Lin Xiaoli

Faculty of Information Technology,
Wuhan College of Foreign Language and Foreign Affairs,
Wuhan 430083, China
thinkview@163.com, aneya@163.com

Abstract. Because any MapReduce job requires a series of complex operations such as task scheduling and resource allocation independently, there are a lot of redundant disk I/O and resource duplicate application operations among multiple MapReduce jobs coordinated by the same algorithm, causing inefficient resource utilization in job computing process. Big data mining algorithms are usually divided into several MapReduce Jobs, taking ItemBased algorithm as an example, this paper has analyzed the resource efficiency of mining algorithm with multi-MapReduce job collaboration scenario. It proposed an ItemBased algorithm based on DistributedCache, which used DistributedCache to cache I/O data between multiple MapReduce Jobs, breaks the defect of independence between jobs, and reduced the waiting delay between Map and Reduce tasks. The experimental results show that, DistributedCache can improve the data reading speed of MapReduce jobs. The algorithm reconstructed by DistributedCache greatly reduces the waiting delay between Map and Reduce tasks, and improves the resource efficiency by more than three times.

Keywords: MapReduce optimization · ItemBased algorithm · Memory file system · I/O efficiency · Resource optimization

1 Introduction

According to the report released by IDC (Internet Data Center) in 2018, the total data generated in the world was close to 20 ZB and this number will be expected to reach 44 ZB in 2020 [1]. The explosive growth of data has brought opportunities and challenges to IT industry. Since Google had published a paper in 2003 that had introduced the distributed storage system GFS [2] and computational model MapReduce [3], MapReduce computing model has gradually become the most common underlying computing framework for big data systems such as Hadoop, Spark, Pig, Hbase and Hive.

When MapReduce job is decomposed into several tasks that don't exist in isolation by scheduling system, the process of collaborative execution between tasks requires a large number of operations in disk read/write and network transmission. The task execution process of MapReduce is shown in Fig. 1. Operations such as Split,

© Springer Nature Switzerland AG 2019
D.-S. Huang et al. (Eds.): ICIC 2019, LNAI 11645, pp. 503–514, 2019.
https://doi.org/10.1007/978-3-030-26766-7_46

RecordReader, Partition and Shuffler have transferred the intermediate calculation results between the network of different machines and saved them to disk as operations after pipeline input. Due to the high complexity of the big data mining algorithm based on MapReduce model, it usually needs to be coordinated by multiple MapReduce jobs. For example, the PageRank algorithm has been decomposed into four jobs, and the Bayes algorithm has been decomposed into ten jobs. But the resource scheduling and management of these multi MapReduce jobs are not coordinated, and the operations of serious redundant disk read/write and repeated I/O resource application have resulted in low resource utilization efficiency [4–6].

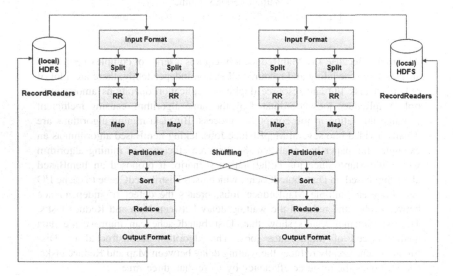

Fig. 1. MapReduce's task execution process

Firstly, this paper has analyzed the resource efficiency problem existed in Item-Based algorithm in MapReduce environment. Secondly, I/O data that coordinated by the same algorithm between multi-MapReduce jobs has been cached in a unified manner based on the problem analysis. It can reduce redundant I/O operations in the process of algorithm calculation overall and optimize the efficiency of resource utilization. Finally, the improvement of resource utilization efficiency has been verified by comparative experiments.

2   Related Research

With the exponential growth of algorithm's processing data volume and the gradually popularization of MapReduce, the algorithm will gradually develop from the stand-alone model to distributed model, and the calculation and resource utilization efficiency of the algorithm will gradually be accepted by academic community.

Literature [7] and literarture [8] had attempted to Improve the computational efficiency of User-Based collaborative filtering algorithm by the parallelism of MapReduce program. The algorithm had been implemented on Hadoop platform, and algorithm's computing efficiency had been further improved by flexibly setting the number of job maps and reduce tasks. Schelter et al. in literature [9] had re-implemented the algorithm based on MapReduce model for the problems of poor scalability when the data would rapid growth in the filtering algorithm with similarity-based neighborhood, and it had improved the computing efficiency by the test of 700 million Yahoo music data. Literature [10] had introduced MinHash clustering, Probabilistic Latent Semantic Indexing (PLSI) and Covisitation counting technology into Google News' collaborative filtering algorithm, which MinHash had used for probability clustering in the scenarios with low clustering accuracy. Then the calculation speed of algorithm would be improved further by using MapReduce, BigTable and other technologies. Literature [11] had divided the core computing steps into four MapReduce jobs in collaborative filtering algorithm, and it had reduced the amount of data transmission during the execution of algorithm by using the data partitioning strategy, the experiments had showed that the utilization efficiency of network resources in ItemBased recommendation algorithm were improved. However, whether it is User-Based or ItemBased algorithm, the collaborative filtering algorithm in MapReduce environment needs multiple jobs to cooperate. There are a lot of redundant I/O and resource re-application operations between MapReduce jobs. There is still a large optimization space for resource utilization in the algorithm. At present, some existing researches select platform porting for solving the problem of MapReduce data mining algorithms with low resource efficiency. For example, the algorithm can be transplanted from Hadoop platform to spark platform [12], and its efficiency will be improved by using the advantages of Spark memory and iterative computing. But the migration of the algorithm platform mainly has the following two problems.

(1) Migration cost problem. The algorithm has been transplanted from the existing Hadoop to Spark platform. Since the development language is from Java to Scala, the business code of the entire implementation must be rewritten according to the new API and syntax [12]. During the transport process, research personnel needs to face higher learning and development costs, while system administrators needs to face higher system migration and deployment costs.

(2) Stability problem. After the algorithm has been transplanted to the new platform which has not undergone long-term stability testing, the algorithm's stability will easily affect the service quality of upper layer application.

Based on the above discussion, this paper has taken the ItemBased recommendation algorithm as an example to analyze the resource efficiency defects in MapReduce platform, and has proposed to unify the I/O data between multi-MapReduce jobs coordinated by the same algorithm. The cache processing can reduce the redundant I/O operations and achieve the purpose of optimizing resource utilization efficiency. Compared with the previous platform migration scheme, it is not necessary to re-develop the service code and can ensure the system's stability when it can save the costs of migration and deployment.

3 Resource Efficiency Deficiencies in Multi-MapReduce Job Collaboration

Because the similarity between items is relatively stable, the Item-Based recommendation algorithm is more suitable for offline calculation than other algorithms (such as popularity, content-based, model-based algorithms, etc.). The core idea of the Item-Based algorithm is that "birds of a feather flock together", it assumes that an Item of interest to user must be similar to an Item with a higher score. The algorithm first calculates the user's preference for the item, then it calculates the similarity between the items, finally it finds the top-N items that are most similar to each Item. Table 1 has shown the input parameters of ItemBased recommendation algorithm in machine learning library Mahout under Hadoop platform.

Table 1. Main parameters of item-based recommendation algorithm

Parameter name	Significance
input	Enter the file address of HDFS, where the line data file format is: userid itemid preference
output	Output path of HDFS file system
numRecommendation	Recommended quantity (Top-N)
usersFile	Recommended userFile
itemsFile	Recommended itemFile
maxPrefsPerUser	Threshold setting: maximum preference value
minPrefsPerUser	Threshold setting: minimum preference value
maxSimilaritiesPerItem	Calculate the maximum number of similar items for each Item
threshold	Remove the item pair that is less than the parameter threshold
similarityClassname	Specifies the method class to be called when performing similarity calculations

The ItemBased algorithm mainly includes the following four stages:

(a) Preparing two matrices of User-Item and Item-Item.
(b) Calculating the similarity between User-Item and Item-Item matrices.
(c) Using the same Item for similarity matrix and aggregating it as a key to prepare for subsequent matrix multiplication.
(d) Calculating the recommended vector.

The above four steps can be subdivided into the following nine MapReduce jobs by further decomposition, and the execution sequence of the specific MapReduce job has shown in Table 2.

Table 2. MapReduce job execution decomposition of item-based recommendation algorithm

Step	MapReduce stage	MapReduce job name
1		ItemIDIndexMapper-Reducer
2	PreparePreferenceMatrixJob	ToItemPrefsMapper-Reducer
3		ToItemVectorsMapper-Reducer
4		CountObservationsMapper-Reducer
5	RowSimilarityJob	VectorNormMapper-Reducer
6		CooccurrencesMapper-Reducer
7		UnsymmetrifyMapper-Reducer
8	partialMultiply	partialMultiply
9	RecommenderJob	PartialMultiplyMapper-Reducer

A single MapReduce job is decomposed into multiple map and reduce tasks by task scheduler. The input data is read from the file system such as HDFS and sent to the corresponding Map task through the setting value of input parameter. When Map calculation is finished, the key-value data <K, V> will be sent to the specified Reduce task by Shuffle and Sort operations, where <key, Iterator <value>> will be used as the data input format in Reduce calculation process, and the Reduce result will be written to the file path that is specified in output parameter. Due to the independence of MapReduce jobs, any MapReduce job requires a series of complex operations such as task scheduling, resource allocation, HDFS data reading, task calculation, data shuffle, and result output. Especially for the case that the ItemBased recommendation algorithm contains nine MapReduce jobs, it needs to execute eighteen reads/writes to HDFS separately in the implementation by taking disk I/O resources as an example. These high-frequency data read/write operations can reduce the utilization efficiency of MapReduce cluster resources and the overall operating efficiency of the algorithm. And because the disk I/O resources are the performance bottleneck of MapReduce cluster, the resource application and release operations are prone to resource competition, which makes the waiting phenomenon between Map and Reduce tasks, so computational efficiency of the algorithm is easy to be reduced. The above analysis shows that the performance optimization of Item-Based algorithm in MapReduce environment cannot be started from optimization itself. The repetitive disk I/O read and write operations that caused by the independence between MapReduce jobs are the main reason for the inefficiency utilization and operation of resources in the execution of ItemBased algorithm.

4 MapReduce Job Resource Optimization in Item-Based Algorithm

4.1 Multi-job Operation Efficiency Analysis in Item-Based Algorithm

It can be shown from Table 2 that one calculation process has divided into four stages in ItemBased recommendation algorithm and included nine MapReduce jobs.

Supposing that the total running time of an algorithm on MapReduce platform is T_{sum}, the computing process is divided into N MapReduce jobs, and the completion time of each MapReduce job is T_i. When a MapReduce is completed, resource preparation will be required before the next MapReduce job is executed. Assuming that the resource preparation time between the two MapReduce jobs i and $i+1$ is P_i^{i+1}, then the overall completion time T_{sum} is shown in Eq. (1).

$$T_{sum} = \sum_{i=1}^{N} T_i + \sum_{i=1}^{N-1} P_i^{i+1} \tag{1}$$

In order to exclude the impact of resource status to the runtime of real MapReduce job, assuming that the data input time T_i^{input}, calculation result output time T_i^{output}, resource waiting time T_i^{wait} during job execution and other variable time are called resource preparation time. Then the resource preparation time P_i^{i+1} of a MapReduce job i is shown in Eq. (2).

$$P_i^{i+1} = T_i^{input} + T_i^{output} + T_i^{wait} \tag{2}$$

After excluding the impact of resource state, the actual running time of a single MapReduce job is accumulated by the running time of each job's sub-stage map, shuffler and reduce.

$$T_i = T_{map} + T_{shuffler} + T_{reduce} \tag{3}$$

From Eqs. (1) to (3), the execution time T_{sum} of entire algorithm can be obtained by Eq. (4).

$$T_{sum} = \sum_{i=1}^{N} \left(T_{map} + T_{shuffler} + T_{reduce} \right) + \sum_{i=1}^{N-1} \left(T_i^{input} + T_i^{output} + T_i^{wait} \right) \tag{4}$$

In Eq. (4), the running time of the job's sub-phases map, shuffler and reduce is closely related to the tasks' number, node computing power, resource status and data volume. If they are the same, the resource status is the main factor that can affect the running time of the job. It can be seen that the resource utilization efficiency (collaboration efficiency) between MapReduce jobs can be improved theoretically and the defect of independence can be broken. Then it can achieve the purpose of improving the computational efficiency of complex MapReduce algorithms.

4.2 Resource Optimization Method of Multi-MapReduce Job

In general, there are two ways to optimize resource efficiency between multi-MapReduce jobs.

(1) Using the distributed cache mechanism which built into the MapReduce framework to break the barrier of resource sharing between MapReduce jobs. The cache system can distribute the shared data to the memory of each compute node and improve the computational efficiency of the job.

(2) In disk-based distributed files (such as HDFS), a memory-centric and virtual distributed storage system has been introduced to load data resources shared between multiple MapReduce jobs from HDFS into distributed shared memory, thus the call times of disk I/O will be reduced.

The first method is more flexible for a specific algorithm level, but part algorithm's code needs to be reconstructed and deployed, so it needs to be partially modified. The second method deploys the new distributed memory file system into the production system, which easily affects the stability of the entire system, but the algorithm's code does not need to be modified. For the single ItemBased algorithm, the first method is used in this paper to improve the efficiency of resource utilization. The steps of ItemBased algorithm based on DistributedCache are as follows.

Algorithm 1. Item-Based algorithm based on Distributed-Cache
INPUT: Parameter1: <userID, itemID, score> user-item matrix,
       Parameter 2: *InputURL*
OUTPUT: Parameter1: <userID, itemIDS> recommended results,
       Parameter 2: *OutputURL*
① **for** *i*=0 to readData(*InputURL*).*blocksize-1* **do**
②   block ← MEMORY.*load(InputURL)*   // Reading file block
③   DistributedCache.*addCacheFile*(block) // Loading file blocks into memory
④ **end for**
⑤ **foreach** job **in** List<PreparePreferenceMatrixJob> *list*
⑥   MapReduce.*start().getJob(list)*
⑦   context.*setCacheFiles*(PreparePreferenceMatrixJob)
⑧   context.*getCacheFiles*()
⑨ **end for**
⑩ **foreach** job **in** List<RowSimilarityJob> *list*
⑪   MapReduce.*start().getJob(list)*
⑫   context.*setCacheFiles*(RowSimilarityJob)
⑬   context.*getCacheFiles*()
⑭ **end for**
⑮ MapReduce.*start().getJob*(partialMultiply)
⑯ context.*setCacheFiles*(partialMultiply)
⑰ context.*getCacheFiles*()
⑱ context.*setCacheFiles*(PartialMultiplyMapper-Reducer)
⑲ context.*getCacheFiles*()
⑳ MapReduce.*start().getJob*(PartialMultiplyMapper-Reducer)
㉑ **return** Job.*addCacheFile*(new File(" *OutputURL* "))

Line ①–④ of ItemBased algorithm has loaded the input data into the distributed cache according to the parameter *InputURL*; Line ⑤–⑨ has executed all MapReduce tasks in the PreparePreferenceMatrixJob stage and shared the data of three jobs that are named ToItemVectorsMapper-Reducer, ItemIDIndexMapper-Reducer and ToItemPrefsMapper-Reducer in memory; Line ⑩–⑭ has executed all MapReduce tasks in the RowSimilarityJob stage, and the DistributedCache has implemented the memory data sharing between the four jobs that are named CountObservationsMapper-Reducer, VectorNormMapper-Reducer, CooccurrencesMapper-Reducer and UnsymmetrifyMapperReducer; Since the partialMultiply and RecommenderJob phases have composed of a single MapReduce job, line ⑮–⑳ has implemented the memory data read and write operations of two jobs that are named partialMultiply and PartialMultiplyMapperReducer. Finally, line ㉑ has written the algorithm's calculation results into the memory file.

5   Experimental Evaluation and Comparison

5.1   Experimental Environment Configuration

In order to compare the execution efficiency and resource utilization of Item-Based algorithm in different test environments, two experiments under the condition of MapReduce native environment and DistributedCache memory cache have been designed in this paper. The first experiment has tested the data throughput, and the second experiment is a comparison test for ItemBased recommendation algorithm. The total number of experimental cluster nodes is 11, in which the number of master nodes (Master) is 1 and the number of job execution nodes (Workers) is 10. The cluster node experimental environment is shown in Table 3.

Table 3. Description of experimental environment

Project	Description
Operation System	Ubuntu 14.04.5 LTS (Trusty Tahr)
JVM Version	1.8 for Linux OS
Hadoop Version	2.7.1
Node CPU	Intel core i5 Skylake 3.2 GHz
Node Memory	8 GB-DDR4-2400 MHz (4 GB*2)
Node Disc	Seagate Barracuda 1 TB 7200 32 MB SATA3 ST31000524AS
Network Information	TP-LINK TG-3269C PCI RJ-45

5.2   Comparison Test for Data Throughput of MapReduce Job

In order to test the difference in data throughput under the condition of MapReduce native environment and DistributedCache memory cache, this experiment has executed ten MapReduce Jobs that can read text files collaboratively, which will control the performance bottleneck of MapReduce jobs to I/O performance. The experimental text file is 40 GB, the data block sizes of HDFS and DistributedCache are set to 512 MB and 1 GB respectively. The distribution strategy of data blocks in HDFS is the default rack-aware mode. Each group has tested three times, and the test results are shown in Table 4 when the data block size is 512 MB. Then the data block size has been set from 512 MB to 1 GB and tested three times again, the test results are shown in Table 5.

Table 4. Test results when data block size is 512 MB

Number	Time consuming/s	HDFS	DistributedCache	Performance improvement
1	Application total time	238	203	14.71%
	Job total time	227	188	17.18%
	Average map running time	8	6	25%
	Shuffler running time	177	153	13.56%
2	Application total time	245	203	17.14%
	Job total time	226	186	17.7%
	Average map running time	8	6	25%
	Shuffler running time	178	153	14.04%
3	Application total time	244	202	17.21%
	Job total time	228	188	17.54%
	Average map running time	8	6	25%
	Shuffler running time	177	151	14.69%

Table 5. Test results when data block size is 1 GB

Number	Time consuming/s	HDFS	DistributedCache	Performance improvement
1	Application total time	204	141	30.88%
	Job total time	176	120	31.82%
	Average map running time	15	9	40%
	Shuffler running time	132	87	34.09%
2	Application total time	187	131	29.95%
	Job total time	169	116	31.36%
	Average map running time	15	10	33.33%
	Shuffler running time	128	86	32.81%
3	Application total time	187	136	27.27%
	Job total time	174	118	32.18%
	Average map running time	15	10	33.33%
	Shuffler running time	133	88	33.83%

It can be seen from Tables 4 and 5 that DistributedCache can improve the data reading speed of MapReduce jobs compared to the original HDFS, and it can be found that the efficiency of 1 GB data block is greater than 512 MB data block. Because when data block becomes larger, the total number of Map tasks in MapReduce jobs has reduced if dataset size of jobs is fixed. Thus the number of resources in the same cluster is fixed and the resource competition pressure becomes smaller, which causes the reading speed of Map task to become larger. Therefore, the larger the data block, the more obvious the performance improvement.

512   Z. Fengli and L. Xiaoli

5.3   Comparison Test of ItemBased Recommendation Algorithm

The implementation version of ItemBased recommendation algorithm used in this experiment is 0.11.1 version of mahout. The entry of running class is PromoterJob class under the package org.apache.mahout.cf.taste.hadoop.item, and the run command that uses native HDFS files as the storage target is as follows.

```
./hadoop jar
/home/hadoop/mahout0.11.1/mahout-examples-0.11.1-job.jar
org.apache.mahout.cf.taste.hadoop.item.RecommenderJob
-i /mahout/itemcf/inputdata
-o /mahout/itemcf/result
-s SIMILARITY_LOGLIKELIHOOD
–tempDir/mahout/itemcf/templ
```

Where inputdata is the input data directory; result is the output file directory; temp is the temporary file directory during the job running procedure. When DistributedCache has been used as storage system, the ItemBased algorithm will be reconstructed according to Algorithm 1. Then it will need to run in the DistributedCache after refactoring, and its running command is as follows.

```
./hadoop jar
/home/hadoop/mahout0.11.1/mahout-examples-0.11.1-job.jar
org.apache.mahout.cf.taste.hadoop.item.RecommenderJob
-i distributedCache://ubuntu201:23456/ratings/ratings.data
-o distributedCache://ubuntu201: 23456/ratings/result
-s SIMILARITY_LOGLIKELIHOOD
–tempDir distributedCache://ubuntu201:19998/ratings/temp
```

Where distributedCache://ubuntu201:23456/ratings/ratings.data is the input data path of algorithm; distributedCache://ubuntu201: 23456/ratings/result is the output data path; distributedCache://ubuntu201:19998/ratings/temp is the directory of temporary file; and SIMILARITY_LOGLIKELIHOOD is the similarity calculation parameter.

In this experiment, the data pair <userID, itemID, score> in the user scoring matrix is 20 million, it runs three times and the average value will be taken. The resource efficiency comparison of HDFS and DistributedCache is shown in Table 6.

Table 6. Comparison of computational efficiency when data is 20 million

Stage segmentation of ItemBased algorithm			Running time/s			
Step	MapReduce stage	Name of MapReduce job	Running time of HDFS job	Waiting delay of HDFS job	Running time of distributed-cache job	Waiting delay of distributed-cache Job
1	PreparePreferenceMatrixJob	ItemIDIndexMapper-Reducer	201	64	199	8
2		ToItemPrefsMapper-Reducer	269	15	268	4
3		ToItemVectorsMapper-Reducer	207	14	207	3

(continued)

Table 6. (*continued*)

Stage segmentation of ItemBased algorithm			Running time/s			
Step	MapReduce stage	Name of MapReduce job	Running time of HDFS job	Waiting delay of HDFS job	Running time of distributed-cache job	Waiting delay of distributed-cache Job
4	RowSimilarityJob	CountObservationsMapper-Reducer	195	12	192	3
5		VectorNormMapper-Reducer	218	16	216	4
6		CooccurrencesMapper-Reducer	436	14	433	4
7		UnsymmetrifyMapper-Reducer	423	13	424	4
8	partialMultiply	partialMultiply	222	17	218	5
9	RecommenderJob	PartialMultiplyMapper-Reducer	826	16	821	4

According to the experimental data in Table 6, the total resource waiting delay of algorithm is 181 s when the storage system is HDFS, and the total resource waiting delay is shortened to 39 s when the storage system is DistributedCache. The algorithm that has been reconstructed by DistributedCache greatly improves the resource utilization efficiency of jobs. The experimental results have verified the analysis of MapReduce jobs' operation efficiency in Sect. 3.1, which proves that DistributedCache has greatly improved the preparation time and utilization efficiency of resources.

6 Conclusion

While MapReduce computing mode has gradually become a parallel computing standard, some problems of resource utilization efficiency have also existed, and multi-MapReduce jobs will be collaborated especially in complex big data mining scenarios. Redundant disk reading/writing and repeated resources application between multiple jobs have made the low efficiency of MapReduce algorithm's execution time, resource utilization and energy consumption. This paper finds that these problems are caused by the independence of MapReduce Job. Therefore, this paper analyzes the resource efficiency flaws of ItemBased algorithm in MapReduce environment and finds the existing problems. Then it proposes that the intermediate data generated by the calculation process of DistributedCache cache can be used to reduce redundant I/O operations and optimize the efficiency of resource utilization. Finally, two sets of experiments have proved that fast memory I/O replaces dense disk I/O in DistributedCache will optimize resource utilization efficiency and shorten resource preparation time during job operation. The experimental results show that the optimized I/O resource efficiency is more than three times higher. And the next research work is to study the usage and scheduling efficiency of DistributedCache resource when the cluster load pressure is large.

Acknowledgment. This work was supported in part by Research Project of Hubei Provincial Department of Education (No. B2017590).

References

1. The digital universe in 2020: big data, bigger digital shadows, and biggest growth in the far east.   http://www.emc.com/collateral/analyst-reports/idc-the-digitaluniverse-in-2020.pdf. Accessed 15 Mar 2018
2. Ghemawat, S., Gobioff, H., Leung, S.T.: The Google file system. In: Proceedings of the 19th ACM Symposium on Operating System Principles, pp. 29–43. ACM Press, New York (2003)
3. Chen, C., Lin, J., Kuo, S.: MapReduce scheduling for deadline-constrained jobs in heterogeneous cloud computing systems. IEEE Trans. Cloud Comput. **6**(1), 127–140 (2018)
4. Liao, B., Zhang, T., Yu, J., et al.: Energy consumption modeling and optimization analysis for MapReduce. J. Comput. Res. Dev. **53**(9), 2107–2131 (2016)
5. Wu, Q., Wang, L.P., Luo, X.Z., et al.: Top-k high utility pattern mining algorithm based on MapReduce. Appl. Res. Comput. **34**(10), 2897–2900 (2017)
6. Liao, B., Zhang, T., Yu, J., et al.: Temperature aware energy-efficient task scheduling strategies for MapReduce. J. Commun. **37**(1), 61–75 (2016)
7. Zhao, Z.D., Shang, M.S.: User-based collaborative-filtering recommendation algorithms on Hadoop. In: Proceedings of International Conference on Knowledge Discovery and Data Mining, pp. 478–481. IEEE Press, Piscataway (2010)
8. Ma, M.M., Wang, S.P.: Research of user-based collaborative filtering recommendation algorithm based on Hadoop. In: Proceedings of International Conference on Computer Information Systems and Industrial Applications, pp. 63–66. Atlantis, New York (2015)
9. Schelter, S., Boden, C., Markl, V.: Scalable similarity-based neighborhood methods with MapReduce. In: Proceedings of ACM Conference on Recommender Systems, pp. 163–170. ACM Press, New York (2012)
10. Das, A.S., Datar, M., Garg, A., et al.: Google news personalization: scalable online collaborative filtering. In: Proceedings of International Conference on World Wide Web, pp. 271–280. ACM Press, New York (2007)
11. Jiang, J., Lu, J., Zhang, G., et al.: Scaling-up item-based collaborative filtering recommendation algorithm based on Hadoop. In: Proceedings of IEEE World Congress on Services, pp. 490–497. IEEE Press, Piscataway (2011)
12. Liao, B., Zhang, T., Guo, B.L., et al.: Performance optimization of ItemBased recommendation algorithm based on spark. J. Comput. Appl. **37**(7), 1900–1905 (2017)

A Fuzzy Constraint Based Outlier Detection Method

Vasudev Sharma[1](✉), Abhinav Nagpal[1],
and Balakrushna Tripathy[2](✉)

[1] School of Computer Science and Engineering, VIT University,
Vellore 632014, India
vasudevsharma74@yahoo.com
[2] School of Information Technology and Engineering, VIT University,
Vellore 632014, India
tripathybk@vit.ac.in

Abstract. With a huge amount of data generated every second, it has become important to remove data anomalies. Outliers are the extreme value that deviates from other observations in data. We propose a novel outlier detection method; FCBODM (Fuzzy Constraint based Outlier Detection Method) that takes into account of fuzzy constraint and background knowledge to discover the outliers in a dataset. Our key idea is to use fuzzy constraint technology wherein we used nearness measure theory in fuzzy mathematics for finding similarities between data objects and background information. It helps in finding more meaningful outliers. Our novel approach can be integrated with traditional outlier detection methods to improve the outlier ranking. In order to validate and demonstrate the effectiveness and scalability of our method we experimented it on real and semantically meaningful datasets.

Keywords: FCBODM · Fuzzy constraints · Outlier detection · Data mining · Background knowledge · Nearness measure

1 Introduction

An outlier or an anomaly is an observation which deviates from the rest of observations significantly based on some measure. They are usually present due errors in measurements or different system conditions and thus, does not abide the with common properties of the system. With the increase in the amount of data, outlier detection has recently become an important data mining job. It is almost impossible to analyze a large dataset manually to detect outliers present in it. Hence, a mechanism that can identify outliers present in the data is essential. Outlier detection finds usage in many applications like fraud detection in credit cards, network security, medicine and public health etc. For high dimensional data, locating the correct outliers is not an easy job as the traditional outlier detection methods are not efficient. Traditional outlier detection techniques are based on a full dimension space, and incapable of detecting outliers hidden in partial dimensions because of the dimensionality curse. The outliers present in a high dimensional dataset remains unidentified due to the presence noise effects of

© Springer Nature Switzerland AG 2019
D.-S. Huang et al. (Eds.): ICIC 2019, LNAI 11645, pp. 515–527, 2019.
https://doi.org/10.1007/978-3-030-26766-7_47

many dimensions in it. However, the number of subspaces increases exponentially when the number of dimension increases, and exhausting all subspaces in high-dimensional data is impossible.

There are many existing clustering algorithms that detect outliers apart from clustering [1]. However, these algorithms detect only those points that are not present in any of the major cluster and call them as an outlier. Thus, the algorithms indirectly believe that outliers are the background noise with clusters embedded in them. [2] defines that outliers are those points that are not a member of any cluster and background noise. They are points which do not follow similar patterns or behavior compared to the other points in the dataset. Distance-based outlier methods take note of the outlier of a data object by its distance distances to other nearby objects and by the number of objects nearby [3, 4]. The angle-based outlier method detects an outlier by checking the difference in the angles formed by the distance vectors of all pair of points with the query point suspected to be an outlier in the dataset. A good example of such an algorithm is ABOD [5].

In Density-based outlier detection methods, density of each point in the dataset is compared w.r.t. the nearby neighborhood [6]. Breunig et al. assigned local outlier factor (LOF), a score, to all objects in the dataset [7]. In this method, similarities of a candidate outlier and its density is calculated according to its distance from the surrounding points. The LOF method has been modified many times. Some example of its other versions are uncertain local outlier factor [8], the flexible kernel density estimates [9], and natural outlier factor [10]. Eskin [11] proposed a statistical method that uses statistical tests and machine learning methods for finding anomalies. Chen et al. [12] presented robust estimation and outlier detection approaches based on their proposed generalized local statistical framework. However, all the above method follows the assumption that some set of fixed features are important for the detecting outliers.

There are methods that try to find outliers in an arbitrarily oriented subspace. Searching for an outlier using all the dimensions is less complex than dealing with a subset of the dimensions. To solve this problem, an algorithm was proposed by Aggarwal et al. [13] based on outlier detection in subspace, which can find outliers in any subspace. Kriegel et al. [14] formulated a local outlier method to find exceptional outliers by the subspace method. Müller et al. [15] propose an outlier ranking, which computes local density deviation by searching relevant subspaces for objects deviating in subspace projections.

The remaining paper is divided into the following sections – Sect. 2 explains the design of the fuzzy constraint based outlier detection method in detail. It elaborates on extension of fuzzy constraint method on the traditional methods of detecting outliers. Section 3 discusses the experimental evaluation of FCBODM on datasets along with the various evaluation measures used for detecting the quality of outlier results. Section 4 provides conclusion and the possibility of future scope towards FCBODM.

2 Design of the Proposed Algorithm

The following section introduces the fuzzy set and fuzzy similarity scale. Figure 1 shows the complete flowchart of the algorithm.

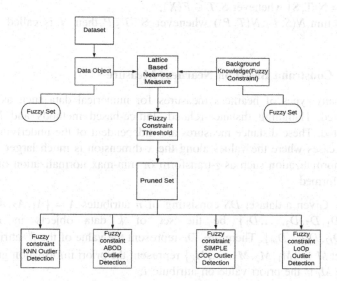

Fig. 1. Flowchart of the proposed algorithm

2.1 Fuzzy Set and Similarity Scale

The notion of fuzzy set was introduced by Zadeh [17] as an extension of the notion of crisp sets in order to model uncertain data. In a crisp set, an element is either a member of the set or not. Fuzzy sets, on the other hand, allow elements to be partially in a set. Each element is given a degree of membership in a set. This membership value can range from 0 (not an element of the set) to 1 (a member of the set). Formally, fuzzy sets can be defined as follows:

Definition 1. Let U be a universe of discourse. F is a fuzzy subset of U if there is a membership function $\mu_F : U \to [0, 1]$, which associates with each element u belonging to U a membership value $\mu_F : U$ in the interval $[0, 1]$. The membership value $\mu_F(u)$ for each u belonging U represents the grade of membership of the element u in the fuzzy set F. Equation 1 gives the notation for a fuzzy set F as proposed by Zadeh.

$$F(u) = \{(u, \mu_F(u)) : u \in U\} \tag{1}$$

Given $\{s_1,, s_2,, s_3,, s_4, \ldots, s_n,\}$ be n fuzzy sets on n standard classes on X. Given $S \in F(X)$, we need to know which class s should belong. To solve this problem, we need to measure the closeness of fuzzy set using nearness measure which is formally defined as follows:

Definition 2. If $N : F(X) \times F(X) \rightarrow [0, 1]$ satisfies that

1. $N(\emptyset, X) = 0$ and $N(S, S) = 1$ whenever $S \in F(X)$,
2. $N(S, T) = N(T, S)$ whenever $S, T \in F(X)$,
3. $N(S, P) \leq \min(N(S, T), N(T, P))$ whenever $S \subseteq T \subseteq P$ then N is called a nearness measure.

2.2   Fuzzy Constraint Based on Nearness Measure

There are many types of nearness measures for numerical data such as Euclidean distance-related, Hamming distance-related, lattice-based methods and Minkowski distance-related. These distance measures are independent of the underlying data distribution. In cases where the values along the x-dimension is much larger than the y-dimension, normalization such as z-transform or min-max normalization of each data object is performed.

Definition 3. Given a dataset DS consisting of n attributes $A = \{A_1, A_2, A_3, \ldots, A_n\}$. Let $D = \{D_1, D_2, D_3, \ldots, D_k\}$ be the set of k data objects in DS where $D_i = \{D_{i1}, D_{i2}, D_{i3} \ldots D_{in}\}$. Therefore, D_{ij} represent the value of the jth attribute of the ith object. Let $M = \{M_1, M_2, M_3, \ldots, M_k\}$ represent the priori information given by the users, where M_i is the priori value on attribute i.

Let $G(X)$ be a fuzzy set, where X is a subset of the attributes and $D_i, M \in F(X)$. Equations 3 and 4 represents inner and outer product of D_i and M_i respectively.

$$D_i \oplus M_i = \vee_{x \in X}(D_i(x) \wedge M_i(x)) \tag{3}$$

$$D_i \otimes M_i = \wedge_{x \in X}(D_i(x) \vee M_i(x)) \tag{4}$$

The lattice based nearness measures Z_L can be defined using x and y as follows:

$$Z_L(O_i, U) = (D_i(x) \oplus M_i(x)) \wedge (1 - D_i(x) \otimes M_i(x)) \tag{5}$$

Let us now describe how lattice based nearness can be used to prune the outliers from the dataset. Given an object D_i, priori knowledge M, and threshold value σ, if $Z_L(D_i, M) \geq \sigma$, then object D_i is called a required object, which matches the constraint condition given by the user. This means the data object is of user's interest. If $Z_L(D_i, M) < \sigma$, then object D_i needs to be pruned as it does not match the constraint condition given by the user. This means the data object is not of user's interest. In this algorithm, threshold value σ also known as nearness-threshold is provided by users. Thus, we calculate the nearness measure between each object in dataset DS and priori knowledge M. This prunes the dataset DS removing data objects of disinterest from DS. This reduced dataset helps in improving the efficiency of outlier detection when further steps are applied on it.

2.3 Pseudocode

```
program FCBODM (Outliers)
  var    DS: given dataset
  const M: priori information
  const di: i^th data object
  const σ: pruning threshold
  const n: number of data objects in DS
  begin:
    normalized_DS:= min_max_normalization(DS)
    repeat:
    inner:= inner_product(di,Mi)
    outer:= outer_product(di,Mi)
    nearness_di:= min(inner,1-outer)
    if nearness_di< σ then
       pruned_DS:= prune(di,normalized_DS)
    until i=n
    outliers:= method(pruned_DS)#method:ABOD,KNN,LoOP,COP
  end
```

2.4 Outlier Detection Methods

FCBODM ABOD. It has been seen that comparing distances between points to identify outliers is not efficient if the dimensionality of the dataset is large. ABOD algorithm proposes uses the distance between points and the direction of the distance vectors. Comparison of the angles between two distance vectors to other points is carried out in the algorithm. This helps to identify outliers in the dataset. If the angles between the distance vectors of an object are relatively large, then the object is inside the cluster. However, if the angles between the distance vectors of an object are relatively small, then the object is outside the cluster. The difference in the direction of the distance vectors of objects is calculated using ABOF - angle based outlier factor [7]. The ABOF(A) is the variance over the angles between difference vectors of all pairs of points in dataset D to a data point A weighted by the distance of the points:

$$ABOF(\overrightarrow{A}) = VAR_{\overrightarrow{B},\overrightarrow{C} \in D}\left(\frac{\langle \overline{AB}, \overline{AC} \rangle}{\|\overline{AB}\|^2 \cdot \|\overline{AC}\|^2}\right) \tag{6}$$

For every object in the dataset, the ABOF value is calculated and the points are ranked on their basis. ABOD algorithm has the advantage of not requiring any parameter.

FCBODM KNN. Developed by Ramaswamy and Shim [4], KNN is a distance-based outlier that calculates the distance of a point from its neighboring points. All points in the dataset are ranked according to their distances from their nearest neighbor. The points with the largest distance are declared as the outlier. Let there be a point p, then DK(p) denotes the distance from the kth nearest neighbor. Let n be the number of outliers that need to be removed. Here, DK(p) also describes the degree of how much outlier is the point p. Points with larger DK(p) value have sparse surroundings and have more chances of being an outlier than points inside a dense cluster that have a lower value of DK(p). Let us say we need to find n outliers. Then, the n points with the maximum DK(p) values are declared as outliers. An advantage of this method is that the user need not have to provide a distance variable to qualify a point as an outlier.

FCBODM Simple COP. In order to find an outlier present in the subspaces of the original attribute space, COP algorithm was made. It considers many combinations of subsets of attributes, to find outliers deviating from their values. The points detected as outliers do not relate to any major correlation in the data. The local correlations within the outlier detection method are considered first. Then, outliers present in the subset of the actual dataset are identified. It then chooses the relevant correlation of attributes to detect the corresponding outlier. An object is considered as an outlier if it does not match the correlations. The objects that are present on a δ-dimensional hyperplane, called correlation hyperplane, show local correlation. Here, d is the dimensionality of the dataset and δ < d. Thus, outliers are the objects that are not present and do not show any correlation in such hyperplanes.

FCBDOM LoOP. Local outlier probabilities (LoOP) is an outlier detection method which is based on local density. It evaluates whether a point is an outlier or not by giving a score from 0 to 1. Let us take a set P containing k objects with d as the distance function. Let $o \in D$ be the probabilistic distance to a context set $S \subseteq P$, referred to as pdist(o,S). This distance has the following property: $\forall s \in S: P[d(o,s) \leq pdist(o,S)]$ φ. A sphere around o with a radius of pdist covers objects with a probability of φ in set S. The probabilistic distance pdist(o, S) between o and S can be calculated as the statistical extent of set S. Based on this, density around an object w.r.t. a context set, the Probabilistic Local Outlier Factor (PLOF) of an object $o \in P$ w.r.t. a significance λ and a context set $S(o) \subseteq P$, is defined as:

$$PLOF_{\lambda,S(o)} = \frac{pdist(\lambda, o, S(o))}{E_{s \in S(o)}[pdsit(\lambda, s, S(s))]} \tag{7}$$

For every object o in the dataset, the PLOF value is calculated which is the ratio of the estimation for the density around o and the expected value of the estimations for the densities around all objects in the context set S(o). The points with the minimum sorted PLOF values are declared outliers.

3   Experimental Evaluation

3.1   Datasets

We evaluated our algorithm on two types of datasets; real datasets, which have been used in the research literature and semantically meaningful datasets, where the semantic interpretation of outliers can be given from the datasets. The algorithm was tested on six UCI datasets. Prior to the evaluation, we performed dataset preparation for the evaluation of outlier detection algorithms. The classification datasets have been used where we assumed class having minority labels as outlier class since outlier detection is tantamount to detect objects belonging to a rare class. The classification datasets need to transformed to outlier datasets hence we performed down-sampling of a class, duplicates removal, min-max normalization and cleaning to deal with categorical and missing attributes.

Real Datasets. To evaluate our outlier detection we selected three UCI machine learning repository datasets [20] which are real world benchmark datasets namely - Shuttle, WPBC, and Ionosphere. Outlier mining is nothing but conceptually similar to detecting objects belonging to a rare class hence we focus on datasets where the class labels feature a clear minority class. We assume this class to contain the outliers in these datasets. In addition, these datasets are pre-processed as illustrated in the above section. Table 1 summarizes the characteristics of these three datasets.

Table 1. Characteristics of real datasets used in literature

Name	Instances	Outliers	Attributes
Shuttle	351	126	32
WPBC	198	47	33
Ionosphere	148	6	19

Semantically Meaningful Datasets. These datasets have certain classes that can be identified with real world scenarios. Data points containing outliers are both rare and digressing, for instance, consider patients suffering from 'Hutchinson-Gilford Progeria'; a rare disease; among a population of patients. But for some scenarios, there might be a possibility that outliers are dominated within a discrimination dataset. To overcome this problem, we down-sampled outlier class (2, 5, 10, 20% of outliers). UCI repository datasets [20] were selected and processed for evaluation of outlier results. These datasets are Cardiotocography, Arrhythmia and Heart Disease as illustrated in Table 2.

Table 2. Characteristics of semantically meaningful datasets.

Name	Instances	Outliers	Attributes
Cardiotocography	2126	471	21
Heart Disease	270	120	13
Arrhythmia	450	206	259

3.2   Evaluation Measures

Outlier detection methods used here yields a complete ranking of the database objects. Data points are given an outlier score upon evaluation by the outlier detections methods. Not every data object is relevant as the user is only interested in finding out the outlier score of say topmost ranked objects of the whole set. One such evaluation criteria are Precision at n (P@n) [16] where the target number of data objects; n is specified well in advance. Precision at n signifies ratio of correct results amid top n ranks [16]. Consider a database(DB) of size N consisting of outliers $O \subseteq DB$ and inliers $I \subseteq DB$ where $DB = I \cup O$. P@n can be formulated as

$$P@n = \frac{|\{o \in O | rank(o) \le n\}|}{n} \tag{8}$$

While P@n is a measure to evaluate the robustness of the outlier detection algorithm, it is unclear on what value of parameter n to choose. When the number of the outlier(n = |O|) is low in comparison to large N, P@n value is marginally small hence not useful enough whereas when the number of outliers(n = |O|) is large enough with respect to N, P@n would he high as small fraction of inliers exist. For an unambiguous measure, P@n should be adjusted for a chance to compare different measures where there is variation in an expected score. Since the maximum number of outliers are O, P@n maximum value is |O| / n provided that O > n else it is 1. The expected value of a completely random ranking is given by |O| / n. Henceforth Adjusted P@n is formulated by the given formula.

$$Adjusted\ P@n = \frac{P@n - |O|/N}{1 - |O|/N} \tag{9}$$

An anomaly with both these measures is the trade-off between the number of outlier and inliers. Generally for an outlier dataset, $|I| \gg |O|$ and $|I| = N$. P@n and Adjusted P@n measures are highly sensitive to n. The same issue of sensitivity toward n occurs with Adjusted P@n. Nonetheless, the other evaluation measures solve this problem by averaging over values of n. On such measure is average precision (AP) used in information retrieval evaluation methods. Instead of evaluation over single n, the values are averaged over ranks of outlier objects.

$$AP = \frac{1}{|O|} \sum_{o \in O} P@rank(o) \tag{10}$$

Similar to Adjusted P@n, an adjusted form of average precision is used for comparing different datasets, having the expected value of random ranking as |O| / n and maximum value as 1.

Another evaluation measure used widely in unsupervised learning is the Receiver Operating Characteristic (ROC). It's obtained by plotting across all n the true positive rate, and the false positive rate. If a ROC curve is close to the diagonal then it may be probably due to random outlier ranking, whereas a perfect ranking would result in a curve where a vertical line is at false positive rate 0 and a horizontal line is at the top of the plot. ROC adjusts for a chance as the normalization of false positives rate by false positive and the normalization of true positive rates by true positive is carried inherently. Therefore, ROC is insensitive to adjustment for a chance. ROC AUC (value varies between 0 and 1), a measure which summarizes a ROC curve by a single value. It can be thought as the average of the recall at n, with n taken over the all the ranks of inlier data objects in |I|. External ground truth labels that are inliers and outlier are required in all the above evaluation measures.

3.3 Evaluation on the Datasets

To assess and validate the quality of our outlier detection algorithm we have used 3 different curves namely PR AUC, ROC AUC and P@n (where we took n = |O|) as indicated in Sect. 3.2. For each of the evaluation measure, 3 plots were produced on the ionosphere (35.9% outliers) and Heart disease (44% outliers) datasets.

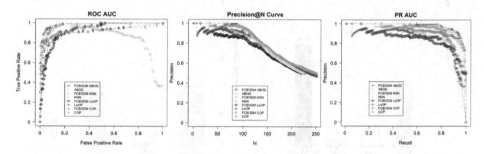

Fig. 2. Results on real dataset (IONOSPHERE), comparing ROC AUC, Precision@n and PR AUC with existing outlier algorithms

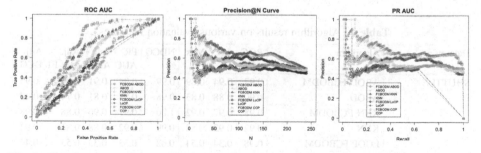

Fig. 3. Results on semantically meaningful dataset (HEART DISEASE), comparing ROC AUC, Precision@n and PR AUC with existing outlier algorithms

On these two groups of datasets we can clearly see the improvement in quality when compared FCBODM with traditional outlier methods. Figure 2 illustrates Precision@n, PR AUC and ROC AUC measures on the ionosphere dataset; a real dataset, with 4 traditional outlier methods as indicated in Sect. 2.3 and four improvised

FCBODM. An equivalent analysis is depicted in Fig. 3 on the Heart Disease dataset; a semantically meaningful dataset. Table 3 enumerates all the evaluation measures suggested in Sect. 3.2 on two real (Shuttle and WPBC) and two semantically meaningful (Heart disease and Arrhythmia) datasets.

We evaluated the runtimes of our algorithm FCBODM with existing outlier detection methods used in literature. Run time evaluation was carried on a real and a semantically meaningful UCI dataset. To our observation, FCBODM proved to excel at runtime when evaluated against earlier outlier methods. Figure 4 (left bar chart) shows the runtime percentage change in performance due to FCBODM when tested against traditional methods on SHUTTLE dataset constituting 1.38% of outliers. Computationally expensive algorithms such as ABOD and simple COP reported a change in the runtime of about 39.31% and 69.6% respectively while algorithms such KNN and LoOP suggested an improvement of about 21.05% and 2% respectively.

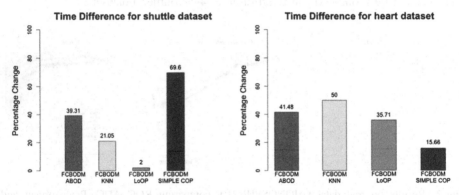

Fig. 4. Measure of percentage change in runtime milliseconds when compared with existing outlier detection techniques on shuttle (left bar chart) and heart dataset (right bar chart).

Table 3. Algorithm results on various evaluation measures.

Datasets	Algorithm	ROC AUC	AP	Max. F1	NDCG	PR AUC	Adj. AP	Adj. max F1	Adj. DCG
SHUTTLE	ABOD FCBODM	0.97	0.94	0.89	0.99	0.94	0.91	0.84	0.95
	ABOD	0.90	0.88	0.83	0.98	0.88	0.81	0.74	0.90
	KNN FCBODM	0.98	0.97	0.92	1.00	0.97	0.96	0.88	0.98
	KNN	0.92	0.92	0.88	0.99	0.92	0.88	0.81	0.94
	LOOP FCBODM	0.98	0.34	0.54	0.62	0.33	0.33	0.53	0.46
	LOOP	0.96	0.20	0.36	0.52	0.19	0.19	0.35	0.32
	SIMPLE COP FCBODM	0.97	0.27	0.37	0.62	0.25	0.26	0.37	0.45
	SIMPLE COP	0.85	0.22	0.39	0.56	0.20	0.21	0.38	0.37

(continued)

Table 3. (*continued*)

Datasets	Algorithm	ROC AUC	AP	Max. F1	NDCG	PR AUC	Adj. AP	Adj. max F1	Adj. DCG
HEART DISEASE	ABOD FCBODM	0.78	0.71	0.76	0.93	0.71	0.48	0.57	0.65
	ABOD	0.72	0.69	0.76	0.92	0.68	0.38	0.53	0.56
	KNN FCBODM	0.68	0.61	0.67	0.90	0.60	0.29	0.40	0.44
	KNN	0.55	0.53	0.66	0.86	0.52	0.08	0.34	0.17
	LOOP FCBODM	0.56	0.50	0.62	0.86	0.42	0.09	0.31	0.23
	LOOP	0.53	0.51	0.66	0.85	0.41	0.05	0.33	0.14
	SIMPLE COP FCBODM	0.64	0.56	0.65	0.87	0.55	0.20	0.38	0.31
	SIMPLE COP	0.54	0.51	0.67	0.84	0.50	0.04	0.36	0.06
WPBC	ABOD FCBODM	0.49	0.29	0.40	0.72	0.26	0.05	0.20	0.16
	ABOD	0.48	0.25	0.39	0.69	0.23	0.02	0.20	0.08
	KNN FCBODM	0.53	0.26	0.40	0.66	0.26	0.03	0.21	-0.01
	KNN	0.47	0.23	0.39	0.63	0.22	-0.02	0.19	-0.09
	LOOP FCBODM	0.58	0.31	0.44	0.71	0.27	0.08	0.25	0.13
	LOOP	0.59	0.28	0.43	0.69	0.25	0.06	0.25	0.08
	SIMPLE COP FCBODM	0.59	0.32	0.47	0.67	0.32	0.09	0.29	0.03
	SIMPLE COP	0.59	0.29	0.42	0.67	0.29	0.08	0.25	0.01
ARRHYTHMIA	ABOD FCBODM	0.74	0.74	0.68	0.95	0.74	0.53	0.40	0.70
	ABOD	0.72	0.71	0.64	0.94	0.71	0.50	0.37	0.69
	KNN FCBODM	0.75	0.76	0.68	0.95	0.75	0.55	0.41	0.72
	KNN	0.73	0.73	0.64	0.94	0.72	0.52	0.38	0.71
	LOOP FCBODM	0.74	0.73	0.68	0.95	0.72	0.51	0.40	0.69
	LOOP	0.72	0.71	0.65	0.94	0.69	0.49	0.39	0.69
	SIMPLE COP FCBODM	0.93	0.86	0.82	0.97	0.86	0.79	0.73	0.88
	SIMPLE COP	0.90	0.85	0.82	0.97	0.84	0.77	0.72	0.86

We tested FCBODM on semantically meaningful Heart Disease dataset which comprised 22% of outliers. Figure 4 (right bar chart) reveals that computationally expensive detection methods resulted in a great improvement in run time; ABOD and Simple COP, when evaluated on FCBODM, indicated an increase in runtime of 41.48% and 15.66% respectively, whereas other outlier detection methods like KNN and LoOP led to improvement of 50% and 35.71% respectively.

4 Conclusion and Future Work

We formulated a novel fuzzy constraint-based outlier detection which can be extended on top of existing outlier detection algorithms to improve not only the various evaluation parameters (ROC AUC, PR AUC, NDCG, F1 score) but also helps us to fathom pertinence of outlier mining results. For improving the relevance of outlier results we relied on lattice-based nearness measure in fuzzy mathematics is where we pruned some existing data objects that do not adhere to constraint condition (background knowledge). Nearness measure technique coupled with constraint condition drastically reduced the size of the dataset. Our algorithm, when combined with existing outlier detection methods (ABOD, Simple COP, LoOP, and KNN), yields an improvement in the performance measures. We validated our results on three real and three semantically meaningful UCI datasets and our novelty proved to be better than existing outlier methods. Due to computational constraint, we were unable to evaluate our results on a large number of datasets and large varieties of algorithms. We aim to use parallel and distributed environments to improve our results and extend to approach on a greater number of datasets and existing outlier detection algorithms.

References

1. Aggarwal, C.C., Yu, P.: Finding generalized projected clusters in high dimensional spaces. In: Proceedings of ACM SIGMOD, pp. 70–81 (2000)
2. Arning, A., Agrawal, R., Raghavan, P.: A linear method for deviation detection in large databases. In: Proceedings of KDD, pp. 164–169 (1996)
3. Knorr, E.M., Ng, R.T.: Algorithms for mining distance-based outliers in large datasets. In: Proceedings of VLDB (1998)
4. Ramaswamy, S., Rastogi, R., Shim, K.: Efficient algorithms for mining outliers from large data sets. In: Proceedings of SIGMOD, pp. 427–438 (2000)
5. Kriegel, H.P., Schubert, M., Zimek, A.: Angle-based outlier detection in high-dimensional data. In: Proceedings of KDD, pp. 444–452 (2008)
6. Jin, W., Tung, A., Han, J.: Mining top-n local outliers in large databases. In: Proceedings of KDD, pp. 293–298 (2001)
7. Breunig, M.M., Kriegel, H.P., Ng, R.T., Sander, J.: LOF: identifying density-based local outliers. In: ACM SIGMOD Record, vol. 29, no. 2, pp. 93–104. ACM (2000)
8. Liu, B., Xiao, Y., Yu, P.S., Hao, Z., Cao, L.: An efficient approach for outlier detection with imperfect data labels. IEEE Trans. Knowl. Data Eng. **26**(7), 1602–1616 (2014)
9. Schubert, E., Zimek, A., Kriegel, H.P.: Generalized outlier detection with flexible kernel density estimates. In: Proceedings of the 14th SIAM International Conference on Data Mining (SDM), Philadelphia, PA, pp. 542–550 (2014)
10. Huang, J., Zhu, Q., Yang, L., Feng, J.: A non-parameter outlier detection algorithm based on natural neighbor. Knowl.-Based Syst. **92**, 71–77 (2016)
11. Eskin, E.: Anomaly detection over noisy data using learned probability distributions. In: International Conference on Machine Learning, pp. 255–262 (2000)
12. Chen, F., Lu, C.T., Boedihardjo, A.P.: GLS-SOD: a generalized local statistical approach for spatial outlier detection. In: ACM SIGKDD International Conference on Knowledge Discovery and Data Mining (2010)
13. Aggarwal, C.C., Philip, S.Y.: An effective and efficient algorithm for high-dimensional outlier detection. Int. J. Very Large Data Bases **14**(2), 211–221 (2005)
14. Kriegel, H.P., Kröger, P., Schubert, E., Zimek, A.: Outlier detection in arbitrarily oriented subspaces. In: IEEE International Conference on Data Mining, pp. 379–388 (2012)
15. Müller, E., Schiffer, M., Seidl, T.: Statistical selection of relevant subspace projections for outlier ranking. In: 2011 IEEE 27th International Conference on Data Engineering (ICDE), pp. 434–445. IEEE (2011)
16. Campos, G.O., et al.: On the evaluation of unsupervised outlier detection: measures, datasets, and an empirical study. Data Mining Knowl. Discov. **30**(4), 891–927 (2016)
17. Zadeh, L.: Fuzzy sets. Inf. Control **8**(3), 338–353 (1965)

18. Wu, D., Mendel, J.M.: A vector similarity measure for linguistic approximation: Interval type-2 and type-1 fuzzy sets. Inf. Sci. **178**(2), 381–402 (2008)
19. Wu, D., Mendel, J.M.: A comparative study of ranking methods, similarity measures and uncertainty measures for interval type-2 fuzzy sets. Inf. Sci. **179**(8), 1169–1192 (2009)
20. UCI machine learning repository. http://archive.ics.uci.edu/ml

Chinese Agricultural Entity Relation Extraction via Deep Learning

Kai Zhang[1,2], Chuan Xia[1,2], Guoliang Liu[1,2], Wenyu Wang[1,2], Yunzhi Wu[1,2], Youhua Zhang[1,2], and Yi Yue[1,2(✉)]

[1] Anhui Provincial Engineering Laboratory of Beidou Precision Agriculture, Anhui Agricultural University, Hefei, China
yyyue@ahau.edu.cn
[2] School of Information and Computer, Anhui Agricultural University, Hefei 230036, Anhui, China

Abstract. With the advent of Deep Learning (DL), Natural Language Processing (NLP) has progressed at a high speed in the past few decades. Some DL models have been established for Relation Extraction and outperform than the traditional Machine Learning (ML) methods. In this paper, we built four DL models: Piecewise Convolutional Neural Network (PCNN), Convolutional Neural Network (CNN), Recurrent Neural network (RNN) and Bidirectional Recurrent Neural Network (Bi-RNN) to tackle the task. Using PCNN, we outperform than other three models, and achieve Area under Curve (AUC) of 0.154 with just 24 epochs. And we use some selector mechanism to improve the model. Our experimental results show that: (1) Attention mechanism got the best compatibility with all models, but in some case, max pooling may perform better than it. (2) Using only word embeddings, the performance of the model will discount a lot.

Keywords: Piecewise Convolutional Neural Network · Relation Extraction · Deep Learning · Neural Network

1 Introduction

Relation Extraction is one of the most important subtasks of NLP. In this paper, we built some DL models to tackle the relation extraction task. Word embedding and position embedding are the crucial step. This step aims to gain the input vector representation for our models. (Zhang et al. 2018) have discussed the use of network algorithms to replace the original frequency-based distribution and achieved great result. Then DL models will learn how to capture the relation information of the vector representation. By selector layer, the network will classify the relations and give the rank of them. Attention mechanism is a great selector mechanism. (Zhang et al. 2018) have increased the speed and reliability of attention network.

D.-S. Huang et al. (Eds.): ICIC 2019, LNAI 11645, pp. 528–534, 2019.
https://doi.org/10.1007/978-3-030-26766-7_48

2   Methodology

Relation extraction is formulated as a classification problem. A form of two entities and a relation will be as the input:

$$entity_1 \#entity_2 relation : sentence$$

For example, the following sentences "Planthopper will destroy rice" contain a relation example "impact" and two entities: "planthopper", "rice", the form will be like that: planthopper#rice impact : Planthopper will destroy rice. We convert the formed data into vectors and then put into the network as input, then the network will give output with scores for each relation and give the predict. Figure 1 shows the overview of our network.

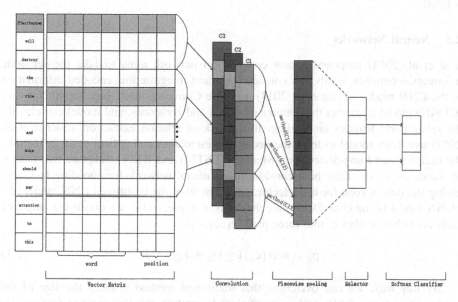

Fig. 1. The Piecewise Convolutional Neural Network for relation extraction

2.1   Word Embedding

Our first layer's job is to transform words into vector representation that hold the basic and semantic information about the word. We use the Continuous Bag of Words (CBOW) to give each word of the words-base a N-dimensional real-valued vector. This approach has been discussed as an effective way to tackle the task. Similar to (Lin et al. 2016), given a sentence S with N words: $x = \{w_1, w_2, w_3, \ldots, w_i, \ldots, w_N\}$, every word is represented by a low-dimensional distributed vector. Then this sentence will be converted into a sequence and will be converted into a real-valued matrix that can be put into the network. By CBOW, we map every word of the dictionary to a low-dimensional space and form a real-valued vector.

2.2   Position Embedding

In relation extraction task, some words are close to the entities for giving information to determine the relation. (Zeng et al. 2014) has used the position features (PFs) to specify entity pairs. We take the similar way (Kumar et al. 2017) to define a PF, entity e_1 and entity e_2 got a distance value. For example, if there are three words between e_1 and e_2, then e_1 will got a value 3, and e_2 will got a value -3. Like the sample shown in Fig. 2:

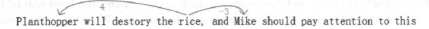

Fig. 2.   Word positions in a sentence, and how to value it.

And we combine the position vector and word vector together as the input of network.

2.3   Neural Networks

(Liu et al. 2013) proposed a new convolution network aims to take the semantic information between words into consideration and integrate this encoded information to the CNN model. (Zeng et al. 2014) used the Convolutional Deep Neural Network (CDNN) model to extract the features of words and sentences, and model can classify the relation via features embedding. In the task of relation extraction, (Socher et al. 2012) use RNN model to learn the vector representation and (Miwa et al. 2016) take the bidirectional Long-Short-Term-Memory (LSTM) and RNN to capture the relation of words sequence. They both consider using neural network for encoding task, and giving the output code for the selector. Different from the traditional CNN model, the PCNN model (Zeng et al. 2015) take the thought of piecewise, we divide the output of each convolution filter c_i into three parts $\{c_1, c_2, c_3\}$.

$$p_{ij} = \max(c_{ij}) \, 1 \leq i \leq n, 1 \leq j \leq 3 \qquad (2.1)$$

By this way, we can overcome that traditional method reduces the size of the hidden layer too quickly, and is insufficient to capture the fine grained features for relation extraction.

2.4   Output Mechanism

In relation extraction task, selector is an important subtask. DL model will learn how to classify the relation between two entities, and that is not enough for the extraction. Similar to (Lin et al. 2016), we build Attention Mechanism (ATT) to improve the performance of relation extraction. We assume relation is a mapping of one sequence to another sequence.

$$r = g(s_1, s_2) \tag{2.2}$$

And we will build an attention matrix to make model classify the relation smarter. Traditional selector: Max pooling (MAX) and Average pooling (AVE) will also be built as comparison.

2.5 Softmax Cross Entropy

Similar to (Liu et al. 2016), softmax function will help models make a better decision.

$$P\left(y^i = j | x^i; \theta\right) = \frac{e^{(\theta_j)^T x_i}}{\sum_{l=1}^{k} e^{(\theta_l)^T x^i}} \tag{2.3}$$

Cross entropy will help softmax function maximize the probability of result. Our relation extraction models use this kind method to calculate the loss, and perform the classification.

3 Experiments

3.1 Experiments Environment and Dataset

All the experiments below are based the following environment (Table 1).

Table 1. Experiment apparatus

Hardware	Software
CPU: Intel Core i5 7200U	Windows 10
GPU: Intel GMA HD 520	Python + Tensor Flow

During our experiments, data is a basic step. We crawled all information about the words that has been listed in the Agricultural pest and disease graphic knowledge base in the baike database as our dataset. Similar to the previous work (Wu et al. 2018), we use Hidden Markov Model (HMM) to finish the Chinese word segmentation task. Then we extracted the entities and formed the entity pairs. About the parameters, we use: word embedding dimensionality of 100. About the other parameters, we use the same as (Zeng et al. 2015) do (shown in Table 2).

Table 2. Parameters in the experiments

Window size	Feature maps	Position dimension	Batch size	Addelta parameter	Dropout probability
W = 3	N = 230	$d_p = 5$	$b_s = 50$	$\rho = 0.95, \varepsilon = 1e^{-6}$	P = 0.5

3.2 Network Comparison

We use some indicators to evaluate our model like F1 score, AUC and some others. To figure out the different kernel function's influence, we build four DL models. We build the PCNN, CNN, RNN, BiRNN and use the ATT, MAX, AVE to do the relation extraction task. We use the ATT as the selector to the four DL models, and the result shown in the Table 3.

Table 3. Indicators of models comparison

	PCNN + ATT	CNN + ATT	RNN + ATT	BiRNN + ATT
AUC	0.153566	0.135109	0.091221	0.115133
F1	0.296651	0.284153	0.229075	0.262857
Average iteration time	2.269197	0.132406	4.802849	8.858684
Test accuracy	0.697350	0.687500	0.656250	0.662500

We can figure out that PCNN performance better than the traditional DL models, all indicators except the average iteration time get a better score. In the Fig. 3, we analyze the Precision-Recall curve. All models tended to have the higher recall score and PCNN also performance the best.

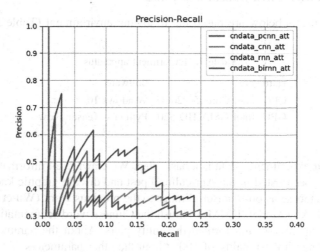

Fig. 3. Models' Precision-Recall curve

3.3 Selector Comparison

Compare to the traditional pooling, we use the ATT to improve the performance of our models, we can see the performance of ATT and traditional selector (MAX, AVE) in Table 4.

Table 4. Selector comparison

	PCNN + ATT	PCNN + MAX	PCNN + AVE
AUC	0.153566	0.160085	0.152178
F1	0.296651	0.316384	0.300000
Average iteration time	2.269197	1.964954	1.090331
Test accuracy	0.697350	0.675000	0.681250

The comparison in Fig. 4 shows that the PCNN model steps further than the predecessor.

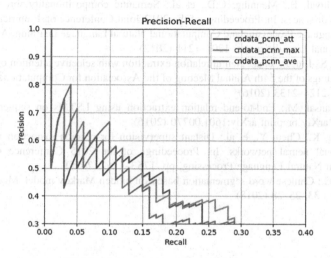

Fig. 4. Selectors' Precision-Recall curve

4   Conclusion

In this work we solve the Chinese agricultural entity relation extraction by using DL models. By giving the classification ranking, we evaluate the performance of our model. The main contribution of this work: (1) the PCNN model we use well solved the Chinese agricultural entity relation extraction task. (2) The use of ATT as the pooling layer shows the effectiveness and the great performance of the predict accuracy. PCNN model steps further than the predecessor.

References

Zhang, Z., Zweigenbaum, P.: GNEG: graph-based negative sampling for word2vec. In: Proceedings of the 56th Annual Meeting of the Association for Computational Linguistics, vol. 2, pp. 566–571 (2018)

Zhang, B., Xiong, D., Su, J.: Accelerating neural transformer via an average attention network. arXiv preprint arXiv:1805.00631 (2018)

Liu, W., Wen, Y., Yu, Z., et al.: Large-margin softmax loss for convolutional neural networks. In: Proceedings of the 33rd International Conference on Machine Learning, pp. 507–516 (2016)

Zeng, D., Liu, K., Lai, S., et al.: Relation classification via convolutional deep neural network (2014)

Kumar, S.: A survey of deep learning methods for relation extraction. arXiv preprint arXiv:1705. 03645 (2017)

Liu, C.Y., Sun, W.B., Chao, W.H., et al.: Convolution neural network for relation extraction. In: Motoda, H., Wu, Z., Cao, L., Zaiane, O., Yao, M., Wang, W. (eds.) ADMA 2013. LNCS, vol. 8347, pp. 231–242. Springer, Heidelberg (2013). https://doi.org/10.1007/978-3-642-53917-6_21

Socher, R., Huval, B., Manning, C.D., et al.: Semantic compositionality through recursive matrix-vector spaces. In: Proceedings of the 2012 Joint Conference on Empirical Methods in Natural Language Processing and Computational Natural Language Learning. Association for Computational Linguistics, pp. 1201–1211 (2012)

Lin, Y., Shen, S., Liu, Z., et al.: Neural relation extraction with selective attention over instances. In: Proceedings of the 54th Annual Meeting of the Association for Computational Linguistics, vol. 1, pp. 2124–2133 (2016)

Miwa, M., Bansal, M.: End-to-end relation extraction using LSTMs on sequences and tree structures. arXiv preprint arXiv:1601.00770 (2016)

Zeng, D., Liu, K., Chen, Y., et al.: Distant supervision for relation extraction via piecewise convolutional neural networks. In: Proceedings of the 2015 Conference on Empirical Methods in Natural Language Processing, pp. 1753–1762 (2015)

Wu, S., Pan, H.: Chinese word segmentation based on hidden Markov model. Modern Comput. (Prof. Edn.) 33, 25–28 (2018)

Dimensions Effect in Word Embeddings of Knowledge Graph

Qinhua Huang[⊠] and Weimin Ouyang

Department of Computer, Shanghai University of Political Science and Law,
Shanghai 201701, China
{hqh, oywm}@shupl.edu.cn

Abstract. Word embedding is one of the basic of knowledge graph. It is designed to represent the entities and relations with vectors or matrix to make knowledge graph model. Recently many related models and methods were proposed, such as translational methods, deep learning based methods, multiplicative approaches. The TransE models take a relation as transition from head entity to tail entity in principle. The further researches noticed that relations and entities might be able to have different representation to be casted into real world relations. Thus it could improve the embedding accuracy of embeddings in some scenarios. To improve model accuracy, the variant algorithms based on TransE adopt strategies like adjusting the loss function, freeing word embedding dimension freedom limitations or increasing other parameters size etc. After carefully investigate these algorithms, we motivated by researches on the effect of embedding dimension size factor. In this paper, we carefully analyzed the factor impact of dimensions on the accuracy and algorithm complexity of word embedding. By comparing some typical word embedding algorithms and methods, we found there are tradeoff problem to deal with between algorithm's simplicity and expressiveness. We carefully designed an experiment to test such kind of effect and give some description and possible measure to adopt.

Keywords: Word embeddings · Dimension size · Knowledge graph

1 Introduction

To deal with problems of natural language information extraction, especially problems with background strong semantic meaning, the knowledge graph model has been proposed. It was first introduced by google to organize and search the huge information more efficiently from text sources [12]. And it has been developed into knowledge representation learning. To represent information in natural language, the entities are abstracted and extracted from text, as well as the relations between entities. So a knowledge graph can be formed in a network. Those entities and their relations are very rich in both structure and content. One of the basic research problem is how to represent the entities and relations using vectors or matrix, which is called the word embedding problem. With the word embedding, the new head or tail entity with a specified relation can be concluded. A KG can be denoted by a triple (head, relation, tail). It is more like the way human processing information. So it has been taken as an

© Springer Nature Switzerland AG 2019
D.-S. Huang et al. (Eds.): ICIC 2019, LNAI 11645, pp. 535–544, 2019.
https://doi.org/10.1007/978-3-030-26766-7_49

artificial intelligence technique. As this technique developed, currently a typical knowledge graph may contain millions of entities and billions of relational facts, but it is far from complete. The task of knowledge graph completion is to find and describe more relations in an existing knowledge graph. It aims at predicting relations between entities under supervision of the existing knowledge graph. The technique to find if there is a relationship between two entities is called link predication.

Link predication is a popular research area with many important applications, including biology, social science, security, and medicine. Traditional approach of link prediction is not capable for knowledge graph completion. Recently, a promising approach for the task is embedding a knowledge graph into a continuous vector space while preserving certain information of the graph. Following this approach, many methods have been further explored, related details can be found by going through Tables 1 and 2.

Among these methods, typical translational methods, starts from TransE [1] and later TransH are simple and effective, achieving the state-of-the-art prediction performance. TransE, inspired by the work of word embedding, learns vector embeddings for both entities and relationships. These vector embeddings are set in R^k. TransH is proposed to enable an entity having different representations when involved in various relations. Both TranH and TransE have the assumption that entities should be in the same space. TransR suggested in some situations, entity and relations should have different aspects due to the complexity of real world. Thus TransR proposed a method that the embeddings has different spaces for entities and relations.

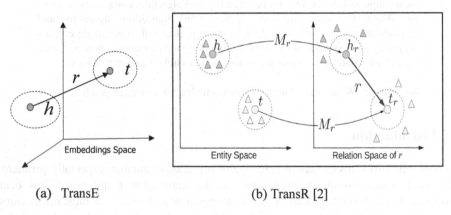

(a) TransE   (b) TransR [2]

Fig. 1. A simple comparison illustration between TransE and TransR

The basic idea of TransE and TransR can be illustrated in Fig. 1 in a simple way. In TransR, entities are looked as locations in a unique embeddings space. A relation is taken as the transition vector, which is the distance vector from head entity to tail entity in the same space. Based the complexity assumption, TransR goes further. It takes relations and entities as in different spaces. For each triple (h, r, t), entities in the entity space are first projected into r-relation space as h_r and t_r with operation Mr, and then

$h_r + r \approx t_r$. The relation-specific projection can make the head/tail entities that actually hold the relation close with each other, and also get far away from those that do not hold the relation. Moreover, under a specific relation, head-tail entity pairs usually exhibit diverse patterns. It is insufficient to build only a single relation vector to perform all translations from head to tail entities [2].

Tensor factorization method is a way to decompose a tensor or matrix. This kind of approaches for KG completion [29], [4], [39, 26] consider word embeddings for each entity and each relation. A typical Tensor factorization model will compute the probability of true triple. The computation function will take some input forms of (h, r, t), for example, h ∘ r ∘ t, where the triple is the word embedding. Details and discussions of these approaches can be found in several recent surveys.

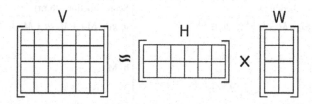

Fig. 2. An illustration of Tensor factorization.

This paper is organized as follows. First part is introduction, in which we shortly give an intuitive description on some word embedding models, including using translational method and tensor factorization. In related work section, to give a full picture, we list some important related embedding models. We analyzed the dimension factor impact on these two method by experiment comparison. Finally, we conclude our comparison and discuss the penitential performance improving direction.

2 Related Works

In this section, we briefly listed the related models in knowledge graph embeddings. Due to the space limitation we mainly discuss two kinds of method. One the translational method. The other is tensor factorization method.

TransE firstly purposed projecting the entities into the same space, where the relation can be taken as a vector from head entity to tail entities. Formally, we have a triple (h,r,t), where $h, r, t \in \mathbb{R}^k$, h is the head entity vector, r is the relation vector and t is the tail entity vector. The TransE model represents the a relationship by a translation from head entity to tail entity, thus it holds $h + r \approx t$. By minimizing the score function $f(h, r, t) = \| h + r - t \|_2^2$, which means h + r is the closest to t in distance. This representation has very clear geometric meaning as it showed in Fig. 1.

TransH was proposed to address the issue of N-to-1, 1-to-N and N-to-N relations. It projected (h, r, t) onto a hyperplane of w_r, where w_r is the hyperplane normal vector of r. TransR noticed both TransE and TransH took the assumption that embeddings of entities and relations are represented in the same space $\mathbb{R}^k$. And relations and entities might have different semantic meaning. So TransE suggest project entities and relations onto different spaces in representation, respectively. The score function will be minimized by translating entity space into relation space.

There are some other models like Unstructured Model, which is a simplified TransE. It suppose that all r = 0; Structured Embedding, it adopted L_1 as its distance measure since it has two relation-specific matrices for head and tail entities; Neural Tensor Network (NTN), which has some complexity that makes it only suit for small knowledge graphs. For the convenience of comparison, we listed the embeddings and score functions of some models in Table 1.

Table 1. Entity and relation embedding models: embeddings and score functions

Model name	Embeddings	Score function s(h,r,t)
Neural Tensor Network (NTN)	$M_{r,1}, M_{r,2} \in \mathbb{R}^{k \times d}$, $b_r \in \mathbb{R}^k$	$u_r^\top g(h^\top M_r t + M_{r,1} h + M_{r,2} t + b_r)$
Latent Factor Model (LFM) [10]		$h^\top M_r t$
Semantic Matching Energy (SME)	M_1, M_2, M_3, M_4 are weight matrices, $\otimes$ is the Hadamard product, b_1, b_2 are bias vectors.	$((M_{1h}) \otimes (M_2 r) + b_1) \top ((M_3 t) \otimes (M_4 r) + b_2)$
TranE [1]	$h, r, t \in \mathbb{R}^k$	$\| h + r - t \|$
TransH [5]	$h, t \in \mathbb{R}^k, w_r, d_r \in \mathbb{R}^k$	$\| (h - w_r^\top h w_r) + d_r - (t - w_r^\top t w_r) \|$
TransD	$\{h, h_p \in \mathbb{R}^k\}$ for entity h, $\{t, t_p \in \mathbb{R}^k\}$, for entity t, $\{r, r_p \in \mathbb{R}^d\}$ for relation r.	$\| (h + h_p^\top h r_p) + r - (t + t_p^\top t r_p)) \|$
TransR [2]	$h, t \in \mathbb{R}^k, r \in \mathbb{R}^d, M_r \in \mathbb{R}^{k \times d}$, M_r is a projection matrix.	$\| M_r h + r - M_r t \|$

In Table 2, the constraints of each models are presented. As we should point out that with the models developed, the embeddings and constraints actually become more complicated. One thing is sure that if the model is more complicated, the computation cost goes higher. This problem should be carefully considered in algorithm design.

Table 2. Entity and relation embedding models: constraints

Model name	Constraints
TranE	$h, r, t \in \mathbb{R}^k$
TransH	$h, t \in \mathbb{R}^k, w_r, d_r \in \mathbb{R}^k$
TransD	$\{h, h_p \in \mathbb{R}^k\}$ for entity h, $\{t, t_p \in \mathbb{R}^k\}$, for entity t, $\{r, r_p \in \mathbb{R}^d\}$ for relation r.
TransR	$h, t \in \mathbb{R}^k, r \in \mathbb{R}^d, M_r \in \mathbb{R}^{k \times d}$, M_r is a projection matrix.

Also there are ways of freeing limitation on the entity embeddings. The main idea is to let head and tail embedding representation independent on each other. We give a possible implementation method here, as showed in Fig. 2.

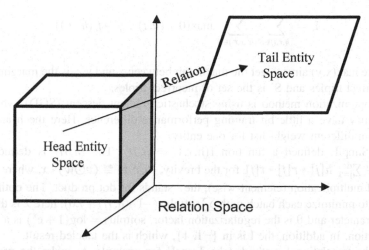

Fig. 3. A simple illustration of entity and relation spaces in embeddings model, where the space dimensions k, l and d might not be the same between any two spaces.

For each triple (h, r, t), $h \in \mathbb{R}^k, t \in \mathbb{R}^l, while\ r \in \mathbb{R}^d$. Here k, l and d can be different. For the sake of calculation, like in TransR and TransE, we project head entities and tail entities into relation space. The projected vectors of head entities and tail entities are defined as

$$h_r = hM_{hr}, t_r = tM_{tr}$$

Where $M_{hr} \in \mathbb{R}^{k \times d}, M_{tr} \in \mathbb{R}^{l \times d}$ are transition matrix.
Routinely the score function is thus defined as

$$f_r(h, t) = \| h_r + r - t_r \|_2^2$$

And there are also constraints on the norms of embeddings h, r, t and the transition matrix. As it showed below

$$\forall h, r, t, \| h \|_2 \leq 1, \| r \|_2 \leq 1, \| t \|_2 \leq 1, \| hM_{hr} \|_2 \leq 1, \| tM_{hr} \|_2 \leq 1$$

The Canonical Polyadic method in link prediction take the head and tail entities by learning independently. After analyzing the negative side of independency, the SimplE model simplified the freedom limitation of Canonical Polyadic decomposition. It let the two kind entities learn dependently, while both of them have the same idea that using two embedding representation for each one entity while it could take head or tail position (Fig. 3).

3 Training Method

The transitional methods generally adopted margin-based score function, as showed in follow equation.

$$L = \sum_{(h,r,t)\in S} \sum_{(h',r,t')\in S'} \max(0, f_r(h,t) + \gamma - f_r(h',t'))$$

Where max(x,y) aims to get the maximum between x and y, γ is the margin, S is the set of correct triples and S' is the set of incorrect triples.

The optimization method is using stochastic gradient descent (SGD), while other method may have a little bit training performance difference. Here the h and t may come from different weight list for one entity.

The SimplE defined a function $f(h,r,t) = \langle h, rt \rangle$. The $\langle h, rt \rangle$ is defined to be $\langle h, r\ t \rangle \overset{\text{def}}{=} \sum_{j=1}^{d} h[j] * r[j] * t[j]$, for the brevity, $\langle h, r, t \rangle \overset{\text{def}}{=} (v \odot w) \cdot x$, where the $\odot$ is Hadamrd multiplication (element-wise), the · stands for dot product. The optimization object is to minimize each batch with $\sum$ softplus$(-1 \cdot \langle h, rt \rangle + \lambda\theta)$, here λ is the model hyper-parameter and θ is the regularization factor, softplus $= \log(1 + e^x)$ is a softened relu function. In addition, the l is in $\{-1, 1\}$, which is the labeled result.

For the translational method, take TransE for example, to make the predication model work, the parameters mainly are $\{h^{k\times d}, r^{n\times d}\}$. If we free the ties between head entity and tail entity for an entity, a $t^{k\times d}$ also is needed to represent the tail entity, for simplicity, assuming the 3 types has same embedding dimension. Let us consider the size of weight parameters in SimplE. We can rewrite parameter spaces here, parameter $= \{h^{k\times d}, r^{n\times d}, r^{n\times d}, t^{k\times d}\}$. As a comparison, the TransE method has a smaller weight spaces in size. Roughly, the SimplE's weight size is about twice as the TransE's if we specify the two embedding size for the same dimension. It could be an important factor that can weaken TransE's expressiveness. If we simply try to increase TransE's embedding size to improve the expressiveness, it might cause problem of overfitting, which is not wanted. Generally there are no perfect theoretical result in deciding how the dimension d should be in this sort of problem. The parameter n, k are characters of the specified training data. Consider if the Knowledge Graph can grow, n, l, k can be very different. Thus it will generate performance impact on a trained model.

4 Experiments and Result

In this section, firstly we describe our dataset. Then we give the experiment result. With the convention of knowledge graph, we use link prediction to test the performance with other algorithms.

4.1 Data Sets and Experiment Setting

In this paper, we evaluate our methods with two typical knowledge graphs, built with Freebase [8] and WordNet [9]. These two datasets are chosen in the same way with

SimplE [13] and other embedding models. The statics of the datasets is showed in Table 3.

Table 3. Datasets statics

Dataset	#Rel	#Ent	#Train	#Valid	#Test
FB15 k	1,345	14,951	483,142	50,000	59,071
WN18	18	40,943	141,442	5,000	5,000

These two datasets have a little difference in semantic meaning layer. We briefly give a short introduction here. Freebase provides general facts of the world. For example, the triple (Steve Jobs, founded, Apple Inc.) builds a relation of founded between the name entity Steve Jobs and the organization entity Apple Inc. In WordNet, each entity is a synset consisting of several words, corresponding to a distinct word sense. Relationships are defined between synsets indicating their lexical relations, such as hypernym, hyponym, meronym and holonym.

4.2 Link Prediction

Link prediction aims to predict the missing h or t for a relation fact triple (h, r, t), used in [1–3]. In this task, for each position of missing entity, the system is asked to rank a set of candidate entities from the knowledge graph, instead of only giving one best result. As set up in [1–3], we conduct experiments using the data sets FB15 K and WN18.

In testing phase, for each test triple (h, r, t), we replace the head/tail entity by all entities in the knowledge graph, and rank these entities in descending order of similarity scores calculated by score function. Following TransR [2], ParaGraphE [3], SimplE [13], we use some measures as our evaluation metric. Details can be checked Table 4. Each measure has 2 settings, raw and filtered. The meanings of each measure are showed in Table 4.

Table 4. The meaning of each measure

Measure	Meaning
mr	The Mean rank of correct entities
mrr	The value of mean reciprocal rank
hits@1	The proportion of ranks list at first
hits@10	The proportion of correct entities in top-10 ranked entities
Raw	The metrics calculated on all corrupted triples
Filter	The metrics calculated on corrupted triples without those already existing in knowledge graph

Generally, if link predictor can have lower mean rank or higher Hits@1, Hits@10, we think it had better performance. In fact, a corrupted triple may also exist in knowledge graphs, which also should be considered as correct. However, the above evaluation may under-estimate those systems that rank these corrupted but correct triples high. Hence, before ranking we may filter out these corrupted triples which have appeared in knowledge graph. We name the first evaluation setting as "Raw" and the latter one as "Filter".

We describe the experiment running parameters here. We implemented our algorithm based on the work of SimplE [13], which using tensorflow framework. The learning rate λ is set to a fixed value of 0.01, the margin γ set 4. In the comparison tests with TransE and SimplE, the dimensions of head entities, tail entities, relations are set start from 50. And the initial dimension of TransE test is set to 50. The batch size is set to 1415. For datasets tests of WN18, each has iterations of 2000. We save five models from 1500 for interval 500 iterations and pick out the best one for the reported result. The SGD method is employed to do optimization. For SimplE, each test has 1000 iterations and the best model was chosen from models of start iteration 800 with interval iterations of 50.

Table 5. Experiment raw results on WN18 of TransE

Embedding dimension size	raw_mrr	raw_hit@1	raw_hit@3	raw_hit@10
50	0.01342	0.0063	0.0129	0.0276
100	0.020955	0.0108	0.0221	0.0386
150	0.024674	0.0104	0.0255	0.051
200	0.030065	0.0109	0.0323	0.0646
300	0.260614	0.017700	0.4519	0.655700
400	0.260597	0.0122	0.4564	0.6587

Table 6. Experiment filtered results on WN18 of TransE

Embedding dimension size	fil_mrr	fil_hit@1	fil_hit@3	fil_hit@10
50	0.013443	0.0063	0.013	0.0276
100	0.021242	0.0112	0.0222	0.0387
150	0.024925	0.0107	0.0257	0.051
200	0.030453	0.0112	0.0327	0.0648
300	0.340896	0.0285	0.6309	0.7432
400	0.339586	0.0243	0.6339	0.7401

From Tables 5 and 6, in this parameters setting, the dimension size affect the performance very significant. While in experiment on SimplE, the performance did not show too much variance. If we just compare performance with dimension size 400 of TransE and 200 of SimplE, We may find they do have a close performance in some degree. These are two different algorithms, but from the view of weights space, these two has nearly close size when dim = 400 in TransE and dim = 200 in SimplE. As we should pointed out, the tensor product definition in SimlE using $\langle h, r, t \rangle \overset{\text{def}}{=} (v \odot w) \cdot x$, the space will be far small than that generated by Cartesian product. Generally SimplE has better performance than TransE on dataset wn18, but if we increase the dimension size parameter of TransE to 200, the performance variance tend to become small. In the mean time we can notice that SimplE is much less sensitive to the dimension size parameter than TransE (Tables 7 and 8).

Table 7. Experiment raw results on WN18 of SimplE

Embedding dimension size	raw_mrr	raw_hit@1	raw_hit@3	raw_hit@10
50	0.559214	0.4345	0.6377	0.7942
100	0.580682	0.4633	0.6566	0.8026
150	0.57687	0.4593	0.6483	0.8008
200	0.575338	0.4578	0.6491	0.7978

Table 8. Experiment filtered results on WN18 of SimplE

Embedding dimension size	fil_mrr	fil_hit@1	fil_hit@3	fil_hit@10
50	0.8154	0.7281	0.9019	0.9372
100	0.9326	0.926	0.9391	0.9392
150	0.93832	0.937	0.9395	0.9398
200	0.938574	0.9373	0.9395	0.9405

5 Conclusion

In this paper, we researched the dimension size factor in the word embedding algorithms. Because for this kind of algorithms, to get better performance such as expressiveness, link prediction accuracy, running speed, scalability, etc., each one may have different optimization, especially with different loss function, which makes the embedding dimension problem more complex. We investigated this problem by comparing two typical algorithms, TransE and SimplE. By the performance comparison under a conversion dimension size, we found embedding dimension size could actually play a key role. While higher expressiveness algorithm can get better performance, but it will pay higher weight spaces size price and it will need to design more complex loss function, which might increase the risk of training complexity. Because all Knowledge Graph need to update, which means actually the number of relation size and entity size will increase. This will make the embedding performance worsen. In the worst case, this kind of change might have bad effect on the chosen of a specified embedding size. So there is a trading balance to deal with in word embedding.

For this work is computation resource intensive, our job in this paper is very preliminary. We hope it can be an initial step for further big evaluation experiments, which will need to test on more embedding algorithms, more kinds of datasets and in more scenarios. As our further work, we are going to investigate more embedding algorithms to make a more comprehensive understanding on word embedding. In addition, a unified theoretical model is in need to describe the dimension size impact on the simplicity and expressiveness.

References

1. Bordes, A., Usunier, N., Garcia-Duran, A., Weston, J., Yakhnenko, O.: Translating embeddings for modeling multi-relational data. In: NIPS, pp. 2787–2795 (2013)
2. Yankai, L., Zhiyuan, L., Maosong, S., Yang, L., Xuan, Z.: Learning entity and relation embeddings for knowledge graph completion. In: AAAI, pp. 2181–2187 (2015)
3. Xiao-Fan, N., Wu-Jun, L.: ParaGraphE: A Library for Parallel Knowledge Graph Embedding arXiv:1703.05614v3 (2017)
4. Recht, B., Re, C., Wright, S., Niu, F.: Hogwild: a lock-free approach to parallelizing stochastic gradient descent. In: NIPS, pp. 693–701 (2011)
5. Wang, Z., Zhang, J., Feng, J., Chen, Z.: Knowledge graph embedding by translating on hyperplanes. In: AAAI, pp. 1112–1119 (2014)
6. Xiao, H., Huang, M., Yu, H., Zhu, X.: From one point to a manifold: knowledge graph embedding for precise link prediction. In: IJCAI, pp. 1315–1321 (2016)
7. Zhao, S.-Y., Zhang, G.-D., Li, W.-J.: Lock-free optimization for nonconvex problems. In: AAAI, pp. 2935–2941 (2017)
8. Miller, G.A.: Wordnet: a lexical database for English. Commun. ACM **38**(11), 39–41 (1995)
9. Bollacker, K., Evans, C., Paritosh, P., Sturge, T., Taylor, J.: Freebase: a collaboratively created graph database for structuring human knowledge. In: Proceedings of KDD, pp. 1247–1250 (2008)
10. Jenatton, R., Roux, N.L., Bordes, A., Obozinski, G.R.: A latent factor model for highly multi-relational data. In: Proceedings of NIPS, pp. 3167–3175 (2012)
11. Ji, G., He, S., Xu, L., Liu, K., Zhao, J.: Knowledge graph embedding via dynamic mapping matrix. In: ACL, pp. 687–696 (2015)
12. Singhal, A.: Introducing the Knowledge Graph: Things, Not Strings. Google Official Blog, 16 May 2012. Accessed 6 Sept 2014
13. Kazemi, S.M., Poole, D.: SimplE embedding for link prediction in knowledge graphs. In: Advances in Neural Information Processing Systems (2018)

Short Text Mapping Based on Fast Clustering Using Minimum Spanning Trees

Pingrong Li[✉]

College of Electronic Commerce, Longnan Teachers College,
Longnan 746000, China
51081937@qq.com

Abstract. Due to short length and limited content, short text representation has the problem of high-dimension and high-sparsity. For the purpose of achieving the goal of reducing the dimension and eliminate the sparseness while preserve the semantics of the information in the text to be represented, a method of short text mapping based on fast clustering using minimum spanning trees is proposed. First, we remove the irrelevant terms, then a clustering method based on minimum spanning tree is adopted to identify the relevant term set and remove the redundant terms to get the short text mapping space. Finally, a matrix mapping method is designed to represent the original short text on a highly correlated and non-redundant short text mapping space. The proposed method not only has low time complexity but also produces higher quality short text mapping space. The experiments prove that our method is feasible and effective.

Keywords: Short text representation · Minimum spanning tree · Semantic feature space · Text mapping

1 Introduction

With the increasing popularity of social platforms, the way we consume and produce information online are changing. Consequently, user-generated textual contents are always in the form of short texts, including search result snippets, forum titles, image or video titles, tweets, and frequently asked questions (FAQs) and so on, and users may become overwhelmed by the massive data. Many researchers have attempted to employ the techniques of traditional text/document classification (referred as long text classification) to reorganize these short texts. However, traditional long text classification methods cannot perform well on short texts, because of the insufficient word occurrences and extreme sparsity of short texts. Besides, short texts are in general much shorter, nosier, and sparser, which calls for a revisit of text classification [1].

To deal with the shortness and sparsity, most approaches proposed for short text classification aim to enrich short text representation by introducing additional semantics. The additional semantics could be derived from the short text data collection [2] or be derived from a much larger external knowledge base with much longer documents in a similar domain as the short texts like Probase, Wikipedia and WordNet [3–5]. Some noticeable works include: In [2], a semi-supervised short text categorization method based on attribute selection was presented. Li et al. [3] proposed an efficient and

D.-S. Huang et al. (Eds.): ICIC 2019, LNAI 11645, pp. 545–554, 2019.
https://doi.org/10.1007/978-3-030-26766-7_50

effective approach for semantic similarity computation using a large scale probabilistic semantic network, known as Probase. In [4], the authors proposed a Wikitop system to automatically generate topic trees from the input text by performing hierarchical classification using the Wikipedia Category Network. Piao et al. [5] employed synsets from WordNet and concepts from DBpedia for representing user interests. In addition, short texts can be mapped to a semantic space learned from a set of common repositories, with the new representation in this semantic space being a combination of source information and external information. However, these external knowledge-enhanced methods are domain-based, which limits short text representation. What is worse, such outer or additional information are not always available, not to mention the high cost for acquiring such information. In the opposite direction of enriching short text representation, another approach known as search and vote is proposed to trim a short text representation to get a few most representative words for topical classification [6].

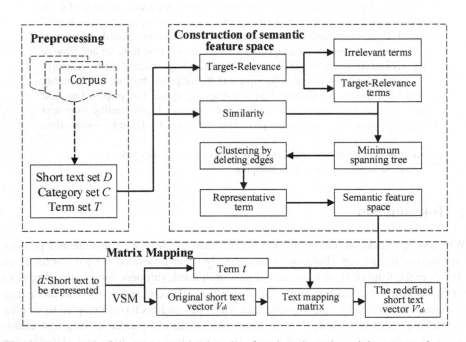

Fig. 1. Framework of short text mapping based on fast clustering using minimum spanning trees

In this work, we propose a short text mapping based on fast clustering using minimum spanning trees algorithm (SFM for short). The SFM algorithm works in two steps. In the first step, term are firstly divided into clusters using minimum spanning tree clustering method and the most representative term that is strongly related to target classes is selected from each cluster to form a subset of terms. In the second step, text mapping matrix is established to map the original short text vectors into the new feature space. Figure 1 summarizes the overview of the proposed short text mapping based on fast clustering using minimum spanning trees.

The main contributions of this paper are as follows:

(1) A core term selecting algorithm is proposed to reduce the representation dimension.
(2) A matrix mapping method is proposed, which can be used to handle the sparseness issue.

The remainder of this paper is organized as follows. The preliminaries for this paper are presented in Sect. 2. Section 3 introduces the proposed core terms selection and short text matrix mapping method. Experimental designs and findings are presented in Sect. 4. Section 5 concludes the proposed work and points out our future work.

2 Problem Preliminaries

In order to more precisely introduce the algorithm, and our proposed feature subset selection framework involves irrelevant feature removal and redundant feature elimination, we firstly provide some definitions.

Definition 1 (SIM). *SIM* (similarity) is adopted to indicate the degree of association between two random variables U and V. Three different normalized correlation measures are defined as follows:

$$SIM_{jac}(U,V) = \frac{I(U,V)}{H(U,V)} \tag{1}$$

$$SIM_{max}(U,V) = \frac{I(U,V)}{\max\{H(U),H(V)\}} \tag{2}$$

$$SIM_{sum}(U,V) = \frac{2 \times I(U,V)}{H(U)+H(V)} \tag{3}$$

Where $H(U)$ and $H(V)$ are the entropy of a discrete random variables U and V, $I(U, V)$ is the mutual information of U and V, $H(U, V)$ is the joint entropy of U and V.

Definition 2 (DIF). *DIS* (distance) is defined to indicate the degree of difference between the two random variables:

$$DIS(U,V) = 1 - SIM(U,V) \tag{4}$$

Obviously, *DIS* is a normalized measure as well.

Definition 3 (Target-Relevance). The relevance between the term t_i and the target concept C is referred to as the Target-Relevance of t_i and C, and denoted as $SIM(t_i, C)$. If $SIM(t_i, C)$ is greater than a predetermined threshold θ, we denote that t_i is a Target-Relevance term.

Definition 4 (Redundant term). Let $T = \{t_1, t_2, \ldots, t_N\}$ be a cluster of terms. if $\exists t_i \in S$, $SIM(t_j, C) \geq SIM(t_i, C) \wedge SIM(t_i, t_j) > SIM(t_j, C)$ is always corrected for each $t_i \in S(i \neq j)$, then t_i is redundant term with respect to the given t_j.

Definition 5 (Representative term). A term $t_i \in T = \{t_1, t_2, \ldots, t_N\}$ is a representative feature of the cluster S if and only if $t_i = \operatorname{argmax}_{t_j \in S} SIM(t_j, C)$.

For the continuous value, the well-known off-the-shelf MDL method [7] is used to discretize the continuous.

3   The Proposed Approach

3.1   Core Terms Selection Using Minimum Spanning Tree Clustering Methods

Feature subset selection can be viewed as the process of identifying and removing as many irrelevant and redundant features as possible. This is because: irrelevant features do not contribute to the predictive accuracy [8], and redundant features do not redound to getting a better predictor for that they provide mostly information which is already present in other feature. In order to obtain a high quality feature subset, we mainly referred to the methods from [9].

According to the above analysis, and definitions in Sect. 2, core terms set selection can be the process that identifies and retains the target-relevance terms and selects representative terms from term clusters. The behind heuristics are that

(1) Irrelevant terms have no/weak correlation with target concept;
(2) Redundant terms are assembled in a cluster and a representative term can be taken out of the cluster.

The proposed method logically consists of three steps: (1) removing irrelevant terms, (2) constructing a minimum spanning tree from relative terms, and (3) partitioning the minimum spanning tree and selecting representative terms.

For a short text set D with N terms $T = \{t_1, t_2, \ldots, t_N\}$ and class C, we compute the $SIM(t_i, C)$ value for each term $t_i(1 \leq i \leq N)$ in the first step. The terms whose $SIM(t_i, C)$ values are greater than a predefined threshold θ comprise the target-relevant term subset $T' = \{t'_1, t'_2, \ldots, t'_k\}$.

In the second step, we first calculate the $SIM\left(t'_i, t'_j\right)$ value for each pair of features t'_i and $t'_j\left(t'_i, t'_j \in T' \wedge i \neq j\right)$, we can easily calculate $DIS\left(t'_i, t'_j\right)$ through formula (4). Then, considering features t'_i and $t'_j(t'_i \neq t'_j)$ as vertices and $DIS\left(t'_i, t'_j\right)$ as the weight of the edge between vertices t'_i and t'_i, Weighted complete graph $G = (V, E)$ is constructed where $V = \{t'_i | t'_i \in t' \wedge i \in [1, k]\}$ and $E = \left\{\left(t'_i, t'_j\right) | \left(t'_i, t'_j \in t' \wedge i, j \in [1, k] \wedge i \neq j\right)\right\}$. G is an undirected graph.

The complete graph G reflects the correlations among all the target-relevant features. Unfortunately, there are k vertices and $k(k-1)/2$ edges of graph G. For high dimensional data, it is heavily dense and the edges with different weights are strongly interweaved. Moreover, the decomposition of complete graph is NP-hard [10]. Thus for graph G, a minimum spanning tree is constructed, connecting all vertices such that the sum of the weights of the edges is the minimum, the well-known Prim algorithm is performed [11]. The difference from the existing method is the weight of edge $\left(t_i', t_j'\right)$ is $DIS\left(t_i', t_j'\right)$.

After building the minimum spanning tree, in the third step, we remove the edges $E = \left\{\left(t_i', t_j'\right) \mid \left(t_i', t_j' \in T' \wedge i, j \in [1, k] \wedge i \neq j\right)\right\}$, whose $SIM\left(t_i', t_j'\right)$ value (note that all $SIM\left(t_i', t_j'\right)$ value has already been acquired in second step) between connected vertices are smaller than both of the target-relevant $SIM(t_i', C)$ and $SIM\left(t_j', C\right)$, from the minimum spanning tree. Each deletion results in two disconnected trees $Tree_1$ and $Tree_2$.

Assuming the set of vertices in any one of the final trees to be $V(Tree)$, then for each pair of vertices $t_i', t_j' \in V(Tree)$, $SIM\left(t_i', t_j'\right) > SIM(t_i', C) \wedge SIM\left(t_i', t_j'\right) > SIM\left(t_j', C\right)$ always holds. From Definition 6 we know that this property guarantees the features in $V(Tree)$ are redundant.

We may take Fig. 2 as an illustrative example. We first traverse all the edges, and then decide to remove the edge $\left(t_1', t_2'\right)$ since its weight $SIM\left(t_1', t_2'\right) = 0.3$ is smaller than both $SIM(t_1', C) = 0.4$ and $SIM(t_2', C) = 0.7$. This makes the minimum spanning tree clustered into two clusters denoted as $V(Tree_1)$ and $V(Tree_2)$ while each cluster is a minimum spanning tree as well. Take $Tree_1$ as an example, we know that $SIM(t_4', C) > SIM(t_2', C) \wedge SIM(t_2', t_4') > SIM(t_2', C)$ We also observed that there is no edge exists between t_3' and t_4'. Considering that $Tree_1$ is a minimum spanning tree, so we conclude that $SIM(t_3', t_4')$ is greater than $SIM(t_2', t_3')$ and $SIM(t_2', t_4')$. Thus $SIM(t_4', C) > SIM(t_2', C) \wedge SIM(t_2', t_4') > SIM(t_2', C)$ and $SIM(t_4', C) > SIM(t_3', C) \wedge SIM(t_3', t_4') > SIM(t_3', C)$ also hold.

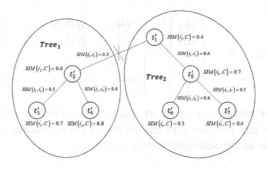

Fig. 2. Example of remove the edges

After removing all the unnecessary edges, a forest $Forest = \{Tree_1, Tree_2, \ldots, Tree_L\}$ is obtained. Each tree $Tree_i \in Forest$ represents a cluster that is denoted as $Tree_i$, which is the vertex set of $V(Tree_i)$ as well. As illustrated above, the terms in each cluster are redundant, so for each cluster $V(Tree_i)$, a representative term t_{c_j} whose Target-Relevance $SIM(t_{c_j}, C)$ is the greatest is selected. All $t_{c_j} (j = 1 \ldots L)$ comprise the final core term subset $T_c = \{t_{c_1}, t_{c_2}, \ldots, t_{c_L}\}$.

3.2   Short Text Representation via Matrix Mapping

The vector space model is adopted to represent short text, and each short text is considered to be a vector, in which each term is associated with a weight indicating its importance. However, the main weakness of this term-vector representation is that different but semantically related terms are not matched, the dimension and sparseness are always high. Our goal is to construct a projection matrix that maps the corresponding term-vectors into a low-dimensional and low-sparseness semantic feature space such that classification performance can be improved.

For an original short text d_i which consists of N terms $T = \{t_1, t_2, \ldots, t_K\}$, we employ the IDF term weighting model, in which the short text d_i is represented as V_{d_i}.

$$V_{d_i} = [idf_1, idf_2, \ldots, idf_K] \tag{5}$$

In the refined approaches, all the terms of d_i and all the core terms are considered to construct a text mapping matrix M. Each entry reflects the semantic proximity between terms t_i and t_j.

$$M = \begin{bmatrix} SIM_{1,1} & SIM_{1,2} & \cdots & SIM_{1,L} \\ SIM_{2,1} & SIM_{2,2} & \cdots & SIM_{2,L} \\ \vdots & \vdots & \ddots & \vdots \\ SIM_{K,1} & SIM_{K,2} & \cdots & SIM_{K,L} \end{bmatrix} \tag{6}$$

Then, the redefined short text semantic mapping vector V'_{d_i} can be obtained via the multiplication of V_{d_i} and the mapping matrix M:

$$V'_{d_i} = V_{d_i} \times M = \begin{bmatrix} idf_1 \\ idf_2 \\ \vdots \\ idf_K \end{bmatrix}^T \times \begin{bmatrix} SIM_{1,1} & SIM_{1,2} & \cdots & SIM_{1,L} \\ SIM_{2,1} & SIM_{2,2} & \cdots & SIM_{2,L} \\ \vdots & \vdots & \ddots & \vdots \\ SIM_{K,1} & SIM_{K,2} & \cdots & SIM_{K,L} \end{bmatrix} \tag{7}$$

Therefore, after mapping, each short text is represented by a less sparse and low dimensional vector that has non-zero entries for all terms that are semantically similar to those that appear in the semantic feature space.

4 Experiments and Results Analysis

In this section, we conduct a series of experiments to evaluate the performance of SFM and analyze these experiments and the results.

4.1 Data Sets

For the purpose of evaluating the performance of our algorithm, we carry out experiments on both Chinese and English test collections, respectively. We adopt 10 classes obtained from DBLP, with 1000 paper titles obtained from CCF recommended list in Rank A and B as English data sets. Besides, we select 10 categories from the Sogou corpus as Chinese data sets. For each category, 1000 news abstracts are selected as experimental data.

We did a series of preprocessing work on both data sets including data denoising, stop words removal, and for Chinese dataset, jieba is utilized as word segmentation tool to obtain the vocabulary.

4.2 Evaluation Metrics

We take advantage of F1-measure and accuracy as the evaluation of metrics [12], and the definitions are as follows:

$$accuracy = \frac{TP + TN}{TP + TN + FP + FN}$$
$$F1\text{-}Measure = \frac{2 \times P \times R}{P + R} \qquad (8)$$

TP, TN, FP, FN represent True Positive (The fact is the positive samples, which were judged as positive samples), True Negative, False Positive, False Negative, respectively. P stands for precision, R is recall, which is defined as follows.

$$P = \frac{TP}{TP + FP} \qquad (9)$$

$$R = \frac{TP}{TP + FN} \qquad (10)$$

4.3 Experimental Results and Analysis

We present experimental results on two tasks. Firstly, we visualize the selection results and evaluate our schemes for short text feature selection and compare the performances with three other selection methods. Secondly, classification performance will be evaluated with different size of feature of our method and compare the performances with other methods. The three other feature selection strategies are SFSM (Effectively Representing Short Text via the Improved Semantic Feature Space Mapping), CDCFS

(Leveraging Term Co-occurrence Distance and Strong Classification Features for Short Text Feature Selection) and TF-IDF method.

Comparison of Different Term Selection Results. We compare feature dictionaries obtained from the above strategies to verify that our method can get a high accuracy for short text feature selection. We select top 20 terms from 1955 terms in the class of Artificial Intelligence and Pattern Recognition using different methods. Table 1 is the comparison of different term selection results.

Table 1. Comparison of Different Selection Results

Method	TF-IDF	CDCFS	SFSM	SFM
1	Algorithm	Algorithm	Algorithm	**Recognition**
2	Model	Model	Network	**Learning**
3	Recognition	Recognition	Recognition	**Cluster**
4	Learning	Feature	Learning	**Classify**
5	Graph	Learning	Cluster	**Feature**
6	Cluster	Cluster	Graph	**Goal**
7	Optimization	Classify	Model	**Tag**
8	Data	Optimization	Data	**Label**
9	Feature	Data	Feature	**Space**
10	Classify	Graph	Classify	**Graph**
11	Detection	Detection	Application	**Rough**
12	Application	Improve	System	**Stripe**
13	Improve	Goal	Detection	**Fuzzy**
14	Goal	Application	Optimization	**Deep**
15	Fuzzy	Theme	Attack	**Local**
16	Network	Fuzzy	Oriented	**Expression**
17	Space	Forecast	Goal	**Algorithm**
18	Mix	Network	Achieve	**Emotion**
19	Object	Space	Solve	**Constraint**
20	Part	Parallel	Analyze	**Mix**

From Table 1 we can easily see that our method has the strongest indication to representation of this category.

Comparison of Different Methods Performance. To compare the performance between our method and other methods, two different types of classification algorithms KNN and SVM are employed to classify data sets. Existing short text classification work suggests that most of the term-weighting schemes shows its best performance in the range of 20–45. Therefore, we parameterize k-NN by choosing different value k in this range and demonstrated the best performance using optimal k. As for the SVM algorithm, we use the linear kernel functions and implement it on the libsvm tool, the other parameters of SVM are set to their default values. We perform 5-fold cross-validation on both DBLP and Sogou data sets. The results of the accuracy and the F1-measure are as follows (Fig. 3):

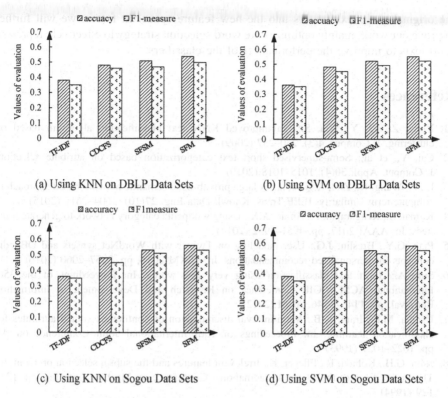

Fig. 3. Performance Comparison between our approaches and baselines

Firstly, in all cases, we can see that our methods performed significantly better than other methods. Furthermore, we find that the experimental results on KNN and SVM have slight difference, which also demonstrates that the proposed representation method is insensitive to the classification algorithm and has good robustness.

Secondly, from the experimental results, we can clearly see the worst performance is the traditional TF-IDF method. Then we can see that the CDSTM and the SFSM are significantly improved compared to TF-IDF. Even in this case, our approaches is still better. In conclusion, the results of these experiments demonstrate that our method is beneficial to improve semantic feature space, thereby improving the short text classification ability.

5 Conclusion

In this paper, we propose a new method for improving the classification of short texts using a combination of relevant feature selection plus a text mapping mechanism. Term are firstly divided into clusters by using minimum spanning tree clustering methods and the most representative term that is strongly related to target classes is selected from each cluster to form a subset of terms. Text mapping matrix is then established to map the original short text vectors into the new feature space. In future, we will further improve our work, mainly optimize core word selection strategy to effectively represent short texts to improve the performances of the classifiers.

References

1. Yong, Z., Li, Y., Xia, S.: An improved KNN text classification algorithm based on clustering. J. Comput. **4**(3), 230–237 (2009)
2. Cai, Y., et al.: Semi-supervised short text categorization based on attribute selection. J. Comput. Appl. **30**(4), 1015–1018 (2010)
3. Li, P., Wang, H., Zhu, K.Q., et al.: A large probabilistic semantic network based approach to compute term similarity. IEEE Trans. Knowl. Data Eng. **27**(10), 2604–2617 (2015)
4. Kumar, S., Rengarajan, P., Annie, A.X.: Using wikipedia category network to generate topic trees. In: AAAI 2017, pp. 4951–4952 (2017)
5. Piao, G.Y., Breslin, J.G.: User modeling on Twitter with WordNet synsets and DBpedia concepts for personalized recommendations. In: CIKM 2016, pp. 2057–2060 (2016)
6. Sun, A.: Short text classification using very few words. In: Proceedings of the 35th International ACM SIGIR Conference on Research and Development in Information Retrieval, pp. 1145–1146. ACM (2012)
7. Usama, M.F., Irani, K.B.: Multi-interval discretization of continuous valued attributes for classification learning. In: Proceedings of 13th International Joint Conference on AI, pp. 1022–1027 (1993)
8. John, G.H., Kohavi, R., Pfleger, K.: Irrelevant features and the subset selection problem. In: The Proceedings of the Eleventh International Conference on Machine Learning, pp. 121–129 (1994)
9. Song, Q., Ni, J., Wang, G.: A fast clustering-based feature subset selection algorithm for high-dimensional data. IEEE Trans. Knowl. Data Eng. **25**(1), 1–14 (2013)
10. Garey, M.R., Johnson, D.S.: Computers and Intractability: A Guide to the Theory of NP-Completeness. W. H. Freeman & Co., New York (1979)
11. Prim, R.C.: Shortest connection networks and some generalizations. Bell Syst. Tech. J. **36**, 1389–1401 (1957)
12. Gao, L., Zhou, S., Guan, J.: Effectively classifying short texts by structured sparse representation with dictionary filtering. Inf. Sci. **323**, 130–142 (2015)

Missing Data Imputation for Operation Data of Transformer Based on Functional Principal Component Analysis and Wavelet Transform

Jiafeng Qin[1], Yi Yang[1], Zijing Hong[2($\boxtimes$)], and Hongyi Du[2]

[1] State Grid Shandong Electric Power Research Institute, Jinan, China
[2] School of Mathematical Sciences, Fudan University, Shanghai, China
15300180051@fudan.edu.cn

Abstract. The operation data of transformer plays an important role in assessing the status of electric devices. However, the real operation data often have missing values, which will result in the unreliability of the following data analysis. In view of this, we propose a method based on functional principal component analysis (FPCA) and wavelet transform to impute the missing data. According to the characteristic of the operation data, we separate the daily data curve to low frequency part and high frequency part. We first investigate the fluctuation patterns over the historical data and use FPCA to estimate the low frequency part. We then estimate the residual function and use wavelet transform to estimate the high frequency part. Combining these two parts, we get the approximation to the original data and impute the missing values. Applications of the proposed method to the real operation data show that the method performs very well and it works well whenever the missing points are randomly distributed or in continuous form.

Keywords: Functional principal component analysis · Wavelet transform · Operation data of transformer · Data imputation

1 Introduction

With the development of information technology in the power system and the accelerating construction of the smart grid, the monitoring and operation data of the power grid are exponentially increasing [1–3]. These data lay the foundation for the following scheduling and decision making, at the same time, they bring new challenges in data processing [4, 5]. As a significant part of the smart grid, transformers generate a huge amount of operation data, which a very important role in the assessment of the equipment's status. However, the real operation data often have missing values due to equipment failures, communication failures, operational errors, etc., which will result in the unreliability of the following data analysis. Therefore, imputing the missing

The research of this work is supported by Science and Technology Commission of Shanghai Municipality 16JC1402600.

D.-S. Huang et al. (Eds.): ICIC 2019, LNAI 11645, pp. 555–567, 2019.
https://doi.org/10.1007/978-3-030-26766-7_51

operation data of transformer is critical to improving the efficiency of the smart grid and improving the reliability and self-healing capabilities of the grid system [6].

There have been a lot of methods on data imputation for grid systems [7–9]. Zhang et al. estimated the characteristic function of electric load data based on the idea of clustering [7]. And then used this characteristic function to impute the missing data. However, this method requires a high degree of similarity across daily curves. Yan et al. considered the power transmission and transformation equipment status data as time series [8]. And they used ARIMA to fit the curve iteratively. This method is not suitable for the situation when missing data are continuously distributed. Liu et al. proposed a historical data mining and recovery method based on Pearson correlation coefficient [9]. This method is essentially doing linear regression. However, the dependent variable and the independent variables may not satisfy the linear relationship. And when the missing points are continuously distributed, the linear model is too simple to achieve satisfactory results.

Compared to the above mentioned data, operation data of transformer have different characteristics. We summarize three characteristics of the operation data as follows:

1. Time dependency. The operation data of the transformer are time dependent. The sensor measures the data at discrete time points. However, the operation data of transformer should be generated continuously in reality. Therefore, we should use methods that can deal with time-continuous data to process the operation data.
2. Periodicity. A certain type of operation data (e.g. high side voltage[1], high side current (See footnote 1), etc.) often has similar patterns on a daily scale. This phenomenon occurs because the operation data of the transformer are related to the power load. And the power load is greatly related to the consumption habits of the households and enterprises served by the transformer, which do not change greatly in a certain period of time. As a result, the daily operation data have some similar patterns. We call this kind of similarity as periodicity with a period one day.
3. Oscillation. The transformer is a complicated system. The voltage and the current on the high and the low voltage sides will be affected by various factors such as the power load, the wire temperature, and the operating state of the transformer. Therefore, the operation data always change rapidly, which results in the oscillation of the curve.

Viewing the weaknesses of the existing methods and considering the three characteristics listed above, we propose a method based on functional principal component analysis and wavelet transform to impute the missing operation data of transformer. Functional principal component analysis (FPCA) is a high-dimensional feature extraction method, which can find some similar fluctuation patterns among curves in the low-dimensional function space [10]. Wavelet transform is a signal processing method which focuses on the local details of the signal. And by analyzing the signal in multi-scale, it can effectively eliminate noise and extract the effective information of the signal [11].

---

[1] This type of operation data measures the voltage on the high voltage side of a transformer.

In our study, due to the periodicity of the transformer operation data, different daily curves of a certain type of data can be regarded as repeated observations. This motivates us to use FPCA to fit the data. To deal with the oscillations in each day, we take advantages of wavelet transform, and use it to extract the effective information in the local oscillation pattern. By combining the low frequency and high frequency parts, we can get the approximation of the original signal and finish the imputation of missing values. Applications of the proposed method to the real operation data of transformer show that the method performs very well and can work well whenever the missing points are randomly distributed or continuously distributed.

2 Imputation Principle Based on FPCA and Wavelet Transform

2.1 Applicability of FPCA to the Operation Data

We consider the voltage, the current, the active power, the reactive power on the high voltage side, the medium voltage side and the low voltage side separately. Let's take high side current as an example. The minimal time interval between two adjacent observed data points is 1 min, thus there should be 1,440 data points within a day (24 h). However, in reality, there are cases where no data points are observed in several minutes or even tens of minutes due to sensor failure, data loss, etc. This paper assumes that most of the observed data points can reflect the true fluctuations, and a small number of outliers exist. Under this assumption, we use FPCA to fit the low frequency signal, which considers the data from two aspects.

Firstly, FPCA considers the longitudinal data, and it can be applied to extract the fluctuation patterns from historical data. Figure 1 shows the high side current of a transformer in three consecutive days. The red arrow can roughly represent the fluctuation tendency. It is clear that, the peaks and the valleys have some correspondence in a certain time period of a day, although the detailed shapes of the peaks or the valleys are not exactly the same.

FPCA utilizes the similarities in consecutive days to extract the important fluctuation patterns, which is called "principal component curve". Theoretically, the number of the "principal component curve" is infinity. According to Karhunen–Loève theorem [12], when the number of selected principal component curves tends to infinity, the original daily curve can be expressed as an infinite linear combination of these principal component curves. In practical applications, we only need to select a small number of principal component curves and then use linear combination to fit the data.

Secondly, FPCA considers the horizontal data, and it adjusts the coefficients of the selected principal component curves according to the data points observed on the current day to achieve the best fitting effect.

By combining the historical patterns and the characteristic of the current day's curve, FPCA can fit the curve reliably even if the value is missing in a long period of time, where normal interpolation methods fail.

Since some outliers inevitably exist in the actual data, it is necessary to discuss the influence of outliers on the FPCA method:

1. The outliers due to short-term strong electromagnetic interference or abnormal transmission tend to appear randomly. The number of this kind of outliers is very small, which has little effect on the estimation of the linear combination by the conditional expectation. Thus the fitting result of FPCA will conform to the normal fluctuation law.
2. When the outliers are caused by a transformer failure in a long time, the outliers tend to appear continuously. The number of this kind of outliers often tends to be large. Under this circumstance, the fluctuations during the transformer failure time interval will be different from that in the same time period of normal days. The principal component curves extracted by FPCA will be impacted under this situation and will include some abnormal fluctuation trends. It is necessary to use other methods to remove this kind of outliers and consider them as missing points before using FPCA to fit the data.

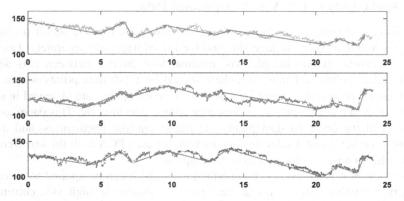

Fig. 1. Comparison of high side current in three consecutive days (Color figure online)

2.2 Model Construction

In this subsection, we take high side current as an example to show the model construction process. Let $Y_i(t)$ be the high side current curve of the ith day, and $T_{ij}, 1 \leq j \leq n_i$ are observed time points. Then the high side current of the ith day at T_{ij} is $Y_i(T_{ij})$, which can be denoted as Y_{ij}. We decompose $Y_i(t)$ to low frequency part and high frequency part as:

$$Y_i(t) = X_i(t) + \varsigma_i(t) \tag{1}$$

And we use FPCA to estimate the low frequency part, and use wavelet transform to estimate the high frequency part. According to Karhunen–Loève theorem [13], $X_i(t)$ could be expressed as:

$$X_i(t) = \mu(t) + \sum_{k=1}^{\infty} \alpha_{ik} \varphi_k(t) \tag{2}$$

where $\mu(t)$ is the mean function, and $\varphi_k(t)$ is the eigenfunction of the covariance surface, or it can be called principal component function. α_{ik} is the score of $X_i(t)$ on the principal component function. Then the model can be written as:

$$Y_i(t) = X_i(t) + \varsigma_i(t) = \mu(t) + \sum_{k=1}^{\infty} \alpha_{ik} \varphi_k(t) + \varsigma_i(t) \tag{3}$$

2.3 Estimate the Low Frequency Part Using FPCA

In this subsection, we use FPCA to estimate the low frequency part of the operation data, which is motivated by Sect. 3 of [13].

First, we use weighted least square to estimate the mean function $\mu(t)$. The model is formulated as the following minimization problem:

$$\min_{\beta_0, \beta_1} \sum_{i=1}^{n} \sum_{j=1}^{n_i} k\left(\frac{T_{ij} - t}{h_\mu}\right) \left[Y_{ij} - \beta_1 \cdot (t - T_{ij}) - \beta_0\right]^2, \tag{4}$$

where $\beta_0(t)$ is the estimation of the mean function $\mu(t) : \hat{\mu}(t) = \hat{\beta}_0(t)$, and $\beta_1(t)$ is the slope of a linear correction function given the time point t. $Y_{ij} - \beta_1 \cdot (t - T_{ij})$ is the correction value for Y_{ij} at time point t. We take Gaussian kernel function $k(\frac{T_{ij}-t}{h_\mu})$ as the weight. The parameter h_μ in Eq. (4) is a constant which represents the bandwidth, which can be chosen using cross-validation.

Then we use weighted least square to estimate the covariance surface. When $s \neq t$, the covariance surface can be estimated by solving the following minimization problem:

$$\min \sum_{i=1}^{n} \sum_{j\neq l} k_2\left(\frac{T_{ij} - s}{h_G}, \frac{T_{il} - t}{h_G}\right) \left[G_i(T_{ij}, T_{il}) - \beta_{11} \cdot (s - T_{ij}) - \beta_{12}(t - T_{il}) - \beta_0\right]^2. \tag{5}$$

where $\beta_0(s,t), \beta_{11}(s,t), \beta_{12}(s,t)$ are all parameters to be estimated given a certain time point (s,t). $\beta_0(s,t)$ is the estimator of the covariance surface $G(s,t) : \hat{G}(s,t) = \hat{\beta}_0(s,t)$. $\beta_{11}(s,t)$ and $\beta_{12}(s,t)$ are the slopes of the linear correction function on the two orthogonal directions. $k_2(\cdot, \cdot)$ is a binary Gaussian kernel function, and h_G is also a bandwidth. For simplicity, one can let $h_G = h_\mu$. And we can also choose the bandwidth with cross-validation.

When $s = t$, since the shape of the covariance surface in the vertical diagonal direction is closer to the quadratic curve [14, 15], we may use quadratic correction function rather than linear correction function in the vertical diagonal direction. By solving a similar minimization problem like Eq. (5), we can get the estimation of the covariance surface when $s = t$ as $\hat{G}(t,t)$.

At last, we estimate the principal component function $\varphi_k(t)$ and the scores of $X_i(t)$ on the principal component function. Since the principal component functions are the eigenfunctions of the covariance surface, we could get the estimation of them by solving the following integral equation:

$$\int_{\Gamma} \hat{G}(s,t)\hat{\varphi}_k(s)ds = \hat{\lambda}_k\hat{\varphi}_k(t) \tag{6}$$

By discretizing the smoothed covariance, we could estimate $\varphi_k(t)$ [16, 17]. We then get K groups of eigenfunction and eigenvalue $\left(\hat{\varphi}_k(t), \hat{\lambda}_k\right), 1 \le k \le K$. And we may select the top g largest eigenvalues and the corresponding eigenfunctions, where g is a given constant.

After that, we estimate the scores of $X_i(t)$ on the principal component functions by using conditional expectation. Let $\hat{\varphi}_{ik} = (\hat{\varphi}_k(T_{i1}), \hat{\varphi}_k(T_{i2}), \ldots, \hat{\varphi}_k(T_{iN_i}))^T$, $\tilde{Y}_i = (Y_{i1}, Y_{i2}, \ldots, Y_{iN_i})^T$, $(\hat{\Sigma}_{Y_i})_{j,l} = \hat{G}(T_{ij}, T_{il}) + \hat{\sigma}^2\delta_{jl}$, $\hat{\mu}_i = (\mu(T_{i1}), \mu(T_{i2}), \ldots, \mu(T_{iN_i}))^T$, where δ_{jl} is a indicator function, i.e. $\delta_{jl} = 1$ when $j = l$, and $\delta_{jl} = 0$ when $j \ne l$. Γ is the domain of t, and $\hat{\sigma}^2$ can be expressed as:

$$\hat{\sigma}^2 = \frac{2}{|\Gamma|}\int_{\frac{|\Gamma|}{4}}^{\frac{3|\Gamma|}{4}} \left[\hat{\beta}_0(t,t) - \hat{G}(t,t)\right]dt \tag{7}$$

And $\hat{\alpha}_{ik}$ is estimated as:

$$\hat{\alpha}_{ik} = E\left(\alpha_{ik}|\tilde{Y}_i\right) = \hat{\lambda}_k\hat{\varphi}_{ik}^T\hat{\Sigma}_{Y_i}^{-1}(\tilde{Y} - \hat{\mu}_i) \tag{8}$$

And we finally get the estimation of the low frequency part of the daily curve:

$$\hat{X}_i(t) = \hat{\mu}(t) + \sum_{j=1}^{k} \hat{\alpha}_{ij}\hat{\varphi}_j(t) \tag{9}$$

2.4   Estimate the High Frequency Part Using Wavelet Transform

After estimating the low frequency part, we can calculate the residual on the observed time points. And we also consider it as a function, which contains both effective signal and noise. Then the residual function can be expressed as:

$$\xi_i(t) = \varsigma_i(t) + e_i(t) \tag{10}$$

Here $\varsigma_i(t)$ is the effective signal, and $e_i(t)$ is the noise. In order to improve the performance of the wavelet transform, we need to estimate the values of the residual function on all time points from 1 to 1440. We denote the set of the observed time points in ith day as $V_i = \{T_{i1}, T_{i2}, \ldots, T_{in_i}|T_{i1} \le T_{i2} \le \ldots \le T_{in_i}\}$. If $t \in V_i$, there exists a number m, s.t. $t = T_{im}$. Then the value of the residual function at t can be expressed as $\xi_i(t) = Y_{im} - \hat{X}_i(t)$. If $t \notin V_i$, define $t_{i0} = 1, t_{i(n_i+1)} = 1440$, then there must exist a number p s.t. $t \in (t_{ip}, t_{i(p+1)})$. We further define $gap = |t_{ip} - t_{i(p+1)}|$. When gap is small, the value of residual function at t can be estimated by moving average. And when *gap* is large, it is meaningless to estimate the value since it may add extra error to the final estimation. In a word, the residual function $\xi_i(t)$ could be estimated as follows:

$$\xi_i(t) = \begin{cases} Y_{im} - \hat{X}_i(t), & t \in V_i \\ 0, & t \notin V_i, gap > 10 \\ \frac{1}{q}\sum_{s=1}^{q} \hat{\xi}_i(T_{i(p+1-s)}), & t \notin V_i, gap \leq 10, t \geq T_{il}, q = \min\{p, 10\} \\ 0, & t \notin V_i, gap \leq 10, t < T_{i1} \end{cases} \tag{11}$$

Then we use wavelet transform to eliminate the noise and estimate the high frequency part $\varsigma_i(t)$. Since $\xi_i(t)$ is a continuous function, whose domain is a bounded and closed interval, we could assume that $\xi_i(t)$ is square integrable. So according to the theory of discrete wavelet transform [18], $\xi_i(t)$ could be expressed as:

$$\xi_i(t) = \sum_k c_{j_0}(k)2^{\frac{j_0}{2}}\Phi(2^{j_0}t - k) + \sum_k \sum_{j=j_0}^{\infty} d_j(k)2^{\frac{j}{2}}\Psi(2^j t - k) \tag{12}$$

Here $\Phi(t)$ is the scaling function and $\Psi(t)$ is the wavelet function. And $c_{j_0}(k)$ and $d_j(k)$ are the coefficients corresponding to the scaling function. Since we only get finite discrete sample points in reality, the scale cannot be subdivided infinitely. Thus, $\xi_i(t)$ can be expressed as:

$$\xi_i(t) = \sum_k c_{j_0}(k)2^{\frac{j_0}{2}}\Phi(2^{j_0}t - k) + \sum_k \sum_{j=j_0}^{\infty} d_j(k)2^{\frac{j}{2}}\Psi(2^j t - k) \tag{13}$$

where $j_0 = 0$ and j_1 is the number of decomposition level, which is a selected constant. After the wavelet function, which includes Daubechies wavelets, Symlets, Coiflets and so on, and the decomposition level are selected, we could calculate the coefficients $c_{j_0}(k)$ and $d_j(k)$ by Mallat algorithm [19]. Typically, a large absolute value of a coefficient of the wavelet function $d_j(k)$ represents an effective signal, while a small absolute value of a coefficient means the energy of the signal is low, which is always the characteristic of noise. So we use some strategies to accept large enough coefficients and set other small coefficients as zero, in the view of absolute value. At last, we use the processed coefficients to reconstruct the signal.

The strategies of processing the coefficients include two parts. We define a threshold at first, and then we use hard thresholding method or soft thresholding method to process the coefficients. Four different definitions of threshold are listed as follows:

1. Adaptive threshold selection based on Stein's Unbiased Risk Estimate. This definition could be divided into three steps. Firstly, assuming that there are K coefficients of the wavelet function, we write it as $d_{j1}, d_{j2}, \ldots, d_{jK}$. And then we write $\vec{d}_j' = (d_{j1}'^2, d_{j2}'^2 \ldots, d_{jK}'^2)$, where $d_{j1}', d_{j2}', \ldots, d_{jK}'$ is a permutation of $d_{j1}, d_{j2}, \ldots, d_{jK}$ s.t. $|d_{j1}'| \leq |d_{j2}'| \leq \ldots \leq |d_{jK}'|$.   Secondly,   we   write

$I = \underset{i}{\operatorname{argmin}}\{[K - 2i + (K - i) \cdot d_{ji}'^2 + \sum_{k=1}^{i} d_{jk}'^2]/K\}$. Thirdly, the threshold could be written as $\lambda_{rigr} = \sqrt{d_{ji}'^2}$.

2. Fixed-form threshold. And the threshold could be written as $\lambda_{sqt} = \sqrt{\log(2K)}$.
3. Heuristic threshold. This definition could be divided into two steps. Firstly, we write $crit = \sqrt{\frac{\log^3 K}{K \log^3 2}}$, $eta = \frac{\|\vec{d_j}'\|_2^2 - K}{K}$. Secondly, the threshold λ_{heur} could be written as $\lambda_{heur} = \lambda_{sqt}$ if $eta \leq crit$. Otherwise, it could be written as $\lambda_{heur} = \min\{\lambda_{rigr}, \lambda_{sqt}\}$.
4. Minimax threshold. If $K > 32$, the threshold is $\lambda_{mima} = 0.3936 + 0.1829 \frac{\log N}{\log 2}$. Otherwise, the threshold is zero.

And two methods of processing the coefficients are as follows:

1. Hard thresholding. The processed coefficient d_{ji}' is equal to d_{ji} if $|d_{ji}| \geq \lambda$, otherwise, $d_{ji}' = 0$.
2. Soft thresholding. The processed coefficient d_{ji}' could be expressed as $d_{ji}' = \operatorname{sgn}(d_{ji}) \cdot (|d_{ji}| - \lambda)^+$.

After getting the processed coefficients, we could use Mallat algorithm [20] to reconstruct the high frequency signal.

Since different wavelet functions, different numbers of decomposition level and different thresholding strategies can be suitable to different kinds of operation data. In order to make the algorithm more general, we use cross-validation to choose the most suitable strategy to estimate the high frequency part. Specifically, we divide the data set to two parts. 90% randomly selected data points are taken as the training set, while the remaining 10% data points are in the validation set. Then we use different strategies to process the residual function by wavelet transform. After that, we could get the estimated high frequency part by using Mallat algorithm to reconstruct the signal. And combing with the low frequency part, we could get the whole estimation of the daily curve $\hat{Y}_i(t)$. We separately calculate the RMSE for every method on the validation set and choose the method having the smallest RMSE on the validation set.

At last, we use the selected method to estimate the low frequency part and the high frequency part on the whole data set to get the final estimation $\hat{Y}_i(t)$.

3   Missing Value Imputation Procedure

In this section, we will give the specific steps of our proposed method. Take the high side current for example, we consider the data of n continuous days.

- Step1: Divide the whole data set into training set and validation set. For the data of each day, we randomly select 90% of the observed points as the training set, and the remaining 10% as the validation set.
- Step 2: Estimate the mean function according to Eq. (4).

- Step 3: Estimate the covariance surface according to Eq. (5).
- Step 4: Given a constant g, estimate the top g largest eigenvalues and the corresponding eigenfunctions $\left(\hat{\lambda}_k, \hat{\varphi}_k(t)\right)$ according to Eq. (6).
- Step 5: Estimate the scores of $X_i(t)$ on the principal component functions according to Eqs. (7) and (8).

After steps 1–5, we get the estimation of all the low frequency parts of n days' curves on the training set.

- Step 6: Estimate the residual function $\xi_i(t)$ according to Eq. (11).
- Step 7: Use cross-validation to choose the most suitable strategy, which includes the wavelet function, the decomposition level and the thresholding strategies, to estimate the high frequency part, as we formerly discussed in Sect. 2.3.
- Step 8: Combine the training set and the validation set and consider the whole data set. And combining with the low frequency part and the high frequency part, we get the final estimation of the daily curve: $\hat{Y}_i(t) = \hat{X}_i(t) + \hat{\varsigma}_i(t)$.

4 Applications to a High Side Current Data Set

We first demonstrate our method in a high side current data set produced by a 220 kV main transformer from Jan. 1st to Jan. 10th, 2016. Since the missing values could be randomly distributed or continuously distributed in reality, we test the performance of the proposed method under these two situations separately.

First, we consider the situation that the missing values appear randomly. In order to evaluate the performance of the proposed method, we randomly select 10% points in the data set to form a test set to evaluate the performance of the proposed method.

After steps 1–5, we get the estimation of the low frequency part, the result of Jan. 1st, 2016 in shown in Fig. 2. The black dotted line shows the observed points. The scattered blue points represent the missing points, which are in the test set only. And we use red solid line to represent the estimated low frequency signal.

It can be shown that the low frequency signal fits the observed curve well. When we focus on the local details, we can see some disparities between the observed curve and the low frequency signal. We then continue to implement steps 6–8 on the basis of steps 1–5. And the final fitting curve is showed in Fig. 3. It is clear that the final fitting curve fits the true value very well.

To further show the reliability of the proposed method, we compared the performance of our method with the performance of the cubic spline interpolation by comparing RMSE on the test set. We repeated the experiment ten times. And every time we randomly select the test set. The comparison of RMSE of two methods is showed in Table 1. The result shows that, RMSE of our method is apparently smaller than that of cubic spline interpolation. The decrease of RMSE is 21.3% by average. And the standard deviation of RMSE of cubic spline interpolation is 0.203, while ours is 0.138. This shows that our method is more stable than cubic spline interpolation.

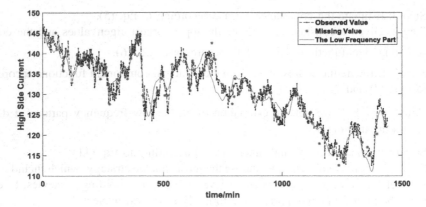

Fig. 2. The low frequency part (Color figure online)

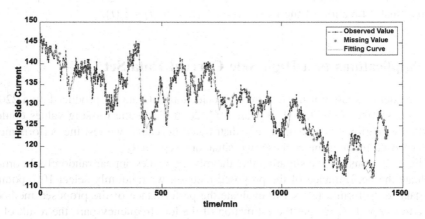

Fig. 3. The final fitting curve when missing points are randomly distributed

Table 1. Comparison of RMSE on high side current

Experiment time	RMSE of our method	RMSE of cubic spline interpolation	Decrease of RMSE
1	1.4368	2.0281	29.2%
2	1.4810	1.9233	23.0%
3	1.2364	1.5931	22.4%
4	1.4714	2.2083	33.4%
5	1.3976	1.6288	14.2%
6	1.6693	2.0508	18.6%
7	1.4434	1.8302	21.1%
8	1.5428	1.8500	16.6%
9	1.6559	2.2288	25.7%
10	1.7191	1.8779	8.5%

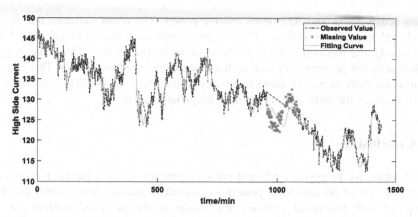

Fig. 4. The final fitting curve when missing points are continuously distributed

When the missing values are distributed continuously, normal interpolation methods like cubic spline interpolation are not available. However, our method still works well. This is because it considers the longitudinal data and the horizontal data at the same time.

To see the results under this situation, we randomly selected 10% continuous observed points on Jan. 1st as the test set and used the proposed method to impute the missing values. The results are shown in Fig. 4. It is shown that the fitting curve and the missing points have similar fluctuation trend. Although there is a certain deviation, the error is within 5%.

In order to test if our method is suitable to other types of operation data besides high side current, we further tested the performance on other 11 types of the operation data. The results are shown in Table 2, where 'U' means voltage, 'I' means current, 'P'

Table 2. Comparison of RMSE on 12 types of data

Types of data	RMSE of our method	RMSE of cubic spline interpolation	Decrease of RMSE
High Side U	0.1523	0.2269	32.9%
High Side I	1.4393	1.902	24.3%
High Side P	0.5989	0.7068	15.3%
High Side Q	0.3073	0.3105	1.0%
Medium Side U	0.0802	0.102	21.4%
Medium Side I	3.0817	4.1444	25.6%
Medium Side P	0.6005	0.7273	17.4%
Medium Side Q	0.2685	0.2832	5.2%
Low Side U	0.008	0.0103	22.3%
Low Side I	8.7064	10.6326	18.1%
Low Side P	0.1154	0.1502	23.2%
Low Side Q	0.0983	0.1434	31.5%

means active power and 'Q' means relative power. The result indicates that the proposed method performs well on different types of operation data of transformer. Except high side Q and medium side Q, the decrease of RMSE is over 15%. The proposed method does not perform very well on high side Q and medium side Q. This is mainly because the daily curves of these two types of data have a low degree of similarity, which interfere the estimation of the low frequency signal.

5   Conclusion

In the paper, we proposed a missing value imputation method for operation data of transformer based on functional principal component analysis and wavelet transform. Compared with traditional methods, the results of the proposed method are more reliable and stable. And the proposed method works whenever the missing points are randomly distributed or continuously distributed. It should be noted that our proposed method can be successfully applied when the historical data have certain similarities.

References

1. Song, Y., Zhou, G., Zhu, Y.: Present status and challenges of big data processing in smart grid. Power Syst. Technol. **37**(04), 927–935 (2013)
2. Zhao, T., Zhang, Y., Zhang, D.: Application technology of big data in smart distribution grid and its prospect analysis. Power Syst. Technol. **38**(12), 3305–3312 (2014)
3. Liu, K., Sheng, W., Zhang, D., et al.: Big data application requirements and scenario analysis in smart distribution network. Proc. CSEE **35**(2), 287–293 (2010)
4. Yan, L., Li, Y., Li, B., Zhao, Z.: Opportunity and challenge of big data for the power industry. Electric Power Inf. Technol. **11**(04), 1–4 (2013)
5. Peng, X., Deng, D., Cheng, S., et al.: Key technologies of electric power big data and its application prospects in smart grid. Proc. CSEE **35**(3), 503–511 (2015)
6. Zhang, D., Miao, X., Liu, L., Zhang, Y., Liu, K.: Research on development strategy for smart grid big data. Proc. CSEE **35**(1), 2–12 (2015)
7. Zhang, X., Cheng, Q., Zhou, Q., et al.: Dynamic intelligent cleaning for dirty electric load data based on data mining. Autom. Electric Power Syst. **29**(8), 60–64 (2005)
8. Yan, Y., Sheng, G., Chen, Y., et al.: Cleaning method for big data of power transmission and transformation equipment state based on times sequence analysis. Autom. Electric Power Syst. **39**(7), 138–144 (2015)
9. Liu, Y., Luan, W., Xu, Y., Wang, P., Guo, C.: Data cleaning method for distribution transformer. Power Syst. Technol. **41**(03), 1008–1014 (2017)
10. James, G.M., Hastie, T.J., Sugar, C.A.: Principal component models for sparse functional data. Biometrika **87**(3), 587–602 (2000)
11. Daubechies, I.: The wavelet transform, time-frequency localization and signal analysis. IEEE Trans. Inf. Theory **36**(5), 961–1005 (1990)
12. Ghanem, R.G., Spanos, P.D.: Stochastic finite elements: a spectral approach, pp. 17–20. Dover Publications (2003)
13. Zeng, X., et al.: Science and Technology Report of National 863 Science and Technology Plan Major Project - Key Technologies for Big Data Application Intelligent Distribution (2017)

14. Yao, F., Müller, H.G., Clifford, A.J., et al.: Shrinkage estimation for functional principal component scores with application to the population kinetics of plasma folate. Biometrics **59** (3), 676–685 (2003)
15. Yao, F., Müller, H.G., Wang, J.L.: Functional data analysis for sparse longitudinal data. J. Am. Stat. Assoc. **100**(470), 577–590 (2005)
16. Rice, J.A., Silverman, B.W.: Estimating the mean and covariance structure nonparametrically when the data are curves. J. Roy. Stat. Soc. Ser. B (Methodol.) **53**(1), 233–243 (1991)
17. Capra, W.B., Müller, H.G.: An accelerated-time model for response curves. J. Am. Stat. Assoc. **92**(437), 72–83 (1997)
18. Burrus, C.S., Gopinath, R.A., Guo, H., et al.: Introduction to Wavelets and Wavelet Transforms: A Primer, p. 17. Prentice hall, New Jersey (1998)
19. Mallat, S.G.: A theory for multiresolution signal decomposition: the wavelet representation. IEEE Trans. Pattern Anal. Mach. Intell. **7**, 674–693 (1989)

Periodic Action Temporal Localization Method Based on Two-Path Architecture for Product Counting in Sewing Video

Jin-Long Huang[1,2], Hong-Bo Zhang[1,2], Ji-Xiang Du[1,2(✉)],
Jian-Feng Zheng[1,2], and Xiao-Xiao Peng[1,2]

[1] Department of Computer Science and Technology, Huaqiao University,
Xiamen, China
jxdu@hqu.edu.cn
[2] Xiamen Key Laboratory of Computer Vision and Pattern Recognition,
Huaqiao University, Xiamen, China

Abstract. Automatically product counting in the handmade process plays a vital role in the manufacturing industry, especially at the sewing industry. Nevertheless, there is currently a few methods to count the product number in the hand sewing process from surveillance video automatically. Due to the sewing procedure is a cyclical action, the product counting in hand sewing process is regarded as periodic action temporal localization and counting problem. In this paper, in order to solve this problem, we propose a novel two-path method, based on pose estimation and region-based convolutional neural network. The pose estimation method is used to obtain the trajectory information of human joint points, and the periodic action is located by detecting the periodic changes in joint trajectory. An effective two-threshold method is proposed to locate each action and count the number of periodic action from the trajectory information. To more accurately localization, we use a convolutional neural network to predict whether the workbench is empty or not. We fuse the results of joint trajectory and the status of the workbench to adjust the final periodic action localization and counting the number. To verify the proposed method, we built a new video database collected in the real sewing industry. The experimental results show that the proposed method is effective and constructive at periodic action localization and counting in the video for the sewing industry.

Keywords: Periodic action temporal localization · Deep learning ·
Two-Path architecture · Product counting · Pose estimation

1 Introduction

Sewing efficiency plays an essential role in the apparel industry, providing an essential contribution to the national economy, while also maintaining the competitiveness of the apparel industry. The proportion of sewing time in the whole process of garment production is particularly massive. In the actual production, the standard time required for predicting operation in sewing industry is calculated by Modular Arrangement of predetermined Time Standard (MOD) method [1], and General Sewing Time

(GST) method [2], or by multiple video recordings and stopwatch timing methods. These methods are very time-consuming and laborious. In this paper, we discuss the problem of automatic counting the number of product in the handmade sewing process from surveillance video.

Obviously, the sewing process is a continuous and repetitive action, as shown in Fig. 1. We can define it as a cyclical action. The problem of counting the number of product can be regarded as the problem of counting the number of periodic human action in the video. Moreover, the key to solving this problem is to achieve the localization and segmentation of the periodic human action in the temporal dimension. Since the time boundary of action is vague and difficult to define, it is tough to locate the time boundary. Periodic action localization is one of the temporal action detection [3] problem, and its target is to locate the time boundary of each action. However, the main difference is that the former is to do the same action repeatedly and continuously, while the latter is to locate the start and end time of different action instances.

Fig. 1. Examples of the sewing process. There are n sewing cycles in this figure. Each sewing cycle is divided into three stages: the beginning, the processing, and the end. The content of the red rectangular box represents the process of production. (Color figure online)

Inspired by the two-stream network [4] architecture, we propose a novel solution to the problem of periodic action localization. As shown in Fig. 2. RGB camera was used to capture video stream, and RGB frames were sent into the pose estimation system. OpenPose [6] is used to obtain joint point trajectory information. We introduce an effective two-method to locate the time boundary of each periodic action and count the number of periodic action from the trajectory information. In order to compensate for the localization error of this method, a convolutional neural network was used to train the classifier which is used to predict the status of the workbench. Finally, we fuse the classification result and trajectory result to get the final output.

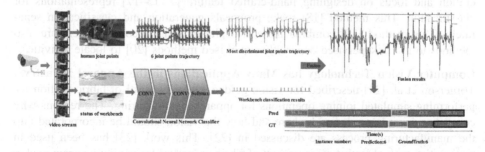

Fig. 2. The pipeline of our proposed method for periodical action localization.

The rest of the paper is organized as follows: In Sect. 2 we review the related work on temporal action detection. In Sect. 3 we introduce the two-path architecture in detail. In Sect. 4 we illustrate the experiment result based on our video dataset. We introduce the Conclusions and directions in Sect. 5.

2 Related Work

Periodic action localization of human body in sewing video is a new application requirement. Temporal action detection [3, 5] has many similarities to our work. In the section, we have reviewed the methods of temporal action detection and application of computer vision technology in the industry. Temporal action detection consists of two parts: action recognition [8] and temporal action localization [7].

Action Recognition. Researchers usually consider action recognition as a classification problem. Early work mainly such as improved Dense Trajectory (iDT) algorithm [9, 10] utilized manual features such as HOG, HOF, and MBH features; Since ILSVRC 2012 [11], the convolutional neural network has made a huge breakthrough in image classification. Some researchers have also been trying to use the deep learning method for video classification in video analysis [4, 12, 13]. Simonyan et al. [29] designed a two-stream CNN network containing spatial and temporal networks. Tran et al. [13] proposed 3D CNN [8] on real and large-scale video datasets.

Temporal action localization is able to identify the category and boundary of each action instance in the untrimmed video. Temporal action localization, like object detection, is a visual detection problem. Therefore, many methods of temporal action localization are inspired and influenced by the progress of object detection. Generally, typical forms of object detection transfer to temporal action localization is to divide temporal action localization into two stages: proposals generation and classification. For object detection, these methods [14, 28] first generate a series of class-agnostic proposals from the whole image, and then classify objects for each proposal. To temporal localization. One typically can adhere to this pattern by first producing segment proposals from the entire video, followed via classifying each proposal. Previous works on temporal action localization mainly use sliding windows as proposals generation and focus on designing hand-crafted feature [7, 15–17] representations for classification. This method [18] make proposals generation and classification separately. This method [19] combine these two phases simultaneously. There are also methods to use unsupervised or weakly supervised methods [20] to locate activities.

Computer Vision Technology has Many Applications in the Apparel Industry. Torgerson et al. [21] described for vision-guided robotic control of fabric motion for performing simulated joining operations for apparel manufacturing. The reasons why computer vision is needed and where and how the vision should be incorporated into the manufacturing process are discussed in [22]. This work [23] has been used in conjunction with objective measurement of fabric properties for predicting seam pucker and sewing machine dynamics for optimization of sewing machinery settings and for the manufacture of lightweight synthetic materials. In this work [24], it describes the application of a low-cost computer vision system to the automatic cutting of lace.

3   The Proposed Approach

Our task is to calculate the number of periodic actions and locate their temporal boundaries in the sewing industry. Obviously, video can be decomposed into spatial and temporal path module. The spatial path carries information about scenes and human described in the video via the form of individual frame. The temporal path module, in the form of motion of consecutive frame, shows the movement of the human activities in video. We devise the two-path method to temporal periodic action localization in sewing video accordingly. As shown in Fig. 3. One path using a deep neural network classifier to represent spatial information. Another path using human joint point trajectory to represent temporal information. Finally, we fuse spatial information and temporal information to get the final output.

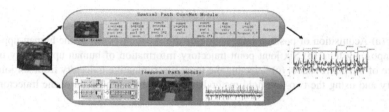

Fig. 3. Two-Path architecture for action localization in video

Temporal Path Module. Reviewed in Fig. 3, the input to the temporal path module is a full video, and its output is the trajectory information of the most discriminant. Since the joint points with obvious periodic human activities in the sewing video are in the upper body, we obtained the trajectories of the six joint points from the upper body. We can calculate the variance of six trajectories, and select the trajectory with the largest variance as the most discriminant trajectory. We find the peak point of the most discriminant joint trajectory by the two-threshold method as the starting point of the period.

Two-Threshold Methods. In order to locate the peak point of the joint point trajectory, we introduce an effective localization method. Joint point trajectory consists of the position of joint points in each frame from the video. Therefore, we can consider the joint point trajectories as a set of discrete points. As shown in Fig. 4(a), we set thresholds on the X and Y coordinates of the coordinate axis respectively to narrow the solution range of peak point and reduce unnecessary computational cost. Specifically, the definitions of the x-axis direction threshold and the y-axis direction threshold are as follows:

$$y_{threshold} = \frac{\max(v_i) + \frac{1}{n}\left(\sum_{i=1}^{n} v_i\right)}{2} \tag{1}$$

$$x_{threshold} = \frac{1}{2}\left(\frac{1}{k1}\sum_{j=1}^{j=k1} l_t^j + \frac{1}{k2}\sum_{j=1}^{j=k2} s_t^j\right) \tag{2}$$

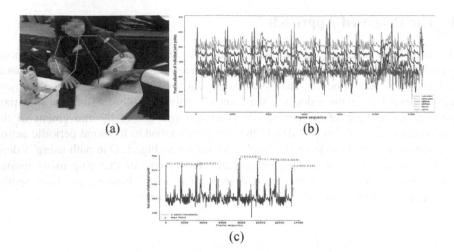

(a)   (b)

(c)

Fig. 4. (a) Acquisition of position information of the six joint points of the human upper body by Openpose system. (b) Six joint point trajectory information of human upper body in frame sequence of the entire video. (c) Selecting the most discriminative trajectory from six joint point trajectory and using the two-threshold methods to find the peak position of the trajectory.

Where i denotes i-th frame in the video. v_i denotes the joint point position at i-th frame. According to the duration of the action cycle, we classify all periodic action in the data set into two categories: long, short. There are $k1$ long videos and $k2$ short videos respectively. l_t^j represents the duration of the j-th video in long videos, and s_t^j represents the duration of the j-th video in short videos.

We first obtain $y_{threshold}$ according to formula (1), and get candidate points set of peak points through $y_{threshold}$. Next, we select those points where the adjacent point difference is greater than the $x_{threshold}$ as the intervals of the candidate point set. We divide the candidate points set into multi-segment local point sets by interval sets. Finally, we get the maximum values from these local segments as the local peak points respectively (Fig. 5).

Spatial Path ConvNet Module. To more accurately localization, we use a region-based convolutional neural network to train a classifier to correct localization errors. The two-threshold algorithm relies too much on two thresholds. Thus, we need more information to alleviate this dependence. Through the study of periodic actions in the sewing video, we find that the workbench will be empty at the end and beginning of the actions. A neural network model is trained to classify whether the workbench is empty or not, to locate the peak points more accurately and correct some wrong peak points.

Fusion Strategy. When the workbench area on the frame is empty, the frame is represented by a high value, and vice versa is a low value. We set up a window with a size of 200 frames. We can count the number of workbench empty frames around the peak point in the window. When the ratio of the workbench empty and non-empty is greater than alpha, where alpha is set to 0.1, the peak point is reserved. Otherwise, it will be eliminated from the peak point set (Fig. 6).

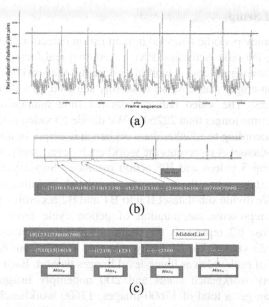

(a)

(b)

(c)

Fig. 5. (a) Two-threshold method. Discrete point trajectory formed by the human joint points on each frame of video. Blue line represents $y_{threshold}$ line. (b) Two-threshold method. The set of peak candidate points selected by low$_{threshold}$ and the set of break points obtained by $x_{threshold}$. Red triangle indicates break point. (c) Two-threshold method. Localizing the peak candidate set points with break points and solve the local maximum as the peak point. (Color figure online)

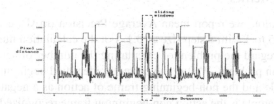

Fig. 6. The Fusion of joint point trajectory and classifier features for more accurate peak localization. The green line represents the joint point trajectory; the red line indicates the classification result of the classifier on the video frame sets. When the workbench is empty, the status of this frame is set to high value. Otherwise, it is set to a low value. (Color figure online)

4 Experimental Results

In this section, we quantitatively evaluate our methods on the tasks of locating periodic sewing time and detecting the number of processes via using video datasets. We demonstrate the effective performance on the task of locating periodic sewing time and detecting the number of processes.

4.1 Dataset and Setup

In order to better study periodic temporal human action detection task, we introduced a new video dataset, Periodic Activities (PA), based on a real factory scenario. In a real factory, when the sewing process is working normally, the workbench state is divided into empty and non-empty states, so we randomly sampled 53 videos from the factory scenario as datasets. The video dataset contains time annotation for 522 action instances and total time longer than 22260 s. We divide 53 video datasets into two sub-datasets A and B according to whether the workbench is empty or not when the process is completed. Sub-dataset A represents the workbench never empty when the process is completed, including 5 videos and 108 action instances. Sub-dataset B represents the empty workbench status when the process is completed, including 48 videos and 414 action instances. We divide sub-dataset B into B1 and B2 according to the time of each action cycle. B1 represents the duration of action cycle more than one minute, including 13 videos. B2 represents the duration of the action cycle less than one minute, including 35 videos. We randomly select 44 videos from dataset B to cut the workbench region of each video as frame-level images dataset. Each video extracts 200 images with empty workbench status and 200 nonempty images with nonempty workbench. We can get a total of 17600 images. 17600 workbench-region images is divided into training, validation sets by the ratio of 4:1. Similarly, the remaining 4 video workbench data can be used as test sets. Finally, PA is divided into Sub-dataset B, including LONG(13) and SHORT(35) dataset, and Sub-dataset A. What is more, we apply the Two-Path method to seven real-time online videos.

4.2 Evaluation Metrics

For action localization, we report mean Average Precision (mAP) using different IOU thresholds from 0.5 to 1 with a step size of 0.1. For the calculation number of an action instance. We can regard the problem of calculating the number of action instances as a binary classification problem. We use the correctly identified beginning frame of action as a positive sample and the non-beginning frame of action as a negative sample. Then the false positive sample is the action non-beginning frame recognized as the beginning frame of action and the false negative sample indicates that the beginning frame of action is recognized as the non-be frame of action. Therefore, we use the mean average accuracy of the beginning frame of action and recall as the Evaluation metrics. Since the beginning frame of action is not well defined, we set a threshold range, [starting frame − 100, starting frame + 100], on the annotated the action starting frame, as long as the frame in this range can be regarded as the beginning frame.

4.3 Implementation Details

To train the classifier, we obtained 17600 images datasets from 44 videos. We make training dataset and validation dataset according to the ratio of 4:1. Similarly, we can get 1600 images from the remaining 4 videos as test dataset. We use different convolutional neural network architectures to train on this dataset and select the optimal neural network model for this task. Base on the Caffe framework, we use different convolutional neural network architecture, including CaffeNet, VGG-16 [25], VGG19 [26], ResNet-50 [27], to iterate 10000 times on the image training set, and We use SGD to learn CNN parameters in our methods, with the batch size of 32 and momentum as 0.9. The initial learning rate is set to 0.001.

In this experiment, we used the OpenPose to detect the position of the human joint point of each frame in the video, so we can construct the temporal sequence of human joint points from the entire video. Since the maximal peak joint point of human joint point temporal position indicates the time of human takes things, it can be regarded as the beginning moment of an action cycle. Next, we introduce a novel two-threshold method to find the maximal peak point from human joint point trajectory. Finally, we fuse the maximum peak point and classifier results to get the localization and the number of action.

4.4 Exploration Study

We first conduct experiments to evaluate the effectiveness of individual components in the Two-Path architecture and investigate how they contribute to performance improvement.

Table 1. Comparison the accuracy under different Convolutional Neural Network

Network architecture	Accuracy on validation set	Accuracy on test set
CaffeNet	0.997	0.9232
VGG-16	0.995	0.9625
VGG-19	0.997	**0.9763**
GoogleNet	0.994	0.9531
ResNet-50	0.97	0.9575

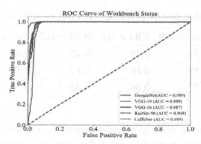

Fig. 7. Different network architecture performance for detecting workbench status. These ROC curves demonstrate the trade-off between the true positive rate and the false-positive rate, as the detection thresholds are varied. The area under the ROC curve (AUC) is an aggregate measure of detection performance and indicates the probability that the model will rank a positive example more highly than a negative example. (A model whose predictions are 100% correct will have an AUC of 1.0)

Table 2. Average precision accuracy of the number of periods of different processes

Process mark	Temporal path module	Two-path method
LONG(13)	0.53	**0.61**
SHORT(35)	0.34	**0.42**
sub-datasets A (5)	0.6	0.6
ONLINE(7)	0.7	0.7

sub-datasets B is divided into two video dataset, LONG and SHORT. LONG indicates that the workbench area in the video is empty and the duration time of periodic action is more than one minute. SHORT indicates the workbench area in the video is empty and the duration time of periodic action is less than one minute. **sub-datasets A** represents the workbench area in the video is never empty. ONLINE(7) represents 7 video data processed online and in real time. In Table 2. We use temporal path module and Two-Path method to predict the number of processes in each video dataset.

$$f_k^i = \begin{cases} 1, & pred = GT \\ 0, & pred \neq GT \end{cases} \tag{3}$$

$$PA_k = \frac{\sum_1^N f_k^i}{N} \tag{4}$$

Where f_k^i represents the result of i-th video in sub-video dataset k. When the number of predictions is same as GroundTruth, the result response of video is 1, and vice versa, it is 0. PA_k represents average precision accuracy of the number of periods on different sub-video datasets k, and the N denotes the number of videos in sub-video dataset k, with values of 13, 35, 5 and 7, respectively.

Table 3. Average precision results of Temporal Path module under different IOU values.

Process	IOU = 0.9	IOU = 0.8	IOU = 0.7	IOU = 0.6	IOU = 0.5
LONG	0.39	0.45	0.48	0.48	**0.48**
SHORT	0.15	0.16	0.17	0.19	**0.21**
sub-datasets A	0.22	0.30	0.35	0.36	**0.38**

Table 4. Comparison of mean average precision (mAP) accuracy of temporal path module of two-path method and two-path method on PA video dataset with different IOU values.

MAP	Temporal path module	Two-path method
IOU = 0.9	0.25	**0.43**
IOU = 0.8	0.30	**0.44**
IOU = 0.7	0.33	**0.46**
IOU = 0.6	0.34	**0.48**
IOU = 0.5	0.36	**0.49**

In Tables 3 and 4, we evaluate the accuracy of Two-Path under different IOU values. IOU is the overlap rate of the duration time of prediction and the duration time of GroundTruth. The IOU ideal value is 1, i.e., the predicted action time coincides with GroundTruth.

$$IOU = \frac{Prediction \cap GroundTruth}{Prediction \cup GroundTruth} \tag{5}$$

$$F_k^i = \begin{cases} 1, & IOU \geq thresold \\ 0, & IOU < thresold \end{cases} \tag{6}$$

$$PA_k = \frac{\sum_1^N F_k^i}{N} \tag{7}$$

$$mAP_k = \frac{\sum_1^N F_k^i}{m * N} \tag{8}$$

Where F_k^i represents the result of i-th video in sub-video dataset k. When the IOU value is greater than or equal to the threshold, the result response of video is 1, and vice versa, it is 0. IOU thresholds from 0.5 to 1 with a step size of 0.1. PA_k denotes average precision results of Temporal Path module under different IOU values on different sub-video datasets k. mPA_k represents mean average precision accuracy of prediction on different sub-video datasets k, and the N denotes the number of videos in sub-video dataset k, with values of 13, 35, 5 respectively. m is 3, representing 3 sub-video dataset.

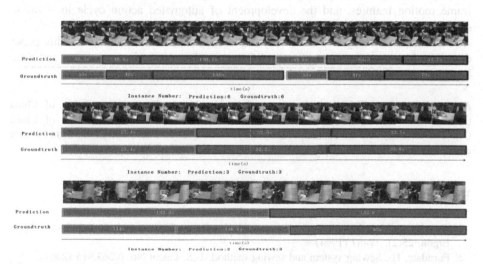

Fig. 8. Examples of quantitative analysis on PA Dataset using Two-Path method. Each box of different colors represents a cycle duration time.

In comparison with the results in Table 1 and Fig. 7, we can realize that VGG-19 model has the highest verification accuracy on the test set and largest AUC area, which indicates that VGG-19 network is more suitable to be used as a classifier in this task. The comparative experiments in Table 2 show that we can calculate the number of action periods in the sewing videos using only the temporal path module in the Two-Path architecture method. However, when we fuse the temporal path module with spatial convolution module, that is, the Two-Path method, the performance of the method will be significantly improved compared to the single temporal path module method on the dataset with empty and non-empty alternating workbench state. And the performance of the Two-Path method is consistent with the temporal module on the sewing video dataset where the workbench state is never empty. Tables 3 and 4 show that the effectiveness and constructiveness of the Two-Path Method in dealing with periodic human action localization (Fig. 8).

5   Conclusions and Discussion

In this work, we proposed a new method of joint human pose estimation and region-based convolution neural network to solve the periodic action localization of the sewing industry in the real world. We demonstrate that our approach achieves effective performance on periodic localization tasks. In addition, the method can be used to evaluate the level of employee competence, standardization of action, capacity allocation and a series of other operations. Future work includes continuing to build on the types of meaningful information that can be automatically extracted from sewing videos, including multi-trajectory fusion features, global features of static images, inter-frame motion features, and the development of automated action cycle localization system.

As a seminal study, we have only explored a little in this direction. This paper describes how to compute automatically the number of sewing processes and time boundary location by computer vision.

Acknowledgement. This work is supported by the Nature Science Foundation of China (No. 61673186, 61871196), the Natural Science Foundation of Fujian Province of China (No. 2019J01082), and the Promotion Program for Young and Middle-aged Teacher in Science and Technology Research of Huaqiao University (ZQN-YX601).

References

1. Genaidy, A.M., Al-Shedi, A.A., Karwowski, W.: Postural stress analysis in industry. Appl. Ergon. **25**(2), 77–87 (1994)
2. Furudate, H.: Sewing system and sewing method. U.S. Patent No. 6,263,815 (2001)
3. Singh, G., Cuzzolin, F.: Untrimmed video classification for activity detection: submission to activitynet challenge, 1 July 2016. https://arxiv.org/abs/1607.01979
4. Feichtenhofer, C., Pinz, A., Zisserman, A.: Convolutional two-stream network fusion for video action recognition. In: Proceedings of the IEEE Conference on Computer Vision and Pattern Recognition, pp. 1933–1941 (2016)

5. Shou, Z., Wang, D., Chang, S.F.: Temporal action localization in untrimmed videos via multi-stage CNNs. In: Proceedings of the IEEE Conference on Computer Vision and Pattern Recognition, pp. 1049–1058 (2016)
6. Cao, Z., Simon, T., Wei, S.E., et al.: Realtime multi-person 2D pose estimation using part affinity fields. In: Proceedings of the IEEE Conference on Computer Vision and Pattern Recognition, pp. 7291–7299 (2017)
7. Gaidon, A., Harchaoui, Z., Schmid, C.F.: Temporal localization of actions with actoms. IEEE Trans. Pattern Anal. Mach. Intell. **35**(11), 2782–2795 (2013)
8. Ji, S., Xu, W., Yang, M., et al.: 3D convolutional neural networks for human action recognition. IEEE Trans. Pattern Anal. Mach. Intell. **35**(1), 221–231 (2013)
9. Wang, H., Kläser, A., Schmid, C., et al.: Action recognition by dense trajectories. In: CVPR 2011-IEEE Conference on Computer Vision and Pattern Recognition, pp. 3169–3176 (2011)
10. Wang, H., Schmid, C.: Action recognition with improved trajectories. In: Proceedings of the IEEE International Conference on Computer Vision, pp. 3551–3558 (2013)
11. Russakovsky, O., Deng, J., Su, H., et al.: Imagenet large scale visual recognition challenge. Int. J. Comput. Vis. **115**(3), 211–252 (2015)
12. Karpathy, A., Toderici, G., Shetty, S., et al.: Large-scale video classification with convolutional neural networks. In: Proceedings of the IEEE Conference on Computer Vision and Pattern Recognition, pp. 1725–1732 (2014)
13. Tran, D., Bourdev, L., Fergus, R., et al.: Learning spatiotemporal features with 3D convolutional networks. In: Proceedings of the IEEE International Conference on Computer Vision, pp. 4489–4497 (2015)
14. Ren, S., He, K., Girshick, R., et al.: Faster R-CNN: towards real-time object detection with region proposal networks. In: Advances in Neural Information Processing Systems, pp. 91–99 (2015)
15. Tang, K., Yao, B., Fei-Fei, L., et al.: Combining the right features for complex event recognition. In: Proceedings of the IEEE International Conference on Computer Vision, pp. 2696–2703 (2013)
16. Mettes, P., Van Gemert, J.C., Cappallo, S., et al.: Bag-of-fragments: selecting and encoding video fragments for event detection and recounting. In: Proceedings of the 5th ACM on International Conference on Multimedia Retrieval, pp. 427–434 (2015)
17. Yuan, J., Ni, B., Yang, X., et al.: Temporal action localization with pyramid of score distribution features. In: Proceedings of the IEEE Conference on Computer Vision and Pattern Recognition, pp. 3093–3102 (2016)
18. Singh, G., Cuzzolin, F.: Untrimmed video classification for activity detection: submission to activitynet challenge, July 2016. https://arxiv.org/abs/1607.01979
19. Lin, T., Zhao, X., Shou, Z.: Single shot temporal action detection. In: Proceedings of the 25th ACM International Conference on Multimedia, pp. 988–996 (2017)
20. Wang, L., et al.: Temporal segment networks: towards good practices for deep action recognition. In: Leibe, B., Matas, J., Sebe, N., Welling, M. (eds.) ECCV 2016. LNCS, vol. 9912, pp. 20–36. Springer, Cham (2016). https://doi.org/10.1007/978-3-319-46484-8_2
21. Torgerson, E., Paul, F.W.: Vision-guided robotic fabric manipulation for apparel manufacturing. IEEE Control Syst. Mag. **8**(1), 14–20 (1988)
22. Norton-Wayne, L.: Automated garment inspection using machine vision. In: IEEE International Conference on Systems Engineering, pp. 374–377 (1990)
23. Stylios, G., Sotomi, J.: Seam pucker in lightweight synthetic fabrics: measurement as an aesthetic property using computer vision. Int. J. Cloth. Sci. Technol. **3**(3), 14–17 (1991)
24. Russel, R.A., Wong, P.: An application of computer vision to lace cutting. Robot. Auton. Syst. **5**(2), 91–96 (1989)

25. Dodge, S., Karam, L.: Understanding how image quality affects deep neural networks. In: 8th International Conference on Quality of Multimedia Experience, pp. 1–6 (2016)
26. Park, E., et al.: Combining multiple sources of knowledge in deep CNNs for action recognition. In: IEEE Winter Conference on Applications of Computer Vision (WACV), pp. 1–8 (2016)
27. Szegedy, C., Ioffe, S., Vanhoucke, V., et al.: Inception-v4, Inception-ResNet and the impact of residual connections on learning. In: Thirty-First AAAI Conference on Artificial Intelligence (2017)
28. Redmon, J., Divvala, S., Girshick, R., et al.: You only look once: unified, real-time object detection. In: Proceedings of the IEEE Conference on Computer Vision and Pattern Recognition, pp. 779–788 (2016)
29. Simonyan, K., Zisserman, F.: Two-stream convolutional networks for action recognition invideos. In: Advances in Neural Information Processing Systems, pp. 568–576 (2014)

Investigating the Capability of Agile Processes to Support Medical Devices Regulations: The Case of XP, Scrum, and FDD with EU MDR Regulations

Mohmood Alsaadi[1,2]([⊠]), Alexei Lisitsa[2], Mohammed Khalaf[1],
and Malik Qasaimeh[3]

[1] Applied Computing Research Group, Al-Maarif University College,
Anbar, Iraq
Mahmood89.ma@gmail.com, M.i.khalaf@acritt.org.uk
[2] Department of Computer Science, University of Liverpool, Liverpool, UK
A.Lisitsa@liverpool.ac.uk
[3] King Hussein Faculty of Computing Sciences,
Princess Sumaya University for Technology, Amman, Jordan
M.qasaimeh@psut.edu.jo

Abstract. Nowadays, medical devices rely on software whether completely such as mobile medical applications or as embedded software into medical chips such as tele surgery systems. Medical device software usually developed by using traditional development methods such as V-model and Waterfall model as these methods are straightforward and has the ability to comply with regulatory requirements such as documentation that ensures traceability. However, these methods could take a long time, could be costly, and these are inappropriate in case of requirements changing. In contrast, agile processes have several benefits such as producing a product of a high-quality with low cost and in short period with the capability to embrace change during development. Therefore, companies that develop medical device software can benefit from adopting agile practices. While the adoption rate of agile practices in software development in different industries is increasing, healthcare industries still have a low rate of agile adoption. This due to the gaps between agile practices and the stringent requirements of healthcare regulations such as documentation, traceability, and formality. The main question of this research is how capable are Agile processes can be in dealing with MDR requirements? This paper will investigate the capability of agile processes to support European Medical Device Regulation (MDR) requirements by extracting the MDR requirements that are related to software development life cycle. Then investigating the robustness of agile processes, to support MDR regulations. These objectives will be conducting by comparing and analysing the most popular agile processes (XP, Scrum, and FDD) to identify the gaps between MDR requirements and agile processes. The analysis revealed that XP is inappropriate for MDR requirements since it lacks the fixed up-front planning and also have insufficient documentations for treatability, Scrum has a model status report can be used for traceability but also

© Springer Nature Switzerland AG 2019
D.-S. Huang et al. (Eds.): ICIC 2019, LNAI 11645, pp. 581–592, 2019.
https://doi.org/10.1007/978-3-030-26766-7_53

it has insufficient documentation for MDR, and FDD is the closest agile practices to satisfy MDR requirements because it has a develop overall model phase which can be considered as semi-fixed up-front planning as well as has more documentations than the XP for traceability purposes such as UML modelling.

Keywords: Agile practices · Healthcare regulations · Adoption of agile · MDR

1 Introduction

Recently, the reliance on software systems has led to an increase in the attention to regulatory compliance [1, 2]. Medical device software development companies are surrounded by regulatory requirements and they must ensure that their products comply with the regulations. Compliance with the healthcare regulations varies depending on the marketing region. For example, medical device software to be marketed in the USA must comply with Food and Drug Administration (FDA) while in the European Union they must comply with Medical Device Regulation (MDR) [3]. On the other hand, the adoption rate of agile software development has been increased in different industries to improve the efficiency of software development. According to the 11th annual state of agile report, the adoption of agile practices across almost all the industries is rising at an accelerated rate [4]. In addition, [5] mention that agile practices such as XP and Scrum are the most widely method used in developing software as well as [6] states that agile principles are mostly used in the IT industry. Agile practices have several benefits that medical device industries can obtain such as producing a product of a high-quality with low cost and in short period [7]. Moreover, agile methods have the capability to embrace change throughout the development process that leads to customers and business satisfaction.

Despite the fact that "agile practices can be successfully adopted to develop regulatory compliant software" as shown in the technical report of the Association for the Advancement for Medical Instrumentation (AAMI) [8]. Yet, the adoption rate of agile practices in medical devices software development is still low as mentioned in the 11th annual state of the agile report only 6% of the respondents have adopted agile practices in healthcare industries [4]. This has been confirmed by a survey that conducted by [9] amongst twenty medical device software development organisations in Ireland. They found that 50% of medical devices companies are using V-model in developing software, 25% of the organisations are using tailored agile practices in developing medical device software, and 25% of organizations are using traditional methods in developing software such as Waterfall model. The reason behind the low rate of agile adoption in healthcare industry is the gaps that exists between agile practices and the stringent requirements of healthcare regulations These include lack of documentation, maintaining traceability, lack of up-front planning, formal level of the requirements gathering, requirement changes and the process of managing multiple releases [5, 10, 11]. The contributions of this research are presenting a literature review of the relationship between agile practices and healthcare industry as well as conducting a study on the MDR requirements to extract the requirements that effect direct or indirect on software development life cycle. Moreover, this paper will investigate the capability of agile practices with MDR requirements by conducting a comparison between agile practices (XP, Scrum, and FDD) and MDR requirements based on the gaps to show how agile practices lack the necessary practices to support MDR regulations.

2 Background and Related Work

2.1 Agile Practice

This section will explain briefly the most common agile practices that have been selected in this research based on literature review:

2.1.1 Agile XP

Agile eXtreme Programming which known as agile XP is most widely used approach to agile software development [12]. It has four framework activities [13]:

Planning: during this phase requirements are gathered from the customer by user story card or face to face communication. New requirements can be added to the story card any time of the development process.

Design: agile XP use the principle of KIS (keep it simple), XP members team try to keep the design as simple as possible. Techniques such as spike solution (operational prototype of that part of the design), refactoring (keep the design clear), and CRC cards (Class Responsibility Collaborator) used to solve the complexity of the design and also to enhance the structure of the design.

Coding: the first step in the coding phase is that the developers write the unit tests that will check the requirements of each story. Coding phase usually achieved by pair programming which is two people work together to create code for a story.

Testing: during this phase different tests conducted such as unit test, smoke test during the integration, a regression testing occurs each time a new module is added, and acceptance tests conducted by the customer.

2.1.2 Scrum

Scrum is an agile software development method that focuses on teamwork activities to produce a quality product in a changing environment [14]. The process of the Scrum follows five framework activities: requirements, analysis, design, evaluation, and delivery [15]. Requirements are gathered from the customer during the planning phase, the development team and customer then create an architecture for the project. After the initial planning, the product manager identifies tasks and capture them in a list called product backlog where items can be added to the backlog at any time. The main Scrum techniques are product backlog, sprint, and daily scrum meeting.

2.1.3 Feature-Driven Development (FDD)

FDD is an iteration software development process consists of five phases that mainly focuses on design and building phases [16]. FDD phases are:

Develop overall model: during this phase requirements are gathered, user requirements gathering process involves developers that familiar with UML, customer, and problem domain experts [17]. The outcome of this phase is UML diagrams such as class diagrams that consist of classes, relationships, methods, and attributes. At the end of this phase, the domain experts create walkthrough document that contains high-level description of the system [18].

Build feature list: based on the outcome or the walkthrough documentation of the previous phase, features list is built. Which is then can be used to track the progress [19]. Features are small pieces of user functional requirements that written in form of <action> <result> <object> [16]. At the end of this phase, features list is reviewed for validation purpose by the customer, domain experts and developers.

Plan by feature: during this phase a high-level description plan is created as well as prioritize feature sets via iteration planning meeting. At the end of this phase features set submitted to the programmers as UML classes [16].

Build by feature: during this phase developer start to build the feature in iterative procedure which each iteration should take from 2 days to 2 weeks. Build by feature phase involved the following steps: coding, unit testing, integration, and code inspection.

2.2 Agile Processes in Healthcare Industry

This section will present a literature review of the relationship between agile processes and healthcare industry. Medical device software usually developed by using plan-driven software development life cycle (SDLC) [20]. There are several reasons for that such as that plan-driven is inflexible and rigid development process, and also it has the capability to comply with the requirements of healthcare regulations. However, medical devices software development companies have suffered from long and costly development life cycles [5]. Using agile practices would bring benefits such as achieving lower development costs, high efficiency, and improved software quality. In addition, there are studies show that using an iterative method such as agile practice in developing medical devices software would enhance the software development process [21]. Several studies have focused on agile practices within medical devices to investigate if agile practices have been used in medical device software development and if so, how have they been adopted and to what success.

Other studies focused on how to modify agile practices in order to be used in developing medical device software. Thus [22] found that agile and lean methods have been tailored for regulated development by integrating them with plan-driven practices. The paper [23] presented a mapping study between agile practices and plan-driven methods. As a result of the mapping study they found that companies that develop medical device can use agile practices but with modification on the agile practices. Based on the mapping study they also found all the successful projects that adopted agile were not pure agile, it was agile practices mixed with plan-driven methods rather than wholly embracing a complete agile methodology such as Scrum or XP. The paper [20] Implemented an agile methods within a medical device software development organization based in Ireland. Their agile method was a combination of plan-driven V-model and agile practice. Two studies by [24] and [25] show that agile development practices are being extensively used in the medical software industry with proper adaptations and with modification to the agile practices. The paper [26] demonstrated a successful implementation of a tailored Scrum approach in a regulated environment and mention that "our findings suggest that agile is highly suitable when tailored to meet the needs of regulated environments and supported with appropriate tools".

Overall, these studies indicates that there is an interest in agile software development from medical device industries. Almost all the studies in this section show that adopting agile alone would not work unless it integrated with some practice of plan-driven methods.

2.2.1 Compliance with MDR

MDR is Medical Device Regulations (was called as Medical Device Directive MDD), which is the new EU regulations that aims to ensure the smooth functioning of the internal market as regards medical devices, taking as a base a high level of protection of health for patients and users, and taking into account the small- and medium-sized enterprises that are active in this sector [14]. MDR Article 2(4) defined "software" as an active device which means any "active device used, whether alone or in combination with other devices, to support, modify, replace or restore biological functions or structures with a view to treatment or alleviation of an illness, injury or disability" [14]. In addition, MDR (Annex VIII, Chapter III) classified software into four categories based on the risk level (see Table 1. MDR software classification). Therefore, any company that dealing software as defined by MDR must confirm to MDR requiems.

On the other hand, failing to comply with healthcare regulations has consequences such as fines and loss of reputation can be costly. There are different reasons for medical device failure. For instance, software design as Medical Device Recall Report FY2003 to FY2012 mentioned that software design is the biggest cause of medical device recalls as shown in Fig. 1 [27]. Other report [28] conducted by Trustworthy Medical Device Software shows that the common software failure reason is that company failed to apply well-known system engineering techniques, especially during specification of requirements and analysis.

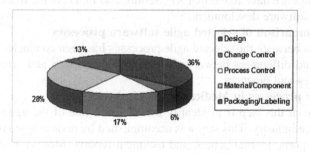

Fig. 1. Recall reasons [27]

Table 1. MDR software classification

Class	Risk level	Definition (which software)
Class I		• All other software is classified as class I.
⇩ Class IIa	L	• All active therapeutic devices intended to administer or exchange energy. • Active devices intended for diagnosis and monitoring. • Software intended to provide information which is used to take decisions with diagnosis or therapeutic purposes. • Software intended to monitor physiological processes.
Class IIb		• All active devices intended to control or monitor the performance of active therapeutic
Class III	H	• All active devices that are intended for controlling, monitoring or directly influencing the performance of active implantable devices are classified as class

3 Methodology

This research methodology consists of five steps (as shown in Fig. 2):

Step 1: Agile processes in the literature review

The objective of this step is to study the agile processes and to understand the differences between agile processes as well as to determine the most widely used agile processes.

This step of the methodology consists of the following sub-step:

Step1.1. Analysis of agile software processes

In this step, we have identified and analyses different agile processes to determine which agile processes are the most common in software development. Based on the literature review we have found that XP, Scrum, and FDD are the most common agile processes in software development.

Step 1.2. Comparison of selected agile software processes

A comparison between the selected agile processes has been conducted to indicate the similarities and differences between selected agile processes and also to understand their principles and practices.

Step 2: Agile processes in Medical devices industry

The objective of this step is to identify the adoption rate of the agile process in the medical device industry. This step was accomplished by reviewing conference papers, experience reports, journal papers, and technical reports. Moreover, in this step, we have studied and analysed the challenges and boundaries that medical device development organizations faced when adopting agile processes.

Step 3: Study of EU Medical Device Regulation (MDR)

The objective of this step is to conduct a study on MDR requirements in order to extract requirements that would affect software development process directly or indirectly.

Step 4: Identify the gaps between MDR requirements and selected agile processes
The objective of this step is to identify the lacks between MDR requirements and
selected agile processes (XP, Scrum, and FDD).

Step 5: Investigate the capability of selected agile processes in MDR environment
The objective of this step is to check the capability of agile (XP, Scrum, and FDD)
with MDR requirements. Based on the previous steps, we have looked from MDR
perspective for what each agile processes have to minimize the lacks that found in
step 4.

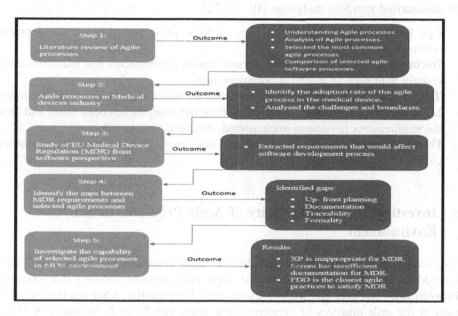

Fig. 2. Research methodology

4 Gaps Between Agile Practices and MDR Requirements

Due to the nature of agile processes and the stringent requirements of MDR, gaps could
occur. MDR has several requirements that related to medical devices regarding design
and manufacture. However, the focus of our research here is on the requirements that
affect directly or indirectly on the software development life cycle. These requirements
outlined in annexes of legislative acts of medical devices that published by official
journal of the European Union [14] such as Annex II Technical documentation. MDR
required the manufacturer to provide technical documentation that contains detailed
information such as device description and specification and software verification and
validation describing the software design and development process and also evidence
of the validation of the software, as well as documentation for traceability purpose. In
contrast, agile methods principle focus on having light documentation as much as
possible for the development process.

A review of the current research literature on the use of agile practices has revealed weaknesses in its capabilities to satisfy the requirements of healthcare regulations. However, there is no research identify the weaknesses of agile processes in supporting MDR rules. MDR is one of the healthcare regulations that have common rules with other healthcare regulations, therefore, the weaknesses that found in the literature review between healthcare regulations and agile processes also can be considered for MDR. In this research, we have selected only the weaknesses that occur between agile processes and MDR requirements. These weaknesses can be attributed to the gaps that exist between the MDR requirements and the agile software processes. These gaps can be summarized into four categories [10, 11, 22]:

Up- front planning: MDR requires to have fixed plan for the development activates.

Documentation: MDR has strict documentation requirements that industries must comply with such as technical documentation.

Traceability: MDR requires documentation for traceability purpose. For example, MDR requires documented evidence on how risk analysis has been done, and how the requirements have been implemented, designed, and tested.

Formality: MDR requires to have formal documents such as formal document for customer requirements.

5 Investigating the Capability of Agile Processes in MDR Environment

This section will investigate the capabilities of agile practices (XP, Scrum, and FDD) with MDR requirements, based on the gaps (see previous section) that occurred between agile practices and MDR requirements. By exploring what has been used in each of the agile practices to overcome these gaps. See Table 2. Summary of Capabilities of Agile practices with MDR.

5.1 Agile XP

Up-front planning: in agile XP there is no fixed up-front planning as the agile XP planning approach focuses on delivering the features of the product rather than project activities. In addition, a customer has the ability to check iteration and to add new story any time of the development process. Agile XP planning phase divided into three phases release planning, iteration planning, and daily planning [13]. During release planning phase, user stories are identified to be included in the first release. Release planning can be updated at the beginning of each iteration where user stories are divided into smaller tasks. At daily planning agile XP team members allocate single tasks to specific individuals and revise their plans one more time. Therefore, to have fixed up-front planning in agile XP, an improvement on XP planning process should happen.

Documentation: XP focused mainly on programming, therefore the main concept behind agile XP is to have simple documentation as possible. The documentation in the agile XP exist in some phases:

In the planning phase there is user story card which is an index card that used by the customer to write the requirements of the software in form of as a <type of user> I want <some goal or objective> so that <benefit, value>. And then the customer assign a value to the index card for prioritization [13]. User story then transformed into acceptance test, the customer write the acceptance test which contains what the story should exactly do.

During the iteration planning, developers use a card called task card that used to convert the stories of the iteration into tasks. It is smaller than the whole story, and sometimes one task will support several stories. Task card consist of story number, software engineer, task estimate, task description, software engineer's notes, and task tracking [13].

Finally, in the design phase there is CRC card (class responsibility collaborator) which is used to identify and organize the classes that are relevant to the system requirements. It is a standard index card that consists of three sections: name of the class, responsibility (attributes and operations that are relevant to the class), and collaborator (the classes that the responsibilities point to) [15]. To sum up, the documentation that exists in agile XP are user story card, acceptance test card, task card, and CRC card. These documentation do not satisfy MDR requirements as MDR requires documentation for each development activities such as documentation for design, implementation, testing, and risk management.

Traceability: since the requirements in agile XP can be change anytime during the development process, is it difficult to trace requirements between planning and testing [18]. In the planning phase user story card and the acceptance test card can be used to trace the requirement by using user story ID, the developer will be able to identify the origin of the requirement. However, this tractability documentation will not satisfy MDR requirements as MDR requires details traceability document.

Formality: requirements are gathered from the customer in agile XP is in an informal way such as face-to-face communication or via the user story card. Therefore, an improvement on the user story is needed to make it more formal in order to satisfy MDR requirements.

We have followed the same investigating method that has been used above for agile XP to investigate the capabilities of Scrum and FDD (see Table 2). As result, Scrum has a model status report can be used for traceability but also it has insufficient documentation for MDR, and FDD is the closest agile practices to satisfy MDR requirements because it has a develop overall model phase which can be considered as semi-fixed up-front planning as well as has more documentations than the XP for traceability purposes such as UML modelling.

Table 2. Summary of capabilities of agile practices with MDR.

Gaps	XP	Scrum	FDD
Up-front planning	No fixed up-front planning: *User story*	Semi-Fixed for specific period: *product backlog*	Semi-Fixed: *Develop overall model* phase
Documentation	Insufficient docs: *user story card, acceptance test card, task card, and CRC card*	Insufficient docs: *product backlog, development sprint backlog, and model status report*	*More documentation: UML modelling, list of features, design documentation, user documentation*
Traceability	*Only CRC card used for traceability*	*Model status report used for traceability*	*Features list, UML modelling, and user documentation are used for traceability*
Formality	Informal	Relatively informal	Semi-formal

6 Conclusion

Although, agile practices could bring benefits to medical devices software development companies, the adoption rate of agile practices in healthcare industry still low. This due to the obstacles that medical devices software development companies faced when complying with healthcare regulations. Therefore, using agile practices in healthcare industry required enhancement on agile practices. As part of our in-progress project, we have discussed the barriers that agile developers faced when developing a medical device to comply with MDR regulatory. Moreover, this paper investigated the capabilities of agile practices (XP, Scrum, and FDD) with MDR requirements, based on the gaps that occur between agile practices and MDR requirements. As result, Agile XP could not satisfy MDR requirements, Agile Scrum has some part that lightly satisfies MDR requirements, and FDD is the closest agile practices to satisfy MDR requirements. For future work, authors will conduct a study to enhance agile XP user story in order to satisfy MDR requirements. Also authors will analysis MDR requirements in details and resolve their ambiguities by using software engineering techniques to make them clearer for developers.

References

1. Heeager, L., Nielsen, P.: A conceptual model of agile software development in a safety-critical context: a systematic literature review. Inf. Softw. Technol. **103**, 22–39 (2018)
2. Mehrfard, H., Pirzadeh, H., Hamou-Lhadj, A.: Investigating the capability of agile processes to support life-science regulations: the case of XP and FDA regulations with a focus on human factor requirements. In: Lee, R., Ormandjieva, O., Abran, A., Constantinides, C. (eds.) Software Engineering Research, Management and Applications 2010. Studies in Computational Intelligence, vol. 296. Springer, Heidelberg (2010). https://doi.org/10.1007/978-3-642-13273-5_16

Investigating the Capability of Agile Processes 591

3. Özcan-Top, Ö., McCaffery, F.: A hybrid assessment approach for medical device software development companies. J. Softw.: Evol. Process **30**, e1929 (2017)
4. VersionOne: The 11th annual State of Agile Report (2017)
5. Özcan-Top, Ö., McCaffery, F.: To what extent the medical device software regulations can be achieved with agile software development methods? XP—DSDM—Scrum. J. Supercomput. (2019)
6. Poth, A., Sasabe, S., Mas, A., Mesquida, A.: Lean and agile software process improvement in traditional and agile environments. J. Softw.: Evol. Process **31**(1), e1986 (2018)
7. Nazir, N., Hasteer, N., Majumdar, R.: Barriers to agile adoption: a developer perspective. In: Kapur, P.K., Klochkov, Y., Verma, A.K., Singh, G. (eds.) System Performance and Management Analytics. AA, pp. 87–95. Springer, Singapore (2019). https://doi.org/10.1007/978-981-10-7323-6_8
8. AAMI, AAMI TIR 45:2012: Guidance on the use of Agile practices in the development of medical device software (2012)
9. Mc Hugh, M., Mc Caffery, F., Casey, V.: Barriers to using agile software development practices within the medical device industry. In: European System, Software & Service Process Improvement & Innovation, Vienna, Austria (2012)
10. Mc Hugh, M., Mc Caffery, F., Casey V.: Barriers to adopting agile practices when developing medical device software. In: Mas, A., Mesquida, A., Rout, T., O'Connor R.V., Dorling, A. (eds.) Software Process Improvement and Capability Determination, SPICE 2012 (2012)
11. Alsaadi, M., Qasaimeh, M., Tedmori, S., Almakadmeh, K.: HIPAA security and privacy rules auditing in extreme programming environments. Int. J. Inf. Syst. Serv. Sect. **9**(1), 1–21 (2017)
12. Demissie, S., Keenan, F., McCaffery, F.: Investigating the suitability of using agile for medical embedded software development. In: Clarke, Paul M., O'Connor, Rory V., Rout, T., Dorling, A. (eds.) SPICE 2016. CCIS, vol. 609, pp. 409–416. Springer, Cham (2016). https://doi.org/10.1007/978-3-319-38980-6_29
13. Beck, K.: Extreme Programming Explained. Addison-Wesley, Boston (2010)
14. European Commission: Regulation (EU) 2017/745 of the European parliament and of the council of 5 April 2017 on medical devices, amending directive 2001/83/EC, regulation (EC) no 178/2002 and regulation (EC) no 1223/2009 and repealing council directives 90/385/EEC and 93/42/EEC. Official Journal of the European Union (2017)
15. Pressman, R., Maxim, B.: Software Engineering, 8th edn, pp. 66–86. McGraw-Hill Education, New York (2015)
16. Palmer, S., Felsing, M.: A Practical Guide to Feature-Driven Development, 1st edn. Pearson Education, London (2002)
17. Khramtchenko, S.: Comparing eXtreme Programming and Feature Driven Development in academic and regulated environments. Final paper for CSCIE-275: Software Architecture and Engineering. Harvard University (2004)
18. Cleland-Huang, J.: Traceability in Agile Projects. In: Cleland-Huang, J., Gotel, O., Zisman, A. (eds.) Software and Systems Traceability, pp. 265–275. Springer, London (2011). https://doi.org/10.1007/978-1-4471-2239-5_12
19. Abrahamsson, P., Salo, O., Ronkainen, J., Warsta, J.: Agile soft-ware development methods: review and analysis, VTT Technical report (2002)
20. Mc Hugh, M., Mc Caffery, F., Coady, G.: An agile implementation within a medical device software organisation. In: Mitasiunas, A., Rout, T., O'Connor, R.V., Dorling, A. (eds.) SPICE 2014. CCIS, vol. 477, pp. 190–201. Springer, Cham (2014). https://doi.org/10.1007/978-3-319-13036-1_17

21. Ge, X., Paige, R., McDermid, J.: An iterative approach for development of safety-critical software and safety arguments. In: 2010 Agile Conference (2010)
22. Cawley, O., Wang, X., Richardson, I.: Lean/Agile software development methodologies in regulated environments – state of the art. In: Abrahamsson, P., Oza, N. (eds.) LESS 2010. LNBIP, vol. 65, pp. 31–36. Springer, Heidelberg (2010). https://doi.org/10.1007/978-3-642-16416-3_4
23. Mc Hugh, M., Cawley, O., Mc Caffery, F., Richardson, I., Wang, X.: An agile v-model for medical device software development to overcome the challenges with plan-driven software development lifecycles. In: 2013 5th International Workshop on Software Engineering in Health Care (SEHC), pp. 12–19. IEEE (2013)
24. Manjunath, K., Jagadeesh, J., Yogeesh, M.: Achieving quality product in a long term software product development in healthcare application using Lean and Agile principles: software engineering and software development. In: 2013 International Mutli-Conference on Automation, Computing, Communication, Control and Compressed Sensing (iMac4s) (2013)
25. Oshana, R.: Software Engineering for Embedded Systems: Methods, Practical Techniques, and Applications. Elsevier, Waltham (2013)
26. Fitzgerald, B., Stol, K., O'Sullivan, R., O'Brien, D.: Scaling agile methods to regulated environments: an industry case study. In: 2013 35th International Conference on Software Engineering (ICSE) (2013)
27. Ferriter, A.: Medical Device Recall Report FY 2003- FY 2012. [ebook] Division of Analysis and Program Operations Office of Compliance Center for Devices and Radiological Health (2014). http://fmdic.org/wp-content/uploads/2014/04/Medical-Device-Recall-Report-amf-2.pdf. Accessed 22 Mar 2019
28. Trustworthy Medical Device Software: Public Health Effectiveness of the FDA 510(k) Clearance Process: Measuring Postmarket Performance and Other Select Topics: Workshop Report. Washington (DC): National Academies Press (US) (2011). https://www.ncbi.nlm.nih.gov/books/NBK209656/. Accessed 22 Mar 2019

Age Group Detection in Stochastic Gas Smart Meter Data Using Decision-Tree Learning

William Hurst(✉) , Casimiro A. Curbelo Montanez ,
and Dhiya Al-Jumeily

Department of Computer Science, Liverpool John Moores University,
Liverpool L3 3AF, UK
{W.Hurst, C.A.CurbeloMontanez, D.Aljumeily}@ljmu.ac.uk

Abstract. Smart meters are the next generation gas and electricity meters where the meter readings are presented digitally and accurately to the consumer via an In-Home Display unit. Access to the data sets generated by smart meters is becoming increasingly prevalent. As such, this paper presents an approach for detecting age groups from aggregated smart meter data. The benefits of achieving this range from healthcare cluster mapping for smart resource allocation and intelligent forecasting, to anomaly detection within age-range groups. The technique proposed and presented in this paper uses a cloud analytics platform for the data processing. Using this approach, the classification is able to achieve a 75.1% AUC prediction accuracy using a two-class decision forest and a 74.6% AUC with a boosted two-class decision tree. A two-class linear regression model, which is able to achieve a 53.7% accuracy, is applied as a benchmark for comparison with the decision tree approach.

Keywords: Machine learning · Smart meter · Variability

1 Introduction

The introduction of the smart meter and the wider smart grid enables the collection and analysis of data from considerably large distributed networks. Researchers are finding progressively ingenious ways to use the data generated by the various smart grid technologies outside their intended use [1]. For example, considerable work has been done in the area of smart meter analytics for remote healthcare purposes, but focusing on electrical smart meter data in particular. This is achieved by detecting appliance usage and correlating it to activities of daily living [2]. In a society where there is a growing widespread need for a better and a scalable instrument for health analytics [3], research in this area offers tremendous benefits. Detecting habits and routines of activity and understanding how these routines change over time, facilitates the detection of anomalous activities that may indicate a decline in health [4].

However, to be effective, home profiling to a granular level requires typically higher data sampling rates than that which is provided by smart meters at their default sampling rate of 30-min intervals. A sub-second or 10-s [2] sampling rate caters for the detection of specific appliances within the home is preferable. Then by monitoring the appliance usage a correlation with activities of daily living can be established. This

© Springer Nature Switzerland AG 2019
D.-S. Huang et al. (Eds.): ICIC 2019, LNAI 11645, pp. 593–605, 2019.
https://doi.org/10.1007/978-3-030-26766-7_54

complex process is not possible with 30-min samples alone. Therefore, most smart meter analytics projects tend to move away from health and are constrained to anomaly detection, forecasting for grid management or bad data countermeasures [5]. Furthermore, gas meter data is often overlooked when considering remote monitoring applications. This is because, there are considerable benefits of being able to detect appliance usage.

However, the big data sets generated by both gas and electricity smart meters alone do offer a means for learning and discovering higher-level human activity patterns. For example, high levels of activity at night time, no or lack of energy/gas consumption may indicate a change in health condition; such as the onset of sundown syndrome in Alzheimer's disease [6]. The construction of other general human activity patterns is also a common source of research [4]. This is true since grouping individuals into similar usage and consumption patterns has an important role in designing clinical trials, improving care delivery, when considering the benefits from a healthcare perspective. However, this also has advantages for load forecasting, tariff/billing design and general resource allocation.

For that reason, this research focuses on the detection of different age groupings by analysing gas smart meter data, which is collected at 30-s samples. To support this research, the Commission for Energy Regulation (CER) Smart Metering Project Gas Customer Behaviour Trial, 2009-2010 ISSDA data set is used [7]. The data is ideal for this experiment, as the gas meter readings are accompanied with survey data which details the age of the customer. The ambition of this research is to provide an approach which can be used for health care applications. Therefore, the focus of the work is on the detection of users over 65. This is due to the fact that this age group would be most likely to contain individuals with self-limiting conditions that would need close remote patient monitoring services. For example, conditions such as dementia are most common within this age group [8]. The benefits of this research are two-fold: (1) Being able to detect the age of the end user would provide insights into healthcare-based cluster mapping. For example, health care services could be allocated based on demographic groupings identified from the smart meter data analytics process. (2) A greater understanding of the consumption traits of the end-users based on their age range will also improve the process for intelligent billing and load forecasting. For example, if the energy provider is aware that the end-user is of retirement age, then it is more likely that their consumption will be higher during day time periods.

The approach put forward in this paper details a two-class and boosted two-class decision tree classification approach for detecting users which are 65 and above in the gas consumption data. The classifiers are compared against a two-class logistic regression approach, which is provided as a benchmark experiment. To the best of our knowledge, this is the first time these techniques have been applied to this dataset. The remainder of this paper is as follows. Section 2 provides a background discussion on related work and details the dataset used for this research. Section 3 outlines the methodology adopted in this paper. The results are presented in Sect. 4 and the paper is concluded in Sect. 5.

2 Background

In this section, a case study on the gas dataset used in this research is put forward. This serves as a demonstration of how rudimentary visualisations can be employed to show trends in behavioural patterns; yet it also shows how advanced data analytics are required to ascertain meaning from the information.

2.1 Gas Meter Data

Gas meter data is comprised of samples/readings collected at 30-min intervals. To serve as a proof of concept, the data used in this research is comprised of 20 users from different residential homes chosen at random from the larger dataset, for a 7-day time period. The data is accompanied by a survey which details the occupants age. This enabled us to split the data into two groups of ten users: (1) all users 65 and over and (2) all customers under the age of 65. A sample consisting of two hours' worth of 30-min samples is presented in Table 1 as a demonstration of the raw data structure. The original time is presented in Julian's Day format. However, in Table 1 the values have been converted to the correct time stamp.

Table 1. Sample of gas consumed during 30-min intervals (kW)

Over 65 User			Other		
ID	Time	Usage	ID	Time	Usage
1221	11:00:00	0.066485	1000	11:00:00	0.928069
1221	11:30:00	0.088646	1000	11:30:00	0.640186
1221	12:00:00	0.077565	1000	12:00:00	0.939141
1221	12:30:00	0.066485	1000	12:30:00	0.629114

The smart meter network is a fully maintained and highly accurate sensing network. Within the data generated by the meters, it is possible to identify patterns and routines in behaviour from the raw data. This detailed insight into routine activities creates an opportunity for a variety of applications that can offer benefits to the end user, as previously discussed. Figure 1(a) displays a stacked line graph of 10 users selected at random under the age of 65. Clear collective peaks in activities are apparent on visual inspection. For example, three peaks in gas consumption in the typical morning, lunch and evening periods are evident for the 24-h block of data displayed.

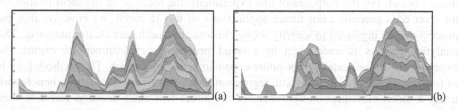

Fig. 1. (a) Stacked line graph for 10 residential gas consumers; (b) stacked line graph for 10 residential gas consumers over the age of 65

Similarly, Fig. 1(b) displays a stacked line graph for 10 customers over the age of 65 (Fig. 1b).

The three higher peaks in gas consumption are once again visible; however, there is a staggered variation for the lunchtime period when compared with the under 65 grouping. These peak consumption times for the customers over 65 are categorised in the tree map diagram visualisations presented in Figs. 2a and b. The diagrams classify the highest gas consumptions periods, with the highest value in the top left corner.

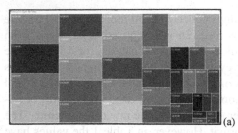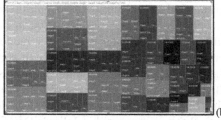
(a)   (b)

Fig. 2. (a) Tree diagram for one customer over the age of 65; (b) Tree diagram for all customers over the age of 65

Clearly, ascertaining any meaning (for example detecting anomalies or energy usage patterns) from a gas smart meter dataset on visual inspection does produce any meaningful insight. The visualisations provide an effective overview of patterns of activity; however, it is a challenge to identify related groups or individuals with similar socio-demographic types without the use of advanced data analytics as a supporting metric. Also, without an intelligent analytics service, learning and understanding these trends in behaviour that are unique to the individual household is a near impossible task. For that reason, and considering the stochastic nature of the data, advanced classification techniques are required to discern patterns in consumption. Therefore, in the following section, the methodology used for the age group classification process is documented.

3   Methodology

In this section, three different classification techniques are employed. The methodology put forward demonstrates a process for identifying age groups within the gas smart mater dataset. For the purposes of this experiment, the focus is on the identification of the over 65's grouping. For future applications of this research, we envision that the process can be deployed to identify social clusters for health care cluster mapping. The analytics process is undertaken in a cloud processing environment to ensure the experiments can be scaled up for future expansions of the research. The methodology is as follows; (1) A two-class logistic regression is applied to the dataset as a benchmark using a direct classification approach. (2) A direct two class decision forest is then applied to the dataset with a normalisation stage; (3) An in-direct classification approach is then adopted using a two-class boosted decision tree, with normalisation, feature extraction phases and with a dimensionality reduction stage prior to classification.

3.1 Direct Classification Benchmark

The benchmark experiment involves a direct classification process using a two-class logistic regression model. The model for the algorithm is displayed in Fig. 3. The raw data is split into a training and test set.

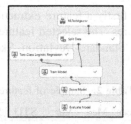

Fig. 3. Two-class logistic regression model

The two-class logistic regression process is calculated using Eq. (1), where x is a d-dimensional vector, p is the logistic distribution and $\beta 0$ relates to the unknown parameters for the logistic distribution [9].

$$p(x; \beta 0, \ldots, \beta D - 1) \tag{1}$$

3.2 Direct Two Class Decision Forest

As a comparison with the benchmark classification, a two-class decision forest technique is employed. However, in this case, prior to the classification a data normalisation process is conducted. The data is normalised using the sliding Zscore. This is calculated using the following:

$$z = \frac{x - mean(x)}{Std(x)} \tag{2}$$

Zscore normalisation is appropriate in this case as it ensures that the raw data conforms to a common scale for the classification. As the visualisations show in Sect. 2, often the peak consumption values are in considerably high, compared to other values in the dataset; which results in spikes in the data visualisation. Whereas, the lower-end values are often a true 0. Therefore, the normalisation process rescales the features so that they have the properties of a standard normalisation. The approach for the two-class decision forest is displayed in Fig. 4.

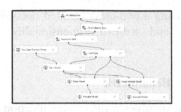

Fig. 4. Two-class decision forest

3.3 In-Direct Classification Approach

For the in-direct classification, firstly a feature extraction process is undertaken using standard time series-based features. The selected features for this research include Min, Max, Mean, Median, Variance, Standard Deviation, Skewness and Kurtosis (Table 2).

Table 2. Sample of features

Min	Max	Mean	Med	Var	STD	Skews	Kurt
1.508	4.850	3.993	4.806	2.745	1.657	−1.996	3.988
6.144	8.266	6.892	6.579	0.995	0.998	1.203	0.484
5.805	6.631	6.157	6.097	0.142	0.377	0.619	−1.913
1.252	6.249	3.886	4.022	5.809	2.410	−0.147	−4.442
0.594	7.454	4.289	4.556	7.941	2.818	−0.559	1.665

The features are calculated at two-hour time blocks. This is due to the selection of skewness and kurtosis as features, as both require minimum three values as input. Variance is calculated using (3) where $\bar{x}$ is the sample mean, and n is the sample size [10].

$$\sigma^2 = \frac{\sum(x - \bar{x})^2}{(n-1)} \tag{3}$$

Similarly, the Standard Deviation calculation takes x for the sample mean and n is the sample size, as displayed in (4) [10].

$$\sigma = \sqrt{\frac{\sum(x - \bar{x})^2}{(n-1)}} \tag{4}$$

The calculation for skewness is outlined line (5), where s is the sample standard deviation and x is the mean value [11].

$$\frac{n}{(n-1)(n-2)} \sum (\frac{x_j - \bar{x}}{s})^3 \tag{5}$$

Similarly, kurtosis, which is a measures of outliers [12], also uses the standard deviation (s) and is calculated in (6) [11].

$$\left\{ \frac{n(n+1)}{(n-1)(n-2)(n-3)} \sum (\frac{x_j - \bar{x}}{s})^4 \right\} - \frac{3(n-1)}{(n-2)(n-3)} \tag{6}$$

The scatter matrix in Fig. 5 displays the positive and negative correlation between the features. In this case, from the visual inspection, most features have a positive correlation (other than Min to Variance and Min to Std).

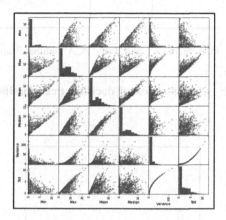

Fig. 5. Scatter matrix feature comparison

Figure 6 displays the correlation between the Skewness and Kurtosis feature values.

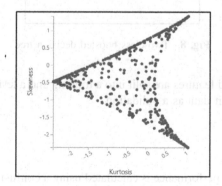

Fig. 6. Skewness vs Kurtosis

The dimensionality reduction process, using Principal Component Analysis (PCA), reduces the features from eight to four. The four newly generated columns contain an approximation of the feature space of the 8 original features. Figure 7 displays scatter plot visualisations of the newly generated features.

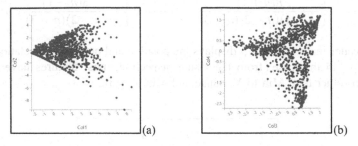

Fig. 7. PCA features 1 and 2 (a), PCA features 3 and 4 (b)

The model for the two-class boosted decision tree classification is displayed in Fig. 8.

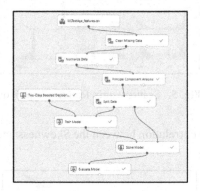

Fig. 8. Two-class boosted decision tree

The PCA-generated features are split into a training and a test set. The classification is scored using the split data as a validation.

4 Results

Each of the classifiers' performance is calculated using a confusion matrix to assess the success of the classification or Area Under the Curve (AUC) and error. The AUC measures the entire two-dimensional area underneath a Receiver Operating Characteristic (ROC) curve. The ROC curve displays the true positive against the false positive predictions. This can be expressed mathematically as shown below, where TP is True Positive, TN is True Negative, FP is False Positive and FN is False Negative:

$$AUC = \frac{(TP + TN)}{(TP + FP + FN + TN)} \tag{7}$$

4.1 Experiment 1 – Two-Class Logistic Regression

The benchmark two-class linear regression approach is able to classify the over 65's age grouping with a 53.7% AUC success rate. This is calculated from the ROC curve displayed in Fig. 9 (Table 3).

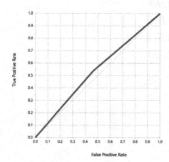

Fig. 9. ROC curve

Table 3. Confusion matrix

Actual class	Predicted class	
	Positive	*Negative*
Positive	**TP = 2021**	**FN = 0**
Negative	**FP = 1625**	**TN = 0**

A full breakdown of the classification scores is presented in Table 4 and the results are visualised in Fig. 10. For all considerations, the classifier is able to achieve over a 53.77% accuracy at minimum, with a maximum success of 60.04%.

Table 4. Two-class logistic regression results

Statistics	Score	Statistics	Score
Mean	0.5696	Accuracy	0.554
Median	0.6004	Precision	0.554
Min	0.5377	Recall	1.000
Max	0.6004	F1 Score	0.713
Std	0.0313	AUC	0.537

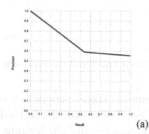
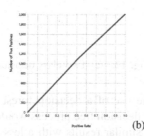

Fig. 10. (a) Two-class logistic regression recall and (b) Precision

4.2   Experiment 2 – Two-Class Decision Forest

The direct two-class decision forest serves as a point of reference comparison with the two-class logistic method, prior to the inclusion of data pre-processing stages. The classification results are displayed in Table 4 and Fig. 11 (Table 5).

Table 5. Two-class decision forest results

Statistics	Score	Statistics	Score
Mean	0.7022	Accuracy	0.698
Median	0.5897	Precision	0.698
Min	0.5092	Recall	1.000
Max	1.00	F1 Score	0.822
Std	0.2071	AUC	0.751

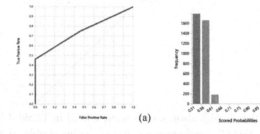

(a)   (b)

Fig. 11 (a) ROC curve and (b) Scored probabilities

The results demonstrate and improvement on the two-class logistic regression approach. The results immediately improve from 53.7% AUC to 75.1% AUC. The precision and recall scores are displayed in Fig. 12.

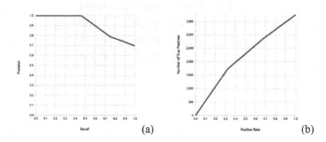

(a)   (b)

Fig. 12. (a) Two-class decision forest recall and (b) Precision

However, the use of a decision forest requires more computationally demanding resources than a decision tree process would. For that reason, the following experiment aims to achieve similar results with a less computationally challenging algorithm, but with a more advanced data pre-processing stage.

4.3 Experiment 3 – Two-Class Boosted Decision Tree

The results for the two-class boosted decision tree are presented in Table 6.

Table 6. Two-class boosted decision tree results

Statistics	Score
Accuracy	0.681
Precision	0.635
Recall	0.816
F1 Score	0.714
AUC	0.746

The results are higher than experiment one and comparable with the decision forest approach. Whilst the results are marginally less successful than the two-class decision forest, the overall distribution of the score probabilities is more optimal than the decision forest approach. A visualisation of the results is presented in Figs. 13 and 14.

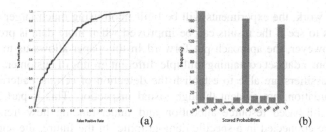

Fig. 13. (a) ROC curve and (b) Scored probabilities

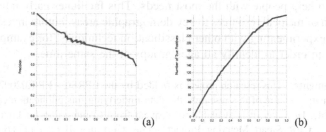

Fig. 14. (a) Two-class boosted decision tree recall and (b) Precision

5 Discussion

Three different algorithms are assessed to demonstrate how they perform using the smart meter 30-min readings. The two-class decision forest achieves the highest AUC. A comparison is detailed in Fig. 15.

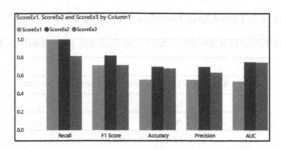

Fig. 15. Experiment comparison

However, the boosted decision tree is comparable when feature extraction and dimensionality reduction are applied to the dataset. As the results display, the scores for the decision tree approaches are higher than that of the logistic regression benchmark experiment.

6 Conclusion and Future Work

In the future work, the experiments will be built upon using much larger dataset. This will allow us to see if the results can be improved when more data is provided to the classifier. However, the approach put forward in this paper allows us to identify one age group from a dataset containing multiple different groups of customers with various ages. The classifiers are able to establish the detection of certain patterns and trends within a population, not evident through visual inspection. This is particularly beneficial for health-based resource allocation and understanding how trends in health conditions are connected in a specific demographic. In the future, the approach could be built on to help understand and visualise the health patterns that can be seen within an urban area. This offers an effective insight into the type of intervention that should be in place to help people with the most needs. This facilitates early intervention and the allocation of medical resources to key demographic areas. Future investigations will also include experimenting with other classification techniques; for example, clustering will be used in order to identify all ages groups at the same time.

Acknowledgements. This research project is funded by the EPRSC - EP/R020922/1. Owing to the ethical sensitive nature of this research, the data underlying this publication cannot be made openly available. However, it is available for request from Commission for Energy Regulation (CER). (2012). CER Smart Metering Project - Gas Customer Behaviour Trial, 2009-2010. [dataset]. 1st Edition. Irish Social Science Data Archive. SN: 0013-00. www.ucd.ie/issda/CER-gas

.

References

1. Song, H., Srinivasan, R., et al.: Health Monitoring Using Smart Systems in Smart Cities: Foundations, Principles, and Applications, pp. 773–792. Wiley, Hoboken (2017)
2. Chalmers, C., Hurst, W., et al.: Smart monitoring: an intelligent system to facilitate health care across an ageing population. In: International Conference on Emerging Networks and Systems Intelligence (2016)
3. Kalogridis, G., Dave, S.: Privacy and eHealth-enabled smart meter informatics. In: International Conference on e-Health Networking, Applications and Services, Natal, pp. 116–121 (2014)
4. Yassine, A., Singh, S., et al.: Mining human activity patterns from smart home big data for health care applications. IEEE Access **5**, 13131–13141 (2017)
5. Wang, Y., Chen, Q., Hong, T., Kang, C.: Review of smart meter data analytics: applications, methodologies, and challenges. IEEE Trans. Smart Grid **10**, 3125–3148 (2018)
6. Yamagata, C., Coppola, J.F., et al.: Mobile app development and usability research to help dementia and Alzheimer patients. In: Long Island Systems, Applications and Technology Conference, pp. 1–6 (2013)
7. CER Smart Metering Project - Gas Customer Behaviour Trial, 2009–2010 and also ISSDA, in the following way: Accessed via the Irish Social Science Data Archive. www.ucd.ie/issda. Accessed: 11 Sept 2018
8. Higuera, V., Ellen Ellis, M.: 10 Early Symptoms of Dementia. Healthline Newsletter, April 2017
9. Yun, W., Kim, D., et al.: Two-dimensional logistic regression. In: International Conference on Tools with Artificial Intelligence, Patras, pp. 349–353 (2007)
10. Wellmer, F.W.: Standard deviation and variance of the mean. In: Wellmer, F.W. (ed.) Statistical Evaluations in Exploration for Mineral Deposits. Springer, Heidelberg (1998). https://doi.org/10.1007/978-3-642-60262-7_6
11. McNeese, B.: Are the Skewness and Kurtosis Useful Statistics? SPC for Excel (2016)
12. Westfall, P.H.: Kurtosis as Peakedness, 1905–2014. R.I.P. Am. Stat. **68**(3), 191–195 (2014)

Non-directional Learning Strategy Particle Swarm Optimization Algorithm

Zhi Ye[1], Cong Li[1(✉)], Yingshi Liang[2], Zhexin Chen[2],
and Lijing Tan[3]

[1] College of Mathematics and Statistics, Shenzhen University, Shezhen, China
frlicong@163.com
[2] College of Management, Shenzhen University, Shenzhen, China
[3] College of Management, Shenzhen Institute of Information Technology,
Shenzhen, China

Abstract. In conventional particle swarm optimization (PSO) algorithm, each particle adjusts their position and velocity to achieve an optimal solution by iteration, but it has the tendency to fall into the local optimum. In order to avoid the classic PSO problem, a new variant of PSO exemplar based on non-directional learning strategy (NLS) is introduced in this paper, which uses random information of partial dimension of personal best experience in every iteration. Initially, the above method randomly extracts dimensional experience from all dimensions of personal best position of particles. Then, the non-directional position is generated by information of random-dimension. Based on the above mechanism, particles are set to obtain information from personal, population and non-directional position, which can enhance particles search ability. Non-directional learning strategy PSO is tested by several benchmark functions, along with some novel PSO algorithms, and the results illustrate that the convergence accuracy is improved significantly.

Keywords: PSO optimization · Non-directional learning strategy ·
Random dimension

1 Introduction

Particle swarm optimization (PSO) is an evolutionary algorithm through simulating social foraging behavior of birds. It was put forward by Kennedy and Eberhart in 1995 [1]. PSO is applied to many domains such as function optimization, artificial neural networks training [2], pattern classification and system control. PSO is one of the effective tools to solve optimization problems. Nevertheless, PSO still exists some drawbacks, for instance, it is easily trapped in a local optimum. Moreover, the convergence rate decreases considerably in the late iteration especially with multimodal functions, for the reason that the algorithm stops optimizing when it's close to the particle optimal solution.

Due to these above-mentioned problems, a great number of attempts have been proposed to ameliorate the algorithm. Many approaches enhance the performance of the algorithm on exploration via utilizing different kinds of learning strategies of PSO,

© Springer Nature Switzerland AG 2019
D.-S. Huang et al. (Eds.): ICIC 2019, LNAI 11645, pp. 606–616, 2019.
https://doi.org/10.1007/978-3-030-26766-7_55

for example, the comprehensive learning PSO (CLPSO) [3], the social learning PSO (SLPSO) [4] and the Unified PSO (UPSO) [5]. CLPSO utilizes other particles i historical best information to update a particles' velocity in order to jump out of the local optima. Unlike PSO, SLPSO enables each group to learn from better groups one by one. Although SLPSO can solve the shortcomings that PSO is easy to fall into local optima, SLPSO has the following defects in solving complex optimization problems. UPSO updates the velocity through studying all the neighbors i experience of the swarm.

A modification of non-directional learning strategy particle swarm optimizer (NLS-PSO) based on randomly-dimension position learning is proposed in this paper, to solve the problem in premature convergence in PSOs. In this algorithm, apart from learning the historic experience of particles and population, particles obtain partially dimensional information from p_{best} stochastically. When updating the position of particles in each iteration, swarms would make use of potential knowledge of dimension instead of all dimensions. In this case, particles in NLS-PSO balance the capability of exploration and exploitation in different stages, which can enhance the performance in optimization problems.

The structure of this research is as follows. In Section TWO, it introduces the standard PSO algorithm. Section THREE proposes the non-directional learning strategy PSO based on dimensional learning. Section FOUR designs a simulation experiment to compare the conventional version of PSO algorithms, CLPSO, UPSO and SLPSO. Section FIVE presents computational experience with a series of test figure and table. Finally, conclusions are given in Section SIX.

2   Non-directional Learning Strategy PSO

2.1   Motivation

According to the updating formulas in basic PSO [5], algorithm uses personal best position and global best position to update the particles in every iteration. Some studies pointed out that classic PSO could easily cause the particles to be trapped into local optima [7]. However, particles would lose potential information of dimensions by conducting two acceleration coefficients. Classic PSO could cause the result that particles wander between the local best and global best location, which reduce the optimization efficiency [8].

To solve the above-mentioned problem, most novel PSOs have been proposed in the following areas: parameter setting, selection of neighborhood topology, learning strategy, and hybridization rules [9, 10]. The study of Liang showed that, when minimizing the fitness value of a function, dimensional learning can improve the performance of particles in the search area [3]. But few research consider the partial information of random-dimension thoroughly. In the following section, the mechanism to obtain promising knowledge in random dimensions is introduced.

2.2 Improvement

In NLS-PSO, the velocities of particles are updated through Eq. (1):

$$V_{i,d}^{t+1} = \omega V_{i,d}^t + C_1 rand_1 \left(p_{best}^t - X_{i,d}^t \right) + C_2 rand_2 \left(g_{best}^t - X_{i,d}^t \right)$$
$$+ C_3 rand_2 \left(p_{d-random}^t - X_{i,d}^t \right) \tag{1}$$

where d-random defines the dimension selected from all dimension. $p_{d-random}^t$ is a corresponding stochastically dimensional position which is called non-directional position in t^{th} iteration. The flowchart of NLPSO is given in Fig. 3, where N is population size of particles, D is dimension size, and T is number of iteration.

According to the updating equation, particles learn from non-directional positions in every acceleration and these positions are different from each particle.

The main process to generate the $p_{d-random}^t$ in every iteration is as follows and shown in Fig. (1):

(1) We randomly choose a integer from 1 to D, where D denotes the dimension size of particles.
(2) We then extract the certain dimensional information from pbest according to the number above.
(3) We generate p^t by copying the chosen dimensional information to all dimension.

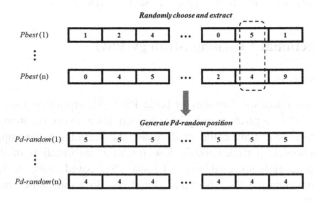

Fig. 1. Process of re-adjust the $p_{d-random}^t$

Based on the updating formula, the movement of particles compared to the basic PSO in Sphere function can be seen in Fig. 2(a) and (b). In Fig. 2(b), particles of NLS-PSO can explore more area in search space, which means particles can obtain more information from non-directional position. And the pseudo-code of NLS-PSO is as follows.

3 Experiment

3.1 Experiment Design and Evaluation Criteria

To compare the effectiveness and convergence feature of NLS-PSO with four PSOs, which include CLPSO, SLPSO, UPSO, PSO, two optimization experiments are presented in this section. Experiment 1 is made on several benchmark functions in 30 dimensions in Table 5, which include 4 unimodal functions $(f_1$-$f_4)$, 4 multimodal functions $(f_5$-$f_8)$, to verify the performance of NLS-PSO. Experiment 2 is performed with two unimodal functions and one multimodal function in 10, 30, 50 dimensions. Each experiment test is tested for 30 times running to show the stability of algorithms and the running time of each PSOs on benchmark functions is recorded.

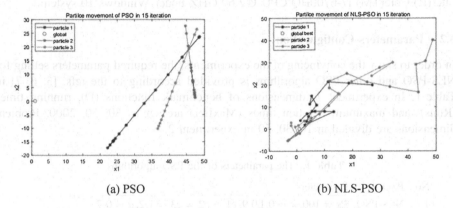

(a) PSO   (b) NLS-PSO

Fig. 2. Comparison results between NLS-PSO and PSO

Algorithm 1: Pseudo-code of NLS-PSO
Input: Population size N; Dimension of space D; Iteration T.
Output: Global optima.
1: Initialize boundary parameter: $X_{max}, X_{min}, V_{max}, V_{min}$;
2: Initialize hyper-parameter: C_1, C_2, C_3;
3: Initialize position and velocity: X_i, V_i;
4: Set $p_{best}^t, p_{best}^{t-1}, g_{best}^t = INF$, and calculate the fitness of X_i
5: for $iter < T$ do
6:   Calculate the finesse of each particle and whole population
7:   if $f(x_i) < f(p_i)$ then
8:     $p_i := x_i$
9:     $p_{best}^{t+1} := f(x_i)$
10:    if $p_{best}^{t+1} < g_{best}^t$ then
11:      $g_i := p_i$
12:      $g_{best}^t := p_{best}^{t+1}$
13:    Generate the non-directional position $p_{d-random}^t$
14:    Update the velocity through Eq. (4)
15:    $iter = iter + 1$

The evaluation criterion include mean (Mean) value, standard deviation (Std) of the best fitness value, running time, and the convergence speed of algorithms. In this 2 experiments, convergence speed is measured with the mean iteration number when algorithms reach their best fitness value. The criterion is as follows: the algorithms Mean value on each functions compared, the smaller the Mean value is, the better the algorithm performance will be; if the algorithms have the same Mean value on each function, the Std values are compared, the smaller the Std value is, the better the stability of the algorithm will be; if the algorithms have the same common Mean and Std values, the convergence of the global optimal curve is compared. The faster the curve converges, the better the search capability of the algorithm will be. If all three criteria are the same, the performance of algorithms are equal.

Besides, all algorithms are implemented in Matlab R2017 (Win 64) of PC with Inter(R) Core(TM) i7-6700HQ CPU @2.60 GHZ under Windows 10 system.

3.2 Parameters Configuration

In order to show the convincing of the experiments, the required parameters setting for NLS-PSO and other PSO algorithms is provided according to the refs. [3, 6, 7] in Table 1. In experiment 1, dimensions of benchmark functions (D), running times (Runs), and maximum iteration times (MaxIter) are set as 30, 30, 2000. Problem dimensions are divided in 10, 30, 50 in experiment 2.

Table 1. The parameters of the PSO algorithms

No.	PSO	Parameters
1	NLS-PSO	$Ps = 100, w = 0.1\,0.9, c1 = c2 = c3 = 1.2, \alpha = 0.7$
2	CLPSO	$Ps = 100, c1 = c2 = 1.49445, t = 0 : \frac{1}{Ps-1} : 1; t = 5.*t,$ $Pc = 0.5.*(expt - expt(1))./(expt(Ps) - expt(1))$
3	SLPSO	$Ps = 100, c3 = Dimension/Ps * 0.01$
4	UPSO	$Ps = 100, c1 = c2 = 1.49445, \omega = 0.729, u = 0.1, \mu = 0, \sigma = 0.01$
5	PSO	$Ps = 100, c1 = c2 = 1.5, \omega = 0.8$

3.3 Comparison and Analysis

Results of Experiment 1. In this section, NLS-PSO is compared with basic PSO, CLPSO, SLPSO, and UPSO with the parameters setting of experiment 1. The results are listed in Table 2 and the convergence curves of different benchmark functions are shown in Fig. 4.

On four unimodal benchmark functions, NLS-PSO performs the best performance on $(f_1–f_3)$ and SLPSO ranks first on f_4. From Table 2, NLS-PSO reaches the global optima value on f1 and f3 with zero standard deviation. For Dixon-Price function (f4), NLS-PSO's Mean and Std value is slightly worse than the SLPSO and UPSOs. However, four unimodal functions illustrate that the proposed PSO converges faster than other PSOs.

On four multimodal benchmark functions, NLS-PSO ranks first on $(f_5–f_8)$ and locates the minimum value of all functions with zero standard deviation. For f_5, UPSO also reaches the optima with zero standard deviation. For f_6, NLS-PSO shows the same performance before 800 iterations approximately but obtains best performance after that, which means that it is difficult to trap in local optimum and has a powerful global searching capability. Besides, four multimodal functions illustrate that the convergence speed of NLS-PSO is faster than other.

Results of Experiment 2. Experiment 2 is presented to compare the stability of PSOs in varied dimensions of 10, 30, and 50 on three benchmark functions with the same parameters setting of experiment 1. Results of experiment 2 are listed in Table 3 and the convergence curves are shown in Fig. 5.

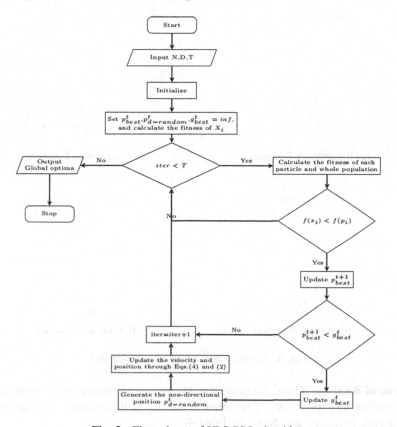

Fig. 3. Flow charts of NLS-PSO algorithm

On two unimodal functions, dimensional test on sphere function (f_1) shows high stability of NLS-PSO. Figure 9 shows that NLS-PSO obtains best mean value on f_1 and maintains the approximate convergence speed in all varied dimension. For rosenbrock function (f_3), NLS-PSO presents a better performance in both dimensional stability and high convergence accuracy compared to other PSO variants.

On multimodal function, results in Fig. 6 illustrate the difficulty of other four PSOs when facing high dimension problems. NLS-PSO ranks first in best mean value and convergence speed with the help of non-directional learning strategy.

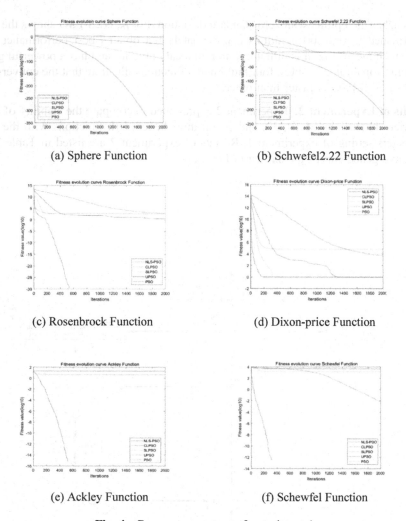

(a) Sphere Function

(b) Schwefel2.22 Function

(c) Rosenbrock Function

(d) Dixon-price Function

(e) Ackley Function

(f) Schewfel Function

Fig. 4. Convergence curves of experiment 1

Analysis of Running Time. To prove the advantages of convergence speed of NLS-PSO, runtime of experiments is recorded to discuss in this section. Average running time of each functions are shown in Table 4.

From Table 4, on average time, convergence speed of PSO is the fastest in three different dimensions among PSO algorithms, and the running time of NLS-PSO in 10, 30, 50 dimensions is about 14 s on f_1, which ranks second compared to others. Run time of NLS-PSO is about 5 s on f_4 and is about 4 s on f_6 Compared to other PSOs, NLS-PSO is more stable in running time.

The reason of NLS-PSO's performance is as follows. Firstly, the updating rule of particles is improved on the basis of basic PSO, and the learning strategy is less complicated than four PSO algorithms in this paper. Secondly, non-directional learning strategy considers the partial information of personal best position in every iteration, which reduces the runtime compared to the one-by-one dimensional learning rules of SLPSO. On the whole, particles in NLS-PSO obtain more dimensional information and the non-directional position illustrates better directivity.

Table 2. Results of PSOs on experiment 1

Fun.		NLS-PSO	CIPSO	SLPSO	UPSO	PSO
f_1	Mean	0	3.3×10^{-4}	4.15×10^{-91}	1.34×10^{-31}	0.79
	Std	0	6.49×10^{-5}	6.83×10^{-91}	7.87×10^{-32}	0.49
f_2	Mean	3.0×10^{-219}	0.0807	0.9412	400.9998	2.01×10^3
	Std	0	0.0139	0	896.6628	1.01×10^3
f_3	Mean	0	621.2431	680.5239	128.6868	394.1351
	Std	0	265.7259	1.14×10^3	128.6868	173.1201
f_4	Mean	0.9768	4.90×10^3	**0.6213**	0.9412	6.17×10^3
	Std	0.0201	1.90×10^3	0	7.23×10^{-13}	9.91×10^4
f_5	Mean	0	3.56×10^{-4}	21.600	0	0.0942
	Std	0	1.46×10^{-4}	32.2017	0	0.0541
f_6	Mean	0	18.6003	0.0525	30.6468	17.1932
	Std	0	0.8051	0.1174	2.1473	4.9639
f_7	Mean	0	20.1693	20.6952	20.6952	4.4414
	Std	0	0.0245	0.209	0.209	1.2599
f_8	Mean	0	0.0079	1.46×10^3	4.531×10^{-3}	5.57×10^3
	Std	0	0.0035	192.6345	187.3992	301.6351

Table 3. Results of PSOs on experiment 2

		Sphere		Schewel		Rosenbrock	
	Dim	Mean	Std	Mean	Std	Mean	Std
NLS-PSO	10	0	0	0	0	0	0
	30	0	0	1.8×10^{-11}	0	0	0
	50	0	0	0	0	0	0
CLPSO	10	9.5×10^{-25}	1.2×10^{-24}	0	0	0.7587	0.5758
	30	1.6×10^{-4}	6.6×10^{-5}	0.0072	0.0098	7.7×10^2	4.2×10^2
	50	1.8619	0.3469	28.6749	22.4668	5.8×10^4	4.4×10^4
SLPSO	10	3.2×10^{-187}	0	4.5×10^2	9.1×10^2	0.2983	0.0449
	30	1.9×10^{-90}	7.4×10^{-90}	1.3×10^3	6.7×10^2	89.8029	3.1×10^2
	50	4.3×10^{-61}	6.0×10^{-61}	2.7×10^3	8.8×10^2	2.0×10^2	7.2×10^2
UPSO	10	6.5×10^{-89}	2.0×10^{-88}	3.2×10^2	3.4×10^2	0.746	1.3649
	30	2.4×10^{-31}	3.7×10^{-31}	4.1×10^3	1.0×10^3	44.6425	1.3×10^2
	50	1.6×10^{-18}	1.2×10^{-18}	8.9×10^3	9.1×10^2	57.5063	77.8526
PSO	10	2.6×10^{-75}	8.9×10^{-75}	1.2×10^3	1.2×10^2	0	0
	30	2.4796	2.9042	4.2×10^3	848.7840	1.0×10^4	2.3×10^4
	50	34.1538	19.0626	9.6×10^3	2.0×10^3	4.7×10^5	1.9×10^6

Table 4. Comparison of running time

Fun.	Dim	NLS-PSO	CLPSO	SLPSO	UPSO	PSO
f_1	10	12.030095	84.771843	9.853811	152.18554	**1.764323**
	30	14.688366	91.233081	29.701747	159.491137	**2.120672**
	50	13.28869	98.95141	87.865642	166.622715	**2.81128**
f_4	10	5.18519	108.227627	4.938399	159.230968	**3.160651**
	30	5.543088	113.221279	21.800833	184.291231	**3.670509**
	50	5.693909	119.720147	45.830091	203.502421	**4.283535**
f_6	10	3.710952	96.215415	7.975474833	127.437492	**1.892317**
	30	4.007363	114.991377	52.984579	197.409223	**2.645699**
	50	4.299971	123.587748	118.48833	190.670511	**3.173249**

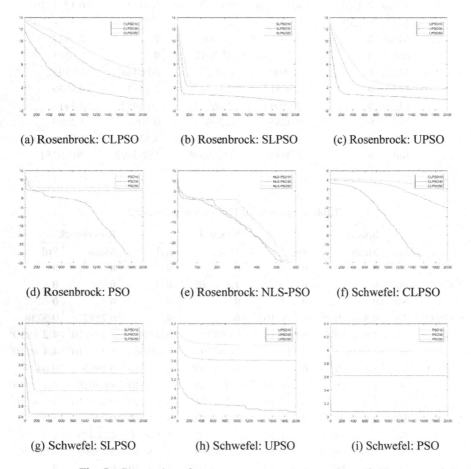

(a) Rosenbrock: CLPSO   (b) Rosenbrock: SLPSO   (c) Rosenbrock: UPSO

(d) Rosenbrock: PSO   (e) Rosenbrock: NLS-PSO   (f) Schwefel: CLPSO

(g) Schwefel: SLPSO   (h) Schwefel: UPSO   (i) Schwefel: PSO

Fig. 5. Comparsion of convergence curves on experiment 2 (1)

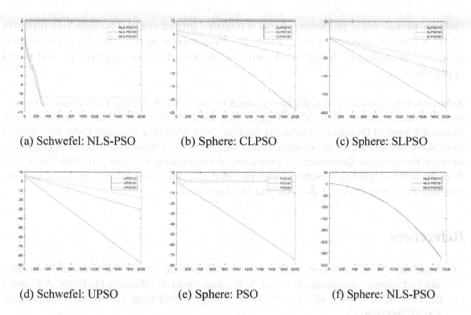

(a) Schwefel: NLS-PSO   (b) Sphere: CLPSO   (c) Sphere: SLPSO

(d) Schwefel: UPSO   (e) Sphere: PSO   (f) Sphere: NLS-PSO

Fig. 6. Comparsion of convergence curves on experiment 2 (2)

Table 5. 8 benchmark functions

No.	Function name	Search range	x*	$f(X^*)$
f_1	Sphere	$x_i \in [-500, 500]^d$	$(0, ..., 0)$	0
f_2	Schwefel 2.22	$x_i \in [-500, 500]^d$	$(0, ..., 0)$	0
f_3	Rosenbrock	$x_i \in [-500, 500]^d$	$(1, ..., 1)$	0
f_4	Dixon-price	$x_i \in [-500, 500]^d$	$x_i = 2^{-\frac{2^i-2}{2^i}}, i = 1, ..., d$	0
f_5	Griewank	$x_i \in [-500, 500]^d$	$(0, ..., 0)$	0
f_6	Weierstrass	$x_i \in [-500, 500]^d$	$(0, ..., 0)$	0
f_7	Ackley	$x_i \in [-500, 500]^d$	$(0, ..., 0)$	0
f_8	Schewfel	$x_i \in [-500, 500]^d$	$(420.9687, ..., 420.9687)$	0

4 Conclusion

This paper introduces a novel dimensional learning based on PSO algorithm which is called non-directional learning strategy PSO (NLS-PSO). In this mechanism, particles learn from stochastically dimensional positions when updating their velocities and positions, which significantly enhances the search ability and avoids premature convergence. Besides, from empirical studies in this paper, NLS-PSO has a better performance in most of benchmark functions than other PSO algorithms with high convergence accuracy and fast convergence speed. However, NLPSO may reduce the stability in real-world problems, so that it requires further study. The attractive property of NLS-PSO is that few studies focus on dimensional information of particles best position. In addition, future work will be expected to be applied in real-world application.

Acknowledge. This work is partially supported by the Natural Science Foundation of Guangdong Province (2016A030310074), Project supported by Innovation and Entrepreneurship Research Center of Guangdong University Student (2018A073825), Research Cultivation Project from Shenzhen Institute of Information Technology (ZY201717) and Innovating and Upgrading Institute Project from Department of Education of Guangdong Province (2017GWTSCX038). And the authors appreciate everyone who provided us with constructive suggestions and discussions, especially Professor Ben Niu and Ms. Lulu Zuo.

References

1. Angeline, P.J.: Evolutionary optimization versus particle swarm optimization: Philosophy and performance differences. In: Porto, V.W., Saravanan, N., Waagen, D., Eiben, A.E. (eds.) EP 1998. LNCS, vol. 1447, pp. 601–610. Springer, Heidelberg (1998). https://doi.org/10.1007/BFb0040811
2. Cao, L.: A collaboration-based particle swarm optimizer with history-guided estimation for optimization in dynamic environments. Expert Syst. Appl. **120**, 1–13 (2019)
3. Cheng, R.: A social learning particle swarm optimization algorithm for scalable optimization. Inf. Sci. **291**, 43–60 (2015)
4. Durán-Rosal, A.M.: A hybrid dynamic exploitation barebones particle swarm optimization algorithm for time series segmentation. Neurocomputing (2019). https://doi.org/10.1016/j.neucom.2018.05.129
5. Kennedy, J.: Particle swarm optimization. In: Proceedings of International Conference on Neural Networks, pp. 1942–1948. IEEE, Perth, The University of Western Australia, Perth, Western Australia (1995)
6. Liang, J.J.: Comprehensive learning particle swarm optimizer for global optimization of multimodal functions. IEEE Trans. Evol. Comput. **10**(3), 281–295 (2006)
7. Parsopoulos, K.E., Vrahatis, M.N.: Unified particle swarm optimization for solving constrained engineering optimization problems. In: Wang, L., Chen, K., Ong, Y.S. (eds.) ICNC 2005. LNCS, vol. 3612, pp. 582–591. Springer, Heidelberg (2005). https://doi.org/10.1007/11539902_71
8. Rosli, A.D.: Application of particle swarm optimization algorithm for optimizing ANN model in recognizing ripeness of citrus. In: Proceedings of IOP Conference Series on Materials Science and Engineering, Phuket, Thailand, vol. 340, pp. 012–015. IOP Publishing (2018)
9. Seo, J.H.: Multimodal function optimization based on particle swarm optimization. IEEE Trans. Mag. **42**(2), 1095–1098 (2006)
10. Tian, D.: MPSO: modified particle swarm optimization and its applications. Swarm Evol. Comput. **41**, 49–68 (2018)

An Improved Social Learning Particle Swarm Optimization Algorithm with Selected Learning

Haoruo Hu[1(✉)], Ming Chen[1], Xi Song[2], E. T. Chia[3], and Lijing Tan[4]

[1] College of Economics, Shenzhen University, Shenzhen 518060, China
huhaoruo2017@email.szu.edu.cn
[2] Greater Bay Area International Institute for Innovation, Shenzhen University,
Shenzhen 518060, China
[3] School of Cyber Engineering, Xidian University, Xian 710071, China
[4] College of Management, Shenzhen Institute of Information Technology,
Shenzhen 518172, China

Abstract. Particle Learning Optimization (PSO) is a novel heuristic algorithm that has undergone decades of evolution. Social learning particle swarm optimization (SL-PSO) proposed by Cheng, Jin et al. in 2016 [1] remarkably improves the PSO algorithm by applying multi-swarm learning strategy. Nevertheless, randomness on setting inertia and choosing learning objects gives rise to an unbalanced emphasis on global search, and thus impairs convergence rate and exploitation ability. The proposed ISL-PSO algorithm strengthens global search capability through modelling selected learning mechanism, in which learning objects are selected through generated learning possibility subjected to Gauss distribution. Furthermore, ISL-PSO algorithm models condition-based attraction process, in which particles are attracted to the center by calculating transformed distance between particles and the center. By applying the strategies, ISL-PSO improves convergence speed and accuracy of the original algorithm.

Keywords: Multi-swarm · PSO · Swarm intelligence ·
Social learning particle swarm learning · ISL-PSO

1 Introduction

Swarm-intelligence-based metaheuristic search methods have become increasingly popular for their power of solving difficult scientific problems. Popular heuristic algorithms include Particle Swarm Optimization algorithm (PSO) [2], Ant Colony Optimization Algorithm (ACO) [3], Bacterial Foraging Optimization Algorithm (BFO) [4] and so forth. These methods simulate socialized populations, in which individuals act with limited cognitive ability and communicate with simple rules, while the swarm behaves in an organized manner without conscious control and ultimately seeks out the global optima [5]. Despite differences among heuristic algorithms, they usually share certain characteristics. For example, communication mechanisms are

© Springer Nature Switzerland AG 2019
D.-S. Huang et al. (Eds.): ICIC 2019, LNAI 11645, pp. 617–628, 2019.
https://doi.org/10.1007/978-3-030-26766-7_56

incorporated to all sorts of heuristic algorithm, and that the balance between exploitation and exploration invariably remains pivotal [6].

Particle Swarm Optimization (PSO), introduced by Kennedy and Eberhart in 1995 [2], is among the most popular swarm-intelligence metaheuristic search methods in solving problems on optimization. In PSO algorithm, each particle is capable of storing the best positions found previously. The best positions found by the swarm of particles and the best position found by every single particle are termed global best and personal best respectively. Particles move towards global optimum through learnings from personal bests as well as global bests.

Admittedly, PSO algorithm has achieved broad applications in feature selection [7], economic dispatch [8], optimal control [9], network design [10] and so forth, but it is still criticized for several dysfunctional problems such as trapping in local optima and converging at low speed. Moreover, the algorithm is considered insufficient in solving problems of high dimensions and inseparability [11]. The following categories of variants have been proposed to enhance its capability and amend the limitations.

The first category focuses on adaptive control over parameters of PSO. Generally, researchers modifies ω, r_1 and c_2 [12]. Apart from the three parameters, other parameters are introduced [13].

The second category, namely Hybrid PSO, is featured by incorporating other search strategies and operators to standard PSO, such as genetic algorithms [14] and differential evolutions [15]. Existing variants of this category includes GA-PSO [16], PSACO [17], BFOA [18], etc.

The third category is termed Multi-Swarm PSO. PSO algorithms of this category are modified for enhancing the diversity of the swarm by communication among different swarms, in which the swarm can ameliorate the deficiency of premature convergence. Paradigms of this category includes Cooperative Multi-Swarm PSO (CPSO) [19] and Dynamic Multi-Swarm PSO (DMS-PSO) [20].

The fourth category of improved PSO introduces structure of topology to enhance the diversity of the swarm, thereby ameliorating the limitations such as premature convergence [21]. Researchers develop various strategies such as mutation and social learning. Social learning strategy is derived from observations that birds of Swaythling opened bottles of milk by learning from the crowd [22]. Inspired by the interaction of bird swarm, researchers introduce the mechanisms of social learning in PSO. Paradigms of such mechanism include local enhancement strategy [23], observational conditioning strategy, and contagion strategy. Of all existing social learning strategies, imitation strategy [24] is generally viewed as a unique mechanism, in which certain particle behaviors spread at the population level by operations through the whole swarm [25].

SL-PSO algorithm is proposed by Cheng et al. in 2015 for addressing limitations of standard PSO [1]. Overall, SL-PSO algorithm is modeled to make each particle learn from better particles (examples) for each dimension as well as the behaviors of the current population. In this paper, we modify the learning strategies adopted in original SL-PSO to reduce the randomness on choosing learning objects, and hence effectively improve the performance of the algorithm.

2   PSO Modification

2.1   Standard PSO

In the standard PSO proposed by Kennedy et al. [2], velocity updating Eq. 1 and position updating Eq. 2 can be written as follows:

$$V_i(t+1) = \omega V_i(t) + c_1 R_1(t)(t_i(t) + X_i(t)) + c_2 R_2(t)(gbest_i(t) + X_i(t)) \tag{1}$$

$$X_i(t+1) = X_i(t) + V_i(t+1) \tag{2}$$

where

　t denotes the generation number
　$V_i(t)$ denotes the velocity of the i particle $X_i(t)$ denotes the position of the i particle
　c_1 and c_2 are the accelerating factors
　$R_1(t)$ and $R_2(t)$ are two vectors generated in the range of $[0, 1]^n$
　$pbest_i(t)$ denotes the personal best of the i particle
　$gbest_i(t)$ denotes the global best in all particles.

2.2   SL-PSO

Social learning particle swarm optimization is proposed for addressing defects of standard PSO. In summary, SL-PSO takes advantages that each particle learns from better particles called Examples for each dimension as well as the mean behavior of the current population. The social learning process is briefly demonstrated in Fig. 1. The swarm would keep iteration until the stopping criteria are met.

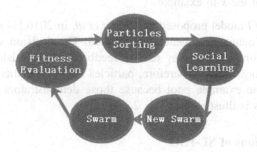

Fig. 1. The process of SL-PSO

According to Cheng et al. [1] in 2015, particle X_i updates itself by learning from a better particle X_k ($1 \leq i \leq N-1$ and $i \leq k \leq N$) through:

$$X_{i,j}(t+1) = X_{i,j}(t) + \Delta X_{i,j}(t+1) \tag{3}$$

$\Delta X_{i,j}$ is the behavior correction as follows:

$$\Delta X_{i,j}(t+1) = r_1 \cdot \Delta X_{i,j}(t) + r_2 \cdot I_{ij}(t) + r_3 \cdot s \cdot C_{i,j}(t) \tag{4}$$

For attaining s as the society influence factor, the following equation is applied:

$$s = \beta * D/100 \tag{5}$$

For attaining $I_{i,j}$ and $C_{i,j}$, the following equation is applied:

$$\begin{cases} I_{i,j}(t) = X_{k,j}(t) - X_{i,j}(t) \\ C_{i,j}(t) = X_j(t) - X_{i,j}(t) \end{cases} \tag{6}$$

$$\bar{X}_j(t) = \frac{\sum_{i=1}^{N} x_i^j}{N} \tag{7}$$

Where

$I_{i,j}$ denotes the distance from the example to $X_{i,j}$

$C_{i,j}$ denotes the distance from the mean of the whole group to $X_{i,j}$ $r_{i=1,2,3}$ is random number distributed uniformly within [0,1]

s is the society influence

β is a predefined parameter

$X_{k,j}$ is the k-th example of the j-th dimension

X_j is the mean behavior of all particles of the current population on the j-th dimension of the k-th example

Original SL-PSO model proposed by Cheng et al. in 2016 [1] generates an m-size swarm, in which t denotes the generation iterated. The decision vector $X_i(t)$ of each particle forms randomly. Once they receive feedback from decision vectors, each of them is given a fitness value. Therefore, particles can amend their activities through demonstrators in the example pool because those demonstrators have better fitness values. The process is illustrated in Fig. 2.

2.3   The Limitations of SL-PSO

Even though SL-PSO has ameliorated limitations of standard PSO in certain aspects, it is still less perfect than Cheng et al. indicated. As Zhang et al. suggested in 2019 [24], SL-PSO algorithm has the following problems.

1. The randomness on generating inertia r_1 and $X_{i,j}$ negatively affects the convergence rate in the later search phase.
2. The randomness brings SL-PSO robust global search ability. Nonetheless, the swarm is relatively easy to fall into a lopsided emphasis on exploration.
3. It is difficult for SL-PSO algorithm to adopt parallel computation since it calculates dimension by dimension, which may constrain the running speed.

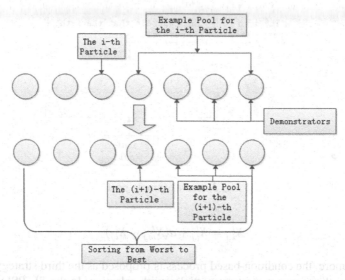

Fig. 2. Example pool

4. SL-PSO does not take full advantage of every particle because the best particle does not update throughout searching process. And therefore the search efficiency is negatively influenced.

3 The Improved SL-PSO

3.1 Novel Strategies

To reduce randomness and achieve balanced emphasis on both exploitation and exploration, the way the sorted particles learn from the example pools will be modified based on following strategies. In the first strategy, instead of learning from example randomly, particles are rearranged randomly and compared to each other in terms of fitness value. Subsequently, the rearranged particles exchange positions with those of lower fitness value on the condition that the exchanged particles attain lower fitness value. This step is aiming to strengthen the particles' capability of global search.

Moreover, instead of choosing learning objects randomly in original SL-PSO algorithm, learning objects of particles are chosen with respective possibility, in which the learning possibility is subjected to Gauss distribution. Experiments have been done to set the most appropriate means and standard deviation while applying the strategy. By ameliorating the randomness when choosing learning objects, particles learn with better quality. In original SL-PSO algorithm, the formula is given as $X_{k,j}^t = X_{i,j}^t + rand(X_{m,j}^t - X_{i,j}^t)$. The improved SL-PSO formula is as follows. The a is a random number generated through normal distribution (Fig. 3).

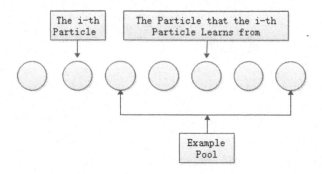

Fig. 3. New example pool

$$X_{k,j}^t = X_{i,j}^t + a(X_{m,j}^t - X_{i,j}^t) \qquad (8)$$

Furthermore, the condition-based process is proposed as the third strategy, in which the center of the swarm can attract particles with selection. In the SL-PSO algorithm, each particle is attracted by the center of the swarm. In condition-based approach, primary, the position of the center of swarm is obtained as $\overline{p_{i,j}}$, and then the distance vector between each particle and the center of the swarm $p_{i,j} - \overline{p_{i,j}}$ is calculated and transformed to Euclidean Distance $d_i(i = [1, 30])$, and hence the mean of the Euclidean Distance $\overline{d_i}$ is calculated. When $d_i > \overline{d_i}$, the particle will be attracted by the center of the swarm, as demonstrated in Eq. (9) behavior correction. Otherwise, the particle will not be attracted because it is close enough to the center of the swarm.

$$\Delta X_{i,j}(t+1) = r_1 \cdot \Delta X_{i,j}(t) + r_2 \cdot I_{i,j}(t) - r_3 \cdot s \cdot C_{i,j}(t) \qquad (9)$$

3.2 Process of the Proposed ISL-PSO

The process of the improved SL-PSO algorithm is demonstrated in as follows. For elucidate the framework of the proposed ISL-PSO algorithm, flowchart is demonstrated in Fig. 4.

Step1. Set parameters and initialize velocities and positions of each particle
Step2. Compute fitness values of particles and attain $P\ best$
Step3. Re-arrange particles from the best to the worst by fitness values
Step4. For i = 1:m, example pool = m − i
Step5. Randomly interrupt the indices of particles and re-evaluate fitness value
Step6. Determine learning object *pwin* through modified learning possibility *s.t.* normal distribution
Step7. Update fitness values through the third strategy
Step8. Update positions by the formula $X_{k,j}^t = X_{i,j}^t + a(X_{m,j}^t - X_{i,j}^t)$
Step9. Stop algorithm when the termination conditions are met

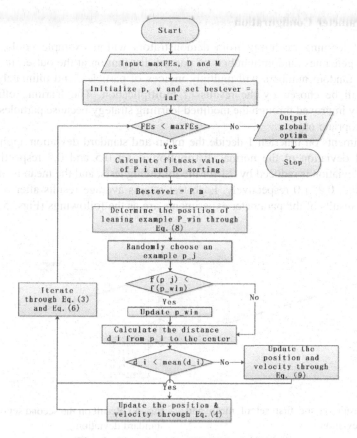

Fig. 4. Flowchart for ISL-PSO algorithm

4 Experiment Design

4.1 Benchmark Functions

Benchmark Functions applied in evaluating original SL-PSO and improved SL-PSO methods are shown in Table 1:

Table 1. No., function names, and dimensions for benchmark functions

No.	Functions	Minimum	Dimensions
f1	Sphere	0	30
f2	Schwefel 2.22	0	30
f3	Rosenbrock	0	30
f4	Schwefel 2.21	0	30
f5	Rastrigin	0	30
f6	Ackley	0	30
f7	Griewank	0	30
f8	Penalized 1	0	30

4.2 Parameter Configuration

Instead of learning randomly from demonstrators within example pools, ISL-PSO algorithm generates random numbers of normal distribution at the outset, in which the generated random numbers will multiply indices of particles, and ultimately learning objects will be chosen by the results of multiplication. The learning efficiency is remarkably improved through the modified learning strategy because particles can learn from appropriate objects.

Experiments on function 1 decide the mean and standard deviation applied. Initial mean and deviation of the normal distribution is set 0.5 and 0.5 respectively. The standard deviation is reduced by 0.1 for the following tests, and the mean is modified to 0.6, 0.7, 0.8, 0.9, 1.0 respectively. Each test attains average results after 30 times of runs. The results of the parameter experiments are as the followings (Figs. 5, 6, 7, 8, 9 and 10).

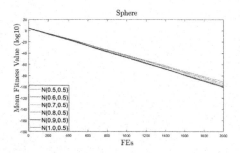

Fig. 5. Result on the first set of mean and standard deviation

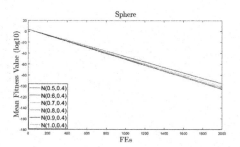

Fig. 6. Result on the second set of mean and standard deviation

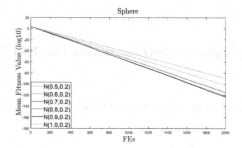

Fig. 7. Result on the third set of mean and standard deviation

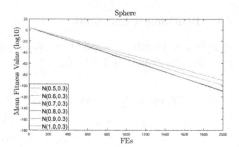

Fig. 8. Result on the fourth set of mean and standard deviation

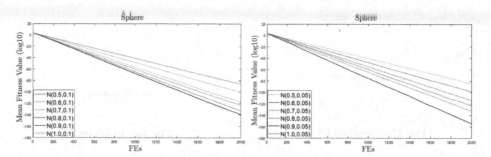

Fig. 9. Result on the fifth set of mean and standard deviation

Fig. 10. Result on the sixth set of mean and standard deviation

The test results indicate that when the standard deviation is set equal, the convergence performance ameliorates if the mean gets close to 1; When the standard deviation reduces gradually, the convergence speed improves. The lowest curves shown in the above figures are all attributed to a mean of 0.9. The results support that the stability is guaranteed in choosing the most appropriate learning objects. Moreover, the experiment suggests that moderate fluctuation of choice over learning objects contributes to the improvement of the global search ability. Therefore, ISL-PSO will adopt normal distribution with mean = 0.9 and standard deviation = 0.05. Note that this is a classical configuration of parameters. Moderate modification of mean and standard deviation is recommended for multi-modal problems.

4.3 Experiment Results

Initial number of the swarm is set as 100 and the dimensions of test functions are set as 30. After 2000 times of evaluation in 30 runs The test results are as follows (Figs. 11, 12, 13, 14, 15, 16, 17 and 18 and Table 2):

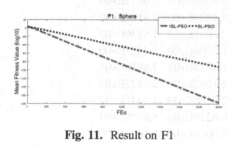

Fig. 11. Result on F1

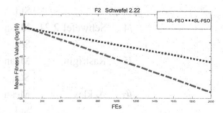

Fig. 12. Result on F2

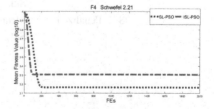

Fig. 13. Result on F3

Fig. 14. Result on F4

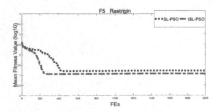

Fig. 15. Result on F5

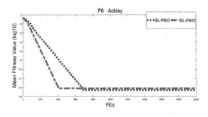

Fig. 16. Result on F6

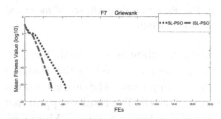

Fig. 17. Result on F7

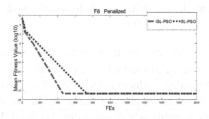

Fig. 18. Result on F8

Table 2. Mean values and standard deviation of the eight algorithm on benchmark functions

No.	Functions	Mean&Std.	ISL-PSO	SL-PSO
f1	Sphere	Mean	4.3621E−87	9.8459E−45
		Std.	1.8933E+52	3.6345E+27
f2	Schwefel 2.22	Mean	3.6819E−45	2.4750E−24
		Std.	9.9686E+26	1.7839E+14
f3	Rosenbrock	Mean	4.8050E+01	2.6672E+00
		Std.	2.4324E+00	2.8183E+00
f4	Schwefel 2.21	Mean	2.1035E+03	1.6136E+03
		Std.	1.2308E+00	1.4425E+00
f5	Rastrigin	Mean	1.4879E+01	2.4026E+01
		Std.	2.1597E+00	2.3772E+00
f6	Ackley	Mean	2.9180E−13	1.7441E−12
		Std.	8.6216E+03	5.8528E+04
f7	Griewank	Mean	0.0000E+00	0.0000E+00
		Std.	–	–
f8	Penalized	Mean	2.2121E−28	4.3002E−26
		Std.	1.3740E+09	3.8557E+10

5   Conclusion

This paper has examined the social-learning particle swarm optimization algorithm and proposed an improved social-learning particle swarm optimization algorithm. Experiments have been implemented to compare the initial SL-PSO algorithm with the ISL-PSO algorithm in eight functions. The result supports that ISL-PSO algorithm effectively improves the performance of SL-PSO algorithm, particularly in classical uni-modal and multimodal functions. Based on the experiment results, the modified SL-PSO algorithm proposed in this paper is expected to work out for an improved particle swarm optimization algorithm.

Acknowledgements. Ming Chen and Haoruo Hu contributed equally to this paper and shared the first authorship. This work is partially supported by the Natural Science Foundation of Guangdong Province (2016A030310074), Project supported by Innovation and Entrepreneurship Research Center of Guangdong University Student (2018A073825), Research Cultivation Project from Shenzhen Institute of Information Technology (ZY201717) and Innovating and Upgrading Institute Project from Department of Education of Guangdong Province (2017GWTSCX038).

References

1. Cheng, R.: A social learning particle swarm optimization algorithm for scalable optimization. Inf. Sci. **291**, 43–60 (2015)
2. Poli, R.: Particle swarm optimization. Swarm Intell. **1**(1), 33–57 (2007)
3. Dorigo, M.: Ant colony optimization: a new meta-heuristic. In: Proceedings of the 1999 Congress on Evolutionary Computation-CEC 1999 (Cat. No. 99TH8406), pp. 1470–1477, Washington, D.C., USA. IEEE (1999)
4. Passino, K.M.: Biomimicry of bacterial foraging for distributed optimization and control. IEEE Control Syst. Mag. **22**(3), 52–67 (2002)
5. Beheshti, Z.: A review of population-based meta-heuristic algorithms. Int. J. Adv. Soft Comput. Appl. **5**(1), 1–35 (2013)
6. Chen, F.: Tradeoff strategy between exploration and exploitation for PSO. In: 2011 Seventh International Conference on Natural Computation, pp. 1216–1222, Shanghai, China. IEEE (2011)
7. Wang, X.: Feature selection based on rough sets and particle swarm optimization. Pattern Recognit. Lett. **28**(4), 459–471 (2007)
8. Gaing, Z.L.: Particle swarm optimization to solving the economic dispatch considering the generator constraints. IEEE Trans. Power Syst. **18**(3), 1187–1195 (2003)
9. Ruiz-Cruz, R.: Particle swarm optimization for discrete-time inverse optimal control of a doubly fed induction generator. IEEE Trans. Cybern. **43**(6), 1698–1709 (2013)
10. Nagesh, R., Raga, S., Mishra, S.: Design of an energy-efficient routing protocol using adaptive PSO technique in wireless sensor networks. In: Sridhar, V., Padma, M.C., Rao, K. A.R. (eds.) Emerging Research in Electronics, Computer Science and Technology. LNEE, vol. 545, pp. 1039–1053. Springer, Singapore (2019). https://doi.org/10.1007/978-981-13-5802-9_90
11. Yang, Y.: A comparative study on feature selection in text categorization. In: ICML, Nashville, Tennessee, USA, no. 412–420, p. 35. Morgan Kaufmann (1997)

12. Bansal, J.C., Singh, P.K.: Inertia weight strategies in particle swarm optimization. In: 2011 Third World Congress on Nature and Biologically Inspired Computing, Salamanca, Spain, pp. 633–640. IEEE (2011)
13. Zhan, Z.H.: Adaptive particle swarm optimization. IEEE Trans. Syst. Man Cybern. Part B (Cybern.) 39(6), 1362–1381 (2009)
14. Robinson, J., Sinton, S., Rahmat-Samii, Y.: Particle swarm, genetic algorithm, and their hybrids: optimization of a profiled corrugated horn antenna. In: IEEE Antennas and Propagation Society International Symposium (IEEE Cat. No. 02CH37313), San Antonio, TX, USA, pp. 314–317. IEEE (2002)
15. Juang, C.F.: A hybrid of genetic algorithm and particle swarm optimization for recurrent network design. IEEE Trans. Syst. Man Cybern. Part B (Cybern.) 34(2), 997–1006 (2004)
16. Kao, Y.T.: A hybrid genetic algorithm and particle swarm optimization for multimodal functions. Appl. Soft Comput. 8(2), 849–857 (2008)
17. Shelokar, P.S.: Particle swarm and ant colony algorithms hybridized for improved continuous optimization. Appl. Math. Comput. 188(1), 129–142 (2007)
18. Abd-Elazim, S.M., Ali, E.S.: A hybrid particle swarm optimization and bacterial foraging for optimal power system stabilizers design. Int. J. Electr. Power Energy Syst. 46, 334–341 (2013)
19. Van den Bergh, F., Engelbrecht, A.P.: A cooperative approach to particle swarm optimization. IEEE Trans. Evol. Comput. 8(3), 225–239 (2004)
20. Liang, J.J.: Dynamic multi-swarm particle swarm optimizer. In: Proceedings 2005 IEEE Swarm Intelligence Symposium, SIS 2005, Pasadena, CA, USA, pp. 124–129. IEEE (2005)
21. Shi, Y.: Particle swarm optimization: developments, applications and resources. In: Proceedings of the 2001 Congress on Evolutionary Computation (IEEE Cat. No. 01TH8546), Seoul, South Korea, pp. 81–86. IEEE (2001)
22. Lefebvre, L.: Culturally-transmitted feeding behavior in primates: evidence for accelerating learning rates. Primates 36(2), 227–239 (1995)
23. Gumaida, B.F.: A hybrid particle swarm optimization with a variable neighborhood searchfor the localization enhancement in wireless sensor networks. Appl. Intell. 49, 1–19 (2019). https://doi.org/10.1007/s10489-019-01467-8
24. Zhang, X.: Differential mutation and novel social learning particle swarm optimization algorithm. Inf. Sci. 480, 109–129 (2019)
25. Zentall, T.R.: Imitation in animals: evidence, function, and mechanisms. Cybern. Syst. 32(1–2), 53–96 (2001)

An Integrated Classification Algorithm Using Forecasting Probability Strategies for Mobile App Statistics

Jingyuan Cao, Hong Wang$^{(\boxtimes)}$, and Mo Pang

College of Management, Shenzhen University, Shenzhen 518060, China
ms.hongwang@gmail.com

Abstract. Classification is an important data mining technique for classifying the items according to a series of associated features. Even so, most of them are not stable in performance, and they may get high classification accuracy rate in some datasets but poor in other issues. To solve this problem, in this paper, an integrated algorithm is proposed to keep balance between the classification accuracy rate and stability. The proposed algorithm integrates the K-Nearest Neighbor (KNN), Naive Bayes (NB), Regression Tree (RT), Random Forest (RF), Bagging, and Discriminant Analysis Classifier (DAC) using forecasting probability strategies. Specifically, the majority voting strategy and weighted voting strategy are presented using the forecasting probability obtained from the classification algorithms. To demonstrate the effectiveness of the proposed algorithm, numerous experiments are conducted by applying the classification algorithms to real mobile APP statistics. Results indicate that it can get a comprehensive and stable classification accuracy rate.

Keywords: Classification algorithm · Forecasting probability · Majority voting strategy · Weighted voting strategy

1 Introduction

With the rapid development of the Internet, the customer data is increasing gradually, and more perspectives of impression evaluation factors are needed to be taken into account. As a result, the dimensions of items become more complex, and it is a great challenge to draw effective conclusions from ordinary scales and manual observation. As a matter of fact, many companies build their own databases, and then use intelligent classification algorithms to find the marketing mix strategy with high customer satisfaction, in order to select the best program [1].

Scholars all over the world have done a lot of researches on various classification algorithms. As we all know, K-Nearest Neighbor (KNN) is a kind of parameterless learning, in other words, it does not "learn" anything when classifying. Instead, this algorithm classifies sample directly through the distance between feature vectors. It is a simple but effective classification algorithm and widely used in visual recognition [2]. However, its high computational complexity and outputting result without interpretable meaning waste a lot of time for people in waiting result and explaining the result [3].

© Springer Nature Switzerland AG 2019
D.-S. Huang et al. (Eds.): ICIC 2019, LNAI 11645, pp. 629–639, 2019.
https://doi.org/10.1007/978-3-030-26766-7_57

In risk prediction, Naive Bayes (NB) [4] and Random Forest (RF) [5] are frequently used, which are referenced in many studies. The basis of the NB classifier comes from the very famous Bayes' theorem, which can calculate the probability of future events from the known various conditions and probabilities. RF is the combination of random and forest. Forest combines a large number of different classifiers, and random reflects the flexibility of this classifier. They can analyze data with high latitude freely, without feature selection, but both of them have weakness. For NB, though study shows that it is effective to increase the accuracy if we release the attribute independence, it often needs the hypothesis of independent distribution in the model [6]. While in real life, no predictor could be completely independent. For RF, if the pretreatment is not rigorous enough, the result will have a distinct difference [7]. Regression Tree (RT) is applied in predicting the cost of project because the model is very lucid to build and can be reused without modification [8]. RT divides the data to the node at each time into two parts. If the data meets certain feature requirements, it is divided into left subtree; if not, it is divided into right subtree. The data are assigned to the appropriate degree, different models can be established according to the characteristics of different data groups for classification, which can greatly improve the classification accuracy. However, this classification algorithm is very sensitive to noise data [9]. What's more, Support Vector Machine (SVM) [10], Bagging [11], and Discriminant Analysis Classifier (DAC) [12] are all widely used in modeling. According to the limited sample information, SVM seeks the best compromise between the complexity of the model and the learning ability in order to obtain the best generalization ability. Bagging is a classification composed of multiple methods, its method part can be easily extracted and substituted into different methods for replacement. According to the previous papers, SVM can show us the most important feature, but it has a problem in dealing with multiple classes. Bagging has an advantage in dealing with unevenly distributed data, but the shortcoming that it often need help from other classification algorithm is tough. DAC is able to show the differences between the samples more clearly, but it is restricted by the type of sample.

Based on the brief analysis above, it can be concluded that each classifier has advantages and disadvantages. However, many of them can complement mutually and progress together. Therefore, the integrated classification algorithm is designed to improve the performance. Nugrahaeni and Mutijarsa presented an integrated classification algorithm of KNN, SVM and RF is applied in facial expression [13]. In the paper, they categorized human facial expression into 7 sorts like angry, happy and sad, and the result reached 80%. Constructed by Wang et al. [14], an integration of SVM, NB, RT and RF is used for detecting Android malicious apps and categorizing benign apps, which is proved to be effective. There is also an integration of SVM, NB, BLR, KNN and several classification algorithms in selective omission in road network [15]. According to this literature, the classification algorithms don't have much statistically significant difference, so they can be integrated smoothly. But the result can be concluded from 70% to 90%, which is not stable. They all proved that integrated classification algorithm is more effective than single classification algorithm since it can adjust more kinds of data. Even so, most of them consume high computational complexity or unstable accuracy rate.

To solve the previous problem appeared in the integrated algorithms mentioned above, this paper proposes an integrated classification algorithm using forecasting probability strategies with majority voting strategy and weighted voting strategy. It contains totally six classical classification algorithms, KNN, NB, RT, RF, Bagging, and DAC.

This paper is organized as follows. Section 2 gives a brief introduction of a classical integrated classification algorithm. Section 3 describes the proposed integrated classification algorithm. Application of Integrated Classification Algorithm to Mobile App Statistics will be conducted in Sect. 4. Finally, the corresponding conclusion will be given in Sect. 5.

2 Integrated Classification Algorithm

The integrated strategies employed in classification algorithms can be generally divided into two categories: majority voting [16] and weighted voting [17]. The majority voting rule is used to find a particular class in any given segmentation which has more votes than all the others. The weighted voting rule is to assign the weight coefficient to the classes based on their influence degree. The sample is divided into the class whose accuracy is highest. The detail process is showed in the following statement.

Assumed that classification algorithms are C_i, and i = [1, 2, ..., n], correspondingly their common output accuracy are Acc(i). Then classifications with better performance are selected and named as C_j, and j = [1, 2, ..., m], correspondingly their accuracy are Acc(j). As a result, the principles of two different strategies are illustrated as follows.

- **Classification based on majority voting strategy**

The majority voting strategy is based on the calculation theory of geometric average. Accuracy rates from classification algorithms whose performance are on the top rank are on average. The formula is in the following:

$$ACC1 = (\Sigma Acc(j))/m \quad j = 1, 2, ..., m \tag{1}$$

- **Classification based on weighted voting strategy**

The weighted voting strategy is based on the calculation theory of weighted average. The weight assigned to accuracy from classification algorithms indicates the performance of individual classification in the optimization process of weighted voting strategy is significant or not. The weight coefficient ω is larger in a more significant classification accuracy. Besides, the ω is decided by the ratio of the accuracy from every classification and the sum of all accuracy rates. The formulas are in the following:

$$ACC2 = \Sigma \omega j \cdot Acc(j) \quad j = 1, 2, ..., m \tag{2}$$

$$\omega = (Acc(j))/(\Sigma Acc(k)) \quad k = 1, 2, ..., m \tag{3}$$

From the formulas above, there will be corresponding final accuracy ACC1 and ACC2. ACC1 is the result output by majority voting strategy, and ACC2 is the result output by weighted voting strategy. Both results coming from majority voting strategy and weighted voting strategy are saved for comparison. Figure 1 shows the structure of integrated classification algorithm.

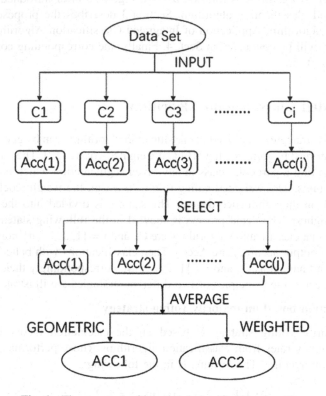

Fig. 1. The structure of integrated classification algorithm

3   The Proposed Integrated Classification Algorithm Using Forecasting Probability Strategies

As we all know, the classification algorithms normally can provide a series of categorical predictor indices for a sample to predict the class labels. Those categorical predictor indices mostly manifest as the forecasting probabilities. In this section, the forecasting probability based voting strategies used in these classification algorithms will be illustrated in details. The basic concept in the proposed integrated classification algorithm is to revaluate the forecasting probability by majority voting strategy and weighted voting strategy, and then direct the samples to segmentation with the largest probability. For example, when one sample can be segmented into 3 classes (Class 0-1-2), each class has one forecasting probability. Assumed that the probability in class 0, 1 and 2 were 0.2, 0.5 and 0.3 separately, so the sample will be classified into class 2.

Assumed that classification algorithms are C_i, and $i = [1,2,...,n]$, correspondingly their output accuracy are Acc(i). Then classifications with better performance are selected and named as C_j, and $j = [1,2,...,m]$. There high quality classifications can give the samples their correspondingly forecasting probability, if 's' is regarded as the quantity of testing samples, The forecasting probability of all samples output by the classifications can be named as P_{js}. If 'a' means the quantity of forecasting probability that the classifications need to output, every sample has their own probability in every class and it's named as P_{jsa}, After that, P_{jsa} in the same class are added together. The principles of two different strategies are illustrated as follows.

- **Classification using forecasting probability based on majority voting strategy**

The majority voting strategy using forecasting probability is based on the calculation theory of geometric average. The corresponding forecasting probability with the same location from classification algorithms are on average. The formula updated forecasting probability of every cell is shown as follows:

$$P_{jsX} = (\Sigma P_{jsa})/m \quad j = 1, 2, ..., m \tag{4}$$

- **Classification using forecasting probability based on weighted voting strategy**

The weighted voting strategy using forecasting probability is based on the calculation theory of weighted average. The weight assigned to each probability in a specific sample indicates the performance of individual class in the optimization process of weighted voting strategy is significant or not. The weight coefficient ω is larger in a more significant class. Besides, the ω is decided by the ratio of the accuracy from every classification and the sum of all accuracy rates. The formula updated forecasting probability of every cell is presented in the following:

$$P_{jsY} = \Sigma \omega_{jsa} \cdot P_{jsa} \quad j = 1, 2, ..., m \tag{5}$$

$$\omega = (Acc(j))/(\Sigma Acc(k)) \quad k = 1, 2, ..., m \tag{6}$$

According to the forecasting probability updated by voting strategies, each sample will be reevaluated. They will be allocated to the significant class with the highest forecasting probabilities. After comparing the updated class with the original segmented class in testing data, there will be corresponding final accuracy ACC3 and ACC4. ACC3 is the accuracy based on integrated classification algorithm using forecasting probability with majority voting strategy, and ACC4 is the accuracy based on integrated classification algorithm using forecasting probability with weighted voting strategy. Figure 2 illustrates the overall framework of the proposed integrated classification algorithm. In the framework, Class (1) is the set of class labels obtained after comparing the forecasting probability calculated by majority voting strategy of samples in each class. And Class (2) the set of class labels obtained after comparing the forecasting probability calculated by weighted voting strategy of samples in each class.

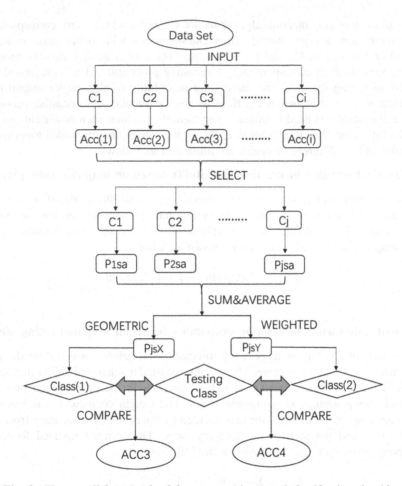

Fig. 2. The overall framework of the proposed integrated classification algorithm

4 Application of Integrated Classification Algorithm to Mobile App Statistics

4.1 Problem Description

With millions of APPs in application market today, one will in the top trending if it gets large download counts. High download counts make more people try to know the product, and they make a higher download counts, so it is an absolutely positive cycle that every company wants. As a result, how to catch customers' eyes is a hot topic for product development companies. They expect to research APP with high customer satisfaction by analyzing users' ratings of the software, and then find out the whole features of these APPs. In this paper, we use a set of software data sets collected in the Apple iOS App Store from the famous data platform Kaggle (https://www.kaggle.com/).

There are 7197 samples with 16 features, including APP's name, APP's size (in Bytes), the price of APP, etc. Table 1 shows the parameter definition of features and their explanation.

Table 1. The parameter definition of features and their explanation

Parameter definition of features	Explanations
id	App ID
track_name	App Name
size_bytes	Size (in Bytes) of APP
currency	Currency Type
price	Price Amount
rating_count_tot	User Rating Counts (for all version)
rating_count_ver	User Rating Counts (for current version)
user_rating	Average User Rating Value (for all version)
user_rating_ver	Average User Rating Value (for current version)
ver	Latest Version Code
cont_rating	Content Rating
prime_genre	Primary Genre
sup_devices.num	Number of Supporting Devices
ipadSc_urls.num	Number of Screenshots Showed for Display
lang.num	Number of Supported Languages
vpp_lic	Vpp Device Based Licensing Enabled

4.2 Data Pre-processing

Before the experiment, features are needed to be screened out the with digital significance. After analysis, it is proved that 10 of 16 features are significant and can be used in later experiment, they are size (in bytes), price, average user rating value (for all version), average user rating value (for current version), latest version code, primary genre, number of supporting devices, number of screenshots showed for display, number of supported languages and VPP device based licensing enabled. Specially, as those categories are named in words in primary genre, such as book and finance, we list them in ascending order of first letters (from A to Z).

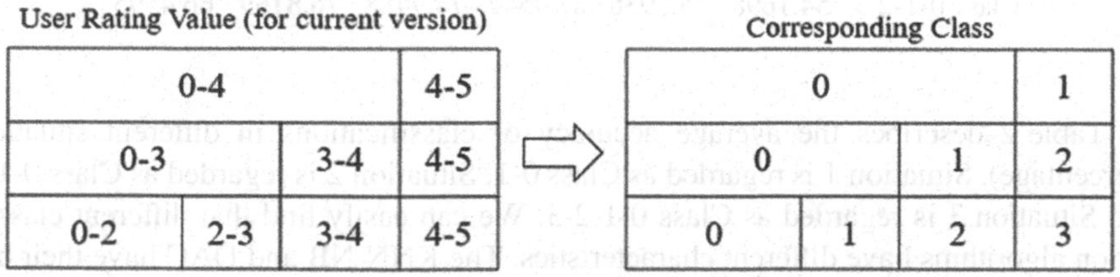

User Rating Value (for current version)				Corresponding Class			
0-4		4-5		0		1	
0-3	3-4	4-5		0	1	2	
0-2	2-3	3-4	4-5	0	1	2	3

Fig. 3. Different intervals defined with different numbers

Then average user rating value (for current version) is highly valued, because it is the most direct reflection of the customer's evaluation of the software. In this section, scores are from 0 to 5, which are existed as integer. When customers prefer the APP more, he/she will give the APP a higher score. As a matter of fact, we naturally regard that the higher score the APP is got, its comprehensive evaluation is better. Therefore, the experiment can be divided into the following three situations according to the grading segmentation, with the interval of 1 unit, and the critical value is reduced to a larger interval. Figure 3 shows different intervals defined with different numbers, which can be more convenient to input later.

- **Situation 1:** Divided into 2 parts, from 0 to 4, from 4 to 5, defined as class 0 and 1 separately.
- **Situation 2**: Divided into 3 parts, from 0 to 3, from 3 to 4, from 4 to 5, defined as class 0, 1 and 2 separately.
- **Situation 3**: Divided into 4 parts, from 0 to 2, from 2 to 3, from 3 to 4, from 4 to 5, defined as class 0, 1, 2, 3 separately.

4.3   Experiment Results and Analysis

In this section, the effectiveness of the proposed integrated classification algorithm will be shown by the classification accuracy rate. The average classification accuracy rates are achieved by classification algorithms based on MATLAB 2018a over 30 repeated times. KNN [2], NB [6], RT [8], RF [7], Bagging [11] and DAC [12] are adopted as classifications for constructing the integrated classification algorithm. On the basis of the results, four classifiers with high accuracy based on data sets will be selected. In the following analysis, two integrated algorithms are compared to demonstrate the effectiveness of the proposed algorithm. The experiment compare the classification algorithms by applying them to the mobile APP statistics with different segmental classes.

As we described above, there are mainly two steps in the experiment. Firstly, result from every single classification algorithm are needed to observe. Secondly, four classification algorithms who have the highest accuracy are selected to construct the integrated classification algorithm.

Table 2. The average accuracy of classification in different situations (percentage)

	KNN	NB	RT	RF	Bagging	DAC
Class 0-1	63.6038	54.1222	66.4732	71.3802	**71.4355**	68.3155
Class 0-1-2	54.8035	39.4522	68.2715	73.7537	**75.7609**	66.6523
Class 0-1-2-3	54.1698	35.3936	67.9542	74.9488	**76.8159**	66.4595

Table 2 describes the average accuracy of classifications in different situations (percentage). Situation 1 is regarded as Class 0-1, Situation 2 is regarded as Class 0-1-2, and Situation 3 is regarded as Class 0-1-2-3. We can easily find that different classification algorithms have different characteristics. The KNN NB and DAC have their best results in Class 0-1. In Class 0-1-2-3, the RF and Bagging achieve the highest classification accuracy rate. Only RT has its optimal performance in situation Class 0-1-2.

From another perspective, when comparing the accuracy rate of classification in the same class, RT, RF, Bagging and DAC are always in the top four, though their ranks may change in different situation. And Bagging always has best performance in all situations, which have been marked in bold type. Furthermore, from the whole table, Bagging in Class 0-1-2-3 has the best accuracy rate, which is 76.8159.

Considering all the situations, three algorithms RT, RF, Bagging and DAC rank the top four places. As a conclusion, these four classifiers are admitted to participate in the integrated classification.

Table 3. The parameter definition of final results from different strategies and their explanation

Parameter definition of final results	Explanation
ACC1	The accuracy based on integrated classification algorithm with majority voting strategy
ACC2	The accuracy based on integrated classification algorithm with weighted voting strategy
ACC3	The accuracy based on integrated classification algorithm using forecasting probability with majority voting strategy
ACC4	The accuracy based on integrated classification algorithm using forecasting probability with weighted voting strategy

Table 3 shows the parameter definition of final results from different strategies and their explanation. And Table 4 reveals the result of integrated classifications in different strategies (percentage). The result is output by RT, RF, Bagging and DAC because they perform better in Table 2. When comparing the results on the same class, ACC4 always perform the best, which have been marked in bold type. From the table, we can also find that the classification accuracy rate is not just influenced by the algorithm and the strategy the accuracy obtained. The number of classes to be classified has largely impacted the classification performance. Though different single classification algorithms have their best results in different segmental ways, both integrated classification algorithms perform best when ratings are segmented into 4 classes (Class 0-1-2-3). From the whole table, ACC4 in Class 0-1-2-3 has the best accuracy rate, which is 76.2631.

It is apparently that results that comes out by new classes updated by forecast probability are better than that comes out by original classes. The more categories there were, the greater the difference. Additionally, when comparing with two different strategies, accuracy related to weighted voting strategy is better than that related to majority voting strategy.

Table 4. The result of integrated classification in different strategies (percentage)

	ACC1	ACC2	ACC3	ACC4
Class0-1	69.4115	69.4763	70.7979	**70.8950**
Class0-1-2	71.1096	71.3084	75.3594	**75.4701**
Class0-1-2-3	71.5446	71.8176	76.1386	**76.2631**

According to the result, it can be concluded that the accuracy comes out by proposed integrated classification algorithm with weighted voting strategy is the best. Though the accuracy in ACC4 is slightly lower than that of Bagging, it is also can be accepted because it reveals a stable and overall consideration from several classical classification algorithms. What's more, the comprehensive method avoids the high computational complexity appeared in some of the classification algorithms. In short, the proposed method can not only meet the best accuracy of single classification, but also show the comprehensive decision.

5   Conclusion and Future Work

The integrated classification algorithm using forecasting probability has been proposed in this paper. The experiment results show that this integrated classification algorithm is effective and efficient. Both results of majority voting strategy and weighted voting strategy coming from forecasting probability are better than that of original integrated classification algorithm. According to the final result, ACC4, which comes from forecasting probability strategy with weighted voting strategy, is 70.8950 in Class 0-1, 75.4701 in Class 0-1-2, and 76.2631 in Class 0-1-2-3 separately. They are all the best in their segmentation. When comparing ACC4 with the result of Bagging, it is apparently that their difference value is less than 1%, which is within acceptable limit. At the same time, we can get a stable and more comprehensive answer. In the conclusion, the proposed integrated classification algorithm using forecasting probability with majority voting strategy and weighted voting strategy take advantage of classifications included to achieve stable accuracy and lower computational complexity.

Even though the improved method can exceed the original method, there is still room for the improvement. In our future study, we want to try more combination of classification algorithms so that it can adjust more kinds of data sets.

Acknowledgement. This work is partially supported by The Natural Science Foundation of Guangdong Province (2018A030310575), and Research Foundation of Shenzhen University (85303/00000155).

References

1. Liu, J.W.: Using big data database to construct new Gfuzzy text mining and decision algorithm for targeting and classifying customers. Comput. Ind. Eng. **128**, 1088–1095 (2018)
2. Liu, Q., Liu, C.: A novel locally linear KNN method with applications to visual recognition. IEEE Trans. Neural Netw. Learn. Syst. **28**(9), 2010–2021 (2016)

3. Sahu, S.K., Kumar, P., Singh, A.P.: Modified K-NN algorithm for classification problems with improved accuracy. Int. J. Inf. Technol. **10**(1), 65–70 (2017)
4. Wolfson, J., Bandyopadhyayy, S., et al.: A naive Bayes machine learning approach to risk prediction using censored, time-to-event data. Stat. Med. **34**(21), 2941–2957 (2015)
5. Hu, C., Steingrimsson, J.A.: Personalized risk prediction in clinical oncology research: applications and practical issues using survival trees and random forests. J. Biopharm. Stat. 1–17 (2017)
6. Arar, Ö.F., Ayan, K.: A feature dependent naive Bayes approach and its application to the software defect prediction problem. Appl. Soft Comput. **59**, 197–209 (2017)
7. Cano, G., Garcia-Rodriguez, J., Garcia-Garcia, A., Perez-Sanchez, H., Benediktsson, J.A., Thapa, A., et al.: Automatic selection of molecular descriptors using random forest: application to drug discovery. Expert Syst. Appl. **72**, 151–159 (2017)
8. Yoonseok, S.: Application of boosting regression trees to preliminary cost estimation in building construction projects. Comput. Intell. Neurosci. 1–9 (2015)
9. Yang, R.M., Zhang, G.L., et al.: Comparison of boosted regression tree and random forest models for mapping topsoil organic carbon concentration in an alpine ecosystem. Ecol. Indic. **60**, 870–878 (2016)
10. Prabhakar Karthikeyan, S., Jacob Raglend, I., Sathish Kumar, K., Kumar Sahoo, S., Priya Esther, B.: Application of SVM as classifier in estimating market power under deregulated electricity market. In: Kamalakannan, C., Suresh, L.P., Dash, S.S., Panigrahi, B.K. (eds.) Power Electronics and Renewable Energy Systems. LNEE, vol. 326, pp. 1309–1317. Springer, New Delhi (2015). https://doi.org/10.1007/978-81-322-2119-7_127
11. Zareapoor, M., Shamsolmoali, P.: Application of credit card fraud detection: based on bagging ensemble classifier. Procedia Comput. Sci. **48**, 679–685 (2015)
12. Gyamfi, K.S., Brusey, J., Hunt, A., Gaura, E.: Linear classifier design under heteroscedasticity in linear discriminant analysis. Expert Syst. Appl. **79**, 44–52 (2017)
13. Nugrahaeni, R.A., Mutijarsa, K.: Comparative analysis of machine learning KNN, SVM, and random forests algorithm for facial expression classification. In: Technology of Information & Communication, pp. 163–168 (2017)
14. Wang, W., Li, Y., Wang, X., Liu, J., Zhang, X.: Detecting android malicious apps and categorizing benign apps with ensemble of classifiers. Futur. Gener. Comput. Syst. **78**, 987–994 (2017)
15. Zhou, Q.: A comparative study of various supervised learning approaches to selective omission in a road network. Cartogr. J. **54**, 1–11 (2016)
16. Gopinath, B., Gupt, D.B.R.: Majority voting based classification of thyroid carcinoma. Procedia Comput. Sci. **2**(2), 265–271 (2010)
17. Zhu, X., Song, Q., Jia, Z.: A weighted voting-based associative classification algorithm. Comput. J. **53**(6), 786–801 (2010)

An Improved PSO Algorithm
with an Area-Oriented Learning Strategy

Tengjun Liu[1], Jinzhuo Chen[1(✉)], Yuelin Rong[2], Yuting Zheng[3],
and Lijing Tan[4]

[1] College of Electronic and Information Engineering, Shenzhen University,
Shenzhen 518000, China
2017133107@email.szu.edu.cn
[2] College of Mathematics and Statistics, Shenzhen University,
Shenzhen 518000, China
[3] College of Fashion Design and Engineering, Shenzhen University,
Shenzhen 518000, China
[4] College of Management, Shenzhen Institute of Information Technology,
Shenzhen 518000, China

Abstract. The classical particle swarm optimization (PSO) trains the particles
to move toward the global best particle in every iteration. So, it has a great
possibility of being trapped into local optima. To deal with this issue, this paper
improves the learning strategy of PSO. Therefore, an area-oriented particle
swarm optimization (AOPSO) is proposed, which contributes to leading the
particles to move toward an area surrounded by some suboptimal particles
besides the best one. 10 test functions are employed to compare the performance
of AOPSO with the classical PSO and 3 other improved PSOs. AOPSO performs the best in 5 test functions and relatively better than some of the other
algorithms in the rest, which sufficiently demonstrates the effectiveness of
AOPSO.

Keywords: PSO · AOPSO · Suboptimal particles · Learning strategy

1 Introduction

The particle swarm optimization (PSO) is one of the stochastic global search algorithms, imitating social behavior like fish schooling and bird blocking by which particles move to search for food with their joint effort. Proposed by Eberhart and
Kennedy in 1995 [1], PSO has been applied in abundant fields to solve problems
related with feature selection and neural network, etc., due to its fast convergence and
implementation ability.

Despite of these advantages above, it's difficult for the classical PSO to avoid being
trapped in local optima completely. As a consequence, variant methods have been
raised to improve the PSO algorithm: changing the inertia weight, like DNCPSO [2];
combining the classical PSO with other searching algorithms, such as HGAPSO [3],
FPSO+FGA [4], PSACO [5] and I-PR-PSO [6]; introducing mathematical theories
such as MOPSO [7] and DTT-PSO [8]; and developing the learning strategy.

© Springer Nature Switzerland AG 2019
D.-S. Huang et al. (Eds.): ICIC 2019, LNAI 11645, pp. 640–650, 2019.
https://doi.org/10.1007/978-3-030-26766-7_58

Variants of PSO focusing on improving the learning strategy attempt to enhance the searching ability and alleviate premature convergence. For instance, J. J. Liang and A. K. Qin illustrated a cooperative approach to particle swarm optimization (CPSO) aiming to achieve better solutions as the dimension increases [9]. Different from the classical PSO, CPSO takes advantages of Genetic Algorithm, by partitioning particles into sub-swarms before searching for the optimization, especially when encountering high-dimensional problems. Combined with the functional-link-based neural fuzzy network (FLNFN), it has shown its good performance in predicting time series problems [10]. In [11], the social learning PSO (SLPSO) was developed. Instead of learning from the p_{bestk} and g_{best}, which represent the best previous position of the kth particle and the best previous position of all particles, each particle is influenced by the demonstrators in the current swarm. By comparison, SLPSO manifests its superiority over low dimensional circumstances. Based on SLPSO, X. M. Zhang et al. introduced a differential mutation and novel social learning PSO algorithm (DSPSO) to solve problems that are more complicated [12]; Apart from CPSO and SLPSO, there introduced a comprehensive optimization algorithm, namely ensemble particle swarm optimizer (ESPO) [13]. It is embedded to deal with heterogeneous problems by making full use of diverse variants of PSO such as self-organizing hierarchical PSO (HPSO-TVAC), comprehensive learning PSO (CLPSO) and distance-based locally informed PSO (LIPS) and then finds out the superior method. However, the processing time of EPSO tends to be longer than a single algorithm.

From what have been discussed above, we may find that though these improved algorithms did creative explorations on improving learning strategies, while they seem to ignore the potential progress if the particles do not just learn from the p_{best} and the g_{best}. Besides they can also learn from a well performed area. Considering the situation that the global optima is always surrounded by several local optima, particles could be confused by the local optima easily so that it might be difficult for them to find the best solution. Our work is leading the particles to move toward a specific area in order to reduce the probability of being trapped in local optima.

The rest of the paper is organized as follows. Section 2 introduces the classical PSO. In Sect. 3, we state the motivation and details of the new area-oriented PSO (AOPSO) algorithm. Section 4 presents the testing results of AOPSO on benchmark functions and, as a consequence, we end this paper with an important conclusion in Sect. 5.

2 Classical Particle Swarm Optimization

Utilizing the theory of swarm intelligence, PSO can imitate the biological group behavior to find the optimal solution. And each individual in PSO has two attributes: x_k and v_k, which represent the current position and velocity of the kth particle respectively. In order to find the best position among all particles, these two attributes are updated by the following equations:

$$v_k^d(t+1) = \omega v_k^d(t) + c_1 r_1 (p_{bestk}^d(t) - x_k^d(t)) + c_2 r_2 (g_{bestj}^d(t) - x_k^d(t)) \qquad (1)$$

$$x_k^d(t+1) = x_k^d(t) + v_k^d(t+1) \qquad (2)$$

where c_1 and c_2 are two acceleration constants controlling how far a particle will move toward its direction, r_1 and r_2 are two uniformly random distribution numbers within the range of $(0,1)$, and ω is the inertia weight. The variables p_{bestk} and g_{bestj}, where $k = 1, 2, 3, \ldots m, j = 1, 2, 3 \ldots, n$, represent the best personal position of the kth particle and the best global position of the jth iteration.

A large inertia weight helps PSO explore the best global position and a small one tends to explore the best local position. And each particle can learn from its $v_k(t+1)$, p_{bestk} and g_{bestj} to converge to the optimal solution.

3 The Proposed Algorithm

3.1 Motivation

In the classical PSO, the particles just focus on the global best particle besides themselves. This learning strategy sometimes will cause them trapped in the local optima. In this paper, we propose another feasible learning strategy to train the particles to find the global optimum.

In some cases, especially the multimodal ones, the suboptimal particles sometimes locate in a relatively far area in every iteration. And the global optimum is always situated in an area where these suboptimal particles lie. Considering these facts, we suppose that the global optimum lies in the area enveloped by three positions of the particles with the currently top three best function values, and near the currently global optimum in every iteration. Therefore, we modify the learning strategy so that the particles move toward the area mentioned above, which can help the particles get away from the local optima to some extent. This newly proposed method is named area-oriented PSO, i.e., AOPSO.

3.2 An Area-Oriented PSO (AOPSO)

With our newly proposed learning strategy, explicitly, the particles move toward the center of the area as defined above with a bias to the current optimum in every iteration ((5) and (6)), and the inertia weight ω changes every iteration according to (4). Apart from these changes, the rest of AOPSO are identical to the classical PSO. The detailed processes are listed as follows:

(1) Initialization. Firstly, we initialize all the parameters, including the number of the particles N, the dimension of every particle d, the number of iterations T, the individual learning factor c_1, the socially area and partner learning factors c_2 and c_3, the boundaries of positions, velocities and the inertia weight $x_{max}, x_{min}, v_{max}, v_{min}, \omega_{max}, \omega_{min}$, the positions of all the particles x, the velocities of all the particles v, the best position of every particle p_{ibest}. And then, we calculate the initial function values of every particle g_{ibest}, so we can get the globally top three best function values g_{gbest}.

(2) Calculate the function values. After the initialization, we calculate the function value of every particle in a new iteration.

(3) Update individual and partner optimal position. Afterward, we update g_{gbest} and g_{ibest}, and calculate the new individual optimal position p_{ibest}, the new socially partner optimal position p_{gbest}, i.e., the position with a best function value among all the particles.

(4) Update socially area optimal position. Using the 3 particles' positions with the 3 best function values, we calculate the p_{area} representing the target area by taking the average position of them as formula (3).

$$p_{area} = E(p_{ibestk}) \quad (k = 1, 2, 3) \tag{3}$$

In formula (3), p_{ibest1} represents the position with the best function value among all the particles, and p_{ibest2} with the second one and p_{ibest3} with the third one.

(5) Update ω, x and v. In line with (4), we decrease ω linearly in every iteration. i is the times of iterations.

$$\Omega = \omega_{min} + \frac{T - i}{T} (\omega_{max} - \omega_{min}) \tag{4}$$

After that, as (5), (6) and (7), we update v and x according to the p_{area}, p_{gbest} and p_{ibest} we get from the above steps, and i is the times of iterations while d is the dimension of every particle.

$$v_{i+1}^d = \omega v_i^d + c_1 r_i^d \Delta p_{ibest}^d + c_2 r_i^d \Delta p_{area}^d + c_3 r_i^d \Delta p_{gbest}^d \tag{5}$$

$$\begin{cases} \Delta p_{ibest}^d = p_{ibesti}^d - x_i^d \\ \Delta p_{area}^d = p_{areai}^d - x_i^d \\ \Delta p_{gbest}^d = p_{gbesti}^d - x_i^d \end{cases} \tag{6}$$

$$x_{i+1}^d = x_i^d + v_{i+1}^d \tag{7}$$

(6) Constrain processing. We constrain x and v within two specific ranges, i.e., x_{min}, x_{max} and v_{min}, v_{max}. Consequently, if x or v exceed these ranges, we will set them back in the ranges randomly.

(7) Iteration. We repeat steps (2) to (6) until meeting the number of iterations.

3.3   Flowchart

The flowchart of the proposed algorithm is shown as follows (Fig. 1):

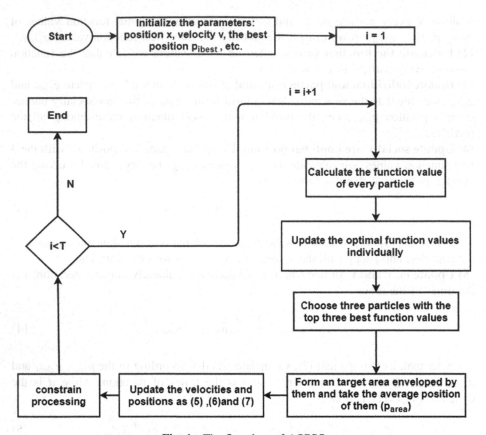

Fig. 1. The flowchart of AOPSO

4 Experiment and Evaluation

4.1 Test Functions

We evaluate the performance of the newly proposed PSO model, i.e., the AOPSO, by ten test benchmark functions, which are shown in Table 1.

All the test functions have the minima 0. Among them, f_1, f_3, f_4, f_7, f_8, f_{10} are unimodal optimization problems while the rest (f_2, f_5, f_6, f_9) are multimodal. And the Noisy Quadric function (f_7) is the sole function with noise. Meanwhile, the Noisy Quadric function and the Step function (f_7, f_8) are two discontinuous optimization problems. Moreover, Rotated hyper-ellipsoid function (f_{10}) is the only rotated function.

Table 1. Test functions

	Test function	f_{min}
Sphere function	$f_1(x) = \sum_{i=1}^{d} x_i^2$	0
Generalized Rastrigrin function	$f_2(x) = \sum_{i=1}^{d} (x_i^2 - 10\cos(2\pi x_i) + 10)$	0
Schwefel function	$f_3(x) = \sum_{i=1}^{d} |x_i| + \prod_{i=1}^{d} |x_i|$	0
Rosenbrock function	$f_4(x) = \sum_{i=1}^{d-1} \left(100\left(x_i^2 - x_{i+1}\right)^2 + (x_i - 1)^2\right)$	0
Ackley function	$f_5(x) = -20\exp\left(-0.2\sqrt{\frac{1}{d}\sum_{i=1}^{d} x_i^2}\right) - \exp(\frac{1}{d}\sum_{i=1}^{d}\cos(2\pi x_i)) + 20 + e$	0
Griewank function	$f_6(x) = \sum_{i=1}^{d} \frac{x_i^2}{4000} - \prod_{i=1}^{d} \cos(\frac{x_i}{\sqrt{i}}) + 1$	0
Noisy Quadraic function	$f_7(x) = \sum_{i=1}^{d} ix_i^4 + rand$	0
Step function	$f_8(x) = \sum_{i=1}^{d} (\lfloor x_i + 0.5 \rfloor)^2$	0
Noncontinuous Rastrigin function	$f_9(x) = \sum_{i=1}^{d} (y_i^2 - 10\cos(2\pi y_i) + 10)$ $y_i = \begin{cases} x_i, & |x_i| < 0.5 \\ \frac{round(2x_i)}{2}, & |x_i| \geq 0.5 \end{cases}$	0
Rotated hyper-ellipsoid function	$f_{10}(x) = \sum_{i=1}^{d}\sum_{j=1}^{i} x_j^2$	0

4.2 Parameters Setting up

In order to verify the performance of AOPSO, the classical PSO and 3 improved PSO (CPSO [9], SLPSO [11], EPSO [13]) were chosen to make a comparison. To analyze the performance among all the algorithms fairly, we set the sharing parameters as the same, including $N = 100$, $d = 30$, $T = 2000$, $x_{max} = 20$, $x_{min} = -20$, $v_{max} = 10$, $v_{min} = -10$. For the classical PSO, c_1 and c_2 were both set as 1.5. For the AOPSO, we set the extra parameter as $c_3 = 0.45$ and $c_2 = 1.5$, while $c_1 = 1.5$. For CPSO, EPSO, SLPSO, the same parameters used in [9, 11, 13] were adopted here.

4.3 Results and Discussion

The experimental results, in terms of the best, worst, mean and standard deviation of each algorithm on each test function with 30-D, are listed in Table 2 and the best among all algorithms on each test function are shown in bold in the table. Meanwhile, for a more intuitive comparison between AOPSO and other PSO variants, the performance of each algorithm on each test function has been shown in the following figures (Fig. 2).

Table 2. Results for test functions.

		SLPSO	PSO	EPSO	CPSO	AOPSO
f_1	Best	4.28E−96	0.078133	5.19E−36	1.35E-311	**0**
	Worst	4.79E−93	6.308784	4.22E−27	4.85E-300	**0**
	Mean	6.84E−94	2.026681	2.51E−28	1.64E-301	**0**
	Std.	1.02E−93	1.62015	8.79E−28	**0**	**0**
f_2	Best	2.84E−48	121.8832	13.92942	**0**	**0**
	Worst	5.32E−47	280.4618	53.72828	**0**	176.58498
	Mean	1.88E−47	193.5951	29.42433	**0**	45.970315
	Std.	1.19E−47	40.09614	9.503923	**0**	46.812668
f_3	Best	1.53E−09	5.844938	5.80E−21	9.30E−157	**4.69E−235**
	Worst	2.00E−07	34.3479	1.57E−16	3.35E−152	**3.71E−217**
	Mean	2.53E−08	14.05269	8.16E−18	3.57E−153	**1.25E−218**
	Std.	4.25E−08	7.615038	2.90E−17	6.98E−153	**0**
f_4	Best	1.55E−26	161.436	3.840092	1.35E-311	**0**
	Worst	**2.59E−25**	1809.511	77.43807	0.521754	28.678638
	Mean	**5.81E−26**	669.0739	23.81214	0.109376	20.963489
	Std.	**5.37E−26**	428.9003	20.70364	0.151841	12.857775
f_5	Best	17.81785	3.486975	7.99E−15	2.22E−14	**8.88E−16**
	Worst	78.70065	7.109438	2.22E−14	5.06E-14	**4.44E−15**
	Mean	22.99561	4.795511	1.45E−14	3.69E-14	**3.02E−15**
	Std.	14.24704	0.861474	3.74E−15	6.24E-15	**1.77E−15**
f_6	Best	**0**	0.19449	**0**	**0**	**0**
	Worst	**0**	0.29833	0.029481	0.360652	0.0877079
	Mean	**0**	0.116442	0.008535	0.120117	0.0134309
	Std.	**0**	0.008186	0.008186	0.092114	0.0255846
f_7	Best	0.00466	2.880936	0.001587	0.000187	**3.26E−06**
	Worst	0.018437	168.0453	0.00681	**0.000877**	0.0232354
	Mean	0.010284	38.98480	0.003894	**0.000393**	0.0014449
	Std.	0.002832	41.09498	0.001186	**0.000162**	0.0043509
f_8	Best	11986.72	8	**0**	**0**	**0**
	Worst	12033.16	53	**0**	**0**	**0**
	Mean	12003.75	27.73333	**0**	**0**	**0**
	Std.	11.02143	12.08286	**0**	**0**	**0**
f_9	Best	7.959672	124.0549	8	**0**	**0**
	Worst	23.879	309.25	28	**0**	114.35156
	Mean	13.73043	211.7355	16.5679	**0**	37.48131
	Std.	3.83481	53.60137	4.995113	**0**	40.069354
f_{10}	Best	2.66E−15	10.18431	2.63E−10	2.37E-44	**4.07E−321**
	Worst	6.22E−15	78.18703	6.89E−08	200	**5.04E−84**
	Mean	5.74E−15	38.95989	1.06E−08	6.666667	**1.68E−85**
	Std.	1.23E−15	17.60305	1.60E−08	36.51484	**9.20E−85**

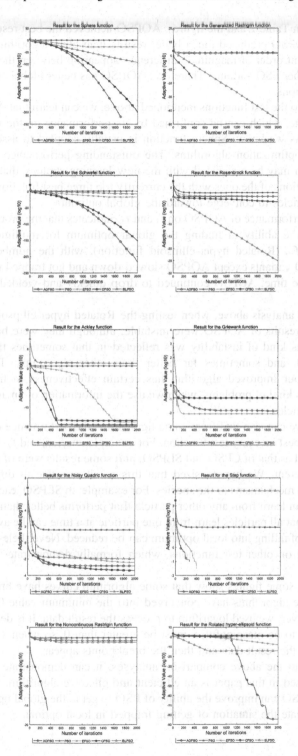

Fig. 2. Convergence characteristics of test functions and 30 independent trials of the results were run.

As shown in Table 2 and the figures, AOPSO achieved the best results on f_1, f_3, f_5, f_8, f_{10}. And AOPSO produced much better results than other algorithms on f_1, f_3, f_{10}, with a significant order of magnitude difference appearing between the best results of AOPSO and other PSO variants. Therefore, AOPSO has better global search ability on these test functions.

According to the test functions mentioned above, we can learn that AOPSO has fast convergence rate, which can be confirmed by its performance on the unimodal functions, like f_1. As we know, Sphere function is usually adopted to assess the convergence rate of optimization algorithms. The outstanding performance of AOPSO on Sphere function may be associated with the new learning strategy that particles learn from three positions of the ones with the currently top three best function values, which leads other particles to converge toward the global optimum.

Also, the performance of AOPSO on f_3 and f_5, indicates that the proposed algorithm has enhanced the ability of finding the global optimum for multimodal functions. Especially on f_{10} (Rotated hyper-ellipsoid function), with the number of iterations increasing, PSO variants except AOPSO slowed down and got trapped in local optima, but, at the same time, AOPSO continued to drop quickly and yielded a much better solution.

Besides the analysis above, when testing the Rotated hyper-ellipsoid function, we found that the results of AOPSO were unstable, although they were better than other algorithms. This kind of instability was reflected in that sometimes the results were relatively good, and sometimes far better than other algorithms. This potentially indicates that our improved algorithm has certain effectiveness in finding the best location for this kind of problems, when we use the information of an area to guide the learning of particles.

However, we can also find from some figures that the performance of AOPSO was not always the best on every test function. For example, on f_1, f_4 and f_9, its performance was not as good as that of CPSO and SLPSO, and some results were of a large order of magnitude different. We hypothesized that this result was due to differences in the communication mechanisms of the species. For example, in SLPSO, each particle, after being sorted, can learn from any other particle that performs better than it. In this way, the limitation that all particles learn from one particle at a time can be avoided and then the possibility of falling into local optimum can be reduced. Nevertheless, AOPSO still performed well on other test functions, which formally demonstrates the validity of AOPSO.

In addition, some figures show that some algorithm curves have breakpoints. This is because these algorithms have converged into the minimum value 0 at the time of iteration. However, we used logarithm to process the result data. It is deserved to know that if $log_m x$ is to be meaningful, x must be greater than 0, so when $x = 0$, logarithm cannot process the result data and then the breakpoints appear.

According to the above comparison analyses, it can demonstrate that our novel AOPSO proposed in this paper is an efficient and effective algorithm. And the results show that AOPSO can improve the ability of PSO to get to the global optimum and can relatively alleviate the situation of getting trapped in local optima.

5 Conclusion

In this paper, besides the linear function to change the inherent weight ω, a new learning strategy was introduced to improve the performance of PSO. In the classical PSO, all the particles move toward the best position. However, in view of the somewhat misleading learning strategy, we proposed the AOPSO to lead each particle to move toward an area defined by three points with the top three best function values.

In order to test the effectiveness of our algorithm, we compared the proposed algorithm with four popular algorithms on 10 test functions. According to the comparison results, the novel AOPSO was the best in 5 test functions, especially in Sphere function and Rotated hyper-ellipsoid function. However, the performance of AOPSO in Griewank function, Rosenbrock function and Generalized Rastrigin function was inferior to some of the other algorithms, such as EPSO, SLPSO and CPSO respectively.

By contrast, we can conclude that, this strategy can help particles search more widely in space and find a potential solution, which can enhance the diversity of population and improve the global searching ability.

However, only ten test functions have been used in this paper and we will use more test functions to test the effectiveness of AOPSO in the future.

Acknowledgments. This work is partially supported by the Natural Science Foundation of Guangdong Province (2016A030310074), Project supported by Innovation and Entrepreneurship Research Center of Guangdong University Student (2018A073825), Research Cultivation Project from Shenzhen Institute of Information Technology (ZY201717) and Innovating and Upgrading Institute Project from Department of Education of Guangdong Province (2017GWTSCX038). And the authors appreciate everyone who provided us with constructive suggestions and discussions, especially Professor Ben Niu and Ms. Churong Zhang.

References

1. Kennedy, J., Eberhart, R.C.: Particle swarm optimization. In: Proceedings of IEEE International Conference on Neural Networks, pp. 1942–1948. IEEE, Perth (1995)
2. Xie, Y., Zhu, Y., Wang, Y., et al.: A novel directional and non-local-convergent particle swarm optimization based workflow scheduling in cloud–edge environment. Future Gener. Comput. Syst. **97**, 361–378 (2019)
3. Li, D., Li, K., Liang, J., et al.: A hybrid particle swarm optimization algorithm for load balancing of MDS on heterogeneous computing systems. Neurocomputing **330**, 380–393 (2019)
4. Valdez, F., Melin, P., Castillo, O.: Modular neural networks architecture optimization with a new nature inspired method using a fuzzy combination of particle swarm optimization and genetic algorithms. Inf. Sci. **270**, 143–153 (2014)
5. Shelokar, P., Siarry, P., Jayaraman, V.K., Kulkarni, B.D.: Particle swarm and ant colony algorithms hybridized for improved continuous optimization. Appl. Math. Comput. **188**, 129–142 (2007)
6. Di Cesare, N., Domaszewski, M.: A new hybrid topology optimization method based on I-PR-PSO and ESO. Application to continuum structural mechanics. Comput. Struct. **212**, 311–326 (2019)

7. Campos Jr., A., Pozo, A.T.R., Duarte Jr., E.P.: Parallel multi-swarm PSO strategies for solving many objective optimization problems. J. Parallel Distrib. Comput. **126**, 13–33 (2019)
8. Wang, L., Yang, B., Orchard, J.: Particle swarm optimization using dynamic tournament topology. Appl. Soft Comput. **48**, 584–596 (2016)
9. Van den Bergh, F., Engelbrecht, A.P.: A cooperative approach to particle swarm optimization. IEEE Trans. Evol. Comput. **8**(3), 225–239 (2004)
10. Lin, C.J., Chen, C.H., Lin, C.T.: A hybrid of cooperative particle swarm optimization and cultural algorithm for neural fuzzy networks and its prediction applications. IEEE Trans. Syst. Man Cybern. Part C (Appl. Rev.) **39**, 55–68 (2009)
11. Cheng, R., Jin, Y.: A social learning particle swarm optimization algorithm for scalable optimization. Inf. Sci. **291**, 43–60 (2015)
12. Zhang, X.M., Wang, X., Kang, Q., Cheng, J.F.: Differential mutation and novel social learning particle swarm optimization algorithm. Inf. Sci. **480**, 109–129 (2019)
13. Lynn, N., Suganthan, P.N.: Ensemble particle swarm optimizer. Appl. Soft Comput. **55**, 533–548 (2017)

An Improved PSO Algorithm with Migration Behavior and Asynchronous Varying Acceleration Coefficient

Shukun Jiang[1], Jingzhou Jiang[2(✉)], Canhua Zheng[3], Yunyu Liang[4], and Lijing Tan[5]

[1] College of Computer and Software, Shenzhen University,
Shenzhen 518000, China
[2] College of Mathematics and Statistics, Shenzhen University,
Shenzhen 518000, China
2018193031@email.szu.edu.com
[3] College of Management, Shenzhen University, Shenzhen 518000, China
[4] College of Economics, Shenzhen University, Shenzhen 518000, China
[5] College of Management, Shenzhen Institute of Information Technology,
Shenzhen 518172, China

Abstract. Particle swarm optimization (PSO) has been developing at a fast pace and has a wide range of applications since it is easy to understand and implement. However, it is greatly limited by the problem of being trapped in local optimum. Inspired by the migration behavior of animals, MBPSO is proposed. In MBPSO, the whole evolutionary cycle is divided into several increasing sub-evolutionary cycles. As the particles complete each sub-evolutionary cycle, they move around from their original position. Moreover, in order to ensure convergence and improve con-vergence speed, acceleration coefficient and inertia weight change asynchronously with time. Standard PSO, SAPSO and SecPSO are selected for comparison with MBPSO. Then the performance of the four algorithms in six functions is tested. It is ultimately proved that the proposed MBPSO algorithm is more effective than the other three PSO algorithms.

Keywords: Particle Swarm Optimization · Migration Mechanism · Asynchronously varying · Acceleration coefficient

1 Introduction

In 1995, Kennedy and Eberhart proposed particle swarm optimization (PSO) [1] a stochastic population based optimization algorithm. It was inspired by observing the natural world, especially the social behavior of flocking birds and fish schools.

Compared with other modern optimization methods, particle swarm optimization algorithm has some obvious advantages: it is easier to understand, requires less parameters to be adjusted, is simple and easy to operate, and has a fast convergence speed. Because of these advantages, PSO has been used in a variety of application areas: function optimization [2, 3], artificial neural network training [4, 5], fuzzy system control [6] and so on.

© Springer Nature Switzerland AG 2019
D.-S. Huang et al. (Eds.): ICIC 2019, LNAI 11645, pp. 651–659, 2019.
https://doi.org/10.1007/978-3-030-26766-7_59

Although PSO performs well when dealing with unimodal functions, it is prone to fall into local optimum and diversity declines prematurely when dealing with complex multimodal functions. In order to deal with the shortcomings of standard PSO algorithm, many improved particle swarm optimization algorithms have been proposed by researchers. Gang proposed LPSO-TVAC for the migration of particles between groups [7]. PSO with edge offset, which is combined with random dynamic topology, is proposed by Mohais [8]. El-Abd put forward a collaborative particle swarm optimization algorithm with heterogeneous probability model migration [9]. Many scholars have also improved the parameters of the formula. Cui [10] and Hu [11] give the PSO algorithm a linear change of inertia weight. Cui used a time-varying acceleration factor to improve the PSO algorithm [12]. Mao introduces an algorithm for adjusting asymmetric learning factors [13]. These methods can cope with the shortcomings of PSO algorithm, effectively improve the population diversity, global exploration ability and local search ability.

Illuminated by the behavior of animals in nature, a particle swarm optimization algorithm with migration behavior is proposed in this paper. When a particle completes one cycle of evolution, it moves randomly around, and then on to the next cycle. This way, the diversity of population is improved and the local optimum is avoided. Additionally, in order to avoid the population moving out of the global optimum, the optimal value of each evolutionary cycle is retained and compared. The particle swarm will always learn the optimal location, and the evolutionary cycle increases with the number of evolutions. Moreover, the use of inertia weights and acceleration coefficient varying with time facilitates the convergence of the algorithm.

Next, the paper will be divided into several sections. The standard PSO is introduced in Sect. 2. In Sect. 3, the revised PSO is explained in detail. Section 4 shows the test functions and the discussion of the experimental results. Conclusions are given in Sect. 5.

2 Standard Particle Swarm Optimization

Through the information transmission among individuals in a group, an optimization method based on the group often finds the best solutions with efficiency. Particle swarm optimization (PSO) is a kind of swarm intelligence algorithm, which was proposed by Dr. Eberhart and Dr. Kennedy in 1995 [1]. It's inspired by the collective behavior of animals when they're looking for food.

Each particle needs to learn its personal best position (Pbest) in history and the global best position (Gbest) is discovered by the entire population. The velocity and position of ith particle is denoted by $V_i = (V_{i1}, V_{i2}, \cdots, V_{iD})$ and $X_i = (X_{i1}, X_{i2}, \cdots, X_{iD})$ in D-dimensional space. The velocity and position renewal of particles are governed by the following formula:

$$V_{ij}(t+1) = wV_{IJ}(t) + c_1 \cdot r_1 \cdot \left(Pbest_{ij} - X_{ij}(t)\right) + c_2 \cdot r_2 \left(\text{Gbest}_{ij} - X_{ij}(t)\right) \quad (1)$$

$$X_{ij}(t+1) = X_{ij}(t) + V_{ij}(t+1) \quad (2)$$

where, w is inertia weight and r_1, r_2 are randomly generated on the interval $[0,1]$. t is the current iteration number and c_1, c_2 are acceleration coefficient. $i = 1, 2, 3, ..., N$, N is the swarm size. We take $w = 0.8$, $c_1 = c_2 = 1.5$.

From Eq. 1, each particle can rapidly converge to the best position by learning its Pbest and Gbest. However, if the best position is the local best position, the particle swarm will lose its diversity and it will be difficult to get rid of the local optimal position.

3   Proposed MBPSO Algorithm

3.1   Motivation

In the actual dynamic optimization problem, there is often more than one peak value in the fitness landscape. When the original particle swarm optimization algorithm is used to find the optimal solution of complex functions, premature convergence will occur, resulting in local minimum.

In this section, we propose a particle swarm optimization algorithm with random migration behavior and dynamic parameters (MBPSO) used to better escape from the local optimum, improving the population diversity and the local searching ability.

In a natural environment, resources are limited. When a bird finds food, other birds gather to get it. However, the amount of food available in that location is limited. When enough birds congregate where food is available, the food run out quickly. At that time, birds tend to randomly disperse in search of other food, rather than always approach the global optimal location. Based on this idea, we endow particle swarm with certain migration ability. We divide the iterative process of particle swarm into several incremental evolutionary cycles. After completing an evolutionary cycle, the swarm needs to find resources again, and each particle randomly disperses in all directions. After completing this step, the particle will continue to search for the optimal value until the next evolution cycle. At the same time, the optimal value of each evolution cycle is recorded, and the particle swarm has enough time to converge to the optimal value in the last evolution cycle, so that the particle swarm will not lose the global optimal value. Although this method will consume more time than standard PSO, it can effectively improve the population and the global exploration ability of particle swarm, so as not to fall into premature convergence when dealing with complex multimodal functions.

The velocity of random particle migration is manipulated by the following formula:

$$X_r = rand_{1D} \cdot (X_{max} - X_{min}) + X_{min} \tag{3}$$

$$V_{ij}(t+1) = w_a \cdot V_{ij}(t) + (rand \cdot (X_{max} - X_{min}) + X_{min}) \cdot (X_r - X_{ij}) \tag{4}$$

The velocity of normal movement is controlled by the following formula:

$$V_{ij}(t+1) = w_a \cdot V_{ij}(t) + c_{1a}(Pbest_{ij} - X_{ij}) + c_{2a}(ttbest_{ij} - X_{ij}) \tag{5}$$

The particle position update formula is:

$$V_{ij}(t+1) = V_{ij}(t) + V_{ij} \qquad (6)$$

where V_{ij} represents the velocity of the particle, X_{ij} represents the position of the particle. X_{max} and X_{min} are the upper and lower bound of the search space w_a stands for inertia weight and represents the inheritance of its initial velocity. c_{1a} and c_{2a} represent acceleration coefficients, which stands for the learning degree of their own optimal historical position and the global optimal historical respectively, which is changed by c_1 and c_2. i represents the number of iterations, and T represents the total number of iterations.

The algorithm steps are as follows:

Step1: Initialize a set of particles (population size), including random positions and velocities. The evolutionary cycle is divided into several sub-cycles;
Step2: Calculate the degree of adaptability of each particle;
Step3: For each particle, its degree of adaptability value is compared with the best position Pbest, and if it is better, it becomes the current best location (Pbest);
Step4: For each particle, its degree of adaptability value is compared with the best position Gbest, and if it is better, make it the current best location (Gbest);
Step5: Update the position and velocity of each particle;
Step6: After a certain number of iterations, particles migrate randomly and then turn back to **step2**;
Step7: If the end condition is not met (usually set to the maximum number of iterations), then return to **step 2**. If so, the calculation will be stopped.

3.2   Inertia Weight Setting

In addition, the greater the inertia weight, the greater the incremental change in velocity per step, which means that new search areas need to be explored for better solutions. On the other hand, a small inertia weight means a small change in velocity and can be searched more finely in a local range. It can be inferred that the system should start with a large inertia weight, and then decrease non-linearly with the increase of the number of iterations so as to conduct more detailed local exploration in future iterations. It is pointed out that the decrement strategy of a linear function is better than that of a convex one, but the decrement strategy of a concave function is better than that of a linear one. We use the nonlinear inertial weight decreasing formula mentioned by Hu [14].

$$w_a = arctan \frac{T-t}{T} w \qquad (7)$$

3.3   Improved Acceleration Coefficient

In the process of optimization iteration, two acceleration coefficients vary with time, which is called as asynchronous change. The sizes of acceleration co-efficient c_1 and c_2 determine the influence of particle self-cognition and social cognition on particle trajectory, reflecting the degree of information exchange between particles in the group. We optimize the learning factors c_1 and c_2. We hope to take a larger value of c_1 and a smaller value of c_2 in the first stage of search, in this way, the particles will learn more from their optimal position and less from the global optimal position. This improvement aims to strengthen the global search ability of the algorithm. In the following step, the situation is just the opposite. c_1 takes a smaller value and c_2 takes a larger value, so that the particles learn more from the global optimal position and enhance the local search ability of the algorithm. We use a trig strategy in which c_1 decreases nonlinearly with concave functions and c_2 increases nonlinearly with concave functions. The change formula of learning factor is

$$c_{1a} = c_1 \cdot arctan \tfrac{T-t}{T} \tag{8}$$

$$c_{2a} = c_2 \cdot sin \tfrac{t}{T} \tag{9}$$

where, c_{1a} and c_{2a} are dynamic acceleration coefficient. The range of c_{1a} is 0 to 1.177 and c_{2a} is 0 to 1.26. So at the beginning of optimization iteration, the social learning ability of particles is weak and the self-learning ability is strong, which is convenient to realize fast search and improve global search capability. At the later stage of optimization iteration, the self-learning ability of particles is weak and the social learning ability is strong, which is conducive for local fine search and converges to the global optimal solution with high accuracy. The graphs functions of change for c_{1a} and c_{2a} is given below (Figs. 1 and 2):

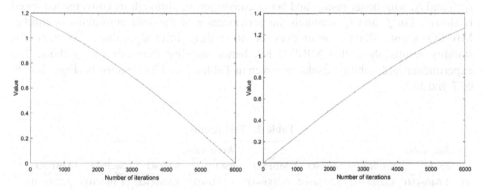

Fig. 1. The changing trend of $c1a$ **Fig. 2.** The changing trend of $c2a$

4 Experiment

4.1 Experiment Design

In our experimental study, we use five common multimodal benchmark functions and one unimodal function to test the performance of our proposed method. In addition, we select standard PSO, second order particle swarm optimization (SecPSO) and PSO algorithm combined with simulated annealing algorithm (SAPSO) to compare with our method. In order to compare the performance of the four algorithms, the optimal value, variance, worst value and average value of the four algorithms in solving, different functions are recorded. The convergence rates of MBPSO and other PSO algorithms are shown in the next section. The characteristics of the six benchmark functions are given in Table 1.

Table 1. Benchmark function characteristic

	Global minimum	Region of search	Dimensionality
$f1$: Ackley function	0	$[-32, 32]$	30
$f2$: Perm function	0	$[-100, 100]$	30
$f3$: Levy function	0	$[-10, 10]$	30
$f4$: Levy function N. 13	0	$[-10, 10]$	30
$f5$: Griewank function	0	$[-600, 600]$	30
$f6$: Rastrigin function	0	$[-5.12, 5.12]$	30

4.2   Experimental Results

In this section, the results of the test functions and the convergence trend graphs of the four algorithms are obtained. It can be seen from the figure that MBPSO is decreasing in step form in solving the convergence process of f_1, f_3, f_5 and f_6. In general, MBPSO performs better than the other three PSO algorithms. Aside from that, MBPSO is poor at tackling f_5. Among them, MBPSO is obviously superior to other PSO algorithms in f_1, f_3 and f_6, with better results and better convergence, although its convergence speed is slower. For f_2 and f_4, although the performance of the four algorithms is similar, MBPSO is only slightly better than the other three PSO algorithms. However, the stability results show that MBPSO has better stability than other algorithms. The experimental and stability results are given in Tables 2 and 3 respectively (Figs. 3, 4, 5, 6, 7 and 8).

Table 2. Test results

	Best value				Worst value			
	PSO	SAPSO	SecPSO	MBPSO	PSO	SAPSO	SecPSO	MBPSO
$f1$:	1.646e+00	1.064e−02	3.354e−01	0.000e+00	5.141e+00	1.804e+00	1.195e+00	3.324e−03
$f2$:	0.000e+00	4.975e−04	6.279e−03	0.000e+00	2.294e−02	2.294e−02	3.475e+01	2.294e−02
$f3$:	9.225e−06	1.000e−01	6.758e−01	0.000e+00	2.995e+00	9.964e−01	2.028e−01	7.950e−11
$f4$:	1.349e−31	1.349e−31	1.539e−03	0.000e+00	1.349e−31	1.349e−31	1.949e−02	1.349e−31
$f5$:	1.223e−05	8.599e−06	1.539e−03	0.000e+00	1.276e−01	6.615e−04	1.949e−02	4.790e−01
$f6$:	9.951e+00	8.112e+00	1.639e+01	0.000e+00	3.780e+01	3.497e+01	6.907e+01	1.913e+01

Table 3. Result stability

	Average				Variance			
	PSO	SAPSO	SecPSO	MBPSO	PSO	SAPSO	SecPSO	MBPSO
$f1$:	3.287e+00	1.325e−01	7.139e−01	1.525e−04	7.353e−01	9.881e−02	1.014e−01	4.183e−07
$f2$:	4.398e−03	6.149e−03	8.301e+00	4.363e−03	3.016e−05	4.762e−05	7.026e+01	2.944e−05
$f3$:	1.180e+00	3.764e−01	4.947e−01	4.958e−12	6.211e−01	3.156e−02	1.440e−02	2.280e−22
$f4$:	1.349784e−31	1.349e−31	9.470e−03	1.124e−31	4.314e−93	1.349e−31	2.612e−09	4.217e−64
$f5$:	4.707e−02	1.338e−04	9.470e−03	1.517e−01	1.439e−03	1.884e−08	1.766e−05	1.252e−02
$f6$:	2.322e+01	1.927e+01	3.896e+01	8.953e+00	4.217e+01	4.203e+01	1.661e+02	1.791e+01

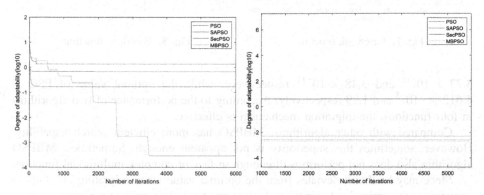

Fig. 3. Ackley function   **Fig. 4.** Perm function

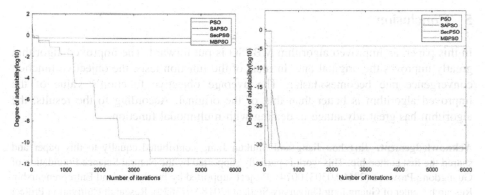

Fig. 5. Levy function   **Fig. 6.** Levy function N. 13

4.3 Discussion

To demonstrate the effectiveness of the migration mechanism, MBPSO without asynchronous varying acceleration is compared with standard PSO. The function Ackley, function Levy and function Rastr are used to compare MBPSO with PSO. In function Ackley, the optimal value of MBPSO is 1.89×10^{-7}, and the optimal value of PSO is 1.42. In functions Levy and Rastr, the optimal value of MBPSO is

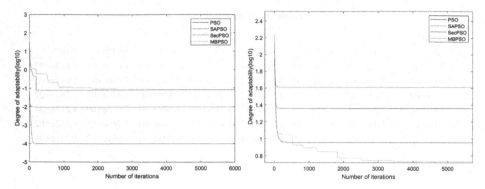

Fig. 7. Griewank function   **Fig. 8.** Rastrigin function

3.77×10^{-16} and 3.48×10^{-11} respectively, while the optimal value of PSO is 4.612×10^{-5} and 1.29 respectively. According to the performance of two algorithms in four functions, the migration mechanism is effective.

Compared with other algorithms, MBPSO has more efficient search capability. However, sometimes this superiority is not apparent enough. Sometimes, MBPSO algorithm also does not perform well enough in face of complex multimodal function.

Since may be that it deviates from the optimal value when migrating and fails to converge to the optimal value in time. At the same time, the running time increases with an increase in the algorithm complexity. So MBPSO has some room for improvement.

5 Conclusion

In this paper, an improved algorithm MBPSO is put forward. The improved algorithm greatly improves the original one. In most of the function tests, the objective function convergence rate becomes faster. The average objective function's value of the improved algorithm is better than that of the original. According to the results, the algorithm has great advantage in dealing with multimodal function.

Acknowledgement. Jingzhou Jiang and Shukun Jiang contributed equally to this paper and shared the first authorship. This work is partially supported by the Natural Science Foundation of Guangdong Province (2016A030310074), Project supported by Innovation and Entrepreneurship Research Center of Guangdong University Student (2018A073825), Research Cultivation Project from Shenzhen Institute of Information Technology (ZY201717) and Innovating and Upgrading Institute Project from Department of Education of Guangdong Province (2017GWTSCX0 38). The author is sincerely grateful to all those who gave advice and discussions, especially Ben Niu. In addition, the author also wants to express sincere thanks to Qianying Liu who provided valuable advice.

References

1. Kennedy, J.: The behavior of particles. In: Porto, V.W., Saravanan, N., Waagen, D., Eiben, A.E. (eds.) EP 1998. LNCS, vol. 1447, pp. 579–589. Springer, Heidelberg (1998). https://doi.org/10.1007/BFb0040809
2. Hu, X.: Recent advances in particle swarm. In: Proceedings of the 2004 Congress on Evolutionary Computation, pp. 90–97. IEEE, Portland (2004)
3. Shi, Y., Liu, H., Fan, M., Huang, J.: Parameter identification of RVM runoff forecasting model based on improved particle swarm optimization. In: Tan, Y., Shi, Y., Mo, H. (eds.) ICSI 2013. LNCS, vol. 7928, pp. 160–167. Springer, Heidelberg (2013). https://doi.org/10.1007/978-3-642-38703-6_19
4. Niu, B., Zhu, Y., Hu, K., Li, S., He, X.: A novel particle swarm optimizer using optimal foraging theory. In: Huang, D.-S., Li, K., Irwin, G.W. (eds.) ICIC 2006. LNCS, vol. 4115, pp. 61–71. Springer, Heidelberg (2006). https://doi.org/10.1007/11816102_7
5. Mendes, R.: Particle swarms for feedforward neural network training. In: Proceedings of the 2002 International Joint Conference on Neural Networks, pp. 1895–1899. IEEE, Honolulu (2002)
6. Venayagamoorthy, G.K.: Navigation of mobile sensors using PSO and embedded PSO in a fuzzy logic controller. In: Conference Record of the 2004 IEEE Industry Applications Conference. 39th IAS Annual Meeting, pp. 1200–1206. IEEE, Seattle (2004)
7. Gang, M.: A novel particle swarm optimization algorithm based on particle migration. Appl. Math. Comput. 218(11), 6620–6626 (2012)
8. Mohais, A.S.: Randomized directed neighborhoods with edge migration in particle swarm optimization. In: Proceedings of the 2004 Congress on Evolutionary Computation, pp. 548–555. IEEE, Portland (2004)
9. El-Abd, M.: A cooperative particle swarm optimizer with migration of heterogeneous probabilistic models. Swarm Intell. 4(1), 57–89 (2010)
10. Cui, H.M.: Convergence analysis and parameter selection in particle swarm optimization. Jisuanji Gongcheng yu Yingyong (Comput. Eng. Appl.) 42(23), 89–91 (2007)
11. Hu, J.: Selection on inertia weight of particle swarm optimization. Comput. Eng. 33(11), 193–195 (2007)
12. Cui, Z.: An improved PSO with time-varying accelerator coefficients. In: 2008 Eighth International Conference on Intelligent Systems Design and Applications, pp. 638–643. IEEE, Kaohsiung (2008)
13. Mao, K.F.: Particle swarm optimization algorithm based on non-symmetric learning factor adjusting. Comput. Eng. 36(19), 182–184 (2010)
14. Hu, J.Z.: Research on particle swarm optimization with dynamic inertia weight. In: 2009 International Conference on Management and Service Science, pp. 1–4. IEEE, Guilin (2009)

Data Clustering Using the Cooperative Search Based Artificial Bee Colony Algorithm

Chen Guo[1], Heng Tang[1], Chang Boon Patrick Lee[1], and Ben Niu[2(✉)]

[1] Faculty of Business Administration, University of Macau, Macau, China
{yb87016,hengtang,cblee}@um.edu.mo
[2] College of Management, Shenzhen University, Shenzhen 518060, China
drniuben@gmail.com

Abstract. Data clustering is a significant and strong data analysis technique which has a broad range of applications in many domains. In this paper, the artificial bee colony algorithm (ABC) is adopted to partition data sets into K clusters. To trade off the global and local searching ability of ABC algorithm, two kinds of cooperative search based ABC algorithms are proposed, that is N2ABC and WCABC. Then, the proposed algorithms are combined with K-means to deal with data clustering. For the purpose of demonstrating the efficiency of two hybrid clustering algorithms (N2ABCC and WCABCC), one artificial data set and six benchmark data sets are selected to test clustering results. Meanwhile, five algorithms, namely K-means, PSOC, ABCC, GABCC and CABCC, are chosen for comparison. The clustering results indicate that the proposed algorithms have better clustering validity than other algorithms.

Keywords: Data clustering · Artificial bee colony algorithm · K-means

1 Introduction

Clustering is a crucial problem that regularly meet and must be addressed as a key link of intricate task in data mining, image segmentation, document retrieval, pattern recognition, market research, web mining and etc. [1, 2]. The current clustering can be simply classified into partitional clustering and hierarchical clustering [2]. Here, it mainly focuses on the former one. K-means is the most welcome and simplest partitional algorithm [3]. However, the K-means has some shortcomings, such as, difficulty selection of initial cluster centers, sensitivity to outliers and easy to be trapped into local optimum [4]. To address those drawbacks, many swarm intelligence algorithms have been adopted to deal with clustering issues.

Artificial bee colony algorithm (ABC) is enlightened by swarm intelligence behavior of honey bees [5]. There are some excellent properties of ABC algorithm, such as simple to use, robust and good at local and global optimization. Due to these advantages, ABC algorithm has been adopted extensively in engineering, neural networks, image processing, transportation problem, data mining and so on [6–10]. However, ABC algorithm also has some shortcomings. Because the ABC algorithm is lack of the ability to memorize the global optimum, it is liable to trap to local optimum. In addition, the searching process is dominated by a random generated neighborhood

© Springer Nature Switzerland AG 2019
D.-S. Huang et al. (Eds.): ICIC 2019, LNAI 11645, pp. 660–671, 2019.
https://doi.org/10.1007/978-3-030-26766-7_60

location, so the exploration process is stochastic enough and exploitation ability is weak [11]. In order to address those problems mentioned above, two kinds of cooperative search based ABC algorithm are proposed here. Whereafter, the proposed ABC algorithms united K-means are applied to deal with clustering issues.

The remainder of the paper is arranged as follows. Section 2 exhibits equations of ABC algorithm. Section 3 describes the proposed ABC algorithms in detail. Section 4 gives the simulation results. Section 5 presents the data clustering steps using proposed algorithms. Data clustering experiments and results are listed in Sect. 6. Finally, Sect. 7 presents the conclusion.

2 Artificial Bee Colony Algorithm

Artificial bee colony algorithm is reviewed in [6], from which it can be seen that ABC algorithm has been widely studied. To avoid redundancy, we assume that the readers have a general understanding of the ABC algorithm and thus do not introduce the algorithm in detail. However, for the convenience of introducing proposed algorithms, the equations of ABC are presented as follows.

- **Equation of generating the location of food sources randomly**

$$x_{i,d} = x_d^{\min} + rand(0,1)(x_d^{\max} - x_d^{\min}) \tag{1}$$

where $i = 1, 2, \ldots, N$ and $d = 1, 2, \ldots, D$. N denotes the size of employed bees. D denotes the dimension of problem. $x_{i,d}$ represents the position of ith employed bee on dth dimension. $x_d^{\max}$ and $x_d^{\min}$ are the maximum and minimum boundary of $x_{i,d}$.
- **Equation of neighborhood search**

$$v_{i,d} = x_{i,d} + \phi_{i,d}(x_{i,d} - x_{j,d}) \tag{2}$$

where $j = 1, 2, \ldots, N$ and $j \neq i$, $\varphi_{i,d}$ is a random number in the scope of $[-1,1]$ to control the generation of neighbor solution around $x_{i,d}$. $v_{i,d}$ denotes the new position of $x_{i,d}$.
- **Equation of greedy selection**

$$\text{fit}_i = \frac{1}{1+f_i} \text{ if } f_i \geq 0; \text{ fit}_i = 1 + \text{abs}(f_i) \text{ if } f_i < 0 \tag{3}$$

where f_i denotes the objective function about the problem, and fit_i is the fitness value of the ith food source.
- **Equation of probability selection**

$$p_i = \frac{\text{fit}_i}{\sum_{i=1}^{N} \text{fit}_i} \tag{4}$$

where p_i is the selection probability of the ith solution.

3 Proposed Cooperative Search Based ABC Algorithms

For the purpose of overcoming the shortcomings associated with ABC algorithms, there are many cooperation-based ABC variants have been proposed [11–18]. Among these ABC variants, [11] proposes GABC which is used for comparison in this paper. Based on those cooperative search ideas, this paper proposes two kinds of cooperative search methods which implement in the stage of employed bee and onlooker bee.

3.1 Two Neighbors Search Method (N2ABC)

Based on the ABC algorithm which searches for better solution under the disturbance of one neighbor $x_{j,d}$, N2ABC adds another neighbor $x_{l,d}$ to do the neighborhood search. Therefore, the neighborhood range is broader than standard ABC and the performance is improved [19]. The new food source is updated using equation below

$$v_{i,d} = x_{i,d} + \varphi_{i,d}(x_{i,d} - x_{j,d}) + \beta_{i,d}(x_{i,d} - x_{l,d}) \tag{5}$$

where $i, j, l = 1, 2, \ldots, N$ and $j \neq i, l \neq i, j \neq l, \varphi_{i,d}$ and $\beta_{i,d}$ are random numbers in the scope of [−1,1]. $\varphi_{i,d}$ and $\beta_{i,d}$ are the parameters to control the production of neighbor solution around $x_{i,d}$.

3.2 Weighted Crossover Operation Based Search Method (WCABC)

Combining the social behavior (*gbest*) of PSO and crossover operation of GA, [20] presents a crossover based global ABC (named CABC here) which shows a better performance than GABC. Specifically, CABC performs neighborhood search first, and then performs crossover operation at a crossover probability *cr*. Based on CABC and weighted PSO [21], this paper proposes a weighted CABC, named WCABC. In particular, a linearly decreasing inertia weight W is added in employed bee and onlooker bee stage, the updating process of WCABC uses Eqs. (6) and (7) first, and then Eq. (8).

$$W = W_{\max} - \frac{t}{Maxcycle}(W_{\max} - W_{\min}) \tag{6}$$

$$v_{i,d} = W * x_{i,d} + \varphi_{i,d}(x_{i,d} - x_{j,d}) \tag{7}$$

$$v_{i,d} = gbest + \eta_{i,d}(gbest - x_{i,d}) \tag{8}$$

where W_{min} is the minimum of W and W_{max} is the maximum of W. t is the number of iteration and *Maxcycle* is the maximum number of iteration. *gbest* is the global optimum, $\varphi_{i,d}$ and $\eta_{i,d}$ are random numbers in the scope of [−1, 1], *cr* is the crossover rate which is 0.8 [16].

4 Simulation Results for Benchmark Functions

Here, six benchmark functions are adopted to test performance of the N2ABC and WCABC. Meanwhile, ABC, GABC and CABC are employed for comparison. The benchmark functions and its global minimum are listed in the Table 1.

Table 1. Benchmark functions.

Function		Definition	Search domain	Global minimum
f_1	Dixon_Price	$f(x) = (x_1 - 1)^2 + \sum_{i=2}^{D} i(2x_i^2 - x_{i-1})^2$	$[-10, 10]$	$f(x^*) = 0$
f_2	Levy	$f(x) = \sin^2(\pi\omega_1) + \sum_{i=1}^{D-1} (\omega_i - 1)^2$ $[1 + 10\sin^2(\pi\omega_i - 1)] + (\omega_D - 1)^2$ $[1 + \sin^2(2\pi\omega_D)]$, where $\omega_i = 1 + \frac{x_i - 1}{4}$, for all $i = 1, 2, \ldots, D$	$[-10, 10]$	$f(x^*) = 0$
f_3	Rastrigin	$f(x) = \sum_{i=1}^{D} (x_i^2 - 10\cos(2\pi x_i) + 10)$	$[-5.12, 5.12]$	$f(x^*) = 0$
f_4	Schaffer2	$f(x) = 0.5 + \frac{\sin^2(x_1^2 - x_2^2) - 0.5}{[1 + 0.001(x_1^2 + x_2^2)]^2}$	$[-100, 100]$	$f(x^*) = 0$
f_5	Schwefel N. 2	$f(x) = 418.9829D - \sum_{i=1}^{D} x_i \sin(\sqrt{|x_i|})$	$[-500, 500]$	$f(x^*) = 0$
f_6	Sphere	$f(x) = \sum_{i=1}^{D} x_i^2$	$[-5.12, 5.12]$	$f(x^*) = 0$

4.1 Parameter Settings

Based on [21, 22], colony size of all algorithms is set 100 and benchmark functions are tested on 10 dimensions. Maximum cycle number is 500,000 and all experiments are repeated for 30 runs. For equal comparison, crossover rate cr is 0.8 in CABC and WCABC [16]; inertia weight W reduces from 0.9 to 0.4 in WCABC and PSO [23]. [5, 24] show that "Limit" is equal to "Number of onlooker bees*Dim". So, limit is equal to 500 in ABC and its variants. Besides, in PSO, c_1 and c_2 are equal to 2 [22].

4.2 Simulation Results

The experiment results are listed in Table 2. From Table 2, comparing the average values and best values, N2ABC has better average value and relatively poor best values than other algorithms on $f_1 \sim f_6$. On the contrary, WCABC has better best values than other algorithms, especially on f_3 and f_6.

Table 2. Simulation results of the algorithms.

F	Criteria	ABC	GABC	CABC	N2ABC	WCABC
f_1	Average	**1.5736**[b]	1.7264	1.9267	**1.2319**[a]	3.9476
	Best	0.0064	2.8684e−09	**5.8737e−13**	2.9851e−05	**5.4571e−10**
	Std	34.4925	53.8389	59.7826	42.1151	66.3743
f_2	Average	0.0017	**0.0011**[a]	0.0020	**0.0011**[a]	**0.0014**[b]
	Best	8.2910e−32	**1.4998e−32**	**1.4988e−32**	1.1862e−10	**1.4998e−32**
	Std	0.0313	0.0294	0.0357	0.0275	0.0226
f_3	Average	0.0169	**0.0121**[b]	0.0202	**0.0095**[a]	0.0180
	Best	**0**	**0**	**0**	4.0762e−08	**0**
	Std	0.1773	0.1562	0.1686	0.1195	0.1482
f_4	Average	9.3421e−05	9.8307e−05	5.5962e−04	**7.5356e−05**[b]	**3.6776e−05**[a]
	Best	4.7110e−06	3.3330e−05	4.8051e−04	9.1340e−06	**1.5725e−07**
	Std	7.5738e−04	4.4420e−04	7.6478e−04	6.1607e−04	3.9567e−04
f_5	Average	1.5768	**0.4775**[a]	4.7008	**0.7530**[b]	2.4413
	Best	**1.2728e−04**	**1.2728e−04**	**1.2728e−04**	**1.4728e−04**	**1.2728e−04**
	Std	6.6107	3.6883	6.8120	4.4186	7.0561
f_6	Average	**0.0013**[a]	0.0020	0.0025	**0.0018**[b]	0.0026
	Best	2.7786e−32	1.0793e−46	**1.3194e−229**	1.8325e−10	**3.0639e−250**
	Std	0.0314	0.0481	0.0397	0.0433	0.0400

a: the first rank
b: the second rank

From Fig. 1, it is clear that the WCABC is superior to other algorithms in searching optimum on f_2, f_3, f_4 and f_6. However, N2ABC has poor ability in searching optimum. As far as convergence speed concerned, WCABC has better performance than other algorithms on f_2, f_3 and f_4. Considering the final stage of iteration in particular, WCABC still maintain outstanding convergence trend on f_1 and f_6, however the other algorithms slow down their convergence speed.

According to the results mentioned above, it is concluded that the proposed ABC algorithms have better performance than ABC, GABC and CABC. So, in the next section, the proposed ABC algorithms are used to deal with data clustering.

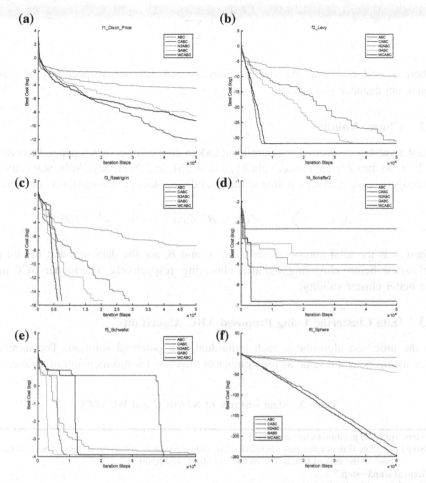

Fig. 1. The average convergence of Cost functions. (a) Dixon_Price function. (b) Levy function. (c) Rastrigin function. (d) Schaffer function N.2. (e) Schwefel function. (f) Sphere function.

5 Data Clustering

5.1 K-means Algorithm

In the k-means algorithm, K clusters are selected randomly, and then assign each data to one class based on the similarity. Euclidian distance is a common way to measure the similarity of clusters. In this paper, sum of the squared error (SSE) is the objective function which is used to measure the similarity of clusters. The smaller objective function value means the better clustering. The equation of SSE and C_j are

$$SSE = \sum_{j=1}^{K} \sum_{s_i \in c_j} dist^2(c_j, s_i) \tag{9}$$

$$c_j = \frac{1}{n_j} \sum_{\forall s_i \in c_j} s_i \qquad (10)$$

where c_j and s_i denote the cluster centroids and data set, respectively. *dist* is the Euclidian distance of c_j and s_i.

5.2 Cluster Validity

Cluster validity evaluation is a significant task in data mining. In this paper, there are two indexes to measure the cluster validity, that is SSE and Accuracy. Accuracy is the ratio of properly classified data points to the total data points [25], equation is defined as

$$ACC = \left(\sum_{i=1}^{n} \left(if(A_i = B_i) \ then \ 1 \ else0 \right) \div n \right) * 100 \qquad (11)$$

where n is the total number of data set. A_i and B_i are the data sets that the *ith* point belongs to before clustering and after clustering, respectively. The higher ACC means the better cluster validity.

5.3 Data Clustering Using Proposed ABC Algorithms

In the proposed algorithms, each individual is a potential solution. The number of dimensions is as the same as the number of clusters. Each dimension of an individual

Table 3. Main structure of N2ABCC and WCABCC.

Step1 Initialize position using equation (1)

Step2 Calculate fitness value using equation (12), and select a half of bees as employed bees on the basis of the rank of fitness, record the best fitness and position

Repeat step3 ~step 7

Step 3 For cycle=1: Maxcycle

Step 4 Employed bee phase

    For each employed bee

        i. Generate new food source using N2ABC or WCABC algorithm, and then implement the K-means

        ii. Compute fitness value using equation (12), and apply greedy selection mechanism

    End

Step 5 Apply for probability selection and calculate p_i

Step 6 Onlooker bee phase

    For each employed bee

        i. Choose a food source according to the p_i

        ii. Search for new food source using N2ABC or WCABC algorithm and implement K-means

        iii. Calculate fitness value using equation (12), and apply greedy selection mechanism

    End

Step 7 Scout bee phase

    If a food source reaches end conditions, the corresponding employed bee transfers to a scout bee

    Initialize new food source to substitute the old one

Step 8 Record the best solution and output the global optimum solution

---

denotes a cluster centroid. Here, the K-means is added in the stage of employed bee and onlooker bee. The main structure of proposed algorithms, namely N2ABCC and WCABCC, is described in Table 3.

In this paper, Eq. (9) is the objective function, the fitness function is the reciprocal of the objective function, that is

$$fit_i = \frac{1}{SSE_i} \tag{12}$$

6 Data Clustering Experiments and Results

6.1 Data Sets

To assess the experimental results of presented algorithms, this paper chooses one artificial data set used in [25] and six benchmark data sets selected from UCI to test the clustering results. The artificial data set is a two-dimensional problem with four classes, and each dimension of the class obeys multivariate normal random distribution, that is $N(\mu = (m_i; 0),\ \Sigma = [0.5\ 0.05;\ 0.05\ 0.5])$, $i = 1, 2, 3, 4$, $m_1 = -3$, $m_2 = 0$, $m_3 = 3$, $m_4 = 6$, μ denotes the average vector and Σ denotes the covariance matrix. The data set is portrayed in Fig. 2. The features of the seven data sets described in Table 4.

Table 4. The features of data sets.

Data set	Class	Feature	Total data points	Points in each class
Artificial	4	2	600	(150, 150, 150, 150)
Iris	3	4	150	(50, 50, 50)
Wine	3	13	178	(59, 71, 48)
Glass	6	9	214	(70, 17, 76, 13, 9, 29)
Breast Cancer	2	9	683	(444,239)
Ionosphere	2	34	351	(126,225)
Zoo	7	16	101	(41,20,5,13,4,8,10)

6.2 Results

Here, the paper assesses the performance of two kinds of data clustering algorithms using proposed ABC algorithms. At the same time, K-means, PSO+K-means (PSOC), ABC+K-means (ABCC), GABC+K-means (GABCC) and CABC+K-means (CABCC) are selected for comparison. For each data set, each algorithm iterates 100 times. The parameter settings are the same as the Sect. 4.1. The SSE and ACC obtained from seven algorithms are concluded in Tables 5 and 6, respectively. The value listed in the table are averages over 10 times simulations.

From Table 5, for artificial data set, WCABCC and K-means have better results than other algorithms. For the Iris and Breast Cancer data sets, the average of SSE for ABCC, GABCC, N2ABCC, CABCC and WCABCC are almost the same as the best

Table 5. Comparison of SSE for the seven clustering algorithms.

Data set	Criteria	ABCC	GABCC	CABCC	WCABCC	N2ABCC	PSOC	K-means
Art	Average	572.2561	592.1628	572.6755	**568.3103**[b]	570.4195	571.7581	**565.1918**[a]
	Best	572.2552	592.1597	572.6729	568.3078	570.4172	567.6718	563.4768
	Std	0.0090	0.0306	0.0258	0.0249	0.0233	12.6855	9.5151
Iris	Average	**6.9981**[a]	**6.9982**[b]	**6.9982**[b]	**6.9982**[b]	**6.9981**[a]	7.0218	10.9171
	Best	6.9981	6.9981	6.9981	6.9981	6.9981	6.9981	10.9083
	Std	2.4123e−04	6.7908e−04	4.5324e−04	6.2868e−04	1.4132e−04	0.1672	0.0870
Wine	Average	48.9559	**48.9551**[a]	48.9556	**48.9555**[b]	**48.9551**[a]	49.1409	49.1078
	Best	48.9540	48.9540	48.9540	48.9540	48.9540	48.9854	49.0154
	Std	0.0161	0.0108	0.0155	0.0135	0.0099	1.1825	0.7421
Glass	Average	18.3438	18.3447	**18.2997**[a]	**18.3225**[b]	18.3687	19.7735	22.1032
	Best	18.3060	18.3117	18.2553	18.2861	18.3353	19.3989	22.0461
	Std	0.1855	0.1925	0.2269	0.2109	0.2006	1.0114	0.3865
Breast Cancer	Average	**238.5578**[b]	**238.5577**[a]	**238.5578**[b]	**238.5578**[b]	**238.5578**[b]	239.0265	238.7740
	Best	238.5577	238.5577	238.5577	238.5577	238.5577	238.5581	238.5577
	Std	5.2068e−04	2.6452e−04	8.1094e−04	6.1668e−04	6.0079e−04	4.1712	1.9785
Ionosphere	Average	628.8981	**628.8968**[a]	628.8979	628.8975	**628.8974**[b]	629.9439	629.8583
	Best	628.8960	628.8960	628.8960	628.8960	628.8960	628.8960	628.9034
	Std	0.0110	0.0078	0.0185	0.0148	0.0135	8.7070	7.1669
Zoo	Average	86.6159	86.2544	86.5711	**85.8479**[a]	**85.4598**[b]	99.9041	92.4360
	Best	86.2197	85.8693	86.1792	85.6324	85.1784	99.6294	92.2650
	Std	0.9828	1.0005	0.9118	0.7987	0.9295	2.0203	1.4471

a: the first rank
b: the second rank

Table 6. Comparison of ACC for the seven clustering algorithms.

Data set	Criteria	ABCC	GABCC	CABCC	WCABCC	N2ABCC	PSOC	K-means
Art	Average	94.7628	93.9635	94.6950	94.9548	93.7083	**97.1800**[a]	**96.7583**[b]
	Best	97.5500	96.7000	98.5333	98.5167	97.3667	97.3500	96.8333
	Std	3.0429	2.9358	3.9558	2.5636	3.0760	0.5269	96.7583
Iris	Average	**87.9967**[b]	86.1673	87.3347	87.3780	85.8793	**88.6420**[a]	57.3393
	Best	89.4000	89.7333	89.3333	89.2667	89.4667	88.8000	57.5333
	Std	0.9208	2.6057	1.6504	1.9031	2.5470	0.1349	0.0316
Wine	Average	**95.3247**[a]	94.9854	**95.2539**[b]	95.1697	94.8994	94.8601	93.2135
	Best	95.7303	95.6742	95.6742	95.6742	95.6742	94.9438	93.2584
	Std	0.3205	0.7849	0.5859	0.6189	0.8917	0.3227	0.3163
Glass	Average	44.4023	43.5794	43.6299	43.6355	**44.4131**[b]	**49.9210**[a]	43.9238
	Best	45.7944	47.2897	48.4112	47.1028	46.8224	50	44.1121
	Std	0.1988	0.5130	0.7291	0.5591	0.4742	0.2056	0.0516
Breast Cancer	Average	96.0097	95.8281	96.0187	96.0092	95.8971	**96.1835**[a]	**96.0363**[b]
	Best	96.1493	96.1054	96.1347	96.3250	96.1493	96.2225	96.0469
	Std	0.0918	0.3772	0.0692	0.2388	0.2439	0.0799	0.0861
Ionosphere	Average	**71.1883**[a]	71.1254	**71.1433**[b]	71.0991	71.1151	71.1917	70.9379
	Best	71.2251	71.2251	71.2536	71.2251	71.2536	71.2251	70.9687
	Std	0.0953	0.1390	0.1666	0.1594	0.1607	0.1429	0.0213
Zoo	Average	**76.7564**[b]	75.8277	**79.5653**[a]	74.5376	73.6842	69.2960	67.3149
	Best	79.6040	77.6238	79.9010	77.8218	76.3366	69.3069	67.3267
	Std	1.1285	0.5388	0.5795	0.7318	0.8587	0.0911	0.1093

a: the first rank
b: the second rank

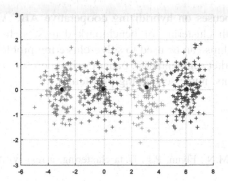

Fig. 2. Distribution and clustering of artificial data set.

distance 6.9981 and 238.5577, respectively. For Wine and Ionosphere data sets, the performance of GABCC and N2ABCC are superior to other algorithms. Take Glass and Zoo data sets into consideration, N2ABCC, CABCC and WCABCC outperform other algorithms. Consequently, for the six real life data sets, N2ABCC and WCABCC have better performance than other algorithms, especially better than PSOC and K-means.

From the Table 6, ABCC and CABCC have higher accuracy than other algorithms for Wine, Ionosphere and Zoo data sets. However, the two proposed algorithms do not compare favorably with PSOC and K-means for Art, Iris and Breast Cancer data sets in terms of accuracy. There is no absolute correlation between SSE and accuracy [25]. But, comparing the accuracy of proposed two types of algorithms, WCABCC outperforms the N2ABCC for Art, Iris, Wine, Breast Cancer and Zoo data sets.

7 Conclusion

Here, on the basis of cooperative ABC variants, two kinds of cooperative search based artificial bee colony algorithms are proposed. For the purpose of assessing the simulation results of the presented algorithms, six test functions are adopted to assess the searching ability and convergence speed of the proposed algorithms. The simulation experiments show that the proposed ABC algorithms have better performance than ABC, GABC and CABC. Then, the paper combines the proposed algorithms with K-means in the stage of employed bee and onlooker bee to solve the clustering problem. The hybrid algorithms are called N2ABCC and WCABCC. For the purpose of evaluating the clustering validity, the hybrid algorithms are tested on one artificial data set and six benchmark data sets acquired from UCI. Meanwhile, K-means, PSOC, ABCC, GABCC and CABCC are selected for comparison. The clustering results indicate that the presented algorithms have better clustering validity than other algorithms. To sum up, it can be concluded that the cooperative search strategy adopted in this paper make the ABC algorithm more efficient, accurate, robust and have certain advantages to compete with other algorithms.

Here, it mainly focuses on hybridizing cooperative ABC variants with K-means algorithm to deal with clustering of benchmark data sets. In the future work, the proposed ABC algorithms can be used to solve clustering problems in healthcare field, such as healthcare market segmentation, patient private profiles clustering, detection of pain patterns in patients, nurse scheduling problem.

References

1. Jain, A.K., Murty, M.N., Flynn, P.J.: Data clustering: a review. ACM Comput. Surv. **31**, 264–323 (1999)
2. Han, J., Pei, J., Kamber, M.: Data Mining: Concepts and Techniques. Elsevier, Amsterdam (2011)
3. Jain, A.K.: Data clustering: 50 years beyond K-means. Pattern Recogn. Lett. **31**, 651–666 (2010)
4. Selim, S.Z., Ismail, M.A.: K-means-type algorithms: a generalized convergence theorem and characterization of local optimality. IEEE Trans. Pattern Anal. Mach. Intell. **1**, 81–87 (1984)
5. Karaboga, D.: An idea based on honey bee swarm for numerical optimization. Technical Report-Tr06, Erciyes University, Engineering Faculty, Computer Engineering Department (2005)
6. Karaboga, D., Gorkemli, B., Ozturk, C., Karaboga, N.: A comprehensive survey: artificial bee colony (ABC) algorithm and applications. Artif. Intell. Rev. **42**, 21–57 (2014)
7. Karaboga, D., Ozturk, C.: A novel clustering approach: artificial bee colony (ABC) algorithm. Appl. Soft Comput. **11**, 652–657 (2011)
8. Boudardara, F., Gorkemli, B.: Application of artificial bee colony programming to two trails of the artificial ant problem. In: 2018 2nd International Symposium on Multidisciplinary Studies and Innovative Technologies (ISMSIT), pp. 1–6. IEEE (2018)
9. Sornam, M., Prabhakaran, M.: Logit-based artificial bee colony optimization (LB-ABC) approach for dental caries classification using a back propagation neural network. In: Krishna, A., Srikantaiah, K., Naveena, C. (eds.) Integrated Intelligent Computing, Communication and Security, vol. 771, pp. 79–91. Springer, Singapore (2019). https://doi.org/10.1007/978-981-10-8797-4_9
10. Gao, H., Shi, Y., Pun, C.-M., Kwong, S.: An improved artificial bee colony algorithm with its application. IEEE Trans. Ind. Inform. **15**, 1853–1865 (2018)
11. Zhu, G., Kwong, S.: Gbest-guided artificial bee colony algorithm for numerical function optimization. Appl. Math. Comput. **217**, 3166–3173 (2010)
12. Liu, H., Gao, L., Kong, X., Zheng, S.: An improved artificial bee colony algorithm. In: 2013 25th Chinese Control and Decision Conference (CCDC), pp. 401–404. IEEE (2013)
13. Jadon, S.S., Bansal, J.C., Tiwari, R., Sharma, H.: Expedited artificial bee colony algorithm. In: Pant, M., Deep, K., Nagar, A., Bansal, J.C. (eds.) Proceedings of the Third International Conference on Soft Computing for Problem Solving. AISC, vol. 259, pp. 787–800. Springer, New Delhi (2014). https://doi.org/10.1007/978-81-322-1768-8_68
14. El-Abd, M.: Local best artificial bee colony algorithm with dynamic sub-populations. In: 2013 IEEE Congress on Evolutionary Computation, pp. 522–528. IEEE (2013)
15. Jadon, S.S., Bansal, J.C., Tiwari, R., Sharma, H.: Artificial bee colony algorithm with global and local neighborhoods. Int. J. Syst. Assur. Eng. Manag. **9**, 589–601 (2018)
16. Sahoo, G.: A two-step artificial bee colony algorithm for clustering. Neural Comput. Appl. **28**, 537–551 (2017)

17. Guo, P., Cheng, W., Liang, J.: Global artificial bee colony search algorithm for numerical function optimization. In: 2011 Seventh International Conference on Natural Computation, pp. 1280–1283. IEEE (2011)
18. Xue, Y., Jiang, J., Zhao, B., Ma, T.: A self-adaptive artificial bee colony algorithm based on global best for global optimization. Soft. Comput. **22**, 1–18 (2018)
19. Zhang, D., Guan, X., Tang, Y., Tang, Y.: Modified artificial bee colony algorithms for numerical optimization. In: 2011 3rd International Workshop on Intelligent Systems and Applications, pp. 1–4. IEEE (2011)
20. Jiang, M., Yuan, D.: Artificial bee colony algorithm and its application. Science Press, Beijing (2014). (in Chinese)
21. Akay, B., Karaboga, D.: Parameter tuning for the artificial bee colony algorithm. In: Nguyen, N.T., Kowalczyk, R., Chen, S.-M. (eds.) ICCCI 2009. LNCS (LNAI), vol. 5796, pp. 608–619. Springer, Heidelberg (2009). https://doi.org/10.1007/978-3-642-04441-0_53
22. Akay, B.: A study on particle swarm optimization and artificial bee colony algorithms for multilevel thresholding. Appl. Soft Comput. **13**, 3066–3091 (2013)
23. Eberhart, R.C., Shi, Y.: Comparing inertia weights and constriction factors in particle swarm optimization. In: Proceedings of the 2000 Congress on Evolutionary Computation. CEC00 (Cat. No. 00TH8512), pp. 84–88. IEEE (2000)
24. Karaboga, D., Akay, B.: A comparative study of artificial bee colony algorithm. Appl. Math. Comput. **214**, 108–132 (2009)
25. Kao, Y.-T., Zahara, E., Kao, I.-W.: A hybridized approach to data clustering. Expert Syst. Appl. **34**, 1754–1762 (2008)

Feature Selection Using a Reinforcement-Behaved Brain Storm Optimization

Ben Niu, Xuesen Yang, and Hong Wang$^{(\boxtimes)}$

College of Management, Shenzhen University, Shenzhen 518060, China
`ms.hongwang@gmail.com`

Abstract. In this era of data explosion, feature selection has received sustained attention to remove the large amounts of meaningless data and improve the classification ac-curacy rate. In this paper, a feature selection method based on reinforcement-behaved strategy is proposed, which identifies the most important features by embedding the Brain Storm Optimization (BSO) algorithm into the classifier. The ideas of the BSO are mapped to feature subsets, and the importance of the feature is evaluated through some indicators, i.e. the validity of the feature migration. In the migration of each feature, the feature is updated to a new feature in the same position between the two generations. The feedback of each action is used as the basis for the ordering of feature importance. An updating strategy is presented to modify the actions based on the current state to improve the feature set. The effectiveness of the proposed algorithm has been demonstrated on six different binary classification datasets (e.g., biometrics, geography, etc.) in comparison to several embedded methods. The results show that our proposed method is superior in high performance, stability and low computing costs.

Keywords: Feature selection · Brain storm optimization · Wrapper method · Binary classification

1 Introduction

Classification plays an important role in pattern recognition, and it has received extensively attention from data mining, artificial intelligence, computer vision and other disciplines [1]. Binary classification is an important branch in the classification task, theirs common application areas include Email spam identification, Cancer detection and Credit management [2].

Real-world binary classification tasks often include big data with high dimensions. However, not all features are important for statistical learning. Irrelevant and redundant features not only reduce the classification performance of the classifier but also occupy storage resources and waste computation time. Therefore, feature selection (FS) has been developed as an important data dimensionality reduction technique [3]. Its objective is to select minimal features to represent useful statistical information.

Most of these methods need to perform two operations: *subset search* and *subset evaluation*. The research on search strategies in the past decade is mainly divided into three categories: *optimal*, *random search* and *heuristic* [4]. Optimal search strategies

© Springer Nature Switzerland AG 2019
D.-S. Huang et al. (Eds.): ICIC 2019, LNAI 11645, pp. 672–681, 2019.
https://doi.org/10.1007/978-3-030-26766-7_61

are the most straightforward methods to find the optimal feature subset, however, they cannot be widely used due to the huge search space. Heuristic search design a rational search method by analyzing the currently selected feature subsets. Since the local properties of these methods, it is easy to be stuck in local optimal solution. In randomized search, combining the feature selection with random search algorithms (such as genetic algorithm) can improve the quality of the subset search. On subset evaluation, feature selection can be divided into three major categories in terms of model building, i.e., *filter*, *wrapper* and *embedded* methods [5]. Filter methods aim to analyze intrinsic properties of data regardless of the model. Wrapper methods assess a given feature subset using a classification model. Embedded methods, which perform feature selection and classification simultaneously with embedded systems.

Since the effectiveness of regarding the classification performance as the evaluation criteria of the importance of feature subsets directly. Wrapper methods with swarm intelligence (SI) algorithms become a hot topic in the field of feature selection within many works [7–15]. The size of the feature subset selected by these methods is relatively small, which is beneficial to the identification of key features. SI algorithms could significantly help search for promising feature subsets.

Among the most SI-based feature selection methods, particle swarm optimization (PSO) [20] is a well-known global search method. A feature selection method based on PSO with mutation operation is proposed, and a new cost matrix is introduced to calculate an appropriate error rate [6]. The recently proposed competitive swarm optimizer feature selection (CSOFS) has proven to be an effective feature selection method, with each particle learn from a pair of randomly selected competitors to promote global search [7]. Some scholars try to reduce the size of feature subsets while selecting promising feature subsets. E.g., a novel 2-D learning framework is proposed to reduce dimension of features. The information about the subset cardinality and the information about the subset location is recorded in 2-D arrays separately [8]. Variable-length PSO representation (VLPSO) has improved the efficiency of feature search through defining learning mechanisms between variable-length particles [9]. Research on feature selection in genetic algorithms (GAs) has always been a popular topic. A hybrid feature selection algorithm (GA-PSOFS) is proposed to integrate the standard update rules of PSO with the selection, crossover and mutation operations of GA [10]. In order to improve the generalization ability of the feature selection model, a novel feature selection method is developed based on integrated some bi-objective genetic algorithm. In each GA agent, attribute dependency of rough set and multivariate mutual information are used be objective functions [11]. In addition, feature selection has also been applied to other SI algorithms, for example, a fixed chemotaxis step was addressed by adaptive one [12]. A novel bacterial population optimization feature selection (BCOFS) based on the life-cycle model was used for cancer detection [13]. In ant colony optimization based feature selection (ACOFS) [14], the features were used to construct a fully connected graph initially. They have two functions: selecting or deselecting others while ants visit all features. Unlike the former, graph clustering based ACOFS adopts a community detection algorithm to allocate search space legitimately [15].

A common drawback has been found in the above FS methods: they are overly dependent on the structure of the SI algorithms and ignore the regular of variation among features. In particular, information that a feature move to another one needs to be recorded and utilized in the wrapper model. It can be said that the reinforcement-behaved framework we propose incorporates the interactive representation of features.

In this work, a novel BSO-based feature selection algorithm is proposed which introduces a new feature ranking measure. The main idea behind the method is to record feedback that features change at the same location and establish a pairwise action relationship between feature distributions. In addition, we modify the specific features of the subset based on whether the paired action brings benefits. The novelty of our approach is to consider the interaction between features, making up for the shortcomings of previous methods based on feature ranking. Clearly, we make full use of the information on population of SI algorithms. In fact, we are bypassing the combinatorial problem in a reasonable way. The reason is that when a feature is paired with other features, the link can be extended to a full link. Therefore the information between the combinations is more fully excavated. Noteworthy, the strategy what we call *reinforcement-behaved* can be applied to other SI versions.

The organization of the paper is as follows. Section 2 describes our feature selection algorithm. The BSO-based subset construction is given in Sect. 2.1, and the reinforcement-behaved strategy is discussed in Sect. 2.2. Section 3 contains experimental evaluations and results. Finally, Sect. 4 gives the conclusion.

2   Reinforcement-Behaved BSOFS

Let $S = \{s_1, s_2, \ldots, s_n\} \in \mathcal{R}^{n \times d}$ denotes the data matrix, where columns are features and rows are samples. Each sample corresponds to a class label from $C = \{c_1, c_2, \ldots, c_n\}$.

In order to perform the proposed feature selection method, the basic BSO is used to search for feature subsets. Next, we record in the action-utility matrix based on the feedback of the feature subsets of the two generations. This $n \times n$ action-utility matrix records the action information between each feature and is used to adjust the value of location with update strategy.

2.1   Brain Storm Optimization

Inspired by the human creative cogitation, brain storm optimization (BSO) [16] has great potential in real applications. The basic BSO algorithm mainly consists of three operators for decision making: *clustering*, *creating* and *selecting operation*. It is supposed that the dimension of search space is D and the population size is N. BSO randomly initializes population with N ideas ($X_i = [x_{i1}, x_{i2}, \ldots, x_{id}]$, $i \in \{1, 2, \ldots, N\}$). Each idea is seen as a feature subset. x_{ij} is a scalar in the boundary range $[x_{j\_min}, x_{j\_max}]$, where x_{j_min} and x_{j_max} are expressed as minimum and maximum boundary. In feature selection issue, $x_{j\_min} = 1$ and $x_{j\_max} =$ total number of features.

Clustering Operation. BSO uses the k-means clustering method to converge ideas into M different groups. Moreover, the cluster center is denoted by the best idea in each group. In order to enhance diversity of swarm, the pre-determined probability

$p\_replace(p\_replace \in [0,1])$ is used to determine the cluster center is replaced by a random idea or not.

Creating Operation. A new idea could be generated based on one cluster or two clusters with control parameter p_one $(p\_one \in [0,1])$. Similarly, control parameters $p\_one\_center(p\_one\_center \in [0,1]))$ and $p\_two\_center(p\_two\_center \in [0,1])$ are used to control the probability based on cluster center or random one.

$$x_{old} = \begin{cases} x_i^d, & one\ cluster \\ \omega_1 \times x_i^d + \omega_2 \times x_j^d, & two\ clusters \end{cases} \tag{1}$$

Where d is the dimension index, x_{old} refers the idea after creating operation. Both ω_1 and ω_2 are random value within range [0, 1], x_i^d and x_j^d come from cluster i and j.

Subsequently, the formulas for generating a new idea with disturbance operation are given as follows:

$$x_{new} = x_{old} + \xi \times G(\mu, \sigma) \tag{2}$$

$$\xi = logsig(\frac{0.5 \times Iter_{max} - Iter_{cur}}{K}) \times rand(\cdot) \tag{3}$$

x_{new} denotes idea after finishing Gauss disturbance $G(\mu, \sigma)$. μ and σ refer mean value and variance value, respectively. Step size ξ is used to control the weight of Gaussian random value. Function logsig is used to control the range of value ξ from 0 to 1. $Iter_{max}$ refers the maximum number of iteration and $Iter_{cur}$ means the value of current iteration. K is a constant used to change the slope of function logsig.

Selecting Operation. The competitive selection strategy is described as follows:

$$x(g+1) = \begin{cases} x_{new}, & if\ fit(x_{new}) \leq fit(x_{old}) \\ x_{old}, & otherwise \end{cases} \tag{4}$$

$$fit(x_i) = \frac{\sum_{n=1}^{k} Err_n}{k} \tag{5}$$

$$Err_n = \frac{FN + FP}{TN + TP + FN + FP} \tag{6}$$

Where $fit(x_{new})$ and $fit(x_{old})$ are the fitness value of the idea x_{new} and x_{old}, respectively. Formula (5) shows that the mean classification error rate after k-fold cross-validation is used to express the fitness value. While TN, TP, FN, FP represent in terms of true negatives, true positives, false negatives and false positives in machine learning [17], respectively.

Specially, both x_{new} and x_{old} are float number vector. A simple repair mechanism is adopted. For example, a feature subset can be processed as $\{1,5,9,11\}$ when $x = \{1.2, 4.5, 9.1, 11.0\}$. Here $\{1, 5, 9, 11\}$ indicates the 1th, 5th, 9th, 11th features are selected. Since the idea are float-number encoding and the feature subsets are integer-version. It will be inevitably appear that same feature appears twice or more within the same feature subset. However, it is not allowed. Defined x_i^j as a recurring feature, A is a set of features that are not in the feature subset. The repair strategy is to replace the x_i^j by randomly selected feature until x_i^j does not exist.

2.2   Reinforcement-Behaved Framework

In a given situation, the rewarded behavior will be strengthened and the punishment one will be weakened. Inspired by this bio-intelligence mode which allows an intelligent individual to gain rewards or penalties from different actions, and then choose the most appropriate actions. The details about the proposed method is described as followed:

Firstly, a reinforcement-behaved framework based on BSO is defined. In our study, for a dataset having a total D number of features. We define the framework as $I = (X^t, R^t, Q^t)$, where t refers t^{th} generation of BSO, and $X_i^t, i \in \{1, 2..., N\}$ represents the population of BSO. Action refers to all possible operations which stored the change of the value at the corresponding position. Specially, each individual has N-dimensional actions, where $N < D$. A matrix of $R_{i,j}^t, i, j \in \{1, 2, ..., D\}$ is created to load all the changes $\left\{ R_i^t \rightarrow R_j^t \right\}$. For example, $x^t = [2, 3, 4]$ and $x^{t+1} = [4, 2, 1]$ are individuals of the previous generation and current generation, respectively. Where $x_1^t = 2$ and $x_1^{t+1} = 4$, thus the action is $\{R_2^t \rightarrow R_4^t\}$. Similarly, the changes of other positions $\{R_3^t \rightarrow R_2^t\}, \{R_4^t \rightarrow R_1^t\}$ could be recorded. Above case is detailed in Fig. 1.

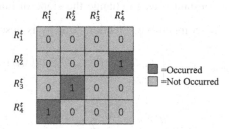

Fig. 1. A 4-dimensional matrix of $R_{i,j}^t$.

An action-utility $D \times D$ matrix Q^t is introduced to record feedback from the previous action, where D is the total number of features. The utility formula is showed as follows:

$$\Delta_i = \begin{cases} + \left| f_i^{t+1} - f_i^t \right|, & \text{if} (f_i^{t+1} - f_i^t) \leq 0 \\ - \left| f_i^{t+1} - f_i^t \right|, & \text{otherwise} \end{cases} \tag{7}$$

$$Q_{m,n}^{t+1} = Q_{m,n}^t + \Delta_i, \quad m, n = 1, 2, ..., N \tag{8}$$

Where i indicates i^{th} individual. t refers the previous generation and $t + 1$ means the current one. m, n is the serial number of feature that recorded the directivity actions.

From the above equations, if the action is deemed beneficial, i.e. $f_i^{t+1} - f_i^t \leq 0$, which means positive encouragement. Otherwise means negative encouragement.

A new strategy based on action-utility matrix Q^t is proposed to improve the feature subset, which could be embedded into the BSO. The update scheme based on the proposed approach is given: Replace the features with other features who have higher returns associated with action-utility matrix Q^t. Since we note that the Q^t could be evaluated the positive weight of each position. Further, the action with higher index should obtain more probability to be applied. In our study, positive actions are defined as the index of the top 10% of all actions. Noteworthy, the adjustment strategy has priority, the higher of the returns, the greater the probability of the position adjustment. In feature adjustment operation, we note that adjusting features will cause redundant feature subset. For example, before adjustment operation, where $Idea_2^t = \{2, 3, 1, 6\}$. However, only action $6 \rightarrow 4$ brings no duplicate features appear in the same feature subset (see Fig. 2).

The details are shown schematically in Fig. 2 which represent the experience gained from state learning.

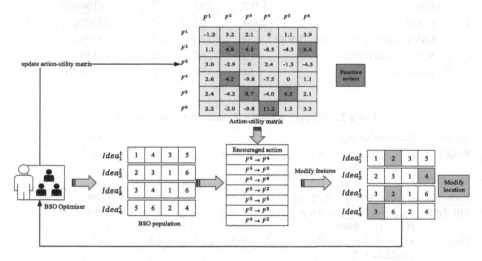

Fig. 2. Reinforcement-Behaved process

3 Experiments and Results

In this section, we utilize 6 benchmark datasets from the UCI Machine Learning Repository to verify the effectiveness of the proposed algorithm [21]. There are total of 4 comparison algorithms from recent literature, including (e.g. basic BSO, MBSO, CLBSO, PSO which has been widely used in optimization research). In all of the methods, the population size of all algorithms is set as 50. The maximum number of iterations is 100. Due to the small size of samples in datasets, 10-fold cross-validation method and k-Nearest Neighbor (k-NN) classifier are implemented. Also, each experiment runs 30 times. Table 1 shows information about datasets and Table 2 describes the parameter setting. '#Features' represents the total number of features, '#Samples' represents the total number of samples (or instances), '#Classes' represents the number of classes. Specially, 'Unbal. ($\pm$)' describes the unbalanced level of datasets, separately (where '+' means the number of positive instances and '$-$' means the number of negative instances). In Table 2, c is constant within formula (3). M represents number of clustering centers. ω means descend range of linear weight, and $c1, c2$ represents individual learning factors and social learning factors are expressed in PSO, respectively. Other parameters are introduced in Sect. 2.

Five evaluation criteria are using for our methods: Accuracy, Precision, Recall, F-measure and AUC (area under curve) score. Accuracy measures classification performance, while Precision, Recall and F-measure are used to assess unbalanced samples. AUC is used to evaluate the generalization of the learner.

Table 1. Datasets description.

Dataset	#Features	#Samples	#Classes	Unbal. ($\pm$)
Sonar	60	208	2	(112/96)
Hill	1212	100	2	(610/602)
Colon	2000	62	2	(40/22)
LSVT voice rehabilitation (LVR)	126	310	2	(42/84)
SPECT Heart (SH)	44	267	2	(212/55)
Ionosphere	34	351	2	(225/126)

Table 2. Wrapper algorithms parameter configurations.

Algorithm	Parameter setting	Reference
BSOFS	$p\_one = 0.2, p\_two = 0.8, p\_one\_center = 0.5, p\_two\_center = 0.5,$ $c = 25, M = 5$	[16]
CLBSOFS	$p\_one = 0.2, p\_two = 0.8, p\_one\_center = 0.5, p\_two\_center = 0.5, M = 5$	[18]
PSOFS	$\omega : 0.9 - 0.4, c1 = 2, c2 = 2$	[19]
MBSOFS	$p\_one = 0.2, p\_two = 0.8, p\_one\_center = 0.5, p\_two\_center = 0.5, c = 25,$ $M = 5, p_r = 0.005$	[20]
Ours	$p\_one = 0.2, p\_two = 0.8, p\_one\_center = 0.5, p\_two\_center = 0.5,$ $c = 25, M = 5$	–

Table 3 shows the best classification accuracy of each wrapper methods among different subset size. Each method contains seven columns, the first five columns are the average performance of the feature selection algorithms with 30 runs and standard deviation are shown in parentheses. The sixth column '#F' indicates the size of selected feature subset, the last column 'Run time(s)' reports total computing time. The highest performance and lowest computing time wherein each dataset are bold.

The result shows that our approach performs better than other SI-based wrapper methods. Although all BSO variants (MBSO, BSO, CLBSO) achieved higher classification accuracy, the proposed method still performed better than theirs on 5 datasets except ionosphere. In binary classification problem, it is unreasonable to evaluate the performance of algorithms only by classification accuracy. Sometimes the Precision, Recall and F-measure are especially important. The specific analysis is as follows:

(1) *Hill dataset*: Our proposed algorithm faintly leads other methods in terms of accuracy. In terms of Precision, Recall and F-measure, the new method outperformed the BSO, MBSO, CLBSO and PSO methods.
(2) *SH dataset:* The proposed method has a slightly higher accuracy, Precision, Recall, F-measure and AUC than other methods.
(3) *LSVT dataset:* In this data set, CLBSO has the worst classification performance and BSO is still leading in all aspects.
(4) *Ionosphere dataset:* For the ionosphere dataset, Table 3 shows that the proposed method performed better than BSO, MBSO, CLBSO and PSO methods in Recall and AUC score, and has a slightly worse Accuracy, precision and F-measure than the MBSO.

Table 3 Average classification results

	Method	Accuracy	Precision	Recall	F-measure	AUC	#F	Run time(s)
Hill	Ours	**0.733(0.014)**	**0.839(0.019)**	**0.577(0.017)**	**0.727(0.014)**	**0.816(0.017)**	15	**1.24e+5**
	MBSOFS	0.729(0.01)	0.827(0.013)	0.574(0.017)	0.719(0.011)	0.806(0.012)	**10**	1.55e+05
	CLBSOFS	0.719(0.01)	0.824(0.011)	0.556(0.018)	0.711(0.011)	0.801(0.01)	20	1.55e+05
	BSOFS	0.727(0.013)	0.833(0.019)	0.569(0.022)	0.72(0.014)	0.805(0.015)	25	1.51e+05
	PSOFS	0.721(0.011)	0.721(0.012)	0.721(0.023)	0.721(0.012)	0.721(0.012)	20	1.47e+05
SH	Ours	**0.868(0.014)**	**0.948(0.011)**	**0.883(0.018)**	**0.816(0.018)**	**0.921(0.013)**	12	**1.02e+5**
	MBSOFS	0.86(0.018)	**0.948(0.013)**	0.871(0.022)	0.807(0.023)	0.916(0.016)	10	1.19e+05
	CLBSOFS	0.84(0.022)	0.922(0.014)	0.872(0.022)	0.773(0.028)	0.887(0.024)	4	1.21e+05
	BSOFS	0.861(0.018)	0.851(0.01)	0.87(0.021)	0.809(0.021)	0.918(0.014)	12	1.20e+05
	PSOFS	0.852(0.02)	0.927(0.015)	0.883(0.019)	0.787(0.028)	0.898(0.022)	4	1.18e+05
LSVT	Ours	**0.814(0.04)**	**0.837(0.064)**	**0.549(0.121)**	**0.763(0.062)**	**0.879(0.043)**	5	**1.01e+5**
	MBSOFS	0.807(0.042)	0.833(0.061)	0.525(0.12)	0.753(0.066)	0.875(0.048)	5	1.19e+05
	CLBSOFS	0.779(0.049)	0.791(0.085)	0.45(0.141)	0.707(0.079)	0.842(0.055)	5	1.18e+05
	BSOFS	0.802(0.036)	0.835(0.058)	0.509(0.107)	0.745(0.057)	0.868(0.04)	5	1.20e+05
	PSOFS	0.807(0.041)	**0.837(0.059)**	0.518(0.116)	0.751(0.064)	0.875(0.043)	5	1.19e+05
Ionosphere	Ours	0.940(0.010)	0.933(0.011)	**0.977(0.011)**	0.934(0.011)	**0.984(0.003)**	8	**1.05e+5**
	MBSOFS	**0.942(0.010)**	**0.936(0.012)**	0.976(0.010)	**0.936(0.011)**	**0.984(0.003)**	6	1.21e+05
	CLBSOFS	0.930(0.008)	0.915(0.010)	0.982(0.006)	0.921(0.010)	0.983(0.002)	12	1.23e+05
	BSOFS	0.938(0.013)	0.935(0.012)	0.972(0.012)	0.932(0.014)	0.983(0.005)	6	1.23e+05
	PSOFS	0.937(0.008)	0.931(0.011)	0.974(0.009)	0.930(0.009)	0.983(0.003)	8	1.20e+05
Colon	Ours	**0.924(0.02)**	**0.963(0.026)**	**0.817(0.053)**	**0.913(0.024)**	**0.962(0.018)**	40	**1.10e+5**
	MBSOFS	0.912(0.037)	0.953(0.037)	0.791(0.092)	0.898(0.049)	0.957(0.025)	50	1.30e+05
	CLBSOFS	0.825(0.061)	0.879(0.086)	0.587(0.168)	0.784(0.089)	0.896(0.05)	30	1.30e+05
	BSOFS	0.9(0.048)	0.934(0.06)	0.772(0.11)	0.885(0.06)	0.946(0.036)	35	1.29e+05
	PSOFS	0.885(0.02)	0.928(0.037)	0.736(0.063)	0.866(0.025)	0.94(0.014)	20	1.28e+05
Sonar	Ours	**0.861(0.015)**	0.929(0.013)	**0.761(0.030)**	**0.858(0.016)**	**0.953(0.007)**	30	**1.01e+05**
	MBSOFS	0.86(0.01)	**0.93(0.02)**	0.75(0.03)	0.85(0.01)	0.95(0.01)	20	1.23e+05
	CLBSOFS	0.836(0.020)	0.920(0.020)	0.710(0.038)	0.830(0.022)	0.937(0.012)	25	1.19e+05
	BSOFS	0.85(0.02)	0.93(0.02)	0.74(0.04)	0.85(0.02)	0.95(0.01)	15	1.20e+05
	PSOFS	0.849(0.02)	0.92(0.02)	0.73(0.04)	0.84(0.02)	0.94(0.01)	20	1.17e+05

(5) *Colon dataset:* For the Colon dataset, Table 3 shows that our proposed method is significantly advanced in all aspects.

(6) *Sonar dataset:* Apart from MBSO within Precision, the proposed method has better classification performance compared to that of the competitor methods.

Specially, there is no evidence that an algorithm tends to pick a smaller size of feature subset. That is worth mentioning, the proposed method runs fastest and saves some computational costs.

4 Conclusion

In this study, a reinforcement-behaved framework combining BSO is proposed. In the proposed method, the population is regarded as a discrete domain, some potential information is extracted from actions between previous and current generation. Each action is regarded as a transform in the different position of the individual, which effectively enhance the learning behavior. Specifically, an action-utility matrix is implemented in which the probabilities of each action are measured whether worth to be exploited, which can effectively reduce the search space and improve search performance. We evaluated the performance of the proposed method in terms of Accuracy, Precision, Recall, F-measure and AUC score on six benchmark datasets. The experimental results show that the proposed method performs better on binary classification task, which outperforms four SI-based feature selection methods. The reinforcement-behaved framework could help search for promising feature subsets. However, the method still has some difficulties in obtaining a satisfactory solution in a short time. In future work, we will investigate the unsupervised strategies to reduce the computational complexity of the proposed feature selection algorithm.

Acknowledgment. This work is partially supported by The National Natural Science Foundation of China (Grants Nos. 71571120, 71271140, 71471158, 71001072, 61472257), Natural Science Foundation of Guangdong Province (2016A030310074, 2018A030310575), Innovation and Entrepreneurship Research Center of Guangdong University Student (2018A073825), Shenzhen Science and Technology Plan (CXZZ20140418182638764), Research Foundation of Shenzhen University (85303/00000155), Research Cultivation Project from Shenzhen Institute of Information Technology (ZY201717), and Innovating and Upgrading Institute Project from Department of Education of Guangdong Province (2017GWTSCX038).

References

1. Bishop, C.: Pattern Recognition and Machine Learning, 1st edn. Springer, New York (2006). https://doi.org/10.1007/978-1-4615-7566-5
2. Friedman, J., Hastie, T., Tibshirani, R.: The Elements of Statistical Learning: Data Mining, Inference, and Prediction, 2nd edn. Springer, New York (2009). https://doi.org/10.1007/BF02985802
3. Guyon, I., Elisseeff, A.: An introduction to variable and feature selection. J. Mach. Learn. Res. **3**, 1157–1182 (2003)

4. Dash, M., Liu, H.: Feature selection for classification. Intell. Data Anal. 1(1–4), 131–156 (1997)
5. Sun, Z., Bebis, G., Miller, R.: Object detection using feature subset selection. Pattern Recogn. 37(11), 2165–2176 (2004)
6. Zhang, Y., Wang, S., Phillips, P., Ji, G.: Binary PSO with mutation operator for feature selection using decision tree applied to spam detection. Knowl.-Based Syst. 64, 22–31 (2014)
7. Gu, S., Cheng, R., Jin, Y.: Feature selection for high-dimensional classification using a competitive swarm optimizer. Soft. Comput. 22(3), 811–822 (2018)
8. Hafiz, F., Swain, A., Patel, N., Naik, C.: A two-dimensional (2-D) learning framework for particle swarm based feature selection. Pattern Recogn. 76, 416–433 (2018)
9. Tran, B., Xue, B., Zhang, M.: Variable-length particle swarm optimisation for feature selection on high-dimensional classification. IEEE Trans. Evol. Comput. 23(3), 473–487 (2018)
10. Ghamisi, P., Benediktsson, J.A.: Feature selection based on hybridization of genetic algorithm and particle swarm optimization. IEEE Geosci. Remote Sens. Lett. 12(2), 309–313 (2015)
11. Das, A.K., Das, S., Ghosh, A.: Ensemble feature selection using bi-objective genetic algorithm. Knowl.-Based Syst. 123, 116–127 (2017)
12. Chen, Y.P., et al.: A novel bacterial foraging optimization algorithm for feature selection. Expert Syst. Appl. 83, 1–17 (2017)
13. Wang, H., Jing, X., Niu, B.: A discrete bacterial algorithm for feature selection in classification of microarray gene expression cancer data. Knowl.-Based Syst. 126, 8–19 (2017)
14. Kashef, S., Nezamabadi-pour, H.: An advanced ACO algorithm for feature subset selection. Neurocomputing 147, 271–279 (2015)
15. Moradi, P., Rostami, M.: Integration of graph clustering with ant colony optimization for feature selection. Knowl.-Based Syst. 84, 144–161 (2015)
16. Shi, Y.: Brain storm optimization algorithm. In: Tan, Y., Shi, Y., Chai, Y., Wang, G. (eds.) ICSI 2011. LNCS, vol. 6728, pp. 303–309. Springer, Heidelberg (2011). https://doi.org/10.1007/978-3-642-21515-5_36
17. Sokolova, M., Lapalme, G.: A systematic analysis of performance measures for classification tasks. Inf. Process. Manag. 45(4), 427–437 (2009)
18. Sun, C., Duan, H., Shi, Y.: Optimal satellite formation reconfiguration based on closed-loop brain storm optimization. IEEE Comput. Intell. Mag. 8(4), 39–51 (2013)
19. Zhan, Z., Zhang, J., Shi, Y., Liu, H.: A modified brain storm optimization. In: IEEE Congress on Evolutionary Computation, pp. 1–8 (2012)
20. Kennedy, J., Eberhart, R.: Particle swarm optimization. Int. Conf. Neural Netw. 4, 1942–1948 (1995)
21. UCI Machine Learning Repository. http://archive.ics.uci.edu/ml/index.php. Accessed 15 May 2019

Shuffle Single Shot Detector

Yangshuo Zhang[✉], Jiabao Wang, Zhuang Miao, Yang Li,
and Jixiao Wang

Army Engineering University of PLA, Nanjing 210007, China
17625944869@163.com, jiabao_1108@163.com,
emiao_beyond@163.com, solarleeon@outlook.com,
jixiao_wang@126.com

Abstract. Real-time object detection is of great significance to embedded mobile platforms. We propose a lightweight object detection network for embedded devices, which we call ShuffleSSD. The ShuffleNet V2 network is used as the backbone, which can effectively reduce the size of the model. In the proposed lightweight detection model, a single shot multi-box detector is adopted and a receptive field block is integrated to obtain high-quality detection results. The evaluation is performed on the public object detection data set (PASCAL VOC) and compared to the most advanced real-time object detection network. The experimental results show that the proposed network has higher detection accuracy than MobileNet-SSD, with smaller network parameters. It is more suitable for real-time object detection on embedded devices.

Keywords: Real-time object detection · ShuffleNet · Single shot detector

1 Introduction

Object detection is a long-standing basic problem in the field of computer vision. It has been an active research field for decades. It is the basis of solving more complex and higher-level visual tasks such as segmentation, object tracking, image description, and event detection and action recognition. Object detection is widely used in many fields of artificial intelligence and information technology, including robot vision, autopilot, human-computer interaction, content-based image retrieval and intelligent video surveillance.

In recent years, with the rise of deep learning technology, significant breakthroughs have been made in the field of object detection. However, a deep learning model usually contains millions or even tens of millions of parameters and dozens or even dozens of layers of networks. A huge number of parameters bring performance improvement, but also bring the disadvantages of huge network parameters and slow computation. For example, the classic VGG network [1] has 140 million parameters, which is not conducive to the application of deep learning models to embedded devices. Therefore, how to efficiently perform real-time image detection on embedded devices has become a major research challenge.

Therefore, in order to run models on the embedded devices while maintaining high detection accuracy, this paper proposes a lightweight object detection model based on

© Springer Nature Switzerland AG 2019
D.-S. Huang et al. (Eds.): ICIC 2019, LNAI 11645, pp. 682–691, 2019.
https://doi.org/10.1007/978-3-030-26766-7_62

ShuffleNet V2 [2]. The backbone of the network performs feature extraction through the ShuffleNet V2 network, which can effectively reduce the size of the model. The single shot multi-box detector is introduced from the SSD network [3], which effectively improves the detection quality. At the same time, the proposed lightweight detection model introduces Receptive Field Block (RFB) [4] to improve the detection accuracy. The model is tested on the public PASCAL VOC dataset [5]. Compared with MobileNet-SSD [6], it has higher detection accuracy with smaller network parameters, and is more suitable for real-time object detection on embedded devices.

2   Related Work

Object detection technology has wide application in intelligent transportation system, intelligent monitoring system and military object detection. Most of the early object detection algorithms were built based on labor-designed features. Due to the lack of effective image feature representation before the birth of deep learning, people have to do their best to design more diversified detection algorithms to compensate for the defects in labor-designed feature expression. Beginning in 2013, the combination of convolutional neural networks and candidate region algorithms made a huge breakthrough in object detection and opened up a craze based on deep learning object detection. These detection frameworks can be divided into two main categories: (1) a two-stage detection framework that includes a pre-processing for the regional proposal, so that the overall process has two stages. (2) A one-stage detection framework, that is, without a regional proposal, the framework does not separate the detection and proposal processes, making the entire process a single-stage.

The most important part of the two-stage detection framework is the work of the R-CNN [7] series. After that, there are many related works to improve the R-CNN series for improving the object detection performance. In 2014, Girshick designed the R-CNN framework to integrate AlexNet [8] with the selective search of regional proposal methods, which made a huge breakthrough in object detection. He et al. proposed that SPP-Net [9] only needs to extract CNN features from the whole graph, which effectively solves the problem of double counting of convolutional layers. Girshick proposed the Fast-R-CNN algorithm [10], which uses a simplified SPP layer-RoI (Region of Interesting) Pooling layer, and introduces multi-task learning to integrate multiple steps into one model. He et al. proposed Faster-RCNN algorithm [11] by introducing the RPN (Region Proposal Networks) network, which enabled regional proposal, classification, and regression to share convolution features, which led to further acceleration.

The one-stage detection framework can obtain faster calculation speeds because there is no intermediate regional proposal process and the prediction results are directly obtained from the images. Therefore, it has become a research hotspot of object detection. At present, one-stage detection networks include YOLO, SSD and their variants. Redmon et al. proposed the YOLO [12] network, adopting the idea of mesh division, directly taking the whole image as the input of the network, and directly obtaining the position and object category of the object bounding box through only one forward propagation. Based on YOLO, Redmon and Farhadi proposed an improved version of YOLO v2 [13], using a smaller network of Darknet19, and many strategies, such as batch standardization, removing the fully connected layer, and more accurate anchor frames with k-means and multi-scale training. YOLOv3 [14] uses a newly designed Darknet53 network by combining multi-scale prediction, reaching SSD accuracy with only one-third of the time. Liu et al. proposed the SSD network in 2016 [3], and the network also adopted the idea of meshing. The difference from the two-stage detection framework is that it integrates all operations into one convolution network. At the same time, prediction is performed on convolutional feature maps of different scales to detect objects of different sizes. In 2017, some of the more advanced single-stage detectors, DSSD [15] and RetinaNet [16] update their original lightweight backbones by the deeper ResNet-101 and apply certain techniques, such as deconvolution [15] or Focal Loss [16], whose scores are comparable and even superior to the ones of state-of-the-art two-stage methods. However, such performance gains largely consume their advantage in speed. In 2018, Zhang et al. proposed a new one-stage detector, RefineDet [17], consisting of two interconnected modules that have high precision while maintaining high efficiency. In 2018, Liu et al. proposed RFB-Net, which can enhance the feature extraction ability of the network by simulating the receptive field of human vision. It introduces Receptive Field Block (RFB) [4], which can achieve fast computation speed while taking into consideration the accuracy.

Although the above-mentioned deep learning methods have achieved good results in object detection, the problems of huge parameters and long time-consuming are not solved, and the models can not be directly deployed to front-end mobile and embedded devices for object detection. In recent years, researchers have focused on exploring lightweight deep neural network that are more suitable for embedded devices. MobileNetV1 [6] uses deep separable convolution to construct a lightweight deep neural network. MobileNetV2 [20] designed a backward residual structure with a linear bottleneck for computing resource constraints, which can reduce model parameters and calculations. SqueezeNet [21] proposed the Fire module to compress model parameters, reducing the depth of the network and the size of the model. ShuffleNet [22] uses group convolution and depth separable convolution to construct lightweight network. Kim et al. [19] developed PVANET by concatenating 3×3 convolutional layer as a building block for the initial feature extraction stage. Wang et al. [18] proposed PeleeNet that uses a combination of parallel multi-size kernel convolutions as a 2-way dense layer. They apply a residual block after feature extraction stage to improve the accuracy. Above networks can be deployed on the embedded devices, to efficiently perform object detection tasks.

3 Method

In this section, the network architecture is presented in detail. The ShuffleNetV2 is used as our backbone, which is designed for mobile object recognition to extract high-level features.

3.1   Network Architecture

In order to improve the efficiency of running the object detection model on embedded devices, we use ShuffleNet V2 as the backbone, to obtain a series of feature maps with reduced resolution and increased receptive field. As shown in Fig. 1, the architecture consists of three parts: the backbone, the RFB module, and the extra layers.

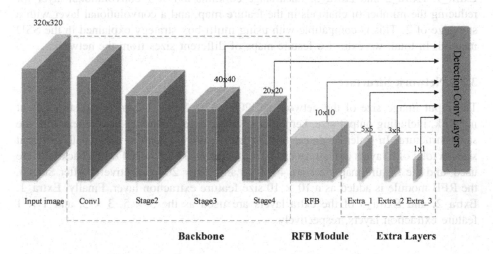

Fig. 1. The architecture of ShuffleSDD. The modified ShuffleNet V2 is used as the backbone, and the RFB module and the extra layers are added to obtain different scale feature maps.

Backbone. It consists of four stages: Conv1, Stage2, Stage3 and Stage4 of the Shuf-fleNet V2 model. We remove the first max-pooling layer from the original network. The reason is that the max-pooling operation destroys the receptive field. In Conv1, a 3 × 3 convolutional layer is applied to the input image with a stride of 2, which reduce the input size by a factor of 2. Stage2, Stage3 and Stage4 are performed with multiple repetitive units. After the Conv1 three stages containing multiple units are performed. The bottleneck is used as a basic unit in ShuffleNet V2 [2]. Stage2 and Stage4 contain 3 bottlenecks respectively while Stage3 is composed of 7 bottlenecks. From Conv1 to Stage4, the size of input image has 16× down-sampling.

RFB Module. It is a multi-branch convolutional block. The inner structure can be divided into two components: the multi-branch convolution layer with different kernels and the dilated convolution. The former part is identical to that of Inception [23], adopts multiple branches with different kernel sizes to capture multi-scale information. To be specific, first, we employ the bottleneck structure in each branch, consisting of a 1 × 1 convolutional layer, to decrease the number of channels in the feature map plus an $n \times n$ convolutional layer. Second, we replace the 5 × 5 convolutional layer by two stacked 3 × 3 convolutional layer to reduce parameters and deeper non-linear layers. The dilated convolution part has rapidly proved competent at semantic segmentation

[22]. The aim of this structure is to generate feature maps of a higher resolution, capturing information at a larger receptive field while keeping the same number of parameters.

Extra Layers. In order to enrich the extracted features from the intermediate layers, we perform extra feature layers in Extra layers. The extra layers include three stages of Extra_1, Extra_2 and Extra_3. Each stage contains a 1 × 1 convolutional layer for reducing the number of channels in the feature map, and a convolutional layer with a step size of 2. This is compatible with using multi-box strategy explained in the SSD method. In total, we extract 6 feature maps of different sizes from the network.

3.2 Network Structure

The input image size of the network is 320 × 320. Table 1 depicts the structure of network, including output size, kernel size, stride, output channels and repeat with the same structure. In order to detect objects from different feature maps with different sizes, a total of 6 layers of features are extracted. Stage3 and Stage4 of the backbone are used, and the feature map sizes are 40 × 40 and 20 × 20, respectively. After Stage4, the RFB module is added as a 10 × 10 size feature extraction layer. Finally, Extra_1, Extra_2, and Extra_3 of the extra layers are used as the 5 × 5, 3 × 3, and 1 × 1 feature extraction layers, respectively.

Table 1. Overall architecture of ShuffleSSD

Layer	Output size	Kernel size	Stride	Repeat	Output channels
Image	320 × 320				3
Conv1	160 × 160	3 × 3	2	1	24
Stage2	80 × 80	3 × 3	2	1	116
			1	3	
Stage3	40 × 40	3 × 3	2	1	232
			1	7	
Stage4	20 × 20	3 × 3	2	1	464
			1	3	
RFB	10 × 10		2	1	512
Extra_1	5 × 5	1 × 1	1	1	128
		3 × 3	2	1	256
Extra_2	3 × 3	1 × 1	1	1	128
		3 × 3	2	1	256
Extra_3	1 × 1	1 × 1	1	1	64
		3 × 3	1	1	128

3.3   Objective Function

Similar to SSD [3], the objective functions includes confidence loss and localization loss, where the localization loss uses the smooth L1 loss. The overall objective function is a weighted sum of the localization loss (loc) and the confidence loss (conf):

$$L(x, c, l, g) = \frac{1}{N}\left(L_{conf}(x, c) + \alpha L_{loc}(x, l, g)\right) \tag{1}$$

N refers the number of default boxes that match the ground truth box. x refers whether the matched default box belongs to class p, and the value belongs to $\{0,1\}$. L_{loc} refers the Smooth L1 loss between the predicted box (l) and the ground truth box (g) parameters. L_{conf} refers the softmax loss over multiple classes confidences (c) and the weight term is set to 1 by cross validation.

In training stage, several strategies are used to improve the effect of training, including data augmentation, hard negative mining, scale and aspect ratios for default boxes.

4   Experiments and Analysis

4.1   Implementation

The experiments are conducted based on Ubuntu 16.04 operating system (2.6 GHz clock speed, 64 GB memory) and Pytorch library. All training and testing are done with NVIDIA GTX 1080TI GPU. We set the batch size at 32 and the initial learning rate at 10^{-3} as in the original SSD. We use a "warmup" strategy that gradually ramps up the learning rate from 10^{-6} to 4×10^{-3} at the first 5 epochs. After the "warmup", it goes back to the original learning rate schedule, divided by 10 at 150 and 200 epochs. The total number of training epochs is 250. A weight decay of 0.0005 and a momentum of 0.9 are adopted in experiments.

4.2   Dataset and Evaluation

We conduct experiments on the Pascal VOC datasets [5], which have 20 object categories. In this experiment, we train our ShuffleSSD with VOC2007 trainval set and VOC2012 trainval set and test our model with VOC2007 test set. The metric to evaluate detection performance is the mean Average Precision (mAP).

4.3   Comparison with Other Architectures

We compare our method with other architectures on the VOC2007 test set. Table 2 shows that our method can achieve competitive performance while having significantly less computation cost compared with the state of the art methods. We re-implement SSD and RFB in the same environment as that of ShuffleSSD. Due to the use of ShuffleNet V2 as the backbone, our ShuffleSSD is significantly smaller than the

VGG-SSD and VGG-RFB models, with an accuracy of mAP of 71.2%. Compared to the lightweight detection network, our method is more accurate than Mobilenet-SSD and ShuffleNetV2-SSD.

Table 2. Comparison of detection methods on the PASCAL VOC 2007 test set. All runtime information is computed on a Graphics card of Geforce GTX 1080TI

Method	MAP	FPS	Model size (MB)
VGG-SSD	77.2	46	100.3
MobileNet-SSD	66.7	83	23.1
ShuffleNet V2-SSD	65.9	76	16.4
VGG-RFB	80.7	37	139.5
MobileNet-RFB	69.7	71	25.6
ShuffleSSD	71.2	56	10.7

In order to more intuitively show the effect of our method. We use the same image to display on the VOC2007 test dataset. In Fig. 2, we show some detection examples on VOC 2007 test with the ShuffleDet model. In Fig. 3, we show some detection examples on VOC 2007 test with the ShuffleNetV2-SSD model. It can be seen from the comparison of the effects of Figs. 2 and 3 that our method has high detection precision, especially for the detection of small targets. For example, in the fourth picture of the first line of Fig. 2, our method detected the edge of the plane and the small target person, but the ShuffleNetV2-SSD model did not detect it. This is because our model uses a 40 × 40 larger resolution feature map.

Fig. 2. Detection examples on VOC 2007 test with ShuffleSSD model.

Fig. 3. Detection examples on VOC 2007 test with ShuffleNetV2-SSD model.

5 Conclusion

In this paper, we propose ShuffleSSD, a new lightweight network detection model based on ShuffleNet V2. It can perform real-time object detection on embedded devices. In this method, the method performs feature extraction through the ShuffleNet V2 network, which can effectively reduce the size of the model. Moreover, we remove the maxpooling layer in shuffleNet V2 to reduce the loss of information while introducing a marginal computation burden. Furthermore, we show RFB module with dilated convolutions are effective modules for enhance the accuracy. Experimental results on the PASCAL VOC datasets indicate that ShuffleSSD outperforms MobileNet-SSD while it is more accurate and computationally efficient. It is more suitable for real-time object detection on embedded devices.

References

1. Simonyan, K., Zisserman, A.: Very deep convolutional networks for large scale image recognition. In: ICLR (2015)
2. Ma, N., Zhang, X., Zheng, H.-T., Sun, J.: ShuffleNet V2: practical guidelines for efficient CNN architecture design. In: Ferrari, V., Hebert, M., Sminchisescu, C., Weiss, Y. (eds.) Computer Vision – ECCV 2018. LNCS, vol. 11218, pp. 122–138. Springer, Cham (2018). https://doi.org/10.1007/978-3-030-01264-9_8

3. Liu, W., et al.: SSD: single shot multibox detector. In: Leibe, B., Matas, J., Sebe, N., Welling, M. (eds.) ECCV 2016. LNCS, vol. 9905, pp. 21–37. Springer, Cham (2016). https://doi.org/10.1007/978-3-319-46448-0_2

4. Liu, S., Huang, D., Wang, Y.: Receptive field block net for accurate and fast object detection. In: Ferrari, V., Hebert, M., Sminchisescu, C., Weiss, Y. (eds.) ECCV 2018. LNCS, vol. 11215, pp. 404–419. Springer, Cham (2018). https://doi.org/10.1007/978-3-030-01252-6_24

5. Everingham, M., Van Gool, L., Williams, C.K., Winn, J., Zisserman, A.: The Pascal Visual Object Classes (VOC) challenge. Int. J. Comput. Vision **88**(2), 303–338 (2010)

6. Howard, A.G., et al.: Mobilenets: efficient convolutional neural networks for mobile vision applications. arXiv preprint arXiv:1704.04861 (2017)

7. Girshick, R., Donahue, J., Darrell, T., Malik, J.: Rich feature hierarchies for accurate object detection and semantic segmentation. In: IEEE Conference on Computer Vision and Pattern Recognition, Columbus, pp. 580–587. IEEE (2014)

8. Krizhevsky, A., Sutskever, I., Hinton, G.E.: ImageNet classification with deep convolutional neural networks. In: Conference and Workshop on Neural Information Processing Systems, Harrahs, pp. 1106–1114 (2012)

9. He, K., Zhang, X., Ren, S., Sun, J.: Spatial pyramid pooling in deep convolutional networks for visual recognition. In: Fleet, D., Pajdla, T., Schiele, B., Tuytelaars, T. (eds.) ECCV 2014. LNCS, vol. 8691, pp. 346–361. Springer, Cham (2014). https://doi.org/10.1007/978-3-319-10578-9_23

10. Girshick, R.: Fast R-CNN. In: IEEE International Conference on Computer Vision, Piscataway, pp. 1440–1448. IEEE (2015)

11. Ren, S., He, K., Girshick, R., Sun, J.: Faster R-CNN: towards real-time object detection with region proposal networks. In: Conference and Workshop on Neural Information Processing Systems, Montreal, pp. 92–99 (2015)

12. Redmon, J., Divvala, S., Girshick, R., Farhadi, A.: You only look once: unified, real-time object detection. In: IEEE Conference on Computer Vision and Pattern Recognition, Las Vegas, pp. 779–788. IEEE (2016)

13. Redmon, J., Farhadi, A.: YOLO9000: better, faster, stronger. In: IEEE Conference on Computer Vision and Pattern Recognition, Piscataway, pp. 6517–6525. IEEE (2017)

14. Redmon, J., Farhadi, A.: YOLOv3: an incremental improvement. arXiv preprint arXiv:1804.02767 (2018)

15. Fu, C.Y., et al.: DSSD: deconvolutional single shot detector. arXiv preprint arXiv:1701.06659 (2017)

16. Lin, T.Y., Goyal, P., Girshick, R., He, K., Dollar, P.: Focal loss for dense object detection. In: International Conference on Computer Vision, Venice, pp. 2999–3007. IEEE (2017)

17. Zhang, S., Wen, L., Bian, X., Lei, Z., Li, S.Z.: Single-shot refinement neural network for object detection. In: IEEE Conference on Computer Vision and Pattern Recognition, Piscataway, pp. 4203–4212. IEEE (2018)

18. Wang, R.J., Li, X., Ao, S., Ling, C.X.: Pelee: a real-time object detection system on mobile devices. arXiv preprint arXiv:1804.06882 (2018)

19. Kim, K.H., Hong, S., Roh, B., Cheon, Y., Park, M.: PVANET: deep but lightweight neural networks for real-time object detection. arXiv preprint arXiv:1608.08021 (2018)

20. Sandler, M., Howard, A., Zhu, M., Zhmoginov, A., Chen, L.C.: Inverted residuals and linear bottlenecks: mobile networks for classification, detection and segmentation. arXiv preprint arXiv:1801.04381 (2018)

21. Iandola, F.N., et al.: Squeezenet: alexnet-level accuracy with 50x fewer parameters and <0.5 MB model size. arXiv preprint arXiv:1602.07360 (2016)

22. Zhang, X., Zhou, X., Lin, M., Sun, J.: Shufflenet: an extremely efficient convolutional neural network for mobile devices. arXiv preprint arXiv:1707.01083 (2017)
23. Szegedy, C., Ioffe, S., Vanhoucke, V., Alemi, A.A.: Inception-v4, Inception-ResNet and the impact of residual connections on learning. In: AAAI, San Francisco, pp. 4278–4284 (2017)
24. Chen, L.C., Papandreou, G., Schro, F., Adam, H.: Rethinking atrous convolution for semantic image segmentation. arXiv preprint arXiv:1706.05587 (2017)

Feasibility of a Non-immersive Virtual Reality Training on Functional Living Skills Applied to Person with Major Neurocognitive Disorder

Simonetta Panerai[1], Valentina Catania[1], Francesco Rundo[1],
Vitoantonio Bevilacqua[2(✉)], Antonio Brunetti[2], Claudio De Meo[2],
Donatella Gelardi[1], Claudio Babiloni[3], and Raffaele Ferri[1]

[1] Oasi Research Institute - IRCCS, Troina, EN, Italy
[2] Department of Electrical and Information Engineering,
Polytechnic University of Bari, Bari, Italy
vitoantonio.bevilacqua@poliba.it
[3] Department of Physiology and Pharmacology "Vittorio Erspamer",
Sapienza University of Rome, Rome, Italy

Abstract. The treatment of Major and Mild NeuroCognitive Disorder (M-NCD, m-NCD) include pharmacological and non-pharmacological therapies, including cognitive training, recreational therapies, exercise, and interventions using technological tools. Proper cognitive training could be effective in preventing the evolution towards more severe cognitive impairment forms. A rehabilitative approach based on the use of Virtual Reality (VR) could be an effective and incentive tool to help prevent and early diagnose.

In this paper, we propose a suite of serious games that stimulates the rehabilitative tasks in order to verify both the feasibility of a non-immersive virtual training on daily living skills for patients with M-NCD and the generalisation of the improved skills in the natural environment.

Results show that virtual training could be a feasible approach for improving the generalisation capabilities of people affected by M-NCD doing activities of daily living. Feasibility, portability, the possibility of customisation and attractivity make Virtual Training an interesting approach for complementing and enriching the existing rehabilitation strategies.

Keywords: Virtual Reality (VR) · Virtual Training (VT) ·
Instrumental Activities of Daily Living (IADL) ·
Major NeuroCognitive Disorder (M-NCD) · Mild Cognitive Disease (MCD) ·
Cognitive rehabilitation

1 Background

The innovative use of technology-based interventions for people with early or prodromal dementia is a growing area of study and offers many possibilities for improving the independence and the quality of life of people with acquired neurocognitive disorders. In the Desktop Visual Display Systems (DVDS) the person interacts with a monitor displaying 3D objects and environments, along with auditory and visual stimuli, making such environments even more similar to the real one. Using such a kind of system is simple, requires limited instrumentation and short training for new users. Only a few studies focusing on functional living skills used Virtual Training (VT) with patients affected by dementia [1–3].

Functional living skills include complex activities of daily life, such as cooking, phoning, cleaning, washing laundry, shopping, taking medicines, and using transport. When these skills deteriorate, the patient loses his/her self-sufficiency and, consequently, his/her self-esteem and well-being. These skills require a great neuropsychological organisation and, in patients with Major NeuroCognitive Disorder (M-NCD), progressively deteriorate as a result of the cognitive decline. Therefore, the interventions for slowing down this decline are of fundamental importance.

In this context, for assessment and rehabilitation purposes, serious games in Augmented (AR) or Virtual Reality (VR) could be a good alternative to the traditional approach to interact with patients, collect statistics, recognise and classify patterns, and design unsupervised machine learning algorithms for clustering their cognitive skills and performance [4–6].

The results of earlier studies on Virtual Reality applied to the empowerment or evaluation of functional living skills are encouraging, and it appears that VT has the same effectiveness of rehabilitation approaches carried out in the natural environment. The study by Hoffmann et al. [1] showed improvements in a shopping task, with decreased errors and times; a follow-up after three weeks showed that people maintained their skills; the participants to the study also appreciated the proposed interactive modality. Van Schaick et al. [2] investigated both facilitators and barriers for a walk in a city centre, highlighting a performance improvement in this task. Foloppe et al. [3] used VR with a patient with Alzheimer's disease to relearn cooking activities, suggesting that VR can produce improvements in the same way as real-life relearning.

2 Objective

Our study aimed to verify the feasibility of a non-immersive VT on daily living skills for patients with M-NCD and to verify the eventual generalisation of improved skills repeating the same tasks in the natural environment, by administering the in vivo tests before (T1) and after (T3) the VT.

3 Materials and Methods

3.1 Participants

Participants were recruited during the years 2017–2018 in the Neurorehabilitation Department of the Oasi Research Institute in Troina (EN, Italy) and randomly assigned to the Experimental Group (EG) or Control Group (CG). Table 1 shows the characteristics of the two groups. All participants benefited from a group cognitive stimulation intervention [7].

Table 1. Characteristics and statistical differences between the Experimental and Control groups (EG, CG) at T1.

	EG	CG	p=
Dementia Etiology, N			
Vascular	2	1	ns[a]
Alzheimer	1	4	
Fronto-temporal	2	3	
Other	1	1	
M-NCD severity level			
Mild	5	5	ns[a]
Moderate	1	4	
Gender, M/F, N	2/4	2/7	ns[a]
Chronological age, median (interquartile range)	68.5 (63.5/70.5)	65 (61/69)	ns[b]
Education years, N	6.5 (5/8)	13 (5/13)	ns[b]
Mini Mental State Examination, age/education corrected score, median (interquartile range)	19.39 (17.72/20.27)	18.86 (15.99/20.46)	ns[b]
Milan Overall Dementia Assessment, corrected scores, median (interquartile range)	69.7 (67.2/70.85)	65.8 (62.63/78.8)	ns[b]
Coloured Progressive Matrices, corrected scores, median (interquartile range)	23.25 (20.88/26)	19 (17.5/24.5)	ns[b]
Visuo-spatial span (Corsi's test), corrected scores, median (interquartile range)	4.15 (3.64/5)	3.75 (3/4)	ns[b]
Digit span, corrected scores, median (interquartile range)	4.25 (3.64/5)	4 (3.75/4.75)	ns[b]
Rey's 15 words, Immediate Recall, corrected scores, median (interquartile range)	29.9 (28.3/32.63)	17.40 (16.8/26.1)	ns[b]
Rey's 15 words, Delayed Recall, corrected scores, median (interquartile range)	6.05 (5.6/6.65)	3.8 (0/4.7)	ns[b]
ADL, raw scores, median (interquartile range)	4 (4/5.5)	5 (4/6)	ns[b]
IADL %, median (interquartile range)	50 (24.38/62.5)	50 (50/87.5)	ns[b]

Legenda: [a]Chi–square test; [b]Mann Whitney's U test.

The inclusion criteria were: (a) DSM-5 diagnostic criteria for M-NCD [8]; (b) score of 1 or 2 at the Clinical Dementia Rating (CDR, Hughes et al. [9]); (c) score between 10 and 23 at the Mini-Mental State Examination (MMSE, Folstein et al. [10]); (d) loss of one or more Instrumental Activities of Daily Living (IADL, Lawton and Brody [11]); (e) sufficient sight, hearing and motor functioning; (f) sufficient communication skills; (g) maintained reading skills.

On the contrary, the criteria for exclusion were: (a) Mild Cognitive Impairment, or severe dementia, or Aphasia diagnosis; (b) score ≤ 0.5 or ≥ 3 at the CDR;

(c) score < 10 or ≥ 24 at MMSE; (d) severe motor, hearing and sight problems; (e) loss of reading skills.

3.2 System and Apps Descriptions

A digital system was set up with a server connecting a database to the apps installed on the touch TV. The database was developed in PostgreSQL with an interface implemented as a C# Dynamic-Link Library (DLL); the system contains a suite of apps created with Unity 3D, a cross-platform real-time game engine, often employed for developing VR scenarios for rehabilitation purposes [12]. Four apps were developed, focusing on four functional skills, and namely:

1. *to provide information* (or *information*): 30 questions in verbal and written form, including general knowledge, personal, family, spatial and temporal orientation, and with multiple-choice answers appearing on the screen in written form;
2. *taking medicines* (or *medicines*): at appropriate times, the scene presents five medicine boxes placed on a kitchen table (Fig. 1D); a set of instructions explains when each drug should be taken; the implemented task consists in choosing the appropriate box by touching one of the shown medicine boxes, for 10 times, as a response to 10 verbal questions, presented randomly during each session;

Fig. 1. App's scenes (A-B-C-D) and examples of the in vivo test (E-F-G-H-I-L).

3. *preparing the suitcase* for a weekend away from home (or *suitcase*): a single scene, with shelves containing clothes to be placed into a suitcase on a sofa (Fig. 1C);

4. *shopping at the supermarket* following a shopping list (or *supermarket*): the shopping list includes five products and remains available on the screen. The first scene is a kitchen, with the shopping list, money and a wallet; the second scene includes a supermarket shelf with different products and a shopping cart (Fig. 1A). The third scene includes the cash counter to pay for products (Fig. 1B); the application also requires the user to pay the right or enough amount due to the cashiers, and check the possible change returned.

In addition, a fifth app, not yet included in the suite of games, has been developed:

5. *Preparing a juice* for two persons (or *juice*): a single scene with a kitchen (Fig. 2) showing several kinds of fruit on a shelf. When the scene starts, the user is asked to choose two types of fruit, randomly selected, and must complete all the steps required for making the juice, serve it and clean the dishes.

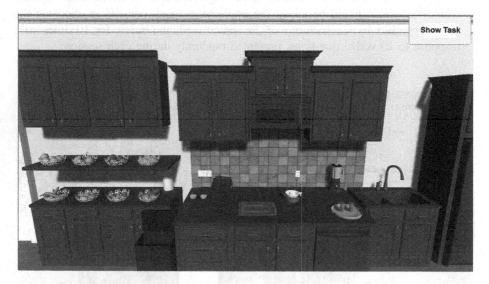

Fig. 2. Scene from *Preparing a juice* app.

Supermarket, *suitcase* and *juice* offered the possibility of a video demonstration before starting the task. The apps were developed based on some principles and procedures of Applied Behaviour Analysis, including:

- verbal reinforcement after the correct response (i.e., good!, ok!, congratulations!, well done!);
- correction after the wrong response, consisting of question repetition and use of the least-to-most prompting (up to a maximum of three prompts);
- task analysis and total task chaining for the *supermarket* and *suitcase* apps.

The first instruction, concerning how to perform the task, was given simultaneously in a verbal and written manner. The application of the procedures previously reported is fundamental for encouraging and stimulating people to perform the rehabilitative sessions in the desired way over time.

3.3 Procedures

This work is a randomised controlled study, including an EG receiving VT for 10 sessions, and a CG without VT treatment. The VT sessions included the performance from the first four apps only (i.e., *information*, *medicines*, *suitcase* and *supermarket*) as the preparing a juice app will be included in the suite of apps only in the next release of the software. The study included the following phases:

– *T1: pre-treatment assessment* for both EG and CG, including the administration of a neuropsychological tests battery as well as the first in vivo test. Neuropsychological battery included MMSE, MODA [13], CPM [14], visuo-spatial Corsi's span and Digit span [15], 15 Rey's words [16], Activities of Daily Living (ADL) [17], and IADL. In vivo tests were performed into adequately arranged environments at the hospital. During the administration, no prompts or reinforcements were provided. Both the number of correct responses and the total execution times were recorded. Table 2 reports the results and the statistical differences between the two groups for the in vivo tests at T1.

Table 2. Results and statistical differences between the Experimental and Control groups for the T1 in vivo tests.

T1 In vivo test	EG	CG	p=
Information, N. of correct answers, median (interquartile range)	26.5 (22.25/28.5)	24 (22/26)	ns[b]
Information, total time execution in seconds, median (interquartile range)	283 (262/316)	349 (289/370)	ns[b]
Suitcase, N. of correct answers, median (interquartile range)	6 (4.5/8.25)	5 (3/9)	ns[b]
Suitcase, total time execution in seconds, median (interquartile range)	194 (142.25/213.75)	303 (192/483)	ns[b]
Medicines, N. of correct answers, median (interquartile range)	7 (6/8)	5 (4/6)	ns[b]
Medicines, total time execution in seconds, median (interquartile range)	190 (177/203)	242 (213/418)	ns[b]
Supermarket, N. of correct answers, median (interquartile range)	10.5 (7/11.75)	10 (5/12)	ns[b]
Supermarket, total time execution in seconds, median (interquartile range)	403 (336/481)	344 (310/381)	ns[b]

Legenda: [a]Chi–square test; [b]Mann Whitney's U test.

- *T2: Virtual Training* lasting 10 sessions, administered only to EG. The following data were recorded: total execution time, time spent for each response, number of correct responses (for the *supermarket* and *suitcase* tasks, the number of correct task steps), number of errors, number of missing responses (participants not answering within 10 s), number of prompts provided to obtain a response.
- *T3: post-treatment in vivo test* including the second administration of in vivo tests for both EG and CG.

The staff involved in the VT consisted of a project coordinator, a neuropsychologist, and electronic engineers. In vivo tests were administered by a different psychologist working in the diagnostic service of the Oasi Research Institute, blind to the aims of the study.

3.4 Statistical Analysis

Most of the variables analysed in this study did not show a normal distribution (asymmetry and kurtosis tests). Thus, non-parametric statistics were used. The Friedman test for repeated measures was used for analysing changes along with the VT sessions; the significance level was set at $p < 0.05$. Comparisons of differences between T1 and T3 in vivo tests administered to EG and CG, and comparison between EG and CG, both in education years and chronological age, were carried out using the Mann-Whitney's U test; the significance level was set at $p < 0.05$. The Chi-Square test was used for comparisons between aetiology, dementia severity level and gender.

3.5 Ethics Committee Approval

Approval was obtained from the Local Committee (2017/05/31/CE-IRCCS-OASI/9). All participants provided informed consent before the onset of the study.

4 Results

At T1, EG and CG showed no significant differences in chronological age, disorder aetiologies and severity, gender, education, results at the neuropsychological battery and in vivo test (Table 1). Statistically significant differences, reported in Table 3 and Fig. 3, were found between the two in vivo tests (T3 vs T1) in the number of correct answers (*suitcase, medicines* and *supermarket*) and in execution time (*medicines*).

During the 10 VT sessions, statistically significant differences were found in the number of correct answers and the number of prompts used for the *suitcase* and *supermarket* tasks, in the number of missing responses for *information, medicines* and *supermarket* tasks and in the total execution time for all the apps. Table 4 and Fig. 4 report the obtained results.

Table 3. Statistical significance of the differences between T1 and T3 in vivo test.

	EG	CG	p=
Information, N of correct answers, median (interquartile range)	1.5 (1/2.75)	−1 (−1/1)	ns
Information, total time execution in seconds, median (interquartile range)	−54 (−58.75/−40.25)	−18 (−62/−4)	ns
Suitcase, N of correct answers, median (interquartile range)	4 (1.75/4.75)	0 (−2/0)	0.008
Suitcase, total time execution in seconds, median (interquartile range)	−0.5 (−2075/28.75)	−92 (−212/−13)	ns
Medicines, N of correct answers, median (interquartile range)	2 (2/3)	0 (0/0)	0.032
Medicines, total time execution in seconds, median (interquartile range)	−57 (−71/−41)	37 (26−58)	0.008
Supermarket, N of correct answers, median (interquartile range)	2.5 (1.25/3.75)	0 (−1/1)	0.05
Supermarket, total time execution in seconds, median (interquartile range)	−84.5 (−186.5/11.5)	−55 (−92/−2)	ns

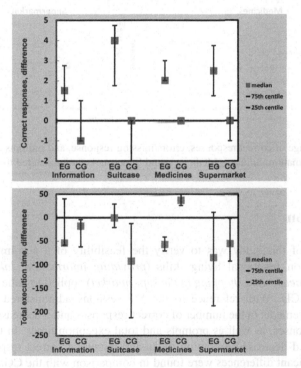

Fig. 3. Top: Difference between correct responses at T1 and T3 in the in vivo test in the Experimental and Control Groups (EG, CG). Bottom: Difference between total execution time at T1 and T3 in the Experimental and Control Groups (EG, CG).

Table 4. Results obtained by the Experimental Group during the 10-VT sessions (Friedman test for repeated measures).

APP	Correct responses p=	Errors p=	Missing responses p=	Prompts p=	Total execution time p=
Information	ns	ns	0.002	ns	<0.001
Suitcase	0.03	ns	ns	0.024	0.037
Medicines	ns	ns	0.005	ns	<0.001
Supermarket	0.001	ns	0.001	0.003	<0.001

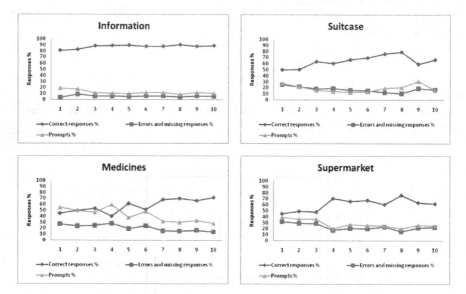

Fig. 4. Percentage of correct responses, errors/missing response, and prompts during the 10 VT sessions of Information, Suitcase, Medicines, and Supermarket administered to the Experimental Group.

5   Discussion

The first aim of this study was to verify the feasibility of a non-immersive virtual training (VT) on functional living skills (*providing information, taking medicines, preparing a suitcase, and shopping to the supermarket*) applied to patients with mild to moderate M-NCD. With reference to the VT sessions administered to EG, results showed improvements in the number of correct responses and a decrease in the number of missed responses, as well as prompts and total execution times. In the in vivo test, EC also showed decreased execution times and increased correct responses, and statistically significant differences were found in comparison with the CG. This last result seems to suggest the need for specific training to improve the functional skills of everyday life; indeed, the CG, despite having benefited from a group cognitive stimulation training, did not improve performances in the in vivo tests at T3 compared to T1.

The second aim was to verify the eventual generalisation in the natural environment of the skills improved during the VT. Results reported in Table 3 showed improvements in three of the four skills performed in the natural environment. This generalisation was spontaneous since it took place without specific naturalistic training. This result encourages the continuation of the study since it will be possible to rehabilitate many different functional living skills, by developing various and appropriate apps, and then administering the VT. This new technological way for rehabilitating and maintaining daily living skills also appears to be able to reduce rehabilitation costs of the healthcare service and might be used as a remote telerehabilitation tool, with decreased rehabilitation costs and saving time for patients and their caregivers.

6   Conclusion and Future of the Study

Despite the still limited number of subjects involved and of VT sessions, the results obtained from this first feasibility study seem to be encouraging, since the improvements were spontaneously generalised to the natural environment. Specific training on daily living skills seems to be essential to promote improvements in this critical domain.

Using VR in teaching functional living skills leads to many advantages, including: (a) the possibility to teach daily living skills in a safe, controlled, visually stimulating environment; (b) the possibility to adapt the task difficulties to the individual characteristics of persons performing it; (c) the possibility to carry out the sessions at any time, also in absence of a human tutor; (d) accuracy and completeness of data collection for monitoring the patient's progress; (e) the possibility to expand the use of virtual tools from the home skills to self-care and community participation domains; (f) the satisfaction of users who enjoy the tool while training.

The VT seems to be an approach that complements and enriches the existing rehabilitation strategies; its flexible use (at the hospital and home) makes it particularly interesting and attractive.

The continuation of the study will include the following objectives: (a) to increase the sample size; (b) to obtain a higher number of VT sessions (20); (c) to develop additional apps on home and self-care skills, as well as community participation; (d) to adapt apps for their use on tablets or more immersive virtual environments.

Acknowledgements. This study was partially supported by a fund from the Italian Ministry of Health "Ricerca Corrente" (RC n. 2634475, Drs. Panerai, Catania, Rundo, Ferri, Gelardi) and by a fund from the INNOLABS program of the Apulia Region "RECALL" (project n. QIRYKE8, Drs. Bevilacqua, Brunetti, and De Meo).

References

1. Hofmann, M., et al.: Interactive computer-training as a therapeutic tool in Alzheimer's disease (2003). https://doi.org/10.1016/S0010-440X(03)00006-3. https://www.sciencedirect.com/science/article/abs/pii/S0010440X03000063
2. Van Schaik, P., Martyr, A., Blackman, T., Robinson, J.: Involving persons with dementia in the evaluation of outdoor environments. Cyberpsychol. Behav. 11, 415–424 (2008). https://doi.org/10.1089/cpb.2007.0105
3. Foloppe, D.A., Richard, P., Yamaguchi, T., Etcharry-Bouyx, F., Allain, P.: The potential of virtual reality-based training to enhance the functional autonomy of Alzheimer's disease patients in cooking activities: a single case study. Neuropsychol. Rehabil. 28, 709–733 (2018). https://doi.org/10.1080/09602011.2015.1094394
4. Bevilacqua, V., et al.: A P300 clustering of mild cognitive impairment patients stimulated in an immersive virtual reality scenario. In: Huang, D.-S., Jo, K.-H., Hussain, A. (eds.) ICIC 2015. LNCS, vol. 9226, pp. 226–236. Springer, Cham (2015). https://doi.org/10.1007/978-3-319-22186-1_23
5. de Tommaso, M., et al.: Testing a novel method for improving wayfinding by means of a P3b virtual reality visual paradigm in normal aging. Springerplus 5 (2016). https://doi.org/10.1186/s40064-016-2978-7
6. Bevilacqua, V., et al.: Design and development of a forearm rehabilitation system based on an augmented reality serious game (2016). https://doi.org/10.1007/978-3-319-32695-5_12
7. Panerai, S., et al.: Group intensive cognitive activation in patients with major or mild neurocognitive disorder. Front. Behav. Neurosci. 10, 34 (2016). https://doi.org/10.3389/fnbeh.2016.00034
8. American Psychiatric Association: DSM-5 Task Force, Diagnostic and statistical manual of mental disorders : DSM-5 (2013)
9. Hughes, C.P., Berg, L., Danziger, W.L., Coben, L.A., Martin, R.L.: A new clinical scale for the staging of dementia. Br. J. Psychiatry 140, 566–572 (1982). https://doi.org/10.1192/bjp.140.6.566
10. Folstein, M.F., Folstein, S.E., McHugh, P.R.: "Mini-mental state": a practical method for grading the cognitive state of patients for the clinician. J. Psychiatr. Res. 12, 189–198 (1975). https://doi.org/10.1016/0022-3956(75)90026-6
11. Lawton, M.P., Brody, E.M.: Assessment of older people: self-maintaining and instrumental activities of daily living. Gerontologist (1969). https://doi.org/10.1093/geront/9.3_Part_1.179
12. Bevilacqua, V., et al.: A RGB-D sensor based tool for assessment and rating of movement disorders. In: Duffy, V., Lightner, N. (eds.) AHFE 2017. AISC, vol. 590, pp. 110–118. Springer, Cham (2018). https://doi.org/10.1007/978-3-319-60483-1_12
13. Brazzelli, M., Capitani, E., Della Sala, S., Spinnler, H., Zuffi, M.: A neuropsychological instrument adding to the description of patients with suspected cortical dementia: the Milan overall dementia assessment. J. Neurol. Neurosurg. Psychiatry 57, 1510-7 (1994). https://doi.org/10.1136/JNNP.57.12.1510
14. Basso, A., Capitani, E., Laiacona, M.: Raven's coloured progressive matrices: normative values on 305 adult normal controls. Funct. Neurol. 2(2), 189–194 (1987)
15. Orsini, A., Grossi, D., Capitani, E., Laiacona, M., Papagno, C., Vallar, G.: Verbal and spatial immediate memory span: normative data from 1355 adults and 1112 children. Ital. J. Neurol. Sci. (1987). https://doi.org/10.1007/BF02333660

16. Carlesimo, G.A., et al.: The mental deterioration battery: normative data, diagnostic reliability and qualitative analyses of cognitive impairment. Eur. Neurol. (1996). https://doi.org/10.1159/000117297

17. Katz, S., Downs, T.D., Cash, H.R., Grotz, R.C.: Progress in development of the index of ADL. Gerontologist **10**, 20–30 (1970)

Design and Development of a Robotic Platform Based on Virtual Reality Scenarios and Wearable Sensors for Upper Limb Rehabilitation and Visuomotor Coordination

Stefano Mazzoleni[1]([✉]), Elena Battini[1], Domenico Buongiorno[2],
Daniele Giansanti[3], Mauro Grigioni[3], Giovanni Maccioni[3],
Federico Posteraro[4], Francesco Draicchio[5],
and Vitoantonio Bevilacqua[2]

[1] The BioRobotics Institute, Scuola Superiore Sant'Anna, Pisa, Italy
stefano.mazzoleni@santannapisa.it
[2] Department of Electrical and Information Engineering (DEI),
Polytechnic University of Bari, Bari, Italy
[3] Istituto Superiore di Sanità, Rome, Italy
[4] UOC Recupero e Rieducazione Funzionale,
Ospedale Versilia, Azienda USL Toscana Nord-Ovest,
Lido di Camaiore, Lucca, Italy
[5] Department of Occupational and Environmental Medicine,
Epidemiology and Hygiene, INAIL, Rome, Italy

Abstract. The work reintegration following shoulder biomechanical overload illness is a multidimensional process, especially for those tasks requiring strength, movement control and arm dexterity. Currently different robotic devices used for upper limb rehabilitation are available on the market, but these devices are not based on activities focused on the work reintegration. Furthermore, the rehabilitation programmes aimed to the work reintegration are insufficiently focused on the recovery of the necessary skills for the re-employment.

In this study the details of the design of an innovative robotic platform integrated with wearable sensors and virtual reality scenarios for upper limbs motor rehabilitation and visuomotor coordination is presented. The design of control strategy will also be introduced. The robotic platform is based on a robotic arm characterized by seven degrees of freedom and by an adaptive control, wearable sensorized insoles, virtual reality (VR) scenarios and the Leap Motion device to track the hand gestures during the rehabilitation training. Future works will address the application of deep learning techniques for the analysis of the acquired big amount of data in order to automatically adapt both the difficulty level of the VR serious games and amount of motor assistance provided by the robot.

Keywords: Robotic devices · Wearable sensors · Virtual reality ·
Upper limb rehabilitation · Pattern recognition · Deep learning

D.-S. Huang et al. (Eds.): ICIC 2019, LNAI 11645, pp. 704–715, 2019.
https://doi.org/10.1007/978-3-030-26766-7_64

1 Introduction

Work reintegration following a trauma or illness is a multidimensional process that involves both physical and psychosocial aspects [1]. The recovery of upper limb abilities is an important factor for work reintegration, especially for those tasks requiring strength, movement control and arm dexterity [2–5]. In the last two decades, different rehabilitation robotic devices have been developed in order to facilitate upper limb recovery following neurological impairment, such as stroke [6, 7]. Recent reviews showed evidence of significant results when providing high-intensity treatments and task-specific goal-oriented exercises combining robot-assisted and conventional rehabilitative programmes [8–11]. Robot-assisted rehabilitation training have shown improvements in terms of muscle strength and motor coordination recovery in neurological patients [12]; their impact in terms of activities of daily living (ADLs) is currently rather limited [12]. The recovery of sensory-motor functions to perform ADLs represents one of the main objectives of rehabilitation for workers, as it can restore greater autonomy and an increased ability to carry out specific tasks [9]. Currently the existing rehabilitation programs aimed to work reintegration are poorly oriented to quantitative and personalized assessment of residual abilities and to the recovery of skills for worker re-employment. To overcome these limits, a new robotic platform based on a robotic arm, wearable sensors and virtual reality scenarios is developed. The aim of this study is to present the technical details of the innovative robotic platform (RoboVir), integrated with wearable sensors and virtual reality scenarios for upper limb rehabilitation and visuomotor coordination in occupational contexts.

2 The Robotic Platform

The RoboVir robotic platform is composed of the following elements (Fig. 1): a 7 degrees of freedom (DoFs) commercial robotic arm (Fig. 1a); wearable customized foot insoles (Fig. 1b); virtual reality (VR) scenarios (Fig. 1c); a commercial hand tracking device (Fig. 1d).

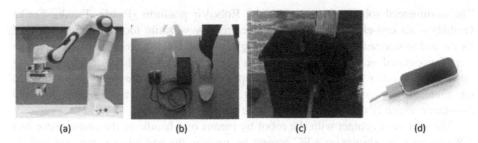

(a)   (b)   (c)   (d)

Fig. 1. Elements of the RoboVir robotic platform.

Figure 2 shows the setup combining the use of the robotic arm, including its workspace, and the commercial hand tracking device.

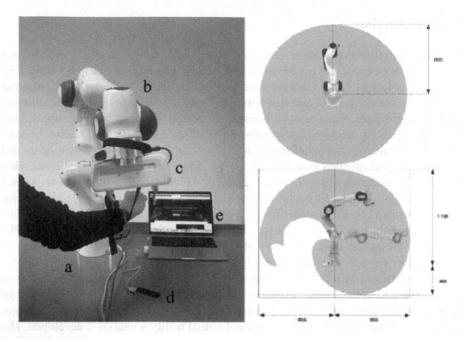

Fig. 2. *Left.* The proposed setup: human arm (a), robot arm (b), end-effector of the robot arm (c), commercial hand tracking device (d) and the PC for the VR (e). *Right-top.* Top view of the Panda robot. *Right-bottom.* Side view of the Panda robot.

During the rehabilitation session, the patient taking on the upright position stays in front of the robot arm and he is also able to see the PC monitor, where the rehabilitation exercises are shown. The commercial hand tracking device is also placed on the same table in order to track the opening/closing movements of the hand. The patient is asked to perform task-oriented exercises in VR while he is physically supported by the robot.

2.1   The Robotic Arm

The commercial robotic arm used in the RoboVir platform (Panda, Franka Emika GmbH) is an end-effector, sensitive and extremely versatile tool characterized by 7 DoFs and torque sensors at each joint (Fig. 1a). The robot is controlled by using an impedance-based technique [13] in order to interact gently with the user. The control is developed in order to implement an adaptive behavior: if the user is not able to complete the requested exercise the robot assist the user by providing the minimal assistance which is needed to complete the task.

The user is in contact with the robot by means of a handle at the end-effector and VR scenarios are shown on a PC screen: by moving the end-effector the user will be able to complete the requested upper limb rehabilitation exercise according to his residual abilities.

2.2 The Wearable Foot Insoles

The wearable foot insoles sensors are used to monitor the subject posture during the execution of the motion task. Figure 3a shows the *subsystem* that includes:

- a *control unit* for converting and sending data to the processing system;
- 4 Force Sensing Resistors (FSRs) sensors that are positioned on the heel and on the plantar of each foot and a push-button trigger;
- an *USB connector.*

The control unit (Fig. 3b and c) is equipped with a conditioning circuit and an analog/digital (A/D) conversion card (National Instruments PN NI-USB-6008). In the front panel the adjustment of the signal amplification (maximum value 5 V) is carried out by means of four knobs. Each knob is associated with an FSR channel:

- White: Right-plantar
- Red: Right-heel
- Green: Left-plantar
- Blue: Left-heel

The *control unit* is integrated with:

- a *kit* composed by a 9-pole female canon panel connector to which a male canon connector is connected with the 4 FSRs and a manual trigger;
- an *USB connector* to connect the system to the PC2 for data exchanging and system power supply. This type of connection also allows to recharge the system, eliminating possible electrical safety problems;
- a further additional trigger by means of a jack-audio type connector

Each connection cable to the FSRs is mapped with the same color indicated on the panel in the adjustment knob. The cables are shielded audio type. Each FSR sensor after being inserted in a pocket provided in Non-Woven Fabric (NWF) hypoallergenic material can be slipped into the shoe through a positioning in a para rubber sole, according to a proposed procedure, based on the color mapping (Fig. 4).

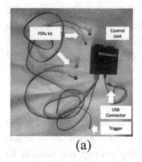
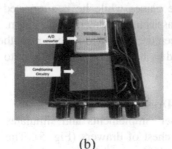

      (a)   (b)   (c)

Fig. 3. The foot insoles subsystem (a), the control unit circuitry (b) and the control unit front panel (c) (Color figure online)

The signals associated with each FSR channel and the trigger can be displayed on the PC2 by the USB connector through a simple user-friendly program developed in Labview, called *Robovirstability*, which also allows data to be saved in Excel. The signals are shown in the four white, red, green and blue mapping colors. The software also allows a post-processing analysis. In particular, it returns:

- % of correct placement
- % placement with rear overload
- % placement with front overload

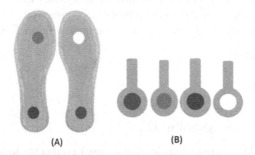

Fig. 4. The design of (A) the insoles and (B) the NWF pockets (Color figure online)

2.3   The Serious Games and the Upper Limb Tracking System

The developed robotic platform considers the integration of the robotic arm with two serious games developed in virtual reality named "Dresser" and "Bookcase". Both virtual games have been implemented using the Unity 3D framework and make use of the Leap Motion device to detect the opening and the closing of the patient's hand. The patient can interact with the virtual objects by controlling the position of a virtual hand that follows the patient's real hand. The serious games have been fully integrated with the robotic arm thus enabling a robotic-aided task-oriented rehabilitation. Following the visual cues shown by the VR, the patient is asked to accomplish some task-oriented exercises proposed by the games while he is supported by the robot that helps the patient according to the assistance-as-needed paradigm. Hence, since the there is a physical connection between the robot and human arm, the position and the orientation of the real hand are tracked by the robotic arm and sent to the virtual game via an UDP connection.

2.3.1   Serious Games "Dresser"
The serious game "Dresser" implements and simulates the interaction between the human upper limb and a chest of drawers (Fig. 5). The aim of the game is to rehabilitate or assess the upper limb motor functions by sequentially asking the patient to (1) reach and grasp objects positioned on the top of a virtual dresser, and (2) open and close the virtual drawers.

The game reproduces, in a virtual environment, a chest of drawers with drawers that can be opened through direct interaction with the user. The chest of drawers consists of three rows of drawers, each row contains three drawers. The difficulty level of the game is modulated by the height of the drawer the patient is asked to interact with (starting from the bottom, the first row represents the base level, the second row the intermediate one and the third the advanced level).

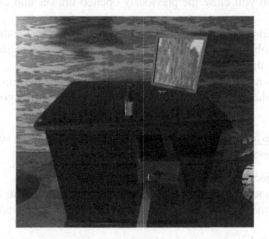

Fig. 5. A screenshot of the "Dresser" serious game.

A rehabilitation session is composed of several independent trials. Each trial considers the interaction with a random drawer positioned at the specific height according to the difficulty level selected by the physical therapist before the session start. The number of trials is chosen by the physical therapist according to the patient's motor capabilities. Each trial is composed of a sequence of sub actions as follows:

1. The patient is asked to position the virtual hand in the rest position, then the random selected drawer starts flashing.
2. The patient moves his hand towards the drawer and grasp the knob of the drawer by closing the hand (the drawer will be 'grabbed' only if the hand closes near the knob).
3. Then, the subject is asked to open the drawer until the drawer is completely overlapped with the fully opened ghost drawer as reported in Fig. 5.
4. Once the drawer is completely opened, the user opens the hand to release the knob and must reach and grasp a random object positioned at the top of the dresser.
5. Then, the patient is asked to put the grabbed object into the previously opened drawer by opening the hand above the drawer.
6. Finally, to complete the trial, the patient must grab the knob, close the drawer, and reach the rest position to trigger a new trial.

The developed game also considers the cases when the patient is not able to complete one or some sub actions of the single trial within a customizable defined amount of time. In particular, we defined the following behaviors:

- if the user fails to reach or completely open the drawer within the time limit the system will automatically open the drawer to ensure continuity of the game;
- if the user is unable to reach or grasp the object on the drawers within the set time limit the system will close the previously opened drawer and consider the trial as finished;
- if the patient fails to place the previously grabbed object inside the drawer the game will make the object disappear from the patient's hand, close the drawer and stop the ongoing trial;
- if the drawer is not completely closed in the set time, the system will automatically close it, invite the patient to bring his hand back to the starting position and start the next trial.

Game Session Performance. At the end of one rehabilitation session, the system computes and shows some performance statistics:

1. Number of objects the patient was able to grasp;
2. The average time in seconds used to reach the object and grab it;
3. The average time in seconds to bring the grasped object towards the drawer and release it;
4. Number of complete drawer openings;
5. Number of complete drawer closures;
6. Average time in seconds used to reach and open the drawer;
7. Average time in seconds used to reach and close the drawer;
8. Number of complete successful trials.

Fig. 6. A screenshot of the "Bookcase" serious game.

2.3.2   Serious Games "Bookcase"

The serious game "Bookcase" implements and simulates the interaction between the human upper limb and a bookcase (Fig. 6). The aim of the game is to rehabilitate or assess the upper limb motor functions by sequential (1) reaching and grasping of books positioned outside the bookcase, and (2) placing the book back into a specific location in the bookcase. The game reproduces, in a virtual environment, a bookcase. The virtual bookcase has five levels composed of three shelves each. The difficulty level of the game is modulated by the height of the shelf the patient is asked to put the book into (starting from the top, the first row represents the advanced level, the second row the intermediate one and the third the base level).

As for the Dresser games, a rehabilitation session is composed of several independent trials. Each trial considers the interaction with a book that has to be moved into a selected shelf according to the difficulty level chosen by the physical therapist before the session starting. The number of trials is defined by the physical therapist according to the patient's motor skills. Each trial is composed of a sequence of sub actions as follows:

1. The patient moves his hand at the starting position, then, a book located at a random position outside the bookcase at the same height of his sternum appears.
2. Then, the patient is asked to reach and grasp the book by closing his hand (the book will be "grabbed" only if the hand closes near the book).
3. Once the book is grabbed, a light spot is activated in correspondence of the selected shelf (that shelf represents the correct place where to place the book).
4. Then, the patient must place the book at the right location and open the hand.
5. Finally, the patient goes back in the rest position to trigger both the end of the current trial and the start of a new one.

The developed game also considers the cases when the patient is not able to complete one or some sub actions of the single trial within a customizable defined amount of time. In particular, we defined the following behaviors:

- if the user fails to reach the book within the time limit the system will automatically position the book into his virtual hand;
- if the user is unable to place the book at the right location within the set time limit the system will automatically position the book consider the trial as finished.

Game Session Performance. At the end of one rehabilitation session, the system computes and reports some performance statistics: (1) Number of books the patient was able to grasp; (2) The average time in seconds used to reach and grab the book; (3) The average time in seconds to bring the grasped book at the right location in the bookcase; (4) Number of complete successful trials.

2.3.3   Hand Movement Quality Metrics

Besides the performances that are strictly related to the goals of each serious game, the robotic platform is equipped with an analysis software that processes the trajectories followed by the patient's hand during the whole session. Such analysis, that is performed offline at the end of the session, extracts a set of metrics that will be used by the medical staff to monitor and validate the rehabilitation plan. In particular, for each trajectory acquired and recorded during each trail sub action, the following metrics are computed:

- *Smoothness (SM):* this metric represents a quantitative measure of the fluidity of movement and is computed as the Spectral Arc Length (SPARC) which returns a dimensionless number [14].
- *Normalized Path Length (NPL):* this parameter is computed as the ratio between two distances: the optimal distance between the initial and the end points of the sub action (estimated as the length of the straight line crossing the two points) and the distance actually traveled by the patient's hand during the game session for the same route.
- *Initial angle error (IAE):* it is a parameter that quantifies, in terms of angle, the initial direction error [15]. The patient initially could move his hand along a direction different from the optimal one to reach his goal, consequently the angle between the line representing the initial direction followed by the patient and the straight line of the optimal path quantifies the angle of error.

2.3.4   Tracking of the Upper Limb Joint Angles

As deeply discussed above, the RoboVir rehabilitation platform is based on an end-effector (EE) robot that is able to apply forces at the patient's wrist. Such EE based configuration allows to track the wrist position and orientation within the robot workspace without considering any information about the full arm pose, i.e. the angle value of the upper limb articulations. This motivated the development of a low-cost tracking system able to track the upper limb movement and extract the trend of the upper limb joint angles over time.

The developed tracking system was designed for the acquisition of the angles of the shoulder, elbow and forearm joints, i.e. the three shoulder rotation angles defined as in the work of Holzbaur et al. [16], the elbow flexion/extension angles and the prono-supination angle. The system is based on the virtual reality HTC Vive platform and, in particular on, the Vive Tracker device that is a battery-powered tracking device that allows the acquisition of the full pose (position and orientation) of a rigid body on which it is fixed.

The tracking system employs four HTC Vive trackers, elastic belts with plastic supports for the trackers and a metallic stick with a known length (Fig. 7). The four markers are positioned at the wrist, the arm, the sternum and above the acromion. The stable positioning of the trackers is ensured by means of elastic bands and appropriate 3D printed plastic supports.

Calibration Procedure. A calibration procedure is needed to individuate the relative position of 8 landmark points of the skeletal system respect to the markers: ulna styloid process, radio styloid process, medial epicondyle, lateral epicondyle, xiphoid process, sternal extremities of the two clavicles and acromion. After positioning the trackers 1, 2 and 3, the tracker 4 is fixed to the stick (see Fig. 7) to identify the relative positions of the landmark points respect to a specific marker. In particular, the ulna styloid process and the radio styloid process are referred respect to the tracker 1; the medial epicondyle and the lateral epicondyle are referred respect to the tracker 2; the xiphoid process and

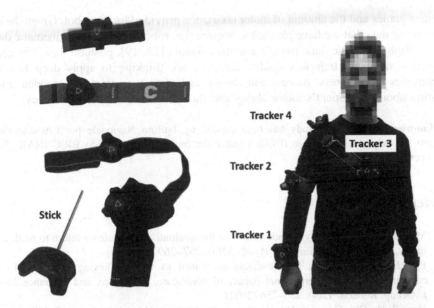

Fig. 7. *Left.* Tracking system accessories. *Right.* A healthy subject wearing the accessories of the tracking system.

the sternal extremities of the two clavicles are referred to the tracker 3. The acquisition of the relative positions is performed by placing the extremity of the stick (that has a known position respect to the tracker 4) above the landmark points. Finally, the tracker 4 is positioned with a plastic support above the acromion at a defined distance.

Joint Angles Reconstruction. During the entire rehabilitative session, a custom-made software is used to record the pose of all trackers. Then, it is possible to reconstruct the trajectory of each landmark point given its relative position to the specific tracker. Finally, the articulation angles are extracted by running an inverse kinematic procedure on a scaled version of the upper limb model developed by Holzbaur et al. [16] using the OpenSim software [17].

3 Future Works: Machine Learning for Data Analysis

We have presented a fully integrated robot-based platform for shoulder rehabilitation. The proposed platform has been designed for VR-based orthopedic rehabilitation of workers that suffered a shoulder biomechanical overload. Experimental tests with patients have already planned. We will run a randomized controlled clinical trial to test the efficacy of the proposed rehabilitation programme and perform a comparison with the traditional orthopedic rehabilitation. Future works will also address the design and implementation of a medical pattern-recognition system able to analyze the acquired and recorded data in order to: (1) objectively assess and rate the motor skill improvement over sessions; (2) automatically adapt both the difficulty level of the serious games and the amount of motor assistance provided by the robot. Given the big amount of data that we have planned to acquire, i.e. robot forces, robot kinematic data, upper limb kinematic data, muscle activities signals [18, 19], postural data, VR game statistics and hand trajectory quality data, we are thinking to apply deep learning techniques to compress, process and cluster such data to extract the encoded information about the patient's motor ability and the rehabilitation process efficacy.

Acknowledgments. This study has been funded by Istituto Nazionale per l'Assicurazione contro gli Infortuni sul Lavoro (INAIL) under the framework of Bando BRiC INAIL 2016 (project ROBOVIR).

References

1. MacEachen, E., et al.: Systematic review of the qualitative literature on return to work after injury. Scand. J. Work Environ. Health **32**(4), 257–269 (2006)
2. Franche, R.-L., Krause, N.: Readiness for return to work following injury or illness: conceptualizing the interpersonal impact of healthcare, workplace, and insurance factor. J. Occup. Rehabil. **12**(4), 233–256 (2002)
3. Hou, W.H., Chi, C.C., Lo, H.L.D., Kuo, K.N., Chuang, H.Y.: Vocational rehabilitation for enhancing return-to-work in workers with traumatic upper limb injuries (2013)
4. Shi, Q., Sinden, K., Macdermid, J.C., Walton, D., Grewal, R.: A systematic review of prognostic factors for return to work following work-related traumatic hand injury. J. Hand Ther. **27**(1), 55–62 (2014)
5. Fadyl, J., McPherson, K.: Return to work after injury: a review of evidence regarding expectations and injury perceptions, and their influence on outcome. J. Occup. Rehabil. **18** (4), 362–374 (2008)
6. Krebs, H.I.: Twenty + years of robotics for upper-extremity rehabilitation following a stroke. In: Rehabilitation Robotics (2018)
7. Buongiorno, D., Sotgiu, E., Leonardis, D., Marcheschi, S., Solazzi, M., Frisoli, A.: WRES: a novel 3 DoF WRist exoskeleton with tendon-driven differential transmission for neurorehabilitation and teleoperation. IEEE Robot. Autom. Lett. **3**(3), 2152–2159 (2018)
8. Krebs, H.I., et al.: Robotic applications in neuromotor rehabilitation. Robotica **21**(1), 3–11 (2003)
9. Lee, S.-S., Park, S.-A., Kwon, O.-Y., Song, J.-E., Son, K.-C.: Measuring range of motion and muscle activation of flower arrangement tasks and application for improving upper limb function. Korean J. Hortic. Sci. Technol. **30**(4), 449–462 (2012)
10. Spreeuwers, D., et al.: Work-related upper extremity disorders: one-year follow-up in an occupational diseases registry. Int. Arch. Occup. Environ. Health **84**(7), 789–796 (2011)
11. Mehrholz, J., Pohl, M., Platz, T., Kugler, J., Elsner, B.: Electromechanical and robot-assisted arm training for improving activities of daily living, arm function, and arm muscle strength after stroke (2018)
12. Lederer, V., Rivard, M., Mechakra-Tahiri, S.D.: Gender differences in personal and work-related determinants of return-to-work following long-term disability: a 5-year cohort study. J. Occup. Rehabil. **22**(4), 522–531 (2012)
13. Siciliano, B., Lorenzo, S., Villani, L., Orilo, G.: Robotics: Modelling, Planning and Control, 2nd edn. Springer, London (2010). https://doi.org/10.1007/978-1-84628-642-1

14. Balasubramanian, S., Melendez-Calderon, A., Roby-Brami, A., Burdet, E.: On the analysis of movement smoothness. J. Neuroeng. Rehabil. **12**, 112 (2015). https://doi.org/10.1186/s12984-015-0090-9
15. Berger, D.J., d'Avella, A.: Effective force control by muscle synergies. Front. Comput. Neurosci. **8**, 46 (2014). https://doi.org/10.3389/fncom.2014.00046
16. Holzbaur, K.R.S., Murray, W.M., Delp, S.L.: A model of the upper extremity for simulating musculoskeletal surgery and analyzing neuromuscular control. Ann. Biomed. Eng. **33**(6), 829–840 (2005). https://doi.org/10.1007/s10439-005-3320-7
17. Delp, S.L., et al.: OpenSim: open-source software to create and analyze dynamic simulations of movement. IEEE Trans. Biomed. Eng. **54**(11), 1940–1950 (2007)
18. Buongiorno, D., et al.: Evaluation of a pose-shared synergy-based isometric model for hand force estimation: towards myocontrol. In: Biosystems and Biorobotics (2017)
19. Buongiorno, D., Barsotti, M., Barone, F., Bevilacqua, V., Frisoli, A.: A linear approach to optimize an EMG-driven neuromusculoskeletal model for movement intention detection in myo-control: a case study on shoulder and elbow joints. Front. Neurorobot. **12**, 74 (2018)

Depth-Awareness in a System for Mixed-Reality Aided Surgical Procedures

Mauro Sylos Labini[1,2,3] (iD), Christina Gsaxner[1,2,4] (iD),
Antonio Pepe[1,2] (iD), Juergen Wallner[2,4] (iD), Jan Egger[1,2,4(✉)] (iD),
and Vitoantonio Bevilacqua[3] (iD)

[1] Institute for Computer Graphics and Vision, Faculty of Computer Science
and Biomedical Engineering, Graz University of Technology, Graz, Austria
egger@tugraz.at
[2] Computer Algorithms for Medicine Laboratory, Graz, Austria
[3] Department of Electrical and Information Engineering,
Polytechnic University of Bari, Bari, Italy
vitoantonio.bevilacqua@poliba.it
[4] Department of Oral and Maxillofacial Surgery,
Medical University of Graz, Graz, Austria

Abstract. Computer-assisted surgery is a trending topic in research, with many different approaches which aim at supporting surgeons in the operating room. Existing surgical planning and navigation solutions are often considered to be distracting, unintuitive or hard to interpret. In this work, we address this issue with an approach based on mixed reality devices like Microsoft HoloLens. We assess the depth sensing capabilities of Microsoft HoloLens, and the potential benefit they could bring to computer-assisted surgery applications.

Keywords: Computer-assisted surgery · Mixed reality · HoloLens ·
Time-of-flight camera · RGB-Depth mapping · Pattern recognition

1 Introduction

Surgical resection of the tumoral mass is one of the primary treatments that patients affected by cancer in the head and neck area must undergo. Due to the high invasiveness of the procedure and the risk of relapses, it is crucial for the surgeon to quickly and precisely evaluate the location and extension of the tumor. As noted by the American National Institutes of Health (NIH) [1], while medical imaging and operating microscopes currently aid the surgeon during this operation, neither of these tools can provide direct visualization of the mass to be removed. This work wants to further investigate the usability of mixed-reality (MR) devices as a visualization aid for a

---

Supported by FWF KLI 678-B31 (enFaced), COMET K-Project 871132 (CAMed) and the TU Graz
Lead Project (Mechanics, Modeling and Simulation of Aortic Dissection).

D.-S. Huang et al. (Eds.): ICIC 2019, LNAI 11645, pp. 716–726, 2019.
https://doi.org/10.1007/978-3-030-26766-7_65

surgeon in the aforementioned scenario [2]. Recent advancements in MR sparked new research effort towards the introduction of this technology in the operating room, with Microsoft HoloLens representing a common hardware choice. An example of this trend is given by Perksin et al. [3], where they evaluate the role of mixed reality during breast cancer surgery, or the work from Pratt et al. [4], on augmented reconstruction surgery. A relevant contribution regarding head and neck surgery is given by Wang et al. [5], who employed a video-see-through AR headset to visualize CT imaging data directly on the patient. Pepe et al. [6] suggested a MR-based application, which features automatic marker-less image registration and a hands-free interface for the facial surgeon. However, their approach is limited by the spatial reasoning capabilities of the device, which only exposes a coarse map of the environment. This is unsuitable for the accuracy required for medical applications. The capabilities of the device have recently been expanded, allowing researchers to access more of its onboard sensors data, therefore we aim at exploiting these new capabilities. In particular Time-of-Flight (ToF) depth sensor, which enabled us to turn the HoloLens into an all-in-one visualization and measuring tool. Furthermore, an approach for improved object-to-patient registration is proposed. Our method, in fact, combines the newly enabled depth-awareness of the headset with insight from pattern recognition algorithms to accurately detect a patient's facial traits and locate them in the user's frame of reference.

2   HoloLens Measuring Capabilities

Following the idea from Pepe et al. [6], we develop a MR image registration system based on the Dlib face recognition library [7–10], which detects a number of the patient's facial landmarks in the frames obtained from the device RGB camera. These landmarks are then identified on the 3D model built from the patient's CT scan so that each key point can be matched to its corresponding point in the camera frame. A major difficulty with this approach lies in the assessment of the patient's position in the camera's optical axis direction. To reconstruct the 3D coordinates of the detected landmark points, Pepe et al. use the combined information of the camera intrinsic parameters and the rough spatial mapping depth estimation. Spatial mapping however, was not designed for fine distance measurements and it is therefore not the ideal candidate for this task.

2.1   Research Mode

In April 2018, Microsoft released a Windows 10 update, which unlocks the so-called Research Mode on the HoloLens headset. Research Mode is a tool aimed at granting developers with an extended access to the data collected by the headset built-in sensors [11]. This provides APIs to three different data sources:

- Four environment tracking cameras used by the system for map building and head tracking.

– Two ToF depth cameras, one for high-frequency (30 FPS) near-depth sensing, commonly used in hand tracking, and the other for lower-frequency (1FPS) far-depth sensing, currently used by the SLAM-based Spatial Mapping.
– Two versions of an IR-reflectivity stream used by the HoloLens to compute depth.

As depth perception was found to represent a major obstacle in previous studies [6], in this work we only considered the short-range depth data, streaming at 30 frames per second (FPS). This choice was also driven by accuracy reasons - ToF sensors performances exponentially degrade with the distance and by the fact that surgeons generally operate within the arm distance from the patient.

2.2   Range Images Un-Projection

Cameras produce 2D images projecting a set of 3D points in the physical world onto a plane, thus reconstructing a 3D model of the camera view from its 2D representation requires some knowledge about the camera projection model. This model depends on physical and optical features of the camera, usually device-specific and hard to retrieve without the manufacturers aid. Microsoft provides a representation of the HoloLens depth camera model in the form of an un-projection mapping, namely a transformation that maps pixel coordinates to a unit-depth plane [12].

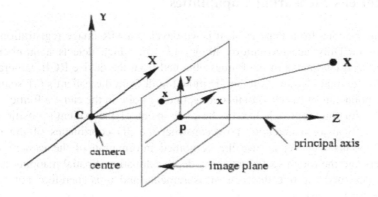

Fig. 1. Illustration of the geometry of the projection process.

With reference to Fig. 1, let us call $[X, Y, Z]$ the 3D coordinates of a point in the real world and $[x, y, 1]$ the coordinates of the point projected on the unit-depth plane. The un-projection mapping specifies, for each point on the projection plane, a pair of $[u, v]$ values such that $[X, Y, Z] = Z * [u, v, 1]$. The mapping values come arranged in two 448×450 matrices – one for u values and one for v values – so that to each pixel in the depth frame correspond a unique pair $[u, v]$. This means that all we need now in order to reconstruct the cameras 3D view is the Z coordinate of the un-projected points. As before mentioned, research mode provides a stream of range images, which define for each pixel a value of distance, measured from the target point to the camera center. However, Fig. 1 clearly shows that this distance does not correspond to the Z coordinate of the target, which we can retrieve through the (u, v) values.

Now, to retrieve the desired 3D coordinates, we apply the mapping over the whole depth frame as follows: if Z_{ij} is the Z value of the pixel on the i-th row and the j-th column and u_{ij} and v_{ij} are the un-projection mapping parameters, we transform the pixel in Z_{11} with the corresponding (u, v) values so that $[X_1, Y_1, Z_1] = Z_{11} * [u_{11}, v_{11}, 1]$. This process theoretically leaves us with $448 * 450 = 201600 - 3D$ points, of which, in practice, more than a half are discarded through background removal.

2.3   Measurements on a Reconstructed 3D Scene

As we propose to use Microsoft HoloLens as a high-precision depth sensing tool, it is crucial that we assess the accuracy performance of the device in different measurement scenarios. Here we consider three scenarios:

- planar surfaces,
- simple 3D objects,
- a complex 3D object.

To perform the measurements, the recordings of the depth sensor were downloaded from the HoloLens, then the relative point clouds were extracted and analyzed in MATLAB, using the built-in point clouds visualizer. As light interferences can negatively affect IR-based depth sensing, the recordings were carried out in an environment as isolated from sunlight as possible, using the on-board IR projector for illumination. For planar surface measurements, we observed a wall in the laboratory, assumed to be perfectly flat. We then employed a RANSAC-based algorithm [13] to fit a plane to the extracted point cloud and we measured the mean squared error over all the inlier points. The depth sensor is located on top of the user's head, and therefore moves together with it. Due to this fact, it is difficult to accurately establish a ground truth for absolute distance measurements. Thus, for the remaining measurements, we decided to maintain a simple setup and to only consider relative distances. The last measurement scenario is worth of particular attention, as this is also a testbed for the actual medical application. The used object was a 3D-printed model built from a high-precision scan of a head-cancer patient; the same employed by Pepe et al. [6] to test the final application. To recover the ground truth for the measurement, we loaded the mesh from which the model was printed into a 3D visualization software, as shown in Fig. 2. Here, we determined the distance of the nose tip from the flat back of the head, to simulate a patient lying face-up on an operating bed. Then, for the actual measurements, we analyzed several reconstructed views of the 3D head, recorded at arm distance. Eventually, we exploited MATLAB plane fitting to determine the parameters of the head's bearing plane, selected the farthest point from such plane and took the distance as our "nose tip - to - plane" measure.

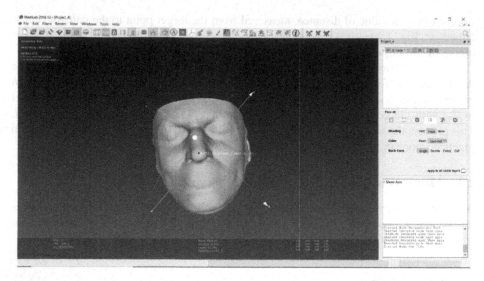

Fig. 2. The 3D model of a patient head visualized in MeshLab to define a ground truth for the measurements.

2.4   RGB-Depth Mapping Pipeline

In order to enhance the RGB camera-based face detection algorithm developed by Pepe et al. [6], it was necessary to map depth values produced by the ToF sensor to the frames shot by the HoloLens front-facing camera. The task was not trivial due to the misalignment between the two camera views and to the differences in field of view (FOV) and resolution, highlighted in Fig. 3.

Fig. 3. A quasi-synchronized shot from the RGB camera (left) and the short-range depth camera (right).

For devices developed with depth measurements in mind, manufacturers usually provide in-house produced calibration results, as the procedure requires rather complex setups and high-precision instrumentation. This was not the case for HoloLens, for which Microsoft only released scarcely documented, partial pieces of calibration data. For this reason, a significant part of the present work was devoted to delineating the RGB-to-Depth mapping pipeline.

2.5 The Mapping Pipeline

In order to map depth information on RGB frames, we need to find a transformation between the two camera views. This transformation can be expressed through a 4×4 roto-translation matrix, composed by a 3×3 matrix and a 1×3 vector, which hold information about the relative rotation and translation, respectively, between two cameras' frames of reference. In practice, as illustrated in Fig. 4, the whole process can be reduced to a series of transformations: from the depth camera 2D projection space to its relative 3D Coordinate System, then to the RGB Coordinate System and finally back to the RGB projection space. One major issue with this approach lies in the temporal misalignment between the recordings of the two sensors. In fact, the HoloLens API does not allow access to the RGB and Depth data streams at the same time [14], leading to a fluctuating mismatch between the frames acquisition time. Because of this, we are forced to consider the sensors as if they were constantly moving with respect to each other. Therefore, the sensors relative position has to be calculated for each pair of frames we want to map between.

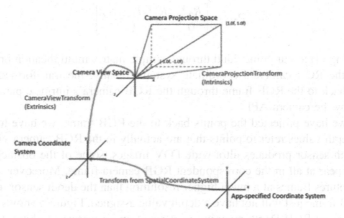

Fig. 4. A scheme of the process of locating frames acquired with HoloLens in the real world (https://docs.microsoft.com/en-us/windows/mixed-reality/locatable-camera).

Initially, we un-project the depth frame pixels to 3D coordinates as discussed in the previous section. Then, we calculate the absolute poses of the sensors. We achieve this by combining the Frame to Origin and the Camera View Transformation 4×4 matrices, accessible through the HoloLens API [15]. Finally, we calculate the transformation between Depth and RGB cameras coordinates, from which the relative pose can be derived as follows:

Let the cameras C1 and C2 have respective camera poses $P_{w1} = \begin{pmatrix} R_1 & t_1 \\ 0 & 1 \end{pmatrix}$ and $P_{w2} = \begin{pmatrix} R_2 & t_2 \\ 0 & 1 \end{pmatrix}$, where W denotes the world's frame of reference, R is a 3×3 rotation matrix and t a 3×1 translation vector. We want to find the transformation matrix P_{21} that defines the transformation from C1 to C2. We can just use P_{W1} and P_{W2} to find this, since they share a similar view. The first step is to convert a point q_1 in the C_1 space to a common world space through P_{W1}:

$$q_w = t_1 + R_1 * q_1 \tag{1}$$

Now, given a point q_w in the world's frame of reference, we can invert the camera pose transformation to write a point q_2 in C_2 as:

$$q_2 = R_1^{-1} * (q_w - t_2) \tag{2}$$

Substituting (1) in (2) we obtain P_{21}, the searched transformation from a source camera space - C_1 - to a target camera space - C_2 - so that:

$$q_2 = P_{21} * q_1 \tag{3}$$

where:

$$P_{21} = \begin{bmatrix} R_2^{-1} & R_2^{-1}(t_2 - t_1) \\ 0 & 1 \end{bmatrix} \tag{4}$$

Applying (4) to our point cloud through simple matrix multiplication brings the 3D points to the RGB camera's coordinate system. Finally, the transformed points are projected back to the RGB frame through the RGB camera's intrinsic parameters [16] provided by the camera API.

Once we have projected the points back to the RGB frame, we have to make sure that our depth values refer to points that are actually in the RGB camera view. In fact, as the depth sensor produces ultra-wide FOV images, some of the obstacles detected will not appear at all in the correspondent RGB camera frame. Moreover, as the RGB camera captures frames at a much higher resolution than the depth sensor, many pixels in the RGB frame just will not have a depth value assigned. Figure 5 shows an example of the proposed RGB-Depth mapping performed on a separate machine with the data recorded on the HoloLens.

2.6   Depth-Enhanced Landmark Detection

This section will discuss the steps taken in order to enhance the hologram to patient registration algorithm proposed in [6] with the acquired depth information. The RGB-Depth mapping process illustrated above, enabled us to assign a "distance" value to pixels in the RGB frames, which we can use to better estimate an object's location in the world. Moreover, through this approach, we could directly build upon the foundation laid by Pepe's team's work, developed around the Dlib RGB-based face detection algorithm.

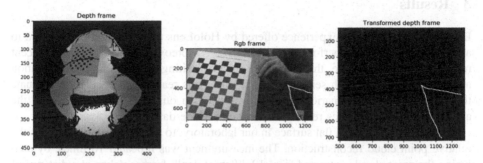

Fig. 5. An example of our RGB-Depth mapping. The white markings highlight the correspondence between the frames.

In particular, we decided to use the depth data to correct the estimated position of the patient's nose tip, one of the landmark points searched by Dlib in order to recognize a human face, like in Fig. 6.

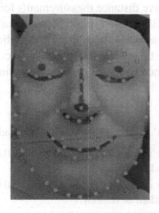

Fig. 6. The Dlib face detection software locating the face landmarks on a 3D-printed head [6].

With the mapped depth values in hand, we moved on in a similar fashion to Pepe's study: We determined the direction of the nose tip landmark point through pixel unprojection. Then, we scaled its position according to the relative depth value, which for a front facing subject is easily found as the point of the patient's face closest to the HoloLens user. This process provides us with an estimate of the patient's head's position but doesn't tell us about its orientation.

The task of recovering an object's orientation with respect to the camera pose from a set of 3D points and their projection on the frame is called a Perspectiven-Point (PnP) problem and can be solved by common computer vision libraries like OpenCV.

3   Results

Due to the nature of the experience offered by HoloLens to the user, it can be hard to assess its visualization performance objectively. Moreover, due to the very recent release of Research Mode, there is, at the time of writing, close to no documentation about the HoloLens depth sensor's accuracy. For this reason, we decided to evaluate the accuracy also in scenarios not exactly related to our particular use-case, to the advantage of future studies relying on Research Mode data.

First, we considered a flat surface in our laboratory, to assess the smoothness of the relative point cloud reconstruction. The measurement was repeated 10 times, by analyzing the point clouds extracted from 10 different depth frames, taken at a distance of about 60 cm from a wall (Table 1).

Table 1. HoloLens' depth sensing accuracy - planar surface smoothness.

RMS error	2.4 ± 0.4 mm
(mean ± standard deviation)	

Next, we performed relative distance measurements for 3D objects, namely a sharp-edged wooden box and the 3D-printed head model already used for testing purposes. For the box, the measure was performed on all the 3 dimensions, while for the 3D-printed head only the nose tip-to-bearing plane distance was evaluated. Each measurement was repeated 5 times and performed at a distance of about 60 cm from the target (Table 2).

Table 2. HoloLens' depth sensing accuracy - wooden box dimensions.

Relative error in the back-front dimension (mean ± standard deviation)	0.033 ± 0.018
Relative error in the up-down dimension (mean ± standard deviation)	0.079 ± 0.028
Relative error in the right-left dimension (mean ± standard deviation)	0.044 ± 0.021

The primary goal of this work is to assess the effects of introducing additional depth information in the hologram-to-patient registration process, reportedly one of the major weak spots in Pepe's study [6] (Table 3).

Table 3. HoloLens' depth sensing accuracy - 3D-printed head facial features.

Relative error in the nose tip - to bearing plane distance (mean ± standard deviation)	0.018 ± 0.011

The technical results, shown in Table 4, are generally in line with what is found in the previous study, but still interesting considering that this preliminary work only exploits a very small portion of the available sensors' data. Again, the measures were repeated 5 times, at a distance of approximately 60 cm.

Table 4. Hologram-to-patient registration error: comparison.

Measured value	Proposed method	Previous method
Error in the back-front dimension (mean ± standard deviation)	3.8 ± 1.7 mm	−4.5 ± 2.9 mm
Error in the up-down dimension (mean ± standard deviation)	−8.6 ± 3.7 mm	3.3 ± 2.3 mm
Error in the right-left dimension (mean ± standard deviation)	−2.2 ± 1.5 mm	−9.3 ± 6.1 mm

4 Conclusion

We evaluated the potential of the Microsoft HoloLens depth sensing capabilities in enabling accurate and seamless imaging data visualization in maxillofacial surgical procedures. In this work, we addressed the most prominent issue of the previous implementations: the bottleneck in hologram-to-patient registration accuracy, which is limited depth perception. Invaluable for this purpose was HoloLens' Research mode, which, only recently released by Microsoft, provided us with a stream of depth data previously unavailable to researchers. First, the data, in the form of a stream of range images, was processed in order to obtain a point cloud representation of the user's view. Then, a conversion pipeline, to map the depth information onto the RGB frames' pixel was established. Given the only recent availability of the data, no information on the sensor's accuracy was found in literature, for this reason, an accuracy evaluation in different measurement scenarios was performed. Ultimately, we proceeded to integrate the mapped depth information into the RGB-based application developed in [6], as pattern recognition techniques like the ones here employed for face detection can be heavily affected by inaccurate spatial perception. Overall, our study found that the rich set of on-board sensors can be exploited beyond its intended use user interface, environment navigation - turning the headset in something more than a mere visualization device. The accuracy evaluation, in fact, demonstrated the device potential for millimeter-accuracy measurement, comparable to other commercially available sensors. With regard to hologram registration, although only slight improvements from the previous study were found, we see huge potential for more spatial aware HoloLens applications. Future developments, in fact, could easily make more extensive use of the headset sensors data, for example considering a 3D-to-3D registration approach, in order to overcome the inherent flaws of two-dimensional pose estimation.

References

1. National Institutes of Health: Technologies Enhance Tumor Surgery: Helping Surgeons Spot and Remove Cancer. News in Health (2016)
2. Chen, X., et al.: Development of a surgical navigation system based on augmented reality using an optical see-through head-mounted display. J. Biomed. Inform. **55**, 124–131 (2015)
3. Perkins, S.L., Lin, M.A., Srinivasan, S., Wheeler, A.J., Hargreaves, B.A., Daniel, B.L.: A mixed reality system for breast surgical planning. In: IEEE International Symposium on Mixed and Augmented Reality (ISMAR-Adjunct), Nantes, pp. 269–274 (2017)
4. Pratt, P., et al.: Through the HoloLens looking glass: augmented reality for extremity reconstruction surgery using 3D vascular models with perforating vessels. Eur. Radiol. Exp. **2**(1), 2 (2018)
5. Wang, J., Suenaga, H., Yang, L., Kobayashi, E., Sakuma, I.: Video see-through augmented reality for oral and maxillofacial surgery. Int. J. Med. Rob. Comput. Assist. Surg. **13**, e1754 (2017)
6. Pepe, A., et al.: Pattern recognition and mixed reality for computer-aided maxillofacial surgery and oncological assessment. In: 2018 11th IEEE Biomedical Engineering International Conference (BMEiCON), pp. 1–5 (2018). https://doi.org/10.1109/BMEiCON.2018.8609921
7. Dlib C++ library. http://www.dlib.net
8. Bevilacqua, V., D'Ambruoso, D., Mandolino, G., Suma, M.: A new tool to support diagnosis of neurological disorders by means of facial expressions. In: IEEE International Symposium on Medical Measurements and Applications, Medical Measurements and Applications Proceedings (MeMeA2011), pp. 544–549 (2011). https://ieeexplore.ieee.org/document/5966766/
9. Bevilacqua, V., Biasi, L., Pepe, A., Mastronardi, G., Caporusso, N.: A computer vision method for the italian finger spelling recognition. In: International Conference on Intelligent Computing (ICIC2015), vol. 9227, pp. 264–274 (2015). https://doi.org/10.1007/978-3-319-22053-6_28
10. Bevilacqua, V., et al.: A comprehensive method for assessing the blepharospasm cases severity. In: Santosh, K.C., Hangarge, M., Bevilacqua, V., Negi, A. (eds.) RTIP2R 2016. CCIS, vol. 709, pp. 369–381. Springer, Singapore (2017). https://doi.org/10.1007/978-981-10-4859-3_33
11. Microsoft: HoloLens Research Mode Tutorial at CVPR (2018). https://docs.microsoft.com/it-it/windows/mixed-reality/cvpr-2018
12. Microsoft: HoloLensForCV C#/C++ library. https://github.com/Microsoft/HoloLensForCV
13. MathWorks: pcfitplane - fit plane to 3D point cloud. https://www.mathworks.com/help/vision/ref/pcfitplane.html
14. Microsoft: Process Media frames with MediaFrameReader. https://docs.microsoft.com/en-us/windows/uwp/audio-video-camera/processmedia-frames-with-mediaframereader#setting-up-your-project
15. Guyman, W., Zeller, M., Cowley, E., Bray, B.: Locatable camera, Microsoft - Windows Dev Center, 21 March 2018. https://docs.microsoft.com/en-us/windows/mixedreality/locatable-camera
16. Microsoft: CameraIntrinsics Class. https://docs.microsoft.com/en-us/uwp/api/windows.media.devices.core.cameraintrinsics

An Innovative Neural Network Framework for Glomerulus Classification Based on Morphological and Texture Features Evaluated in Histological Images of Kidney Biopsy

Giacomo Donato Cascarano[1], Francesco Saverio Debitonto[1],
Ruggero Lemma[1], Antonio Brunetti[1], Domenico Buongiorno[1],
Irio De Feudis[1], Andrea Guerriero[1], Michele Rossini[2],
Francesco Pesce[2], Loreto Gesualdo[2], and Vitoantonio Bevilacqua[1(✉)]

[1] Department of Electrical and Information Engineering (DEI),
Polytechnic University of Bari, Bari, Italy
vitoantonio.bevilacqua@poliba.it
[2] Department of Emergency and Organ Transplantation, Nephrology Unit,
University of Bari Aldo Moro, Bari, Italy

Abstract. Medical Imaging Computer Aided Diagnosis (CAD) systems could support physicians in several fields and recently are also applied in histopathology. The goal of this work is to design and test a novel CAD system module for the discrimination between glomeruli with a sclerotic and non-sclerotic condition, through the elaboration of histological images. The dataset was constituted by 26 kidney biopsies coming from 19 donors with Periodic Acid Schiff (PAS) staining. Preparation, digital acquisition and glomeruli annotations have been conducted by experts from the Department of Emergency and Organ Transplantation (DETO) of the University of Bari Aldo Moro (Italy). Starting from the annotated Regions Of Interest (ROIs), several feature extraction techniques were evaluated. Feature reduction and shallow artificial neural network were used for discriminating between the glomeruli classes. The mean and the best performances of the best ANN architecture were evaluated on an independent dataset. Metric comparison and analysis were performed to face the unbalanced dataset problem. Results on the test set asses that the proposed workflow, from the feature extraction to the supervised ANN approach, is consistent and reveals good performance in discriminating sclerotic and non-sclerotic glomeruli.

Keywords: CKD · Kidney · Glomerulus classification ·
Morphological features · Texture features · ANN

1 Introduction/Background

Chronic Kidney Disease (CKD) is a pathological condition consisting of a functional degeneration of the kidney. Kidney transplantation is the primary therapy for patients affected by CKD, more effective than dialysis treatment in terms of long-term mortality

© Springer Nature Switzerland AG 2019
D.-S. Huang et al. (Eds.): ICIC 2019, LNAI 11645, pp. 727–738, 2019.
https://doi.org/10.1007/978-3-030-26766-7_66

risk [1, 2], having, at the same time, a smaller impact on the public health system. Due to the increasing necessity of kidney transplants [3], different studies tried to widen the criteria of inclusion [4, 5]. In [6] was performed a comparison between dual kidney transplantation (DKT) from expanded criteria donors (ECDs) and single kidney transplantation (SKT) from concurrent ECDs and standard criteria donors. The authors assessed that the use of dual kidney transplantation from marginal donors is a viable option and that renal function can be achieved with older, low nephron mass donors provided that both kidneys are transplanted into a single recipient. In [7], techniques to assess the kidney condition by histological biopsy is proposed. The evaluation criterion, called Karpinski score, is based on the percentage evaluation of a pathological condition of four main functional areas: glomerulosclerosis, tubular atrophy, interstitial fibrosis and arterial sclerosis. The score ranges from 0 to 12, and a higher number means a worse condition [7–9]. Kidneys with a Karpinski score from 0 to 3 and from 4 to 6 are considered suitable for single and dual transplant, respectively. The computation of the score needs the analysis of kidney biopsies by pathologists, that is usually a time-consuming, prone to error and subjective procedure. Due to the reasons mentioned above, developing a clinical support system based on the tissue image analysis for supporting the computation of the score is a desirable headway.

This work is focused on the automatic evaluation of kidney biopsies, dealing with one of the four pathological conditions evaluated in the Karpinski score: glomerulosclerosis. It consists in detecting and discriminating the sclerotic condition affecting the glomeruli from those non-sclerotic. A glomerulus is part of the nephron, the functional renal unit involved in blood filtration, and performing this discrimination is a challenging task, due to their wide intensity variation and inconsistency in terms of shape and size.

In a previous work [10], a Computer Aided Diagnosis (CAD) system for segmentation and discrimination of blood vessels versus tubules from biopsies in the kidney tissue has been designed and tested. Histological images with PAS staining have been used to segment Regions of Interest (ROIs) and extract Haralick features allowing a subsequent classification procedure based on Artificial Neural Network (ANN) algorithms. Test results determined that the supervised ANN approach is consistent and reveals good performance.

In this work, a combination of different feature extraction algorithms has been designed and evaluated, starting from Whole Slide Images (WSI) with Periodic acid–Schiff (PAS) staining, for discriminating two glomerulus conditions: sclerotic and non-sclerotic. The set of extracted features come from a collection of two wide-used, well-known and general purpose features extractor algorithms families; in particular, morphological and texture features have been computed. The set of features was then reduced by means of feature reduction algorithms and then used as input to a shallow Artificial Neural Network.

2 Materials

Whole Slide Images were collected between 07/2011 and 02/2015 by physicians from the Department of Emergency and Organ Transplantations (DETO) of the Bari University Hospital. All the kidney biopsies with PAS staining were scanned by using the Aperio ScanScope CS at 20x with a resolution of 0.50 μm/pixel. The WSIs were collected from a total of 26 kidney biopsies coming from 19 donors and stored at full resolution in SVS file format (an Aperio file format consisting of pyramidal tiled TIFF with non-standard metadata and compression). The collected images presented wide differences in colour and saturation, even if all treated with PAS staining.

Two medical graduands manually annotated the glomeruli independently; the annotation were subsequently validated by a renal pathologist. The manual annotation was performed by outlining the real glomerulus region using the Aperio ImageScope tool; at the same time, the glomeruli were labelled as sclerotic and non-sclerotic.

The obtained initial dataset was composed of 428 sclerotic glomeruli and 2.344 non-sclerotic glomeruli, with a ratio between the two classes of 1/5.5.

The dataset was subsequently divided into two subsets called train set and test set. In particular about 20% of the original dataset has been used as test subset and the information of the target in the test-set has been used to assess final performances only; furthermore, the selection has been achieved randomly with the constraint that if a glomerulus appear in the test-set, all the other glomeruli belonging to the same biopsy must appear in the test-set only. This is equivalent to asses that the train/test division has been performed at biopsy level. The constrained division avoids that particular context information could be present in both the dataset leading to an unfair dataset split. The latest dataset configuration is reported in Table 1.

Table 1. Dataset configuration.

Dataset	Sclerotic glomeruli	Non-sclerotic glomeruli	Total
Train set	341	1852	2193
Test set	87	492	579
Total	428	2344	2772

3 Methods

The main goal of this research was to design a CAD module able to classify the glomerulus condition using a feature-based approach. In detail, the designed solution works with image processing and machine learning techniques to assess the class of the each glomerulus: sclerotic or non-sclerotic.

A detailed representation of the full workflow, described in the following paragraphs, is reported in Fig. 1.

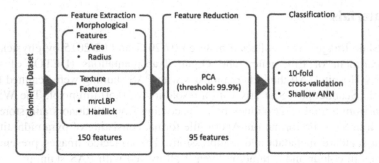

Fig. 1. Full features extraction and classification workflow.

As depicted in Fig. 1, the discrimination process could be divided into three main steps. The first two allow the extraction of several features and the reduction of their space by means of feature reduction algorithm; the last one leads to the assignment of the label.

3.1 Features Extraction

The features extraction is the first step of the workflow [11, 12], allowing the definition of a set of characteristics able to define and discriminate between the two different types of glomeruli. Based on the human reasoning used by the physicians able to address the problem, the best features to face the problem are those related to two main image processing techniques: morphological and texture based features.

As reported by the pathologist involved in this study, the main distinctions between sclerotic and non-sclerotic glomeruli are the shape of the Bowman's capsule, different dimension and a different texture related to blood vessels. All the evaluations, the thresholding values and the decision regarding the best algorithms configuration have been done on train set only.

Morphological Features
Regarding the morphological characteristics, two features are related to the Bowman's capsule and the Bowman's space.

The first feature is computed as the sum of the areas related to the Bowman's capsule, the blood vessels areas and the inter-capillary spaces. Due to the PAS staining, these structures are characterized by a whiteness colouration and the detection of the mask describing the region is based on three parallel image processing procedures. Each process took into account the channels of three different colour space: RGB, CMYK and Lab. In detail:

- Green channel of RGB colour space, as it is the most representative of the glomerulus structure [13];
- complementary of Magenta from the CMYK colour model has been chosen due to the detectable empirical significance of this colour component;
- *a* and *b* components of Lab colour space due to the link with the human colour vision.

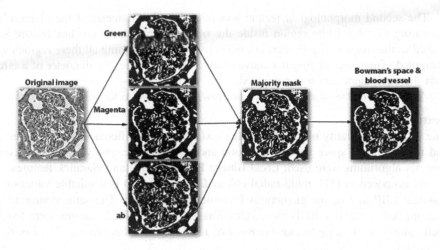

Fig. 2. Workflow of Bowman's space segmentation.

The extraction of the masks of green and magenta channels follows the same steps:

1. *binarisation*: to keep the pixels related to white regions a threshold value has been empirically set to 190 [14];
2. *morphological operators*: to clean the image obtained from the previous step, erosion, dilation and median filtering have been used with a disk of radius ranging from 1 to 3 as structuring element;
3. *active contour*: to clean the shape of the obtained mask, active contour algorithm [15] has been used with 200 iterations (the chosen number of iterations avoid an extreme smoothing of the glomerulus shape).

The three previous steps led to the computation of two masks, one for green channel and one for magenta one; the last mask was computed from a and b components of Lab colour space. The ab matrix has been used as input to k-means clustering algorithm [16]; the number of clusters was empirically set to 5, and the number of repetitions of the clustering process using new initial cluster centroid positions to avoid local minima was set to 3. The mask was computed subsequently by retaining just the pixels belonging to the cluster with the greatest mean grey-scale intensity value. Then the steps 2 and 3 of the Green-Magenta segmentation process were applied.

The final mask was the composition of the resulting three masks computed with a majority criterion. The obtained mask was processed to remove artefact and not interesting regions; in detail, too small regions (lesser than 1000 pixels), and a logical AND with a circle of radius equal to the smaller dimension of the image subtracted by 1/8 of its value was performed.

Figure 2 shows the overview of the Bowman's space segmentation workflow.

Starting from the final mask, the feature of interest was the sum of Bowman's space, blood vessels and the inter-capillary region of the glomerulus, that is, in our workflow, the sum of white region. This value was finally normalised considering the image area.

The second morphological feature was related to the diameter of the glomerulus. Assuming that the white region inside the mask computed for the last feature was related to the shape of the glomerulus, the convex hull containing all these regions was computed. Then, considering the convex hull ROI as a circle, the diameter of a circle with the equivalent area was computed.

As a result of the morphological workflow, a total of two features were computed.

Texture Features

Due to the particularity of the glomerulus texture and the differences in blood vessels and inter-capillary space between sclerotic and non-sclerotic, two well-known texture analysis algorithms were used: Local Binary Pattern (LBP) and Haralick features.

As proposed in [17], multi-radial colour LBP (mrcLBP) is a suitable variation of classical LBP to face the glomerulus identification problem. The same configuration was applied to the raw RGB glomerulus images. The obtained features were ten for each radius, thus leading to a total number of 120 (10 features per radius, 4 radius, three channels).

The second set of texture-based features was obtained from the extraction of Haralick features. The four Grey-Level Co-occurrence Matrix (GLCM), one for each direction, has been computed; then, the 14 Haralick indexes were computed, leading to 56 features. To reduce this number, the mean and the range among the four directions was calculated. The final features were 28 (14 mean and 14 range, one for each Haralick feature).

As a result of the texture features extraction, a total of 148 features were computed.

3.2   Feature Reduction

The created set of features is the union of both morphologic and texture-based features. An overall number of 150 features was achieved. Due to the possibility of correlation among the different subsets of features, and to reduce the total number of inputs to the classification step, Principal Component Analysis (PCA) was applied as feature reduction algorithm. Prior to PCA, each feature was z-score normalized.

As stated in Sect. 2, the dataset was previously split into train and test set with the aim to fairly take all the image pre-processing and classification decisions on the train set only. The feature reduction algorithm, instead, doesn't need or use the label information; for this reason, the application of PCA could be done on the whole dataset or on the training dataset only. Due to the small differences between the number of the two reduced dataset, to take into account all the information inside the dataset, and to avoid the necessity to preserve the transformation matrix for the test phase, the first approach has been chosen.

As a result of the PCA as feature reduction algorithm, a total number of 95 features were computed and will be used for the classification phase.

3.3   Glomeruli Classification

The glomeruli classification steps are based on Artificial Neural Network (ANN), specifically on shallow ANN (Fig. 3).

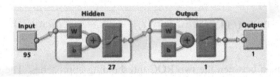

Fig. 3. Shallow artificial neural network.

All the decisions about the ANN architecture and the tuning of its parameters were taken considering the train set only, whereas all the reported results and performance discussions refer to the test set. To generalise, to avoid overfitting and to obtain a classifier independent from the input dataset, k-fold was used as cross-validation technique. Several network initialisation inside each fold and hard voting among the folds was used both to obtain independency from a particular network initialisation and to compute the overall fold class label.

The input of the classifier was the features set obtained from the image processing algorithm and the subsequent PCA feature reduction; the number of input features was 95, and 10-fold cross validation was used. The fixed training parameters were the following: one hidden layer, tansig and softmax as activation functions for the hidden and output layer, respectively, crossentropy as loss function and scaled conjugate gradient as backpropagation algorithm. The stop criterion was based on the validation set and, an early stop of the training was implemented to promote generalisation and to avoid overfitting, stopping it if the performance on validation set did not decrease inside a sliding window of 6 epochs. The last relevant parameter is the number of neurons for the hidden layer, and the choice of the right value is afterwards reported.

To face the heavy problem of unbalanced distribution between sclerotic and non-sclerotic glomeruli, we selected the Matthews Correlation Coefficient (MCC) [18] as a general performance comparison among the folds. MCC (Eq. 5) takes into account false negative and false positive and computes a correlation coefficient between predicted and target classes. As stated in [19], among the usual performance scores, MCC is the only one that takes into account the ratio of the confusion matrix size, and it revealed to be a better index of performance than accuracy or F1 score on unbalanced datasets.

Subsequently, we used the Receiving Operating Characteristic (ROC) curve to choose the correct classification threshold value. Two approaches were analysed. The first one (Approach A) assumes the optimal value as the first intersection point between the ROC curve and a line with slope equal to the ratio between the total number of negative and positive samples and sliding from the upper left corner of the ROC plot ((FPR, TPR) = (0, 1)). The second approach (Approach B), pro-posed in [20], evaluates the point of minimum distance from the point (0, 1) of the ROC plot. The equation is reported in Eq. 1.

The comparison of the two methods in terms of different performance indexes (Eqs. 2, 3, 4 and 5) is reported in Table 2. Due to the medical domain of the work, a higher recall is preferred, thus the second method was chosen.

$$\min_i \sqrt{(1 - sensitivity(i))^2 + (1 - specificity(i))^2} \qquad (1)$$

Table 2. Comparison between the two ROC thresholding approaches. The reported values are the mean among the 10-fold.

	Approach A	Approach B
Accuracy	0.9898	0.9865
Precision	0.9775	0.9332
Recall	**0.9575**	**0.9880**
MCC	0.9612	0.9520

The architecture of the shallow ANN chosen for glomeruli classification was fixed to one hidden layer. To choose the right number of neurons per layer, the performance of 95 networks were compared. In detail, several networks with the number of neurons for the hidden layer ranging from 1 to 95 were trained (95 is the fixed number of the input features). Based on the best MCC value computed as the mean MCC of the folds, the final number of neurons for the hidden layer was set to 27.

4 Results

In this section, the results of the proposed glomerulus classification workflow are reported. In particular, we reported the performance obtained considering the reduced set of features by means of PCA and then classified by using cross-validated shallow ANN. As reported in Table 1 the test set was constituted by 579 glomeruli images (87 sclerotics, 492 non-sclerotics).

Table 3. Confusion matrix for metrics computation.

		True condition	
		Positive (sclerotic)	Negative (non-sclerotic)
Predicted condition	Positive (sclerotic)	True positive TP	False positive FP
	Negative (non-sclerotic)	False negative FN	True negative TN

Several metrics were considered for the evaluation. In particular, Accuracy (Eq. 2) and Matthews Correlation Coefficient (MCC) (Eq. 5) were evaluated on the test set, according to the confusion matrix reported in Table 3.

$$Accuracy = \frac{TP + TN}{TP + FP + FN + TN} \qquad (2)$$

$$Precision = \frac{TP}{TP + FP} \qquad (3)$$

$$Recall = \frac{TP}{TP + FN} \qquad (4)$$

$$MCC = \frac{TP * TN - FP * FN}{\sqrt{(TP + FP) * (TP + FN) * (TN + FP) * (TN + FN)}} \qquad (5)$$

To evaluate the workflow stability, 10 runs of the whole process were performed, and the corresponding results are summarized in Table 4; the results are reported in terms of mean and standard deviation and the corresponding equations are reported in Eqs. 2, 3, 4 and 5. The best result, instead, is reported in Table 5 and the corresponding confusion matrix is reported in Table 6. An example of misclassified glomeruli is reported in Fig. 4, mainly due to the presence of artefacts on the images; the pathologist confirmed this.

Table 4. Metrics comparison of 10 network initialization.

	Mean ± std
Accuracy	0.9874 ± 0.0018
Precision	0.9844 ± 0.0111
Recall	0.9310 ± 0.0153
MCC	0.9501 ± 0.0074

Table 5. Metrics comparison of the best network.

Accuracy	0.9914
Precision	1.0000
Recall	0.9425
MCC	0.9659

Table 6. Confusion matrix of the best network.

		True condition	
		Positive (sclerotic)	Negative (non-sclerotic)
Predicted condition	Positive (sclerotic)	82	0
	Negative (non-sclerotic)	5	492

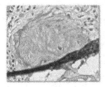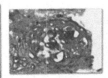

Fig. 4. False negative misclassified by the best model.

As reported in Table 4, the workflow allows the classification of sclerotic and non-sclerotic glomeruli with good performances and low variability. Precision and Recall are equal to 0.98 and 0.93, respectively, showing a better performance in the non-sclerotic evaluation. The results show that the proposed workflow is a valid solution to detect glomeruli and, as assessed by the domain expert, the misclassified images are a challenging detection problem for pathologist too and are usually discarded.

5 Discussion and Conclusion

In this work, a complete workflow for sclerotic and non-sclerotic glomeruli classification has been designed and developed. Several features extraction algorithms were analysed and tested, and two feature typologies were selected among them. Both morphological and texture feature extraction algorithms were tuned to achieve good performance on the training set. We collected 150 feature: 2 morphological features and 148 texture ones extracted by means of mrcLBP and Haralick algorithms. Then, a PCA was performed to reduce the number of features to 95. A cross-validated artificial neural network was trained, and unbalanced dataset and network tuning problems were faced. The final results were computed on an independent test set. The classification workflow achieve a mean MCC and Accuracy of 0.9501 and 0.9874, respectively, and low variability over 10 independent iterations. Good precision and recall were obtained too. The reported results suggest that the proposed workflow set-up is reliable for the investigated domain, supporting the clinical practice of discriminating the two classes of glomeruli. However, there are still common glomeruli misclassification in images affected by artefacts, which are usually discarded by pathologists even in the clinical practice.

In the future works, we will investigate how to reduce the number of empirical assumptions on the feature extraction process and to introduce a weighted classification among the folds; furthermore, a feature analysis, to recognise the better ones, will be conducted. Different classification techniques, such as deep learning [21, 22], will be also evaluated and the results will be compared with the approach proposed in this paper. Finally, the presented workflow will be integrated into a complete CAD tool for kidney biopsies analysis.

Acknowledgment. This study has been partially funded from the INNONETWORK program of the Apulia Region, project SOS – Smart Operating Shelter (project n. 9757YR7).

The authors would like to thank Federica Albanese, Davide Mallardi and Dr Michele Rossini for the medical domain support and the manual annotation of the glomeruli whole slide images for the input dataset creation.

References

1. Wolfe, R.A., Ashby, V.B., Milford, E.L., Ojo, A.O., Ettenger, R.E., Agodoa, L.Y.C., et al.: Comparison of mortality in all patients on dialysis, patients on dialysis awaiting transplantation, and recipients of a first cadaveric transplant. N. Engl. J. Med. (2002). https://doi.org/10.1056/nejm199912023412303
2. Port, F.K., Wolfe, R.A., Mauger, E.A., Berling, D.P., Jiang, K.: Comparison of survival probabilities for dialysis patients vs cadaveric renal transplant recipients. JAMA J. Am. Med. Assoc. (1993). https://doi.org/10.1001/jama.1993.03510110079036
3. Cecka, J.M.: The UNOS Scientific Renal Transplant Registry–ten years of kidney transplants. Clin. Transpl. 1–14 (1997)
4. United Network for Organ Sharing: United Network for Organ Sharing. Annual Report (2004)
5. Perico, N., Ruggenenti, P., Scalamogna, M., Remuzzi, G.: Tackling the shortage of donor kidneys: how to use the best that we have (2003). https://doi.org/10.1159/000072055
6. Moore, P.S., Farney, A.C., Sundberg, A.K., Rohr, M.S., Hartmann, E.L., Iskandar, S.S., et al.: Dual kidney transplantation: a case-control comparison with single kidney transplantation from standard and expanded criteria donors (2007). https://doi.org/10.1097/01.tp.0000266579.11595.95
7. Remuzzi, G., Grinyò, J., Ruggenenti, P., Beatini, M., Cole, E.H., Milford, E.L., et al.: Early experience with dual kidney transplantation in adults using expanded donor criteria. Double Kidney Transplant Group (DKG). J. Am. Soc. Nephrol. 10, 2591–2598 (1999)
8. Karpinski, J., Lajoie, G., Cattran, D., Fenton, S., Zaltzman, J., Cardella, C., et al.: Outcome of kidney transplantation from high-risk donors is determined by both structure and function. Transplantation (1999). https://doi.org/10.1097/00007890-199904270-00013
9. Remuzzi, G., Ruggenenti, P.: Renal transplantation: single or dual for donors aging ≥ 60 years? Transplantation (2003). https://doi.org/10.1097/00007890-200005270-00002
10. Bevilacqua, V., Pietroleonardo, N., Triggiani, V., Brunetti, A., Di Palma, A.M., Rossini, M., et al.: An innovative neural network framework to classify blood vessels and tubules based on Haralick features evaluated in histological images of kidney biopsy. Neurocomputing 228, 143–153 (2017). https://doi.org/10.1016/j.neucom.2016.09.091
11. Bevilacqua, V., Buongiorno, D., Carlucci, P., Giglio, F., Tattoli, G., Guarini, A., et al.: A supervised CAD to support telemedicine in hematology. In: Proceedings of the International Joint Conference on Neural Networks (2015). https://doi.org/10.1109/ijcnn.2015.7280464
12. Bevilacqua, V., et al.: Retinal fundus biometric analysis for personal identifications. In: Huang, D.-S., Wunsch, D.C., Levine, D.S., Jo, K.-H. (eds.) ICIC 2008. LNCS (LNAI), vol. 5227, pp. 1229–1237. Springer, Heidelberg (2008). https://doi.org/10.1007/978-3-540-85984-0_147
13. Kotyk, T., Dey, N., Ashour, A.S., Balas-Timar, D., Chakraborty, S., Ashour, A.S., et al.: Measurement of glomerulus diameter and Bowman's space width of renal albino rats. Comput. Methods Programs Biomed. 126, 143–153 (2016). https://doi.org/10.1016/J.CMPB.2015.10.023
14. Zhao, Y., Black, E.F., Marini, L., McHenry, K., Kenyon, N., Patil, R., et al.: Automatic glomerulus extraction in whole slide images towards computer aided diagnosis. In: 2016 IEEE 12th International Conference on e-Science (e-Science), pp. 165–174. IEEE (2016). https://doi.org/10.1109/escience.2016.7870897
15. Chan, T.F., Vese, L.A.: Active contours without edges. IEEE Trans. Image Process. (2001). https://doi.org/10.1109/83.902291
16. Lloyd, S.: Least squares quantization in PCM. IEEE Trans. Inf. Theory 28, 129–137 (1982)

17. Simon, O., Yacoub, R., Jain, S., Tomaszewski, J.E., Sarder, P.: Multi-radial LBP features as a tool for rapid glomerular detection and assessment in whole slide histopathology images OPEN, **8**, 2032 (2018). https://doi.org/10.1038/s41598-018-20453-7
18. Matthews, B.W.: Comparison of the predicted and observed secondary structure of T4 phage lysozyme. Biochim. Biophys. Acta (BBA)-Protein Struct. **405**, 442–451 (1975)
19. Chicco, D.: Ten quick tips for machine learning in computational biology (2017). https://doi.org/10.1186/s13040-017-0155-3
20. Song, B., Zhang, G., Zhu, W., Liang, Z.: ROC operating point selection for classification of imbalanced data with application to computer-aided polyp detection in CT colonography. Int. J. Comput. Assist. Radiol. Surg. (2014). https://doi.org/10.1007/s11548-013-0913-8
21. Bevilacqua, V., et al.: A novel deep learning approach in haematology for classification of leucocytes. In: Esposito, A., Faundez-Zanuy, M., Morabito, F.C., Pasero, E. (eds.) WIRN 2017 2017. SIST, vol. 103, pp. 265–274. Springer, Cham (2019). https://doi.org/10.1007/978-3-319-95095-2_25
22. Brunetti, A., Buongiorno, D., Trotta, G.F., Bevilacqua, V.: Computer vision and deep learning techniques for pedestrian detection and tracking: a survey. Neurocomputing (2018). https://doi.org/10.1016/j.neucom.2018.01.092

Evaluating Generalization Capability of Bio-inspired Models for a Myoelectric Control: A Pilot Study

Cristian Camardella[1]([⊠]) [iD], Michele Barsotti[1] [iD],
Luis Pelaez Murciego[1] [iD], Domenico Buongiorno[2] [iD],
Vitoantonio Bevilacqua[2] [iD], and Antonio Frisoli[1] [iD]

[1] PercRo Laboratory, Scuola Superiore Sant'Anna, Pisa, Italy
{cristian.camardella,michele.barsotti,
luispelaez.murciego,antonio.frisoli}@santannapisa.it
[2] DEI Department, Polytechnic University of Bari, Bari, Italy
{domenico.buongiorno,
vitoantonio.bevilacqua}@poliba.it

Abstract. The rapid growing interest in the field of wearable robots opens the challenge for the development of intuitive and natural control strategies for establishing an effective human-robot interaction. The myoelectric control could be a valid solution for achieving this goal, since it is a strategy based on decoding the human motor intentions from surface electromyographic signals (sEMG) and mapping them into the control output.

In this work we propose a bio-inspired myocontrol approach able to generalize the hand force estimation in the central point of a rectangle shaped workspace, after being trained and interpolated using data acquired on the vertexes only.

We compared performance of the proposed approach (featuring factorization and clustering techniques for building muscles patterns valid in the whole workspace) versus the ones obtained using the classical muscle synergies extraction strategy and without the use of muscle synergies.

Obtained results show that all the three tested approaches were able to generalize. Moreover, with the proposed approach, we were able to closely estimate the muscle patterns in the testing point, suggesting the possibility of a better generalization capability (compared to other approaches) when increasing the size of the workspace.

Keywords: EMG · Muscle synergies · Blind source separation · Myoelectric-control · Regression · Clustering

---

This work has been partially founded by the PRIN-2015 ModuLimb - Prot.2015HFWRYY and supported by the Italian project RoboVir within the BRIC INAIL-2016 program.

© Springer Nature Switzerland AG 2019
D.-S. Huang et al. (Eds.): ICIC 2019, LNAI 11645, pp. 739–750, 2019.
https://doi.org/10.1007/978-3-030-26766-7_67

1 Introduction

In the last decades, wearable robotics became an important part of an independent, growing line of research in which researchers put their efforts in order to develop effective and reliable systems for assisting or increasing the user's physical performance [19]. In this context, Human Machine Interfaces (HMIs) represents a crucial factor for maximizing these systems' efficacy [8] by developing intuitive and personalized schemes of interaction [2]. Myo-control schemes, thanks to their ability of decoding the human motor intention from non-invasive electromyographic signals (EMG), allow for the establishment of natural and intuitive HMIs. Moreover, they are suitable for being coupled with a human-complied robotic interface (i.e. exoskeletons) [14, 15] even when severe impairments occur [18].

Although the neuro-muscular theory behind the EMG signals generation ensures the existence of applicable schemes [11], still no robust and reliable solutions appeared among both scientific literature and commercial devices [2]. In order to achieve an accurate simultaneous and proportional myocontrol scheme (SPC), the complexity underneath the relation between limbs force generation and the driving neural strategy of the Central Nervous System (CNS) is currently being investigated. As for many engineering problems, biological phenomena in spired many elegant solutions. In the myocontrol field two bio-inspired main approaches have been developed: the model-based approach and the synergistic approach. The model-based approaches, based on mathematical models of the neuro-musculoskeletal system, are particularly suitable for estimating the articulation torques of humans, but their use is limited by both the long lasting calibration phases and the high computational cost [7]. On the other hand, the synergistic approach for myoelectric control aims at minimizing the computational cost of the system by mimicking the CNS through the identification of those muscle activation patterns exploited during task-related movements [4]. D'Avella et al. [10] showed that the spatiotemporal characteristic of the phasic muscle patterns generated during point-to-point reaching movements in vertical planes can be explained by the combinations of a reduced set of muscle synergies. Following this line, Berger et al. [3] demonstrated that muscle synergies can be used to generate target forces in multiple directions with the same level of accuracy achieved using individual muscles in a defined upper-limb pose. Jiang et al. [12] successfully developed a strategy for achieving an accurate simultaneous and proportional control of 2-Degrees of Freedom (DOF) myoelectric prostheses by extracting synergy-based control signals from each independent DoF simultaneously. However, they documented substantial decrease of the performance when introducing the 3rd DoF. Although the previous works successfully achieved the control of a virtual cursor in a 2-DoF task, still an exhaustive exploration of the human upper-limb workspace is missing. In fact, for being applicable in real scenarios (i.e., control of an exoskeleton), a myocontrol scheme should be valid for a large part of the workspace and, at the same time, should feature a short (or even null) training time. The main challenge is thus to understand how to take into account the variation of muscle synergies due to the upper limb pose. In particular, it is interesting understanding how new models can be built, using a reduced training set and how these new models perform in non-trained upper-limb poses (generalization capabilities). In [1], the feasibility of predicting untrained combinations of multiple DoF based on linear regressor models trained with minimal amount data has been proven. In [17], it has been showed how, with linear interpolation, new combined datasets can be generated for training data augmentation purposes. In this work, we want to show how adapting the linear interpolation concept to the upper-limb workspace exploration problem, could eventually lead to a model capable of estimating the force-targets in an untrained point of the workspace. More in detail, we tested the generalization capability of models extracted by spatially interpolating the EMG-to-Force matrices obtained at the vertexes of a predefined planar workspace (approximately A4 size paper). On the other hand, we compared two synergy-extraction strategies: "Pose-Shared", which assumes that muscle patterns can be factorized using data coming from different points, and "Pose-Related" which assumes that each point has its own set of muscle primitives that can be clustered together with the synergies extracted in the other points [6]. We believe that the combined use of the "Pose-Related" muscle synergies extraction method, coupled with the interpolation of the mapping matrices, could reduce the training session of a myocontrol system able to generalize outside the trained points.

2   Materials and Methods

2.1   Participants

In this study three right-handed healthy subjects have participated, aged between 24–31 years, with no previous experience in virtual environments applications. All the experiments have been conducted following the WMA Declaration of Helsinki and all subjects provided written consent.

2.2   Experimental Setup

Participants sat on a firm fixed-height chair at a variable distance from the anthropomorphic robot UR5 (Universal Robots, Denmark). The UR5 is a 6 DoF manipulandum with high stiffness and a maximum payload of 50 N. Joint angles are configured using the API provided for setting the different positions of the robot during the experiment. The distance between the subject and the robot is computed with respect to the base joint of the robot in such a way that the testing point was reachable with an elbow angle of 90°. The UR5 robot end effector is provided with a cylindrical handle along with a triaxial force sensor able to monitor linear forces and torques. Additionally, the upper limb weight is compensated with a 2 DoF passive structure used to support the subject arm through two adjustable pulleys. Subjects were asked to grasp the handle with the dominant hand, while wearing an Oculus Rift HMD (Oculus, USA) used for the immersion in a peaceful countryside virtual environment (VE). This calm environment help novel subjects to relax and accomplish the tasks following the visual cues.

EMG signals were recorded using two biosignals amplifiers gUSBamp (gTec, Austria) of 8 bipolar channels each, at a sampling frequency of 1000 Hz and pre-processed using the amplifier build-in filters of 50 Hz notch and 5–250 Hz band-pass.

Ground electrode was positioned on the right elbow. The amplifiers were configured in a master-slave architecture in order to synchronize the 13 EMG channels used in the experiment. The skin was carefully cleaned before placing disposable passive Ag/AgCl surface electrodes. Muscular activations were acquired from 13 different muscles involving elbow and shoulder joints. The muscles recorded were: biceps short head, biceps long head, brachioradial, triceps long head, triceps lateral head, deltoid anterior head, deltoid posterior head, trapezius, pectoralis major, teres major, infraspinatus, latissimus dorsi and rhomboid.

2.3   Data Acquisition Protocol

The acquisition consists of a set of isometric contractions in four points (training, A B C D) distributed in a rectangular shape with 30 cm base and 20 cm height, and one point (testing, E) located in the centroid of the square (see Fig. 1). Each point corresponds to a different upper-limb pose. At the beginning of the experiment, the robot is controlled up to the first position (i.e. Point A), ensuring the feasibility of isometric contractions, thanks to its high stiffness. The VE shows to the subject the cues and the cursor that moves during the experiment. The position of the cursor is computed using a spring model $P_c = KF_{EE}$, where P_c is the 3D cursor position, F_{EE} is the isometric force vector applied to the sensor and K is the elastic constant.

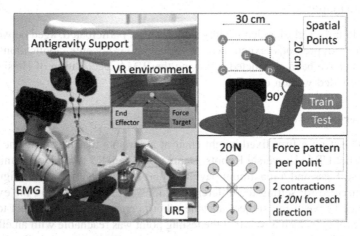

Fig. 1. Experimental setup. (Left) Subject 1 wearing VE glasses and the arm support. (Right-up) Five points acquisition grid. (Right-bottom) Diagram of the tasks performed (Color figure online)

The cues are represented as a green ball indicating the target position for each direction. The subject is then asked to apply force in the direction of the green ball to guide the cursor towards it. The cursor, represented as a small yellow ball, must be placed inside the green ball and remain there for 2 s to complete the task. The force needed to get to the target position is equivalent to 20 N. For each point the subject has to perform 2 not consecutive repetitions per direction, along a total of 8 directions equally spaced at 45° in a circumference. For each direction the task consists in three steps: (1) move the cursor into the target position, (2) keep the cursor inside the target position for 2 s and (3) return the cursor to the rest position, staying still for 2 s. The diameter of the target sphere admits a tolerance of 3 N in the force exerted. Once all the directions are completed for one point, the robot moves to the next point, and the tasks are repeated again until all points are completed. The train and test datasets have been built collecting one contraction for training and one for testing for all the 256 possible combinations (i.e. 2 repetitions per 8 directions). Raw EMG signals were rectified and filtered using a 4 Hz 2nd order Butterworth low-pass filter. All the signals have been normalized using the maximum voluntary contraction (MVC) among all the channels within the entire dataset. The data processing and the conducted analysis are explained in the next sections and summarized in Fig. 2.

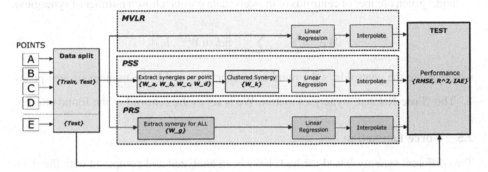

Fig. 2. Block diagram of the investigated control schemes.

2.4   Synergies Extraction

Computing muscle synergies means exploiting a factorization algorithm to separate fundamental components of the input signal. A negative muscle activation has no physiological meaning, thus, the Non-Negative Matrix Factorization (NMF) [13] has become more popular when dealing with electromyographic signals, versus other commonly used statistical methods such as Principal Component Analysis (PCA) or Independent Component Analysis (ICA). Using the NMF algorithm, muscle synergies are computed as following:

$$m = Wc + e_m \qquad (1)$$

where m is the input signal (M × N matrix, being M the number of muscles and N the number of samples), W is the relative muscle activation level (M × s matrix, being s the number of synergies), c is the synergies activations matrix (s × N matrix) and e_m is the muscle activations factorization residuals. Considering the application of the factorization algorithm NMF on different data, two different approaches, namely "Pose-Shared" and "Pose-Related", have been analyzed. The "Pose-Shared" synergies are extracted running the NMF once on the EMG data acquired at all arm poses (i.e. Points A, B, C and D) as follows:

$$m_T = W_g c_T + e_m \tag{2}$$

where m_T is the union of the signals at the boundary points $m_T = [m_A \cup m_B \cup m_C \cup m_D]$ and W_g is the "Pose-Shared" synergy matrix. Instead, the "Pose-Related" synergies are computed with a three steps procedure:

1. The NMF factorization is independently applied on the EMG data recorded at each arm pose producing M synergies matrices (W_i with i = 1:P, where P is the number of arm poses) with (1).
2. All the extracted synergies vectors are clustered using the K-means algorithm with cosine distance. K-means is a simple unsupervised algorithm that minimizes the distance between a set of data points (i.e. the synergies extracted in the training points) and a given number of centroids of the same data points (chosen number of synergies):

$$W_C = \sum_{j=1}^{s} \sum_{i=1}^{P} \left\| dist(W_i^{(j)}, b_i) \right\| \tag{3}$$

where b_i are the centroids, s the number of synergies and P the number of arm poses.

3. The "Pose-Related" synergies matrix, Wc is set as the centroid of the found clusters.

2.5   Force Estimation

Two different synergy-based methods have been analyzed and compared with the state-of-art most common algorithm (i.e. linear regression using all individual muscle activations). In this section is presented an overview of the three approaches.

A. Multi-variate linear regression (MVLR). The main assumption underneath the linear regression computation is that, if the muscle activation is lower enough than the subject MVC and the arm has a fixed pose, the relation between the force exerted at the hand and the EMG measured on the elbow and shoulder muscles is approximately linear [3, 5, 7]. For this reason, it is possible to estimate the forces at the hand position using the following method:

$$F_m(t) = H_m m(t) \tag{4}$$

where $F_m(t)$ is the estimated 2-dimensional force, $m(t)$ is the EMG signal matrix (M × N where N is the number of samples) and H m is the regression matrix (2 × M) calculated as:

$$H_m(t) = reg(m_t(t), F_t) \tag{5}$$

where m_t and F_t correspond to the training data of EMGs and forces respectively.

B. Linear regression and "Pose-Shared" synergies (PSS). The goal idea is to be able to estimate the arm force direction even in spatial points where the model has not been trained at. Multiple synergies extracted in different upper-limb positions can be used to estimate the force direction in that particular point, but the same synergy set may not be suitable when the arm moves to another position. In order to overcome the signal variability introduced by the joint angles difference in the new position, a new pose-shared synergy matrix is calculated as stated above. Using this method, the force estimation proceeds as follows:

$$F_g(t) = H_g c_g(t) \tag{6}$$

where $F_g(t)$ is the estimated 2-dimensional force, H_g is the regression matrix ($2 \times s$ matrix, where s is the number of synergies) and $c_g(t)$ is the synergies activations vector computed using (1) and neglecting the muscle residuals:

$$c_g(t) = W_g^+ m(t) \tag{7}$$

where W_g^+ is the pseudo-inverse of W_g pose-shared synergy matrix seen in (2) and $m(t)$ is the filtered EMG signal matrix.

C. Linear regression and "Pose-Related" synergies (PRS). It has been observed that in a small region of the upper-limb workspace the muscle synergies do not suffer a high variation of the relative muscle weights. We performed a cluster analysis (k-means of all the points-specific synergies matrices, as done in Sect. 2.4) in order to obtain a resultant matrix.

Which is supposed to have a better generalization capability with respect to the previous ones. Eventually the estimation could be computed using the following formula:

$$F_c(t) = H_c c_c(t) \tag{8}$$

where $F_c(t)$ is the computed 2-dimensional force, H_c is the regression matrix ($2 \times s$ matrix, where s is the number of synergies) and $c_c(t)$ is the synergies activations vector computed using the inverse formula of (1) and neglecting the muscle residuals:

$$c_c(t) = W_c^+ m(t) \tag{9}$$

where W_c^+ is the pseudo-inverse of W_c clustered synergies matrix as stated in (3) and $m(t)$ is the training/validation-set filtered EMG signals matrix.

2.6   Test Phase: Regression Matrices Interpolation

Once evaluating the validation performance, the investigation is focused on understanding the generalization properties of each algorithm, in a non-trained point. This means building a model starting from A-B-C-D points and test it in the point E. The force estimation model in E has been generated, for each method, interpolating the regression matrices trained in A-B-C-D in a linear space. Consequently, the new interpolated regression matrix has been used to estimate the forces using EMG signals in the test set.

2.7 Performance Indices

Three different indices have been used for assessing the performance of each method.

Root Mean Square Error (RMSE) is used to measure the difference between the measured and the estimated forces and it is calculated as follows:

$$RMSE = \sqrt{\frac{\sum_{n=1}^{N}(x_i^2 - \hat{x}_i^2)}{N}} \tag{10}$$

where x_i is the measured force sample, $\hat{x}_i$ i is the estimated output sample and N is the total of samples.

The Initial Angle Error (IAE) index computes the difference between the measured and estimated force angles using the last 70% of the rising edge of the signal after the contraction activation. The coefficient of determination (R^2) is also used to highlight a signal total variation explained by the estimates. The R^2 is computed as follows:

$$R^2 = 1 - \frac{SS_{res}}{SS_{tot}} = 1 - \frac{\sum_{n=1}^{N}\left(x_i^2 - \hat{x}_i^2\right)}{\sum_{n=1}^{N}(x_i^2 - \bar{x}_i^2)} \tag{11}$$

where x_i is the measured force sample and $\hat{x}_i$ is the estimated output sample, N is the total of samples. The index ranges from minus infinite to 1 (equal to 1in case of perfect estimate with zero error).

3 Results

As showed in previous works [3, 10], a small number of synergies well reconstructs all the muscles activations channels, building a functional subset of grouped muscles. In order to fix the number of synergies for consistent analysis of the performances, the evolution of the EMG reconstruction rate (i.e. R^2), increasing the number of synergies out of NMF, has been computed. Considering the variance between subjects for each method and a reconstruction rate threshold of 90% (see Fig. 3), 5 synergies have been selected for the training-test phases, since they also build a plausible neural subset in the muscles space [3].

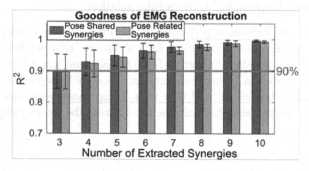

Fig. 3. EMG reconstruction rate varying the synergies number out of NMF algorithm.

The upper panels of Fig. 4 shows the extracted synergies for points A, B, C and D for a representative subject. The medium panel compares the actual synergies extracted in point E with the "Pose-Shared" synergies and the "Pose-Related" synergies. It can be noted that, qualitatively, "Pose-Related" synergies are more similar to the synergies in point E than the "Pose-Shared" synergies.

The bottom panel of Fig. 4, gives a representation of how the interpolated EMG-to-Force matrices extracted using the "Pose-Related" synergies are suitable for covering all the force directions in the explored workspace. Figure 5 shows the performance metrics for the three investigated methods both in the vertex points (A, B, C and D, where the models were also trained) and in the central point (E, where the models were interpolated). It can be noted that the linear regression algorithm without synergies

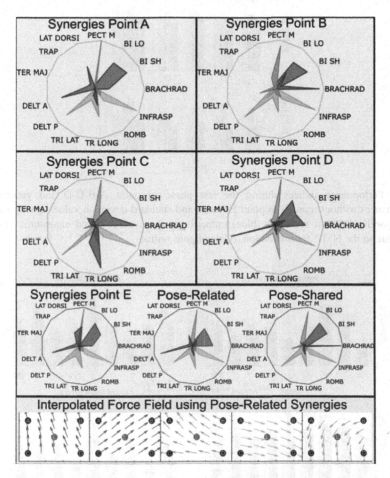

Fig. 4. Synergistic investigation for a representative subject. Synergies extracted in the training points (A, B, C and D) are shown in the upper panel. "Pose-Shared" and "Pose-Related" synergies are reported in the medium panel together with synergies extracted in the point E. The last panel shows the interpolated regression matrix for the "Pose-Related" synergies.

outperformed the synergy-based ones in all the points. Among the two synergistic methods, it is possible to note that using the "Pose-Related" synergies (clustered synergies) leads to better results than using "Pose-Shared" synergies (extracted from the merged training dataset) especially for the initial angle errors (IAE). The

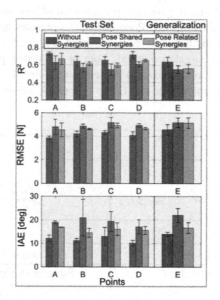

Fig. 5. Performance indices during the test phase in point A-B-C-D and generalization performances without training in point E (mean and standard deviation values among subjects). The algorithm without synergies (blue) outperforms the synergy-based algorithms in the test phase due to the NMF approximation. (Color figure online)

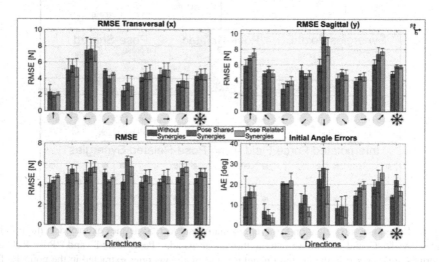

Fig. 6. Performances per direction extracted in the test point E (mean and standard deviation values among subjects) for each method.

generalization capabilities of the three methods can be evaluated by comparing the performance obtained in point E with the performance obtained in the other points.

This analysis reveals a small loss of performances in generalization, thus, the interpolation technique could be eligible for the training-set reduction purpose. In Fig. 6, the errors obtained for the three methods in the generalization set (point E) are shown separately for each direction. It is possible to note how errors vary with the target direction. In particular, the IAE is acceptable (less than 20°) for most of the directions except for the backward direction.

4 Discussion and Conclusion

This work presents a pilot study aimed at investigating the possibility to reduce the training time of a myo-control scheme by building a regression model that could reliably estimate the upper-limb forces in a rectangular region of the horizontal plane after being trained with a minimal amount data acquired in the vertexes only. In particular, we investigated how the modularity of muscles synergies can be exploited for generating a single set of muscles patterns suitable for the force estimation in the explored workspace. For this purpose, we compared two synergies extraction strategies, namely "Pose-Related" and "Pose-Shared" synergies. Even though the "Pose-Shared" synergies could reconstruct the EMG better than "Pose-Related" synergies do (see Fig. 3), performance results were slightly higher when using the "Pose-Related" synergies. This fact suggests that, clustering the synergies obtained in different points of the workspace, could be a valid solution for estimating hand forces in another point of the workspace (see Fig. 4), thus, that the cluster centroids are more representative than the "Pose-Shared" synergies. However, performance obtained with both methods that map synergies to hand-force were slightly lower than the performance obtained by mapping the individual muscles into control signals. Thus, obtained performance results do not confirm the initial hypothesis under which, outside the points where the models were trained on, synergistic approaches performed better than using individual muscles. This fact may be due to the small size of the explored workspace that does not exploit a large portion of the upper-limb range of motion. Further studies are needed for investigating whether, in a larger workspace, the introduction of synergies in the control scheme, could lead to better generalization capabilities. In this way, it will possible to understand the effects of the joints angles variation on the synergies modularity and models interpolation. Another interesting aspect concerns the comparison between a training based on unconstrained upper-limb movements [16] and a training based on isometric contractions [3, 9]. In fact, as shown in Fig. 6, the IAE is deeply influenced by the target force directions. The study has two main disadvantages: a limited amount of test data, as only three participants were involved in this pilot study and they performed only one session, and a lack of limb-impaired subjects among the tested population. In order to conclude whether the performance of the proposed method differs when considering patient population, additional investigation is needed. In future work, we will focus on extending the investigation on a wider workspace and testing the proposed system in the real-time control of a robotic device (upper limb exoskeleton).

References

1. Barsotti, M., Dupan, S., Vujaklija, I., Došen, S., Frisoli, A., Farina, D.: Online finger control using high-density EMG and minimal training data for robotic applications. IEEE Robot. Autom. Lett. **4**(2), 217–223 (2019)
2. Beckerle, P., et al.: A human–robot interaction perspective on assistive and rehabilitation robotics. Front. Neurorobot. **11**, 24 (2017)
3. Berger, D.J., D'Avella, A.: Effective force control by muscle synergies. Front. Comput. Neurosci. **8**, 46 (2014)
4. Bizzi, E., D'Avella, A., Saltiel, P., Tresch, M.: Book review: modular organization of spinal motor systems. Neurosci. **8**(5), 437–442 (2002)
5. Buchanan, T.S., Lloyd, D.G., Manal, K., Besier, T.F.: Neuromusculoskeletal modeling: estimation of muscle forces and joint moments and movements from measurements of neural command. J. Appl. Biomech. **20**(4), 367–395 (2004)
6. Buongiorno, D., et al.: Evaluation of a pose-shared synergy-based isometric model for hand force estimation: towards myocontrol. In: Ibáñez, J., González-Vargas, J., Azorín, J., Akay, M., Pons, J. (eds.) Converging Clinical and Engineering Research on Neurorehabilitation II. BIOSYSROB, vol. 15, pp. 953–958. Springer, Cham (2017). https://doi.org/10.1007/978-3-319-46669-9_154
7. Buongiorno, D., Barsotti, M., Barone, F., Bevilacqua, V., Frisoli, A.: A linear approach to optimize an EMG-driven neuromusculoskeletal model for movement intention detection in myo-control: a case study on shoulder and elbow joints. Front. Neurorobot. **12**, 74 (2018)
8. Clabaugh, C., Matarić, M.: Robots for the people, by the people: personalizing human-machine interaction. Sci. Robot. **3** (2018)
9. D'Avella, A., Lacquaniti, F.: Control of reaching movements by muscle synergy combinations. Front. Comput. Neurosci. **7**, 42 (2013)
10. D'Avella, A., Portone, A., Fernandez, L., Lacquaniti, F.: Control of fast-reaching movements by muscle synergy combinations. J. Neurosci. **26**(30), 7791–7810 (2006)
11. Farina, D., et al.: Man/machine interface based on the discharge timings of spinal motor neurons after targeted muscle reinnervation. Nat. Biomed. Eng. **1**(2), 0025 (2017)
12. Jiang, N., Englehart, K.B., Parker, P.A.: Extracting simultaneous and proportional neural control information for multiple-DOF prostheses from the surface electromyographic signal. IEEE Trans. Biomed. Eng. **56**(4), 1070–1080 (2009)
13. Lee, D.D., Seung, H.S.: Learning the parts of objects by non-negative matrix factorization. Nature **401**(6755), 788 (1999)
14. Leonardis, D., et al.: An EMG-controlled robotic hand exoskeleton for bilateral rehabilitation. IEEE Trans. Haptics **8**(2), 140–151 (2015)
15. Loconsole, C., Dettori, S., Frisoli, A., Avizzano, C.A., Bergamasco, M.: An EMG-based approach for on-line predicted torque control in robotic-assisted rehabilitation. In: 2014 IEEE Haptics Symposium (HAPTICS), pp. 181–186. IEEE (2014)
16. Muceli, S., Jiang, N., Farina, D.: Extracting signals robust to electrode number and shift for online simultaneous and proportional myoelectric control by factorization algorithms. IEEE Trans. Neural Syst. Rehabil. Eng. **22**(3), 623–633 (2014)
17. Nowak, M., Castellini, C.: The LET procedure for prosthetic myocontrol:.towards multi-DOF control using single-DOF activations. PLoS ONE **11**(9), e0161678 (2016)
18. Taborri, J., et al.: Feasibility of muscle synergy outcomes in clinics, robotics, and sports: a systematic review. Appl. Bionics Biomech. **2018** (2018)
19. Buongiorno, D., Sotgiu, E., Leonardis, D., Marcheschi, S., Solazzi, M., Frisoli, A.: WRES: a novel 3 DoF WRist ExoSkeleton with tendon-driven differential transmission for neuro-rehabilitation and teleoperation. IEEE Robot. Autom. Lett. **3**(3), 2152–2159 (2018)

A Survey on Deep Learning in Electromyographic Signal Analysis

Domenico Buongiorno[1,2], Giacomo Donato Cascarano[1,2],
Antonio Brunetti[1,2], Irio De Feudis[1,2],
and Vitoantonio Bevilacqua[1,2(✉)]

[1] Department of Electrical and Information Engineering,
Polytechnic University of Bari, Bari, Italy
vitoantonio.bevilacqua@poliba.it
[2] Apulian Bioengineering s.r.l., Via delle Violette n814, Modugno, BA, Italy

Abstract. In the recent past Deep Learning (DL) has been used to develop intelligent systems that perform surprisingly well in a large variety of tasks, e.g. image recognition, machine translation, and self-driving cars. The huge improvement of the elaboration hardware and the growing need of big data processing have boosted the DL research in several fields. Recently, physiological signal processing has taking advantage of deep learning as well. In particular, the number of studies concerning the analysis of electromyographic (EMG) signals with DL methods is exponentially raising. This phenomenon is mainly explained by both the existing limitation of the myoelectric controlled prostheses and the recent publication of big datasets of EMG recordings, e.g. Ninapro. Such increasing trend motivated us to search and review recent papers that focus on the processing of EMG signals with DL methods. A comprehensive literature search of papers published between January 2014 and March 2019 was performed referring to the Scopus database. After a full text analysis, 65 papers were selected for the review. The bibliometric research shows four distinct clusters focused on different applications: Hand Gesture Classification; Speech and Emotion Classification; Sleep Stage Classification; Other Applications. As expected, the review process revealed that most of the papers related to DL and EMG signal processing concerns the hand gesture classification, and the convolutional neural network is the most used technique.

Keywords: Deep learning · Convolutional neural network · Auto-encoder · Deep belief network · Recurrent neural network · Physiological signal

1 Introduction

The electromyographic (EMG) signal is a representation of the electric potential field generated by the depolarization of the outer muscle fiber membrane. Its detection involves the use of intramuscular or surface electrodes which are placed at a certain distance from the sources. The properties of the tissues that separate the source and the recording electrodes heavily determine the features of the acquired signals (e.g. the signal amplitude and the frequency content). Such property explains both the high intra-subject and inter-subject variability of the acquired EMG signals that encouraged

© Springer Nature Switzerland AG 2019
D.-S. Huang et al. (Eds.): ICIC 2019, LNAI 11645, pp. 751–761, 2019.
https://doi.org/10.1007/978-3-030-26766-7_68

many researchers to develop novel signal processing techniques able to robustly decode the core information encoded into the EMG signals [1].

The surface electromyography (sEMG) is widely used in many fields such as [2]: neurophysiology [3], ergonomics and occupational medicine [4], posture analysis [5], movement and gait analysis [6, 7], EMG-based biofeedback [8], exercise physiology and sports [9], and man-machine interfaces [10, 11]. In applications for man–machine interfacing, muscles are used as biological amplifiers of efferent nerve activity because of the direct association between the action potentials generated by the motor neurons and the electrical activity generated in the innervated muscle fibers. In fact, sEMG signals can be used to trigger or continuously generate control commands for external "devices" including virtual games, powered orthoses, prostheses and exoskeletons [12, 13]. The mapping between the EMG signals and the specific device can be achieved with two main EMG-processing methods: (1) data-driven approach that is based on machine learning techniques [14], and (2) model-driven approach that makes use of the sEMG as the input to subject specific physical models of the musculotendon system [10, 11, 15, 16]. In the last decades, the research on data-driven approaches has been boosted by the existing limitation of the myoelectric prosthetic hands offered by the market [17]. Many researchers are indeed focused on the development of new machine-learning based methods for the detection of the intended hand gesture from the forearm muscle activations to correctly actuate prosthetic devices [18]. However, despite several improvements, the robust natural control of dexterous prosthetic hands is not fully achieved [17].

The well-known limitation of shallow approaches and the increasing amount of multimodal physiologic data in biomedical applications encouraged researchers to adopt Deep Learning (DL) for physiological data processing [17, 19, 20]. Deep learning methods have recently revolutionized several fields of machine learning [21], including speech recognition [22], computer vision [23], image processing [24–26]. Thus, researchers have started investigating the DL ability to process and decode sEMG data, also thank to the recent publication of several benchmark databases of EMG recordings, e.g. NinaPro, csl-hdemg, BioPatRec and CapgMyo.

Such increasing research trend also concerns other physiological signals, e.g. Electroencephalogram (EEG), Electrocardiogram (ECG), and Electrooculogram (EOG), as it has been discussed by Faust, Ganapathy and their colleagues in two surveys about deep learning applied on physiological signals [27, 28]. However, since these two surveys discuss papers published until December 2017, the number of papers related to the processing of EMG signals with deep learning techniques is limited to fifteen [27, 28]. Hence, an updated review on this topic is needed.

2   Paper Search Methodology and Classification

In this section, we explain the methodology used to select the papers and to classify the relevant papers according to the final application of the EMG signal processing.

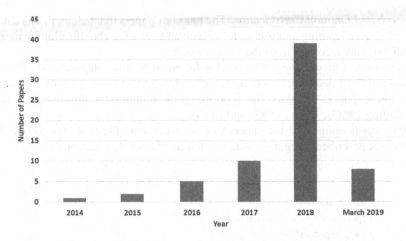

Fig. 1. Number of published papers related to Deep Learning and EMG signals per year.

2.1 Paper Selection

The literature search was performed for articles published on Scopus until March 2019 (Fig. 1). The search focused on retrieving articles that were written in English and included a set of keywords involving the two main topics "Deep Learning Methods" and "Electromyographic Signals" (the specific used query string is *"(TITLE-ABS-KEY ("deep learning*" OR "deep neur*" OR *cnn* OR autoencoder OR *lstm* OR *rnn* OR "recurrent neur*" OR dbn OR rbm OR "deep belief network" OR "restricted Boltzmann machine") AND TITLE-ABS-KEY (electromyo* OR *emg* OR myoelectric)) AND PUBYEAR > 2013 AND (LIMIT-TO (LANGUAGE, "English"))"*. With the above keywords, we originally obtained 203 papers. We then excluded the results either based on topologies composed of only one hidden layer or that actually considered the processing of other kind of signals, e.g. EEG, images or videos. In total, we were able to exclude 138 results, therefore 65 papers (published in journals or conference proceedings) were selected for the survey.

2.2 Paper Classification

Two papers out of 65 were survey/review papers [27, 28]. Then, the remaining 63 papers were classified according to the final goal of the EMG signal processing. In particular, we have grouped the papers in four main categories (see Fig. 2): (1) Hand Gesture Classification; (2) Speech and Emotion Classification; (3) Sleep Stage Classification; (4) Other Applications:

1. *Hand Gesture Classification.* The papers that fall into this category investigate deep learning techniques for the recognition and the classification of hand/finger gestures by the analysis of EMG signals acquired from the main upper limb muscles. Such papers represent the majority of the selected articles.

2. *Speech and Emotion Classification.* The research papers that belong to this category employ deep learning approaches for speech and emotion classification by means of EMG signals recorded from the facial muscles.
3. *Sleep Stage Classification.* The articles falling into this category study the application of deep neural networks for the classification and rating of the sleep stages. All these papers consider the processing of multimodal physiological signals including EMGs, EEGs, EOGs and ECGs.
4. *Other Applications.* All the papers that we were not able to assign to one of the previous described categories were labeled as "Other Applications".

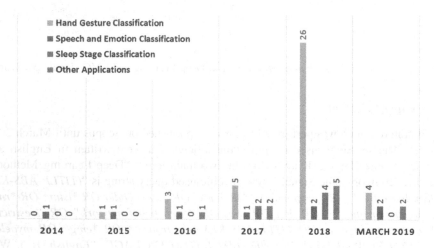

Fig. 2. Number of papers of each individuated category per year.

3 Content Review

The bibliometric research clearly revealed the presence of four distinct paper clusters created according different final applications: (1) Hand Gesture Classification; (2) Speech and Emotion Classification; (3) Sleep Stage Classification; (4) Other Applications.

3.1 Hand Gesture Classification

As clearly shown in Fig. 2, most of the selected papers concern the processing of EMG signals acquired from the upper limb muscles. Such finding is strictly related to the recent increasing interest of many researchers in myoelectric prosthetic control and advanced natural human-machine interaction interfaces [14]. After a deep review of the papers falling into this category, it emerged that the CNN is the most used technique, followed by the AE, the RNN and the DBN. According to the growing popularity of the CNNs in several research domains due to its demonstrated high performance, several authors have proposed classifiers based on just CNNs [17, 29], or CNN-RNN

[30–33], or CNN-AE [34]. Some authors, instead, have developed multi-class classifiers that are completely based on deep AE [35, 36], RNN [37, 38] or DBN [39, 40].

3.2 Speech and Emotion Classification

We have collected in this category all the papers related to the analysis of the facial electromyography (fEMG). Even though the number of found papers is not considerable (see Fig. 2), in our opinion a specific category to such papers is needed due to the important impact that speech and emotion classification have in clinical applications. In detail, four papers out seven investigate the ability of deep learning techniques for the fEMG-based speech classification [41, 42], whereas three selected papers out seven are focused on the emotion detection and classification [43, 44]. Regarding the used technique, the selected papers employ standard deep MLP neural, CNN, and DBN.

3.3 Sleep Stage Classification

Besides the two categories described above, there are also some papers that refer to the classification and rating of the sleep stages. Current sleep medicine relies on the supervised analysis of polysomnographic recordings, which comprise, among others, electroencephalogram, electromyogram, and electrooculogram signals. This explains why these particular types of paper consider the elaboration of multimodal data that include physiological signals, e.g. EEGs, EMGs, EOGs and ECGs. It is however important to remark that, in sleep stage classification, the role of the EMG signals is not as primary as the EEG signal contribution. Nevertheless, the results of the selected and reviewed studies demonstrated that the classification performance benefits from the inclusion of the EMG signals. Likewise for the "Hand Gesture Classification" papers, most research studies of this category are based on CNN [45, 46]. Only one work relies on the integration of a DBN with a RNN [47].

3.4 Other Applications

Ten papers out 63 were not included into any of the three categories that have been presented above. As example, here we report the description of some of the uncategorized papers. Su et al. have proposed a DBN to predict onset of muscle fatigue that occurs while holding a load with the upper limbs [48]. Xia et al. proposed a CNN integrated with a RNN for the estimation of hand trajectory [49]. Said et al. presented a stacked autoencoder for the compression of multimodal biosignals, i.e. EMG and EEG [50]. Bakiya et al. proposed a DNN to discriminate healthy subjects from patients affected by the amyotrophic lateral sclerosis or myopathy [51]. Sengur et al. presented a CNN for efficient classification of amyotrophic lateral sclerosis and normal electromyogram signals [52]. Chen et al. implemented a DBN to extract EMG features for the estimation of the human lower limb flexion/extension joint angles [53]. Rane et al. developed a CNN for lower limb muscle force estimation during gait [54].

4 Discussion

Deep learning has already established itself as a robust and effective method for data processing in several research fields, e.g. image processing, video processing and audio processing. As demonstrated by this work, in the last few years there has been a growing interest of the scientific community in applying deep learning methods for EMG signal analysis and processing, thus confirming the same trend that involving the analysis of other physiological signals, e.g. EEG and ECG [27, 28].

The paper search and analysis conducted in this work revealed that the major contribution in the field of deep learning applied to EMG signals is coming from the researches focused on the myoelectric control of prostheses due to both the limitation of the current products on the market and the availability of many published databases. Besides the papers concerning the hand gesture classification, DL techniques have been applied in other fields that consider the processing of big amount of data, e.g. the researches dealing with the sleep stage analysis and classification, and the speech and emotion classification.

Regarding the used deep learning methods, it resulted that the convolutional neural network is the most used technique among the five investigated DL methods: i.e. the deep neural networks, the convolutional neural networks, the auto-encoders, the deep belief networks and the deep recurrent neural networks (see Fig. 3). In some other cases, these individuated techniques are mixed to take advantage of the pros of several DL methods.

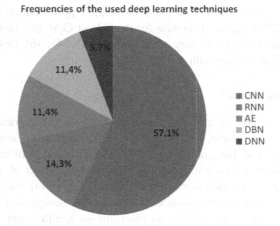

Fig. 3. Frequencies of the used deep learning techniques among the reviewed papers. Papers that investigate mixed approaches are accounted more times.

Concerning the used datasets in the selected papers, it resulted that many papers consider the processing existing datasets. In particular, we found that (a) the Ninapro, the csl-hdemg, and the CapgMyo are the most used datasets among the Hand Gesture Classification papers; (b) the DEAP dataset is the most used in works related to the emotion classification; and (c) the PhysioNet and the MASS are the most used databases for the investigation of the DL techniques in sleep stage classification. We then can state that the presence of existing databases has actually boosted the research in this field. As it is well known, the additional main advantage of employing datasets concerns the possibility to compare the results of different studies. In fact, as it mainly emerged from the analysis of the "hand gesture classification" papers, the large variability of the used setups, i.e. number of acquired EMG signals, the adoption of different kinds of electrodes that can be superficial or intramuscular, the employment of bipolar configurations of electrode arrays, and the number of classes to be classified, makes difficult the comparison of the study results.

Finally, it is important to remark that even if many researchers are mainly focus on the continuous improvement of the data processing performances, some others are studying and investigated compact deep topology in order to shorten both the training and the execution time while maintaining high levels of performance.

5 Conclusion

Concluding, in this work we have searched and reviewed recent articles concerning the processing of EMG signals with DL methods that were published between January 2014 and March 2019. After a full text analysis, 65 papers were selected for the review. The bibliometric research showed that the selected papers could be sorted in four distinct categories focused on different applications, respectively: (1) Hand Gesture Classification; (2) Speech and Emotion Classification; (3) Sleep Stage Classification; (4) Other Applications. As expected, the review process revealed that (a) most of the papers related to DL and EMG signal processing concerns the hand gesture classification, and (b) the convolutional neural network is the most used technique. Given the impact of both the "hand gesture classification" category and the CNN, future works could deeply study the CNN topologies used to classify hand/finger gestures by EMG signals, including a systematic comparison among the several papers.

Acknowledgments. This work has been supported by the Italian project RoboVir within the BRIC INAIL-2016 program.

References

1. Lv, B., Sheng, X., Zhu, X.: Improving myoelectric pattern recognition robustness to electrode shift by autoencoder. In: Proceedings of the Annual International Conference of the IEEE Engineering in Medicine and Biology Society, EMBS, pp. 5652–5655. Institute of Electrical and Electronics Engineers Inc. (2018). https://doi.org/10.1109/EMBC.2018. 8513525
2. Merletti, R., Farina, D.: Surface Electromyography: Physiology, Engineering and Applications (2016). https://doi.org/10.1002/9781119082934
3. Farina, D., Falla, D.: Effect of muscle-fiber velocity recovery function on motor unit action potential properties in voluntary contractions. Muscle Nerve (2008). https://doi.org/10.1002/mus.20948

4. Peppoloni, L., Filippeschi, A., Ruffaldi, E., Avizzano, C.A.: (WMSDs issue) a novel wearable system for the online assessment of risk for biomechanical load in repetitive efforts. Int. J. Ind. Ergon. (2016). https://doi.org/10.1016/j.ergon.2015.07.002

5. Casadio, M., Morasso, P.G., Sanguineti, V.: Direct measurement of ankle stiffness during quiet standing: implications for control modelling and clinical application. Gait Posture (2005). https://doi.org/10.1016/j.gaitpost.2004.05.005

6. Monaco, V., Ghionzoli, A., Micera, S.: Age-related modifications of muscle synergies and spinal cord activity during locomotion. J. Neurophysiol. (2010). https://doi.org/10.1152/jn.00525.2009

7. Buongiorno, D., et al.: Assessment and rating of movement impairment in Parkinson's disease using a low-cost vision-based system. In: Huang, D.-S., Gromiha, M.M., Han, K., Hussain, A. (eds.) Intelligent Computing Methodologies, pp. 777–788. Springer International Publishing, Cham (2018)

8. Cram, J.R.: Biofeedback applications. In: Electromyography (2005). https://doi.org/10.1002/0471678384.ch17

9. Besier, T.F., Lloyd, D.G., Ackland, T.R., Cochrane, J.L.: Anticipatory effects on knee joint loading during running and cutting maneuvers. Med. Sci. Sports Exerc. (2001). https://doi.org/10.1097/00005768-200107000-00015

10. Buongiorno, D., Barsotti, M., Barone, F., Bevilacqua, V., Frisoli, A.: A linear approach to optimize an EMG-driven neuromusculoskeletal model for movement intention detection in myo-control: a case study on shoulder and elbow joints. Front. Neurorobot. (2018). https://doi.org/10.3389/fnbot.2018.00074

11. Buongiorno, D., et al.: A neuromusculoskeletal model of the human upper limb for a myoelectric exoskeleton control using a reduced number of muscles. In: IEEE World Haptics Conference, WHC 2015 (2015). https://doi.org/10.1109/WHC.2015.7177725

12. Buongiorno, D., Sotgiu, E., Leonardis, D., Marcheschi, S., Solazzi, M., Frisoli, A.: WRES: a novel 3 DoF WRist ExoSkeleton with tendon-driven differential transmission for neuro-rehabilitation and teleoperation. IEEE Robot. Autom. Lett. (2018). https://doi.org/10.1109/LRA.2018.2810943

13. Stroppa, F., et al.: Real-time 3D tracker in robot-based neurorehabilitation. In: Computer Vision for Assistive Healthcare (2018). https://doi.org/10.1016/B978-0-12-813445-0.00003-4

14. Vujaklija, I., Shalchyan, V., Kamavuako, E.N., Jiang, N., Marateb, H.R., Farina, D.: Online mapping of EMG signals into kinematics by autoencoding. J. Neuroeng. Rehabil. 15 (2018). https://doi.org/10.1186/s12984-018-0363-1

15. Buongiorno, D., Barone, F., Solazzi, M., Bevilacqua, V., Frisoli, A.: A linear optimization procedure for an EMG-driven neuromusculoskeletal model parameters adjusting: Validation through a myoelectric exoskeleton control. In: Lecture Notes in Computer Science (Including Subseries Lecture Notes in Artificial Intelligence and Lecture Notes in Bioinformatics) (2016). https://doi.org/10.1007/978-3-319-42324-1_22

16. Buongiorno, D., et al.: Evaluation of a pose-shared synergy-based isometric model for hand force estimation: towards myocontrol. In: Biosystems and Biorobotics (2017). https://doi.org/10.1007/978-3-319-46669-9_154

17. Atzori, M., Cognolato, M., Müller, H.: Deep learning with convolutional neural networks applied to electromyography data: a resource for the classification of movements for prosthetic hands. Front. Neurorobot. 10 (2016). https://doi.org/10.3389/fnbot.2016.00009

18. Geethanjali, P.: Myoelectric control of prosthetic hands: state-of-the-art review (2016). https://doi.org/10.2147/MDER.S91102

19. Bevilacqua, V., et al.: A novel BCI-SSVEP based approach for control of walking in virtual environment using a convolutional neural network. In: Proceedings of the International Joint Conference on Neural Networks (2014). https://doi.org/10.1109/IJCNN.2014.6889955

20. Bevilacqua, V., et al.: Advanced classification of Alzheimer's disease and healthy subjects based on EEG markers. In: Proceedings of the International Joint Conference on Neural Networks (2015). https://doi.org/10.1109/IJCNN.2015.7280463

21. Liu, W., Wang, Z., Liu, X., Zeng, N., Liu, Y., Alsaadi, F.E.: A survey of deep neural network architectures and their applications. Neurocomputing **234**, 11–26 (2017)

22. Hinton, G., et al.: Deep neural networks for acoustic modeling in speech recognition. IEEE Sig. Process. Mag. **29**, 82–97 (2012)

23. Brunetti, A., Buongiorno, D., Trotta, G.F., Bevilacqua, V.: Computer vision and deep learning techniques for Pedestrian detection and tracking: a survey. Neurocomputing (2018). https://doi.org/10.1016/j.neucom.2018.01.092

24. Bevilacqua, V., et al.: A Novel deep learning approach in haematology for classification of leucocytes. In: Esposito, A., Faundez-Zanuy, M., Morabito, F.C., Pasero, E. (eds.) Quantifying and Processing Biomedical and Behavioral Signals, pp. 265–274. Springer International Publishing, Cham (2019). https://doi.org/10.1007/978-3-319-95095-2_25

25. Bevilacqua, V., et al.: Retinal fundus biometric analysis for personal identifications. In: Lecture Notes in Computer Science (Including Subseries Lecture Notes in Artificial Intelligence and Lecture Notes in Bioinformatics) (2008). https://doi.org/10.1007/978-3-540-85984-0_147

26. Bevilacqua, V., et al.: A supervised CAD to support telemedicine in hematology. In: Proceedings of the International Joint Conference on Neural Networks (2015). https://doi.org/10.1109/IJCNN.2015.7280464

27. Ganapathy, N., Swaminathan, R., Deserno, T.M.: Deep learning on 1-D biosignals: a taxonomy-based survey. Yearb. Med. Inform. **27**, 98–109 (2018). https://doi.org/10.1055/s-0038-1667083

28. Faust, O., Hagiwara, Y., Hong, T.J., Lih, O.S., Acharya, U.R.: Deep learning for healthcare applications based on physiological signals: a review. Comput. Methods Programs Biomed. **161**, 1–13 (2018)

29. Park, K.-H., Lee, S.-W.: Movement intention decoding based on deep learning for multiuser myoelectric interfaces. In: 4th International Winter Conference on Brain-Computer Interface, BCI 2016. Institute of Electrical and Electronics Engineers Inc. (2016)

30. Geng, W., Hu, Y., Wong, Y., Wei, W., Du, Y., Kankanhalli, M.: A novel attention-based hybrid CNN-RNN architecture for sEMG-based gesture recognition. PLoS ONE **13**, e0206049 (2018)

31. Xu, L., Chen, X., Cao, S., Zhang, X., Chen, X.: Feasibility study of advanced neural networks applied to sEMG-based force estimation. Sensors (Switzerland) **18**, 3226 (2018). https://doi.org/10.3390/s18103226

32. Xie, B., Li, B., Harland, A.: movement and gesture recognition using deep learning and wearable-sensor technology. In: ACM International Conference Proceeding Series, pp. 26–31. Association for Computing Machinery (2018). https://doi.org/10.1145/3268866.3268890

33. Wangshow, W., Chen, B., Xia, P., Hu, J., Peng, Y.: Sensor fusion for myoelectric control based on deep learning with recurrent convolutional neural networks. Artif. Organs **42**, E272–E282 (2018)

34. Zhengyi, L., Hui, Z., Dandan, Y., Shuiqing, X.: Multimodal deep learning network based hand ADLs tasks classification for prosthetics control. In: Proceedings of 2017 International Conference on Progress in Informatics and Computing, PIC 2017, pp. 91–95. Institute of Electrical and Electronics Engineers Inc. (2017). https://doi.org/10.1109/PIC.2017.8359521

35. Zia Ur Rehman, M., Gilani, S.O., Waris, A., Niazi, I.K., Kamavuako, E.N.: A novel approach for classification of hand movements using surface EMG signals. In: 2017 IEEE International Symposium on Signal Processing and Information Technology, ISSPIT 2017, pp. 265–269. Institute of Electrical and Electronics Engineers Inc. (2018)

36. Ibrahim, M.F.I., Al-Jumaily, A.A.: Auto-encoder based deep learning for surface electromyography signal processing. Adv. Sci. Technol. Eng. Syst. **3**, 94–102 (2018). https://doi.org/10.25046/aj030111

37. Sosin, I., Kudenko, D., Shpilman, A.: Continuous gesture recognition from sEMG sensor data with recurrent neural networks and adversarial domain adaptation. In: 2018 15th International Conference on Control, Automation, Robotics and Vision, ICARCV 2018. pp. 1436–1441. Institute of Electrical and Electronics Engineers Inc. (2018)

38. He, Y., Fukuda, O., Bu, N., Okumura, H., Yamaguchi, N.: Surface EMG pattern recognition using long short-term memory combined with multilayer perceptron. In: 2018 40th Annual International Conference of the IEEE Engineering in Medicine and Biology Society (EMBC), pp. 5636–5639 (2018)

39. Shim, H.-M., Lee, S.: Multi-channel electromyography pattern classification using deep belief networks for enhanced user experience. J. Cent. South Univ. **22**, 1801–1808 (2015). https://doi.org/10.1007/s11771-015-2698-0

40. Shim, H.-M., An, H., Lee, S., Lee, E.H., Min, H.-K., Lee, S.: EMG pattern classification by split and merge deep belief network. Symmetry (Basel) **8** (2016). https://doi.org/10.3390/sym8120148

41. Wand, M., Schmidhuber, J.: Deep neural network frontend for continuous EMG-based speech recognition. In: Morgan, N., Georgiou, P., Morgan, N., Narayanan S., Metze, F. (ed.) Proceedings of the Annual Conference of the International Speech Communication Association, INTERSPEECH, pp. 3032–3036. International Speech and Communication Association (2016)

42. Morikawa, S., Ito, S.-I., Ito, M., Fukumi, M.: Personal authentication by lips EMG using dry electrode and CNN. In: Proceedings - 2018 IEEE International Conference on Internet of Things and Intelligence System, IOTAIS 2018, pp. 180–183. Institute of Electrical and Electronics Engineers Inc. (2019). https://doi.org/10.1109/IOTAIS.2018.8600859

43. Abtahi, F., Ro, T., Li, W., Zhu, Z.: Emotion analysis using audio/video, EMG and EEG: a dataset and comparison study. In: Proceedings - 2018 IEEE Winter Conference on Applications of Computer Vision, WACV 2018, pp. 10–19. Institute of Electrical and Electronics Engineers Inc. (2018)

44. Hassan, M.M., Alam, M.G.R., Uddin, M.Z., Huda, S., Almogren, A., Fortino, G.: Human emotion recognition using deep belief network architecture. Inf. Fusion **51**, 10–18 (2019). https://doi.org/10.1016/j.inffus.2018.10.009

45. Chambon, S., Galtier, M.N., Arnal, P.J., Wainrib, G., Gramfort, A.: A deep learning architecture for temporal sleep stage classification using multivariate and multimodal time series. IEEE Trans. Neural Syst. Rehabil. Eng. **26**, 758–769 (2018). https://doi.org/10.1109/TNSRE.2018.2813138

46. Andreotti, F., Phan, H., Cooray, N., Lo, C., Hu, M.T.M., De Vos, M.: Multichannel sleep stage classification and transfer learning using convolutional neural networks. In: Proceedings of the Annual International Conference of the IEEE Engineering in Medicine and Biology Society, EMBS, pp. 171–174. Institute of Electrical and Electronics Engineers Inc. (2018). https://doi.org/10.1109/EMBC.2018.8512214

47. Yulita, I.N., Fanany, M.I., Arymurthy, A.M.: Combining deep belief networks and bidirectional long short-term memory case study: sleep stage classification. In: Rahmawan H. Facta M., R.M.A.S.D. (ed.) International Conference on Electrical Engineering, Computer Science and Informatics (EECSI). Institute of Advanced Engineering and Science (2017)

48. Su, Y., Sun, S., Ozturk, Y., Tian, M.: Measurement of upper limb muscle fatigue using deep belief networks. J. Mech. Med. Biol. **16**, 1640032 (2016)

49. Xia, P., Hu, J., Peng, Y.: EMG-based estimation of limb movement using deep learning with recurrent convolutional neural networks. Artif. Organs **42**, E67–E77 (2018). https://doi.org/10.1111/aor.13004

50. Ben Said, A., Mohamed, A., Elfouly, T., Harras, K., Wang, Z.J.: Multimodal deep learning approach for joint EEG-EMG data compression and classification. In: IEEE Wireless Communications and Networking Conference, WCNC. Institute of Electrical and Electronics Engineers Inc. (2017). https://doi.org/10.1109/WCNC.2017.7925709

51. Bakiya, A., Kamalanand, K., Rajinikanth, V., Nayak, R.S., Kadry, S.: Deep neural network assisted diagnosis of time-frequency transformed electromyograms. Multimed. Tools Appl. **2018**, 1–17 (2018)

52. Sengur, A., Gedikpinar, M., Akbulut, Y., Deniz, E., Bajaj, V., Guo, Y.: DeepEMGNet: an application for efficient discrimination of ALS and normal EMG signals. Adv. Intell. Syst. Comput. **644**, 619–625 (2018)

53. Chen, J., Zhang, X., Cheng, Y., Xi, N.: Surface EMG based continuous estimation of human lower limb joint angles by using deep belief networks. Biomed. Sig. Process. Control **40**, 335–342 (2018)

54. Rane, L., Ding, Z., McGregor, A.H., Bull, A.M.J.: Deep learning for musculoskeletal force prediction. Ann. Biomed. Eng. **47**, 778–789 (2019)

Occluded Object Classification with Assistant Unit

Qing Tang, Youlkyeong Lee, and Kanghyun Jo$^{(\boxtimes)}$

University of Ulsan, Ulsan 44610, Korea
{tangqing,ykleeg}@islab.ulsan.ac.kr,
acejo@ulsan.ac.kr

Abstract. This paper presents a new convolutional neural network (CNN) architecture which improve performance of occluded-object classification by adding an assistant unit. The classification architecture is OCC-VGG19. The OCC-VGG19 outputs two parts classification information. First information is occluded state of target, and second information is the objectness information of target. To access the performance of the proposed architecture, we generate a new dataset that referred to as OCC-CIFAR10 based on CIFAR-10. The OCC-CIFAR10 include 40,000 original images and 10,000 generated image that are occluded by noise, and the OCC-CIFAR10 samples are RGB color images with size 32×32. The OCC-CIFAR10 is used in both of training and testing step. Experimental results show that the proposed assistant unit enhance network robustness in occluded-objects classification task.

Keywords: Classification · Assistant unit · Occluded objects

1 Introduction

Object Classification is a process in which the objects are differentiated. Computer vision applies a classification process by simulating human thinking process to make judgments accurately and quickly. The classification method first extracts features of object then utilizes those similar or unique features of same class objects to classify them from other class objects. In a wide variety of traditional computer vision methods, computer vision first extracts hand-crafted object feature, such as color channels value, corner points, edges and shapes. Secondly, computer vision utilizes the extracted feature to differentiate object's category using supervised learning algorithm such as, support vector machine (SVM) [1] and fully-connect layer [2]. Hand-crafted feature extraction methods, such as, corner detection, line detection, optical flow and template matching appeared in many previous computer vision researches. Hand-crafted features mean the features are presented in the image, such as, edges and corners. It is proved that the features extracted by Convolutional neural network (CNN) easy to classify by classifier, therefore the CNN-based classification methods outperform in object classification field [2–4].

Convolutional neural networks have become indispensable technology in computer vision since AlexNet [5] won the ImageNet Challenge: ILSVRC 2012 [6]. The object classification in some real-world applications has yielded superhuman performance

© Springer Nature Switzerland AG 2019
D.-S. Huang et al. (Eds.): ICIC 2019, LNAI 11645, pp. 762–770, 2019.
https://doi.org/10.1007/978-3-030-26766-7_69

such as traffic sign classification [7]. To improve the performance of classification network, most of methods tend to deepen the network and make the network more complicated, because the deeper and more complicate network learn more high-level features to classify object. But the deeper network requires more calculated resources. Training and testing those deeper networks are energy consuming. Also, the VGG-16 and VGG-19 proves that adding more convolution layer do not always improve the network performance [3]. Both of the efficiency and accuracy of network are important in many real-world applications, such as, robotics, self-driving car and face verification. Many researches of trade-off between efficiency and accuracy has been proposed and developed.

Weak generalization ability is a main shortcoming of CNN. When an image is translated or rotated by a few pixels, the performance of CNNs (e.g. VGG16, ResNet50, and InceptionResNetV2) will drastically decrease [8]. This phenomenon proves that the CNN generalization ability is weak. A set of images of a real-word object looks different because of various light conditions, angles of view and occluded situation. Because of the shortcoming of CNN, CNN need a large number of variety sample for training. A popular method to improving the CNN generalization performance without adding convolution layers is data augmentation [10]. Data augmentation algorithm increases the numbers of images by such as, rotating, translating and cropping original images, so one image can generate different sub-samples. We use these generated samples to train the CNN to robust performance of CNN. Data augmentation algorithm shows there are still many information in one image can be utilized. This paper proposes a method which improve occluded-object classification performance by utilizing the occlusion information.

In this work, an assistant unit is added in output layer to output object occlusion state. The assistant unit is trained using objects' occlusion state that labelled by 0 or 1, 0 refers to non-occluded object, and 1 refers to occluded object. The occlusion state labels were generated by ourselves when we generate the OCC-CIFAR10. The generation method will be introduced in Sect. 2.1. The critical attributions of this paper are that (1) we generated a dataset for occluded-object classification based on CIFAR10 dataset (2) an assistant unit has been proposed that can improve occluded-object classification accuracy.

The paper introduces an efficient network architecture which can improve occluded-object classification accuracy by adding an assistant unit in the output layer. The assistant unit push the network to learn the knowledge about how to classify the occluded and non-occluded objects, and the normal classification network push the network to learn the knowledge about how to classify the object class. After training step, the testing network outputs the occluded state and the class of object. The assistant unit and normal classification network were trained simultaneously, therefore two units share the front convolution layer weights. Compared with VGG19, OCC-VGG19 adds a small number of parameters but apparently improve classification accuracy. Section 2 describes the network architecture and the assistant unit in detail. Section 3 describes experiments on Canadian Institute for Advanced Research (CIFAR-10 [9] and OCC-CIFAR10 dataset. Also, the performance of proposed architectures is shown in Sect. 3. Section 4 describes the problem of the experiment and our future work. Section 5 summarize the work.

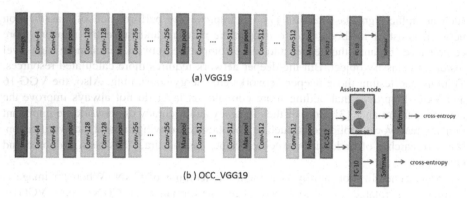

Fig. 1. The standard VGG19 architecture and the proposed OCC-VGG19 architecture.

2   Proposed Methodology

In this section, first the detail of the OCC-CIFAR10 dataset generation will be described, then network architecture will be described in detail. In the last subsection, we introduce the assistant unit. Figure 1 is the comparison of standard VGG19 architecture and OCC-VGG19 architecture.

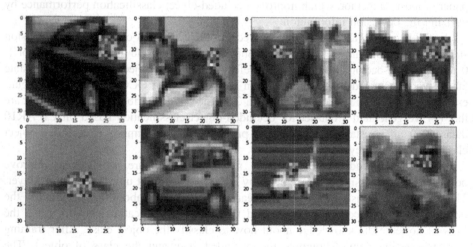

Fig. 2. The OCC-CIFAR10 dataset samples.

2.1   Dataset

The dataset we used in the paper is OCC-CIFAR10, which is generated by ourselves using CIFAR-10 images. CIFAR-10 is a collection of images that commonly used in training and test object classification algorithm. The CIFAR-10 contains 50,000 RGB color training images and 10,000 RGB color test images with size 32 × 32. The 10 different classes represent as airplanes, cars, birds, cats, deer, dogs, frogs, horses, ships, and trucks. In CIFAR-10, each class includes 6000 images.

Table 1. Comparison of CIFAR-10 and OCC-CIFAR10 dataset.

Dataset	CIFAR-10	OCC-CIFAR10
Image size	32 × 32	32 × 32
Class	10	10
Number of images per class in training	5,000	non-occluded: 4,000 occluded: 1,000
Number of images per class in test	1,000	non-occluded: 800 occluded: 200
Training set	non-occluded: 50,000	non-occluded: 40,000 occluded: 10,000
Test set	non-occluded: 10,000	non-occluded: 8,000; occluded: 2,000

The paper aims to prove that the assistant unit enhance the network robustness in occluded-object classification task. Because all the objects in the CIFAR-10 dataset image are not been occluded. Generating occluded objects is needful to prove the capability of assistant unit. For both of training and test sets, we replaced 20% image of each sets with their corresponding occluded image. Figure 1 shows 8 occluded images samples. We randomly set the center location of noise in threshold [6, 27] pixel in horizontal axis and vertical axis of image, respectively. Both of the width and length of noise region are range in [8, 13] pixels. In Fig. 2, almost all the objects locate around the image center, therefore generating noise part which can overlap with object means generating noise part around the image center. We add the occluded information 0 (non-occluded state) and 1 (occluded state) to each images label's matrix to represent the object occluded state.

Table 2. Accuracy of VGG19 that was trained and tested on different dataset.

Architecture	Training dataset	Test dataset	Accuracy
VGG19	CIFAR-10	CIFAR-10	85.12%
VGG19	*OCC-CIFAR-10	*OCC-CIFAR-10	54.75%
VGG19	OCC-CIFAR-10	OCC-CIFAR-10	**44.93%**

Same with CIFAR-10, the OCC-CIFAR10 dataset has 10 classes. Each image is 3 channel RGB color image with size 32 × 32. In the experiment, we used 10,000 occluded images and 40,000 non-occluded images in training step, and we use 2,000 occluded images and 8,000 non-occluded images for test. The comparison of CIFAR-10 and OCC-CIFAR10 dataset is shown in Table 1. We test VGG19 in CIFAR-10 and OCC-CIFAR10 dataset, the experimental result is shown in Table 2. Table 2 shows that the classification performance of the occluded object marked decline from 85.12%

to 54.75% and 44.93%. The *OCC-CIFAR-10 randomly generate occluded sample without considering the class, which means the number of occluded samples in each class are different.

2.2   Network Architecture

Figure 1(top) shows VGG-19 architecture. The images are passed through a stack of convolution layers and max-pooling layer. In the convolutional layer, kernels with a very small receptive field 3×3 were used, because the input images size are 32×32 which is relative small size. The two architectures in Fig. 2 VGG 19, OCC-VGG includes 2 Conv-64 modules, 2 Conv-128 modules, 4 Conv-256 modules, and 8 Conv-516 modules, the number after 'Conv-' means the number of convolution kernels used in the Convolution layer. For one image, the size of input images is $3 \times 32 \times 32$.

These stack of convolutional layer of VGG-19 is followed by one fully connected layer that has 10 dimensions. The outputs of the last Conv-516 module is $512 \times 1 \times 1$, the 512 nodes fully connect with 10 nodes, because CIFAR-10 have 10 classes.

We use VGG-19 Network to extract feature in the paper. Compared with all of the VGG network, includes VGG-11, VGG-11 (LRN), VGG-13, VGG-16 (Conv1), VGG-16 and VGG-19, VGG-16 achieves highest accuracy in the ILSVRC (ImageNet Large Scale Visual Recognition Competition) 2014 in the classification task. VGG-19 obtains the second highest accuracy in the ILSVRC which means the deeper networks do not always improve accuracy. But compared with the common classification task, the classification network with an assistant unit and OCC-CIFAR10 dataset need to learn more information, because the proposed network (VGG with the assistant unit) classify object state(occluded or non-occluded) and object's category in the meantime, therefore we use VGG19 in this paper.

2.3   Assistant Unit

The Assistant Unit architecture which consists two assistant nodes is shown in Fig. 1 (bottom). The Assistant Unit and the objectness classification unit are parallel at the output layer. The OCC-VGG19 is shown in Fig. 2(bottom), the output FC-512 has 512 dimensions features, we fully connect 512 features with 2 dimensions assistant node. Paralleled with the assistant unit, FC-512 layer fully connected with FC-10 output the 10 object's confidence scores. The Cross-Entropy loss function is utilized in here as Eq. (1).

$$occ\_loss = \frac{\sum_{i=1}^{n} H(occ\_target, target)}{n} \tag{1}$$

$$con\_loss = \frac{\sum_{i=1}^{n} H(objectness\_score, target)}{n} \tag{2}$$

$$loss = occ\_loss + \lambda con\_loss \tag{3}$$

The occ_loss denoted as the loss value that the assistant unit generated. The n denoted as the batch size, and $n = 128$ and $n = 100$ have been used in training and test step, respectively. Batch size defines how many samples were inputted to the model for backpropagating the loss one time. occ_target is the binary value 0 (non-occluded object) and 1 (occluded object), which defined as the object occlusion state. The occ_target is the ground truth of the assistant unit, therefore we train the output of assistant node close to the occ_target. The con_loss donates the objectness loss of input object.

The cross-entropy loss function is utilized in occ_loss and con_loss as Eqs. 1 and 2. The occlusion state scores and object's objectness scores for each class are the output of the softmax layer. The occ_loss and con_loss are normalized by n batch size. Equation 3 shows the calculation methods of the total loss, con_loss is weighted by a balancing parameter λ. In our current experiments we set $\lambda = 4$. These two architectures connect with a softmax module.

3 Experiment

In this section CIFAR-10 dataset with VGG19 has been trained and tested, we then show the experiment result of OCC-CIFAR10 using OCC-VGG19 architecture. OCC-VGG19 architecture adds two node which connect with fully connection layer (512 dimensions). Compared with VGG19, OCC-VGG19 have 1024(512*2) parameters more than VGG19, which is relative tiny increasing that compared with the size of VGG19 architecture. The experiment train and test use i5-6600 CPU and one Nvidia GeForce GTX 1080 Ti GPU.

3.1 Network Architecture

Our training used asynchronous stochastic gradient descent with 0.9 momentum. Fixed learning rate is 0.01 and 200 epochs are used. As the epoch increasing, the validation accuracy was decreased slightly which is caused by overfitting. Because we tend to prove the assistant unit is helpful for classifying the object which in occluded case only, we did not use dropout and other regularization method in experiment. Also, the experimental results that are shown in Tables 2 and 3 did not use any data augmentation like rotation, translation and cropping.

Table 3. The accuracy of the VGG19 and OCC_VGG19 architecture

Architecture	Training dataset	Test dataset	Accuracy	Assistance unit accuracy
VGG19	OCC-CIFAR10	OCC-CIFAR10	44.93%	-
OCC_VGG19	OCC-CIFAR-10	OCC-CIFAR-10	**47.75%**	99.19%

3.2   Experimental Result

There are 10,000 test images in CIFAR10 and OCC-CIFAR10. Each image is associated with one ground truth label. Each OCC-CIFAR10 image labeled by class label and occluded state label. For OCC-CIFAR10, the performance (accuracy in experimental result table) is measured only based on objectness output, we did not consider the accuracy of assistant unit which used to predict occluded state.

Table 2 (top) shows VGG-19 results when trained and test using CIFAR-10. We did not use any augmentation method, the top 1 accuracy is 84.12%. The VGG-19 that train and test on OCC-CIFAR10 achieve 54.75%. Table 2 (bottom), OCC-CIFAR10 which switch 10,000 occluded images in training and 2,000 occluded image in test decrease the accuracy about 30%. The OCC-CIFAR10 is more Challenging because of lack of non-occluded training image and extra occluded test image. Table 3 shows the result of experiment using OCC-CIFAR10 for three architecture VGG-19, OCC VGG19 v1 and OCC VGG19 v2. Using assistant unit for occluded object classification without combining with the first FC-10 layer, the top 1 accuracy (55.23%) and top 2 accuracy (73.79%) slightly higher than the VGG19 baseline. It proves that, during the training processing, the network constricted by occluded information enhance network performance. Table 3 (last row) shows the OCC VGG19 v2 performance. Compared with VGG19, the top 1 accuracy (55.56%) and top 2 accuracy (73.86%) higher than the VGG19 baseline. Top 1 accuracy of OCC VGG19 v2 shows OCC VGG19 v2 correctly predict 80 images more than VGG19. Compared with OCC VGG19 v1, OCC VGG19 v2 improve top 1 accuracy .33%. The result of OCC VGG19 v2 proves combine the assistant unit and objectness output to get the finally objectness output improve architecture performance.

3.3   Ablation Experiment

In light of our observation it seems that training the assistance unit is easier than training the objectness classification network. The OCC_VGG19_v1 connect the assistance unit with the first Conv-64 layer of VGG19. Compared with the OCC_VGG19 architecture, The OCC_VGG19_v1 only share the first Conv-64 layer with objectness classification network. The experimental result which is shown in Table 4 proves the OCC_VGG19_v1 outperforms in objectness classification accuracy though, the assistance unit accuracy less than the OCC_VGG19 (Fig. 3).

Table 4. The accuracy of the OCC_VGG19 and OCC_VGG19_v1 architecture

Architecture	Training dataset	Test dataset	Accuracy	Assistance unit accuracy
OCC_VGG19	OCC-CIFAR-10	OCC-CIFAR-10	47.75%	99.19%
OCC_VGG19_v1	OCC-CIFAR-10	OCC-CIFAR-10	**49.98%**	98.03%

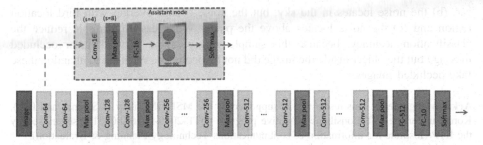

Fig. 3. The architecture of OCC_VGG19_v1

4   Conclusion

This paper presents two architectures for object classification using a convolutional neural network(CNN) with an assistant unit. By sharing feature with the fully connection layer, assistant unit is nearly cost-free. The critical attributions of this paper are that (1) a new dataset OCC-CIFAR10 based on CIFAR-10 has been generated to test the performance of proposed method (2) we propose an assistant unit that improve classification accuracy in object-occluded case.

The assistant unit also adjust learned layer of whole VGG19 architecture, so the assistant unit improves classification quality. The OCC-VGG19 network utilize the object class and object occlusion state together to train the network converge. The experimental results show that using occluded information enhance robustness of network.

5   Discussion

There are some error messages in generated dataset might cause the improvement of proposed network is slight. Figure 4 shows three sample of generated OCC-CIFAR10 dataset. The square noise part does not overlap with the object, but we still label it occluded state. Figure 4(a) the noise locates at the image background and above the

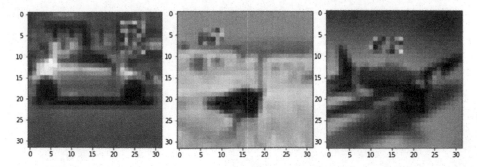

Fig. 4. Drawback of generated dataset.

car, (b) the noise locates in the sky, but the noise should overlap with bird location region and (c) the noise locates above the plane. We guess this sample reduce the classification accuracy, because this sample will give the network error occluded message but the object inside the image did not be occluded. We need to abandon these fake occluded images.

Acknowledgement. This research was supported by the MSIT (Ministry of Science and ICT), Korea, under the ICT Consilience Creative program (IITP-2019-2016-0-00318) supervised by the IITP (Institute for Information & communications Technology Planning & Evaluation).

References

1. Hearst, M.A.: Support vector machines. IEEE Intell. Syst. **13**, 18–28 (1998)
2. LeCun, Y.: LeNet-5, convolutional neural networks (2015). http://yann.lecun.com/exdb/lenet
3. Simonyan, K., Zisserman, A.: Very deep convolutional networks for large-scale image recognition. arXiv:1409.1556, September 2014
4. Szegedyet, C., et al.: Going deeper with convolutions. In: Proceedings of IEEE Conference Computer Vision Pattern Recognition, pp. 1–9 (2015)
5. Krizhevsky, A., Sutskever, I., Hinton, G.E.: ImageNet classification with deep convolutional neural networks. In: Proceedings of Advances in Neural Information Processing Systems, pp. 1097–1105 (2012)
6. Russakovsky, O., et al.: ImageNet large scale visual recognition challenge. Int. J. Comput. Vis. **115**(3), 211–252 (2015)
7. Ciresan, D.C., Meier, U., Masci, J., Schmidhuber, J.: Multi-column deep neural network for traffic sign classification. Neural Netw. **32**, 333–338 (2012)
8. Azulay, A., Weiss, Y.: Why do deep convolutional networks generalize so poorly to small image transformations? arXiv preprint arXiv:1805.12177, abs/1805.12177 (2018)
9. Krizhevsky, A.: Learning Multiple layers of features from tiny images (2009)
10. Wang, J., Luis, P.: The effectiveness of data augmentation in image classification using deep learning. Stanford University research report (2017)

Water Wave Optimization for Flow-Shop Scheduling

Jia-Yu Wu, Xue Wu, Xue-Qin Lu, Yi-Chen Du,
and Min-Xia Zhang[(✉)]

College of Computer Science and Technology,
Zhejiang University of Technology, Hangzhou 310023, China
minxia_zhang@yeah.net

Abstract. Flow-shop scheduling problem (FSP) is a well-known combinatorial optimization problem which has a wide range of practical applications. However, FSP is known to be *NP*-hard when there are more than two machines, for which traditional exact algorithms can only solve small-size problem instances, and many metaheuristic algorithms are mostly suitable for solving large-size instances. Water wave optimization (WWO) is a novel metaheuristic evolutionary algorithm that draws inspiration from shallow water wave model for optimization problems. In this paper, we propose two WWO algorithms for FSP. The first algorithm adapts the original evolutionary operators of the basic WWO according to the solution space of FSP. The second algorithm further improves the first algorithm with a self-adaptive local search procedure. Experimental results on test instances show that the proposed strategies are effective for solving FSP, and the WWO algorithm with self-adaptive local search exhibits significant performance advantages over many other well-known metaheuristic algorithms.

Keywords: Flow-shop scheduling problem (FSP) ·
Water wave optimization (WWO) · Combinatorial optimization · Local search

1 Introduction

Flow-shop scheduling problem (FSP) is one of the most common scheduling problems in combinatorial optimization. It has a wide range of practical applications in science and engineering. However, FSP is known to be *NP*-hard [6] when there are more than two machines, and it is typically hard to derive sharp bounds on the optimal solution of FSP. Consequently, it is difficult to use traditional exact algorithms to solve large-size problem instances. In recent decades, numerous studies have been devoted to solving FSP using metaheuristic evolutionary algorithms, which cannot guarantee the optimal solution but can often quickly obtain a good (near-optimal) solution. The typical metaheuristic algorithms for FSP include genetic algorithm (GA) [4, 15, 20], differential evolution (DE) [12, 13, 17], particle swarm optimization (PSO) [7, 8], biogeography-based optimization (BBO) [9], cuckoo search algorithm (CSA) [1, 10], ant colony optimization (ACO) [14], and so on.

© Springer Nature Switzerland AG 2019
D.-S. Huang et al. (Eds.): ICIC 2019, LNAI 11645, pp. 771–783, 2019.
https://doi.org/10.1007/978-3-030-26766-7_70

Water wave optimization (WWO) is a recently proposed metaheuristic optimization algorithm [21] which takes inspiration from the shallow water wave theory to global optimization problems. Owing to its simple algorithm framework, fewer control parameters, high efficiency in exploring high-dimensional space, and good extensibility, WWO has aroused great research interest and has been applied to a wide range of problems.

In this paper, we adapt the WWO to solve the discrete FSP. We propose two WWO versions. The first version adapts the original evolutionary operators of the basic WWO according to the solution space of FSP. The second version further improves the first one by introducing a self-adaptive local search procedure. Experimental results on test instances show that the proposed strategies are effective for solving FSP, and the WWO algorithm with self-adaptive local search exhibits significant performance advantages over many well-known metaheuristic algorithms.

The remainder of the paper is organized as follows. Section 2 introduces the problem formulation of FSP. Section 3 first describes the original WWO algorithm, and then proposes the two WWO versions for FSP. Section 4 presents the numerical experiments, and finally Sect. 5 concludes.

2 Problem Description

FSP is to schedule n independent jobs $\{J_1, J_2, ..., J_n\}$ on m machines $\{M_1, M_2, ..., M_m\}$. Each job contains exactly m operations. The j-th operation of each job J_i must be executed on the j-th machine, where the execution time t_{ij} is specified ($1 \leq i \leq n$; $1 \leq j \leq m$). No machine can perform more than one operation simultaneously.

The problem needs to decide an execution sequence (permutation) $\pi = \{\pi_1, \pi_2, ..., \pi_n\}$ of the n jobs. Let $C(\pi_i, j)$ denote the completion time of job π_i on machine M_j. For the first machine M_1 we have:

$$C(\pi_1, 1) = t_{\pi_1, 1} \tag{1}$$

$$C(\pi_i, 1) = C(\pi_{i-1}, 1) + t_{\pi_i, 1}, i = 2, ..., n \tag{2}$$

The first job π_1 can be processed on machine M_j immediately after it is completed on machine M_{j-1}:

$$C(\pi_1, j) = C(\pi_1, j-1) + t_{\pi_1, j}, j = 2, ..., n \tag{3}$$

Each subsequent job π_i can be processed on machine M_j only when the following two conditions are satisfied: (1) The job π_i has been completed on machine M_{j-1}; (2) Its previous job π_{i-1} has been completed on machine M_j. So we have:

$$C(\pi_i, j) = \max\{C(\pi_{i-1}, j), C(\pi_i, j-1)\} + t_{\pi_i, j}, \quad i = 2, ..., n; j = 2, ..., m \tag{4}$$

Thus the makespan of π is:

$$C_{\max}(\pi) = C(\pi_n, m) \tag{5}$$

The goal of the problem is to find an optimal π^* in the set Π of all possible sequences to minimize the makespan:

$$C_{\max}(\pi^*) = \min_{\pi \in \Pi} C_{\max}(\pi) \tag{6}$$

To improve the practicability of FSP, we can pose a due time $d(\pi_i)$ for each job J_i:

$$C(\pi_i, m) \leq d(\pi_i), i = 1, 2, \ldots, n \tag{7}$$

Using the penalty function method, the objective function (6) can be transformed as follows (where M is a large positive number):

$$f(\pi) = \min\left(C_{\max}(\pi) + M \sum_{i=1}^{n} \max(C(\pi_i, m) - d(\pi_i), 0)\right) \tag{8}$$

3   Water Wave Optimization for FSP

3.1   The Original Water Wave Optimization

WWO [21] is a novel metaheuristic evolutionary algorithm that draws inspiration from shallow water wave models to solve optimization problems. In WWO, each solution is analogous to a wave, the solution space is analogous to the seabed area, and the fitness of a solution is measured inversely by its seabed depth: the higher the energy (i.e., fitness), the smaller the wavelength, and the smaller range the wave propagates, as shown in Fig. 1. It should be noted that, for a high-dimensional optimization problem, the 2-D or 3-D space of the seabed should to extended to a hyperspace space.

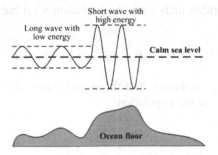

Fig. 1. Illustration of the wavelength model in WWO [21]

Given an optimization problem, WWO first initializes a population of solutions to the problem, and then iteratively evolves the solutions by operators including propagation, refraction, and breaking. At each generation, each solution $\mathbf{x}$ propagates to a new position $\mathbf{x}'$ by shifting each dimension d in a range proportional to its wavelength $\lambda_{\mathbf{x}}$ as follows:

$$x'(d) = x(d) + \lambda_{\mathbf{x}} \cdot rand(-1, 1) \cdot L(d) \tag{8}$$

where the function $rand()$ produces a random number uniformly distributed in the specified range, and $L(d)$ is the d-th dimension of the search space.

The wavelength is updated based on solution fitness $f(\mathbf{x})$ at each generation as follows:

$$\lambda_{\mathbf{x}} = \lambda_{\mathbf{x}} \cdot \alpha^{-(f(\mathbf{x}) - f_{\min} + \epsilon)/(f_{\max} - f_{\min} + \epsilon)} \tag{9}$$

where $f_{\max}$ and $f_{\min}$ are the maximum and minimum fitness among the population, respectively, α is the wavelength reduction coefficient which is set up 1.0026, and ϵ is a very small number to avoid division by zero. This makes high-fitness solutions explore large spaces for global search, and make low-fitness solutions exploit small regions.

If the new $\mathbf{x}'$ is better than the original $\mathbf{x}$, then $\mathbf{x}$ is replaced by $\mathbf{x}'$; otherwise $\mathbf{x}$ remains in the population. However, if a solution $\mathbf{x}$ remains after $h_{\max}$ generations (where $h_{\max}$ is a predefined limit typical set to 6), it will be replaced by a new wave $\mathbf{x}'$ generated by refraction as follows:

$$x'(d) = norm\left(\frac{x^*(d) + x(d)}{2}, \frac{|x^*(d) - x(d)|}{2}\right) \tag{10}$$

where $\mathbf{x}^*$ is the current best solution found by the algorithm, and the function $norm()$ produces a Gaussian random number with the specified mean and standard deviation.

After refraction, the wavelength is updated as follows:

$$\lambda_{\mathbf{x}'} = \lambda_{\mathbf{x}} \frac{f(\mathbf{x})}{f(\mathbf{x}')} \tag{11}$$

Whenever the algorithm finds a new best solution $\mathbf{x}^*$, it breaks into several solitary waves:

$$x'(d) = x^*(d) + norm(0, 1) \cdot \beta \cdot L(d) \tag{12}$$

where β is the breaking coefficient. The fittest one among the solitary waves, if better than $\mathbf{x}^*$, will replace $\mathbf{x}^*$ in the population.

3.2   A Discrete WWO Algorithm for FSP

The original WWO algorithms is for continuous optimization problems. FSP is a combinatorial optimization problem, for which we use permutation-based representation, such that each solution x to the problem is encoded as a permutation π of the given n jobs.

In order to adapt WWO to FSP, we must redefine its operators for evolving permutation-based solutions in the discrete search space. Central to WWO is its propagation operator, the principle of which is that, the higher (lower) the solution fitness, the more (less) the changes will be made to the solution. For FSP, propagation on a solution x can be made by performing one or several local search steps on x, where the number of steps is proportional to its wavelength λ_x. Here we define a local search step as reversing a (random) subsequence of the permutation. Consequently, the larger the wavelength λ_x, the more the solution will be disturbed.

Using this approach, λ_x should be an integer instead of a real number in [0, 1]. Thus we replace the wavelength calculation Eq. (10) with the following one:

$$\lambda_x = \lambda_{max}^{(f_{max} - f(x) + \epsilon)/(f_{max} - f_{min} + \epsilon)} \tag{13}$$

where λ_{max} is the maximum allowable wavelength.

The breaking operator is to search solutions very close to the current best solution x^*. For FSP, we generate each solitary wave solution by randomly selecting a job j whose waiting time is larger than the average waiting time of all jobs in x^*, and then reinserting it into another random position in the permutation.

We follow the work of [23] to discard the refraction operator, and employ a population size reduction strategy as follows to remove stagnant or low-quality solutions from the population and to accelerate the convergence speed:

$$NP = NP_{min} + (NP_{max} - NP_{min}) \frac{g}{g_{max}} \tag{14}$$

where g is the current number of generations (or function evaluations), g_{max} is the maximum allowable number of the generations (or function evaluations), and NP_{max} and NP_{min} are the upper and lower limits of the population size. Whenever the size in the current generation is less than that in the previous generation, the worst solution in the population is removed.

Algorithm 1 presents the WWO algorithm for FSP.

Algorithm 1: The WWO algorithm for FSP.

1 Randomly initialize a population of NP solutions to the problem;
2 Let $\mathbf{x}^*$ be the best among the solutions;
3 **while** *the stopping condition is not met* **do**
4   Calculate the wavelengths based on Eq. (14);
5   **foreach** $\mathbf{x}$ *in the population* **do**
6   Let $k = rand(1, \lambda_\mathbf{x})$;
7   Propagate $\mathbf{x}$ to a new $\mathbf{x}'$ by performing k random reversals;
8   **if** $f(\mathbf{x}')$ *is better than* $f(\mathbf{x})$ **then**
9   Replace $\mathbf{x}$ with $\mathbf{x}'$ in the population;
10   **if** $f(\mathbf{x})$ *is better than* $f(\mathbf{x}^*)$ **then**
11   **for** $k = 1$ *to* $rand(1, k_{\max})$ **do**
12   Produce a neighbor $\mathbf{x}'$ by randomly reinserting a job in $\mathbf{x}^*$;
13   **if** $f(\mathbf{x}')$ *is better than* $f(\mathbf{x}^*)$ **then**
14   Replace $\mathbf{x}^*$ with $\mathbf{x}'$;

15   Update the population size based on Eq. (15);
16   **if** *the population size is decreased* **then**
17   Remove the worst solution from the population;

18 **return** $\mathbf{x}^*$.

3.3 A Self-adaptive WWO Version for FSP

The second version further improves the algorithm by considering multiple local search operators in breaking. In addition to the reinsertion-based operator, we add two operators, the first randomly swapping two adjacent jobs in the permutation, and the second using the NEH heuristic [11].

The self-adaptive WWO (denoted by WWO-M) dynamically choose a local search operator from the three operators for breaking based on their past performance during problem solving. Initially, the three local search operators have the same probability of being selected. After the first LP generations (where LP is a parameter for controlling the learning period), the probability of each operator is adjusted at each generation based on its performance during the previous LP generations:

$$\rho_l = \frac{c_l n_l^I / n_l}{\sum_{l=1}^{3} c_l n_l^I / n_l}. \tag{15}$$

where n_l is the number of invocations of the l-th operator, n_l^I is the number of invocations of the operator that produce better solutions, and c_1 is a coefficient representing the computational cost of an invocation of the operator ($1 \leq l \leq 3$). Here, we set $c_1 = 1$, $c_2 = 1$, and $c_3 = n/2$.

Algorithm 2 presents the self-adaptive WWO version for FSP.

Algorithm 2: The self-adaptive WWO (WWO-M) algorithm for FSP.

1   Randomly initialize a population of NP solutions to the problem;
2   Let $\mathbf{x}^*$ be the best among the solutions;
3   **while** *the stopping condition is not met* **do**
4   Calculate the wavelengths based on Eq. (14);
5   **foreach** $\mathbf{x}$ *in the population* **do**
6   Let $k = rand(1, \lambda_{\mathbf{x}})$;
7   Propagate $\mathbf{x}$ to a new $\mathbf{x}'$ by performing k random reversals;
8   **if** $f(\mathbf{x}')$ *is better than* $f(\mathbf{x})$ **then**
9   Replace $\mathbf{x}$ with $\mathbf{x}'$ in the population;
10   **if** $f(\mathbf{x})$ *is better than* $f(\mathbf{x}^*)$ **then**
11   **for** $k = 1$ *to* $rand(1, k_{\max})$ **do**
12   Select a local search operator l according to the probability ρ_l;
13   Use the operator to produce a neighbor $\mathbf{x}'$;
14   **if** $f(\mathbf{x}')$ *is better than* $f(\mathbf{x}^*)$ **then**
15   Replace $\mathbf{x}^*$ with $\mathbf{x}'$;
16                          $n_l^{\mathrm{I}} \leftarrow n_l^{\mathrm{I}} + 1$;
17   Update the selection probabilities of the breaking operators based on Eq. (16);
18   Update the population size based on Eq. (15);
19   **if** *the population size is decreased* **then**
20   Remove the worst solution from the population;
21 **return** $\mathbf{x}^*$.

---

4   Computational Experiments

We select 12 FSP test instances from [19], and compare the proposed two WWO versions for FSP with the following nine metaheuristic algorithms for FSP:

- An ordered-based GA [20].
- A discrete PSO algorithm [8].
- A discrete DE algorithm [13], for which we implement two versions, a basic version without enhanced local search and a version enhanced with NEH and referenced local search (denoted by DE-M).
- A discrete cuckoo search algorithm (CSA) [1].
- A teaching-learning-based optimization (TLBO) algorithm [18], for which we implement two versions, a basic version without enhanced local search and a version enhanced with insert local search based on simulated annealing (denoted by TLBO-M).

- A BBO algorithm using ecogeography-based migration [22] which outperforms other BBO variants [2].
- A hybrid ACO (HACO) algorithm [3].

For WWO and WWO-M, we set $NP_{max} = \min(3n, 100)$, $NP_{min} = 18$, and $\lambda_{max} = 0.8$ n. For WWO-M, we set the learning period $LP = 30$. The control parameters of the other algorithms are set as suggested in the literature. For the sake of fairness, all algorithms use the same stopping condition that NFE reaches 100 mn, and each algorithm is run 30 times on each instance. For convenience, the resulting objective function values obtained by the algorithms are transformed into the relative percentage deviation (RPD) from the best known values.

Table 2 presents the medians and standard deviations of the results of DE-M, TLBO-M, HACO, and WWO-M, all of which are equipped with dedicated local search procedures. The results show that WWO-M also obtains the best median RPD on all 12 instances. On instances 4 and 7, the four algorithms all obtain the exact optimal solutions. On instances 1, 2, and 3, three algorithms except DE-M obtain the exact optimal solutions. On the remaining seven instances, WWO-M obtains significantly better results than DE-M and TLBO-M. The results of WWO-M are also significantly better than those of HACO on three instances, and no worse than those of HACO on all instances. Such results demonstrate that WWO-M is the most efficient metaheuristic algorithm in solving the test FSP instances.

Table 1 presents the medians and standard deviations of the results of GA, PSO, DE, CSA, TLBO, EBO, and WWO, all of which do not use adaptive local search. On each instance, the minimum median value among the seven algorithms is shown in bold, and a superscript [†] before a median value indicates that the result of the algorithm is significantly different from that of WWO (at a confidence level of 95%, according to the nonparametric Wilcoxon rank sum test). As we can observe from the results, WWO achieves the best median PRD on all 12 instances. On the smallest-size instance 1, all algorithms except DE can obtain the exact optimal solution, as indicated by that PRD = 0. Among the remaining 11 instances, the results of WWO are significantly better than GA on all instances, better than PSO, DE, and CSA on 10 instances, better than TLBO on 8 instances, and better than BBO on 6 instances. On the contrary, none of the other algorithms can obtain significantly better results than WWO on any instance. Such results show WWO has significant performance improvements over the other six algorithms on the test set.

Table 2 presents the medians and standard deviations of the results of DE-M, TLBO-M, HACO, and WWO-M, all of which are equipped with dedicated local search procedures. The results show that WWO-M also obtains the best median RPD on all 12 instances. On instances 4 and 7, the four algorithms all obtain the exact optimal solutions. On instances 1, 2, and 3, three algorithms except DE-M obtain the exact optimal solutions. On the remaining seven instances, WWO-M obtains significantly better results than DE-M and TLBO-M. The results of WWO-M are also significantly better than those of HACO on three instances, and no worse than those of HACO on all instances. Such results demonstrate that WWO-M is the most efficient metaheuristic algorithm in solving the test FSP instances.

Table 1. The results of basic metaheuristic algorithms (without adaptive local search) on FSP test instances.

Instance	Size($n \times m$)	Metric	GA	PSO	DE	CSA	TLBO	BBO	WWO
1	20 × 5	Median	**0**	**0**	†0.03	**0**	**0**	**0**	**0**
		Std	0	0	0.02	0	0	0	0
2	20 × 10	Median	†0.03	†0.03	†0.02	†0.03	0.01	0.01	0.01
		Std	0.02	0.02	0.01	0.02	0.01	0.01	0.01
3	20 × 20	Median	†0.08	†0.05	†0.03	†0.03	†0.03	0.02	0.02
		Std	0.02	0.03	0.02	0.03	0.02	0.01	0.01
4	50 × 5	Median	†0.01	**0**	†0.01	†0.01	**0**	**0**	**0**
		Std	0.02	0.01	0.01	0.01	0.01	0.00	0
5	50 × 10	Median	†0.47	†0.39	†0.43	†0.31	0.31	0.29	0.19
		Std	0.30	0.26	0.13	0.14	0.27	0.16	0.03
6	50 × 20	Median	†0.81	†0.61	†0.64	†0.48	†0.47	†0.48	0.28
		Std	0.29	0.25	0.23	0.08	0.31	0.13	0.15
7	100 × 5	Median	†0.23	†0.12	**0**	**0**	†0.05	**0**	**0**
		Std	0.2	0.07	0.02	0.02	0.08	0	0
8	100 × 10	Median	†0.39	†0.45	†0.35	†0.42	†0.37	†0.27	0.21
		Std	0.21	0.18	0.09	0.19	0.15	0.13	0.09
9	100 × 20	Median	†1.61	†1.82	†2.78	†1.54	†1.65	†1.23	0.86
		Std	0.49	0.55	0.35	0.49	0.32	0.31	0.16
10	200 × 10	Median	†0.45	†0.67	†0.21	†0.18	†0.22	†0.14	0.08
		Std	0.24	0.22	0.27	0.18	0.16	0.11	0.06
11	200 × 20	Median	†3.52	†4.16	†5.11	†3.13	†4.25	†2.96	2.36
		Std	1.05	0.68	2.26	1.18	0.89	1.06	1.39
12	500 × 20	Median	†3.96	†4.02	†3.56	†2.95	†3.46	†2.77	2.08
		Std	1.79	1.35	0.91	1.20	1.44	1.63	1.12

Table 2. The results of metaheuristic algorithms with dedicated local search procedures on FSP test instances.

Instances	Size($n \times m$)	Metric	DE-M	TLBO-M	HACO	WWO-M
1	20×5	Median	†0.03	**0**	**0**	**0**
		Std	0.02	0	0	0
2	20×10	Median	†0.01	**0**	**0**	**0**
		Std	0.01	0.00	0	0
3	20×20	Median	†0.02	**0**	**0**	**0**
		Std	0.01	0.00	0.01	0.00
4	50×5	Median	**0**	**0**	**0**	**0**
		Std	0.00	0	0	0
5	50×10	Median	†0.39	†0.15	**0**	**0**
		Std	0.12	0.11	0.07	0.04
6	50×20	Median	†0.55	†0.28	†0.10	**0**
		Std	0.09	0.11	0.05	0.01
7	100×5	Median	**0**	**0**	**0**	**0**
		Std	0	0	0	0
8	100×10	Median	†0.21	†0.05	**0**	**0**
		Std	0.11	0.02	0.02	0.01
9	100×20	Median	†1.08	†0.27	0.27	**0.18**
		Std	0.22	0.37	0.19	0.41
10	200×10	Median	†0.09	†0.09	0.06	**0**
		Std	0.1	0.07	0.05	0.08
11	200×20	Median	†3.01	†2.45	†1.93	**0.83**
		Std	1.35	1.15	1.28	1.23
12	500×20	Median	†2.21	†2.06	†1.93	**0.69**
		Std	1.26	1.12	1.35	0.98

Figure 2 presents the median, maximum, minimum, the first quartile (Q1) and the third quartile (Q3) values of all eleven algorithms. In general, the results of the algorithms with dedicated local search procedures are significantly better than those of the other algorithms, which demonstrates that the integration of dedicated local search procedures can effectively improve the performance of metaheuristic algorithms.

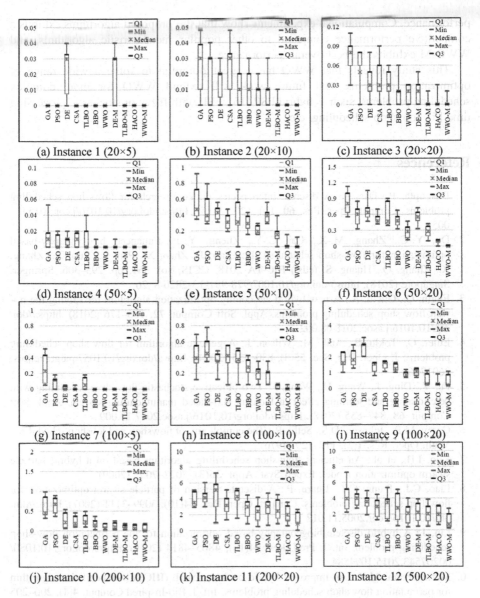

(a) Instance 1 (20×5)   (b) Instance 2 (20×10)   (c) Instance 3 (20×20)

(d) Instance 4 (50×5)   (e) Instance 5 (50×10)   (f) Instance 6 (50×20)

(g) Instance 7 (100×5)   (h) Instance 8 (100×10)   (i) Instance 9 (100×20)

(j) Instance 10 (200×10)   (k) Instance 11 (200×20)   (l) Instance 12 (500×20)

Fig. 2. Box plots of the results of obtained by the eleven algorithms on FSP test instances.

5 Conclusion

This paper proposes two versions of WWO for efficiently solving FSP. We design discrete propagation and breaking operators that can efficiently evolve FSP solutions in the discrete search space. The improved WWO-M version uses a self-adaptive strategy to selecting among four different local search operators to enhance the algorithm performance. Computational experiments show that the proposed algorithms exhibits a competitive performance compared to other popular metaheuristic algorithms, and WWO-M exhibits the best overall performance.

This study reveals that WWO has a great potential to solve a wide range of optimization problems. Our future work will extend WWO for more complex scheduling problems, such as hybrid FSP [16], FSP with fuzzy or stochastic processing time [5], job shop scheduling, etc.

References

1. Dasgupta, P., Das, S.: A discrete inter-species cuckoo search for flow-shop scheduling problems. Comput. Oper. Res. **60**, 111–120 (2015). https://doi.org/10.1016/j.cor.2015.01.005

2. Du, Y.-C., Zhang, M.-X., Cai, C.-Y., Zheng, Y.-J.: Enhanced biogeography-based optimization for flow-shop scheduling. In: Qiao, J., Zhao, X., Pan, L., Zuo, X., Zhang, X., Zhang, Q., Huang, S. (eds.) BIC-TA 2018. CCIS, vol. 951, pp. 295–306. Springer, Singapore (2018). https://doi.org/10.1007/978-981-13-2826-8_26

3. Engin, O., Güçlü, A.: A new hybrid ant colony optimization algorithm for solving the no-wait flow shop scheduling problems. Appl. Soft Comput. **72**, 166–176 (2018). https://doi.org/10.1016/j.asoc.2018.08.002

4. Etiler, O., Toklu, B., Atak, M., Wilson, J.: A genetic algorithm for flow shop scheduling problems. J. Oper. Res. Soc. **55**(8), 830–835 (2004). https://doi.org/10.1057/palgrave.jors.2601766

5. Fu, Y., Ding, J., Wang, H., Wang, J.: Two-objective stochastic flow-shop scheduling with deteriorating and learning effect in Industry 4.0-based manufacturing system. Appl. Soft Comput. **68**, 847–855 (2018). https://doi.org/10.1016/j.asoc.2017.12.009

6. Garey, M.R., Johnson, D.S., Sethi, R.: The complexity of flowshop and jobshop scheduling. Math. Oper. Res. **1**(2), 117–129 (1976)

7. Kuo, I.H., et al.: An efficient flow-shop scheduling algorithm based on a hybrid particle swarm optimization model. Expert Syst. Appl. **36**(3), 7027–7032 (2009)

8. Liao, C.J., Tseng, C.T., Luarn, P.: A discrete version of particle swarm optimization for flowshop scheduling problems. Comput. Oper. Res. **34**(10), 3099–3111 (2007). https://doi.org/10.1016/j.cor.2005.11.017

9. Lin, J.: A hybrid discrete biogeography-based optimization for the permutation flow-shop scheduling problem. Int. J. Prod. Res. **54**(16), 4805–4814 (2016). https://doi.org/10.1080/00207543.2015.1094584

10. Marichelvam, M.K.: An improved hybrid cuckoo search (IHCS) metaheuristics algorithm for permutation flow shop scheduling problems. Int. J. Bio-Inspired Comput. **4**(4), 200–205 (2012). https://doi.org/10.1504/IJBIC.2012.048061

11. Nawaz, M., Enscore, E.E., Ham, I.: A heuristic algorithm for the m-machine, n-job flow-shop sequencing problem. Omega **11**(1), 91–95 (1983). https://doi.org/10.1016/0305-0483(83)90088-9

12. Onwubolu, G., Davendra, D.: Scheduling flow shops using differential evolution algorithm. Eur. J. Oper. Res. **171**(2), 674–692 (2006). https://doi.org/10.1016/j.ejor.2004.08.043

13. Pan, Q.K., Tasgetiren, M.F., Liang, Y.C.: A discrete differential evolution algorithm for the permutation flowshop scheduling problem. Comput. Ind. Eng. **55**(4), 795–816 (2008). https://doi.org/10.1016/j.cie.2008.03.003

14. Rajendran, C., Ziegler, H.: Ant-colony algorithms for permutation flowshop scheduling to minimize makespan/total flowtime of jobs. Eur. J. Oper. Res. **155**(2), 426–438 (2004). https://doi.org/10.1016/S0377-2217(02)00908-6

15. Reeves, C.R.: A genetic algorithm for flowshop sequencing. Comput. Oper. Res. **22**(1), 5–13 (1995). https://doi.org/10.1016/0305-0548(93)E0014-K

16. Ruiz, R., Vázquez-Rodríguez, J.A.: The hybrid flow shop scheduling problem. Eur. J. Oper. Res. **205**(1), 1–18 (2010). https://doi.org/10.1016/j.ejor.2009.09.024

17. Santucci, V., Baioletti, M., Milani, A.: Algebraic differential evolution algorithm for the permutation flowshop scheduling problem with total flowtime criterion. IEEE Trans. Evol. Comput. **20**(5), 682–694 (2016). https://doi.org/10.1109/TEVC.2015.2507785

18. Shao, W., Pi, D., Shao, Z.: An extended teaching-learning based optimization algorithm for solving no-wait flow shop scheduling problem. Appl. Soft Comput. **61**, 193–210 (2017). https://doi.org/10.1016/j.asoc.2017.08.020

19. Taillard, E.: Benchmarks for basic scheduling problems. Eur. J. Oper. Res. **64**(2), 278–285 (1993). https://doi.org/10.1016/0377-2217(93)90182-M

20. Wang, L., Zhang, L., Zheng, D.Z.: A class of order-based genetic algorithm for flow shop scheduling. Int. J. Adv. Manuf. Technol. **22**(11), 828–835 (2003). https://doi.org/10.1007/s00170-003-1689-8

21. Zheng, Y.J.: Water wave optimization: a new nature-inspired metaheuristic. Comput. Oper. Res. **55**(1), 1–11 (2015). https://doi.org/10.1016/j.cor.2014.10.008

22. Zheng, Y.J., Ling, H.F., Shi, H.H., Chen, H.S., Chen, S.Y.: Emergency railway wagon scheduling by hybrid biogeography-based optimization. Comput. Oper. Res. **43**(3), 1–8 (2014). https://doi.org/10.1016/j.cor.2013.09.002

23. Zheng, Y.J., Zhang, B.: A simplified water wave optimization algorithm. In: IEEE Congress on Evolutionary Computation, pp. 807–813 (2015).https://doi.org/10.1109/CEC.2015.7256974

The Implementation of Pretrained AlexNet on PCG Classification

Haya Alaskar[1]([⊠]), Nada Alzhrani[1], Abir Hussain[2],
and Fatma Almarshed[1]

[1] Computer Science Department, College of Computer Engineering and Science,
Prince Sattam Bin Abdulaziz University, Alkharj, Saudi Arabia
{h.alaskar, ny.alzhrani, f.almarshad}@psau.edu.sa
[2] Liverpool John Moores University, Liverpool, UK
A.husssain@ljmu.edu.uk

Abstract. Heart sounds are essential components of cardiac diagnosis, in which heart conditions can be detected using phonocardiogram (PCG) signals. PCG signals provide useful information and can help in the early detection and diagnosis of heart diseases. Many studies have attempted to discover automated tools that analyse heart sounds by applying machine learning algorithms. Although tremendous efforts have been made in this area, no successful framework currently exists to detect pathology in signals because of issues with background noise or poor quality. One part of the evolution of machine learning is the development of deep learning networks, which are designed to exploit the compositional structure of data.

This paper investigates the performance of a convolutional neural network called AlexNet, focusing on two approaches for distinguishing abnormality in PCG signals with data collected from the 2016 PhysioNet/CinC Challenge dataset. This dataset contains heart sound recordings collected from clinical and non-clinical environments. Our extensive simulation results indicated that using AlexNet as feature extractor and Support Vector Machine as classifier a 87% recognition accuracy was achieved, this is an improvement of 85% accuracy obtained by end-to-end learning AlexNet in comparison to the benchmarked techniques.

Keywords: Convolutional neural network · PCG · AlexNet · Classification · End-to-end learning · Feature extraction

1 Introduction

Cardiovascular diseases (CVDs) are the major health problem and cause of death globally [1].Statistics indicated that more than 48% of deaths in Europe [2], in 2016 more than 75% of deaths in low- and middle-income countries [1, 3] are caused by CVDs. Therefore, efforts are underway to discover and develop automated diagnosis methods for CVDs that could drastically reduce the risk factors that can lead to CVD-related death [4].

Different methods are used to detect CVDs. The most common method is listening to the heart sounds as the blood pumps through the body. These sounds are known as

© Springer Nature Switzerland AG 2019
D.-S. Huang et al. (Eds.): ICIC 2019, LNAI 11645, pp. 784–794, 2019.
https://doi.org/10.1007/978-3-030-26766-7_71

phonocardiogram (PCG) signals. PCG signals can be listened to and recorded with a electronic stethoscope [5]. Doctors can detect CVDs based on PCG signals; however, this method is not always sufficient for the diagnosis of CVDs because the likelihood of obtaining an accurate diagnosis usually depends on the cardiologist's expertise. According to Mangione and Nieman [6], medical students and primary care physicians are able to diagnosis CVDs with 40% accuracy, while expert cardiologists are able to detect CVDs with 80% accuracy. For this reason, further investigation into analysing PCG signals is needed to help doctors make accurate diagnoses [3, 7, 8], and discover a suitable tool to detect heartbeat abnormalities.

Over the last several decades, automated heart sound signal analysis has been widely studied. According to Octavian et al. [9], CVD pathologies in a variety of clinical applications can be efficiently detected using automated heart sound classification technique. Therefore, the importance of processing, analysing and recognising abnormalities in these signals are essential. Consequently, PCG signals have been analysed extensively. Typically, four steps are used to recognize and analyse PCG signals: pre-processing, segmentation, feature engineering and classification.

In the literature, many comparative analyses of machine learning algorithms have been performed to study automated PCG classification [3, 10]. However, the data on time series signals consist of very complex high dimensional datasets. It is necessary to utilise some feature extraction methods to transform large time series signals into a small number of features that optimally able to distinguish the data into right group [11].

Most research has concentrated on understanding and analysing PCG signals by applying a number of feature engineering methods [12–16]. However, using feature engineering for PCG signals is a challenging task (Prasad and Kumar, 2015). For example, there are a number of feature sets that can be related to different domains, such as the temporal domain and the frequency domain. Furthermore, feature engineering is done manually, and the features must be examined several times to determine the optimal feature set. Therefore, determining the ideal features for characterising the properties of the PCG signals requires a substantial amount of time and effort [12–15].

Lately, automated intelligent models have been generated directly from PCG signals using deep learning. Deep learning has transformed the machine learning field; it is most frequently used in image categorizing, object or handwriting recognition and speech recognition [17–19]. Furthermore, deep learning plays an important role in the biomedical field for automated disease analysis [20–25]. Deep learning networks can successfully replace the traditional feature engineering-based approach. In the last few years, the most popular type of deep learning model is convolutional neural network (CNN). CNN was initially developed by Fukushima in 1980 [26]. However, in the last decade, its first evolution appeared when AlexNet won the ImageNet competition [27]. With their high levels of qualified information processing, CNNs offer effective representations of complex high dimensional datasets. Many studies using CNNs have achieved greater efficiency and performance in real-time classifications [17, 20, 23, 24].

This paper aims to classify normal and abnormal PCG signals using AlexNet in two ways: as a feature extraction tool and as an end-to-end learning classifier. The first approach will implement Support vector machine as classifier. To accomplish this, the

PCG signals are filtered and segmented; lastly, scalogram images are created from each signal using wavelet analysis.

2   Literature Review

PCG signals linking other biomedical signals suffer from high dimensionality, which leads to difficulties in their analysis. Hence, transforming these signals into subsets is crucial. Feature engineering step is the process of converting signals to features, and these features can embed the main characteristics and properties of each signal [12, 22]. Feature engineering-based methods have dominated the PCG signal recognition domain for the last several decades, including statistical feature extraction [15], frequency-based feature extraction and time-frequency feature [16].

The existence of a variety of feature sets is vital to improving the discriminative performance of classifiers. In some feature engineering cases, the feature vectors are high dimensional, which will require an increased computational effort and might lead to the appearance of unrelated and noisy features, affecting the classification performance. For a high dimensional dataset, the compotation time for measuring and calculating similarities between data becomes more complex and can lead to misunderstandings about the data structure. Therefore, the high dimensionality of the extracted feature vectors must be reduced using feature selection techniques. These steps should be taken into consideration when dealing with any type of data. However, more time and effort would be required from the specialist to decide which tools and techniques should be selected.

The step follows the feature engineering is classification. Studies have used various machine learning tools to classify PCG signals, such as Support Vector Machines [16], and Artificial Neural Networks [28, 29].

Each feature provides a different classification power for different diagnoses. Furthermore, some diseases can share some similar propriety and features with other diseases, which might affect the network performance [30]. Therefore, complex feature extraction procedures and high sampling rates are mandatory.

There are some major limitations in the existing PCG signal analysis approaches; they are extremely reliant on pre-extracted features from the training dataset and are ineffective in dealing with new PCG records. It is worth noting that the investigation of the efficiency of each feature is a time-consuming, difficult process, and it generally requires field expertise [18, 20, 21, 31].

2.1   Application of CNN with PCG Signals

CNN is a promising method for large-scale data analytics [32]. During the last few decades, the application of CNN to identify the properties of biomedical signals has improved significantly. According to the literature, CNN have been employed to analyse biomedical signals, such as electroencephalogram EEGs [21], electrocardiogram (ECGs) [22, 24, 33, 34], and electromyography (EMGs) [20, 35, 36].

Rubin et al. [37] used deep learning network for the automatic recognition of abnormal heartbeats and created a two-dimensional heat map from PCG signals. Their results for sensitivity and accuracy were 0.7278 and 0.8399, respectively.

Other study developed a method based on combined outputs from AdaBoost and CNN to classify heart sounds as normal or abnormal [38]. Aykanat et al. [39] used two types of machine learning algorithms to classify heart sounds: a support vector machine (SVM) with a Mel Frequency Cepstral Coefficient (MFCC) and spectrogram images in a CNN [39]. CNN has been widely used for the automated identification of abnormal heart sound PCG arrhythmia. For example, Nilanon et al. [25] showed that CNN is able to detect heart disease from PCG signals with 82% accuracy.

A deep learning algorithm was used by Renna et al. [40] to segment heart sounds into their main components using CNN. The authors tested their approach on heart sound signals for PCG segmentation and found that CNN outperformed the current segmentation methods with 93.4% sensitivity in detecting the first S1 and secound S2 heard sounds. Renna et al. [40] developed systems to detect abnormal heart sounds using CNN. In the former, the authors segmented the PCG signals into equal lengths and then extracted a scalogram image from each signal using a wavelet transformation. In [8], the authors addressed the classification of PCG signals using Continuous Wavelet Transform (CWT) and CNN. Then, the resulted wavelet was passed to CNN, SVM and the k-nearest neighbours (kNN) classifiers. They concluded that using CWT and CNN in heart sound classification is promising for distinguishing between the first and second heart sound with 86% accuracy. However, using multiple features can improve the accuracy of the classifier. [11] extracted MFCCs and the DWT for machine learning methods, they used SVM, a deep learning network and a centroid displacement-based KNN. The dataset includes five categories of PCG signals with one normal and four abnormal groups. The proposed system achieved an accuracy of 97%.

In [41] used DWT to create PCG scalogram images. The authors run two model of CNNs. The first model was with two conventional layers and the second model was with three conventional layers. Their experiment showed that the second model achieved the highest accuracy with 75%. In their study, the features were extracted automatically using the CNN classifier. In contrast, in [16], the authors extracted 515 features from multiple domains, such as time interval, state amplitude, energy and entropy. Correlation analysis was used to show the effects of each feature. The analysis presented that "frequency spectrum of state", "energy", and "entropy" are best domains of the optimal features. They concluded that SVM classifier with the top 400 features performed better with 88% sensitivity.

Deep learning had huge impact on detecting abnormalities in heart sound images. The system proposed by [42] examined the classification of a patient-recorded audio segment of heartbeats as healthy or pathological. They compared AlexNet, GoogleNet, VGG16, ResNet and long short-term memory (LSTM) networks and showed that CNNs outperformed LSTM. The best result was achieved using 3164 readings resulting in 85811 files.

3   Methodology

As noted previously, two types methods are used in the present study: end-to-end AlexNet learning and CNN feature extraction.

The procedure steps are shown in Fig. 1. First, the raw PCG signals of each subject were pre-processed. Since recording signals can be affected by background noises, which will badly impact the signals, filtering signals were applied to reduce the noise. Then, the segmentation method was used to detect the component. Next, PCG scalo-gram images were created by applying CWT to each segmented signal. Lastly, the two studied approaches were run: first, the pre-trained AlexNet was used to extracted features then based to SVM classifier, and, second, AlexNet was used as an end-to-end learning classifier.

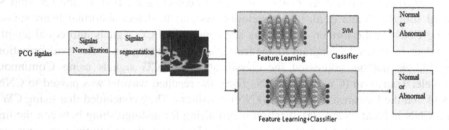

Fig. 1. The procedure steps

3.1   Heart Sound PCG Dataset

The 2016 PhysioNet/Computing in Cardiology (CinC) Challenge is the largest public heart sound database. It has 4,430 recordings taken from 1,072 subjects and collected from eight sources around the world. The signals were recorded using heterogeneous tools in both clinical and non-clinical, e.g. in-home visit, settings.

3.2   Data Preparation

The PCG signals had to be filtered because some of the signals were recorded in unrestrained environments, and affected by some noise. Therefore, the signals were passed through a high-pass filter with a cut-off frequency of 10 Hz to eliminate high-frequency noise besides artefacts, such as baseline wandering. The Schmidt spike removal technique was used to remove signal spikes [43]. Each signal was normalized to a zero mean and variance of 1.

3.3   Segmentation

The segmentation method was used to guarantee that incoming signal inputs were suitably aligned before each classification task. Each signal was segmented into shorter segments to detect the beginning of each heartbeat, i.e. the first S1 heart sound and second S2 heart sound. In the present study, Springer's segmentation algorithm, which is a state-of-the-art solution for heart beat segmentation, was used [44]. This algorithm, which is based on a logistic regression hidden semi-Markov mode, was used to identify the most probable sequence of heart sound states (S1→Systole→S2→Diastole) by incorporating information about expected state durations.

3.4   Time-Frequency Representation (Scalogram)

Continuous Wavelet Transform (CWT) was utilised as it is considered the most suitable method for signal analysis. It measures the time delay between the frequency component of normal and abnormal PCG signals, and it can perform multi-resolution analyses for both time and frequency. According to [45], CWT is the optimal method for non-stationary and transient signals.

In this step, the original one-dimensional PCG signals were transformed into two-dimensional time-frequency representations, so each PCG segment could be processed as a separate image as illustrated on Fig. 2.

(a)   **(b)**

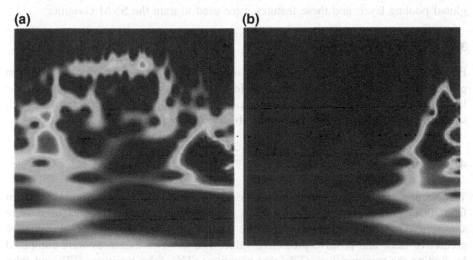

Fig. 2.   (a) Abnormal PCG signal. (b) Normal PCG signal

3.5   Pre-trained CNN Architectures

Each CNN has one or more convolutional and max pooling layers followed by one or more fully connected layers.

The CNN layers are characterized as a block that has number of layers. Convolutions and pooling procedures are implemented on the input data. These layers acts as filter to produce an optimal feature map. The feature maps are then combined as the final output of the convolution layers, and the classification process is achieved in the fully connected layers.

In the literature, a number of CNNs have been examined, such as AlexNet, VGGNet, GoogLeNet and ResNet [46–48]. However, AlexNet is regularly used for recognition tasks, attaining very good results. After training on millions of images to attain low error rates, the pre-trained AlexNet can classify 1,000 possible classes. In the present study, a pre-trained AlexNet with appropriate fine-tuning was used. The implementation of a pre-trained network has been recommended by a number of studies in which all layers of the pre-trained CNN except the last one are fine-tuned [20].

3.6   Feature Extraction Using AlexNet and SVM

In this step, each scalogram PCG image was passed into the pre-trained AlexNet. Each AlexNet layer created a response from an input image, although only a few layers were suitable for image feature extraction. For example, the first layers of the network captured the basic image features, such as edges and blobs, which were then processed by deeper network layers. The deeper layers held the higher-level features, which were created using the simple features of the first layers. These higher-level features transformed all of the basic features into a richer image representation that was suitable for recognition. In total, 9216 features were extracted from one of the deeper layers, the global pooling layer, and these features were used to train the SVM classifier.

3.7   End-to-End Learning

End-to-end learning methods were applied to directly deal with the classification problem. Therefore, AlexNet was used as feature extractor as well as a classifier. In the last fully connected layers, the classification layers defined the right class for each scalogram PCG image if it is normal or abnormal class.

4   Discussion

This section explains the experiments carried out on the pre-trained AlexNet to learn from PCG time-frequency representation. The evaluation measurements used were sensitivity, specificity, accuracy, positive likelihood, negative likelihood, negative predictive value and positive predictive value. The first two measures were computed by finding the true positives (TP), true negatives (TN), false positives (FP) and false negatives (FN). The TPs were correctly identified as abnormal. The TNs were correctly identified as normal. The FP and FN were the number of uncorrected abnormal and normal detections, respectively. Table 1 shows the confusion matrix used to calculate the estimation measurements.

Table 1. The confusion matrix

	Predicted Positive	Predicted Negative
Actual Positive	TP (True Positive)	FN (False Negative)
Actual Negative	FN (False Positive)	TN (True Negative)

The training options were modified to fit the problem. The size of the mini-batch, which refers to subset of the training set used in each iteration of the experiment, was set at 15. The max epochs, which refers to the maximum number of epochs used in training, was set at 10.

This experiment used MATLAB 2018. For the purposes of reproducibility, the network was trained in a stand-alone system with an Intel Core Processor i7-7500U CPU that had 2.70 GHz, 2904 MHz, two cores and 64 GB of RAM.

Initially, the pre-processing steps were implemented, and then continuous wavelet transform (CWT) was used to extract the scalogram images. Each image was fed into the AlexNet input layer. The training testing rate was randomly selected as 80% to 20%.

Two approaches were applied to classify the patterns of the PCG signals as abnormal or normal.

Each scalogram image was fed into the AlexNet as an input. As shown in Table 2, the pre-trained AlexNet and SVM outperformed the end-to-end learning approach performance, achieving higher levels of accuracy in sensitivity and specificity.

Table 2. The results

	AlexNet+SVM	Pre-trained AlexNet
Accuracy	87.65%	85%
Sensitivity	83.71%	76.88%
Specificity	89.99%	92.23%
Positive predictive value	83.22%	88.17%
Negative predictive value	90.30%	84.12%
Positive likelihood	8.36	9.90
Negative likelihood	0.18	0.25

When comparing the methods with AlexNet, it is clear that adapting the pre-trained AlexNet is more effective than the methods described in the literature, which are presented in Table 3. Moreover, in general, the SVM classification of features extracted from the pre-trained AlexNet performed better than the AlexNet end-to-end learning

Table 3. Results from the literature

Literature source	Machine learning	Feature extraction	Results
[7]	• CNN	• Wavelet transformation	76.8% sensitivity
[37]	• CNN	• MFCC heat maps	0.95, 0.73 and 0.84 ratings in specificity, sensitivity and overall accuracy, respectively
[8]	• CNN • SVM • kNN	• CWT	CNN achieved an 85.9% accuracy rating

classifier. The reason could be due in part to the SVM classifier being better to suit to the relatively smaller amount of training data available in the PhysioNet/CinC dataset than the soft-max classifiers in AlexNet. Finally, the strongest performance, 87% accuracy, was obtained on the test set using the AlexNet+SVM' method.

5 Conclusion

Each experiment in the present study demonstrated that deep learning using CNN can automatically extract the best features from PCG signals and learn to discriminate among different classes.

The application of deep learning in the medical field will require extra investigation as it has showed outstanding results in some medical data analysis, and more precisely, in biomedical signal detection and diagnosis. The CNNs have showed success achievements on various of medical applications compared with traditional machine learning algorithms. Consequently, CNN, and more generally, deep learning, will positively increase the efficiency and quality of healthcare, reduce the risk of mis-classification and increase the early diagnosis of serious diseases.

References

1. World Health Organization: Cardiovascular diseases (CVDs) fact sheet. World Health Organization (2017)
2. Lloyd-Jones, D., et al.: Executive summary: heart disease and stroke statistics—2010 update: a report from the American Heart Association. Circulation **121**(7), 948–954 (2010)
3. Latif, S., Usman, M., Rana, R., Qadir, J.: Phonocardiographic sensing using deep learning for abnormal heartbeat detection. IEEE Sens. J. **18**(22), 9393–9400 (2018)
4. Yang, Z.-J., Liu, J., Ge, J.-P., Chen, L., Zhao, Z.-G., Yang, W.-Y.: Prevalence of cardiovascular disease risk factor in the Chinese population: the 2007–2008 China National Diabetes and Metabolic Disorders Study. Eur. Heart J. **33**(2), 213–220 (2011)
5. Randhawa, S.K., Singh, M.: Classification of heart sound signals using multi-modal features. Procedia Comput. Sci. **58**, 165–171 (2015)
6. Mangione, S., Nieman, L.Z.: Cardiac auscultatory skills of internal medicine and family practice trainees: a comparison of diagnostic proficiency. JAMA **278**(9), 717–722 (1997)
7. Ren, Z., Cummins, N., Pandit, V., Han, J., Qian, K., Schuller, B.: Learning image-based representations for heart sound classification. In: Proceedings of the 2018 International Conference on Digital Health, pp. 143–147 (2018)
8. Meintjes, A., Lowe, A., Legget, M.: Fundamental heart sound classification using the continuous wavelet transform and convolutional neural networks. In: 2018 40th Annual International Conference of the IEEE Engineering in Medicine and Biology Society (EMBC), pp. 409–412 (2018)
9. Mukhopadhyay, S.C., Jayasundera, K.P., Postolache, O.: Modern Sensing Technologies. Springer, Cham (2019). https://doi.org/10.1007/978-3-319-99540-3
10. Latif, S., et al.: Mobile technologies for managing non-communicable diseases in developing countries. In: Mobile Applications and Solutions for Social Inclusion. IGI Global, pp. 261–287 (2018)

11. Son, G.-Y., Kwon, S.: Classification of heart sound signal using multiple features. Appl. Sci. **8**(12), 2344 (2018)
12. Babaei, S., Geranmayeh, A.: Heart sound reproduction based on neural network classification of cardiac valve disorders using wavelet transforms of PCG signals. Comput. Biol. Med. **39**(1), 8–15 (2009)
13. Grzegorczyk, I., et al.: PCG classification using a neural network approach. In: 2016 Computing in Cardiology Conference (CinC), pp. 1129–1132 (2016)
14. Lubaib, P., Muneer, K.A.: The heart defect analysis based on PCG signals using pattern recognition techniques. Procedia Technol. **24**, 1024–1031 (2016)
15. Prasad, G.V.H., Kumar, P.R.: Performance analysis of feature selection methods for feature extracted PCG signals. In: 2015 13th International Conference on Electromagnetic Interference and Compatibility (INCEMIC), pp. 225–231 (2015)
16. Tang, H., Dai, Z., Jiang, Y., Li, T., Liu, C.: PCG classification using multidomain features and SVM classifier. BioMed Res. Int. **2018** (2018)
17. Shin, H.-C., et al.: Deep convolutional neural networks for computer-aided detection: CNN architectures, dataset characteristics and transfer learning. IEEE Trans. Med. Imaging **35**(5), 1285–1298 (2016)
18. Dong, Y., et al.: Evaluations of deep convolutional neural networks for automatic identification of malaria infected cells. In: 2017 IEEE EMBS International Conference on Biomedical & Health Informatics (BHI), pp. 101–104 (2017)
19. Hinton, G., et al.: Deep neural networks for acoustic modeling in speech recognition: the shared views of four research groups. IEEE Signal Process. Mag. **29**(6), 82–97 (2012)
20. Alaskar, H.: Deep learning of EMG time frequency representations for identifying normal and aggressive action. IJCSNS Int. J. Comput. Sci. Netw. Secur. **18**(12) (2018)
21. Acharya, U.R., Oh, S.L., Hagiwara, Y., Tan, J.H., Adeli, H.: Deep convolutional neural network for the automated detection and diagnosis of seizure using EEG signals. Comput. Biol. Med. **100**, 270–278 (2018)
22. Acharya, U.R., Fujita, H., Lih, O.S., Adam, M., Tan, J.H., Chua, C.K.: Automated detection of coronary artery disease using different durations of ECG segments with convolutional neural network. Knowl.-Based Syst. **132**, 62–71 (2017)
23. Alaskar, H.: Deep learning-based model architecture for time-frequency images analysis. Int. J. Adv. Comput. Sci. Appl. **9**(12) (2018)
24. Kim, K.: Arrhythmia classification in multi-channel ECG signals using deep neural networks (2018)
25. Nilanon, T., Yao, J., Hao, J., Purushotham, S., Liu, Y.: Normal/abnormal heart sound recordings classification using convolutional neural network. In: Computing in Cardiology Conference (CinC), pp. 585–588 (2016)
26. Desai, U., Martis, R.J., Acharya, U.R., Nayak, C.G., Seshikala, G., Shetty, K.: Diagnosis of multiclass tachycardia beats using recurrence quantification analysis and ensemble classifiers. J. Mech. Med. Biol. **16**(01), 1640005 (2016)
27. Russakovsky, O., et al.: Imagenet large scale visual recognition challenge. Int. J. Comput. Vis. **115**(3), 211–252 (2015)
28. Uğuz, H.: A biomedical system based on artificial neural network and principal component analysis for diagnosis of the heart valve diseases. J. Med. Syst. **36**(1), 61–72 (2012)
29. Ölmez, T., Dokur, Z.: Classification of heart sounds using an artificial neural network. Pattern Recognit. Lett. **24**(1–3), 617–629 (2003)
30. Savalia, S., Emamian, V.: Cardiac arrhythmia classification by multi-layer perceptron and convolution neural networks. Bioengineering **5**(2), 35 (2018)

31. Alaskar, H., Jaafar Hussain, A.: Data mining to support the discrimination of amyotrophic lateral sclerosis diseases based on gait analysis. In: Huang, D.-S., Gromiha, M.M., Han, K., Hussain, A. (eds.) ICIC 2018. LNCS (LNAI), vol. 10956, pp. 760–766. Springer, Cham (2018). https://doi.org/10.1007/978-3-319-95957-3_80

32. Längkvist, M., Karlsson, L., Loutfi, A.: A review of unsupervised feature learning and deep learning for time-series modeling. Pattern Recognit. Lett. **42**, 11–24 (2014)

33. Andreotti, F., Carr, O., Pimentel, M.A., Mahdi, A., De Vos, M.: Comparing feature-based classifiers and convolutional neural networks to detect arrhythmia from short segments of ECG. Computing **44**, 1 (2017)

34. Kiranyaz, S., Ince, T., Hamila, R., Gabbouj, M.: Convolutional neural networks for patient-specific ECG classification. In: 2015 37th Annual International Conference of the IEEE Engineering in Medicine and Biology Society (EMBC), pp. 2608–2611 (2015)

35. Biagetti, G., Crippa, P., Orcioni, S., Turchetti, C.: Surface EMG fatigue analysis by means of homomorphic deconvolution. In: Conti, M., Martínez Madrid, N., Seepold, R., Orcioni, S. (eds.) Mobile Networks for Biometric Data Analysis. LNEE, vol. 392, pp. 173–188. Springer, Cham (2016). https://doi.org/10.1007/978-3-319-39700-9_14

36. Xia, P., Hu, J., Peng, Y.: EMG-based estimation of limb movement using deep learning with recurrent convolutional neural networks. Artif. Organs **42**(5), E67–E77 (2018)

37. Rubin, J., Abreu, R., Ganguli, A., Nelaturi, S., Matei, I., Sricharan, K.: Recognizing abnormal heart sounds using deep learning. arXiv preprint arXiv:1707.04642 (2017)

38. Potes, C., Parvaneh, S., Rahman, A., Conroy, B.: Ensemble of feature-based and deep learning-based classifiers for detection of abnormal heart sounds. In: 2016 Computing in Cardiology Conference (CinC), pp. 621–624 (2016)

39. Aykanat, M., Kılıç, O., Kurt, B., Saryal, S.: Classification of lung sounds using convolutional neural networks. EURASIP J. Image Video Process. **2017**(1), 65 (2017)

40. Renna, F., Oliveira, J., Coimbra, M.T.: Convolutional neural networks for heart sound segmentation. In: 2018 26th European Signal Processing Conference (EUSIPCO), pp. 757–761 (2018)

41. Low, J.X., Choo, K.: Automatic classification of periodic heart sounds using convolutional neural network. World Acad. Sci. Eng. Technol. Int. J. Electr. Comput. Eng. **5**, 100–105 (2018)

42. Andersson, G.: Classification of heart sounds with deep learning (2018)

43. Schmidt, S.E., Holst-Hansen, C., Graff, C., Toft, E., Struijk, J.J.: Segmentation of heart sound recordings by a duration-dependent hidden Markov model. Physiol. Meas. **31**(4), 513 (2010)

44. Springer, B., Tarassenko, L., Clifford, G.D.: Logistic regression-HSMM-based heart sound segmentation. IEEE Trans. Biomed. Eng. **63**(4), 822–832 (2016)

45. Shaker, M.M.: EEG waves classifier using wavelet transform and Fourier transform. Brain **2**, 3 (2006)

46. Krizhevsky, A., Sutskever, I., Hinton, G.E.: Imagenet classification with deep convolutional neural networks. In: Advances in Neural Information Processing Systems, pp. 1097–1105 (2012)

47. Simonyan, K., Zisserman, A.: Very deep convolutional networks for large-scale image recognition. arXiv preprint arXiv:1409.1556 (2014)

48. Szegedy, C., et al.: Going deeper with convolutions (2015)

Power Supply and Its Expert System for Cold Welding of Aluminum and Magnesium Sheet Metal

Liling Zhang$^{(\boxtimes)}$, Bing Li, and Jianxiong Ye

Jiangxi Province Key Laboratory of Precision Drive & Control,
Nanchang Institute of Technology, Nanchang, China
zlljdx@163.com

Abstract. According to the welding characteristics of aluminum and magnesium plates, the overall design scheme of cold welding power supply was discussed. The main control circuit board with Digital Signal Processor (DSP) as the core is designed on the hardware. In order to improve the dynamic characteristics of the power supply, the voltage and current waveform dual polarity phase shift soft switch control system is adopted. On the software, in order to realize the self-adjustment of welding process parameters, a special welding expert system is designed, which adopts Newton interpolation algorithm, with the characteristics of parameter unification, parameter automatic generation, parameter storage and self-learning. The welding experiment results show that the welding process is stable, and the weld joint is well formed. Overall, the welding quality is good. The validity of the cold welding power supply based on expert system for aluminum and magnesium sheet metal is proved.

Keywords: Aluminum magnesium sheet metal · Cold welding power supply ·
Newton interpolation · Expert system

1 Introduction

1.1 A Subsection Sample

Lightweight materials such as aluminum, magnesium and titanium, as important materials for aerospace, rail traffic and automobile products, have become new materials that China focuses on [1, 2]. As weld is an important energy consumption process of the manufacturing process, the manufacturing and development of welding technology and equipment all put energy saving in an important position. The cold welding equipment developed in recent years as energy-saving and environment-friendly welding equipment are getting more and more attention. Its energy saving effect is not only reflected in the welding process of small energy consumption, but also embodied in the small weld seam, so less solders and less consumable material is used to weld the finished product [3, 4].

Aluminum, Magnesium and their alloys have physical properties such as low resistivity, low density, high linear expansion coefficient, high thermal conductivity, high metal atomic activity, etc. Their physical properties determine that slag inclusion,

D.-S. Huang et al. (Eds.): ICIC 2019, LNAI 11645, pp. 795–804, 2019.
https://doi.org/10.1007/978-3-030-26766-7_72

incomplete fusion, incomplete welding, shrinkage hole, hot crack and hydrogen porosity are easily formed during welding. However, hydrogen porosity is the most common defect [5, 6]. In addition, Aluminum and Magnesium are mainly used in aerospace, transportation and other industrial fields as thin plates (within 3 mm). These require welding to reduce heat input, reduce deformation, improve the weld ability, and ensure the stability of the welding process.

The CMT method introduced by Austria's Fronius Company adopts the wire feeding method of push and pull. When there is a short circuit, the wire feeding machine reverses and pulls back the welding wire, so that the welding wire is separated from the melting droplet and the melting droplet is transited in a state of almost no current, which fundamentally eliminates the factor of spatter. However, this method has the disadvantage that the blowback of the welding gun easily leads to broken wire, and it requires high synchronization and collaborative control. The wire feeding system has complex control circuit, so its cost is high. The application range of the welding machine is relatively single for the poor compatibility of the welding gun [7, 8]. The Cold Arc welding technology is introduced by EWM Germany. It adopts inverter technology combined with digital control system, which rapidly reduces the power output energy when the Arc is reignited after short circuit, making the welding process energy input lower. However, this method still adopts the method of short circuit transition in principle. In the case of short circuit, a higher short circuit current is adopted to complete the droplet transition through the "small bridge" blasting off, which will inevitably generate splash and is not suitable for welding aluminum alloy. Moreover, it adopts special Zn base welding wire, which has a narrow application range and is not easy to be popularized in actual production [9, 10].

In order to ensure the low heat input, A cold welding inverter power supply – a kind of GMAW welding is designed, which can precisely control and adjust the arc voltage and current in the welding process, so that the arc is stable and the welding seam is uniform.

2   The Hardware Design

DSP digital signal processor is used to control the voltage and current waveform. In order to improve the dynamic characteristics of the power supply, the phase-shift soft switch control is adopted, and the main switch frequency is designed as 20–30 kHz. Figure 1 shows the block diagram of hardware circuit structure, including the main inverter circuit, control circuit, wire feeding drive, human-computer interaction system and other parts composed of rectifier filter circuit and IGBT full bridge.

Basic design principle: DSP chip is adopted to realize the limited bipolar soft switch control, and the output PWM is isolated and amplified by an optical coupler to drive the IGBT module. As the process control core, DSP mainly completes the protection of over current, over voltage, over temperature, initial signal setting, process logic sequence control, CAN bus and man-machine communication, data storage, expert system generation and management and other functions.

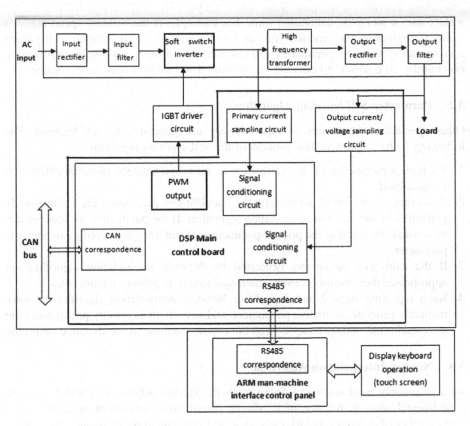

Fig. 1. System hardware structure diagram

3 Software Architecture

Cold welding digital power supply system needs to realize the synchronous control of droplet transition and wire feeding, and the setting and self-adjustment of process parameters. The "self-adjustment" function of expert system is realized by control software.

Expert system is one of the most active and accomplished research fields in artificial intelligence. The function of the expert system mainly depends on a large number of knowledge, facts, rules, assumptions and enhanced intelligent functions, also known as knowledge engineering system. In order to precisely control the input energy, the expert system should have parameter self- adjustment ability, that means the attributes of expert database could be changed from "read-only" to flexible "self-learning".

3.1 Calibration of Typical Parameters

In order to determine the initial value of welding machine parameters, it is necessary to calibrate the typical parameters. To simplify the parameter calibration, the speed of the welder uses a isometric calibration form. For example, if the welding speed range is from 20 cm/min to 100 cm/min, you can set the interval as 10 cm/min, so 9 groups corresponding welding parameters (20 cm/min, 30 cm/min, 40 m/min, 50 cm/min, 60 cm/min, 70 cm/min, 80 cm/min, 90 cm/min, 100 cm/min) need be calibrated.

3.2 Parameter Self-learning Algorithm

Other specialized parameters of expert library are generated by self-learning. The following is the implementation process of the self-learning algorithm:

1. Set typical parameters of step calibration as initial parameters, so initial calibration is completed.
2. According to the initial parameters, the uncalibrated parameters are automatically generated by the Newton interpolation algorithm. If the parameters are appropriate, they would be saved as the priority parameters; If not, they are saved as an alternate parameter.
3. If the calibrated parameters generated by Newton interpolation algorithm are appropriate, they would be overwritten and saved as priority parameters.
4. Keep repeating steps 2 and 3, and use Newton interpolation algorithm to automatically generate alternative parameters, and save them as priority parameters after reasonable fine-tuning, so as to keep the expert database in the dynamic optimum.

3.3 Newton Interpolation Algorithm

As a commonly used numerical fitting method, Newton difference is widely used in experimental analysis because of its simple calculation, convenient calculation of a large number of difference points, clear logic and convenient programming calculation. Using interpolation basis function, it is easy to get the Lagrange interpolation polynomial. It has compact structure and is convenient in the theoretical analysis. But when increasing or decreasing the interpolation node, all interpolation basis functions will change, and the whole formula will also change, which leads to inconveniences in the actual calculation. In order to overcome this defect, Newton interpolation is put forward. It is an adaptive algorithm and easy to estimate error.

Given $(x_i, y_i)_{i=0,1,\cdots,n}$, where $x_i \neq x_j (i \neq j)$, and $y_i = f(x_i)$, then

$$f[x_{i-k}, x_{i-k+1}, \cdots, x_i] = \frac{f[x_{i-k+1}, \cdots, xi] - f[x_{i-k}, \cdots, x_{i-1}]}{x_i - x_{i-k}}, (i = k, k+1, \cdots, n)$$

(1)

is the k order difference quotient of $f(x)$ about the nodes $x_{i-k}, x_{i-k+1}, \cdots, x_i$.

Set $(x_i, y_i)_{i=0,1,\cdots n}$, $x_i \neq x_j (i \neq j)$, x is any point on the interpolation interval $[a, b]$. Find the first order, second order, ..., No. $n+1$ order difference quotient:

$$f[x, x_0] = \frac{f(x_0) - f(x)}{x_0 - x},$$

$$f[x, x_0, x_1] = \frac{f[x_0, x_1] - f[x - x_0]}{x_1 - x},$$

$$f[x, x_0, x_1, x_2] = \frac{f[x_0, x_1, x_2] - f[x, x_0, x_1]}{x_2 - x},$$

$$\vdots$$

$$f[x, x_0, x_1, \cdots, x_n] = \frac{f[x_0, x_1, \cdots, x_n] - f[x, x_0, \cdots, x_{n-1}]}{x_n - x}$$

Solve from the above in order,

$$f(x) = f(x_0) + f[x, x_0](x - x_0),$$
$$f[x, x_0] = f[x_0, x1] + f[x, x_0, x1](x - x_1),$$
$$f[x, x_0, x_1] = f[x_0, x_1, x_2] + f[x, x_0, x_1, x_2](x - x_2),$$
$$\vdots$$
$$f[x, x_0, \cdots, x_{n-1}] = f[x_0, \cdots, x_{n-1}] + f[x, x_0, x_1, \cdots, x_n](x - x_n).$$

Substitute the latter formula into the former one successively,

$$f(x) = f(x_0) + f[x_0, x_1](x - x_0) + f[x_0, x_1, x_2](x - x_0)(x - x_1)$$
$$+ \cdots + f[x_0, x_1, \cdots, x_n](x - x_0)(x - x_1) \cdots (x - x_{n-1})$$
$$+ f[x, x_0, x_1, \cdots x_n](x - x_0)(x - x_1) \cdots (x - x_n). \tag{2}$$

Set

$$\omega_k(x) = (x - x_0) \cdots (x - x_{k-1}), (k = 1, 2, \cdots, n+1) \tag{3}$$

$$N_n(x) = \sum_{k=0}^{n} f[x_0, x_1, \cdots x_k] \omega_k(x), \tag{4}$$

$$R_n(x) = f[x, x_0, x_1, \cdots, x_k] \omega_{n+1}(x), \tag{5}$$

So

$$f(x) = N_n(x) + R_n(x). \tag{6}$$

$N_n(x)$ is called Newton interpolation polynomial, and $R_n(x)$ is the remainder [11].

In the parameter automatic generation of intelligent welding power supply, the balance between interpolation efficiency and generation speed should be considered. In this design, priority parameters, initial parameters and alternative parameters are used to carry out three points quadratic local Newton interpolation [12]. The algorithm has high efficiency, high computing speed and meets the design requirements.

In practical programming, Newton interpolation algorithm is implemented through iteration. Assuming that the value of n points: (x_1, y_1), (x_2, y_2), ..., (x_n, y_n) is known, the corresponding y value of x can be solved. This can be described using the flowchart shown in Fig. 2.

4 Test Results and Analysis

The self-made GMAW welding inverter power supply was used in the test after implanting the expert system program. Test conditions and process parameters are as follows. Automatic walking mechanism is used. Welding current range is 50–180 A. The specimens are pure aluminum plate with a size of 2–6 mm. Weld wire model is 4043 Φ 1.2 mm, whose extended length is 12 mm. The shielding gas is 99.99% high purity argon gas, whose flow rate is 15 L/min. The welding method is flat welding.

The welding arc dynamic wavelet analyzer was used to collect the voltage and current waveforms during the welding process, and the U-I diagram and the instantaneous energy diagram were displayed after the wavelet filtering. Figures 3 and 4 show the instantaneous voltage and current waveform, instantaneous energy waveform and appearance of welding seam obtained under the condition of the same average voltage value and the different average current value.

Several experiments with different currents were carried out under this scheme. The results show that the welding current is regular and stable, the arc is concentrated, the stretching rhythm is stable, the low frequency noise is soft, and there is no arc breaking phenomenon. It can be seen from Figs. 3(a), (b) and 4(a), (b) that the welding voltage and current peak is smooth, the peak input energy is low, and the welding arc transition is smooth. It can be seen from Figs. 3(c) and 4(c) that there is basically no spatter or undercut in the weld, and the weld seam present a uniform and beautiful fish-scale pattern with clear lines.

A large number of tests were done, the current is gradually adjusted from small to large, and the pulse welding current can cover all the currents from 50 to 180 A. The tests proved that, this expert database system not only ensures that the welding wire of model 4043 with a diameter of 1.2 mm provides a wide range of welding work, but also ensures that the accuracy of this database meets the actual production requirements.

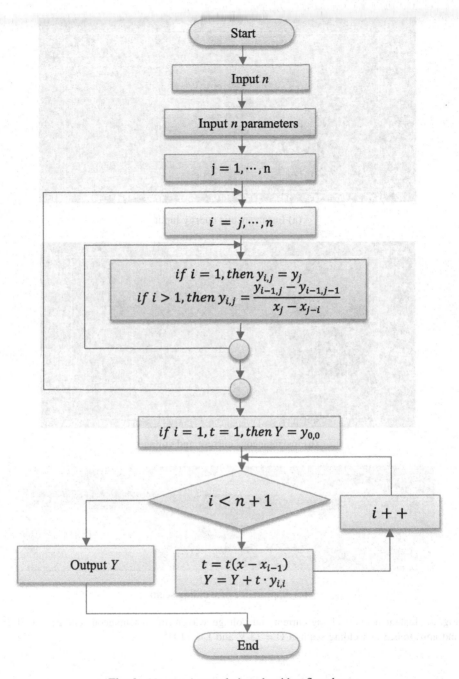

Fig. 2. Newton interpolation algorithm flowchart

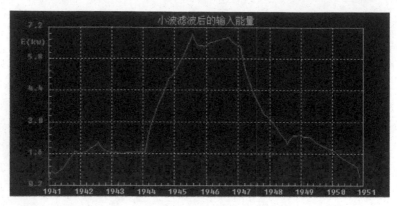

(a) Instantaneous energy input

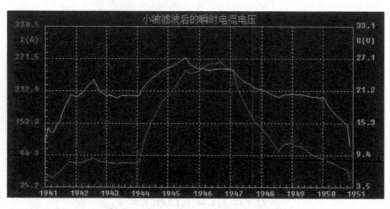

(b) Instantaneous current and voltage

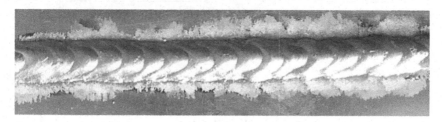

(c) Appearance of welding seam

Fig. 3. Instantaneous welding current and voltage waveform, instantaneous energy waveform and appearance of welding seam at U = 25 V and I = 110 A

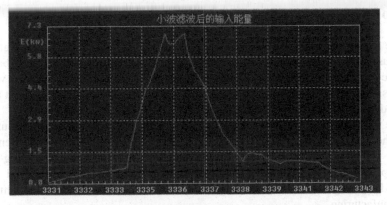

(a) Instantaneous energy input

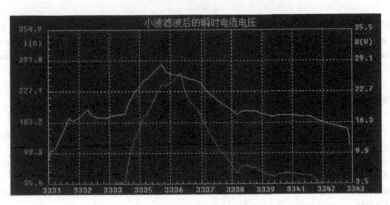

(b) Instantaneous current and voltage

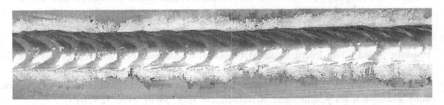

(c) Appearance of welding seam

Fig. 4. Instantaneous welding current and voltage waveform, instantaneous energy waveform and appearance of welding seam at U = 25 V and I = 80 A

5   Conclusion

• The parameters can be automatically generated and reset through the parameter self-adjustment algorithm of Newton interpolation, so as to keep the expert database in a state of dynamic optimization with good parameter effectiveness. Through welding tests, the validity of the method is verified. Stable welding can be carried out and the weld seam is well formed.

• The low heat input of cold welding arc can greatly reduce the energy consumption in the welding process of Magnesium or aluminum sheet, realize high efficiency and low consumption, and improve the welding process. The cold welding power supply with the expert system of parameter self-adjustment can be used for the welding of light aluminum and magnesium materials so as to realize the lightweight manufacturing.

Acknowledgment. This work was supported by Industrial Support Project of Jiangxi province under Grant 20151BBE50040 and Open Foundation of Jiangxi Province Key Laboratory of precision drive & control under Grant PLPDC-KFKT-201621, which is gratefully appreciated.

References

1. Lin, B., Peng, C., Wang, R., et al.: Recent development and prospects for giant plane aluminum alloy. Chin. J. Nonferr. Met. **20**(9), 1705–1715 (2010)
2. Zhang, T., Li, X.: Research progress of magnesium alloy welding technology. Shandong Sci. **10**(5), 39–46 (2012)
3. Bruckner, J.: Cold metal transfer has a future joining steel to aluminum. Weld. J. **84**(6), 38–40 (2005)
4. Cong, B., Ouyang, R., Qi, B., Ding, J.: Influence of cold metal transfer process and its heat input on weld bead geometry and porosity of aluminum-copper alloy welds. Rare Met. Mater. Eng. **45**(3), 606–611 (2016)
5. Verhaeghe, G., Hilton, P.: Laser welding of low-porosity aerospace aluminum alloy. In: Hinduja, S. (ed.) 34th International MATADOR Conference on Proceedings, pp. 241–246. Springer, London (2004). https://doi.org/10.1007/978-1-4471-0647-0_36
6. Ma, M., Bi, X., You, J., et al.: Research progress of property and its application of aluminum alloy aluminium alloy auto sheet. Mater. Mech. Eng. **34**(6), 1–5, 32 (2010)
7. Furukawa, K.: New CMT arc welding process-welding of steel to aluminium dissimilar metals and welding of super-thin aluminium sheets. Weld. Int. **20**(6), 440–445 (2006)
8. Johne, P.: CMT welding: sputter-free welding thanks to controlled energy and material transfer. Aluminium **82**(10), 938–943 (2006)
9. Yang, X.-R.: A revolution in GMAW welding technology. Cailiao Kexue yu Gongyi/Mater. Sci. Technol. **14**(12), 236–238 (2006)
10. Kou, S., Firouzdor, V., Haygood, I.W.: Hot cracking in welds of aluminum and magnesium alloys. In: Böllinghaus, T., Lippold, J., Cross, C. (eds.) Hot Cracking Phenomena in Welds III, pp. 3–23. Springer, Heidelberg (2011). https://doi.org/10.1007/978-3-642-16864-2_1
11. Faires, J.D., Burden, R.L.: Numerical Methods. PNS Publishing Company, Boston (1993)
12. Jamens, J.D., Smith, G.M., Wolford, L.C.: Applied Numerical Methods for Digital Computation. Harper Collins College Publishing, New York (1993)

Author Index

Printed in the United States
By Bookmasters

Printed in the United States
By Bookmasters